Health-promoting Properties of Fruit and Vegetables

To
Joanne, Oscar and Alice –
for your patience and love

Health-promoting Properties of Fruit and Vegetables

Edited by

Leon A. Terry

of

Cranfield University, UK

www.cabi.org

CABI is a trading name of CAB International

CABI Head Office	CABI North American Office
Nosworthy Way	38 Chauncey St
Wallingford	Suite 1002
Oxfordshire OX10 8DE	Boston, MA 02111
UK	USA
Tel: +44 (0)1491 832111	Tel: +1 617 395 4056
Fax: +44 (0)1491 833508	Fax: +1 617 354 6875
Email: cabi@cabi.org	Email: cabi-nao@cabi.org
Web site: www.cabi.org	

A catalogue record for this book is available from the British Library, London, UK.

Library of Congress Cataloging-in-Publication Data

Health-promoting properties of fruits and vegetables / edited by Leon A. Terry.
     p. cm.
  ISBN 978-1-84593-528-3 (alk.paper)
  1. Fruit in human nutrition. 2. Vegetables in human nutrition. I. Terry, Leon A. (Leon Alexander) II. Title.

QP144.F78H43 2011
613.2—dc22

                                                                    2011000833

ISBN-13: 978-1-84593-528-3 (HB)
ISBN-13: 978-1-78064-422-6 (PB)

First published (HB) 2011
First paperback edition 2013

Printed and bound in the UK by Berforts Information Press Ltd.

Contents

Contributors

Jeffrey K. Brecht, Horticultural Sciences Department, Institute of Food and Agricultural Sciences, University of Florida, POB 110690, Gainesville, FL 32611-0690, USA. E-mail: jkbrecht@ufl.edu

Vicky Buchanan-Wollaston, School of Life Sciences, Wellesbourne Campus, The University of Warwick, Wellesbourne, Warwick CV35 9EF, UK. E-mail: vicky.b-wollaston@warwick.ac.uk

Christian Chervin, INP/ENSAT, GBF, BP 32607, Université de Toulouse, 31326 Castanet-Tolosan, France. E-mail: chervin@ensat.fr

Gemma A. Chope, Plant Science Laboratory, Vincent Building, Cranfield University, Bedfordshire MK43 0AL, UK. E-mail: g.a.chope@cranfield.ac.uk

Luis Cisneros-Zevallos, Department of Horticultural Sciences, Food Science Program, Horticulture/Forest Science Building, Texas A&M University, College Station, Texas 77843-2133, USA. E-mail: lcisnero@ag.tamu.edu

Katherine Cools, Plant Science Laboratory, Vincent Building, Cranfield University, Bedfordshire MK43 0AL, UK. E-mail: k.cools.s06@cranfield.ac.uk

Carlos H. Crisosto, Department of Plant Sciences, University of California, Davis, One Shields Avenue, Davis, California 95616, USA. E-mail: chcrisosto@ucdavis.edu

Manuela Donetti, Plant Science Laboratory, Vincent Building, Cranfield University, Bedfordshire MK43 0AL, UK. E-mail: m.donetti.s06@cranfield.ac.uk

Charles F. Forney, Atlantic Food and Horticulture Research Centre, Agriculture and Agri-Food Canada, 32 Main Street, Kentville, Nova Scotia B4N 1J5, Canada. E-mail: Charles.Forney@agr.gc.ca

Jordi Giné Bordonaba, Plant Science Laboratory, Vincent Building, Cranfield University, Bedfordshire MK43 0AL, UK. E-mail: j.ginebordonaba.s05@cranfield.ac.uk

D. Mark Hodges, Agriculture and Agri-Food Canada, Atlantic Food and Horticulture Research Centre, Kentville, Nova Scotia, Canada. E-mail: Mark.Hodges@agr.gc.ca

Wilhelmina Kalt, Atlantic Food and Horticulture Research Centre, Agriculture and Agri-Food Canada, 32 Main Street, Kentville, Nova Scotia B4N 1J5, Canada. E-mail: Wilhelmina.Kalt@agr.gc.ca

Sandra Landahl, Plant Science Laboratory, Vincent Building, Cranfield University, Bedfordshire MK43 0AL, UK. E-mail: s.landahl@cranfield.ac.uk

Gene E. Lester, Food Quality Laboratory, USDA Agricultural Research Service, 10300 Baltimore Avenue, Building 002, Room 103, Beltsville, MD 20705-2350, USA. E-mail: Gene.Lester@ARS. USDA.GOV

Rui Hai Liu, Department of Food Science, Cornell University, New York 14853, USA. E-mail: rl23@cornell.edu

George A. Manganaris, Department of Agricultural Production and Food Science and Technology, Cyprus University of Technology, Athinon & Anexartisias 57 Corner, 3603 Lemessos, Cyprus. E-mail: george.manganaris@cut.ac.cy

Marjolaine D. Meyer, Plant Science Laboratory, Vincent Building, Cranfield University, Bedfordshire MK43 0AL, UK. E-mail: marjomeyer@hotmail.fr

Cecilia do Nascimento Nunes, Food Quality Laboratory, College of Human and Social Sciences, University of South Florida Polytechnic, 4100 South Frontage Road, Building 100, Suite 112, Lakeland, Florida 33815, USA. E-mail: mariacecilia@poly.usf.edu

Anne Pihlanto, MTT, Biotechnology and Food Research, 1600 Jokioinen, Finland. E-mail: anne.pihlanto@mtt.fi

Naila Rabbani, Clinical Sciences Research Institute, Warwick Medical School, University of Warwick, Coventry, UK. E-mail: N.Rabbani@warwick.ac.uk

Mark A. Ritenour, Indian River Research and Education Center, Institute of Food and Agricultural Sciences, 2199 S. Rock Road Fort Pierce, FL 34945-3138, USA. E-mail: ritenour@ufl.edu

Amarat H. Simonne, Family, Youth and Community Sciences Department, Institute of Food and Agricultural Sciences, 3025 McCarty Hall, PO Box 110310, University of Florida, Gainesville, FL 32611-0310, USA. E-mail: asim@ufl.edu

Nettra Somboonkaew, Postharvest and Processing Research and Development Office, Department of Agriculture, Chatuchak, Bangkok 10900. Thailand. E-mail: nettra_s@yahoo.com

Pierre-Louis Teissedre, Faculté d'Oenologie - ISVV, Université Victor Segalen Bordeaux 2, UMR 1219 Oenologie, 210 Chemin de Leysotte, CS 50008, 33882 Villenave d'Ornon Cedex, France. E-mail: pierrelouis.teissedre@u-bordeaux2.fr

Leon A. Terry, Head of Plant Science Laboratory, Head of Food Security and Environmental Health, Vincent Building, Cranfield University, Bedfordshire, MK43 0AL, UK. E-mail: l.a.terry@cranfield.ac.uk

Thiruchelvam Thanaraj, Plant Science Laboratory, Vincent Building, Cranfield University, Bedfordshire MK43 0AL, UK. E-mail: tthanaraj29@gmail.com

A. Keith Thompson, Plant Science Laboratory, Cranfield University, Bedfordshire, UK. E-mail: keiththompson28@yahoo.com

Paul J. Thornalley, Clinical Sciences Research Institute, Warwick Medical School, University of Warwick, Coventry, UK. E-mail: P.J.Thornalley@warwick.ac.uk

Peter M.A. Toivonen, Agriculture and Agri-Food Canada, Pacific Agri-Food Research Centre, Summerland, British Columbia, Canada. E-mail: Peter.Toivonen@agr.gc.ca

Ariel R. Vicente, Facultad de Ciencias Agrarias y Forestales, University of La Plata, Calle 60 y 119 s/n, CP 1900 La Plata, Argentina. E-mail: arielvicente@quimica.unlp.edu.ar

Peter Glen Walley, School of Life Sciences, Wellesbourne Campus, The University of Warwick, Wellesbourne, Warwick CV35 9EF, UK. E-mail: Peter.G.Walley@warwick.ac.uk

Chris B. Watkins, Department of Horticulture, Cornell University, New York 14853, USA. E-mail: cbw3@cornell.edu

Mingzhan Xue, Clinical Sciences Research Institute, Warwick Medical School, University of Warwick, Coventry, UK. E-mail: Mingzhan.Xue@warwick.ac.uk

About the Editor

Leon A. Terry, BSc (Hons) ARCS MSc PhD (Reader in Plant Science at Cranfield University, UK) received his BSc from Imperial College, London, and both his MSc and PhD from Cranfield University. Dr Terry established the Plant Science Laboratory at Cranfield University in 2001, which is now one of the largest groups dedicated to research, consultancy and education in postharvest science of fresh produce in the EU. As well as being Head of the Plant Science Laboratory, Dr Terry currently heads Food Security and Environmental Health at Cranfield University, with responsibility over all staff and students in the department.

Acknowledgements

———————

The Editor is grateful for the contributions from all authors. Without their knowledge and expertise, the completion of this book would have been impossible.

All current and past staff and students of the Plant Science Laboratory at Cranfield University (UK) are thanked for their enthusiasm for all things postharvest.

1 Introduction

Leon A. Terry and A. Keith Thompson

1.1 Introduction

Fruit and vegetables have been an important constituent of the human diet from time immemorial. Indeed, some 3500 years ago, when the children of Israel were wandering in the desert from Egypt to Canaan, they were reported to complain to Moses, 'And wherefore have ye made us to come up out of Egypt... it is no place of seed or of fig or of vine or of pomegranate...' (Numbers 20:5) and 'We remember the fish, which we did eat in Egypt freely; the cucumbers, and the melons, and the leeks, and the onions, and the garlick' (Numbers 11:5). Hippocrates (c.460–370 BC) later wrote 'Let food be thy medicine and medicine be thy food'.

Fruit and vegetables are doubtless eaten for their unique taste and flavour, yet their health and healing properties have been known for centuries. In the 18th century a French pharmacist, Antoine-Augustin Parmentier, demonstrated, for several years by his own diet, that all the nutrients required to sustain a healthy life were found in potatoes (Block, 2008). Yet potatoes are specifically excluded from the modern '5-a-day' fruit and vegetable recommendations. The Chinese have long used plant parts, including fruit and vegetables, to prevent or control diseases. In the West, specific plants have been used; for example, in 1753 a Royal Naval surgeon (James Lind) showed that scurvy could be prevented with citrus juice. However, it was not until 1932 that Albert Szent-Györgyi identified vitamin C as the curative chemical in the juice, for which he received a Nobel Prize. However, Carr and Frei (1999) suggest that the amount of vitamin C required to prevent scurvy is not sufficient to protect against other chronic diseases and have argued that requirements should be increased to take these effects into account.

The type of fruit or vegetable consumed is influenced, among other things, by availability, weather, culture, price and promotion. With the introduction of modern postharvest storage technologies and greater efficiencies in international trade, the availability and thus opportunity for consumers to eat more fresh produce of greater diversity has increased dramatically. Improvements in breeding, crop management, transport, storage and packaging have resulted in seasonality becoming less important, such that the opportunity for consumers to choose from a large array of different fresh produce types and thus expose their bodies to a greater diversity of phytochemicals is welcome.

Much myth surrounds the supposed health-giving properties of fresh fruit and vegetables. For instance, the term 'fresh' is ill defined. The notion that a stored product is universally less healthy is false. Most consumers

© CAB International 2011. *Health-promoting Properties of Fruit and Vegetables* (ed. L.A. Terry)

eat products that have been stored for variable periods. This fact is often overlooked, and indeed the effect of storage on health-promoting properties of fruit and vegetables has been overshadowed largely by the drive towards reducing waste and maintaining eating quality and storage life. With more urbanization, the greater reliance on logistics and supply chain management will serve to make stored fruit and vegetables more accessible, yet only to those who can afford it.

The incidence of chronic non-communicable diseases, namely coronary heart disease, diabetes, hypertension and obesity, has become a major public health problem in developing countries, and these diseases are expected to account for about 70% of deaths by 2020 (WHO, 2002). However, deaths from chronic non-communicable diseases will escalate with the greater consumption of animal-based food products, but will decrease with higher consumption of fruit and vegetables. It is clear that these worrying trends are now starting to be reflected in the emerging economies.

The increased dietary intake of antioxidant vitamins (vitamins A, C and E) and fibre from vegetables and fruit is highly recommended in order to reduce the risk of cardiovascular disease, stroke and cancer (mouth, oesophagus, stomach and colon). Fruit is generally a good source of dietary antioxidants such as vitamins, phenolic compounds and carotenoids (β-carotene, lycopene, lutein and zeaxanthin). Fruit antioxidants play an important role in reducing the risk of degenerative diseases, in particular cardiovascular diseases, diabetes and several types of cancer. The antioxidant activity of fruit is also believed to reduce the progress of senescence. However, antioxidant capacity varies with genetic differences, harvest maturity, harvesting season, postharvest storage and processing. The antioxidant activities of fruit like apples and berries have been studied extensively, since they are rich in polyphenolics and ascorbic acid (see Chapters 11, 14 and 15 of this volume). Generally, most research on elucidating the health-promoting effects of fruit and vegetables has been conducted on products that are important to Western consumers. Fruits that are consumed more commonly in the developing world are not as well studied, even though they can make up a larger proportion of the global diet.

The contribution of fruit and vegetables to human health has been increasingly recognized. However, low fruit and vegetable intake was identified as an important risk factor for chronic diseases in the WHO World Health Report 2002. Overall, it was estimated that up to 2.7 million lives potentially could be saved each year if fruit and vegetable consumption was sufficiently increased (Keller and Tukuitonga, 2007). In 1988 the '5-a-day' portion of fruit and vegetables campaign was initiated in the USA, and thereafter in Europe and elsewhere. A portion was defined as 80 g of fruit (or vegetable) or 150 ml of unsweetened fruit juice, although juice should contribute only one portion. The WCRF/AICR (1997) recommended that the inclusion of 400–800 g/day of a variety of fruit and vegetables should be included in the diet to reduce the risk of cancer. In their subsequent review, the WCRF/AICR (2007) still recommended to include 'at least 5 portions/servings (at least 400 g per day) of a variety of non-starchy vegetables and of fruit each day', but this consumption might not be effective in reducing the risk of all cancers. However, in a UK survey more than 60% of adults had intakes below the recommended level of 400 g/day (Ashfield-Watt et al., 2003) and the situation has not radically improved. Although the '5-a-day' message is flawed in that it equates or assumes that one portion by weight of one product is as 'valuable' as another, the '5-a-day' recommendation can only be a generalization to give a simple indication and message to consumers to eat more fruit and vegetables. It is likely that the body mass, gender, ethnicity and age of the person affect their requirements, as does the percentage of each portion that consists of peel and other parts that are not eaten.

A study in the USA showed that the concentration of flavonoids in fruit and vegetables consumed by the public was highly variable. Harnly et al. (2006) suggested that this was most likely due to different cultivars, local growing conditions and processing methods. There is a range of phytochemicals in individual species and cultivars of fruit and vegetables; for example, the bioactive phytochemicals in citrus fruit include limonoids, vitamin C, β-carotene, flavonoids, folic acid and dietary fibre (see Chapter 6 of this volume). Taking phenolics as an example, Tomás-Barberán and Gil (2008) showed that the intake of phenolics from selected fruit and vegetables, depended on the method of preparation and the parts that were consumed (Table 1.1).

Besides the chemicals required for human nutrition, fruit and vegetables contain what have been referred to as phytochemicals. Of the thousands of phytochemicals (known and unknown) in fruit and vegetables, only a limited number have been evaluated in terms of their effects on human health. Rice-Evans and Miller (1995), reviewing the literature on antioxidants, found that there was a strong case that they could contribute to the prevention or delaying of the onset of cancer and heart disease. Silalahi (2002) also found strong evidence that there was a low risk of degenerative diseases, cardiovascular disease, hypertension, cataract, stroke and cancers in people with a high intake of fruit and vegetables.

The effect of different phytochemicals on human health is influenced not only by the identity and abundance of a single target analyte or cluster of chemicals, but also by the intake, bioavailability and metabolism of these bioactive compounds in the individual. The abundance and profile of bioactive compounds is affected by many factors; yet many of the main features that influence health-promoting properties often have been overlooked or disregarded in most epidemiological studies to date. Fruit and vegetables increasingly are

sourced from different locations with varied soil type and climate. In addition, fruit and vegetables are harvested at different horticultural maturities and may be stored and transported for long distances before they are sold and then eventually eaten. Fresh produce continues to undergo physiological, mechanical and biochemical changes after harvest and it is these temporal and even spatial changes, as defined by genotype and the way the product is grown, harvested, stored, processed or cooked, that should be taken into account in any epidemiological study, as all these changes and factors can impact on intake, bioavailability and the interaction between the phytochemical(s) and the individual. For instance, postharvest changes can be influenced by implementing appropriate (or inappropriate) technologies, such as controlling temperature, relative humidity, gaseous composition and so on, and these are known to have profound effects on phytochemicals. Many fruit and vegetables are processed or cooked before being consumed and often are not eaten in isolation; that is, they are commonly eaten with other foodstuffs in combination (i.e. in a meal).

Classification of phytochemicals is complicated. Jaganath and Crozier (2008) defined phytochemicals as 'bioactive non-nutrient plant compounds that have been linked to reductions in the risk of major chronic diseases'. They classified them into four major groups: nitrogen-containing alkaloids, phenolics and polyphenolics, sulfur-containing compounds, and terpenoids. The premise of this book is not, however, to reassess this classification or to expand upon it, but rather

Table 1.1. Amount of phenolic antioxidants from 150 g of selected fruit and vegetables (modified from Tomás-Barberán and Gil, 2008).

Fresh fruit or vegetable	Phenolic antioxidant intake	
Peach	110 mg not peeled for cv. Snow King	15 mg peeled for cv. Flavor Crest
Grape cv. Napoleon	150 mg with seeds and peel	5 mg without seeds and peel
Orange cv. Navel	400 mg with part of albedo	100 mg without albedo
Lettuce	300 mg cv. Lollo Rosso leaving external leaves	10 mg cv. Iceberg white midribs
Spinach	120 mg steam cooked	50 mg boiled and water removed

to classify the health-promoting properties of fruit and vegetables on a species-specific basis. Consequently, each chapter provides a review on the current understanding of the identity, abundance, variance, bioavailability and efficacy of individual bioactive phyto-chemicals specific to that fresh produce type(s) and describes the measures and research needs that will be required in the future to preserve 'bioactive life'.

References

Ashfield-Watt, A., Welch, A., Day, N.E. and Bingham, S.A. (2003) Is 'five-a-day' an effective way of increasing fruit and vegetable intakes? *Public Health Nutrition* 7, 257–261.

Block, B.P. (2008) Antoine-Augustin Parmentier: pharmacist extrodinaire. *Pharmaceutical Historian* 38, 6–14.

Carr, A.C. and Frei, B. (1999) Toward a new recommended dietary allowance for vitamin C based on antioxidant and health effects in humans. *American Journal of Clinical Nutrition* 69, 1086–1087.

Harnly, J.M., Doherty, R.F., Beecher, G.R., Holden, J.M., Haytowitz, D.B., Bhagwat, S.A. and Gebhardt, S.E. (2006) Flavonoid content of US fruits, vegetables, and nuts. *Journal of Agricultural and Food Chemistry* 54, 9966–9977.

Jaganath, I.B. and Crozier, A. (2008) Overview of health promoting compounds in fruit and vegetables. In: Tomás-Barberán, F.A. and Gil, M.I. (eds) *Improving the Health-promoting Properties of Fruit and Vegetable Products.* Woodhead Publishing, Cambridge, UK, pp. 3–37.

Keller, I. and Tukuitonga, C. (2007) The WHO/FAO fruit and vegetable promotion initiative. *Acta Horticulturae* 744, 27–37.

Rice-Evans, C. and Miller, N.J. (1995) Antioxidants – the case for fruit and vegetables in the diet. *British Food Journal* 97, 35–40.

Silalahi, J. (2002) Anticancer and health protective properties of citrus fruit components. *Asia Pacific Journal of Clinical Nutrition* 11, 79–84.

Tomás-Barberán, F.A. and Gil, M.I. (2008) *Improving the Health-promoting Properties of Fruit and Vegetable Products.* Woodhead Publishing, Cambridge, UK.

WCRF/AICR (1997 and 2007) www.dietandcancerreport.org (accessed July 2008).

WHO (2002) http://www.who.int/whr/2002/en/whr02_en.pdf (accessed 23 February 2011).

2 Alliums
[Onion, Garlic, Leek and Shallot]

Gemma A. Chope, Katherine Cools and Leon A. Terry

2.1 Introduction

The *Allium* genus is very large and consists of many wild edible species; however, only a small selection is cultivated commercially and these include onion (*Allium cepa* L), garlic (*A. sativum* L), leek (*A. ampeloprasum* leek group) and shallot (*A. cepa* Aggregatum group). Garlic and leek are part of the subgenus *Allium*, which comprises 37 species in 15 sections, while onion and shallot fall into the subgenus *Cepa*, which has 30 species in five sections (Fritsch and Keusgen, 2006). In South–east (SE) Asia, species such as Chinese chives (*A. tuberosum* Rottl.) and Rakkyo (*A. chinense* G. Don), which are often used for pickles, are cultivated commercially (Brewster, 1994). The Japanese bunching onion (*A. fistulosum* L.), also known as the spring onion, salad onion, Welsh onion or scallion, is also historically from eastern Asia but is now cultivated and consumed in numerous countries (Brewster, 1994). Onion production extends from the tropics to temperate regions, in countries ranging from the equator to Scandinavia and South Africa. Onion plants are sensitive to photoperiod; therefore, a very wide range of cultivars exists, from 'short day' to 'very long day' types, which are adapted to cover all latitudes. Therefore, onions grown in one zone may not be readily transferable to another (Currah *et al.*, unpublished data).

Shallots are genetically very close to the common onion and are used in SE Asia as a substitute for onion. Onions are the most commercially important member of the *Allium* genus, with a worldwide production of almost 64.5 million tonnes (Mt), followed by garlic (15.5 Mt) (FAOSTAT, 2008). Shallots are less commercially important, with a worldwide production of green onions, including shallot, of more than 3.5 Mt.

Allium vegetables have been cultivated and eaten for centuries by many cultures all over the world. Many health benefits historically have been attributed to the consumption of *Allium* vegetables since the Egyptians in 1550 BC (El-Bayoumy *et al.*, 2006). Since then, their health-promoting properties, which have been shown or suggested, include anticancer, antimicrobial, antiplatelet, antithrombotic, antihyperlipidaemic, antihypertensive, anti-asthmatic and immunostimulatory (Block, 2005; Corzo-Martínez *et al.*, 2007).

2.2 Identity and Role of Bioactives

2.2.1 Organic acids

The literature on the organic acid content of onion bulbs is very limited. Malic and citric acid, determined by gas chromatography (GC) prior to storage, were found to be the major organic acids in onion bulbs cv. Sentinel, comprising 2.8 and 1.8% dry weight (DW) (c.280 and 180 mg/100 g fresh weight (FW)), respectively. Both fumaric and succinic acids were also present; however, levels were quoted as very low, with no data given (Salama *et al.*, 1990). Benkeblia and Varoquaux (2003) also agreed that succinic and fumaric acids were found at very low concentrations, of 0.2 and 0.75 mg/100 g FW, respectively. Both malic and citric acid concentrations were lower than those stated by Salama *et al.* (1990), at 102 and 20 mg/100 g FW, and an additional organic acid, namely oxalic acid, was measured at a concentration higher than citric acid: 37 mg/100 g FW (Benkeblia and Varoquaux, 2003). More recently, an extended range of organic acids, including tartaric and glutamic acids, has been measured, using HPLC, in six onion cultivars from Tenerife stored for 1 week at 20–25°C. Glutamic acid was found in much higher concentrations than malic and citric acids, at 192–433 mg/100 g FW, whereas tartaric acid was c.8.9–25.2 mg/100 g FW (Rodríguez Galdón *et al.*, 2008). Discrepancies between these published results may be due to variations between cultivars and growing conditions, combined with the use of different techniques to extract and measure the organic acids. Salama *et al.* (1990) used 80% ethanol (v/v) for extraction and GC coupled with flame ionization detection (FID) for analysis, whereas Benkeblia and Varoquaux (2003) and Rodríguez Galdón *et al.* (2008) used water to extract and as a mobile phase for HPLC analysis coupled to a diode array detector (DAD). Differences between the two latter studies could be explained, as only Rodríguez Galdón *et al.* (2008) used acidified water for both the extraction and HPLC mobile phase, which would have helped to preserve antioxidants.

Ascorbic acid is the most abundant vitamin in onion bulbs, with concentrations around 1 mg/g DW (Breu, 1996). According to Gorinstein *et al.* (2008), ascorbic acid content (measured by the cupric-reducing antioxidant capacity (CUPRAC) assay) was 735, 1994 and 1385 µg/g DW in garlic, red onion and white onion, respectively (cultivar and storage time not stated).

2.2.2 Phenolics

Phenolics are secondary metabolites characterized by hydroxylated aromatic rings which include flavonoids, hydroxycinnamic acids and phenolic acids (Velioglu *et al.*, 1998). Phenolics have been found to contribute to the antioxidant properties of onions and have been correlated positively with radical scavenging activity and antioxidant activity (Nuutila *et al.*, 2003). The phenolic acids, ferulic, gallic and protocatechuic acid, were highest in red onions at 21.4, 263 and 138 µg/g DW, respectively, and were more concentrated in the outer scales (Prakash *et al.*, 2007). These recent data support work from the early 20th century on the antifungal properties of onion skin. It was found that red onion skins contained protocatechuic acid and catechol (not found in the outer scales of brown onions) and were toxic to the fungus, *Colletotrichum circinans*, responsible for the disease, smudge (Walker *et al.*, 1929; Link and Walker, 1933). The literature on the phenolic acid content of leek is scarce, but Schmidtlein and Herrmann (1975) quoted that the most abundant phenolic acids were ferulic acid and *p*-coumaric acid.

The total phenolic content of *Allium* species extracted using various solvents and analysed using the standard Folin–Ciocalteu reagent has been investigated. Lin and Tang (2007) reported that the total phenolic concentration of red and white onions (cultivars not stated) were, respectively, c.310 and 216.7 mg gallic acid equivalent (GAE)/100 g FW, as determined using lyophilized onion powder dissolved in deionized water. Garlic and shallot both contained c.55 mg GAE/100 g FW, as determined using a mixture of juice filtrate and hexane extract, which was then hydrolysed in acidified 50% methanol (v/v) (Leelarungrayub *et al.*, 2006). Leek contained the lowest concentration of phenolics at 27.7 mg GAE/100 g FW, extracted from lyophilized powder with 80% methanol (v/v) (Marinova *et al.*, 2005). Discrepancies exist between values reported in the literature; these are possibly due to variations in cultivar, growing site, tissue choice and extraction method. Nuutila *et al.* (2003) agreed that red onion (cultivar not stated) had a higher total phenolic content than yellow onion, but the concentrations in the scales were much lower than those reported by Lin and Tang (2007), at *c.*20.8 and 15.5 mg GAE/100 g FW, respectively. However, Nuutila *et al.* (2003) investigated spatial differences in total phenolic content and found that concentrations (GAE) in skins from red and yellow onions (cultivars not stated) were 38- and 17-fold higher than in the bulb scales, respectively. Similarly, garlic total phenolic concentration was 2.3-fold higher in the skin than the bulb, and leek concentrations were 1.3-fold higher in the leaves than the stem. The increased levels of total phenolics in the outer layers of these *Alliums* may be due to their UV-B exposure. Onion tissue perceives UV-B as a stress signal, which results in enhanced synthesis of enzymes, such as phenylalanine ammonia-lyase (PAL) and chalcone synthases, that catalyse the biosynthesis of phenolic compounds (Mogren *et al.*, 2006). Onion scales (second or fourth) cv. Mansang irradiated with white fluorescent light (3000 lux, 25°C) had 1.2- to 1.6-fold higher quercetin and isorhamnetin glucoside concentrations, after just 24 h, than those held in the dark (Lee *et al.*, 2008b).

2.2.3 Flavonoids

Flavonoids consist of two subgroups, flavonols and anthocyanins (Leighton *et al.*, 1992), which are thought to protect against cancer and cardiovascular disease by inhibiting tumour growth and microbial cells (Griffiths *et al.*, 2002). In the main, this link has been shown only by *in vitro* assays. The major flavonols found in onion are quercetin, kaempferol and isorhamnetin, which can exist as aglycons (quercetin) or as sugar conjugates (glycosides), with 25 different flavonol derivatives characterized in onion bulbs to date (Slimestad *et al.*, 2007). The most abundant flavonol, quercetin, and its glycosides are found mainly in the outer scales of the onion bulbs, specifically in the abaxial epidermis (Hirota *et al.*, 1998). Unlike those in onion, leek-derived flavonols are comprised mainly of kaempferol derivatives, of which Fattorusso *et al.* (2001) identified and isolated five: three known and two previously unknown.

Quercetin in the bulb scales of onion cv. Sherpa was mainly in the form of quercetin 3,4-diglucoside (4.68 mg/g FW) and quercetin 4-glucoside (2.87 mg/g FW), whereas in the skin it was comprised mainly of quercetin aglycon (10.29 mg/g FW) (Figs 2.1 and 2.2; Downes *et al.*, 2009, 2010). In rats, the bioavailability of quercetin glucosides was found to be 50% that of quercetin aglycon, possibly due to the hydrophilic sugar group hindering passive diffusion across the intestine lining (Scalbert and Williamson, 2000; Wiczkowski *et al.*, 2003). Onion skin is not consumed, although recent work by Roldán *et al.* (2008) showed that onion by-products could be processed into a paste, followed by mild pasteurization to form a stable product with high antioxidant capacity for possible use as a food ingredient.

The pigment in red onions is predominantly due to anthocyanins, which are comprised mainly of cyanidin derivatives, although peonidin, delphinidin, petunidin and 5-carboxypyranocyanidin derivatives also exist (Fig. 2.3; Slimestad *et al.*, 2007).

Lee *et al.* (2008b) investigated the effect of baking (5 min), boiling (5 min), frying (2 min), microwaving (1 min), sautéing (3 min) and steaming (5 min) on onion cv. Tubo flavonoid content extracted using 80% aqueous ethanol (v/v). The largest percentage loss (32.8%) of total flavonoids occurred after frying, whereas steaming and microwaving caused the least total flavonoid reduction at 5.7 and 4.4%, and baking resulted in a 1.1% increase, although this was not significant.

2.2.4 Organosulfur and organoselenium compounds

Onions and other *Allium* vegetables are eaten for their unique taste and the medicinal properties of their flavour compounds (Griffiths *et al.*, 2002). The majority of the compounds that contribute to flavour and taste are secondary metabolites whose biosynthesis involves the metabolism of cysteine and glutathione, which are essential pathways for uptake of sulfur and detoxification (Jones *et al.*, 2004). In intact *Alliums*, the major organosulfur compounds are γ-glutamyl-*S*-allyl-L-cysteines

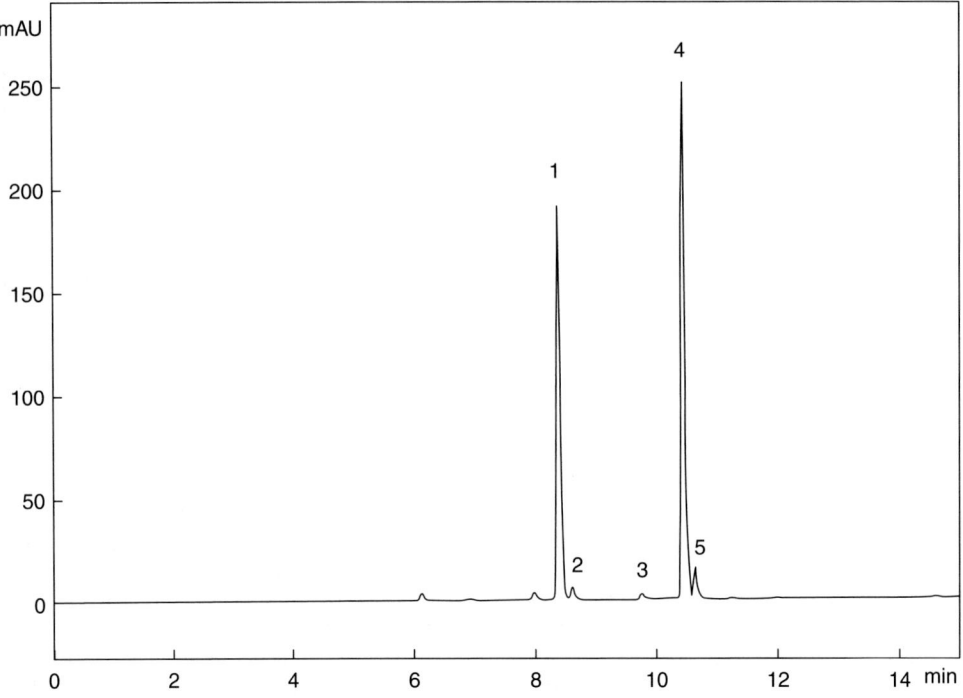

Fig. 2.1. HPLC-DAD chromatographic profile of major flavonols from the fleshy bulb scales of onion cv. Wellington (Downes *et al.*, 2010). 1, quercetin 3,4-diglucoside; 2, isorhamnetin 3,4-diglucoside; 3, quercetin 3-glucoside; 4, quercetin 4-glucoside; 5, isorhamnetin 4-glucoside.

and *S*-allyl-ʟ-cysteine sulfoxides (ACSOs). In an intact cell, the enzyme alliinase [EC 4.4.1.4] is located in the vacuole, and the ACSOs in the cytoplasm. When the tissue is disrupted (for example, during mastication or homogenization), alliinase hydrolyses the ACSOs, yielding pyruvate, ammonia and transient and unstable sulfenic acids (Uddin and MacTavish, 2003), which then condense spontaneously in pairs to form volatile thiosulfinates that contribute to perceived flavour (Briggs and Goldman, 2002). Four different ACSOs are present in *Alliums*; *S*-2-propenyl-(alliin), *S*-1-propenyl-(isoalliin, or 1-PrenCSO), *S*-methyl-(methiin, or MCSO) and *S*-propyl-(propiin, or PSCO) ʟ-cysteine *S*-oxides (Fig. 2.4). The composition and concentration of these compounds are responsible, in part, for imparting the characteristic taste and odour of individual *Alliums*. Methiin is found in garlic, onion, leek and shallot (Fritsch and Keusgen, 2006). The predominant ACSO in garlic

is alliin, while in onion and shallot isoalliin predominates. Isoalliin gives rise to the lachrymatory factor, thiopropanal *S*-oxide, via an enzyme known as lachrymatory factor synthase (Imai *et al.*, 2002). Shallot has a higher relative propiin content (more than 10%) and lower isoalliin than other onion-type *Alliums* and, in general, leek and shallot have relatively low amounts of ACSOs. Allicin is a thiosulfinate formed from alliin, and many of the reputed health benefits of garlic are attributed to this compound (Cavagnaro *et al.*, 2007).

The thiosulfinates are themselves unstable and decompose rapidly into a variety of strong-smelling volatile sulfur compounds, such as polysulfides, cepaenes/ajoenes and zwiebelanes. Cepaenes and ajoenes are α-sulfinyl disulfides and zwiebelanes are cyclic S-S compounds. The compounds produced as a result of this reaction are highly dependent on conditions such as: the initial concentration and ratio of sulfenic acids, pH,

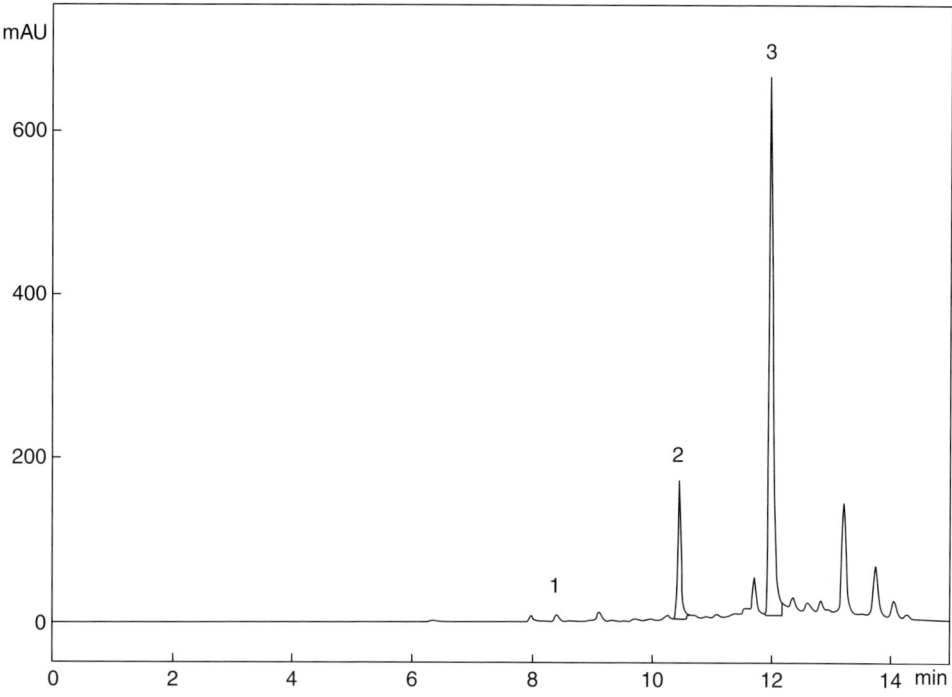

Fig. 2.2. HPLC-DAD chromatographic profile of major flavonols from the skins of onion cv. Wellington (Downes *et al.*, 2009). 1, quercetin 3,4-diglucoside; 2, quercetin 4-glucoside; 3, quercetin.

temperature, polarity of extraction solvent, etc. For example, in garlic, ajoenes (Fig. 2.4) are formed in ethanolic extracts and dithiins are formed in oil extracts (Keusgen, 2002), and diallyl disulfide is formed during steam distillations (Brewster, 2008).

The organosulfur compounds are the main active antimicrobial, antifungal and antibacterial agents in *Allium* vegetables (Corzo-Martínez *et al.*, 2007). Ajoene is a stable rearrangement product of allicin, which is reputed to demonstrate a range of biological activities such as antithrombotic, antimicrobial, antifungal and anticancer (Hunter *et al.*, 2008). It is thought that the organosulfur compounds produced by *Alliums* have a role in chemical defence against grazing animals and some fungi and bacteria (Brewster, 2008). The organosulfur compounds and their γ-glutamyl derivatives contribute a significant amount to the dry weight of *Allium* plants, constituting between 1 and 5%; therefore, it is also likely that these compounds play a role in nitrogen,

sulfur and carbon turnover, storage and transport.

Organoselenium compounds also exist in the *Alliums*, which are analogous to the organosulfur compounds in that selenium is substituted for sulfur in these molecules. Selenium can be incorporated in the plant metabolism in place of sulfur where it is available in the growing medium. The major organoselenium compound in onion bulbs is γ-glutamyl-*Se*-methyl selenocysteine, whereas in onion leaves it is *Se*-methyl selenocysteine (Arnault and Auger, 2006). Selenium can be incorporated further into *Allium* chemistry when *Alliums* are grown in a selenium-rich environment (Wróbel *et al.*, 2004; Block, 2005). The major organoselenium compound in selenium-enriched onion and garlic is *Se*-methyl selenocysteine, accompanied by other compounds including γ-glutamyl-*Se*-methyl selenocysteine, selenocysteine and selenomethionine (Arnault and Auger, 2006). Selenosulfur compounds have been reported to have greater

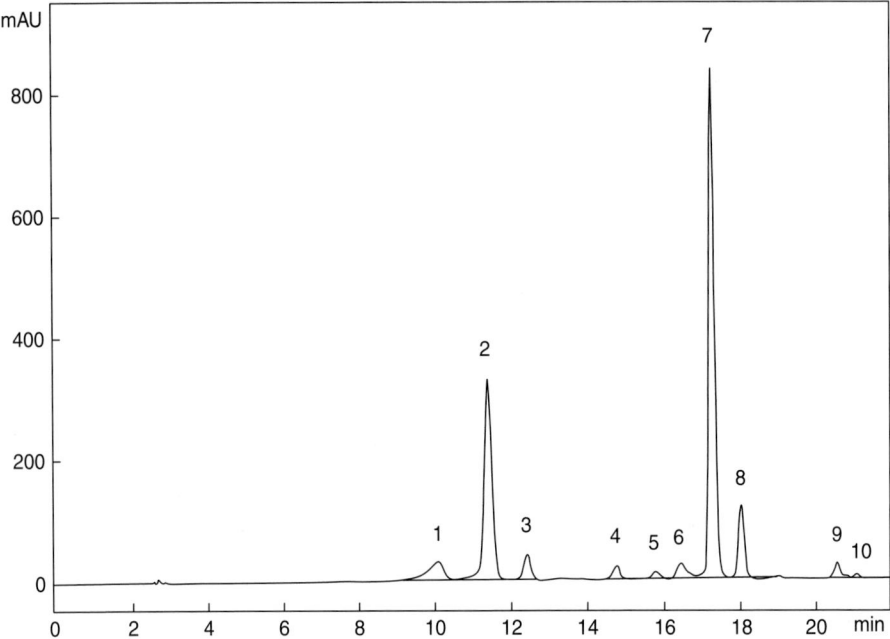

Fig. 2.3. HPLC-DAD chromatographic profile and chemical structure of major anthocyanins from the skin of onion cv. Red Baron (Downes *et al.*, 2009). 1, cyanidin 3-(malonoyl)-glucose-5-glucose; 2, cyanidin 3-glucose; 3, cyanidin 3-laminariboside; 4, cyanidin 3-(3″-malonoylglucoside); 5, peonidin 3-glucose; 6, cyanidin 3-(3″-acetoyl)glucoside; 7, cyanidin 3-(6″-malonoylglucoside); 8, cyanidin 3-(6″-malonoyl-laminariboside); 9, peonidin 3-(malonoyl)glucoside; 10, cyanidin 3-(malonoyl)(acetoyl)glucoside.

anticancer activity than their sulfurous counterparts demonstrated by *in vivo* studies on mice and rats (El-Bayoumy *et al.*, 2006). *Se*-methyl selenocysteine is unstable in water extracts at room temperature, but its stability can be maintained by freeze-drying (Arnault and Auger, 2006).

2.2.5 Fructans

Fructans are the principal storage carbohydrates in *Alliums*. In a study of 60 vegetables, onion, garlic, shallot and leek were among the top six in terms of fructan concentration (a range of 1.8–17.4 g/100 g FW edible portion) (Muir *et al.*, 2007). These polysaccharides are not digested in the upper intestine and are a source of energy for bacteria-producing β-fructosidases in the caeco-colon. Fructan has a reported prebiotic effect, whereby it is believed to promote proliferation of beneficial bacteria like *Bifidobaceteria* (Bielecka *et al.*, 2002) and *Lactobacilli* (probiotic bacteria), which in turn results in a decrease in the population of potentially harmful bacteria (Roberford, 2007). The beneficial colonic microbiota produce short-chain fatty acids (e.g. lactate and butyrate) which lower pH, thus favouring increased absorption of mineral cations (such as Ca and Mg) from the gut into the bloodstream. Changes in colonic bacteria may reduce carcinogen activation in the colon and stimulate the immune system. Animal studies have also shown benefits for glucose metabolism (increased insulin secretion and changes to hormone metabolism) (Brewster, 2008). It has been suggested that fructans with a higher degree of polymerization (such as those present in garlic and leek) are less likely to induce undesirable gastrointestinal side effects (Muir *et al.*, 2007). Fructan profiles (Fig. 2.5) vary with cultivar for onion, garlic,

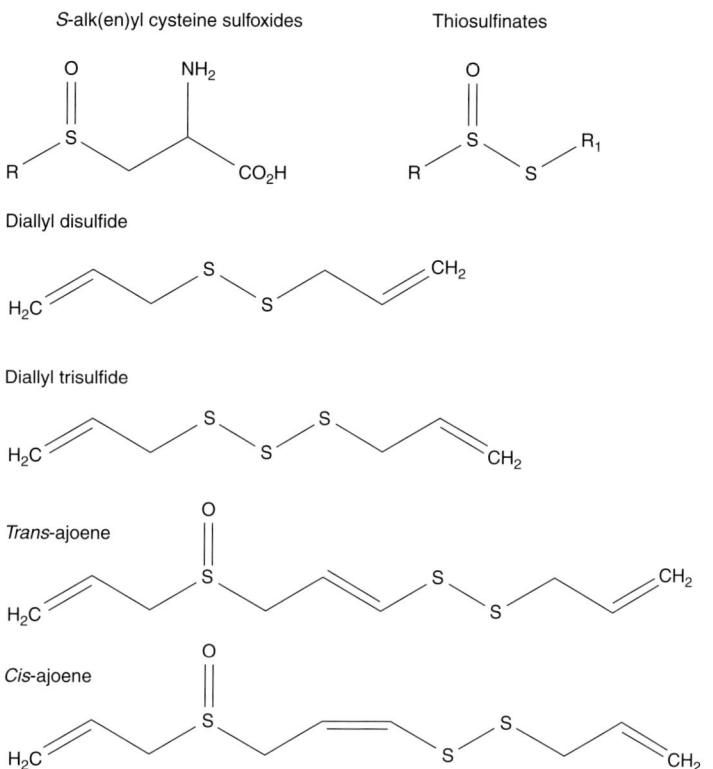

Fig. 2.4. Structures of some of the widely studied organosulfur compounds found in *Allium* species.

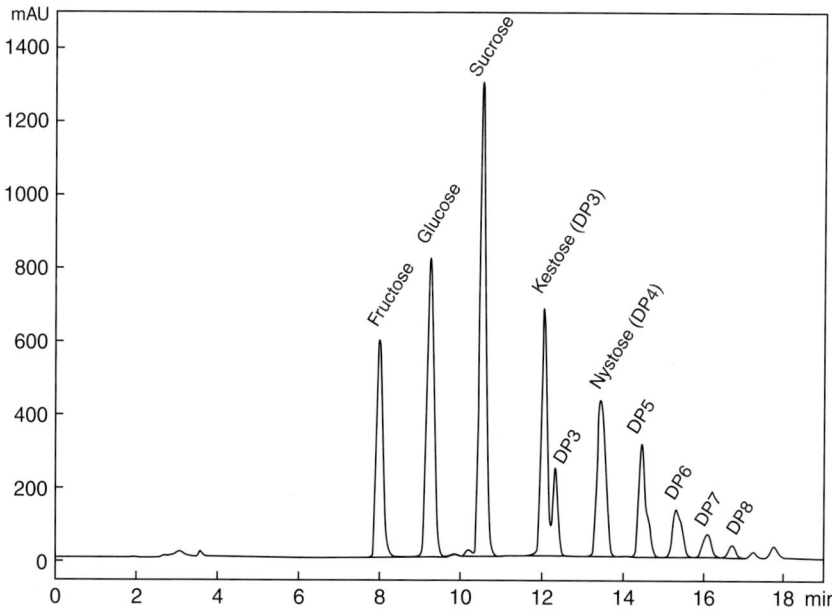

Fig. 2.5. HPLC-evaporative light scattering detector (ELSD) chromatographic profile of non-structural carbohydrates from the fleshy bulb scales of onion cv. Wellington.

leek and shallot; however, there are characteristic patterns for each species, with onion and shallot containing fructo-oligosaccharides with a degree of polymerization (DP) of up to *c*.18, garlic containing high concentrations of high DP fructans, and leek containing high DP fructans as well as smaller fructo-oligosaccharides (Ernst *et al.*, 1998).

2.2.6 Saponins

Allium vegetables are a source of steroid saponins (Carotenuto *et al.*, 1999; Lanzotti, 2006). Steroid saponins have haemolytic, antiparasitic and antifungal activities, and a bitter taste. Steroidal saponins can be divided into two groups: the spirostanol glycosides and the furostanol glycosides. Sapogenins are the aglycones of the saponins. β-Chlorogenin is a characteristic steroid sapogenin of garlic and is bioavailable *in vivo*. Plant saponins are thought to prevent the absorption of cholesterol in the intestine (Amagase, 2006).

2.3 Chemopreventive Activity and Bioavailability

Many of the studies conducted on the potential health benefits of *Allium* vegetables are made on known doses of a particular purified chemical or concentrated extract. Therefore, the benefits of dietary intake of the fresh or cooked product depend on the bioavailability of the bioactive constituents, and whether the required concentration for clinical effectiveness can be reached through consumption alone rather than with dietary supplements (Powolny and Singh, 2008). Early studies suggest that organosulfur compounds are readily available, but further investigation is required. It has been suggested that chewing garlic, as is traditional practice for folk medicine, could increase the bioavailability of the active ingredients (Borrelli and Izzo, 2008). The bioavailability of the allyl thiosulfinates found in garlic can be assessed by measuring the amount of allyl methyl sulfide on the breath following consumption of garlic preparations (Lawson and Gardner, 2005).

2.3.1 Onion

Many studies have been conducted since the turn of the century to help identify the mechanisms by which onion intake can help sustain or improve human health (Table 2.1). These investigations involved mainly human studies which took into account the individual's lifestyle choices, such as diet, body mass index (BMI), education, smoking and alcohol intake, to reduce bias. These human studies, plus those using animal models and cell culture, rarely take into account onion genotype, which accounts for huge variability in concentrations of sulfur compounds and antioxidant capacity. Of the authors listed in Table 2.1, most specified a region in which the onions were sourced but did not state the cultivar. Apart from cultivar, onion biochemistry varies with postharvest treatment, i.e. fresh, cured or stored, and these details are also rarely specified.

Cancer studies

Phenolics have been found to contribute to the antioxidant properties of onions and have been correlated positively with radical scavenging activity and antioxidant activity contributing to anticarcinogenic actions (Nuutila *et al.*, 2003). Alkyl sulfides and diallyl disulfides have also been suggested to protect against cancer by the metabolism of carcinogenic compounds (Griffiths *et al.*, 2002). A comprehensive review on the mode of action of these organosulfur compounds in cancer chemoprevention/chemotherapy found the mechanisms were based on interactions with cellular proteins, DNA or oxidative stressors. Cell death is caused by apoptosis or inhabitation of proliferation. *Allium* vegetables can prevent cancers such as prostate, skin, lung, etc., showing that topical application is not necessary, but systemic effects occur. Some organosulfur compounds have selective activity against cancer cells and can modulate drug resistance of cancer cells (Scherer *et al.*, 2009). Specifically, diallyl disulfide induces apoptosis in human colon cancer cells (COLO 205) by a mechanism associated with an increase in the production of reactive oxygen species (ROS) (Yang *et al.*, 2009).

Table 2.1. Health-promoting action of *Allium cepa*.

Activity	Action	System	Dose	Extract type	Reference
Anticancer	Reduced risk of colorectal, laryngeal, ovarian, oral cavity and oesophagus cancer	Human	At least seven portions of onion per week	Fresh/cooked	Galeone et al. (2006)
Anticancer	Reduced risk of stomach cancer	Human	'High' onion intake. High/low cut-off point based on the median distribution of the controls	Fresh/cooked	Setiawan et al. (2005)
Anticancer	Antiproliferation of HL-60 leukaemia cells and induction of differentiation into granulocytic lineage	HL-60 cell cultures	20 µg/ml	Oil	Seki et al. (2000)
Anticancer	Decreased CYP 2E1 activity which metabolizes low molecular weight carcinogens and nitrosamines	Rat	200 g/kg diet for 9 days	Powder	Teyssier et al. (2001)
Anticancer	The antioxidant effect of disulfides and thiols found in onion oil caused an inverse correlation between antioxidant enzymes and lipid peroxidation in nicotine-treated rats	Rat	100 mg/kg body weight	Oil	Helen et al. (2000)
Anticardiovascular disease	Reduced cholesterol in the liver of high-fat-fed rats resulting in decreased glutamic pyruvate transaminase activity, indicating improved liver function	Rat	50 g/kg diet (+ 40 g/kg lard and 10 g/kg cholesterol)	Powder	Lee et al. (2008a)
Antithrombotic	Antiplatelet, antithrombosis and thrombolytic activity. Mechanism inconclusive as no correlation was found between thrombosis and quercetin content	Mouse and rat	Thrombosis experiment: mouse oral administration 3.85 ml/kg. Platelet reactivity and dynamic coagulation: rat blood mixed with onion juice 9:1	Juice[a]	Yamada et al. (2004)
Antiplatelet	Mechanism may involve inhibition of arachidonic acid, thromboxane A_2 (TXA_2) synthase and TXA_2/PGH_2 receptor blockage decreasing TXA_2 production, which is a potent platelet aggregator	Rat	0.1–1 g/ml	Aqueous extraction reduced to powder	Moon et al. (2000)
Antidiabetic	Decrease in urea and creatinine, markers of renal function, in alloxan-induced diabetic rats. In addition, markers for hepatic dysfunction were reduced in onion-fed diabetic rats, as well as a significant decrease in blood glucose levels	Rat	1 ml onion juice/100 g body weight/day	Juice	El-Demerdash et al. (2005)

Note: [a]From onion cvs. Kitamiko27, Toyohira, Kitawase3, Tsukisappu, K83211, 2935A, Superkitamomiji, CS3-12, Tsukiko22 and Rantaro.

Cardiovascular disease

Cardiovascular diseases can be caused by many factors, including increased blood cholesterol and triglycerol levels, increased blood platelets and homocysteine levels, leading to heart disease and clotting, hypertension, diabetes and obesity (Corzo-Martínez *et al.*, 2007). An Italian case–control study consisting of 760 patients with non-fatal acute myocardial infarction (MI), and 682 controls, found that decreased risk of MI was associated with increased intake of onion but not garlic (Galeone *et al.*, 2009). Moon *et al.* (2000) studied the effect of aqueous onion extract on animal models and suggested that it inhibited the production of thromboxane A_2 (TXA$_2$), an eiconasoid platelet aggregating agent. TXA$_2$ is synthesized from membrane phospholipids, which are converted into arachidonic acid (AA) by phospholipase A_2 (PLA$_2$) and then to TXA$_2$. Onion was found to lower the concentration of intracellular Ca^{2+} ([Ca^{2+}]$_i$) and, as PLA$_2$ is [Ca^{2+}]$_i$ dependent; this might lead to the inhibition of PLA$_2$, reducing the downstream production of AA and TXA$_2$. However, a second pathway involved in the production of TXA$_2$ via cyclooxygenase (COX) was not affected by aqueous onion extract; therefore, it was suggested that onion might also block the TXA$_2$ receptor. Since the anti-aggregatory effect of flavonoids has not been replicated *in vivo*, it is generally considered that organosulfur compounds are responsible for the antiplatelet effect of onion (Griffiths *et al.*, 2002). However, although onions are well known for their positive health benefits, it has been demonstrated in rats that toxic sulfide compounds may be formed from *S*-allyl-L-cysteine sulfoxides and, after chronic ingestion, these can cause haemolytic anaemia (Munday and Manns, 1994).

Antibiotic effects

Onions contain an antifungal peptide, allicepin, which is distinct from the antimicrobial peptide (Ace-AMP1) contained in the onion seed. Allicepin is a chitinase and is active against a variety of fungi, including *Botrytis cinerea*, *Fusarium oxysporum* and *Mycosphaerella arachidicola* (Wang and Ng, 2004). Onion extracts have been found to inhibit oral bacteria, as well as many yeast species and Gram-positives (Griffiths *et al.*, 2002). Proteins, saponins, phenolics and ACSOs are thought to contribute to the antimicrobial activity. Ramos *et al.* (2006) investigated the antibacterial effects of yellow onion skin compounds and found that quercetin had a mild inhibitory effect on multi-resistant *Staphylococcus aureus* (MRSA) and *Helicobacter pylori*. However, the most potent antibacterial compound was a previously unknown quercetin-derived oxidation product, 3-(quercetin-8-yl)-2,3-epoxyflavanone, which showed high antibacterial activity against two strains of MRSA (MRSA#5 and MRSACOL) and *H. pylori* and increased activity in the presence of β-lactam (part of the structure of several antibiotic families). ACSOs are broken down enzymatically into thiosulfinates and their derivatives, which posses a –S(O)-S- group that disrupts the essential proteins of microorganisms by reacting with –SH groups (Kyung and Lee, 2001).

Metabolic diseases

Diabetes increases levels of urea and creatinine in the blood, leading to renal dysfunction. Treatment of alloxin-induced diabetic rats with onion decreased the levels of urea in the plasma by 16% compared with the untreated group (El-Demerdash *et al.*, 2005). Diabetic rats also had higher levels of liver enzymes in the bloodstream; however, these were reduced significantly in those fed with onion. It was suggested that onion was able to reduce the leakage of liver enzymes into the bloodstream by inhibiting liver damage. This was supported by the high bilirubin concentrations found in the alloxin-induced diabetic rats, thought to be due to reduced liver uptake; those fed an onion diet had significantly reduced bilirubin levels.

Other beneficial effects

Moon *et al.* (2000) found that aqueous onion extracts had no effect on COX activity; however, compounds isolated from onion, such as thiosulfoxides and cepaenes, have been shown to inhibit sheep seminal microsomal COX and porcine leukocyte 5-lipogenase activity, resulting in anti-inflammatory and antiasthmatic effects (Wagner *et al.*, 1990). Quercetin has been shown to have antiasthmatic properties in guinea pigs treated with aerosolized ovalbumin to stimulate specific airway resistance and immediate- and late-phase asthmatic response. Quercetin was found to inhibit these responses as well as leukocyte recruitment at a similar level to the potent anti-inflammatory drug, dexamethasone (Jung *et al.*, 2007).

The cell walls (dietary fibre) of onions consist mostly of cellulose and non-cellulosic polysaccharides such as pectin and xyloglucan. Sun-Waterhouse *et al.* (2008) found that dietary fibre had a protective effect on ascorbic acid *in vitro*. It was suggested that the polygalacturonic acid portion of pectin might form a complex with ascorbic acid via calcium ions or a complex with multivalent metal ions, catalysts for the oxidation of ascorbic acid.

2.3.2 Garlic

The therapeutic and medicinal values of garlic have been reviewed by Keusgen (2002). Therefore, studies published in the past 5–10 years are mainly considered here. The bioactive components depend on the method of preparation, and Tripathi (2009) reviewed the effects of various garlic preparations on extract composition. For example, the major sulfur compound present in raw garlic and garlic powder is alliin, while allicin is the major sulfur compound in crushed garlic. Solvent-extracted garlic and garlic oils contain mainly allyl and methyl sulfides (Rahman, 2007). Allicin from garlic is unstable and is metabolized rapidly by both blood and liver cells, and therefore is not present for long periods in the body. In addition, allicin is not formed in the acidic conditions of the stomach, as the enzyme alliinase is inactivated at the low pH (> pH 3), but is converted by the liver to other compounds such as diallyl disulfide (DADS) (Amagase, 2006). There is a paucity of information on the bioavailability of the bioactive compounds in garlic. Examples of various studies demonstrating the health-promoting properties of garlic are summarized in Table 2.2.

Cancer studies

Epidemiological evidence suggests that dietary consumption of *Allium* vegetables is correlated inversely with the occurrence of colorectal, laryngeal, ovarian, oral, oesophageal, prostate, stomach and renal cell cancers (Galeone *et al.*, 2006; Stan *et al.*, 2008; Chan *et al.*, 2009). A population-based case–control study conducted in Shanghai showed that men who consumed more than 10 g of *Allium* vegetables per day had a reduced risk of prostate cancer compared with those who consumed less than 2.2 g/day. Among individual *Allium* vegetables, garlic had a pronounced effect on the reduction of prostate cancer risk (Hsing *et al.*, 2002).

The anticancer effect of garlic is thought to be due partly to the following pathway – enzymes such as glutathione transferases (phase 2) inactivate the carcinogenic intermediates that are activated by cytochrome p450-dependent monooxygenases (phase 1). The sulfur compounds can both inhibit phase 1 enzymes and increase expression of phase 2 enzymes. They can also halt cell cycle progression in neoplastic cells. There are also reports of organosulfur compound-mediated apoptosis in cancer cell lines. Diallyl sulfide, diallyl disulfide and diallyl trisulfide are compounds found in garlic and have been shown to induce apoptosis in T98G and U87MG cells. These cells are human glioblastoma cells and are the most malignant type among primary brain tumours. The effect is brought about by a mechanism involving the production of ROS, which are thought to signal the activation of stress kinases and cysteine proteases (Das *et al.*, 2007). Compounds present in garlic have long been reputed to have a selective antiproliferative effect on tumour cells, mediated in a variety of ways including inhibition of metabolism, inhibition of DNA adduct formation, free-radical scavenging, antiproliferative activities and induction of apoptosis. Other modes of action include histone modification and inhibition of angiogenesis. For in-depth reviews on the anticancer effects of garlic, see Shukla and Kalra (2007) for *in vivo*, *in vitro* and epidemiological studies and Powolny and Singh (2008) for mechanisms of action.

Table 2.2. Health-promoting action of *Allium sativum*.

Activity	Action	System	Dose	Reference
Antithrombotic	Antiplatelet activity	Human blood	30 µl garlic juice/ml blood	Cavagnaro et al. (2007)
Antiheavy metal poisoning	Protect against gonadotoxic and spermiotoxic effects of Cd poisoning	Rats	0.5–1.0 ml aqueous extract/100 g body weight/day	Ola-Mudathir et al. (2008)
Antidiabetic, cardiovascular effects	Reduced serum triglycerides and reduced serum fructosamine levels	Human study, patients with type 2 diabetes mellitis	Time release garlic powder tablets, 600 mg/day for 4 weeks	Sobenin et al. (2008)
Antidiabetic	Improved glucose tolerance and decreased fasting blood glucose	Fructose-fed male albino Wistar rats	Daily injection of aqueous garlic extract for 8 weeks	Jalal et al. (2007)
Antidiabetic	Decreased serum glucose, triglycerides and total cholesterol, and increased serum insulin levels	Streptozotocin-induced diabetic male Wistar rats	Daily oral administration of 0.25 or 0.5 g ethanolic extract of Iranian garlic/kg body weight for 14 days	Eidi et al. (2006)
Anticancer	Growth inhibition of cancer cells	Human promyelocytic leukaemia cell line (HL-60)	Ethanolic extract	Nishida et al. (2008)
Anticancer	Not stated	Population based case–control study in Shanghai	*Allium* vegetables	Hsing et al. (2002)
Anticancer	Reduction in size and number of colorectal adenomas	Human subjects with colorectal adenomas	High dose 2.4 ml AGE/day; low dose 0.16 ml/day for 12 months.	Tanaka et al. (2006)
Anticancer	Reduced incidence of papilloma-bearing mice and decreased size and number of papillomas	Female 'Swiss albino mice' with DMBA-induced skin carcinoma	0.5 ml of c.10.4 mg/ml aqueous garlic paste/day orally for 8 weeks	Das and Saha (2009)
Anticardiovascular disease	Prolonged bleeding and thrombin time, enhanced anticoagulation factor activity	Sprague–Dawleys rats	5–50 mg garlic oil/kg body weight	Chan et al. (2007)
Anticardiovascular disease	Reduced blood pressure in hypertensive rats	Sprague–Dawleys rats with/without two-kidney, one-clip induced hypertension	500 mg aqueous garlic extract/kg body weight/day for 2 weeks as 0.5 ml intraperitoneal injection	Al-Qattan et al. (2006)
Anticardiovascular disease	Reduced the rise in total cholesterol and LDL cholesterol, accompanied by decrease in plasma fibrinogen in cholesterol-fed rats	Wistar rats fed on diets +/– cholesterol	25 mg lyophilized garlic/kg body weight	Jastrzebski et al. (2007)
Anticardiovascular disease	Reduced systolic blood pressure in subjects with marginal hypertension	Human study ($n = 25$)	2500 mg garlic powder/day	Mousa and Mousa (2007)

Cardiovascular disease

Extracts of garlic have been shown to have an antihypertensive effect, and also to inhibit platelet aggregation *in vitro*. However, others have shown raw garlic and extracts of garlic (garlic powder tablets and aged garlic powder) to have no effect on the plasma low-density lipoprotein concentrations in patients with moderate hypercholesterolaemia, with the study lasting 6 months (Gardner *et al.*, 2007). The authors recommend that more studies with a wider range of dosage levels and situations are carried out. Ried *et al.* (2008) conducted a systematic review and meta-analysis of the currently available data, considering only those studies that included a placebo control and garlic only supplements, and concluded that garlic supplementation reduced both systolic and diastolic blood pressure in hypertensive adults, and that a high starting blood pressure was a significant predictor of a positive effect. The benefit for patients in this group was likened to the hypotensive effects of conventional drugs.

Garlic extract has been shown to double nitric oxide (NO) production *in vitro* by human umbilical vein endothelial cells (Mousa and Mousa, 2007). Nitrous oxide is a well-known vasodilator, and the hypotensive effects of garlic have been attributed, in part, to this effect. Benavides *et al.* (2007) hypothesized that hydrogen sulfide mediated the vasodilatory effect of garlic. They showed that the garlic-derived compounds, DADS (diallyl disulfide) and DATS (diallyl trisulfide), were converted to the endogenous cardioprotective vascular cell signalling molecule hydrogen sulfide by human red blood cells *in vitro*. It is likely that a combination of these effects occurs.

Antibiotic effects

The antibacterial activity of garlic extract has been shown to decline in concert with the decline in concentration of allicin. Therefore, it has been concluded that allicin is responsible for most of the antibacterial activity of garlic, although there is a difference in the chemical and biological half-lives of the extract, indicating that other compounds or breakdown products of allicin also contribute (Fujisawa *et al.*, 2008; Portz *et al.*, 2008). Aqueous fresh garlic extract is effective against *Candida albicans* (Low *et al.*, 2008).

Metabolic diseases

Garlic has been reputed to have an antidiabetic effect and has been used traditionally in Indian medicine for treatment of this condition. Liu *et al.* (2007) have reviewed research concerning the effect of garlic in the treatment of diabetes. They conclude that there is evidence for the antidiabetic effect of garlic, but the studies have been undertaken mainly on animals with drug- or fructose-induced diabetes (Eidi *et al.*, 2006; Jalal *et al.*, 2007), with results comparable to those obtained by administration of commonly used antidiabetic drugs. A study has also shown a beneficial effect of garlic powder in human subjects with diabetes (Sobenin *et al.*, 2008), although the component of garlic responsible for the effect is yet to be identified.

Negative effects

When garlic is consumed to an excessive degree, it can cause unpleasant odours on both the breath and the skin, and has been reported to cause allergic reactions. Allicin has been identified as the major irritant in raw garlic, and the oil-soluble components are generally more toxic than the water-soluble ones. Aged garlic extract is left for up to 20 months, by which time most of the harsh, odorous, irritating compounds are metabolized into naturally stable sulfur compounds (Rahman, 2007). Direct contact of the skin with raw garlic can cause contact dermatitis (Borrelli *et al.*, 2007). High dietary intake of garlic has been implicated in interactions with drugs used to control blood pressure (e.g. warfarin) and diabetes. These interactions are thought to be due to the antiplatelet and hypoglycaemic activity of garlic and have not been proven (Borrelli *et al.*, 2007).

2.3.3 Leek and shallot

There have been few studies into the anticarcinogenic properties of leek in isolation, although population-based studies involving the category of '*Allium* vegetable intake' have found inverse links between such intake and risk of stomach, colon, oesophagus, breast and prostate cancer (Sengupta *et al.*, 2004). Leeks contain both spirostanol-type saponins and a rare cholestanol-type saponin not found in onion, garlic or shallot (Lanzotti, 2005). There is little information on the health-benefiting properties of cholestanol-type saponins, although the spirostanol saponins have been linked with cholesterol-lowering and antitumour effects (Lanzotti, 2005). Leeks contain antifungal chitinases (isoforms APC-Dr, -D and -F) that are structurally distinct from those of other *Alliums*, yet all share a high proportion of cysteine residues. The chitinase of onion showed antifungal activity, determined using bioassays against *F. oxysporum* and *B. allii*, yet the chitinase of leek did not (Wang and Ng, 2004). Many antifungal compounds found in *Alliums* are also bioactives.

Shallot has an antioxidant activity similar to that reported in fresh garlic, and this is highest when extracted using hexane as apposed to water or pressing (Leelarungrayub *et al.*, 2006). Shallot is the only *Allium* to contain the flavonoid isoliquiritigenin (ISL). ISL is a potent antioxidant and has been shown to inhibit cell proliferation in HepG2 (human liver cancer cells) and A549 (human lung cancer cells) (Kuo *et al.*, 2005). Further *in vitro* studies showed that ISL caused apoptosis in human prostate cancer cells and reduced metastatic potential. Aqueous extracts of shallot were compared with aqueous garlic extracts for hypoglycaemic properties. Shallot was found to improve intraperitoneal glucose tolerance and reduced, to a greater degree than garlic extract, the fasting insulin resistance index (FIRI) in Wistar rats fed a high-fructose diet for 8 weeks (Jalal *et al.*, 2007). Shallot bulbs contain a novel antifungal peptide, ascalin, which is similar to chitinases from other *Allium* species (>30 kDa), but much smaller at only 9.5 kDa. Ascalin not only inhibited *B. cinerea* mycelial growth, but also inhibited human immunodeficiency virus type 1 (HIV-1) reverse transcriptase activity at a very low IC_{50} of 10 μM (normal range 100–300 μM) (Wang and Ng, 2002).

2.4 Effect of Preharvest and Postharvest Continuum

Preharvest treatment and conditions in the field, including nutrition, temperature during the growing season, crop maturity at harvest and the harvesting process, can affect the quality of field vegetables. In addition, many biochemical and physical changes occur in stored fresh produce, including changes in sensory quality perception by the consumer and variations in the concentration of bioactive compounds. A better understanding of how changes in health-promoting compounds vary with preharvest factors and postharvest storage conditions and time will allow optimization of the health-promoting properties in these products.

2.4.1 Onion

Preharvest consideration must include cultivar selection, as the health-benefiting properties of onions are not uniform throughout the vast array of cultivars available. As already mentioned, red onions tend to have a higher antioxidant capacity due to the presence of anthocyanins, although this does not always hold true. Vågen and Slimestad (2008) analysed 15 cultivars of red, brown and low pungency onion varieties for flavonol content using HPLC and total antioxidants using trolox equivalent antioxidant capacity (TEAC). Six individual flavonols were measured and their total concentrations ranged from 35 to 159 mg/100 g FW, for cvs. Domenica Supersweet and Powell Brown, respectively. Total antioxidant capacity ranged from 72 to 509 μmol/100 g FW, for cvs. Colossus and Powell Brown, respectively.

The Department of Crop Science in Sweden has published a great deal of literature on the effects of preharvest factors on quercetin content. Mogren *et al.* (2007) investigated the effect of low N (72 kg/ha) compared to levels similar to those used by many commercial growers, of 80 kg/ha extra. The authors found that low N had no deleterious effects on the quercetin glycoside concentrations of onion cvs. Barito and Summit F_1 as determined immediately after harvest and after 5 months storage at 1°C. Quercetin concentrations in the fleshy scales also were not affected by the application method (harrowing or rotary cultivation), the type of N fertilizer (organic or non-organic) (Mogren *et al.*, 2008) or lifting time (Mogren *et al.*, 2007). Atmospheric temperature was found to have no clear effect on quercetin, although global radiation had a significant effect, especially during August, when irradiation levels were 4937 W/m^2 in 2005 and 6059 W/m^2 in 2002, with corresponding quercetin concentrations of 175 and 564 mg/kg FW, respectively (Mogren *et al.*, 2006).

Although onion skin is currently not consumed, there is increasing research into how it can be utilized as a food additive. Anthocyanin content of onion bulb skin was influenced by curing temperature in onion cv. Red Baron. Immediately after curing, onions cured at 28°C contained less than half the cyanidin 3-(3''-malonoylglucoside), cyanidin 3-(6''-malonoyl-laminaribioside) and peonidin 3-(malonoyl)glucoside than those cured at 20 or 24°C (Downes *et al.*, 2009). It has been suggested that anthocyanins do not remain stable throughout storage. Red onions cv. Tropea stored for 2 and 4 weeks at 5°C and 30% RH showed less cyanidin 3-(6''-malonylglucoside) degradation in the whole bulb than those stored at 25°C and 66% RH, or 30°C and 50% RH; however, after 6 weeks the cyanidin 3-(6''-malonylglucoside) content had reduced to *c*.8 mg/kg FW for all treatments (Gennaro *et al.*, 2002).

The literature on changes in the quercetin concentrations of the fleshy scales during curing is conflicting. Field curing (mean 16.7°C) of onion cvs. Barito and Summit F1 increased levels of quercetin glycosides significantly in the onion fleshy scales from 10 to 40 mg/100g FW (Mogren *et al.*, 2006). Other authors (Price *et al.*, 1997) have reported that curing onions cv. Cross Bow for 10 days at 28°C reduced flavonol concentrations in the flesh, due mainly to reductions in quercetin monoglycosides. However, Downes *et al.* (2010) found quercetin and isorhamnetin glucoside concentrations increased during curing at 28°C for 6 weeks in onions cv. Sherpa. Quercetin biosynthesis has been linked to UV-B radiation, with exposure resulting in an increase in soluble flavonoids (Mogren *et al.*, 2006). As Mogren *et al.* (2006) cured their onions in the field, they may have been exposed to more UV-B radiation, causing an increase in quercetin levels, whereas artificial curing usually takes place in the dark. Price *et al.* (1997) did not detail their specific method of curing. The role of UV-B radiation in quercetin biosynthesis was also discussed by Hirota *et al.* (1998), who found higher concentrations of quercetins in the outer and top section of onion cv. Takanishiki bulbs, where light exposure was at a maximum.

Controlled atmosphere (CA) is used to extend the storage life of onions. Onions cv. Hysam held in 2 kPa O_2 and 2 kPa CO_2 (80% RH) at 0.5°C for 9 weeks had decreased concentrations of three ACSOs: MCSO, PrenCSO and PCSO. Increasing CO_2 concentration to 8% resulted in a further decrease in ACSO concentration. The concentrations of MCSO and PrenCSO in onions cv. Hysam held in regular atmosphere storage (21 kPa O_2 and 0.1 kPa CO_2) increased, causing an overall increase in total ACSOs (Uddin and MacTavish, 2003). However, only one cultivar of onion was considered in this study. Chope *et al.* (2007) investigated the effect of CA on pyruvate concentration in three onion cvs., Renate, Ailsa Craig and SS1. The enzyme alliinase hydrolyses pyruvate from ACSOs after cell disruption and is a reliable indicator of pungency. Onion cvs. Renate and SS1 showed a 1.9- and a 1.2-fold increase in pyruvate after storage in 3.03 kPa CO_2 and 5.05 kPa O_2 at 2°C, for 230 and 81 days, respectively. Pyruvate concentration of cv. Ailsa Craig decreased 1.9-fold after 129 days CA storage. These studies suggest that consideration of cultivar choice and storage regime may provide onions with increased organosulfur compound content, and possibly greater health-benefiting properties. Higher sulfur compounds result in a more pungent onion, and with a growing market for sweet, low pungency onions, there must be a compromise between consumer taste preference and nutritional value.

2.4.2 Garlic

Garlic is stored in a number of formats and storage conditions for varying durations, and such factors are likely to affect the concentration, nature and bioactivity of endogenous health-promoting bioactives. As with many other cultivated crops, a wide range of genetic diversity exists in the form of different cultivars, which in turn results in a range of phenotypes that vary according to morphology and biochemical composition. Baghalian *et al.* (2005) studied 24 Iranian garlic ecotypes and found that the allicin content was significantly higher in some varieties than in others. They concluded that this difference was more likely to be due to genetic variation than to geographical origin.

Typically, garlic cloves can be stored at room temperature for *c.*2 months, at an intermediate temperature (15–18°C) for *c.*4 months, or in the cold (–1 to 0°C) for an extended period (Hughes *et al.*, 2006). During storage of intact garlic cv. Fukuchi-howaito cloves for 150 days, the concentrations of γ-glutamyl peptides decreased and the concentrations of the ACSOs increased, with this change being most pronounced in those cloves stored at 4°C, as compared with –3 and 23°C (Ichikawa *et al.*, 2006). This was also the case in garlic cv. Printanor bulbs (Hughes *et al.*, 2006). Similarly, selenium compounds in aqueous garlic extracts from garlic grown on naturally selenium-rich soils were stable during storage for 1 month at 0°C, but were degraded at 4°C (Auger *et al.*, 2004).

A hot water dip at 55°C for 10 min can control sprouting and rooting of garlic cloves during subsequent storage for 4 weeks at 10°C. This treatment did not affect the thiosulfinate concentration (measured by colorimetric assay) of garlic; however, thiosulfinates did increase during the storage period (Cantwell *et al.*, 2003). Gamma irradiation can also be used to control sprout growth of garlic. Garlic cloves of a Korean cultivar were irradiated with 0.1 kGy and stored at 3 ± 1°C, $80 \pm 5\%$ RH, for 10 months. In both irradiated and control cloves, total sulfur content decreased after 6 months of storage (Kwon *et al.*, 1989). If boiled for 20 min at 100°C garlic retained its bioactivity in reducing lipid

levels in rats, but did not do so after 40 or 60 min of boiling (Jastrzebski *et al.*, 2007).

There are many different garlic preparations available on the market and these have been shown to contain different concentrations of the supposed bioactive compounds, both between one another and compared with that shown on the label (Arnault *et al.*, 2005).

The mode of preparation of the extract affects the amount and relative composition of organosulfur compounds. For example, the content of allicin decreases with extraction temperature and time, and an increased proportion of monosulfur compounds is observed with exposure to high temperatures (100–130°C) over a period of 1–3 h (Woo *et al.*, 2007). Antioxidant activity of garlic extract increased on heating at 110–130°C for 2–3 h (Woo *et al.*, 2007). The antimicrobial activity of garlic juice is diminished on heating and storage (Al-Waili *et al.*, 2007).

Garlic is often dried to be taken as a supplement. The method of drying affects the allicin content of the final product. Hot air drying at moderate temperatures results in an allicin content similar to that of fresh garlic, but allicin concentration is decreased at 60°C (Ratti *et al.*, 2007). The best method of garlic drying, in terms of allicin concentration, is to freeze dry whole cloves at 20°C. Ajoene is formed when garlic is heated (e.g. frying and sautéing) and the concentration of ajoene in garlic oil depends on the garlic's country of origin. The concentration of ajoene in fresh garlic oil decreased by *c.*5-fold during 6 months storage at –20°C (Naznin *et al.*, 2008).

2.4.3 Leek and shallot

The effects of preharvest and postharvest conditions on the health-benefiting properties of shallot are limited. As shallots are considered to be conspecific with bulb onions, changes in health-benefiting properties in response to growing and storage conditions may be similar.

Sørensen *et al.* (1995) investigated the effects of preharvest factors, namely, nitrogen and water supply and harvest maturity, on leek nutrition. High nitrogen supply (280 kg/ha) increased leek nitrate concentrations to 307 mg/kg, compared with 5 mg/kg in those grown at low nitrogen supply (100 kg/ha). Leeks grown under water stress conditions (deficit of 29% below plant available water) had higher concentrations of vitamin C, protein, magnesium and manganese in both the stem and trimmed leaves. Additionally, when harvested later in October or November rather than September, vitamin C content was higher, although this was significant for only one out of two growing seasons. Results are conflicting concerning whether nitrates are beneficial or detrimental to human health. It has been postulated that, when swallowed, salivary nitrate is converted into HNO_2, which may nitrosate secondary amines to produce nitrosamines, some of which have been demonstrated as carcinogenic in animal studies (McKnight *et al.*, 1999). However, more recently it has been suggested that a reduction in endogenous nitric oxide can cause cardiovascular disease such as atherosclerosis, hypertension and ischaemic heart disease, but that nitric oxide concentrations can be replaced by nitrate ingestion (Lundburg *et al.*, 2006). Other suggested benefits of nitrate include a non-immune antimicrobial effect against ingested pathogens in the gastrointestinal (GI) tract, plus regulation of recirculation, platelet activity and GI mobility (McKnight *et al.*, 1999). High nitrogen supply may therefore improve the health-benefiting properties of leeks by increasing nitrate concentrations, as well as improving on yield. There is limited literature on the effect of CA storage on the health-benefiting properties of leek, especially in recent years. Kurki (1979) stored leeks in air at 0°C (100% RH) or optimum CA conditions (1% O_2 and 10% CO_2) at 0°C (100% RH) and found that vitamin A content remained higher in leeks stored in CA conditions. A recent study compared the effects of heat treatment (55°C, 17.5 min) with those of the removal of 2 cm from the base on the nutritional quality of leeks tray-packaged, wrapped with 16 µm stretch film, then stored at 10°C for 7 days. Although both treatments controlled leaf growth, total thiosulfinates decreased from 0.857 to 0.305 µmol/g FW, ascorbic acid content reduced from 51.7 to 24.2 µg/g FW, total soluble phenols from 0.369 to 0.247 mg/g FW and antioxidant capacity from 49.5 to 32.9 µg/g AEAC (ascorbic acid equivalents antioxidant capacity) FW (Tsouvaltzis *et al.*, 2007).

2.5 Conclusions

The future for research in the health-promoting properties of fruit and vegetables in general, and *Allium* vegetables in particular, should include detailed analysis of the concentrations and bioavailability of the suspected bioactive components and how this is affected by both pre- and postharvest factors, including cooking and processing. There should be a combination of using pure compounds in studies, and intervention studies where the whole product is consumed. *Allium* vegetables are rarely eaten in isolation, but form the basis of many meals; therefore, the interactions with other foodstuff should be taken into consideration. It is yet to be proven that the concentrations of individual compounds required to produce the desirable effects *in vivo* can feasibly be reproduced by dietary intake or supplementation.

References

Al-Qattan, K.K., Thomson, M., Al-Mutawa'a, S., Al-Hajeri, D., Drobiova, H. and Ali, M. (2006) Nitric oxide mediates the blood-pressure lowering effect of garlic in the rat two-kidney, one-clip model of hypertension. *Journal of Nutrition* 136, 774S–776S.

Al-Waili, N.S., Saloom, K.Y., Akmal, M., Al-Waili, T.N., Al-Waili, A.N., Al-Waili, H., Ali, A. and Al-Sahlani, K. (2007) Effects of heating, storage, and ultraviolet exposure on antimicrobial activity of garlic juice. *Journal of Medicinal Food* 10, 202–212.

Amagase, H. (2006) Significance of garlic and its constituents in cancer and cardiovascular disease. *Journal of Nutrition* 136, 716S–723S.

Arnault, I. and Auger, J. (2006) Seleno-compounds in garlic and onion. *Journal of Chromatography A* 1112, 23–30.

Arnault, I., Haffner, T., Siess, M.H., Vollmar, A., Kahane, R. and Auger, J. (2005) Analytical method for appreciation of garlic therapeutic potential and for validation of a new formulation. *Journal of Pharmaceutical and Biochemical Analysis* 37, 963–970.

Auger, J., Yang, W., Arnault, I., Pannier, F. and Potin-Gautier, M. (2004) High-performance liquid chromatographic-inductively coupled plasma mass spectrometric evidence for Se-'alliins' in garlic and onion grown in Se-rich soil. *Journal of Chromatography A* 1032, 103–107.

Baghalian, K., Ziai, S.A., Naghavi, M.R., Badi, H.N. and Khalighi, A. (2005) Evaluation of allicin content and botanical traits in Iranian garlic (*Allium sativum* L.) ecotypes. *Scientia Horticulturae* 103, 155–166.

Benavides, G.A., Squadrito, G.L., Mills, R.W., Patel, H.D., Isbell, T.S., Patel, R.P., Darley-Usmar, V.M., Doeller, J.E. and Kraus, D.W. (2007) Hydrogen sulphide mediates the vasoactivity of garlic. *Proceedings of the National Academy of Sciences of the United States of America* 104, 17977–17982.

Benkeblia, N. and Varoquaux, P. (2003) Effect of nitrous oxide (N_2O) on respiration rate, soluble sugars and quality attributes of onion bulbs *Allium cepa* cv. Rouge Amposta during storage. *Postharvest Biology and Technology* 30, 161–168.

Bielecka, M., Biedrzycka, E. and Majkowska, A. (2002) Selection of probiotics and prebiotics for synbiotics and confirmation of their *in vivo* effectiveness. *Food Research International* 35, 125–131.

Block, E. (2005) Biological activity of *Allium* compounds: recent results. *Acta Horticulturae* 688, 41–58.

Borrelli, F. and Izzo, A.A. (2008) Could chewing garlic reveal clinical efficacy. *International Journal of Cardiology* 130, 117.

Borrelli, F., Capasso, R. and Izzo, A.A. (2007) Garlic (*Allium sativum* L.): adverse effects and drug interactions in humans. *Molecular Nutrition and Food Research* 51, 1386–1397.

Breu, W. (1996) *Allium cepa* L. (Onion). Part 1: Chemistry and analysis. *Phytomedicine* 3, 293–306.

Brewster, J.L. (1994) *Onions and Other Vegetable Alliums*, 1st edn. CAB International, Wallingford, UK.

Brewster, J.L. (2008) *Onions and Other Vegetable Alliums*, 2nd edn. CAB International, Wallingford, UK.

Briggs, W.H. and Goldman, I.L. (2002) Variation in economically and ecologically important traits in onion plant organs during reproductive development. *Plant Cell and Environment* 25, 1031–1037.

Cantwell, M.I., Kang, J. and Hong, G. (2003) Heat treatments control sprouting and rooting of garlic cloves. *Postharvest Biology and Technology* 30, 57–65.

Carotenuto, A., Fattorusso, E., Lanzotti, V. and Magno, S. (1999) Spirostanol saponins of *Allium porrum* L. *Phytochemistry* 51, 1077–1082.

Cavagnaro, P.F., Camargo, A., Galmarini, C.R. and Simon, P.W. (2007) Effect of cooking on garlic (*Allium sativum* L.) antiplatelet activity and thiosulfinates content. *Journal of Agricultural and Food Chemistry* 55, 1280–1288.

Chan, K.C., Yin, M.C. and Chao, W.J. (2007) Effect of diallyl trisulfide-rich garlic oil on blood coagulation and plasma activity of anticoagulation factors in rats. *Food and Chemical Toxicology* 45, 502–507.

Chan, R., Lok, K. and Woo, J. (2009) Prostate cancer and vegetable consumption. *Molecular Nutrition and Food Research* 53, 201–216.

Chope, G.A., Terry, L.A. and White, P.J. (2007) The effect of the transition between controlled atmosphere and regular atmosphere storage on bulbs of onion cultivars SS1, Carlos and Renate. *Postharvest Biology Technology* 44, 228–239.

Corzo-Martínez, M., Corzo, N. and Villamiel, M. (2007) Biological properties of onions and garlic. *Trends in Food Science and Technology* 18, 609–625.

Das, A., Banik, N.L. and Ray, S.K. (2007) Garlic compounds generate reactive oxygen species leading to activation of stress kinases and cysteine proteases for apoptosis in human glioblastoma T98G and U87MG cells. *Cancer* 110, 1083–1095.

Das, I. and Saha, T. (2009) Effect of garlic on lipid peroxidation and antioxidation enzymes in DBMA-induced skin carcinoma. *Nutrition* 25, 459–471.

Downes, K., Chope, G.A. and Terry, L.A. (2009) Effect of curing temperature on the biochemical composition of skin from three onion (*Allium cepa* L.) cultivars. *Postharvest Biology and Technology* 54, 80–86.

Downes, K., Chope, G.A. and Terry, L.A. (2010) Effects of postharvest application of ethylene and 1-methylcyclopropene either before or after curing on onion (*Allium cepa* L.) bulb quality during long-term cold storage. *Postharvest Biology and Technology* 55, 36–44.

Eidi, A., Eidi, M. and Esmaeili, E. (2006) Antidiabetic effect of garlic (*Allium sativum* L.) in normal and streptozotocin-induced diabetic rats. *Phytomedicine* 13, 624–629.

El-Bayoumy, K., Sinha, R., Pinto, J.T. and Rivlin, R.S. (2006) Cancer chemoprevention by garlic and garlic-containing sulphur and selenium compounds. *American Society for Nutrition* 136, 864S–869S.

El-Demerdash, F.M., Yousef, M.I. and Abou El-Naga, N.I. (2005) Biochemical study on the hypoglycaemic effects of onion and garlic in alloxan-induced diabetic rats. *Food and Chemical Toxicology* 43, 57–63.

Ernst, M.K., Chatterton, N.J., Harrison, P.A. and Matitschka, G. (1998) Characterization of fructan oligomers from species of the genus *Allium* L. *Journal of Plant Physiology* 153, 53–60.

FAOSTAT (2008) Food and Agriculture Statistics Division (http://faostat.fao.org/site/567/Desktop-Default.aspx?PageID=567#ancor, accessed 10 September 2008).

Fattorusso, E., Lanzotti, V., Taglialatela-Scafati, O. and Cicala, C. (2001) The flavonoids of leek, *Allium porrum*. *Phytochemistry* 57, 565–569.

Fritsch, R.M. and Keusgen, M. (2006) Occurrence and taxonomic significance of cysteine sulphoxides in the genus *Allium* L. (*Alliaceae*). *Phytochemistry* 67, 1127–1135.

Fujisawa, H., Suma, K., Origuchi, K., Kumagai, H., Seki, T. and Ariga, T. (2008) Biological and chemical stability of garlic-derived allicin. *Journal of Agricultural and Food Chemistry* 56, 4229–4235.

Galeone, C., Pelucchi, C., Levi, F., Negri, E., Franceschi, S., Talamini, R., Giacosa, A. and La Vecchia, C. (2006) Onion and garlic use and human cancer. *American Journal of Clinical Nutrition* 84, 1027–1032.

Galeone, C., Tavani, A., Pelucchi, C., Negri, E. and La Vecchia, C. (2009) *Allium* vegetable intake and risk of acute myocardial infarction in Italy. *European Journal of Nutrition* 48, 120–123.

Gardner, C.D., Lawson, L.D., Block, E., Chatterjee, L.M., Kiazand, A., Balise, R.R. and Kraemer, H.C. (2007) Effect of raw garlic vs commercial garlic supplements on plasma lipid concentrations in adults with moderate hypercholesterolemia. *Archives on Internal Medicine* 167, 346–353.

Gennaro, L., Leonardi, C., Esposito, F., Salucci, M., Maiani, G., Quaglia, G. and Fogliano, V. (2002) Flavonoid and carbohydrate contents in Tropea red onions: effects of homelike peeling and storage. *Journal of Agricultural and Food Chemistry* 50, 1904–1910.

Gorinstein, S., Leontowicz, H., Leontowicz, M., Namiesnik, J., Najman, K., Drzewiecki, J., Cvikrová, M., Martincová, O., Katrich, E. and Trakhtenberg, S. (2008) Comparison of the main bioactive compounds and antioxidant activities in garlic and white and red onions after treatment protocols. *Journal of Agricultural and Food Chemistry* 56, 4418–4426.

Griffiths, G., Trueman, L., Crowther, T., Thomas, B. and Smith, B. (2002) Onions – a global benefit to health. *Phytotherapy Research* 16, 603–615.

Helen, A., Krishnakumar, K., Vijayammal, P.L. and Augusti, T. (2000) Antioxidant effect of onion oil (*Allium cepa*. Linn) on the damages induced by nicotine in rats as compared to alpha-tocopherol. *Toxicology Letters* 116, 61–68.

Hirota, S., Shimoda, T. and Takahama, U. (1998) Tissue and spatial distribution of flavonoid and peroxidase in onion bulbs and stability of flavonoid glucosides during boiling of the scales. *Journal of Agricultural and Food Chemistry* 46, 3497–3502.

Hsing, A.W., Chokkalingam, A.P., Gao, Y.-T., Madigan, M.P., Deng, J., Gridley, G. and Fraumeni, J.F. (2002) *Allium* vegetables and risk of prostate cancer: a population-based study. *Journal of the National Cancer Institute* 94, 1648–1651.

Hughes, J., Collin, H.A., Tregova, A., Tomsett, A.B., Cosstick, R. and Jones, M.G. (2006) Effect of low temperature storage on some of the flavour precursors in garlic (*Allium sativum*). *Plant Foods for Human Nutrition* 61, 81–85.

Hunter, R., Kaschula, C.H., Parker, I.M., Caira, M.R., Richards, P., Travis, S., Taute, F. and Qwebani, T. (2008) Substituted ajoenes as novel anti-cancer agents. *Bioorganic and Medicinal Chemistry Letters* 18, 5277–5279.

Ichikawa, M., Nagatoshi, I. and Ono, K. (2006) Changes in organosulfur compounds in garlic cloves during storage. *Journal of Agricultural and Food Chemistry* 54, 4849–4854.

Imai, S., Tsuge, N., Tomotake, M., Nagatome, Y., Sawada, H., Nagata, T. and Kumagai, H. (2002) Plant biochemistry: an onion enzyme that makes the eyes water. *Nature* 419, 685.

Jalal, R., Bagheri, S.M., Moghimi, A. and Rasuli, M.B. (2007) Hypoglycemic effect of aqueous shallot and garlic extracts in rats with fructose-induced insulin resistance. *Journal of Clinical Biochemistry and Nutrition* 41, 218–223.

Jastrzebski, Z., Leontowicz, H., Leontowicz, M., Namiesnik, J., Zachwieja, Z., Barton, H., Pawelzik, E., Arancibia-Avila, P., Toledo, F. and Gorinstein, S. (2007) The bioactivity of processed garlic (*Allium sativum* L.) as shown by *in vitro* and *in vivo* studies on rats. *Food and Chemical Toxicology* 45, 1626–1633.

Jones, M.G., Hughes, J., Tregova, A., Milne, J., Tomsett, A.B. and Collin, H.A. (2004) Biosynthesis of the flavour precursors of onion and garlic. *Journal of Experimental Botany* 55, 1903–1918.

Jung, C.H., Lee, J.Y., Cho, C.H. and Kim, C.J. (2007) Anti-asthmatic action of quercetin and rutin in conscious guinea-pigs challenged with aerosolized ovalbumin. *Archives of Pharmacal Research* 30, 1599–1607.

Keusgen, M. (2002) Health and Alliums. In: Rabinowitch, H.D. and Currah, L. (eds) *Allium Crop Science: Recent Advances.* CAB International, Wallingford, UK, pp. 357–373.

Kuo, P.-L., Hsu, Y.-L. and Lin, C.-C. (2005) The chemopreventive effects of natural products against human cancer cells. *International Journal of Applied Science and Engineering* 3, 203–214.

Kurki, L. (1979) Leek quality changes in CA-storage. *Acta Horticulturae* 93, 85–90.

Kwon, J.-H., Choi, J.-U. and Yoon, H.-S. (1989) Sulfur-containing components of gamma-irradiated garlic bulbs. *Radiation Physics and Chemistry* 34, 969–972.

Kyung, K.H. and Lee, Y.C. (2001) Antimicrobial activities of sulfur compounds derived from *S*-alk(en)yl-L-cysteine sulfoxides in *Allium* and *Brassica*. *Food Reviews International* 17, 183–198.

Lanzotti, V. (2005) Bioactive saponins from *Allium* and *Aster* plants. *Phytochemistry Reviews* 4, 95–110.

Lanzotti, V. (2006) The analysis of onion and garlic. *Journal of Chromatography A* 112, 3–22.

Lawson, L.D. and Gardner, C.D. (2005) Composition, stability, and bioavailability of garlic products used in a clinical trial. *Journal of Agricultural and Food Chemistry* 53, 6254–6261.

Lee, K.-H., Kim, Y., Park, E. and Hwang, H.-J. (2008a) Effect of onion powder supplementation on lipid metabolism in high fat-cholesterol fed SD rats. *Journal of Food Science and Nutrition* 13, 71–76.

Lee, S.U., Lee, J.H., Choi, S.H., Lee, J.S., Ohnisi-Kameyama, M., Kozukue, N., Levin, C.E. and Friedman, M. (2008b) Flavonoid content in fresh, home-processed, and light-exposed onions and in dehydrated commercial onion products. *Journal of Agricultural and Food Chemistry* 56, 8541–8548.

Leelarungrayub, N., Rattanapanone, V., Chanarat, N. and Gebicki, J.M. (2006) Quantitative evaluation of the antioxidant properties of garlic and shallot preparations. *Nutrition* 22, 266–274.

Leighton, T., Ginther, C., Fluss, L., Harter, W.K., Cansado, J. and Notorio, V. (1992) Molecular characterization of quercetin and quercetin glycosides in *Allium* vegetables. In: Huang, M.T., Ho, C.T. and Lee, C.Y. (eds) *Phenolic Compounds in Food and Their Effects on Health II*. American Chemical Society, Washington, DC, pp. 220–238.

Lin, J.-Y. and Tang, C.-Y. (2007) Determination of total phenolic and flavonoid contents in selected fruits and vegetables, as well as their stimulatory effects on mouse splenocyte proliferation. *Food Chemistry* 101, 140–147.

Link, K.P. and Walker, J.C. (1933) The isolation of catechol from pigmented onion scales and its significance in relation to disease resistance in onions. *Journal of Biological Chemistry* 100, 379–383.

Liu, C.-Z., Sheen, L.-Y. and Lii, C.-K. (2007) Does garlic have a role as an antidiabetic agent? *Molecular Nutrition and Food Research* 51, 1353–1364.

Low, C.F., Chong, P.P., Yong, P.V.C., Lim, C.S.Y., Ahmad, Z. and Othman, F. (2008) Inhibition of hyphae formation and SIR2 expression in *Candida albicans* treated with fresh *Allium sativum* (garlic) extract. *Journal of Applied Microbiology* 105, 2169–2177.

Lundburg, J.O., Feelisch, M., Björne, H., Jansson, E.Å. and Weitzberg, E. (2006) Cardioprotective effects of vegetables: is nitrate the answer? *Nitric Oxide* 15, 359–362.

McKnight, G.M., Duncan, C.W., Leifert, C. and Golden, M.H. (1999) Dietary nitrate in man: friend or foe? *British Journal of Nutrition* 81, 349–358.

Marinova, D., Ribarova, F. and Atanassova, M. (2005) Total phenolics and total flavonoids in Bulgarian fruits and vegetables. *Journal of the University of Chemical Technology and Metallurgy* 40, 255–260.

Mogren, L.M., Olsson, M.E. and Gertsson, U.E. (2006) Quercetin content in field-cured onion (*Allium cepa* L.): effects of cultivar, lifting time, and nitrogen fertilizer level. *Journal of Agricultural and Food Chemistry* 54, 6185–6191.

Mogren, L.M., Olsson, M.E. and Gertsson, U.E. (2007) Quercetin content in stored onions (*Allium cepa* L.): effects of storage conditions, cultivar, lifting time and nitrogen fertiliser level. *Journal of the Science of Food and Agriculture* 87, 1595–1602.

Mogren, L.M., Caspersen, S., Olsson, M.E. and Gertsson, U.E. (2008) Organically fertilized onions (*Allium cepa* L.): effects of the fertilizer placement method on quercetin content and soil nitrogen dynamics. *Journal of Agricultural and Food Chemistry* 56, 361–367.

Moon, C.H., Jung, Y.S., Kim, M.H., Lee, S.H., Baik, E.J. and Park, S.W. (2000) Mechanism for antiplatelet effect of onion: AA release inhibition, thromboxane A_2 synthase inhibition and TXA_2/PGH_2 receptor blockade. *Prostaglandins, Leukotrienes and Essential Fatty Acids* 62, 277–283.

Mousa, A.S. and Mousa, S.A. (2007) Cellular effects of garlic supplements and antioxidant vitamins in lowering marginally high blood pressure in humans: pilot study. *Nutrition Research* 27, 119–123.

Muir, J.G., Shepherd, S.J., Rosella, O., Rose, R., Barett, J.S. and Gibson, P.R. (2007) Fructan and free fructose content of some common Australian vegetables and fruit. *Journal of Agricultural and Food Chemistry* 55, 6619–6627.

Munday, R. and Manns, E. (1994) Comparative toxicity of prop(en)yl disulfides derived from Alliaceae: possible involvement of 1-propenyl disulfides in onion-induced haemolytic anemia. *Journal of Agricultural and Food Chemistry* 42, 959–962.

Naznin, M.T., Akagawa, M., Okukawa, K., Maeda, T. and Morita, N. (2008) Characterization of *E*- and *Z*-ajoene obtained from different varieties of garlics. *Food Chemistry* 106, 1113–1119.

Nishida, M., Hada, T., Kuramochi, K., Yoshida, H., Yonezawa, Y., Kuriyama, I., Sugawara, F., Yoshida, H. and Mizushina, Y. (2008) Diallyl sulfides: selective inhibitors of family X DNA polymerases from garlic (*Allium sativum* L.). *Food Chemistry* 108, 551–560.

Nuutila, A.M., Puupponen-Pimiä, R. and Aarni, M. (2003) Comparison of antioxidant activities of onion and garlic extracts by inhibition of lipid peroxidation and radical scavenging activity. *Food Chemistry* 81, 485–493.

Ola-Mudathir, K.F., Suru, S.M., Fafunso, M.A., Obioha, U.E. and Faremi, T.F. (2008) Protective roles of onion and garlic extracts on cadmium-induced changes in sperm characteristics and testicular oxidative damage in rats. *Food and Chemical Toxicology* 46, 3604–3611.

Portz, D., Koch, E. and Slusarenko, A.J. (2008) Effects of garlic (*Allium sativum*) juice containing allicin on *Phytophthora infestans* and downy mildew of cucumber caused by *Pseudoperonospora cubensis*. *European Journal of Plant Pathology* 122, 197–206.

Powolny, A.A. and Singh, S.V. (2008) Multitargeted prevention and therapy of cancer by diallyl trisulfide and related *Allium* vegetable-derived organosulfur compounds. *Cancer Letters* 269, 205–314.

Prakash, D., Singh, B.N. and Upadhyay, G. (2007) Antioxidant and free radical scavenging activates of phenols from onion (*Allium cepa* L.) *Food Chemistry* 102, 1389–1393.

Price, K.R., Bacon, J.R. and Rhodes, M.J.C. (1997) Effect of storage and domestic processing on the content and composition of flavonol glucosides in onion (*Allium cepa* L.). *Journal of Agricultural and Food Chemistry* 45, 938–942.

Rahman, M.S. (2007) Allicin and other functional active components in garlic: health benefits and bioavailability. *International Journal of Food Properties* 10, 245–268.

Ramos, F.A., Takaishi, Y., Shirotori, M., Kawaguchi, Y., Tsuchiya, K., Shibata, H., Higuti, T., Tadokoro, T. and Takeuchi, M. (2006) Antibacterial and antioxidant activities of quercetin oxidation products from yellow onion (*Allium cepa*) skin. *Journal of Agricultural and Food Chemistry* 54, 3551–3557.

Ratti, C., Araya-Farias, M., Mendez-Lagunas, L. and Makhlouf, J. (2007) Drying of garlic (*Allium sativum*) and its effect on allicin retention. *Drying Technology* 25, 349–356.

Ried, K., Frank, O.R., Stocks, N.P., Fakler, P. and Suulivan, T. (2008) Effect of garlic on blood pressure: a systematic review and meta-analysis. *BMC Cardiovascular Disorders* 8, 13.

Roberford, M.B. (2007) Inulin-type fructans: functional food ingredients. *Journal of Nutrition* 137, 2493S–2502S.

Rodríguez Galdón, B., Rodríguez Rodríguez, E.M. and Díaz Romero, C. (2008) Flavonoids in onion cultivars (*Allium cepa* L.). *Journal of Food Science* 73, 599–605.

Roldán, E., Sánchez-Moreno, C., de Ancos, B. and Cano, M.P. (2008) Characterization of onion (*Allium cepa* L.) by-products as food ingredients with antioxidant and antibrowning properties. *Food Chemistry* 108, 907–916.

Salama, A.M., Hicks, J.R. and Nock, J.F. (1990) Sugar and organic acid changes in stored onion bulbs treated with maleic hydrazide. *Hortscience* 25, 1625–1628.

Scalbert, A. and Williamson, G. (2000) Dietary intake and bioavailability of polyphenols. *Journal of Nutrition* 130, 2073S–2085S.

Scherer, C., Jacob, C., Dicato, M. and Diederich, M. (2009) Potential role of organic sulfur compounds from *Allium* species in cancer prevention and therapy. *Phytochemistry Reviews* 8, 349–368.

Schmidtlein, H. and Herrmann, K. (1975) Quantitative analysis from phenolic acids by thin-layer chromatography. *Journal of Chromatography* 115, 123–128.

Seki, T., Tsuji, K., Hayato, Y., Moritomo, T. and Ariga, T. (2000) Garlic and onion oils inhibit proliferation and induce differentiation of HL-60 cells. *Cancer Letters* 160, 29–35.

Sengupta, A., Ghosh, S. and Bhattacharjee, S. (2004) *Allium* vegetables in cancer prevention: an overview. *Asian Pacific Journal of Cancer Prevention* 5, 237–245.

Setiawan, V.W., Yu, G.-P., Lu, Q.-Y., Lu, M.-L., Yu, S.-Z., Zhang, J.G., Kurtz, R.C., Cai, L., Hsieh, C.-C. and Zhang, Z.-F. (2005) *Allium* vegetables and stomach cancer risk in China. *Asian Pacific Journal of Cancer Prevention* 6, 387–395.

Shukla, Y. and Kalra, N. (2007) Cancer chemoprevention with garlic and its constituents. *Cancer Letters* 247, 167–181.

Slimestad, R., Fossen, T. and Vågen, I.M. (2007) Onions: a source of unique dietary flavonoids. *Journal of Agricultural Food Science* 55, 10067–10080.

Sobenin, I.A., Nedosugova, L.V., Filatova, L.V., Balabolkin, M.I., Gorchakova, T.V. and Orekov, A.N. (2008) Metabolic effects of time-released garlic powder tablets in type 2 diabetes mellitus: the results of double-blinded placebo-controlled study. *Acta Diabetologica* 45, 1–6.

Sørensen, J.N., Johansen, A.S., Kaack, K. (1995) Marketable and nutritional quality of leeks as affected by water and nitrogen supply and plant age at harvest. *Science of Food and Agriculture* 68, 367–373.

Stan, S.D., Kar, S., Stoner, G.D. and Singhy, S.V. (2008) Bioactive food components and cancer risk reduction. *Journal of Cellular Risk Reduction* 104, 339–356.

Sun-Waterhouse, D., Smith, B.G., O'Connor, C.J. and Melton, L.D. (2008) Effect of raw and cooked onion dietary fibre on the antioxidant activity of ascorbic acid and quercetin. *Food Chemistry* 111, 580–585.

Tanaka, S., Haruma, K., Yoshihara, M., Kajiyama, G., Kira, K., Amagase, H. and Chayama, K. (2006) Aged garlic extract has potential suppressive effect on colorectal adenomas in humans. *Journal of Nutrition* 136, 821S–826S.

Teyssier, C., Amiot, M.-J., Mondy, N., Auger, J., Kahane, R. and Siess, M.-H. (2001) Effect of onion consumption by rats on hepatic drug-metabolizing enzymes. *Food and Chemical Toxicology* 39, 981–987.

Tripathi, K. (2009) A review – garlic, the spice of life – (Part 1). *Asian Journal of Research* Chemistry 2, 8–13.

Tsouvaltzis, P., Gerasopoulos, D. and Siomos, A.S. (2007) Effects of base removal and heat treatment on visual and nutritional quality of minimally processed leeks. *Postharvest Biology and Technology* 43, 158–164.

Uddin, M.D. and MacTavish, H.S. (2003) Controlled atmosphere and regular storage-induced changes in S-alk(en)yl-L-cysteine sulphoxides and alliinase activity in onion bulbs (*Allium cepa* L. *cv.* Hysam). *Postharvest Biology and Technology* 28, 239–245.

Vågen, I.M. and Slimestad, R. (2008) Amount of characteristic compounds in 15 cultivars of onion (*Allium cepa* L.) in controlled field trials. *Journal of the Science of Food and Agriculture* 88, 404–411.

Velioglu, S., Mazza, G., Gao, L. and Oomah, B.D. (1998) Antioxidant activity and total phenolics in selected fruits, vegetables, and grain products. *Journal of Agricultural and Food Chemistry* 46, 4113–4117.

Wagner, H., Dorsch, W., Bayer, Th., Breu, W. and Willer, F. (1990) Antiasthmatic effects of onions: inhibition of 5-lipoxygenase and cyclooxygenase *in vitro* by thiosulfinates and 'cepaenes'. *Prostaglandins Leukotrienes and Essential Fatty Acids* 39, 59–62.

Walker, J.C., Link, K.P. and Angell, H.R. (1929) Chemical aspects of disease resistance in the onion. *Proceedings of the National Academy of Sciences of the United States of America* 15, 845–850.

Wang, H.X. and Ng, T.B. (2002) Ascalin, a new anti-fungal peptide with human immunodeficiency virus type 1 reverse transcriptase-inhibiting activity from shallot bulbs. *Peptides* 23, 1025–1029.

Wang, H.X. and Ng, T.B. (2004) Isolation of Allicepin, a novel antifungal peptide from onion (*Allium cepa*) bulbs. *Journal of Peptide Science* 10, 173–177.

Wiczkowski, W., Nemeth, K., Bucinski, A. and Piskula, M.K. (2003) Bioavailability of quercetin from flesh scales and dry skin of onion in rats. *Polish Journal of Food and Nutrition Sciences* 12, 95–99.

Woo, K.S., Yoon, H.-S., Lee, Y.R., Lee, J., Kij, D.J., Hong, J.T. and Jeong, H.S. (2007) Characteristics and antioxidative activity of volatile compounds in heated garlic (*Allium sativum*). *Food Science and Biotechnology* 16, 822–827.

Wróbel, K., Wróbel, K., Kannamkumarath, S.S., Caruso, J.A., Wysocka, I.A., Bulska, E., Świątek, J. and Wierzbicka, M. (2004) HPLC-ICP-MS speciation of selenium in enriched onion leaves – a potential dietary source of Se-methylslenocysteine. *Food Chemistry* 86, 617–623.

Yamada, K., Naemura, A., Sawashita, N., Noguchi, Y. and Yamamoto, J. (2004) An onion variety has natural antithrombotic effect as assessed by thrombosis/thrombolysis models in rodents. *Thrombosis Research* 114, 213–220.

Yang, J.-S., Chen, G.-W., Hsia, T.-C., Ho, H.-C., Ho, C.-C., Lin, M.-W., Lin, S.-S., Yeh, R.-D., Ip, S.-W., Lu, H.-Fi and Chung, J.-G. (2009) Diallyl disulfide induces apoptosis in human colon cancer cell line (COLO 205) through the induction of reactive oxygen species, endoplasmic reticulum stress, caspases cascade and mitochondrial-dependent pathways. *Food and Chemical Toxicology* 47, 171–179.

3 Avocado

Marjolaine D. Meyer, Sandra Landahl,
Manuela Donetti and Leon A. Terry

3.1 Introduction

Avocado fruit (*Persea americana* Mill.) originates from Meso-America, where it has been
consumed for no less than 9000 years (Smith,
1966). Avocados are now cultivated in numerous regions at tropical and subtropical latitudes, including Mexico, Chile, Indonesia, the
Dominican Republic, Brazil, Peru, the USA,
Israel, South Africa, Australia, New Zealand
and Spain, among others (Table 3.1). Avocado
flesh is eaten raw, either alone or incorporated into salads or guacamole. The oil
extracted from the pulp of the fruit is also
used as a food ingredient or cooking oil, but
has also been identified for its potential use in
cosmetics and skin-care products (Athar and
Nasir, 2005). Although relatively new in international commerce terms compared with
many other fruit, avocado has gained popularity worldwide and volumes traded inter-
nationally have increased significantly in the
past decades (Table 3.1), with the main
importers being Europe and North America.
Avocado ranks as the second most commonly
consumed raw fruit in the USA (Electronic
Code of the Federal Regulations, 2006).

The fruit is appreciated not only for its
unique taste and flavour – it is also highly
nutritious and constitutes a good supply of
minerals, vitamins and fibre (Pursglove, 1968;
Bergh, 1992; Table 3.2). A detailed publication

by Slater *et al.* (1975) indicated that one half of
an avocado cv. Fuerte (about 80 g edible fruit)
supplied a substantial percentage of the daily
nutritional needs in magnesium (13%), iron
(11%), vitamin B_6 (pyridoxine; 15%), vitamin
B_3 (niacin; 8%), vitamin B_9 (folacin; 16%) and
vitamins A (12%), C (12%) and E (19%) of a
child aged 7–10 years.

Avocado fruit is a naturally rich dietary
source of health-beneficial bioactive substances, with reported medicinal effects
toward many diseases, including prevention
against cardiovascular risk and potential anticancer activity (Ding *et al.*, 2007) (Table 3.3).
These compounds include monounsaturated
and polyunsaturated lipids, carotenoids, vitamins B, C and E, terpenoids, D-mannohepru-
lose, β-sitosterols, persenone A and B and
phenols. Studies have reported on the antioxidant activity (Leong and Shui, 2002; Soong
and Barlow, 2004; Bertling *et al.*, 2007), radical
suppressing (Kim, O.K. *et al.*, 1998, 2000;
Vinson *et al.*, 2001), acetylCoA carboxylase
inhibitory (Hashimura *et al.*, 2001), antifungal
(Prusky *et al.*, 1991; Domergue *et al.*, 2000)
and chemopreventive (Lu *et al.*, 2005; Ding
et al., 2007, 2009) activities of the bioactive
compounds present in the avocado fruit and
its extracts. Additionally, avocado has been
shown to assist in the uptake of nutrients
from other foodstuffs (Unlu *et al.*, 2005).
However, the high caloric value of avocado

Table 3.1. Values in t/year for main avocado producers and importers (from FAOSTAT, 2007).

Country	Production (t/year)	Country	Import (t/year)
Mexico	1,142,892	USA	348,858
Chile	250,000	France	110,632
Indonesia	201,635	Netherlands	63,211
Colombia	193,996	UK	44,526
Dominican Republic	183,468	Japan	26,511
USA	175,177	Canada	23,252
Spain	120,000		
Israel	85,913		
South Africa	65,203		

associated with its high fat content may dissuade consumers. Nevertheless, contrary to popular assumptions, including avocado in the diet could be part of a successful reduced energy intake diet (Walker and O'Dea, 2001) and scientific evidence has shown that eating avocado does not compromise good weight control (Pieterse *et al.*, 2003).

This chapter reviews the identity, abundance and effect of different bioactive compounds found in avocado fruit and presents the latest scientific evidence on the health-promoting properties of avocado fruit and its extracts.

3.2 Origin of Avocado (*Persea americana* Mill.)

The commercial avocado tree belongs to the large tropical family of *Lauraceae* and to the genus *Persea*. Other known members of the genus exist but have not been recognized as being commercially important. The crop originated in a large geographic area extending from the eastern and central highlands of Mexico, through Guatemala and up to the Pacific coast of Central America (Popenoe, 1920; Smith, 1966; Storey *et al.*, 1986). The fruit was appreciated by both Mayan and Aztec civilizations, which were believed to have semi-domesticated the crop and selected for larger fruit size with improved eating quality (Smith, 1966; Storey *et al.*, 1986). Three distinct, ecologically separate races are generally recognized – Mexican, Guatemalan and West Indian, or Lowland – based on morphological differences and their respective climatic adaptations around 90 years ago (Popenoe, 1920). The Mexican race is adapted to elevated and cool habitats with a 6–8 months winter–spring dry period (Wolstenholme and Whiley, 1999). The Guatemalan race is native to tropical highlands, with year-round cool conditions, although it can also be found in warmer subtropical areas. The West Indian race, in contrast, is adapted more to a hot and humid tropical, lowland climate with a short dry season. The flesh has a lower oil percentage than the other two types and a different flavour. It is possible that fruit properties such as ripening, quality and abscission may be related to the climatic conditions of the respective areas of origin (Praloran, 1970). The three races are compatible and their hybrids represent the varieties that dominate the international market (Scora *et al.*, 2002). The cultivar Hass is predominantly Guatemalan with some Mexican germplasm, and is the cultivar grown most widely today. This cultivar is particularly appreciated for its postharvest qualities since it shows better storability, essential for long transit times, than other cultivars. The skin is thick and confers protection from pests and disease. Unlike many other green-skin cultivars, the ripening process is accompanied by a distinct skin colour change from green to purplish-black. The cv. Fuerte, another economically important cultivar, is a Guatemalan × Mexican hybrid showing good resistance to cold conditions, while other commercially important varieties include cvs. Ryan, Lula, Booth8,

Table 3.2. Avocado nutrient composition.

Nutrients	Amounts per 100 g fresh weight
Water (g)	73.3
Energy (kcal)	160
Energy (kJ)	670
Protein (g)	2.00
Total lipid (fat) (g)	14.66
Total carbohydrate (g)	8.53
Total sugars (g)	0.66
Sucrose	0.06
Glucose (dextrose)	0.37
Fructose	0.12
Total dietary fibre (g)	6.7
Ash (g)	1.58
Calcium, Ca (mg)	12
Magnesium, Mg (mg)	29
Phosphorus, P (mg)	152
Potassium, K (mg)	485
Vitamin C, total ascorbic acid (mg)	10.0
Vitamin A (IU)	146
α-Carotene (μg)	24
β-Carotene (μg)	62
β-Cryptoxanthin (μg)	28
Vitamin E (α-tocopherol) (mg)	2.07
Lutein + zeaxanthin (μg)	271
Vitamin K, phylloquinone (μg)	21.0
Fatty acids, total saturated (g)	2.126
16:0	2.075
18.0	0.049
Fatty acids, total monounsaturated (g)	9.799
16:1 (g)	0.698
18:1 (g)	9.066
Fatty acids, total polyunsaturated (g)	1.816
18:2 (g)	1.674
18:3 (g)	0.125
Stigmasterol (mg)	2
Campesterol (mg)	5
β-Sitosterol (mg)	76

Source: USDA National Nutrient Database (2010).

Walden, Pollock, Pinkerton, Bacon, Lamb Hass and Zuton.

From a botanical point of view, the avocado fruit is classified as a berry (Storey *et al.*, 1973/1974) comprising a single seed and a pericarp. The pericarp is further divided into exocarp (skin), mesocarp (flesh, edible portion) and the thin layer around the seed coat, the endocarp (Fig. 3.1). Fruit characteristics

including size and skin texture vary considerably among the races (Knight, 1980).

3.3 Identity and Role of Bioactives

3.3.1 Phenolics and polyphenolic compounds

Phenolic compounds are plant secondary metabolites including a variety of compounds such as phenolic acids, flavonoids, stilbenes, coumarins and tannins. Avocado is not valued as good a source of phenolics as some other fruit (e.g. soft fruit), probably explaining why investigation into the polyphenolic profile of avocado has remained scarce. Most information on phenolic content has been carried out principally to understand better the economic significance and mechanism of the browning reaction when fruit are cut or where internal disorders occur (Golan *et al.*, 1977; Kahn, 1983). Phenolic acids generally are found in the cell vacuole or in special tissues, and are precursors of many other phytochemicals. The concentration of these compounds and the activity of the enzyme polyphenol oxidase (PPO) have long been known to be related to the process of browning. When membrane integrity is lost, the phenols are released and oxidized to quinines by PPO (Torres *et al.*, 1987). PPO induces the conversion of polyphenolic compounds in *o*-quinones, which in turn form melanins, responsible for the brown coloration (Kahn, 1983; Bower and Cutting, 1988). This process is influenced by different factors, such as concentration of H_2O_2 in cells, the structure of phenolic compounds and the rapidity of the main reaction (Amiot *et al.*, 1997).

The phenolic content in avocado fruit varies according to cultivar, tissue type and ripening stage (Golan *et al.*, 1977; Torres *et al.*, 1987). For instance, Golan *et al.* (1977) reported that the mesocarp of cv. Fuerte contained significantly more phenolics than the mesocarp of the Lerman variety (0.29 versus 0.03 mg/g fresh weight (FW) of chlorogenic acid equivalent, respectively). The same authors found a difference in phenolic content between the proximal and distal end of the mesocarp of

Table 3.3. Health-promoting action of avocado (*Persea americana* Mill.).

Activity	Action	System	Dose	Extract type	Reference
Cholesterol					
Lower cholesterol levels in human coronary system	Unsaturated FAs lower LDL (but also tend to lower protective HDL). FA composition showed seasonality	Review	67% monounsaturated FA found in the study presented, which compares favourably with data for olive oil	Blanched, homogenized, dissolved in methylene chloride–methanol. Organic base-catalysed technique to obtain FAMEs of cv. Hass avocado flesh from two different sites (cool, warm)	Kaiser and Wolstenholme (1994)
Lower cholesterol levels in human coronary system	Fall of *c*.12% total cholesterol and LDL in human plasma. Monounsaturated (MU) most important dietary FA. MUFA diet decreases oxidative stress of LDL	Human	Diets enriched with olive oil, avocado, almonds (total cholesterol consumption was const. around 310 mg/day)	Blood samples (enzymatic, phosphotungstic acid, GLC analysis on erythrocyte-FA)	Berry *et al.* (1995)
Lower cholesterol levels in human coronary system	Dietary phytosterol (β-sitosterol) anticholesterolaemic	Analyses of pooled batches of avocado flesh at different times (human)	Average of 76.4 mg/100 g β-sitosterol (ingestion)	Not specified (raw cv. Hass avocado fruit)	Duester (2001)
Cancer prevention					
Inhibition of tumour growth	Inhibition of glucose uptake by the tumour cell: 'energy starving'	Human tumour in experimental animals	Suggested: 1.7 mg/g daily for 5 days (in a drink)	Mannoheptulose purified from avocados	Board *et al.* (1995)
Inhibition of tumour growth	Inhibition of the proliferation of human prostate cancer cell lines. Cell cycle arrest resulting from downregulation of p27	Prostate cancer cell lines *in vitro*	Androgen-independent cancer proliferation was 60% inhibited by 300 µg/ml dose. Androgen-dependent cancer: 100 µg/ml dose	Acetone extract of California cv. Hass avocado	Lu *et al.* (2005)
Inhibition of tumour growth	Targeting of multiple signalling pathways and increasing intracellular reactive oxygen leading to apoptosis	Human oral cell lines	GI$_{50}$ for malignant cell line 14 µg/ml	Chloroform extract from California cv. Hass avocado	Ding *et al.* (2007)

Inhibition of tumour growth	Inhibitory effect on ^{3}H-leucine incorporation for protein synthesis in Ehrlich ascites tumour cells	Mice	Perseitol K$^+$ complex (molar ratio 20:1)	Methanol extract of *Scurrula fusca* (Bl.) leaves	Ishizu et al. (2002)
Nutraceutical					
Nutraceutical	Increase essential fatty acids in human breast milk	Human	Partially breast-fed (mother eats avocado c.twice a week)	Milk	Rocquelin et al. (1998)
Excipient in cosmetics and medicine	Cosmetic: lipids act as emollient, protector, regenerator. Oleic acid is an excipient, improves biovailability of poorly water-soluble drugs and is raw material for ointments. Linoleic and linolenic acid activate metabolic processes in the skin, promote vitamins A and E and help recovery of stratum corneum. Pharmaceutical: FAs prevent cardiovascular disease by reducing total and LDL cholesterol levels. Linoleic has antiarrhythmic effect on heart, lowers cholesterol, prevents clots in arteries	Human	Ingestion or application on skin	Avocado oil	Rabasco Alvarez and González Rodríguez (2000)
Antioxidant	Lowering incidence of human degenerative disease	Review	143 mg/100 g L-ascorbic acid equivalent antioxidant capacity (high)	50% aq. ethanol extract of homogenized Thai avocado flesh. ABTS free-radical decolourization assay, DPPH radical-scavenging assay, RP-HPLC	Leong and Shui (2002)
Antioxidant	Highest antioxidant activity and phenol content in avocado leaf compared to 47 other medical plants. Treatment of respiratory infection	Correlation to traditional medicine	Equivalent ascorbid acid: 0.157 g/g, equivalent gallic acid: 0.061 g/g	Aqueous extract from Portuguese avocado leaves. Antioxidant: modified ABTS free-radical decolourization assay. Phenol content: Folin–Ciocalteu reagent	Giao et al. (2007)

(Continued)

Table 3.3. *Continued*

Activity	Action	System	Dose	Extract type	Reference
Treatment of osteoathritis	Superior to placebo with osteoarthritis of hip and knee	Human	Ingestion of components	Unsaponifiable fraction of avocado oil after hydrolysis (1:2 avocado:soybean extract)	Curatolo and Bogduk (2001)
Treatment of osteoathritis	Symptom relief for osteoarthritis patients	Human	300 mg once per day	Unsaponifiable fraction of avocado oil after hydrolysis (1:2 avocado:soybean extract)	Ameye and Chee (2006)
Allergy NEG: latex-associated avocado allergy	Endochitinase class 1 containing hevein domain	Human serum	Serum samples were investigated for hevein-specific IgE antibodies: >0.35 kU/l	Ingestion of raw or stewed avocados, or skin-prick test	Posch *et al.* (1999)

Note: FA = fatty acids; LDL = low-density lipoproteins; HDL = high-density lipoproteins; FAMEs = fatty acid methyl esters; ABTS = 2,2′-azino-*bis*-(3-ethylbenzthiazoline-6-sulfonic acid); DPPH = 2,2-diphenyl-1-picrylhydrazyl; RP-HPLC = reversed phase HPLC; NEG = negative effect.

Fig. 3.1. Structure of avocado cv. Hass fruit.

the same fruit. A total phenolic content of 0.24 mg/g FW of gallic acid equivalent (GAE) was reported in the mesocarp of avocado fruit, but the cultivar was not specified (Luximon-Ramma *et al.*, 2003). Haiyan *et al.* (2007), using the Folin–Ciocalteu method, measured a total phenol content of 0.50 mg/g FW (as caffeic acid) from avocado flesh, while total phenol content of the seed was sixfold greater than in the corresponding flesh.

A wide range of phenolic compounds has been measured in avocado. Previous studies have found mainly caffeic and *p*-coumaric acids in cv. Fuerte (Ramirez-Martinez and Luh, 1973; Golan *et al.*, 1977) and in an unnamed variety (Prabha and Patwardhan, 1980). Ramirez-Martinez and Luh (1973) also identified, in the pulp of cv. Fuerte, epicatechin, isoflavone, chlorogenic acid, leucoanthocyanidine, *p*-coumarylquinic acid and, in much lower amounts, catechin, while Golan *et al.* (1977) reported leucoanthocyanidine and catechin. These findings were supported by Torres *et al.* (1987), who identified, by means of HPLC, benzoic acid derivatives (*p*-coumaric, ferulic, *p*-hydroxi, protocatechic, vanillic and synergic acids) and cinnamic derivatives (caffeic and sinapic acids) as being the main compounds. More recent investigation identified epicatechins (0.06–0.08 μg/g FW) in the edible part of avocado (Arts *et al.*, 2000; de Pascual-Teresa *et al.*, 2000).

The avocado seed has been found to have a phenolic content almost 80-fold higher than that in the edible portion of the fruit (Soong and Barlow, 2004). The seed is rich in polyphenolic compounds such as (+)-catechin and (–)-epicatechin, but also highly polymeric substances (Geissman and Dittmar, 1965). The peel of avocado fruit contains the flavonoids (+)-catechin and (–)-epicatechin (Terasawa *et al.*, 2006) and anthocyanins (in purple-coloured ripe fruit). One anthocyanin in particular, cyanidin 3-*O*-glucoside, has been shown to be responsible for the increase in total anthocyanin concentration in the skin of avocado cv. Hass during ripening, resulting in the purpling development (Cox *et al.*, 2004; Ashton *et al.*, 2006). Levels of cyanidin 3-*O*-galactoside, another anthocyanin detected at harvest (Prabha *et al.*, 1980; Cox *et al.*, 2004), remained stable as skin colour changed.

Availability of phenolic compounds is influenced by their chemical structure and stability. Generally, phenolic acids are absorbed as aglycon at the gastrointestinal and stomach level (Lafay *et al.*, 2006). The metabolic absorption of some of the main phenolics found in avocado fruit has been studied in detail. For instance, work on chlorogenic acid showed that the incorporation of this compound in the stomach could be operated by bilitranslocase or a system of unidentified anion transporters (Lafay *et al.*, 2006). Caffeic acid, which has been reported to have antioxidant activity, is absorbed in the small intestine and a reduction in its absorption is detected when esterificated in chlorogenic acid (Olthof *et al.*, 2001). Ferulic acid has been suggested to have beneficial effects on cardiovascular disease, diabetes and Alzheimer's disease (Zhao and Moghadasian, 2008), and is usually present in a bound form with carbohydrates or acids and seems to be absorbed in the gut (Adam *et al.*, 2002; Zhao and Moghadasian, 2008). *p*-Coumaric acid may have protective action by acting as a free-radical scavenger and antioxidant in the eye tissues of rabbit, reducing the effect of UV-B radiation (Lodovici *et al.*, 2008). Some classes of polyphenols can be present in their acetylated form, such as epicatechins, or glycosilated, such as anthocyanins. The presence of glycosilation seems to limit passive absorption in the small intestine (Scalbert and Williamson, 2000).

3.3.2 Carotenoids

Five main carotenoids are present in avocado fruit: lutein, β-cryptoxanthin, zeaxanthin, α-carotene and β-carotene. Lutein, in particular, is thought to be beneficial by reducing the risk of age-related macular degeneration (Koh *et al.*, 2004; Richer *et al.*, 2004) and can account for up to 70% of total carotenoids in avocado cv. Hass fruit (Lu *et al.*, 2005). The concentration of carotenoids reported in the literature for avocado varies according to the cultivar studied: concentrations on a FW basis of 2.93 µg/g lutein, 0.11 µg/g zeaxanthin, 0.25 µg/g β-cryptoxanthin and α-carotene, and 0.60 µg/g β-carotene in ripe flesh of Californian cv. Hass harvested from January to May were found (Lu *et al.*, 2009), while a Finnish study by Heinonen *et al.* (1989) reported 3.20 µg/g FW conjugated lutein and zeaxanthin, 0.38 µg/g FW β-cryptoxanthin, 0.19 µg/g FW α-carotene and 0.34 µg/g FW β-carotene in an unidentified cultivar. In cv. Nabal avocados, total carotenoids in the flesh ranged from 10 to 14 µg/g FW (Gross *et al.*, 1973). More recently, other compounds have been identified by HPLC in Californian-grown avocado cv. Hass, namely all-*trans*-neoxanthin, violaxanthin, neochrome and chrysanthemaxanthin (Lu *et al.*, 2009).

Carotenoid content in the fruit is influenced by factors such as maturity stage and time of harvest (Lu *et al.*, 2005, 2009). For example, the content of lutein in Californian avocado cv. Hass varied from 2.32 to 3.36 µg/g FW in samples harvested in two different seasons, and from 2.67 to 3.62 µg/g in samples from the same harvest (Lu *et al.*, 2009). In a more recent study, Lu *et al.* (2009) found a significant increase in carotenoid concentrations as fruit were harvested later in the year. This increase was correlated positively with that of total fat content in the fruit. Avocados additionally may vary in their carotenoid content depending on tissue type and ripening stage: the concentration of total carotenoids in the peel is greater than in the fleshy mesocarp, with lutein the most abundant carotenoid in both tissues (Gross *et al.*, 1973; Ashton *et al.*, 2006). The concentration of total carotenoids was found to be greatest in the dark-green flesh just under the skin after harvest and during 10 days ripening of avocado cv. Hass grown in New Zealand (Ashton *et al.*, 2006), these results being confirmed later by Lu *et al.* (2009). This nutrient-rich outer section of the avocado should therefore not be discarded in order for the consumer to benefit from health properties of carotenoids. As fruit ripen, all carotenoids (lutein as well as minor carotenoid compounds, neoxanthin, violaxanthin, α-carotene) present in the peel tend to decline after harvest, while the carotenoid composition in the pulp does not change considerably during the ripening process of cv. Hass fruit (Ashton *et al.*, 2006).

Some carotenoids can be assimilated better when ingested in combination with lipids, which stimulate the production of bile acids and increase their bioavailability (Roodenburg *et al.*, 2000). In that sense, absorption of the carotenoids present in salads and other vegetables can be enhanced when these are consumed in combination with avocado (Unlu *et al.*, 2005).

3.3.3 Phytosterols

Phytosterols (or plant sterols) have a chemical structure similar to that of cholesterol, but a different side-chain configuration. Prominent phytosterols include β-sitosterol, campesterol and stigmasterol, which have anticholesteraemic properties (Moghadasian and Frohlich, 1999). Dietary intake of phytosterols has been associated with a reduction in serum cholesterol levels, contributing to the prevention of cardiovascular diseases, as well as potential protection in the development of several cancers (Awad and Fink, 2000; Piironen et al., 2000; Kritchevsky and Chen, 2005).

Analysis of avocado undertaken in the past decade has provided new information showing that this fruit represents a significant source of dietary phytosterols, with 0.75 mg/g FW compared with 0.20 and 0.23 mg/g FW in grape and orange, respectively (Piironen et al., 2003). Similarly to many other fruit, the major sterol is β-sitosterol, being present at 0.76 mg/g in raw edible fruit (Duester, 2001). The avocado provides more β-sitosterol than other commonly eaten fruit such as strawberries, grapes or banana and up to fourfold the amount present in orange (Duester, 2001). The next most abundant phytosterol is campesterol (0.05 mg/g FW) (Duester, 2001; Piironen et al., 2003) and, in much lesser amounts, stigmasterol (less than 0.03 mg/g FW) (Duester, 2001).

3.3.4 Fatty acids

Avocado is an important oleaginous fruit with a lipid content that can reach over 20% FW (Mazliak, 1970; Biale and Young, 1971; Lewis, 1978), depending on the cultivar. Generally, the oil content is lower in cultivars from the West Indian horticultural race (from 2.5 to 8%; Hatton et al., 1964) than in fruit from the Guatemalan (10–13%) and Mexican (15–22%) races (Knight, 2002). The lipid fraction is predominantly monounsaturated, with oleic acid (C18:1) consistently representing the most abundant fatty acid (around 60% of total fatty acids). Other reported fatty acids are, in order of abundance, the saturated palmitic acid (C16:0), representing 20% of total fatty acids, and the unsaturated linoleic (C18:2), palmitoleic (C16:1) and linolenic (C18:3) acids, constituting c.12%, 8% and 1% or less of total fatty acids, respectively. Trace amounts of stearic, myristic and arachidic acids can be found in the pulp of avocado fruit (Ahmed and Barmore, 1980; Ozdemir and Topuz, 2004; Vekiari et al., 2004).

Avocado ranks among the most important natural dietary sources of food-derived monounsaturates and essential fatty acids, with amounts of unsaturated fatty acids up to fivefold those of saturated fatty acids (Slater et al., 1975; Vekiari et al., 2004). For comparison, in olive oil, which is recognized widely for its health benefits, oleic acid represents 83% of total fatty acids, linoleic 7%, palmitic 6% and stearic 4% (Brown, 1975). High dietary intake of monounsaturates has been associated with potential cardiovascular benefits including effects on serum lipids (Alvizouri-Muñoz et al., 1992; Colquhoun et al., 1992; Carranza et al., 1995; Ledesma et al., 1996; Carranza-Madrigal et al., 1997). It has been shown that the dietary intake of avocado oil increases the percentage of serum high-density lipoprotein (HDL) cholesterol when compared with corn or coconut oils. The reported beneficial effects of avocado oil on atherogenicity are comparable to those of corn and olive oil (Kritchevsky et al., 2003).

Although the fatty acid composition remains generally consistent, with the predominance of oleic acid followed by palmitic and linoleic acids, the concentration of each fatty acid varies with cultivar. For instance, Ozdemir and Topuz (2004) found differences in the fatty acid profile of avocado cvs. Fuerte and Hass grown in Turkey. Vekiari et al. (2004) also reported different fatty acid contents in avocado cvs. Ettinger, Hass and Fuerte. However, the ratios of unsaturated and polyunsaturated to saturated fatty acids were not statistically different for all cultivars studied. Luza et al. (1990), when investigating the fatty acid composition of avocado cvs. Fuerte, Negra La Cruz and Ampolleta Grande grown in Chile, found that cv. Ampolleta had the greatest value of palmitic and palmitoleic acids, while Fuerte had the lowest, and Negra La Cruz had the lowest amount of polyunsaturated fatty acids.

Fatty acid composition also varies with harvest time. For instance, Ozdemir and Topuz (2004) reported an increase in oleic acid and a decrease in palmitic, palmitoleic, stearic, linolenic and arachidic acids between November and January. In agreement with these data, Vekiari *et al.* (2004) also found an increase in oleic acid and a decline in palmitic, palmitoleic, but also linoleic acid content of avocado cvs. Ettinger, Fuerte and Hass sampled over the commercial harvest season. Lu *et al.* (2009), on the other hand, reported no changes in the fatty acid composition of Californian avocado cv. Hass sampled between January and September. So, total amounts of mono- and polyunsaturated fatty acids in the fruit increase with later harvest in the year, as does lipid content.

A recent detailed study showed that the fatty acid profile of avocado cv. Hass differed with different growing regions, hence with different agricultural practices and agroenvironmental conditions, and varied in different parts of the fruit (Landahl *et al.*, 2009). The authors investigated the spatial distribution of fatty acids in avocado cv. Hass originating from three different growing regions (namely, Peru, Chile and Spain) in three different seasons. Palmitic acid and palmitoleic acid contents were lowest in the basal part of Spanish and Chilean fruit, while in Peruvian fruit, palmitic was lowest in the apical region of the fruit. Abundance of oleic acid decreased toward the basal end in Peruvian avocado, whereas it tended to be distributed equally across the vertical axis of fruit mesocarp in Spanish and Chilean fruit. Linolenic and linoleic acid concentrations were found to be highest in the middle and basal regions of fruit, independent of origin. These findings are of nutritional importance since not all parts of the fruit provide the same amount of health-related unsaturated fatty acids.

3.3.5 Vitamins

Vitamins present in avocado fruit are the liposoluble vitamin A (present in the form of its precursor, β-carotene), vitamin E (α-tocopherol) and the water-soluble vitamin C (ascorbic acid). The amount of vitamins varies with cultivar. For instance, cv. Fuerte has been reported to have 484 IU (for 100 g) of vitamin A, 2.4 IU of vitamin E and 0.058 mg/g of vitamin C, while cv. Hass contained 740 IU, 1.6 IU and 0.1 mg/g of the respective vitamins (Slater *et al.*, 1975). Smith *et al.* (1983) found a concentration of vitamin A higher in avocado fruit than in other commonly consumed fruit such as peach, apple, banana and grape. The free form of vitamin A can be toxic for the human cell if in excess, but the β-carotene form present in avocado prevents this eventuality (Bergh, 1992). The presence of γ- and α-tocopherols has been recorded at concentrations of 3.34 and 28.71 μg/g FW, respectively, in Californian cv. Hass fruit (Lu *et al.*, 2005). Tocopherol content may vary according to cultivars and from season to season, while in cv. Hass fruit the concentration of α-tocopherols tends to increase with fruit maturity (Lu *et al.*, 2009). The opposite was observed for cv. Fuerte (Slater *et al.*, 1975). The season with the highest level of α-tocopherol coincided with that of highest total carotenoid content (Lu *et al.*, 2009).

Ascorbic acid (vitamin C) is one of the most important water-soluble vitamins mainly supplemented through fruit and vegetables (Naidu, 2003). Although avocado fruit is not recognized as a good source of vitamin C, as in some others fruit (e.g. blackcurrant, see Chapter 14 of this volume), the concentration of vitamin C in the flesh has been detected between 0.058 and 0.1 mg/g, depending on the cultivar (Slater *et al.*, 1975). Recently, Bertling *et al.* (2007) found ascorbic acid, as measured using the colour reaction with 2,4-dinitrophenylhydrazine (DNPH), to be the main antioxidant present in the seed (accounting for half of the seed total antioxidant activity) and the rind tissue, while ascorbic acid content was lowest in mesocarp tissues. Other important vitamins present in avocado are riboflavine (vitamin B$_2$), thiamine (vitamin B$_1$) and folacin (Slater *et al.*, 1975).

3.3.6 Protein and ashes

Avocado pulp is a better source of protein than many other fruit, although not as protein-rich as meat, milk and some pulses. Higher amounts of free amino acids than in other fruit have been measured in avocado, the major ones being asparagine, aspartic acid, glutamine and glutamic acid. Amino acids found in minor quantities in the flesh of Fuerte were serine threonine, alanine, valine and cystine. Avocado was found to contain relatively high amounts of ash (1.0–1.4% recorded for cvs. Fuerte, Hass and Anaheim), which was relatively rich in iron that was physiologically available to experimental rats, and thus was considered to be of potential value in preventing or curing anaemia (Ahmed and Barmore, 1980).

3.3.7 Seven-carbon sugars

The peculiarity of avocado fruit, besides its high lipid content and ripening physiology, lies in its non-structural carbohydrates composition: the avocado fruit contains large amounts of the uncommon seven-carbon (C7) sugar, D-mannoheptulose, and its reduced form polyol, perseitol (Fig. 3.2). These soluble sugars have been measured in various part of the avocado, such as the leaves, shoots, trunk, roots and fruit mesocarp (Liu et al., 1999a), in equal or greater amounts to that of starch. The concentration of the C7 sugars is influenced by seasonality, with reduced concentrations of mannoheptulose present in fruit harvested later in the season (Liu et al., 1999b; Meyer and Terry, 2010). The studies available differ in the reported C7 sugar concentrations, especially for mannoheptulose, as the variation between harvest dates can be very high. Research in California has reported c.30 mg/g DM of mannoheptulose and perseitol in the mesocarp of mature unripe avocado cv. Hass harvested midseason (Liu et al.,

1999b), and such results have also been found by others in the flesh of early season cv. Hass grown in Spain (Meyer and Terry, 2008, 2010). However, in another Californian study, Liu et al. (2002) recorded 10-fold less mannoheptulose and Meyer and Terry (2010) detected almost no mannoheptulose in late season fruit. As fruit ripen, amounts of C7 sugars decline substantially (Liu et al., 1999b; Meyer and Terry, 2008, 2010; Landahl et al., 2009).

Most studies have used cv. Hass as the material of investigation and have not specified the part of the mesocarp used when quantifying C7 sugars in the flesh. Hence, the available studies may differ greatly in the tissue region (basal, apical or combined regions). Bertling and Bower (2005) found spatial disparities in the distribution of these compounds among different cultivars. Specifically, mannoheptulose in cv. Hass was found to have the highest concentration in the mesocarp tissue, while in cvs. Pinkerton and Fuerte, it appeared to have the highest levels in the rind tissue. In a more recent study, Landahl et al. (2009) examined the spatial distribution of non-structural carbohydrates in avocado fruit (cv. Hass) and found a trend that perseitol concentrations were lower in the middle region. Mannoheptulose concentrations in the fruit tissue from stem end to base were highly heterogeneous, but there was a trend toward greater concentration in the apical region, near the stem.

The mechanism(s) for biosynthesis and metabolism of heptose sugars, as well as their function in avocado fruit remain, to date, largely unknown. It has been suggested that the decline in C7 sugars as fruit ripen may

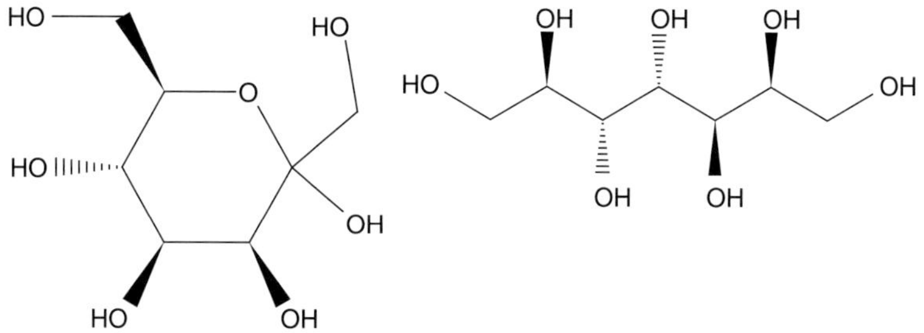

Fig. 3.2. Structures of D-mannoheptulose (left) and perseitol (right; both drawn with ChemDraw Ultra v 11.0, CambridgeSoft©).

indicate a major role in controlling flesh softening and that these carbohydrates could act as ripening inhibitors (Liu *et al.*, 2002). However, more recent work has shown that fruit softening could be delayed, in spite of very low mannoheptulose concentrations in the mesocarp of late season fruit (Meyer and Terry, 2010). It remains unknown whether the decline in C7 initiates ripening or whether it is an artifact of fruit softening. Cowan (2004) proposed various important potential functions for mannoheptulose activity, including protection from damage by reactive oxygen species (ROS) of certain key enzymes that are essential for fruit growth and development, recently confirmed by Bertling *et al.* (2007). Thus, the antioxidative properties of C7 sugars may also carry health benefits for the consumer. Mannoheptulose (Board *et al.*, 1995) and perseitol (Ishizu *et al.*, 2002) have been reported to have anticancer activity; mannoheptulose has been associated with an insulin secretion inhibitory effect (Ferrer *et al.*, 1993).

3.4 Health Benefits

Most of the health benefits associated with the dietary intake of avocado have been attributable largely to its remarkable content in mono- (MUFA) and polyunsaturated (PUFA) fatty acid. Such monounsaturates have been investigated for their potential cardiovascular benefits, including effects on serum lipids (Alvizouri-Muñoz *et al.*, 1992; Colquhoun *et al.*, 1992; Carranza *et al.*, 1995; Carranza-Madrigal *et al.*, 1997). The oil extracted from avocado pulp has been reported to decrease the risk of coronary heart disease (CHD), cataracts, diabetes, prostate cancer and age-related macular diseases (Bendich, 1993; Birkbeck, 2002; Semba and Dagnelie, 2003; Lu *et al.*, 2005). In contrast, less attention has been given to other bioactive substances with potential health-enhancement properties, such as carotenoids, vitamins B, C and E, terpenoids, D-mannoheptulose, β-sisterol, persenone A and B, phenols and phytosterols present in avocado fruit. Carotenoids are known to have antioxidant and anticarcinogenic effects, as well as other

potential mechanisms in the chemoprevention of cancer (Nishino *et al.*, 2000, 2005; Khachik *et al.*, 2004). Lutein, the carotenoid partly responsible for the yellow-green colour of avocado, is known to have antiproliferative and antitumour properties (Chew *et al.*, 1996; Kim, J.M. *et al.*, 1998; Park *et al.*, 1998; Kozuki *et al.*, 2000). Tocopherols (vitamin E), found in substantial amounts in avocado fruit, have been associated with antiproliferative effects on certain types of cancer (Awad and Fink, 2000). Studies have also shown that bioactive substances present in avocado and in its extract have antioxidant (Leong and Shui, 2002; Soong and Barlow, 2004; Bertling *et al.*, 2007), radical suppressing (Kim, O.K. *et al.*, 1998, 2000; Vinson *et al.*, 2001), acetylCoA carboxylase inhibitory (Hashimura *et al.*, 2001), antifungal (Prusky *et al.*, 1991; Domergue *et al.*, 2000) and chemopreventive (Lu *et al.*, 2005; Ding *et al.*, 2007, 2009) activities. Besides containing beneficial bioactive substances, avocado has also been shown to enhance the uptake of nutrients from other food (Unlu *et al.*, 2005).

The following sections provide scientific evidence of the effects of avocado fruit on various diseases and cancer (summarized in Table 3.3).

3.4.1 Antioxidant activity

Several studies have examined the antioxidant activity (principally hydrophilic) of avocado using different methods such as DPPH, FRAP, ORAC or ABTS+ decolourization assays. Depending on the respective study, authors found low (Luximon-Ramma *et al.*, 2003; Garcia-Alonso *et al.*, 2004; Plaza *et al.*, 2009), medium (Soong and Barlow, 2004) or high (Leong and Shui, 2002) activity in avocado. Leong and Shui (2002), when measuring a range of fruit for their general antioxidant capacity based on their ability to scavenge 2,2'-azino-*bis*-(3-ethylbenzthiazoline-6-sulfonic acid) (ABTS), found that avocado fruit contained considerable amounts of antioxidants, while the nature of the compounds contributing to that high antioxidant capacity were not determined. Soong and Barlow

(2004) and Bertling *et al.* (2007) found that the antioxidant activity of avocado was considerably higher in the seed than in the edible portion of the fruit.

3.4.2 Cancer studies

Several recent reports have focused on the chemopreventive activity of avocado fruit and results have shown that chloroform or acetone extracts prepared from avocado pulp have the ability to inhibit cancerous cell growth selectively (Lu *et al.*, 2005; Ding *et al.*, 2007, 2009). Cellular and molecular mechanisms of the phytochemicals present in the avocado extracts and involved in cancer prevention are not yet well understood. However, it has been shown that avocado-derived compounds target multiple signalling pathways and selectively induce cell cycle arrest, growth inhibition and apoptosis in precancerous and cancer cell lines (Lu *et al.*, 2005).

A chloroform extract prepared from mesocarp of Californian cv. Hass avocado and containing phytonutrients was shown to inhibit premalignant and malignant human oral epithelial cell growth selectively and induced apoptosis (Ding *et al.*, 2007). One of the anticancer mechanisms of this extract (Code D003) was believed to involve the targeting of the cell cycle regulatory proteins, cyclin and cyclin-dependent kinase (cdk), which are normally required for the progression and proliferation of cells (Ding *et al.*, 2007). Apoptosis appeared to be another important target for avocado phytochemicals in eliminating cancer cells selectively from normal tissues (Oberlies *et al.*, 1998; Lu *et al.*, 2005; Ding *et al.*, 2007). Apoptosis was attributed to a perturbation of the normal balance of ROS in the cancer-derived cell lines induced by D003, while normal cell lines were not affected. A more recent study by the same group confirmed that the D003 extract initiated apoptosis by acting on ROS levels, activating both the intrinsic and extrinsic pathways (Ding *et al.*, 2009).

Previously, Lu *et al.* (2005) had shown that an acetone extract of avocado containing lipid-soluble carotenoids (lutein and, in minor quantities, the related carotenoids, zeaxanthin,

α-carotene and β-carotene) and tocopherols (vitamin E) inhibited the growth of both androgen-dependent (LNCaP) and androgen-independent (PC-3) prostate cancer cells *in vitro*. Incubation of PC-3 cells with the avocado extract resulted in G2/M cell cycle arrest, with a concomitant increase in the cdk inhibitor (CDKI) p27 protein expression. When researchers assessed the biological activity of lutein alone, the anticancer effect could not be reproduced, substantiating a synergistic rather than an individual effect of the lipid-soluble bioactive substances present in the extract. This observation supports prior evidence that different carotenoids may work synergistically to reduce cancer risk (Zhang *et al.*, 2007). Also, avocado contains large amounts of monounsaturated fat, which is likely to facilitate absorption of bioactive carotenoids into the bloodstream where, mixed with other diet-derived phytochemicals, they may exhibit biological activity against cancer. Yeum and Russell (2002) had reported previously that fats helped increase the absorption of fat-soluble vitamins, including pro-vitamin A carotenoids. Likewise, Unlu *et al.* (2005) demonstrated that the addition of avocado to salsa improved lycopene, lutein and carotene absorption significantly in healthy human subjects (Unlu *et al.*, 2005).

Other potential anticancer phytochemicals present in avocado tissues with cell cycle targeting properties have been identified and are summarized by Ding *et al.* (2007). For instance, persin was shown to suppress progression through the cell cycle in certain breast cancer cell lines (Butt *et al.*, 2006), while quercetin induced a G2/M arrest in a number of cell types, including U937, lung cancer, prostatic carcinoma (PC-3) cell lines and normal tumour fibroblast cells (Vijayababu *et al.*, 2005; Lee *et al.*, 2006). The anticarcinogenic effects of these compounds were attributed to a modulation of the level of expression of cell cycle regulatory proteins (Vijayababu *et al.*, 2005, 2006a,b). Other alkanols isolated from unripe fruit demonstrated moderate cytotoxicity toward selected cancer cell lines (Oberlies *et al.*, 1998). Avocado also contains apigenin (Ding *et al.*, 2007), a very similar flavonoid to genistein, an antioxidant isoflavone with known anticancer properties (Messina *et al.*, 1994). Dietary intake of phytosterols, another class of compounds found in avocado fruit, has been shown to have protective effects in the development of certain cancers (Awad and Fink, 2000). The fruit is not the only source of chemicals with biological anticancer activity: Ye *et al.* (1996) showed that persealide, extracted from the bark of the avocado tree, was moderately cytotoxic towards selected cancer lines, including lung, breast and colon cancer.

Mannoheptulose, which is found in substantial amounts in the mesocarp of the fruit, is thought to be a glucokinase specific inhibitor, and, in a research study by Board *et al.* (1995), mannoheptulose caused inhibition of glucose intake by tumour cells (25–75% at 12 mM mannoheptulose) and inhibited the growth rate of cultured tumour cell lines (I50, 21.4 mM) *in vitro*. However, there exists no study on the effects of mannoheptulose on tumours *in vivo*. A Japanese study, when investigating Indonesian medicinal plants, found that a complex of perseitol and K$^+$ ions isolated from the leaves of *Scurrula fusca* (*Loranthaceae*), traditionally used for the treatment of cancer in Sulawesi Island, had potential cancer-prevention effects (Ishizu *et al.*, 2002).

3.4.3 Osteoarthritis

Avocado combined with soybean (avocado/ soybean unsaponifiables, ASU) demonstrated potential benefits in the treatment of the symptoms of osteoarthritis, in several human trials (Henrotin, 2008). The combination is thought to exert anti-inflammatory and stimulatory effects in chondrocytes. This is supported by *in vitro* studies that have demonstrated that ASU can increase the basal synthesis of aggrecan and reverse the reduction of aggrecan by human chondrocytes in alginate beads. It also decreased the spontaneous production of stromelysin-1, interleukin-6 and -8, prostaglandin E$_2$, macrophage inflammatory protein and finally it stopped fully the inhibitory effects of osteoblasts on cartilage production (Henrotin, 2008).

3.4.4 Lipid-lowering effects

During recent years, research has examined the effects of the consumption of different kinds of lipids and their relationship to obesity, cardiovascular disease and several types of breast and colon cancer (Steven and Bruce, 1998; Rose and Connolly, 1999). Elevated blood concentrations of total cholesterol and low-density lipoprotein cholesterol (LDL-C) increase the risk of cardiovascular disease (CVD), while higher concentrations of high-density lipoprotein cholesterol (HDL-C) have the opposite effect.

Avocado oil seemed appropriate to replace saturated fats, in order to lower total cholesterol and LDL plasma levels, and it could be a good alimentary aid, similar to olive oil, to ameliorate atherosclerosis (Alvizouri-Muñoz *et al.*, 1992). Avocado fruit has inherently high concentrations of MUFA (mainly oleic acid) and research has shown that diets enriched with avocado pulp have a cholesterol-lowering effect (Lerman-Garber *et al.*, 1994; Carranza *et al.*, 1995; Ledesma *et al.*, 1996; Carranza-Madrigal *et al.*, 1997). There is epidemiological evidence that a diet enriched in monounsaturated fatty oils may decrease the risk of coronary heart disease (CHD), mainly via the positive effects of the oils on serum lipids. Indeed, MUFA have the ability to reduce LDL and cholesterol levels without increasing the triglyceride (TG) levels (Amunziata *et al.*, 1999). It has been found more recently that MUFA may be more protective in the early stages of atherogenesis, when inflammatory processes and LDL oxidation take place (Kritchevsky *et al.*, 2003). Polyunsaturated fatty acids, on the other hand, may be even more effective, by not providing aortic cholesterol with its preferred substrate for esterification.

In a randomized parallel controlled study conducted in Mexico, a diet enriched with MUFA derived from avocado decreased the concentrations of total cholesterol and LDL-C significantly, without affecting the HDL-C level in healthy and hypercholesterolaemic subjects (Ledesma *et al.*, 1996). There was also a decrease in TG levels in moderately hypercholesterolaemic patients. Avocado has also been shown to improve lipid levels in a study where patients with phenotype II or phenotype IV dyslipidaemias have received either a diet rich in MUFA, using avocado as the principal source, or a low-saturated fat diet without avocado (Carranza-Madrigal *et al.*, 1997). In the diet enriched with avocado, total cholesterol and LDL-C concentration decreased significantly in the patients with phenotype II dyslipidaemia, and HDL-C increased significantly in all of the patients with (phenotype II or phenotype IV) dyslipidaemias. However, Pieterse *et al.* (2003) found contradicting results, whereby no significant changes in plasma lipid levels were observed, when investigating the effects of avocado in an energy-restricting diet on weight loss and serum lipids in overweight and obese subjects. The difference in patients used for the trials (namely, patients with normal to high serum lipid levels versus dyslipidaemic patients) could have accounted for the discrepancies in results.

Kinetic studies have demonstrated that the lipid-lowering (or hypocholesterolaemic) effect of MUFA probably occurs via an alteration of the small, very low-density lipoprotein (VLDL) particle production rate, on which plasma LDL concentration is dependent. It has been suggested that the LDL-lowering action of dietary MUFA is mediated either by an upregulation of LDL clearance or by decreased conversion of intermediate-density lipoprotein into LDL (Sanderson *et al.*, 2002). High MUFA diets have also been shown to decrease plasma TG levels, but the underlying mechanism for the hypotriacylglycerolaemic (triacylglycerol-lowering) effect of MUFA remains unclear (Kris-Etherton *et al.*, 1999) and requires more research. Not only monounsaturated fatty acids but also phytosterols may contribute to the lipid-lowering effect of avocado. Moghadasian and Frohlich (1999), when reviewing 16 published studies, reported averages of a 13% reduction in LDL concentration and a 10% reduction in total cholesterol concentration in response to various dietary phytosterol mixtures (1–6 g/day).

3.4.5 Diabetes

Diabetes mellitus shows hyperglycaemia as a symptom induced by decreased cellular glucose metabolism and uptake, and it is divided into type 1 and 2, where individuals with type 1 require external insulin and type 2 individuals first are prescribed a diet and lifestyle change and then additional drug therapy (Gallagher *et al.*, 2003). Non-insulin-dependent diabetes mellitus (NIDDM) patients benefit from a regulated nutrient scheme. Studies with animal models have shown the efficacy of certain plant extracts in glycaemic control (Gallagher *et al.*, 2003).

It has been shown that the replacement of certain carbohydrates with MUFA (from avocado) in the diet of NIDDM patients improves lipid profile and maintains good glycaemic control (Lerman-Garber *et al.*, 1994). In NIDDM patients with poor glycaemic control, a high monounsaturated fatty acid (HMUFA) diet showed no benefit over a diet high in complex carbohydrates (HCHO); thus, in these patients dietary treatment alone was not sufficient to ameliorate glucose and lipid metabolism (Lerman-Garber *et al.*, 1995). However, in the same study, patients with the poorest metabolic control had a worse glycaemic profile with the HCHO diet, but not with the HMUFA diet.

According to official sources, 70% of the world population uses traditional medicine to improve their life with diabetes (Gondwe *et al.*, 2008). The effectiveness of many plants to ameliorate diabetes symptoms has been tested scientifically with animal models (Swanston-Flatt *et al.*, 1989); for example, crude methanolic extracts of African mistletoe (*Loranthus micranthus*), a hemiparasitic plant growing on avocado trees, showed a dose-dependent hypoglycaemic effect on alloxan-induced diabetic rats, which was comparable to that seen with the standard drug at 24 h after administration (Osadebe *et al.*, 2004). Extracts from the avocado plant itself have been of interest to researchers: aqueous extract of dried avocado material has been shown to inhibit glucose diffusion across dialysis membrane *in vitro* (Gallagher *et al.*, 2003). Aqueous extracts of dried avocado leaves given to alloxan-induced diabetic rats produced a significant dose-dependent reduction in blood glucose levels (Antia *et al.*, 2005). It has been reported that the active ingredients in avocado leaves are sterols and flavonoids (Andrade-Cetto and Heinrich, 2005). Ethanolic leaf extracts of avocado had a blood glucose lowering effect in non-diabetic and streptozotocin diabetic rats (Gondwe *et al.*, 2008). Possibly, such extracts alter the glucose metabolism in the liver, thus reducing blood glucose concentration by increasing glycogenesis; they have been recommended as a complementary remedy in diabetes (Gondwe *et al.*, 2008). It is unknown whether the fruit would have the same effect.

3.5 Preharvest and Postharvest Continuum

3.5.1 Preharvest

There has been a significant amount of research on preharvest factors affecting postharvest quality of avocado. However, most research understandably has focused on quality parameters such as size, eating quality, ripening and occurrence of disorders rather than on phytochemical content. Some orchard management relating to harvest date has been reported to induce variations in some phytochemicals. Since the avocado, unlike many other fruit, can remain on the tree without ripening when fully mature, it is not unusual that, for economic reasons, part of the production is left hanging on the tree during the picking season, while some fruit are harvested early in the season. These differential harvest dates in the year have been reported to induce some variations in the fruit phytochemical content, such as increased carotenoid content (Lu *et al.*, 2009) and a decrease in C7 sugar concentrations (Liu *et al.*, 1999b; Meyer and Terry, 2010) in later harvested fruit. The increase in mesocarp carotenoid content has been shown to be correlated intimately and positively with the temporal increase in lipid content.

Additionally, growing location may affect the biochemical profile of fruit, since both fatty acid and sugar profiles in avocado cv. Hass vary significantly according to production site (Landahl *et al.*, 2009).

3.5.2 Postharvest continuum

With increasing distances between growing location and consumers, avocado can be stored for long periods. A range of technologies, such as refrigerated and controlled atmosphere (CA), as well as the ethylene inhibitor, 1-methylcyclopropene (1-MCP) (Watkins, 2006, 2008), is used commercially to maintain fruit quality and minimize losses due to premature ripening. While there has been a significant amount of research on the physical attributes of avocado fruit, less attention has been given to investigate changes in the health-related components in avocado fruit after harvest. Moreover, investigation of the effects of different storage techniques on avocado phytochemicals remains limited.

Phenolics have been investigated mostly for their role as substrates for browning enzymes (Golan *et al.*, 1977; Kahn, 1983). Total phenolics content in avocado increases with maturity, as does mesocarp discoloration following cold storage (Cutting *et al.*, 1992). The same authors found no difference in phenolics concentration between cold and non-cold stored fruit. Concentration of a particular anthocyanin, cyanidin 3-*O*-glucoside, can be affected by ripening temperature (Cox *et al.*, 2004).

Changes in fatty acid profile during fruit growth and development have long been known. However, knowledge of such changes in the fatty acid composition of the lipid fraction after harvest is more recent. Some changes occur in the fatty acid composition during the fruit ripening process, with an increase in unsaturated and a decrease in saturated fatty acids (Ozdemir and Topuz, 2006; Meyer and Terry, 2008, 2010). However, these changes remain relatively small and a more important factor that would affect the amount of monounsaturates delivered per fruit would be the lipid content, intimately related to fruit maturity (harvest date). Research with cvs. Fuerte and Hass found that the fatty acid profile remained relatively constant during cold storage (Eaks, 1990; Luza *et al.*, 1990; De la Plaza *et al.*, 2003; Meyer and Terry, 2010).

Other compounds such as heptose sugars show more drastic changes after harvest, with a consistent decrease in mannoheptulose and perseitol during postharvest cold storage and ripening (Liu *et al.*, 1999b, 2002; Meyer and Terry, 2010). These results have suggested that C7 contribute to the carbon reserve necessary during the respiratory process, which requires energy. Mannoheptulose pools within the fruit also tend to decline with increasing maturation on the tree.

The use of CA is effective at extending shelf life and reducing CI in avocado (Hatton and Reeder, 1972; Truter and Eksteen, 1987; Meir *et al.*, 1995; Burdon *et al.*, 2008). However, there is no information available on the effects of CA on the health-related compounds present in avocado.

The effects of 1-MCP on avocado physical attributes have been reviewed extensively (see Watkins, 2006, 2008), but studies on the effects of the ethylene antagonist on the bioactive compounds of avocado fruit remain limited. Recently, the effects of 1-MCP on the fatty acids and sugars content of avocado cv. Hass during cold storage and ripening have been reported (Meyer and Terry, 2010). The authors found that the fatty acid profile differed between 1-MCP-treated and non-treated fruit, although these differences remained small. 1-MCP treatment also had an effect on mannoheptulose and perseitol content, with more of the heptose sugars found in the firmer 1-MCP-treated fruit than in the untreated fruit.

Ethylene scavengers can be used to reduce ethylene in the environment and effectively delay climacteric-induced ripening. While their application is not common for avocado fruit, some significant experimental responses to ethylene removal have been recorded using a palladium (Pd)-promoted ethylene scrubber (Terry *et al.*, 2007; Meyer and Terry, 2010). These studies observed better firmness and greenness maintenance in fruit held in the presence of the scavenger. When ripening (i.e. softening and colour change) was delayed in response to the scavenger, higher C7 content in fruit mesocarp as compared with non-treated fruit was observed. On the other hand, ethylene removal in the storage atmosphere had no effect on the fatty acid composition of the oil, as found previously (De la Plaza *et al.*, 2003).

Processing is becoming increasingly valued in the avocado industry due to the growing popularity of avocado, suitable processing technology and increasing recognition of the health benefits of avocado (Elez-Martínez, 2005; Jacobo-Velázquez and Hernández-Brenes, 2010). Avocado can be processed into guacamole, minimally processed (MP) avocado halves and slices and avocado oil (although the latter is still used predominantly in cosmetics). The impact of minimal processing on health-promoting attributes was investigated on avocado cut in slices or in halves and packaged in plastic bags under nitrogen, air or vacuum (Plaza *et al.*, 2009). Refrigerated storage (8°C) of slices or halves of avocado fruit induced a decrease in fatty acid content. However, halves kept under vacuum had reduced loss of fatty acids. Phytosterols presented no significant changes during refrigerated storage. Antioxidant activity, as measured using DPPH, increased toward the end of refrigerated storage for tissues held under vacuum. Minimal processing under vacuum could hence contribute to better preservation of the health-related attributes of avocado fruit.

3.6 Conclusion

The avocado is a rich source of health-related phytochemicals, although research reporting on bioactive compounds and their health properties is only relatively recent and remains limited in comparison with that on other fruit, berries for instance. The evidence from epidemiological studies suggests that consumption of avocado may help decrease risk of developing chronic diseases such as cardiovascular disease and cancer. However, it is likely that the health benefits of avocado result from the synergistic rather than the individual activity of the complex mixture of phytochemicals and bioactive compounds with different functions. Additionally, the lipid present in the mesocarp plays an important role in the assimilation of various lipid-soluble vitamins and components present in the fruit. Therefore, avocado is an excellent candidate to improve nutrient assimilation of other foodstuffs. More research is necessary to understand the mechanisms underlying the efficacy of avocado toward reduction of various chronic diseases.

Information on the effects of preharvest and postharvest continuum on the biochemical profile remains scant for avocado fruit. Phenolics, carotenoids, vitamin E, phytosterols and heptose sugars have not been quantified sufficiently for different cultivars, maturity (that is, early versus late season) and location sites. The effects that different storage conditions/postharvest techniques have on bioactive compounds have not been defined adequately for avocado fruit. A better understanding of the biochemical response of avocado to different storage conditions would certainly lead to improved pre- and postharvest management and practices to maintain nutritional quality and ultimately enhance the health benefits delivered to consumers.

References

Adam, A., Crespy, V., Levrat-Verny, M.A., Leenhardt, F., Leuillet, M., Demigne, C. and Remesy, C. (2002) The bioavailability of ferulic acid is governed primarily by the food matrix rather than its metabolism in intestine and liver in rats. *Journal of Nutrition* 132, 1962–1968.

Ahmed, E.M. and Barmore, C.R. (1980) Avocado. In: Nagy, S. and Shaw, P.E. (eds) *Tropical and Subtropical Fruits: Composition, Properties and Uses*. AVI, Westport, Connecticut, pp. 121–156.

Alvizouri-Muñoz, M., Carranza, M.J., Herrera-Abarca, J.E., Chavez-Carbajal, F. and Amezcua-Gastelum, J.L. (1992) Effects of avocado as a source of monounsaturated fatty acids on plasma lipid levels. *Archives of Medical Research* 23, 163–167.

Ameye, L.G. and Chee, W.S.S. (2006) Osteoarthritis and nutrition. From neutraceuticals to functional foods: a systematic review of the scientific evidence. *Arthritis Research and Therapy* 8, R127, doi:10.1186/ar2016.

Amiot, M.J., Fleuriet, A., Cheynier, V. and Nicolas, J. (1997) Phenolic compounds and oxidative mechanism in fruit and vegetable. In: Tomas-Barberan, F.A. (ed.) *Phytochemistry of Fruits and Vegetables*. Oxford University Press, New York, pp. 51–85.

Amunziata, C.M., Massaro, M. and Siculella, L. (1999) Oleic acid inhibits endothelial activation. *Arteriosclerosis, Thrombosis and Vascular Biology* 19, 220–228.

Andrade-Cetto, A. and Heinrich, M. (2005) Mexican plants with hypoglycaemic effect used in the treatment of diabetes. *Journal of Ethnopharmacology* 99, 325–348.

Antia, B.S., Okokon, J.E. and Okon, P.A. (2005) Hypoglycemic activity of aqueous leaf extract of *Persea americana* Mill. *Indian Journal Pharmacology* 37, 325–326.

Arts, I.C.W., van de Putte, B. and Hollman, P.C.H. (2000) Catechin contents of foods commonly consumed in The Netherlands. 1. Fruits, vegetables, staple foods, and processed foods. *Journal of Agricultural and Food Chemistry* 48, 1746–1751.

Ashton, O.B.O., Wong, M., McGhie, T.K., Vather, R., Wang, Y., Requejo-Jackman, C., Ramankutty, P. and Woolf, A.B. (2006) Pigments in avocado tissue and oil. *Journal of Agricultural and Food Chemistry* 54, 10151–10158.

Athar, M. and Nasir, S.M. (2005) Taxonomic perspective of plant species yielding vegetable oils used in cosmetics and skin care products. *African Journal of Biotechnology* 4, 36–44.

Awad, A.B. and Fink, C.S. (2000) Phytosterols as anticancer dietary components: evidence and mechanism of action. *Journal of Nutrition* 130, 2127–2130.

Bendich, A. (1993) Biological functions of dietary carotenoids. In: Canfield, L.M., Krinsky, N.I. and Olsen, J.A. (eds) *Carotenoids in Health*. New York Academy of Sciences, New York, pp. 61–67.

Bergh, B. (1992) Nutritious value of avocado. *California Avocado Society Yearbook* 76, 123–135.

Berry, E.M., Eisenberg, S., Friedlander, Y., Harats, D., Kaufmann, N.A., Norman, Y. and Stein, Y. (1995) Effects of diets rich in monounsaturated fatty-acids on plasma-lipoproteins – the Jerusalem nutrition study – monounsaturated vs saturated fatty-acids. *Nutrition Metabolism and Cardiovascular Diseases* 5, 55–62.

Bertling, I. and Bower, J.P. (2005) Sugars as energy sources – is there a link to avocado fruit quality? *South African Avocado Growers' Association Yearbook* 28, 24–27.

Bertling, I., Tesfay, S.Z. and Bower, J.P. (2007) Antioxidants in 'Hass' avocado. *South African Avocado Growers' Association Yearbook* 30, 17–19.

Biale, J.B. and Young, R.E. (1971) The avocado pear. In: Hulme, A.C. (ed.) *The Biochemistry of Fruits and Their Products, Volume 2*. Academic Press, London, pp. 1–63.

Birkbeck, J. (2002) Health benefits of avocado oil. *Food N.Z.* April/May, 40–42.

Board, M., Colquhoun, A. and Newsholme, E.A. (1995) High K_m glucose-phosphorylating (glucokinase) activities in a range of tumour cell lines and inhibition of rates of tumour growth by the specific enzyme inhibitor mannoheptulose. *Cancer Research* 55, 3278–3285.

Bower, J.P. and Cutting, J.G. (1988) Avocado fruit development and ripening physiology. *Horticultural Review* 10, 229–271.

Brown, W.H. (1975) *Introduction to Organic Chemistry*. Willard Grant Press, Boston, Massachusetts, 468 pp.

Burdon, J., Lallu, N., Haynes, G., McDermott, K. and Billing, D. (2008) The effect of delays in establishment of a static or dynamic controlled atmosphere on the quality of 'Hass' avocado fruit. *Postharvest Biology and Technology* 49, 61–68.

Butt, A.J., Roberts, C.G., Seawright, A.A., Oelrichs, P.B., MacLeod, J.K., Liaw, T.Y.E., Kavallaris, M., Somers-Edgar, T.J., Lehrbach, G.M., Watts, C.K. and Sutherland, R.L. (2006) A novel plant toxin, persin, with *in vivo* activity in the mammary gland, induces Bim-dependent apoptosis in human breast cancer cells. *Molecular Cancer Therapeutics* 5, 2300–2309.

Carranza, J., Alvizouri, M., Alvarado, M.R., Chávez, F., Gómez, M. and Herrera, J.E. (1995) Effects of avocado on the level of blood lipids in patients with phenotype II and IV dyslipidemias. *Archivos del Instituto de Cardiologia de Mexico* 65, 342–348.

Carranza-Madrigal, J., Herrera-Abarca, J.E., Alvizouri-Muñoz, M., Alvarado-Jimenez, M.D.R. and Chavez-Carbajal, F. (1997) Effects of a vegetarian diet vs. a vegetarian diet enriched with avocado in hypercholesterolemic patients. *Archives of Medical Research* 28, 537–541.

Chew, B.P., Wong, M.W. and Wong, T.S. (1996) Effects of lutein from marigold extract on immunity and growth of mammary tumours in mice. *Anticancer Research* 16, 3689–3694.

Colquhoun, D.M., Moores, D., Somerset, S.M. and Humphries, J.A. (1992) Comparison of the effects on lipoproteins and apolipoproteins of a diet high in monounsaturated fatty acids, enriched with avocado, and a high-carbohydrate diet. *American Journal of Clinical Nutrition* 56, 671–677.

Cowan, A.K. (2004) Metabolic control of avocado fruit growth: 3-hydroxy-3-methylglutaryl coenzyme a reductase, active oxygen species and the role of C7 sugars. *South African Journal of Botany* 70, 75–82.

Cox, K.A., McGhie, T.K., White, A. and Woolf, A.B. (2004) Skin colour and pigment changes during ripening of 'Hass' avocado fruit. *Postharvest Biology and Technology* 31, 287–294.

Curatolo, M. and Bogduk, N. (2001) Pharmacologic pain treatment of musculoskeletal disorders: current perspectives and future prospects. *Clinical Journal of Pain* 17, 25–32.

Cutting, J.G.M., Wolstenholme, B.N. and Hardy, J. (1992) Increasing relative maturity alters the base mineral composition and phenolic concentration in avocado (*Persea americana* Mill) fruit. *South African Avocado Growers' Association Yearbook* 15, 64–67.

De La Plaza, J.L. Rupérez, P. and Montoya, M.M. (2003) Fatty acids distribution in 'Hass' avocado during storage with ethylene absorber at subcritical temperature. *Acta Horticulturae* 600, 457–460.

de Pascual-Teresa, S., Santos-Buelga, C. and Rivas-Gonzalo, J. (2000) Quantitative analysis of flavan-3-ols in Spanish foodstuffs and beverages. *Journal of Agricultural and Food Chemistry* 48, 5331–5337.

Ding, H., Chin, Y.-W., Kinghorn, A.D. and D'Ambrosio, S.M. (2007) Chemopreventive characteristics of avocado fruit. *Seminars in Cancer Biology* 17, 386–394.

Ding, H., Han, C., Guo, D., Chin, Y.-W., Ding, Y., Kinghorn, A.D. and D'Ambrosio, S.M. (2009) Selective induction of apoptosis of human oral cancer cell lines by avocado extracts via a ROS-mediated mechanism. *Nutrition and Cancer* 61, 348–356.

Domergue, F., Helms, G.L., Prusky, D. and Browse, J. (2000) Antifungal compounds from idioblast cells isolated from avocado fruits. *Phytochemistry* 54, 183–189.

Duester, K.C. (2001) Avocado fruit is a rich source of beta-sitosterol. *Journal of the American Dietetic Association* 101, 404–405.

Eaks, I.L. (1990) Change in the fatty acid composition of avocado fruit during ontogeny, cold storage and ripening. *Acta Horticulturae* 269, 141–152.

Electronic Code of the Federal Regulations I CFR101.44. Identification of the 20 most frequently consumed raw fruit, vegetables, and fish in the United States (Title 21: Food and Drugs. Part 101: Food Labeling) (http://ecfr.gpoaccess.gov/cgi/t/text/text-idx?c=ecfr;rgn=div5;view=text;node=21%3A2.0.1.1.2;id no=21;sid=64fa2cdef4ee54e80414c2f297540271;cc=ecfr#21:2.0.1.1.2.3.1.4, accessed 22 March 2011).

Elez-Martínez, P., Soliva-Fortuny, R.C., Gorinstein, S. and Martín-Belloso, O. (2005) Natural antioxidants preserve the lipid oxidative stability of minimally processed avocado purée. *Journal of Food Science* 70, 325–329.

FAOSTAT (2007) http://faostat.fao.org/site/339/default.aspx (accessed 22 March 2011).

Ferrer, J., Gomis, R., Fernandez Alvarez, J., Casamitjana, R. and Vilardell, E. (1993) Signals derived from glucose metabolism are required for glucose regulation of pancreatic islet GLUT2 mRNA and protein. *Diabetes* 42, 1273–1280.

Gallagher, A.M., Flatt, P.R., Duffy, G. and Abdel-Wahab, Y.H.A. (2003) The effects of traditional antidiabetic plants on *in vitro* glucose diffusion. *Nutrition Research* 23, 413–424.

Garcia-Alonso, M., de Pascual-Teresa, S., Santos-Buelga, C. and Rivas-Gonzalo, J.C. (2004) Evaluation of the antioxidant properties of fruits. *Food Chemistry* 84, 13–18.

Geissman, T.A. and Dittmar, H.F.K. (1965) A proanthocyanidin from avocado seed. *Phytochemistry* 4, 359–368.

Giao, M.S., Gonzalez-Sanjose, M.L., Rivero-Perez, M.D., Pereira, C.I., Pintado, M.E. and Malcata, F.X. (2007) Infusion of Portuguese medicinal plants: dependence of final antioxidant capacity and phenol content on extraction features. *Journal of the Science of Food and Agriculture* 87, 2638–2647.

Golan, A., Kahn, V. and Sadovski, A.Y. (1977) Relationship between polyphenols and browning in avocado mesocarp. Comparison between the Fuerte and Lerman cultivars. *Journal of Agricultural and Food Chemistry* 25, 1253–1260.

Gondwe, M., Kamadyaapa, D.R., Tufts, M.A., Chuturgoon, A.A., Ojewole, J.A.O. and Musabayane, C.T. (2008) Effects of *Persea americana* Mill (Lauraceae) ['Avocado'] ethanolic leaf extract on blood glucose and kidney function in streptozotocin-induced diabetic rats and on kidney cell lines of the proximal (LLC-PK1) and distal tubules (MDBK). *Methods and Findings in Experimental and Clinical Pharmacology* 30, 25–35.

Gross, J., Gabai, M., Lifshitz, A. and Sklarz, B. (1973) Carotenoids in pulp, peel and leaves of *Persea americana*. *Phytochemistry* 12, 2259–2263.

Haiyan, Z., Bedgood, D.B. Jr, Bishop, A., Prenzler, P.D. and Robards, K. (2007) Endogenous biophenol, fatty acid and volatile profiles of selected oils. *Food Chemistry* 100, 1544–1551.

Hashimura, H., Ueda, C., Kawabata, J. and Kasai, T. (2001) Acetyl-CoA carboxylase inhibitors from avocado (*Persea americana* Mill.) fruits. *Bioscience, Biotechnology and Biochemistry* 65, 1656–1658.

Hatton, T.T. Jr and Reeder, W.F. (1972) Quality of 'Lula' avocados stored in controlled atmospheres with or without ethylene. *Journal of the American Society for Horticultural Science* 97, 339–341.

Hatton, T.T. Jr, Harding, P.L. and Reeder, W.F. (1964) Seasonal changes in Florida avocados. *USDA Technical Bulletin*, 1310.

Heinonen, M.I., Ollilainen, V., Linkola, E.K., Varo, P.T. and Koivistoinen, P.E. (1989) Carotenoids in Finnish foods: vegetables, fruits, and berries. *Journal of Agricultural and Food Chemistry* 37, 655–659.

Henrotin, Y. (2008) Avocado/soybean unsaponifiable (ASU) to treat osteoarthritis: a clarification. *Osteoarthritis and Cartilage* 16, 1118–1119.

Ishizu, T., Winarno, H., Tsujino, E., Morita, T. and Shibuya, H. (2002) Indonesian medicinal plants. XXIV. Stereochemical structure of perseitol·K+ complex isolated from the leaves of *Scurrula fusca* (Loranthaceae). *Chemical and Pharmaceutical Bulletin* 50, 489–492.

Jacobo-Velázquez, D.A. and Hernández-Brenes, C. (2010) Biochemical changes during the storage of high hydrostatic pressure processed avocado paste. *Journal of Food Science* 75, 264–270.

Kahn, V. (1983) Multiple effects of hydrogen peroxide on the activity of avocado polyphenol oxidase. *Phytochemistry* 22, 2155–2159.

Kaiser, C. and Wolstenholme, B.N. (1994) Aspects of delayed harvest of Hass avocado (*Persea americana* Mill) fruit in a cool subtropical climate. 1. Fruit lipid and fatty-acid accumulation. *Journal of Horticultural Science* 69, 437–445.

Khachik, F., Beecher, G.R. and Smith, J.C. Jr (2004) Lutein, lycopene, and their oxidative metabolites in chemoprevention of cancer. *Journal of Cellular Biochemistry* 59, 236–246.

Kim, J.M., Araki, S., Kim, D.J., Park, C.B., Takasuka, N., Baba-Toriyama, H., Ota, T., Nir, Z., Khachik, F., Shimidzu, N., Tanaka, Y., Osawa, T., Uraji, T., Murakoshi, M., Nishino, H. and Tsuda, H. (1998) Chemopreventive effects of carotenoids and curcumins on mouse colon carcinogenesis after 1,2-dimethylhydrazine initiation. *Carcinogenesis* 19, 81–85.

Kim, O.K., Murakami, A., Nakamura, Y. and Ohigashi, H. (1998) Screening of edible Japanese plants for nitric oxide generation inhibitory activities in RAW 264.7 cells. *Cancer Letters* 125, 199–207.

Kim, O.K., Murakami, A., Nakamura,Y., Takeda, N., Yoshizumi, H. and Ohigashi, H. (2000) Novel nitric oxide and superoxide generation inhibitors, persenone A and B, from avocado fruit. *Journal of Agricultural and Food Chemistry* 48, 1557–1563.

Knight, R. Jr (1980) Origin and world importance of tropical and subtropical fruit crops. In: Nagy, S. and Shaw, P.E. (eds) *Tropical and Subtropical Fruits*. AVI Publishing Company, Westport, Connecticut, pp. 1–106.

Knight, R.J. (2002) Chapter 1. In: *The Avocado: Botany, Production and Uses*. CABI International, New York, p. 1.

Koh, H.H., Murray, I.J., Nolan, D., Carden, D., Feather, J. and Beatty, S. (2004) Plasma and macular responses to lutein supplement in subjects with and without age-related maculopathy: a pilot study. *Experimental Eye Research* 79, 21–27.

Kozuki, Y., Miura, Y. and Yagasaki, K. (2000) Inhibitory effects of carotenoids on the invasion of rat ascites hepatoma cells in culture. *Cancer Letters* 151, 111–116.

Kris-Etherton, P.M., Pearson, T.A. and Wan, Y. (1999) High–monounsaturated fatty acid diets lower both plasma cholesterol and triacylglycerol concentrations. *American Journal of Clinical Nutrition* 70, 1009–1015.

Kritchevsky, D. and Chen, S.C. (2005) Phytosterols health benefits and potential concerns: a review. *Nutritional Research (NY)* 25, 413–428.

Kritchevsky, D., Tepper, S.A., Wright, S., Czarnecki, S.K., Wilson, T.A. and Nicolosi, R.J. (2003) Cholesterol vehicle in experimental atherosclerosis 24: avocado oil. *Journal of the American College of Nutrition* 22(1), 52–55.

Lafay, S., Gil-Izquierdo, A., Manach, C., Morand, C., Besson, C. and Scalbert, A. (2006) Chlorogenic acid is absorbed in its intact form in the stomach of rats. *Journal of Nutrition* 136, 1192–1197.

Landahl, S., Meyer, M.D. and Terry, L.A. (2009) Spatial and temporal analysis of textural and biochemical changes of imported avocado cv. Hass during fruit ripening. *Journal of Agricultural and Food Chemistry* 57, 7039–7047.

Ledesma, R.L., Munari, A.C.F., Dominguez, B.C.H., Montalvo, S.C., Luna, M.H.H., Juarez C. and Lira, S.M. (1996) Monounsaturated fatty acid (avocado) rich diet for mild hypercholesterolemia. *Archives of Medical Research* 27, 519–523.

Lee, T.J., Kim, O.H., Kim, Y.H., Lim, J.H., Kim, S., Park, J.W. and Kwon, T.K. (2006) Quercetin arrests G2/M phase and induces caspase-dependent cell death in U937 cells. *Cancer Letters* 240, 234–242.

Leong, L.P. and Shui, G. (2002) An investigation of antioxidant capacity of fruits in Singapore markets. *Food Chemistry* 76, 69–75.

Lerman-Garber, I., Ichazo-Cerro, S., Zamora-Gonzalez, J., Cardoso-Saldana, G. and Posadas-Romero, C. (1994) Effect of a high-monounsaturated fat diet enriched with avocado in NIDDM patients. *Diabetes Care* 17, 311–315.

Lerman-Garber, I., Gulias-Herrero, A., Palma, M.E., Valles, V.E., Guerrero, L.A., Garcia, E.G., Gomez-Perez, F.J. and Rull, J.A. (1995) Response to high carbohydrate and high monounsaturated fat diets in hypertriglyceridemic non-insulin dependent diabetic patients with poor glycemic control. *Diabetes Nutrition and Metabolism* 8, 339–345.

Lewis, C.E. (1978) The maturity of avocados – a general review. *Journal of Science Food and Agriculture* 29, 857–866.

Liu, X., Robinson, P.W., Madore, M.A., Witney, G.W., Lu, A.M. and Witney, G.W. (1999a) 'Hass' avocado carbohydrate fluctuations. I. Growth and phenology. *Journal of the American Society for Horticultural Science* 124, 671–675.

Liu, X., Robinson, P.W., Madore, M.A., Witney, G.W. and Arpaia, M.L. (1999b) 'Hass' avocado carbohydrate fluctuations. II. Fruit growth and ripening. *Journal of the American Society for Horticultural Science* 124, 676–681.

Liu, X., Sievert, J., Arpaia, M.L. and Madore, M.A. (2002) Postulated physiological roles of the seven-carbon sugars, mannoheptulose, and perseitol in avocado. *Journal of the American Society for Horticultural Science* 127, 108–114.

Lodovici, M., Caldini, S., Morbidelli, L., Akpan, V., Ziche, M. and Dolora, P. (2008) Protective effect of 4-coumaric acid from UVB ray damage in the rabbit eye. *Toxicology* 225, 1–5.

Lu, Q.-Y., Arteaga, J.R., Zhang, Q., Huerta, S., Go, V.L.W. and Heber, D. (2005) Inhibition of prostate cancer cell growth by an avocado extract: role of lipid-soluble bioactive substances. *Journal of Nutritional Biochemistry* 16, 23–30.

Lu, Q.Z.Y., Wang, Y., Wang, D., Lee, R., Gao, K., Byrns, R. and Heber, R.D. (2009) California Hass avocado: profiling of carotenoids, tocopherol, fatty acid, and fat content during maturation and from different growing areas. *Journal of Agricultural and Food Chemistry* 57, 10408–10413.

Luximon-Ramma, A., Bahorun, T. and Crozier, A. (2003) Antioxidant actions and phenolic and vitamin C contents of common Mauritian exotic fruits. *Journal of the Science of Food and Agriculture* 83, 496–502.

Luza, J.G., Lizana, L.A. and Masson, L. (1990) Comparative lipids evolution during cold storage of three avocado cultivars. *Acta Horticulturae* 269, 153–160.

Mazliak, P. (1970) Lipids. In: Hulme, A.C. (ed.) *Biochemistry of Fruits and Their Products, Volume 1.* Academic Press, London, pp. 209–238.

Meir, S., Akerman, Y., Fuchs, M. and Zauberman, G. (1995) Further studies on the controlled atmosphere storage of avocados. *Postharvest Biology and Technology* 5, 323–330.

Messina, M.J., Parsky, V., Stetchell, K.D. and Barnes, S. (1994) Soy intake and cancer risk: a review of the *in vitro* and *in vivo* data. *Nutrition and Cancer* 21, 113–131.

Meyer, M.D. and Terry, L.A. (2008) Development of a rapid method for the sequential extraction and subsequent quantification of fatty acids and sugars from avocado mesocarp tissue. *Journal of Agricultural and Food Chemistry* 56, 7439–7445.

Meyer, M.D. and Terry, L.A. (2010) Fatty acid and sugar composition of avocado, cv. Hass, in response to treatment with an ethylene scavenger or 1-methylcyclopropene to extend storage life. *Food Chemistry* 121, 1203–1210.

Moghadasian, M.H. and Frohlich, J.J. (1999) Effects of dietary phytosterols on cholesterol metabolism and atherosclerosis: clinical and experimental evidence. *American Journal of Medicine* 107, 588–594.

Naidu, K.A. (2003) Vitamin C in human health and disease is still a mystery? An overview. *Nutrition Journal* 2, 1–10.

Nishino, H., Tokuda, H., Murakoshi, M., Satomi, Y., Masuda, M., Onozuka, M., Yamaguchi, S., Takayasu, J., Tsuruta, J., Okuda, M., Khachik, F., Narisawa, T., Takasuka, N. and Yano, M. (2000) Cancer prevention by natural carotenoids. *BioFactors* 13, 89–94.

Nishino, H., Murakoshia, M., Xiao, Y.M., Wada, S., Masuda, M., Ohsaka, Y., Satomi, Y. and Jinno, K. (2005) Cancer prevention by phytochemicals. *Oncology* 69, 38–40.

Oberlies, N.H., Rogers, L.L., Martin, J.M. and McLaughlin, J.L. (1998) Cytotoxic and insecticidal constituents of the unripe fruit of *Persea americana*. *Journal of Natural Products* 61, 781–785.

Olthof, M.R., Hollman, P.C.H. and Katan, M.B. (2001) Chlorogenic acid and caffeic acid are absorbed in humans. *Journal of Nutrition* 131, 66–71.

Osadebe, P.O., Okide, G.B. and Akabogu, I.C. (2004) Study on anti-diabetic activities of crude methanolic extracts of *Loranthus micranthus* (Linn.) sourced from five different host trees. *Journal of Ethnopharmacology* 95, 133–138.

Ozdemir, F. and Topuz, A. (2004) Changes in dry matter, oil content and fatty acids composition of avocado during harvesting time and post-harvesting ripening period. *Food Chemistry* 86, 79–83.

Park, J.S., Chew, B.P. and Wong, T.S. (1998) Dietary lutein from marigold extract inhibits mammary tumour development in BALB/c mice. *Journal of Nutrition* 128, 1650–1656.

Pieterse, Z., Jerling, J. and Oosthuizen, W. (2003) Avocados (monounsaturated fatty acids), weight loss and serum lipids. *South African Avocado Growers' Association Yearbook* 26, 65–71.

Piironen, V., Lindsay, D.G., Miettinen, T.A., Toivo, J. and Lampi, A.M. (2000) Plant sterols: biosynthesis, biological function and their importance to human nutrition. *Journal of the Science of Food and Agriculture* 80, 939–966.

Piironen, V., Toivo J., Puupponen-Pimia, R. and Lampi, A.M. (2003) Plant sterols in vegetables, fruits and berries. *Journal of the Science of Food and Agriculture* 83, 330–337.

Plaza, L., Sanchez-Moreno, C., de Pascual-Teresa, S., De Ancos, B. and Cano, M.P. (2009) Fatty acids, sterols, and antioxidant activity in minimally processed avocados during refrigerated storage. *Journal of Agricultural and Food Chemistry* 57, 3204–3209.

Popenoe, W. (1920) *Manual of Tropical and Subtropical Fruits.* Macmillan, London, 524 pp.

Posch, A., Wheeler, C.H., Chen, Z., Flagge, A., Dunn, M.J., Papenfuss, F., Raulf-Heimsoth, M. and Baur, X. (1999) Class I endochitinase containing a hevein domain is the causative allergen in latex-associated avocado allergy. *Clinical and Experimental Allergy* 29, 667–672.

Prabha, T.N. and Patwardhan, M.V. (1980) Polyphenols of avocado (*Persea americana*) and their endogenous oxidation. *Journal of Food Science and Technology* 17, 215–217.

Prabha, T.N., Ravindranath, B. and Patwardhan, M.V. (1980) Anthocyanins of avocado (*Persea americana*) peel. *Journal of Food Science and Technology – India* 17, 241–242.

Praloran, J.C. (1970) Le climat des aires d'origine des avocatiers. *Fruits* 25, 543–557.

Prusky, D., Kobiler, I., Fishman, Y., Sims, J.J., Midland, S.L. and Keen, N.T. (1991) Identification of an antifungal compound in unripe avocado fruits and its possible involvement in the quiescent infections of *Colletotrichum gloeosporioides*. *Journal of Phytopathology* 132, 319–327.

Pursglove, J.W. (1968) *Persea americana* Mill. In: *Tropical Crops: Dicotyledons.* 1. Longmans, London, pp. 192–198.

Rabasco Alvarez, A.M. and González Rodríguez, M.L. (2000) Lipids in pharmaceutical and cosmetic preparations. *Grasas y Aceites* 51, 74–96.

Ramirez-Martinez, J.R. and Luh, B.S. (1973) Phenolic compounds in frozen avocados. *Journal of the Science of Food and Agriculture* 24, 219–225.

Richer, S., Stiles, W., Statkute, L., Pulido, J., Frankowski, J., Rudy, D., Pei, K., Tsipursky, M. and Nyland, J. (2004) Double-masked, placebo-controlled, randomized trial of lutein and antioxidant supplementation in the intervention of atrophic age-related macular degeneration: the Veterans LAST study (Lutein Antioxidant Supplementation Trial). *Optometry* 75, 216–229.

Rocquelin, G., Tapsoba, S., Dop, M.C., Mbemba, F., Traissac, P. and Martin-Prevel, Y. (1998) Lipid content and essential fatty acid (EFA) composition of mature Congolese breast milk are influenced by mothers' nutritional status: impact on infants' EFA supply. *European Journal of Clinical Nutrition* 52, 164–171.

Roodenburg, A., Leenen, R., van het Hof, K., Weststrate, J. and Tijburg, L. (2000) Amount of fat in the diet affects bioavailability of lutein esters but not of a-carotene, b-carotene, and vitamin E in humans. *American Journal of Clinical Nutrition* 71, 1187–1193.

Rose, D.P. and Connolly, J.M. (1999) Omega-3 fatty acids as cancer chemopreventive agents. *Pharmacology and Therapeutics* 83, 217–244.

Sanderson, P., Finnegan, Y.E., Williams, C.M., Calder, P.C., Burdge, G.C., Wootton, S.A., Griffin, B.A., Millward, D.J., Pegge, N.C. and Bemelmans, W.J.E. (2002) UK Food Standards Agency alpha-linolenic acid workshop report. *British Journal of Nutrition* 88, 573–579.

Scalbert, A. and Williamson, G. (2000) Dietary intake and bioavailability of polyphenols. *Journal of Nutrition* 130, 2073–2085.

Scora, W.R., Wolstenholme, N.B. and Lav, U. (2002) Taxonomy and botany. In: Whiley, W.A., Scaffer, B. and Wolstenholme, B.N. (eds) *The Avocado: Botany, Production and Uses.* CAB International, Wallingford, UK, pp. 15–37.

Semba, R.D. and Dagnelie, G. (2003) Are lutein and zeaxanthin conditionally essential nutrients for eye health? *Medical Hypotheses* 61, 465–472.

Slater, G.C., Shankman, S., Shepherd, J.S. and Alfin-Slater, R.B. (1975) Seasonal variation in the composition of California avocados. *Journal of Agricultural and Food Chemistry* 23, 468–474.

Smith, C.E. Jr (1966) Archaeological evidence for selection in avocado. *Economic Botany* 20, 169–175.

Smith, J., Goldweber, S., Lamberts, M., Tyson R. and Reynolds, J.S. (1983) Utilization potential for semi-tropical and tropical fruits and vegetables in therapeutic and family diets. *Proceedings of Florida State Horticultural Society* 96, 241–244.

Soong, Y.-Y. and Barlow, P.J. (2004) Antioxidant activity and phenolic content of selected fruit seeds. *Food Chemistry* 88, 411–417.

Steven, M.W. and Bruce, G.J. (1998) Omega fatty acids. In: Akoh, C.C. and Min, D.B. (eds) *Food Lipids: Chemistry, Nutrition and Biotechnology.* Dekker, New York, pp. 463–493.

Storey, W.B., Bergh, B. and Whitsell, R. H. (1973/1974) Factors affecting the marketability of avocado fruit. *California Avocado Society Yearbook* 57, 33–39.

Storey, W.B., Bergh, B. and Zentmyer, G.A. (1986) The origin, indigenous range and dissemination of the avocado. *California Avocado Society Yearbook* 70, 127–133.

Swanston-Flatt, S.K., Day, C., Flatt, P.R., Gould, B.J. and Bailey, C.J. (1989) Glycaemic effects of traditional European plant treatments for diabetes. Studies in normal and streptozotocin diabetic mice. *Diabetes Research* 10, 69–73.

Terasawa, N., Sakakibara, M. and Murata, M. (2006) Antioxidative activity of avocado epicarp hot water extract. *Food Science and Technology Research* 12, 55–58.

Terry, L.A., Ilkenhans, T., Poulston, S., Rowsell, L. and Smith, A.W.J. (2007) Development of new palladium-promoted ethylene scavenger. *Postharvest Biology and Technology* 45, 214–220.

Torres, A.M., Mau-Lastovicka, T. and Rezaaiyan, R. (1987) Total phenolics and high-performance liquid chromatography of phenolic acids of avocado. *Journal of Agricultural and Food Chemistry* 35, 921–925.

Truter, A.B. and Eksteen, G.J. (1987) Controlled and modified atmospheres to extend storage life of avocados. *South African Avocado Growers' Association Yearbook* 10, 151–153.

Unlu, N.Z., Bohn, T., Clinton, S.K. and Schwartz, S.J. (2005) Carotenoid absorption from salad and salsa by humans is enhanced by the addition of avocado or avocado oil. *Journal of Nutrition* 135, 431–436.

USDA National Nutrient Database for Standard Reference, Release 23 (2010) (http://www.nal.usda.gov/fnic/foodcomp/cgi-bin/list_nut_edit.pl, accessed 22 March 2011).

Vekiari, S.A., Papadopoulou, P.P., Lionakis, S. and Krystallis, A. (2004) Variation in the composition of Cretan avocado cultivars during ripening. *Journal of the Science of Food and Agriculture* 84, 485–492.

Vijayababu, M.R., Kanagaraj, P., Arunkumar, A., Ilangovan, R., Aruldhas, M.M. and Arunakaran, J. (2005) Quercetin-induced growth inhibition and cell death in prostatic carcinoma cells (PC-3) are associated

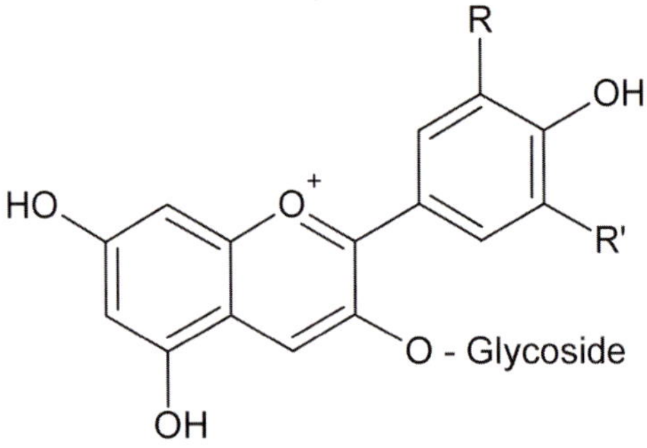

Anthocyanidin	R	R'
Cyanidin	OH	H
Delphinidin	OH	OH
Malvidin	OCH_3	OCH_3
Peonidin	OCH_3	H
Petunidin	OH	OCH_3

Fig. 4.1. Structure of common anthocyanins found in blueberry and cranberry fruit.

referred to as condensed tannins, and can occur as procyanidins or prodelphinidins (Fig. 4.2). These molecules are made up of oligomers and polymers of variously linked flavan-3-ols and flavan-3,4-diols and are found primarily in the peel and seeds of the fruit. Commonly found flavan-3-ols are catechin, epicatechin and galloylated catechins. These monomers typically are linked by single C–C bonds between adjacent flavanol monomers at the 4-6 or 4-8 positions (B-type). *Vaccinium* fruit are notable for the presence of PACs whose monomers are linked by two bonds at the 4-8 and 2-O-7 positions (A-type), which is thought to confer structural rigidity and unique properties (Prior *et al.*, 2001; Gu *et al.*, 2003; Schmidt *et al.*, 2004). This structural feature is not widespread in the plant kingdom. Cranberry PACs are composed primarily of epicatechin units that include A-type linkages, while blueberry PACs are composed primarily of catechin and epicatechin units and have few A-type linkages (Schmidt *et al.*, 2004).

PACs are notoriously difficult to analyse because of their structural diversity and propensity to rearrange, particularly after extraction and processing. A major structural feature that can be analysed is their degree of polymerization. HPLC analysis is conducted using normal phase and fluorescent detection, but cannot characterize heterogeneous and A-linked PACs adequately. MALDI-TOF, a sophisticated and costly technique which requires a relatively purified sample, is currently the most informative approach in the analysis of cranberry PACs (Howell, 2007).

The PAC concentration in blueberries averages 0.65 for lowbush and 0.44 mg/g FW for highbush fruit (Neto, 2007). Cranberry fruit have the highest concentration of PACs among fruit listed in the USDA PAC database

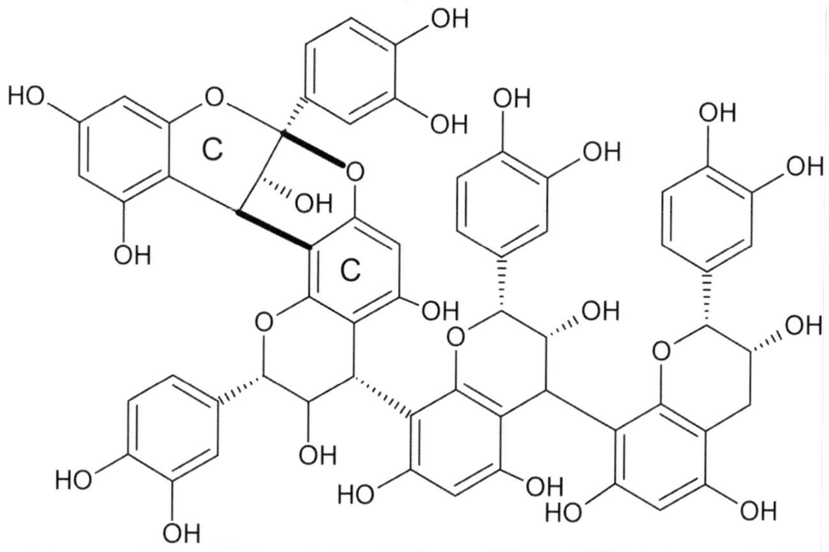

Fig. 4.2. The structure of an A-type proanthocyanin commonly found in cranberry fruit. Note the A-type linkage in bold.

(http://www.nal.usda.gov/fnic/foodcomp). They contain an average of about 1.8 mg/g FW of PACs, with 0.45 mg/g FW as dimers or trimers and 1.33 mg/g FW as oligomers (Neto, 2007). In a survey of 88 different foods including 29 fruit, only cranberry fruit and groundnuts had PACs with A-type linkages at both terminal and extension units (Gu *et al.*, 2003).

4.2.3 Other phenolics

Flavonols are plentiful in both blueberry and cranberry fruit and are found primarily in the skin (Riihinen *et al.*, 2008). Total flavonol concentration of blueberry cvs. Northblue and Northcountry was reported to be 1.2 and 0.7 mg/g DW, respectively (Häkkinen *et al.*, 1999), and the concentration in cranberry fruit ranged from 0.2 to 0.4 mg/g FW (Vvedenskaya *et al.*, 2004; Neto, 2007). The primary flavonols found in blueberries and cranberries are quercetin and myricetin, which exist in several gycosidic forms. Quercetin glycosides comprise about 75–85% of the total flavonols in blueberry and cranberry fruit (Häkkinen *et al.*, 1999; Taruscio *et al.*, 2004; Neto, 2007).

Hydroxycinnamic acids and their esters are abundant in blueberry fruit peel and flesh, of which chlorogenic acid is predominant (Fig. 4.3). Fruit in the highbush blueberry cv. Sierra contained 0.65 mg/g FW chlorogenic acid, which comprised about 29% of the total phenolic compounds in the fruit (Zheng and Wang, 2003). Taruscio *et al.* (2004) found the concentration of phenolic acids of several highbush blueberry cultivars to average 1.3, 0.18, 0.044 and 0.005 mg/g FW for chlorogenic, caffeic, ferulic and *p*-coumaric acids, respectively. In cranberry fruit, Zuo *et al.* (2002) reported the most abundant phenolic acids to be coumaric, sinapic, caffeic, *o*-hydroxycinnamic, ferulic and 2,4-dihydroxybenzoic, having concentrations of 0.25, 0.21, 0.16, 0.089, 0.088 and 0.043 mg/g FW, respectively.

4.2.4 Other bioactive phytochemicals

Blueberry and cranberry fruit contain triterpenes. Blueberry fruit were reported to contain ursolic acid, pomolic acid and β-amyrin (Neto, 2007). Ursolic acid was found in cranberry fruit at a concentration of 0.6–1.1 mg/g

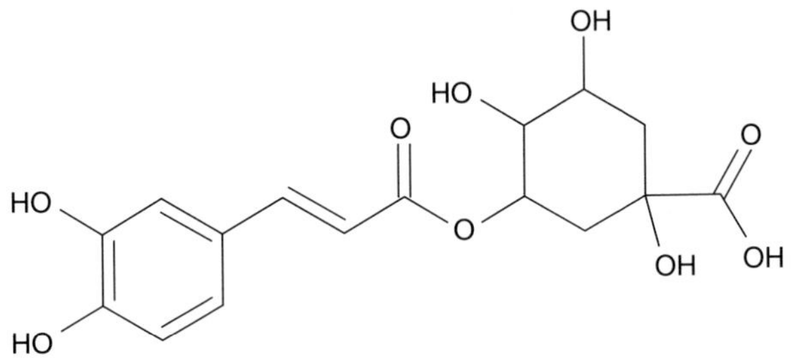

Fig. 4.3.   The structure of chlorogenic acid, a prominent hydroxycinnamic acid in blueberry fruit.

FW and its hydroxycinnamate ester averaged 0.15 mg/g FW (Neto, 2007).

Both blueberries and cranberries contain small amounts of stilbenes. Resveratrol concentrations as high as 1.7 µg/g DW were reported for blueberry fruit and 0.9 µg/g DW for cranberry fruit (Rimando *et al.*, 2004). Pterostilbene and piceatannol, analogues of resveratrol, were also found in rabbiteye and highbush blueberry fruit, respectively, at a concentration ranging from 0.1 to 0.42 µg/g DW.

4.2.5 Antioxidants

Blueberries and cranberries rank highly among fruit for their antioxidant capacity, which has been attributed to their high phenolic content (Prior *et al.*, 1998; Vinson *et al.*, 2001; Moyer *et al.*, 2002). Total phenol antioxidant index (PAOXI) values for blueberry and cranberry fruit were 40.5×10^3 and 31.2×10^3, which ranked third and sixth, respectively, among 20 fruit assayed (Vinson *et al.*, 2001). The antioxidant capacity of blueberry fruit, measured by the oxygen radical absorbance capacity (ORAC) assay, ranged between 19 and 130 µmol Trolox (a water-soluble vitamin E analogue) equivalents per g FW among 20 blueberry genotypes representing lowbush, highbush and rabbiteye fruit (Moyer *et al.*, 2002). When antioxidant capacity was measured using the ferric reducing antioxidant power (FRAP) assay, for these same 20 blueberry genotypes, values ranged from 19 to 161 µmol/g FW and results were similar to those of the ORAC assay (Moyer *et al.*, 2002). It was estimated that anthocyanin content

accounted for about 50% of the antioxidant capacity of blueberry and cranberry fruit (Zheng and Wang, 2003). Chlorogenic acid and peonidin 3-galactoside were reported to be the most important antioxidants in blueberry and cranberry fruit, respectively (Zheng and Wang, 2003). Prior *et al.* (2001) reported that PACs accounted for 32 and 54% of the total ORAC measured in blueberry and cranberry fruit, respectively. Ascorbate contributed <5% of the total ORAC antioxidant capacity in blueberry fruit (Prior *et al.*, 1998; Kalt *et al.*, 1999a).

4.3 Health Benefits of Blueberry and Cranberries

In order to interpret the potential significance of results, it is important to discuss the experimental approaches taken in food and health research. Research conducted using *in vitro* assays is designed to simulate parameters of a specific physiological state (e.g. disease) using appropriate biochemicals and often specific cell types. In *in vitro* testing, the user defines the concentration of, for example, *Vaccinium* extract to be added to an assay, which may or may not be cell based. When tested *in vitro*, components of interest bypass the normal processes of digestive absorption and other factors that determine their bioavailability. While *in vitro* results demonstrate biochemical interactions among test components and may suggest a possible mechanistic basis for possible health effect(s), they provide limited information on the effects that may be expected in a whole living organism. Alternatively, effects shown *in vivo* demonstrate that component(s) in the dietary intervention are adequately bioavailable and in a form capable of producing such effects. Whether *in vitro* or *in vivo*, studies employ 'models' which include specific cell types (*in vitro*) or animals (*in vivo*, often rodents) that may be genetically defined, or treated in such a way to induce a particular condition, often in an attempt to simulate processes related to disease. Along with the important distinctions between *in vitro* and *in vivo* evidence, it is also important to recognize that significant differences exist between the physiology of rodents and humans, as well as the myriad processes that may contribute to disease outcomes. Rigorously designed human clinical trials constitute the current 'gold standard' for evidence of relationships between plant components and human health outcomes. For the reasons outlined above, the following discussion of the health functionality of *Vaccinium* berries focuses primarily on *in vivo* evidence, while the vast abundance of *in vitro* evidence suggestive of the beneficial health effects of *Vaccinium* berries will not be discussed in depth.

4.3.1 Blueberries may protect the brain

Blueberry consumption may protect the brain during normal ageing and when neurodegenerative disease is present. The interest in blueberries as neuroprotective agents arose from epidemiological research supporting a role for fruit and vegetable consumption in reducing the risk of various degenerative diseases (Ames *et al.*, 1993), including neurodegenerative disease (see references in Lau *et al.*, 2007). Concurrently, extensive *in vitro* research suggested that specific fruit and vegetable components might confer health protection via their action as antioxidative agents in biological systems. Since the late 1990s, research on the neuroprotective effects of blueberries has expanded dramatically, with evidence for their benefits extending beyond their activity as antioxidants. Importantly, research supports a role of blueberries as anti-inflammatory agents, due to their ability to affect cell signal transduction pathways to reduce inflammatory processes that can influence

various site-specific aspects of brain ageing and neuropathologies (for review see Lau *et al.*, 2007). Importantly, the beneficial effects of blueberries in the brain have been demonstrated in various *in vivo* rodent models (for review see Lau *et al.*, 2007). These *in vivo* results have been complemented with further *ex vivo* and *in vitro* characterization. Typically, these studies involve the imposition of a physiological stress (e.g. genetic predisposition, ageing, inflammatory insult, oxidative stress), after which cognitive function (e.g. memory, motor abilities) is tested and subsequently neuronal physiology and various biomarkers are examined. Together, results from *in vivo*, *ex vivo* and *in vitro* studies are correlated with specific dietary interventions (e.g. blueberry feeding).

Neuroprotection by various plant and food components has been reported (Ramassamy, 2006). Among the berries, blueberries appear generally to be more potent and distinctive in their effects. In an early study by Joseph *et al.* (1999), rats that demonstrated age-related cognitive impairment were fed diets enriched in spinach, strawberry, or blueberry extract and this resulted in memory improvement in all treatment groups compared with controls, based on the rats' performance in a water maze test. However, only the blueberry-fed rats demonstrated significant improvement in their motor abilities. In another study, rats were fed diets enriched in either strawberries or blueberries before receiving whole-body exposure to high-energy ^{56}Fe irradiation that disrupted dopamine-sensitive neuronal systems, which normally decline during ageing (Rabin *et al.*, 2005). Both blueberry and strawberry feeding preserved the capacity for release of the neurotransmitter, dopamine, compared with the rats fed non-supplemented diets. Also, both strawberry- and blueberry-enriched diets protected rats against cognitive deficits (water maze test performance). However, differences in water maze performance suggested that strawberry supplementation had a greater effect on the hippocampus, which controlled spatial memory, while blueberry supplementation had a greater effect on the striatum, which influenced relearning. The results suggested that not only might the total concentration of dietary phenolics be important in neuroprotection, but also the specific types of phenolics present. Blueberries are notable for their high anthocyanin concentration, while they contain little, if any, ellagitannins. Strawberries contain ellagitannins, but a significantly lower concentration of anthocyanins. These noted differences may aid in the development of products containing optimized blends of phytochemicals to elicit specific health benefits.

The beneficial effects of blueberry supplementation in various models of brain ageing and disease have been reviewed recently (Joseph *et al.*, 2007; Lau *et al.*, 2007). These reviews describe the fundamental role played by inflammation and oxidative stress in the degenerative processes underlying brain ageing and neuropathologies like Alzheimer's disease and Parkinson's disease. Inflammation is a highly regulated process involving numerous intermediates (especially cytokines) that work either to amplify or to dampen the cellular processes involved in inflammation. The immune system is finely tuned to launch rapid responses of an appropriate magnitude, which is critical to the beneficial role of inflammation in acute conditions such as infection and trauma. However, it is widely recognized that chronic low-level inflammation, and the oxidative stress arising from inflammation, are both the hallmark of, and a damaging element in, degenerative conditions such as cardiovascular disease, diabetes, neurodegenerative disease and the ageing process itself. Phytochemicals that can reduce the activity of proinflammatory signal transduction pathways constitute a potentially powerful means to mitigate the degenerative processes of ageing and disease. Numerous lines of evidence suggest that blueberries possess this characteristic.

In the brain, glial cells mediate immune responses and their activation is a definitive marker of neuroinflammation. Activated glial cells produce proinflammatory cytokines (e.g. tumour necrosis factor, interleukins 1 and 6), various growth factors and other proteins that lead to further glial cell activation. Some rodent models designed to examine inflammation involve administration into the brain of the proinflammatory compounds, lipopolysaccharide or kainic acid. Responses

are thought to mimic the effects of inflammatory and oxidative metabolism on cognition and motor function that are evident in disease and ageing (Lau *et al.*, 2007).

Transgenic mice that are predisposed to the development of symptoms of Alzheimer's disease, including amyloid plaque development in several brain regions, and cognitive deficits at middle age, were used to examine the effect of blueberry supplementation on parameters related to the development of the disease (Joseph *et al.*, 2003). After 8 months, blueberry-supplemented transgenic mice performed as well as non-transgenic mice in a Y-maze test, and significantly better than the control transgenic mice. Compared with control transgenic mice, blueberry-fed transgenic mice had a higher activity of specific kinases that played a role in cognitive function, especially in conversion of short- to long-term memory (Joseph *et al.*, 2003).

Old rats that were equivalent to 70-year-old humans received blueberries in their diet and were reported to perform better in an object recognition test than their age-matched counterparts that did not receive berry extract. The blueberry-supplemented rats also had a lower level of NF-κB, a biomarker whose expression was correlated with inflammation and oxidative stress. The effects of blueberries on NF-κB expression varied among brain regions, suggesting additional specificity in the scope of neuroprotective responses to blueberry feeding (Goyarzu *et al.*, 2004).

Different lines of evidence suggest that blueberry components may stimulate the regenerative capacity of the brain. The hippocampus, a region of the brain that functions in learning and spatial recognition, undergoes new cell formation, although the rate of this neurogenesis declines with ageing. Rats that received blueberries for 8 weeks, followed by bromodeoxyuridine, an analogue of uridine that could be detected histochemically, had a significantly higher density of new cells compared with rats that did not receive blueberries (Casadesus *et al.*, 2004). These rats also had higher levels of an insulin-like growth factor that was a key modulator of hippocampal neurogenesis, and higher activity of a kinase that was critical in neuronal signal transduction.

Transplantation of embryonic dopamine neurons can mitigate the progression of Parkinson's disease by restoring dopamine production. However, this approach suffers from poor survival of transplants, in spite of anti-inflammatory and antioxidant therapy. A rat model that simulated this type of human cell transplantation therapy was employed in a study designed to improve the environment of the host tissue prior to transplantation (McGuire *et al.*, 2006). Using this model, diet supplementation with rabbiteye blueberry cv. Tifblue increased the survival of dopamine neurons significantly and reduced behavioural responses associated with poor survival. In another study involving transplantation of neuronal tissue, portions of fetal hippocampus tissue were transplanted into the eye of middle-aged (4-month-old) rats (Willis *et al.*, 2005). The survival, growth and cellular organization of the chimeric grafts were improved significantly in the rats that received blueberries in the diet for 1 week prior to and 6 weeks after the transplantation procedure.

Another study that suggested that blueberry consumption might improve the cellular viability and regenerative capacity of brain tissues employed a rat model of ischaemic stroke (Sweeney *et al.*, 2002). During ischaemic stroke, affected regions of the brain are subjected to hypoxia and then reperfusion when the blockage is relieved. During reperfusion, extremely high levels of reactive oxygen species are released, contributing further to cellular damage arising from hypoxia. Rats that received a lowbush blueberry-supplemented diet for 6 weeks prior to a surgically induced ischaemic stroke were compared with non-supplemented rats 1 week after surgical treatment. When specific regions of the hippocampus were compared histologically for the presence of damaged cells, tissues from blueberry-supplemented rats had significantly less damage in two of the three regions of the hippocampus.

The neuroscience studies summarized above suggest that blueberry consumption may protect the brain in a variety of ways. These observations prompt the question of what is distinctive about the phytochemical profile of blueberries that may account for these benefits. Clearly, blueberries contain a high concentration of anthocyanins compared with other fruit, and even many berries, such as strawberries and grapes. Whether anthocyanins are contributing significantly to these benefits remains to be elucidated, however. In this regard, it is interesting to note a study involving blueberries and longevity (Wilson *et al.*, 2006). The lifespan of a soil nematode (*Caenorhabditis elegans*) is approximately 1 month and is therefore a convenient *in vivo* model for longevity. Including blueberries in the diet of these nematodes extended their lifespan by more than 25%. However, when individual blueberry components were fed, it was found that PACs, and not anthocyanins, were responsible for these effects. The authors also showed that PACs contributed not only to lifespan extension, but also to thermotolerance in these organisms.

4.3.2 Blueberries may benefit vision

Bilberry (*V. myrtillus* L.) anthocyanins have a long history in folk medicine for their purported benefits in night vision. European research conducted mainly between the 1960s and 1980s examined various aspects of anthocyanins in vision physiology both *in vitro* and *in vivo*. One suggestion, which has been supported recently (Matsumoto *et al.*, 2003), is that anthocyanins increase the rate of rhodopsin regeneration. Interestingly, cyanidin, but not delphinidin, glycosides were found to be effective. Rhodopsin is the photoreceptor primarily responsible for vision in low light and darkness. Most recently, molecular modelling studies have been employed to understand the mechanistic basis for the rhodopsin interaction and if and how anthocyanins may be involved (Tirupula *et al.*, 2009).

Clinical benefits to vision arising from the consumption of *Vaccinium* berries with a high anthocyanin concentration (i.e. blueberry species) were reviewed in 2004 by Canter and Ernst (2004). Their meta-analysis of placebo-controlled trials of *V. myrtillus* (i.e. bilberry) effects on night vision concluded that rigorous trials did not support a beneficial role for bilberry in night vision. Of 12 placebo-controlled trials, four trials that used a randomized controlled (RC) design had a negative outcome. A fifth RC trial and seven non-RC trials reported positive effects of bilberry on outcome measures relevant to night vision. Negative outcomes in these trials were associated with more rigorous methodologies and with lower anthocyanin dosage levels. Factors such as the age of subjects and the duration of intervention should also be considered in such trials, since human night vision begins to decline during middle age. Also, the time dependency of the possible effect(s) of anthocyanins on vision parameters is not known.

4.3.3 Cranberries support urinary tract health

The effect of cranberries in maintaining urinary tract health is arguably the most widely recognized and thoroughly studied human health benefit of any fruit. While the benefit of cranberries in urinary tract health has been recognized for over 100 years, conclusive clinical evidence, and the characterization of the effects and their mechanistic basis, is only now becoming clear.

It was thought originally that the high acid concentration of cranberries was sufficient to acidify urine and confer a bacteriostatic effect; however, such a low pH is achieved rarely with cranberry consumption. Instead, research results from the past 20 years favour an antibacterial adhesion mechanism as the basis for their effect. Numerous *in vitro* and *ex vivo* studies support the notion that specific cranberry components possess an antiadhesion property that can reduce or prevent the adhesion of uropathogenic bacteria to the wall of the bladder and urinary tract, which is otherwise essential to the initiation and progression of a urinary tract infection (UTI) (Howell, 2002).

Escherichia coli is the predominant bacterium involved in UTI. Interaction of the proteinaceous fimbriae on the surface of specific *E. coli* with receptor types on the mucosal surface of uroepithelium is the mechanism by which these bacteria adhere, multiply and colonize the urinary tract (Howell, 2002). Most uropathogenic *E. coli* express type-1

fimbriae. However, *E. coli* possessing the P-type fimbriae, which are correlated with cystitis (bacteria in the bladder) and pyelonephritus (infection of the kidneys), appear to be affected specifically by cranberry components. Based on recent *in vitro* studies, cranberry components appear to induce conformational changes in P-type fimbriae, including a reduction in their length and density (Liu *et al.*, 2006). Numerous lines of evidence suggest that cranberry phenolics, specifically their PACs possessing A-linkages, are most important in affecting the action of P-fimbriated *E. coli* in UTIs.

The impact of UTI in society is significant. It is estimated that in the USA 11 million women are affected annually and that 25% of these women suffer from recurring UTI (Howell, 2007). While the typical treatment is to eliminate uropathogenic bacteria using antibiotic therapy, there is growing concern that the widespread use of antibiotics is contributing to the development of antibiotic-resistant bacteria. Cranberries may provide an alternative means to reduce the risk of recurring infections, as well as to treat infections by a non-bacteriostatic means. Therefore, the clinical benefit of cranberry consumption for various types of UTI is actively being investigated. Jepson and Craig (2007) have systematically reviewed clinical trials focused on this topic. Using rigorous criteria, which included only randomized controlled trials (RCT), Jepson and Craig (2008) used meta-analysis to examine selected types of RCT, participants, interventions and outcome measures. On the basis of these criteria, nine clinical trials were included in the meta-analysis and included the outcomes of just over 1000 participants, with interventions using cranberry juice, cocktail and capsules, tablets and a lingonberry–cranberry juice blend. The meta-analysis examined three groups that included women with recurring UTIs, an elderly population and a population requiring intermittent or continuous catheterization (e.g. spinal cord injury patients). This rigorous meta-analysis concluded that cranberry consumption could be beneficial in certain subpopulations of women (uropathogen-specific) with recurring UTIs. This outcome is supported by the notion that the cranberry A-linked PACs confer protection only against the adhesion and further colonization of P-fimbriated *E. coli.*

The antibacterial adhesion mechanisms of cranberry components including A-linked PAC in urinary tract health, or cranberry non-dialysable material in oral health (see below), constitute a novel and potentially powerful means to reduce risk and promote wellness through the consumption of an innocuous food product instead of a pharmaceutical product. A secondary and possibly even more significant beneficial outcome of the antiadhesion effects of cranberries is that their use may reduce the requirement for antibiotic therapy and mitigate the selection for survival of antibiotic-resistant bacteria.

4.3.4 Cranberries may benefit oral health

Dialysis (12,000–15,000 MW cut-off) of cranberry extracts against water gives rise to a fraction called simply non-dialysable material (NDM), which has various bioactivities that may promote oral health (Weiss *et al.*, 1998). The development of dental caries and periodontal disease arises from the effects of dental plaque, which is a dynamic and complex matrix of biotic and abiotic components attached strongly to the teeth. Bacteria are also associated with epithelial cells of the mouth and the saliva. Most significant perhaps is the abundant diversity of proliferating microorganisms that adhere to each other to form biofilms via interspecific cell-to-cell interactions. Biofilms are being reformed constantly after their physical disruption; however, without ongoing intervention through dental hygiene, biofilms become mature stable entities that are resistant to removal and create microconditions for the development of dental caries and periodontal disease.

Various lines of *in vitro* evidence support the notion that cranberry NDM may benefit oral health. An early *in vitro* approach was to examine the extent of bacterial coaggregation, a critical event in the formation of biofilms, among numerous pairs of oral bacteria following preincubation with NDM (Weiss *et al.*, 1998). Cranberry NDM prevented the coaggregation of bacterial pairs at concentrations as low as 0.04 mg/ml. When a non-dialysable

fraction was prepared from various other fruit juices, none was found to prevent coaggregation of pairs of bacteria *in vitro*, with the exception of blueberry, which possessed only a weak activity (Bodet *et al.*, 2008). Once coaggregated, bacterial pairs could be dissociated by an approximately five times greater concentration of NDM. Since *in vivo* NDM will encounter already formed bacterial biofilms, it is notable that NDM could also reverse the aggregation of some bacteria in biofilms. When NDM was tested at the concentration level that occurred in cranberry juice (2.5 mg/ml), a reversal of coaggregation occurred in more that 50% of the bacterial combinations examined, while 90% of the remaining bacterial pairs were reversed completely at a concentration four times greater.

When the effects of saliva and NDM were examined individually *in vitro*, neither saliva nor NDM led to coaggregation of selected bacterial species. However, in combination, saliva and NDM gave rise to significant bacterial aggregation. It was concluded from this result that NDM would improve the ability to remove bacteria from the mouth via the saliva (Weiss *et al.*, 2002).

Hydroxyapatite has been used experimentally to simulate dental enamel, and saliva-coated hydroxyapatite beads were employed as an *in vitro* model to examine factors influencing oral biofilm formation. Yamanaka *et al.* (2004) tested bacterial adherence to these beads and found that cranberry juice inhibited the adherence of several important oral bacteria by up to 95%. It is important to note that, at the concentration employed, NDM did not appear to be affecting bacterial viability, but simply their coaggregation capacity (Duarte *et al.*, 2006). Therefore, the floral profile may not shift to more resistant populations, which is considered a beneficial outcome.

An early stage in the development of dental biofilms is the formation of an underlying pellicle. The pellicle is attached firmly to the dental enamel, and is composed of cell-free and saliva components, including polysaccharides. Polyglucan- and polyfructan-mediated binding of oral bacteria is considered an important mechanism in bacterial biofilm formation (Bodet *et al.*, 2008). These polysaccharides use sucrose as a substrate and are formed through the activity of gluco-syl and fructosyl transferase. Cranberry NDM has been shown to inhibit the activity of these enzymes, but is less effective when they are immobilized on hydroxyapatite, since they then become less accessible to NDM. Cranberry juice reduced bacterial adhesion to glucan binding sites on the pelli-cle by 40–85% and reduced the final biofilm mass (Bodet *et al.*, 2008). Since the formation of the polysaccharide layer by glucosyl and fructosyl transferase will be influenced by the availability of sugar substrates, NDM, and not cranberry cocktails that are rich in simple sugars, will be the more logical material from which to develop cranberry products to sup-port oral health. Dental caries arise due to acid-mediated demineralization of tooth enamel. This occurs when cariogenic bacteria, particularly *Streptococci mutans*, form lactic acid, which occurs when the pH is less than about 5.5. Cranberry extract was found to increase the pH of biofilms and mitigated the formation of lactic acid, which might other-wise have demineralized dental enamel.

The composition of cranberry NDM is not well described. However, in their recent review, Bodet *et al.* (2008) compared cranberry NDM with cranberry PACs, which possess antiadhesin effects on uropathogenic P-fibri-ated *E. coli*. The NDM and PAC fractions dif-fer in their solubility properties, as well as their NMR and MALDI-TOF spectra. While cranberry PACs possess an astringent taste, NDM does not. The mass of NDM and PACs in cranberry cocktail differs. Their concentra-tion required to affect bioactivities in relation to oral pathogens and uropathogenic P-fibri-ated *E. coli* also differs. Cranberry NDM con-tains about 65% phenolic-like materials (based on colorimetric measurements) and is devoid of sugars or acids.

4.3.5 Evidence for cardioprotection by *Vaccinium* fruit

Neto (2007) has reviewed evidence recently for the ability of *Vaccinium* fruit to reduce the risk and mitigate the symptoms of cardiovascular disease. Also included in the review is an over-view of the phytochemical composition of major commercial *Vaccinium* species, includ-ing blueberries and cranberries. A major risk factor in cardiovascular disease is the pres-ence of atherosclerotic plaque in coronary arteries; plaque can restrict blood flow, contrib-ute to hypertension and damage the heart. Development of atherosclerotic plaque involves the uptake of oxidatively modified low-density lipoprotein (LDL) into the lining of the vascu-lar endothelium. Plant phenolics possess potent antioxidant properties. Therefore, pro-vision of dietary phenolics, such as those in *Vaccinium* berries, has been considered an important mechanism to protect plasma LDL from oxidation. However, it should be con-sidered that there are abundant endogenous antioxidants already present in the plasma. This factor, along with research reporting low plasma bioavailability of phenolics, has tem-pered current opinion on the potential impor-tance of dietary phenolic antioxidants in the protection of LDL against oxidative modifica-tion. Other factors considered to be of impor-tance to cardioprotection by *Vaccinia* are their antiplatelet effects, which reduce the propen-sity of blood to clot and create vascular occlu-sion. While antiplatelet effects by phenolics have been demonstrated *in vitro* (Keevil *et al.*, 2000), they have not been shown for blueber-ries *in vivo* (Kalt *et al.*, 2008a), owing perhaps to the low plasma bioavailability of the phe-nolics. It should be noted, however, that in a clinical study involving the consumption of progressively higher amounts of low-calorie cranberry juice, Ruel *et al.* (2008) reported a lower concentration of oxidized plasma LDL, and favourable effects on intercellular and vas-cular cellular adhesion molecules, which are involved in atherosclerotic plaque deposition.

The phenolics, stilbene and pterostil-bene, which are found in blueberries at low concentrations (Rimando *et al.*, 2004, 2005), have been reported to reduce LDL concentra-tion significantly and improve the LDL/HDL ratio *in vivo* in hypercholesterolaemic ham-sters (Rimando *et al.*, 2004). The mechanism underlying this benefit was the induction of peroxisome proliferator activator receptor (PPARα). It is interesting to note that the drug, ciprofibrate, which works via PPARα, was deemed less effective in this respect than pterostilbene (Rimando *et al.*, 2005).

Blueberry components can bind bile acids, resulting in reduced serum cholesterol. Indeed, freeze-dried blueberry was shown *in vitro* to bind almost half the amount of bile acids as the cholesterol-lowering drug, cholestyramine, which works via this mechanism (Kahlon and Smith, 2007). Binding of bile acids favours cholesterol excretion to affect plasma lipid balance beneficially and confer a cardioprotective benefit.

Blueberry feeding was shown recently to reduce total cholesterol and affect the LDL/HDL ratio of pigs favourably (Kalt *et al.*, 2008a). Pigs are considered a good model for cardiovascular studies since several markers of disease, which vary among species (e.g. LDL metabolism, plaque deposition), are similar between pigs and humans. Also, pigs are omnivores and have similar body weight to humans. Plasma cholesterol reduction was dose-dependent among diets containing 0, 1 and 2% blueberries, and levelled off between 2 and 4%. The effects of blueberries on cholesterol lowering was greater when the pig's basal diet was rich in other plant foods (70% soy, oats and barley) than when less (20%) plant foods made up the basal diet. Together, the results suggested that blueberries might affect plasma lipids beneficially when consumed in reasonable doses and as part of a diet rich in plant-based foods.

Ischaemic stroke, another adverse outcome of cardiovascular disease, arises from transient occlusion of blood vessel(s) in the brain. As reported in the discussion of the neuroprotective effects of *Vaccinium* fruit, long-term blueberry feeding mitigated damage to specific regions of the hippocampus in a rat model of ischaemic stroke (Sweeney *et al.*, 2002). Results supporting a beneficial role for blueberries in reducing ischaemic stroke damage were also later reported by Wang *et al.* (2005).

4.3.6 Evidence for cancer chemoprevention

Neto (2007) has recently reviewed the literature on *Vaccinium* fruit in cancer chemoprevention. In this review, the anticancer bioactivity of *Vaccinium* (mainly blueberries and cranberries) phytochemicals (mainly phenolics) is illustrated in a wide range of *in vitro* cancer

cell models. Such *in vitro* models can provide valuable mechanistic information regarding the initiation and progression of cancerous cellular metabolism. To date, anticancer mechanisms suggested for *Vaccinium* phytochemicals include: (i) induction of enzymes involved in the detoxification of carcinogens (e.g. quinone reductase); (ii) promotion of programmed cell death (i.e. apoptosis); and (iii) inhibition of enzymes involved in metastasis (i.e. matrix metalloproteinases). The effect of *Vaccinium* phytochemicals on apoptosis is the most widely cited.

While *in vitro* studies, whether related to cancer or other diseases, can provide valuable information regarding possible mechanisms of action, significant questions remain regarding the ability of specific phytochemicals to provide protection *in vivo*. Indeed, in order to be considered beneficial even *in vitro*, effects must be specific to the cells that manifest the disease (i.e. cancer) instead of affecting both diseased and normal cells. For *in vitro* effects to translate into significant human health benefits, their effects must be demonstrable *in vivo*. Implicit in the demonstration of *in vivo* effects is that putative beneficial phytochemicals are sufficiently bioavailable and effective even after their metabolism by the body. Since it is often difficult to purify sufficient quantities of specific target compounds to conduct long-term *in vivo* studies, beneficial effects are demonstrated most often after feeding these compounds in a complex food source.

The low *in vivo* bioavailability of phenolics may limit their potential benefit in tissues, with one possible exception. Since the concentration of phenolics remains high in the gastrointestinal tract, these compounds (and their gut microfloral catabolites) may protect gastrointestinal tissues against dietary prooxidant and carcinogenic molecules (e.g. nitrosamines) (Halliwell, 2007).

4.4 Effect of Preharvest and Postharvest Continuum

Many factors can affect the composition of blueberry and cranberry fruit and their products, which in turn may impact their potential health benefits. While the mechanisms by which these fruit affect human health and well-being continue to be pursued, the effects of pre- and postharvest factors on their health benefits have not been determined. However, research has been conducted to determine the effects of these factors on the chemical composition and properties of the fruit. Therefore, we will discuss factors that alter the phytochemical profile and properties that have been implicated to contribute to the biological effects of these fruit.

4.4.1 Cultivar

Genotype has a large effect on determining the chemical composition of blueberry and cranberry fruit, including the content of phenolic and other health-promoting compounds. Numerous studies have demonstrated a wide variation in the total phenolic content and the antioxidant capacity of different blueberry genotypes (Ehlenfeldt and Prior, 2001; Kalt *et al.*, 2001; Connor *et al.*, 2002b; Howard *et al.*, 2003). In a survey of 80 highbush cultivars and 155 lowbush clones, chemical composition varied, especially for the lowbush clones, where total anthocyanins, total phenolics and antioxidant capacity varied 5-, 1.6- and 3.3-fold, respectively (Kalt *et al.*, 2001). In a different survey of 87 highbush and species-introgressed highbush blueberry cultivars, similar variation in anthocyanins, phenolics and antioxidant capacity was reported (Ehlenfeldt and Prior, 2001). Extremes in anthocyanin and total phenolic concentration and ORAC antioxidant capacity varied 3-, 10- and 6.8-fold, respectively. In a study with 52 genotypes, Connor *et al.* (2002b) found that fruit from genotypes with ancestry from *V. myrtilloides* and *V. constablaei* × *V. ashei* had the highest antioxidant capacity, while those from *V. corymbosum* and *V. angustifolium* had similar but lower capacities.

In addition to total phenolic concentration and antioxidant capacity, blueberry genotypes can vary substantially in their phenolic composition. Mi *et al.* (2004) compared the flavonoid glycoside content of five blueberry genotypes comprised of a northern and a southern highbush cultivar and three advanced

interspecies hybrid selections. Total anthocyanin concentration ranged from 1.4 to 8.2 mg/g FW, but the relative distribution of anthocyanidins among the genotypes was similar. The blueberry cv. Bluecrop and the selection A-98 were the only genotypes that had a significant concentration of acylated anthocyanins, and Bluecrop had about half the concentration of anthocyanin galactosides and 7- to 13-fold more glucosides than the other genotypes. The predominant flavonol glycoside also varied among genotypes, being quercetin 3-galactoside in the three advanced selections, quercetin 3-glucoside + rutinoside in the blueberry cv. Bluecrop and quercetin 3-acetylrhamnoside in blueberry cv. Ozarkblue. In a comparison of three rabbiteye and three highbush blueberry cultivars, total anthocyanin concentration was greater in the rabbiteye cultivars, ranging from 10.1 to 13.7 compared with 5.8 to 9.6 g cyanidin 3-glucoside equivalents/kg DW for the highbush cultivars (Lohachoompol *et al.*, 2008). Anthocyanin concentration among these cultivars was similar, but proportions of each compound were cultivar dependent. Concentrations of chlorogenic acid also varied among 38 blueberry cultivars, ranging from 0.24 to 1.11 mg/g FW (Giongo *et al.*, 2006).

The heritability of anthocyanin and phenolic content as well as antioxidant capacity in blueberry fruit has been assessed to determine the feasibility of breeding new cultivars that produce fruit with enhanced healthful properties. In a preliminary study, Ehlenfeldt and Prior (2001) suggested that inheritance of the antioxidant capacity of fruit may be additive, based on the evaluation of 11 highbush blueberry cultivar pedigrees. However, data from the Rubel × Duke family suggested that antioxidant content was controlled by epistatic gene action in the blueberry cv. Rubel, which was broken up when it was used as a parent. This might explain why the high antioxidant capacity of the blueberry cv. Rubel was not transferred effectively to progeny. In a heritability study using 20 crosses of *V. corymbosum*, *V. angustifolium* and hybrids between these species, Connor *et al.* (2002c) found narrow-sense heritability estimates of 0.43, 0.46 and 0.56 for antioxidant capacity, total phenolics and total anthocyanins, respectively.

The authors suggested that these moderate values indicated that reasonable progress could be made to improve these traits through breeding.

Anthocyanin content of cranberry fruit also varies among cultivars. The average total anthocyanin concentration of dark red fruit from six cultivars ranged from 0.30 to 0.63 mg/g FW, with smaller fruit having higher concentrations than larger fruit (Sapers *et al.*, 1986a). Similarly, when the total anthocyanins of 12 cultivars were measured over five seasons, concentrations ranged from 0.43 to 0.95 mg/g FW (Schmid, 1977). However, the anthocyanin content within a cultivar can be highly variable among individual fruit, with some having up to fourfold greater anthocyanin concentrations than the mean (Sapers *et al.*, 1986b).

Flavonol composition of cranberry fruit has also been shown to differ among cultivars. The concentration of quercetin and myricetin in six cranberry cultivars ranged from 0.07 to 0.25 and 0.004 to 0.027 mg/g FW, respectively (Bilyk and Sapers, 1986). Within a cultivar, the concentration of each of these flavonols was greater in dark red fruit than in medium red fruit. Kaempferol was detected in only three of the six cultivars at concentrations ranging from 0.001 to 0.003 mg/g FW.

4.4.2 Other preharvest factors

While genetics play a major role in fruit phenolic composition, variation within cultivars is also influenced by the growing environment. Flavonoid biosynthesis can be enhanced by a variety of environmental conditions including light, UV-radiation, water stress, temperature, ozone and pathogen infection (Treutter, 2005). When fruit were harvested over 2 years from 16 highbush and interspecific hybrid blueberry cultivars grown in three locations (Minnesota, Michigan and Oregon), their phenolic and anthocyanin content and antioxidant capacity varied significantly due to both year and location (Connor *et al.*, 2002d). Differences in total phenolic and total anthocyanin content among fruit of the same cultivar harvested from different locations and years ranged from 1.1- to 2.0-fold and 1.2- to 2.0-fold, respectively. Similarly, variation in

antioxidant capacity was observed to range from 1.2- to 2.6-fold. Total anthocyanin and hydroxycinnamic acid concentration and antioxidant capacity of southern highbush blueberry fruit also varied significantly between growing seasons (Howard *et al.*, 2003).

A variety of cultural factors may influence fruit composition. Blueberry fruit produced using organic cultural practices had higher phenolic content than fruit produced using conventional culture, which might have been a result of higher levels of stress arising as a result of the organic cultural methods (Wang, S.Y. *et al.*, 2008). The total phenolics, total anthocyanins and antioxidant capacity of organic-grown blueberries cv. Bluecrop from five commercial farms averaged 1.7-, 1.6- and 1.5-fold higher, respectively, than fruit produced conventionally. Treatment of rabbiteye blueberry fruit with ethephon, an ethylene-releasing compound, stimulated anthocyanin formation 1.9- and 2.2-fold 4 and 8 days after application, respectively (Ban *et al.*, 2007). Treatment of lowbush blueberries cv. Fundy with methyl jasmonate 3 weeks prior to harvest caused a slight increase in total phenolic and anthocyanin concentration by reducing fruit size, but treatments with ReTain®, riboflavin or abscisic acid (ABA) had no effect (Percival and MacKenzie, 2007). Light exposure of cranberry fruit stimulates anthocyanin synthesis both on and off the plant. On the plant, red light was shown to be most effective in stimulating anthocyanin production when compared with white, far-red, green or UV light (Zhou and Singh, 2002). Stimulation of the production of specific anthocyanins was also affected by light quality. Red and far-red light was more effective in stimulating production of cyanidin 3-glucoside than natural sunlight, but less effective in stimulating the production of cyanidin 3-galactoside, cyanidin 3-arabinoside, peonidin 3-galactoside and peonidin 3-glucoside (Zhou and Singh, 2004).

Ripeness affects the phenolic composition of *Vaccinium* fruit significantly, particularly when anthocyanin concentration increases dramatically as berry surface colour changes from green to red and blue. In highbush blueberry fruit, total phenolics, hydroxycinnamic acids and flavonols all decreased as fruit ripened from green to blue, while anthocyanins increased during blue colour formation (Kalt *et al.*, 2003; Castrejón *et al.*, 2008). The antioxidant capacity of the fruit decreased as the fruit matured and ripened but, depending on cultivar, it could also increase slightly as ripe fruit continued to accumulate anthocyanins (Castrejón *et al.*, 2008). Among different ripeness stages of lowbush blueberries, total anthocyanin concentration increased from 0 to about 11 mg/g DW in fully ripe fruit, while levels of chlorogenic acid did not differ among slightly unripe, ripe and overripe fruit (Kalt and McDonald, 1996). Late harvested rabbiteye blueberry fruit had greater anthocyanin and total phenolic content than fruit harvested during a normal commercial harvest (Prior *et al.*, 1998). Similar changes in fruit phenolic content occurred during ripening of cranberry fruit. In cranberry fruit cvs. Ben Lear and Stevens, anthocyanins increased dramatically as fruit ripened, while flavonols and PACs tended to decline during fruit growth but then remained constant or increased during ripening (Vvedenskaya *et al.*, 2004). Ascorbic acid declined during cranberry cv. Pilgram fruit ripening, going from 2.1 to 0.7 mg/ml in green and dark red fruit, respectively (Çelik *et al.*, 2008).

4.4.3 Postharvest handling and storage

When ripe blueberry fruit are stored, changes in anthocyanins, phenolics and antioxidant capacity are minimal. Total phenolic and anthocyanin concentration and antioxidant capacity of fruit representing eight cultivars of highbush blueberries did not change significantly during storage at 5°C for up to 7 weeks, with the exception of blueberry cv. MSU-58, which had a 29% increase in antioxidant capacity (Connor *et al.*, 2002a). Storing fully ripe fruit of the highbush blueberry cv. Bluecrop and a 20-clone mixture of lowbush blueberry fruit for up to 8 days at 0, 10, 20 or 30°C had no significant effect on total phenolic content or ORAC antioxidant capacity (Kalt *et al.*, 1999a). However, in highbush blueberry fruit, total anthocyanins increased 1.2-fold after 8 days at 20°C. In a different study, total anthocyanin content of lowbush blueberry clones increased an average of 18% during 2 weeks of storage at 1°C (Kalt and McDonald, 1996).

When blueberry fruit are harvested underripe, greater postharvest changes in anthocyanin content are seen. Highbush blueberry fruit harvested partially pink or partially blue were able to produce anthocyanins when held at 20°C for up to 8 days (Kalt *et al.*, 2003). If fruit had some blue at harvest, they were able to reach an anthocyanin concentration similar to fruit harvested fully blue. In addition, total phenolic content of these underripe fruit increased during storage. Similar results were reported by Connor *et al.* (2002a) where fruit of the blueberry cv. Elliot harvested before becoming fully blue increased in anthocyanin and phenolic content and antioxidant capacity during storage at 5°C for 3 or more weeks.

Controlled atmosphere storage, comprised of O_2 and CO_2 concentrations ranging from 1 to 15 and 0 to 15 kPa, respectively, of three cultivars of highbush blueberries at 0°C had no beneficial effects on preserving phenolics, anthocyanins, or antioxidant capacity of the fruit (Forney *et al.*, 2008). After 9 weeks of storage, total phenolics, total anthocyanins and antioxidant capacity decreased 5–16%, 8–18% and 6–14%, respectively, depending on cultivar, but this decline was not affected by storage atmosphere composition.

Other postharvest treatments of blueberry fruit have had variable effects on fruit composition. Irradiation of blueberries with 2 or 4 kJ/ m^2 UV-C irradiation increased the anthocyanin content of fruit of the blueberry cv. Bluecrop by 10% following storage for 7 days at 5°C plus 2 days at 20°C (Perkins-Veazie *et al.*, 2008). In addition, 4 kJ/m^2 UV-C irradiation increased the antioxidant capacity of Bluecrop fruit. However 2 kJ/m^2 UV-C radiation reduced total anthocyanin and phenolic concentration and antioxidant capacity in blueberry cv. Collins fruit following storage. Treatment with any of three naturally occurring essential oils, carvacrol, anethole and perillaldehyde, increased total anthocyanin and total phenolic concentration of blueberry cv. Duke fruit about 1.1- to 1.2-fold and antioxidant capacity 1.4- to 1.6-fold (Wang, C.Y. *et al.*, 2008).

Changes in the phenolic composition of cranberry fruit appear to be greater than those of blueberry fruit during storage. Wang and Stretch (2001) reported that total anthocyanin and total phenolic content as well as antioxidant capacity increased during 3 months of storage at temperatures ranging from 0 to 20°C in ten cultivars. Increases were greatest at 15°C, where total anthocyanins, total phenolics and antioxidant capacity increased by an average of 2.3-, 1.5- and 1.5-fold, respectively. However, storing cranberry fruit in controlled atmospheres of 0–70 kPa O_2 with 0–30 kPa CO_2 at 3°C had no effect on total phenolic or total flavonoid content and inhibited the increase in antioxidant capacity that occurred in air-stored fruit (Gunes et al., 2002). Postharvest treatment with ethylene and exposure to light increased anthocyanin content (Craker, 1971). Postharvest light exposure also increased the total phenolic content and antioxidant capacity of fruit, but ABA treatments had no effect (Forney et al., 2009).

4.4.4 Processing

Processing can reduce the phenolic content and antioxidant capacity of blueberry and cranberry fruit. Maceration, heat and various processes can cause oxidation, thermal degradation, leaching and other events that can reduce the fruit's healthful properties (Kalt, 2005). In a survey of processed blueberry products, the greater the processing the lower the antioxidant capacity, with fresh and frozen fruit having the highest levels and products subjected to extensive heat or drying having the lowest (Kalt et al., 2000). When compared with fresh or frozen fruit, drying blueberries reduced their anthocyanin concentration 41%, which was reduced further by 49% if preceded by osmotic dehydration (Lohachoompol et al., 2004). Processing results in a loss of monomeric anthocyanins due to their enzymatic polymerization or degradation, which continues during storage (Brownmiller et al., 2008). Heating and storage conditions reduced phenolic concentration and biological activity, measured as inhibition of cancer cell proliferation, in blueberry extracts stored

in glass bottles following several months of storage (Srivastava et al., 2007). In canned products, significant leakage of anthocyanins can occur from the fruit into the liquid canning medium.

Approaches to reduce the impact of processing on degradation of fruit composition have been identified. More gentle processing methods such as freeze-drying and hot air-drying/microwave vacuum-drying retained higher concentrations of phenolics and anthocyanins than hot air-drying or microwave vacuum-drying alone (Mejia-Meza et al., 2008). Blanching frozen fruit at 95°C for 2 min or adding 50 µl/l SO_2 increased the yield of anthocyanins in extracted blueberry juice (Lee et al., 2002). Low pH and exclusion of oxygen also helped to preserve anthocyanins, phenolics and antioxidant capacity and prevent brown colour formation in blueberry juice due to polymeric phenolics (Kalt et al., 2000).

Cranberry composition may also be altered during processing. Cranberry juice had primarily monomer, dimer and A-type trimer PACs and lacked the higher oligomers observed in whole fruit (Prior et al., 2001). In addition, cranberry fruit processed into juice or powder contains the flavonol aglycones, myricetin and quercetin, as well as quercetin-3-O-(6″-benzoyl)-β-galactoside, which were not found in whole fruit (Vvedenskaya et al., 2004). In a study comparing the stability of phenolics and antioxidant capacity of six fruit juices, including blueberry and cranberry, cranberry juice was found to be the most stable, having the least loss of antioxidant capacity after 29 days of storage (Piljac-Zegarac et al., 2009).

4.5 Conclusions

Evidence continues to accumulate to support a role for both blueberries and cranberries in human health. Based on the nature and strength of the evidence described above, one can conclude that blueberries are most notable for their ability to protect the brain during ageing and under the stresses imposed by neurodegenerative disease. Cranberries are most noted for their beneficial effects in urinary tract health and, due to related properties, may also confer benefits in oral health. Both blueberries and cranberries are generally considered to protect cardiovascular health, and have been found to provide benefits in various models of cancer.

The preponderance of evidence suggests that the phenolic compounds, and particularly the flavonoids, are the principal bioactive components in these berries. *Vaccinium* berries, compared with other fruit and even other berries, have a high concentration of phenolics. Indeed, the high concentration of specific flavonoids is apparent in both these berries. The deep blue coloration of blueberries is due to their high concentration of anthocyanins, and the astringency of cranberries is due to their high concentration of PACs. Since all phenolics possess potent antioxidant properties, both cranberries and blueberries are notable for their high antioxidant capacity compared with other plant foods. Typically, the concentration of total phenolics correlates very well with antioxidant capacity. Although oxidative stress plays a significant role as an underlying element in the degenerative processes of ageing and disease, a decade of research has shown that dietary phenolics per se do not contribute significantly to the antioxidant defences of the body. The reason for their limited role as biological antioxidants is that they are poorly absorbed by the body, and therefore do not bolster the already abundant antioxidant defence machinery present. This finding is an excellent example of how *in vitro* results cannot be interpreted as strong evidence for *in vivo* benefits. In spite of this new knowledge, research continues to demonstrate that *Vaccinium* fruit do provide benefits *in vivo*; however, not likely due to their antioxidant effects. The property of flavonoids as anti-inflammatory agents is highly significant and we can expect that this functionality will continue to be explored in various models of disease and ageing.

Finally, biomedical research is also showing that the specific structure of flavonoids influences their health functionality strongly, whether it is their ability to be absorbed by the body, or their activity in a particular physiological condition. This structure/function specificity creates many exciting opportunities for the fields of horticulture, food chemistry and engineering as specific phytochemical targets are revealed through biomedical research.

Concentrations of the phytochemicals in blueberry and cranberry fruit that have been associated with health benefits can be affected by numerous factors. Genetic manipulation of fruit, particularly through the development of new cultivars, holds the greatest potential for enhancement of specific phytochemicals of interest. Other environmental factors that are important during production or postharvest handling of the fruit can influence the concentration of these phytochemicals and could be used to enhance further the gains obtained through genetics. Processing of fruit can degrade phytochemical content and therefore innovative new methods are needed to preserve fruit health functionality in processed products. As specific modes of action are identified, more directed methods to enhance and preserve their properties can be developed.

References

Ames, B.N., Shigenaga, M.K. and Hagen, T.M. (1993) Oxidants, antioxidants, and the degenerative diseases of aging. *Proceedings of the National Academy of Sciences of the United States of America* 90, 7915–7922.

Ban, T., Kugishima, M., Ogata, T., Shiozaki, S., Horiuchi, S. and Ueda, H. (2007) Effect of ethephon (2-chloroethylphosphonic acid) on the fruit ripening characters of rabbiteye blueberry. *Scientia Horticulturae* 112, 278–281.

Bilyk, A. and Sapers, G.M. (1986) Varietal differences in the quercetin, kaempferol, and myricetin contents of highbush blueberry, cranberry, and thornless blackberry fruits. *Journal of Agricultural and Food Chemistry* 34, 585–588.

Bodet, C., Grenier, D., Chandad, F., Ofek, I., Steinberg, D. and Weiss, E.I. (2008) Potential oral health benefits of cranberry. *Critical Reviews in Food Science and Nutrition* 48, 672–680.

Brownmiller, C., Howard, L.R. and Prior, R.L. (2008) Processing and storage effects on monomeric anthocyanins, percent polymeric color, and antioxidant capacity of processed blueberry products. *Journal of Food Science* 73, H72–H79.

Canter, P.H. and Ernst, E. (2004) Anthocyanosides of *Vaccinium myrtillus* (bilberry) for night vision – a systematic review of placebo-controlled trials. *Survey of Ophthalmology* 49, 38–50.

Carew, R., Florkowski, W.J. and He, S. (2006) Contribution of health attributes, research investment and innovation to developments in the blueberry industry: a Canada–US comparison. *International Journal of Fruit Science* 6, 23–46.

Casadesus, G., Shukitt-Hale, B., Stellwagen, H.M., Zhu, X., Lee, H.G., Smith, M.A. and Joseph, J.A. (2004) Modulation of hippocampal plasticity and cognitive behavior by short-term blueberry supplementation in aged rats. *Nutritional Neuroscience* 7, 309–316.

Castrejón, A.D.R., Eichholz, I., Rohn, S., Kroh, L.W. and Huyskens-Keil, S. (2008) Phenolic profile and antioxidant activity of highbush blueberry (*Vaccinium corymbosum* L.) during fruit maturation and ripening. *Food Chemistry* 109, 564–572.

CCCGA (Cape Cod Cranberry Grower's Association) (2009) Cranberries (http://www.cranberries.org/ accessed 29 January 2009).

Çelik, H., Özgen, M., Serçe, S. and Kaya, C. (2008) Phytochemical accumulation and antioxidant capacity at four maturity stages of cranberry fruit. *Scientia Horticulturae* 117, 345–348.

Chandler, R.F., Freeman, L. and Hooper, S.N. (1979) Herbal remedies of the maritime Indians. *Journal of Ethnopharmacology* 1, 49–68.

Connor, A.M., Luby, J.J., Hancock, J.F., Berkheimer, S. and Hanson, E.J. (2002a) Changes in fruit antioxidant activity among blueberry cultivars during cold-temperature storage. *Journal of Agricultural and Food Chemistry* 50, 893–898.

Connor, A.M., Luby, J.J. and Tong, C.B.S. (2002b) Variability in antioxidant activity in blueberry and correlations among different antioxidant activity assays. *Journal of the American Society for Horticultural Science* 127, 238–244.

Connor, A.M., Luby, J.J. and Tong, C.B.S. (2002c) Variation and heritability estimates for antioxidant activity, total phenolic content, and anthocyanin content in blueberry progenies. *Journal of the American Society for Horticultural Science* 127, 82–88.

Connor, A.M., Luby, J.J., Tong, C.B.S., Finn, C.E. and Hancock, J.F. (2002d) Genotypic and environmental variation in antioxidant activity, total phenolic content, and anthocyanin content among blueberry cultivars. *Journal of the American Society for Horticultural Science* 127, 89–97.

Craker, L.E. (1971) Postharvest color promotion in cranberry with ethylene. *HortScience* 6, 137–139.

Duarte, S., Gregoire, S., Singh, A.P., Vorsa, N., Schaich, K., Bowen, W.H. and Koo, H. (2006) Inhibitory effects of cranberry polyphenols on formation and acidogenicity of *Streptococcus mutans* biofilms. *FEMS Microbiology Letters* 257, 50–56.

Ehlenfeldt, M.K. and Prior, R.L. (2001) Oxygen radical absorbance capacity (ORAC) and phenolic and anthocyanin concentrations in fruit and leaf tissues of highbush blueberry. *Journal of Agricultural and Food Chemistry* 49, 2222–2227.

Forney, C.F., Kalt, W., Jordan, M.A., Vinquist-Tymchuk, M.R. and Fillmore, S.A.E. (2008) Effects of controlled atmosphere storage on antioxidant capacity and phenolics in three highbush blueberry cultivars. *HortScience* 43, 1169.

Forney, C.F., Kalt, W., Abrams, S.R. and Owens, S.J. (2009) Effects of postharvest light and ABA treatments on the composition of white cranberry fruit. *Acta Horticulturae* 810, 799–806.

Giongo, L., Ieri, F., Vrhovsek, U., Grisenti, M., Mattivi, F. and Eccher, M. (2006) Characterization of *Vaccinium* cultivars: horticultural and antioxidant profile. *Acta Horticulturae* 715, 147–151.

Goyarzu, P., Malin, D.H., Lau, F.C., Taglialatela, G., Moon, W.D., Jennings, R., Moy, E., Moy, D., Lippold, S., Shukitt-Hale, B. and Joseph, J.A. (2004) Blueberry supplemented diet: effects on object recognition memory and nuclear factor-kappa B levels in aged rats. *Nutritional Neuroscience* 7, 75–83.

Gu, L., Kelm, M.A., Hammerstone, J.F., Beecher, G., Holden, J., Haytowitz, D. and Prior, R.L. (2003) Screening of foods containing proanthocyanidins and their structural characterization using LC-MS/MS and thiolytic degradation. *Journal of Agricultural and Food Chemistry* 51, 7513–7521.

Gunes, G., Liu, R.H. and Watkins, C.B. (2002) Controlled-atmosphere effects on postharvest quality and antioxidant activity of cranberry fruits. *Journal of Agricultural and Food Chemistry* 50, 5932–5938.

Häkkinen, S., Heinonen, M., Kärenlampi, S., Mykkänen, H., Ruuskanen, J. and Törrönen, R. (1999) Screening of selected flavonoids and phenolic acids in 19 berries. *Food Research International* 32, 345–353.

Halliwell, B. (2007) Dietary polyphenols: good, bad, or indifferent for your health? *Cardiovascular Research* 73, 341–347.

Howard, L.R., Clark, J.R. and Brownmiller, C. (2003) Antioxidant capacity and phenolic content in blueberries as affected by genotype and growing season. *Journal of the Science of Food and Agriculture.* 83, 1238-1247.

Howell, A.B. (2002) Cranberry proanthocyanidins and the maintenance of urinary tract health. *Critical Reviews in Food Science and Nutrition* 42, 273–278.

Howell, A.B. (2007) Bioactive compounds in cranberries and their role in prevention of urinary tract infections. *Molecular Nutrition and Food Research* 51, 732–737.

Jepson, R.G. and Craig, J.C. (2007) A systematic review of the evidence for cranberries and blueberries in UTI prevention. *Molecular Nutrition and Food Research* 51, 738–745.

Jepson, R.G. and Craig, J.C. (2008) Cranberries for preventing urinary tract infections. *Cochrane Database of Systematic Reviews* Issue 1, Art No: CD001321, DOI: 10.1002/14651858.CD001321.pub4 (http://www2.cochrane.org/reviews/en/ab001321.html, accessed January 2009).

Joseph, J.A., Shukitt-Hale, B., Denisova, N.A., Bielinski, D., Martin, A., McEwen, J.J. and Bickford, P.C. (1999) Reversals of age-related declines in neuronal signal transduction, cognitive, and motor behavioral deficits with blueberry, spinach, or strawberry dietary supplementation. *Journal of Neuroscience* 19, 8114–8121.

Joseph, J.A., Denisova, N.A., Arendash, G., Gordon, M., Diamond, D., Shukitt-Hale, B. and Morgan, D. (2003) Blueberry supplementation enhances signaling and prevents behavioral deficits in an Alzheimer disease model. *Nutritional Neuroscience* 6, 153–162.

Joseph, J.A., Shukitt-Hale, B. and Lau, F.C. (2007) Fruit polyphenols and their effects on neuronal signaling and behavior in senescence. *Annals of the New York Academy of Sciences* 1100, 470–485.

Kahlon, T.S. and Smith, G.E. (2007) *In vitro* binding of bile acids by blueberries (*Vaccinium* spp.), plums (*Prunus* spp.), prunes (*Prunus* spp.), strawberries (*Fragaria* X *ananassa*), cherries (*Malpighia punicifolia*), cranberries (*Vaccinium macrocarpon*) and apples (*Malus sylvestris*). *Food Chemistry* 100, 1182–1187.

Kalt, W. (2005) Effects of production and processing factors on major fruit and vegetable antioxidants. *Journal of Food Science* 70, R11–R19.

Kalt, W. and Dufour, D. (1997) Health functionality of blueberries. *HortTechnology* 7, 216–221.

Kalt, W. and McDonald, J.E. (1996) Chemical composition of lowbush blueberry cultivars. *Journal of the American Society for Horticultural Science* 121, 142–146.

Kalt, W., Forney, C.F., Martin, A. and Prior, R.L. (1999a) Antioxidant capacity, vitamin C, phenolics, and anthocyanins after fresh storage of small fruits. *Journal of Agricultural and Food Chemistry* 47, 4638–4644.

Kalt, W., McDonald, J.E., Ricker, R.D. and Lu, X. (1999b) Anthocyanin content and profile within and among blueberry species. *Canadian Journal of Plant Science* 79, 617–623.

Kalt, W., McDonald, J.E. and Donner, H. (2000) Anthocyanins, phenolics, and antioxidant capacity of processed lowbush blueberry products. *Journal of Food Science* 65, 390–393.

Kalt, W., Ryan, D.A.J., Duy, J.C., Prior, R.L., Ehlenfeldt, M.K. and Vander Kloet, S.P. (2001) Interspecific variation in anthocyanins, phenolics, and antioxidant capacity among genotypes of highbush and lowbush blueberries (*Vaccinium* section *cyanococcus* spp.). *Journal of Agricultural and Food Chemistry* 49, 4761–4767.

Kalt, W., Lawand, C., Ryan, D.A.J., McDonald, J.E., Donner, H. and Forney, C.F. (2003) Oxygen radical absorbing capacity, anthocyanin and phenolic content of highbush blueberries (*Vaccinium corymbosum* L.) during ripening and storage. *Journal of the American Society for Horticultural Science* 128, 917–923.

Kalt, W., Foote, K., Fillmore, S.A.E., Lyon, M., Van Lunen, T.A. and McRae, K.B. (2008a) Effect of blueberry feeding on plasma lipids in pigs. *British Journal of Nutrition* 100, 70–78.

Kalt, W., MacKinnon, S., McDonald, J., Vinqvist, M., Craft, C. and Howell, A. (2008b) Phenolics of *Vaccinium* berries and other fruit crops. *Journal of the Science of Food and Agriculture* 88, 68–76.

Keevil, J.G., Osman, H.E., Reed, J.D. and Folts, J.D. (2000) Grape juice, but not orange juice or grapefruit juice, inhibits human platelet aggregation. *Journal of Nutrition* 130, 53–56.

Lau, F.C., Bielinski, D.F. and Joseph, J.A. (2007) Inhibitory effects of blueberry extract on the production of inflammatory mediators in lipopolysaccharide-activated BV2 microglia. *Journal of Neuroscience Research* 85, 1010–1017.

Lee, J., Durst, R.W. and Wrolstad, R.E. (2002) Impact of juice processing on blueberry anthocyanins and polyphenolics: comparison of two pretreatments. *Journal of Food Science* 67, 1660–1667.

Liu, Y., Black, M.A., Caron, L. and Camesano, T.A. (2006) Role of cranberry juice on molecular-scale surface characteristics and adhesion behavior of *Escherichia coli*. *Biotechnology and Bioengineering* 93, 297–305.

Lohachoompol, V., Srzednicki, G. and Craske, J. (2004) The change of total anthocyanins in blueberries and their antioxidant effect after drying and freezing. *Journal of Biomedicine and Biotechnology* 2004(5), 248–252.

Lohachoompol, V., Mulholland, M., Srzednicki, G. and Craske, J. (2008) Determination of anthocyanins in various cultivars of highbush and rabbiteye blueberries. *Food Chemistry* 111, 249–254.

McGuire, S.O., Sortwell, C.E., Shukitt-Hale, B., Joseph, J.A., Hejna, M.J. and Collier, T.J. (2006) Dietary supplementation with blueberry extract improves survival of transplanted dopamine neurons. *Nutritional Neuroscience* 9, 251–258.

Matsumoto, H., Nakamura, Y., Tachibanaki, S., Kawamura, S. and Hirayama, M. (2003) Stimulatory effect of cyanidin 3-glycosides on the regeneration of rhodopsin. *Journal of Agricultural and Food Chemistry* 51, 3560–3563.

Mejia-Meza, E.I., Yanez, J.A., Davies, N.M., Rasco, B., Younce, F., Remsberg, C.M. and Clary, C. (2008) Improving nutritional value of dried blueberries (*Vaccinium corymbosum* L.) combining microwave-vacuum, hot-air drying and freeze drying technologies. *International Journal of Food Engineering* 4(5), Article 5, DOI: 10.2202/1556-3758.1364 (http://www.bepress.com/ijfe/vol4/iss5/art5).

Mi, J.C., Howard, L.R., Prior, R.L. and Clark, J.R. (2004) Flavonoid glycosides and antioxidant capacity of various blackberry, blueberry and red grape genotypes determined by high-performance liquid chromatography/mass spectrometry. *Journal of the Science of Food and Agriculture* 84, 1771–1782.

Moyer, R.A., Hummer, K.E., Finn, C.E., Frei, B. and Wrolstad, R.E. (2002) Anthocyanins, phenolics, and antioxidant capacity in diverse small fruits: *Vaccinium*, *Rubus*, and *Ribes*. *Journal of Agricultural and Food Chemistry* 50, 519–525.

Neto, C.C. (2007) Cranberry and blueberry: evidence for protective effects against cancer and vascular diseases. *Molecular Nutrition and Food Research* 51, 652–664.

Percival, D. and MacKenzie, J.L. (2007) Use of plant growth regulators to increase polyphenolic compounds in the wild blueberry. *Canadian Journal of Plant Science* 87, 333–336.

Perkins-Veazie, P., Collins, J.K. and Howard, L. (2008) Blueberry fruit response to postharvest application of ultraviolet radiation. *Postharvest Biology and Technology* 47, 280–285.

Piljac-Zegarac, J., Valek, L., Martinez, S. and Belscak, A. (2009) Fluctuations in the phenolic content and antioxidant capacity of dark fruit juices in refrigerated storage. *Food Chemistry* 113, 394–400.

Pollack, S. and Perez, A. (2008) *Fruit and Tree Nuts Situation and Outlook Yearbook 2008* (http://www.ers.usda.gov/Publications/FTS/Yearbook08/FTS2008.pdf, accessed 10 December 2008).

Prior, R.L., Cao, G., Martin, A., Sofic, E., McEwen, J., O'Brien, C., Lischner, N., Ehlenfeldt, M., Kalt, W., Krewer, G. and Mainland, C.M. (1998) Antioxidant capacity as influenced by total phenolic and anthocyanin content, maturity, and variety of *Vaccinium* species. *Journal of Agricultural and Food Chemistry* 46, 2686–2693.

Prior, R.L., Lazarus, S.A., Cao, G., Muccitelli, H. and Hammerstone, J.F. (2001) Identification of procyanidins and anthocyanins in blueberries and cranberries (*Vaccinium* spp.) using high-performance liquid chromatography/mass spectrometry. *Journal of Agricultural and Food Chemistry* 49, 1270–1276.

Rabin, B.M., Joseph, J.A. and Shukitt-Hale, B. (2005) Effects of age and diet on the heavy particle-induced disruption of operant responding produced by a ground-based model for exposure to cosmic rays. *Brain Research* 1036, 122–129.

Ramassamy, C. (2006) Emerging role of polyphenolic compounds in the treatment of neurodegenerative diseases: a review of their intracellular targets. *European Journal of Pharmacology* 545, 51–64.

Riihinen, K., Jaakola, L., Kärenlampi, S. and Hohtola, A. (2008) Organ-specific distribution of phenolic compounds in bilberry (*Vaccinium myrtillus*) and 'northblue' blueberry (*Vaccinium corymbosum* × *V. angustifolium*). *Food Chemistry* 110, 156–160.

Rimando, A.M., Kalt, W., Magee, J.B., Dewey, J. and Ballington, J.R. (2004) Resveratrol, pterostilbene, and piceatannol in *Vaccinium* berries. *Journal of Agricultural and Food Chemistry* 52, 4713–4719.

Rimando, A.M., Nagmani, R., Feller, D.R. and Yokoyama, W. (2005) Pterostilbene, a new agonist for the peroxisome proliferator-activated receptor alpha-isoform, lowers plasma lipoproteins and cholesterol in hypercholesterolemic hamsters. *Journal of Agricultural and Food Chemistry* 53, 3403–3407.

Ruel, G., Pomerleau, S., Couture, P., Lemieux, S., Lamarche, B. and Couillard, C. (2008) Low-calorie cranberry juice supplementation reduces plasma oxidized LDL and cell adhesion molecule concentrations in men. *British Journal of Nutrition* 99, 352–359.

Sapers, G.M., Graff, G.R., Phillips, J.G. and Deubert, K.H. (1986a) Factors affecting the anthocyanin content of cranberry. *Journal of the American Society for Horticultural Science* 111, 612–617.

Sapers, G.M., Jones, S.B. and Kelley, M.J. (1986b) Breeding strategies for increasing the anthocyanin content of cranberries. *Journal of the American Society for Horticultural Science* 111, 618–622.

Schmid, P. (1977) Long-term investigation with regard to the constituents of various cranberry varieties (*Vaccinium macrocarpon* Ait.). *Acta Horticulturae* 61, 241–254.

Schmidt, B.M., Howell, A.B., McEniry, B., Knight, C.T., Seigler, D., Erdman, J.W. Jr and Lila, M.A. (2004) Effective separation of potent antiproliferation and antiadhesion components from wild blueberry (*Vaccinium angustifolium* Ait.) fruits. *Journal of Agricultural and Food Chemistry* 52, 6433–6442.

Srivastava, A., Akoh, C.C., Yi, W., Fischer, J. and Krewer, G. (2007) Effect of storage conditions on the biological activity of phenolic compounds of blueberry extract packed in glass bottles. *Journal of Agricultural and Food Chemistry* 55, 2705–2713.

Sweeney, M.I., Kalt, W., MacKinnon, S.L., Ashby, J. and Gottschall-Pass, K.T. (2002) Feeding rats diets enriched in lowbush blueberries for six weeks decreases ischemia-induced brain damage. *Nutritional Neuroscience* 5, 427–431.

Taruscio, T.G., Barney, D.L. and Exon, J. (2004) Content and profile of flavanoid and phenolic acid compounds in conjunction with the antioxidant capacity for a variety of northwest *Vaccinium* berries. *Journal of Agricultural and Food Chemistry* 52, 3169–3176.

Tirupula, K.C., Balem, F., Yanamala, N. and Klein-Seetharaman, J. (2009) pH-dependent interaction of rhodopsin with cyanidin-3-glucoside. 2. Functional aspects. *Photochemistry and Photobiology* 85, 463–470.

Trehane, J. (2004) Cranberries. In: *Blueberries, Cranberries, and Other Vacciniums*. Timber Press, Portland, Oregon, pp. 29–73.

Treutter, D. (2005) Significance of flavonoids in plant resistance and enhancement of their biosynthesis. *Plant Biology* 7, 581–591.

USDA (2008a) *Non-citrus Fruits and Nuts 2008 Preliminary Summary* (http://usda.mannlib.cornell.edu/usda/current/NoncFruiNu/NoncFruiNu-01-23-2009.pdf, accessed 26 January 2009).

USDA (2008b) *US Blueberry Industry* (http://usda.mannlib.cornell.edu/MannUsda/viewDocumentInfo.do?documentID=1765, accessed 26 January 2009).

USHBC (United States Highbush Blueberry Commission) (2006) *An All-American Fruit Goes International* (http://www.blueberry.org/publications/bluespapers/Bluespaper%20April%202006%20Health%20Wave.pdf, accessed 26 January 2009).

Vander Kloet, S.P. (2004) *Vaccinia gloriosa*. In: Forney, C.F. and Eaton, L.J. (eds) *Proceedings of the Ninth North American Blueberry Research and Extension Workers Conference*. Food Products Press, Binghamton, New York, pp. 221–227.

Vinson, J.A., Su, X., Zubik, L. and Bose, P. (2001) Phenol antioxidant quantity and quality in foods: fruits. *Journal of Agricultural and Food Chemistry* 49, 5315–5321.

Vvedenskaya, I.O., Rosen, R.T., Guido, J.E., Russell, D.J., Mills, K.A. and Vorsa, N. (2004) Characterization of flavonols in cranberry (*Vaccinium macrocarpon*) powder. *Journal of Agricultural and Food Chemistry* 52, 188–195.

Wang, C.Y., Wang, S.Y. and Chen, C. (2008) Increasing antioxidant activity and reducing decay of blueberries by essential oils. *Journal of Agricultural and Food Chemistry* 56, 3587–3592.

Wang, S.Y. and Stretch, A.W. (2001) Antioxidant capacity in cranberry is influenced by cultivar and storage temperature. *Journal of Agricultural and Food Chemistry* 49, 969–974.

Wang, S.Y., Chen, C.T., Sciarappa, W., Wang, C.Y. and Camp, M.J. (2008) Fruit quality, antioxidant capacity, and flavonoid content of organically and conventionally grown blueberries. *Journal of Agricultural and Food Chemistry* 56, 5788–5794.

Wang, Y., Chang, C.F., Chou, J., Chen, H.L., Deng, X.L., Harvey, B.K., Cadet, J.L. and Bickford, P.C. (2005) Dietary supplementation with blueberries, spinach, or spirulina reduces ischemic brain damage. *Experimental Neurology* 193, 75–84.

Weiss, E.I., Lev-Dor, R., Kashamn, Y., Goldhar, J., Sharon, N. and Ofek, I. (1998) Inhibiting interspecies coaggregation of plaque bacteria with a cranberry juice constituent. *Journal of the American Dental Association* 129, 1719–1723.

Weiss, E.I., Lev-Dor, R., Sharon, N. and Ofek, I. (2002) Inhibitory effect of a high-molecular-weight constituent of cranberry on adhesion of oral bacteria. *Critical Reviews in Food Science and Nutrition* 42, 285–292.

Willis, L., Bickford, P., Zaman, V., Moore, A. and Granholm, A.C. (2005) Blueberry extract enhances survival of intraocular hippocampal transplants. *Cell Transplantation* 14, 213–223.

Wilson, M.A., Shukitt-Hale, B., Kalt, W., Ingram, D.K., Joseph, J.A. and Wolkow, C.A. (2006) Blueberry polyphenols increase lifespan and thermotolerance in *Caenorhabditis elegans*. *Aging Cell* 5, 59–68.

Wood, G.W. (2004) The wild blueberry industry – past. In: Forney, C.F. and Eaton, L.J. (eds) *Proceedings of the Ninth North American Blueberry Research and Extension Workers Conference*. Food Products Press, Binghamton, New York, pp. 11–18.

Wu, X. and Prior, R.L. (2005) Systematic identification and characterization of anthocyanins by HPLC-ESI-MS/MS in common foods in the United States: fruits and berries. *Journal of Agricultural and Food Chemistry* 53, 2589–2599.

Yamanaka, A., Kimizuka, R., Kato, T. and Okuda, K. (2004) Inhibitory effects of cranberry juice on attachment of oral streptococci and biofilm formation. *Oral Microbiology and Immunology* 19, 150–154.

Zheng, W. and Wang, S.Y. (2003) Oxygen radical absorbing capacity of phenolics in blueberries, cranberries, chokeberries, and lingonberries. *Journal of Agricultural and Food Chemistry* 51, 502–509.

Zhou, Y. and Singh, B.R. (2002) Red light stimulates flowering and anthocyanin biosynthesis in American cranberry. *Plant Growth Regulation* 38, 165–171.

Zhou, Y. and Singh, B.R. (2004) Effect of light on anthocyanin levels in submerged, harvested cranberry fruit. *Journal of Biomedicine and Biotechnology* 5, 259–263.

Zuo, Y., Wang, C. and Zhan, J. (2002) Separation, characterization, and quantitation of benzoic and phenolic antioxidants in American cranberry fruit by GC-MS. *Journal of Agricultural and Food Chemistry* 50, 3789–3794.

5 Brassicas

Peter Glen Walley and Vicky Buchanan–Wollaston

5.1 Introduction

Brassica vegetables have received a great deal of attention in the past 5–10 years as consumers have developed an increased awareness of the nutritional content of foodstuffs in their diet, coupled with government-led strategies to promote the consumption of at least five fruit or vegetable portions per day as part of a healthy diet in order to derive the maximum benefit from the anticarcinogenic properties, for example, that their diet can provide.

Brassica is a genus within the family Brassicaceae (Cruciferae) belonging to the order Brassicales (Hall et al., 2002; Al Shehbaz et al., 2006; Warwick et al., 2006). The Brassicaceae contains more than 3000 species in 370 genera. Several domesticated species in the genus Brassica comprise the crops related most closely to Arabidopsis thaliana (mouse-ear or thale cress), which can thus function as model eudicot crops for comparative genomics (King, 2006; Paterson, 2006). Relationships among the genomes of the Brassica spp. were first delineated by U (1935). In this work, U outlined relationships between the three main Brassica diploid crop species and their amphidiploid species (Fig. 5.1). The underlying taxonomy of the Brassicaceae is complex, with species- and genus level classification often open to question (Spooner et al., 2003).

Indeed, it has been found that Raphanus sativus (radish), although being classified as a different genus, can be crossed, although with difficulty, with B. oleracea (Karpechenko, 1927), B. rapa (Lange et al., 1989) and, with great difficulty, B. napus (Metz, 1995) (see Fig. 5.1). Lysak et al. (2005) suggest that R. sativus is related more closely to B. oleracea than is B. nigra – an observation that has been both supported and disputed (see Flannery et al., 2006; Koch et al., 2007).

5.1.1 Overview of the crop and harvested products

The growing season for broccoli in the UK is between early June to late October. Traditionally, other Brassica such as Brussels sprouts are supplied during November to April; however, increasingly the UK supply of fresh broccoli and other Brassica is supplemented by imports, for example from Spain and Portugal, throughout the year. Environmental and genetic variation in products from internal and external suppliers has a considerable effect on the shelf-life quality of the products (Jeffery et al., 2003). In terms of quantity produced worldwide, cabbage and other Brassica exceed over 69.2 million tones (Mt), with the UK producing over 250 kt of this value (FAO-STAT, 2008).

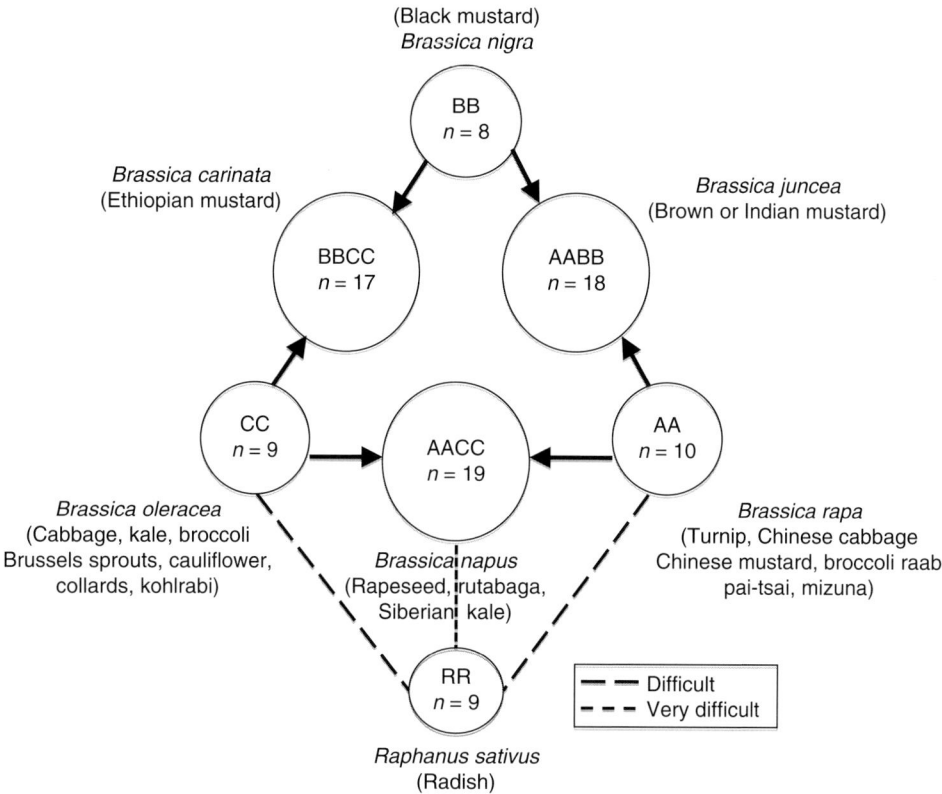

Fig. 5.1. Relationships between genomes of *Brassica* species proposed by U (1935), modified to include radish (RR), as described by Myers (2006). The lines that link radish to the *Brassica* represent the ease with which the species can be crossed. Genomes are represented by letters (A, B, C or R), with haploid chromosome numbers indicated (*n*).

The majority of *Brassica* vegetables are purchased and consumed as fresh vegetables, mostly following cooking. The time and methods of cooking have a considerable influence on nutrient availability and content (see below). There is also considerable consumption of raw or pickled cabbage in products such as coleslaw and sauerkraut. Also, the *Brassica* components of prepared salad bags (e.g. mizuna chard, kale, mustard and turnip greens, and Asian *Brassica* such as red Chinese mustard, tat soi and napa cabbage) are also eaten raw (see Chapter 10 of this volume). In addition, there is a market for frozen *Brassica* vegetables, in particular broccoli, Brussels sprouts and cauliflower.

5.2 Phytochemical Composition

5.2.1 Health-promoting compounds in *Brassica*

Epidemiological studies have shown that increased consumption of vegetables and fruit is associated with a lower risk of degenerative diseases such as cancer, cardiovascular disease, cataracts and brain and immune dysfunction (Block *et al.*, 1992; Hu, 2003). *Brassica* vegetables have been identified as important components of a healthy diet because of their high levels of constituents that may have a beneficial health-promoting role (Van Poppel *et al.*, 1999; Lampe and

Peterson, 2002; Finley, 2003b; Jeffery and Araya, 2009). These vegetables are also known to be beneficial in the prevention of other major illnesses such as Alzheimer's disease, cataracts and some of the functional declines associated with ageing (Verhoeven *et al.*, 1997). The main health-providing properties identified in *Brassica* are dietary flavonoids, essential vitamins and minerals and glucosinolates (and their breakdown products) (Heber, 2004a,b; Moreno *et al.*, 2006).

Brassica vegetables are an excellent source of a variety of vitamins, minerals and dietary fibre (Table 5.1). Vitamin and provitamin antioxidants such as ascorbic acid (vitamin C), tocopherols (vitamin E) and carotenoids are compounds present at high levels in the vegetable *Brassica* and are likely to contribute to the beneficial effects of these vegetables in the diet (Kurilich *et al.*, 1999; Jeffery and Araya, 2009). In addition, *Brassica* vegetables provide significant levels of vitamin A, B_2 (riboflavin), B_6 (pyridoxine), K and folic acid (McKillop *et al.*, 2002). Folic acid is particularly important during pregnancy. Folate supplementation prior to conception can reduce the incidence of neural tube defects significantly (Bailey and Gregory, 1999). Folate deficiencies have also been implicated in the aetiology of megaloblastic anaemia, *Spina bifida*, neuropsychiatric disorders and various forms of cancer.

Important minerals supplied by *Brassica* vegetables include calcium, potassium, iron, zinc, magnesium and selenium (Farnham *et al.*, 2000; De Pascale *et al.*, 2005; Moreno *et al.*, 2006; Broadley *et al.*, 2008). The calcium content of certain *Brassica* vegetables, including broccoli, has good bioavailability, making it a good source of calcium for lactose-intolerant people (Heaney *et al.*, 1993). Supplementation studies with high-selenium broccoli have demonstrated the efficacy of selenium for prevention of colon cancer (Finley *et al.*, 2000). The metabolism of selenium depends on its chemical form, and that which occurs in broccoli appears to be particularly effective at protecting laboratory animals against cancer (Finley, 2003, 2003b).

Flavonoids and hydroxycinnamic acids are phenolic compounds, found in many vegetables, which have been implicated as having a role in reducing the risk of heart disease (Hertog *et al.*, 1993; Knekt *et al.*, 1996) and which may act as both antioxidants and antimutagenic agents. *Brassica* vegetables such as broccoli are a significant source of such compounds (Vinson *et al.*, 1998; De Rijke *et al.*, 2006) (Table 5.2). Flavonoids such as quercetin and kaempferol occur in relatively high concentration in some *Brassica* compared with other fruit and vegetables. Also, many *Brassica* contain other phenolics, such as flavonol glycosides and hydroxycinnamic acid esters. In a comparison between ten common vegetables, broccoli was found to have the highest levels of phenolic-derived antioxidant activity (Chu *et al.*, 2002). The various *Brassica* species contain differing profiles of phenolic compounds (Heimler *et al.*, 2006), which may be relevant to their health-conferring quality.

5.2.2 Produce type specific bioactives

The anticarcinogenic properties of *Brassica* vegetables have been attributed mainly to their relatively high content of glucosinolates and other sulfur-containing compounds (Verhoeven *et al.*, 1997; Traka and Mithen, 2009). The glucosinolate core structure is made up of a β-D-thioglucose group linked to a sulfonated aldoxime moiety and a variable side-chain-derived amino acid. Glucosinolates can be grouped into three different classes: aliphatic, from aliphatic amino acids (methionine, alanine, valine, leucine, isoleucine); aromatic (tyrosine, phenylalanine); and indole (tryptophan). Structural diversity is achieved by amino acid side-chain modifications (elongations), followed by a host of specific secondary modifications to the glucosinolate side-chain (thiol oxidation, desaturation, esterification, hydroxylation) and/or the glucose moiety (Rosa, 1999; Mithen, 2001; Bones and Rossiter, 2006).

The most important glucosinolates in vegetable *Brassica* are the methionine-derived glucosinolates (Table 5.2; see Halkier and Gershenzon, 2006). Glucosinolates are hydrolysed to unstable aglucones, which rearrange into biologically active compounds, typically isothiocyanates and indoles, catalysed by myrosinase, an enzyme

Table 5.1. Nutrients in *Brassica* vegetables (per 100 g) (from USDA nutrition data, http://www.nutritiondata.com).

Produce and form tested	Kale raw	Kale frozen	Kale boiled	Kale frozen boiled	Broccoli raw	Broccoli frozen	Broccoli boiled	Broccoli frozen boiled	Brussels sprouts raw	Brussels sprouts frozen	Brussels sprouts boiled	Brussels sprouts frozen boiled
Calories	50	28	28	30	34	29	35	28	43	41	36	42
Dietary fibre (g)	2	2	2	2	2.6	3	3.3	3	3.8	3.8	2.6	4.1
Protein (g)	3.3	2.7	1.9	2.8	2.8	3.1	2.4	3.1	3.4	3.8	2.5	3.6
Vitamin A (IU)	15,376	6253	13,623	14,705	623	1138	1548	1118	754	617	775	926
Retinol activity (µg)	769	313	681	735	31	57	77	56	38	31	39	46
β-Carotene (µg)	9226	–	8174	8824	361	675	929	663	450	370	465	555
Lutein + zeaxanthin (µg)	39,551	–	18,248	19,698	1403	1525	1080	1498	1590	–	1290	1541
Vitamin C (mg)	120	39.3	41	25.2	89.2	68.3	64.9	40.1	85	74.1	62	45.7
Vitamin E (mg)	–	–	–	–	–	–	1.5	1.3	0.9	–	0.4	0.5
(α-Tocopherol) (mg)	–	–	0.9	0.9	0.8	1.3	–	–	–	–	–	–
Vitamin K (mg)	817	–	817	882	102	101	141	99.5	177	–	140	194
Thiamin (mg)	0.1	0.1	0.1	0	0.1	0.1	0.1	0.1	0.1	0.1	0.1	0.1
Riboflavin (mg)	0.1	0.1	0.1	0.1	0.1	0.1	0.1	0.1	0.1	0.1	0.1	0.1
Niacin (mg)	1	0.7	0.5	0.7	0.6	0.5	0.6	0.5	0.7	0.6	0.6	0.5
Vitamin B$_6$ (mg)	0.3	0.1	0.1	0.1	0.2	0.2	0.2	0.1	0.2	0.2	0.2	0.3
Folate (µg)	29	17	13	14	63	94	108	30	61	123	60	101
Calcium (mg)	135	136	72	138	47	41	40	51	42	26	36	26
Iron (mg)	1.7	0.9	0.9	0.9	0.7	0.7	0.7	0.6	1.4	0.9	1.2	0.5
Magnesium (mg)	34	18	18	18	21	16	21	20	23	20	20	18
Phosphorus (mg)	56	29	28	28	66	59	67	55	69	62	56	56
Potassium (mg)	447	333	228	321	316	250	293	180	389	370	317	290
Sodium (mg)	43	15	23	15	33	17	41	24	25	10	21	15
Zinc (mg)	0.4	0.2	0.2	0.2	0.4	0.3	0.5	0.3	0.4	0.3	0.3	0.2
Copper (mg)	0.3	0	0.2	0	0	0	0.1	0	0.1	0	0.1	0
Manganese (mg)	0.8	0.4	0.4	0.5	0.2	0.3	0.2	0.3	0.3	0.3	0.2	0.2
Selenium (µg)	0.9	0.9	0.9	0.9	2.5	1.9	1.6	1.9	1.6	1.5	1.5	0.6
Water (g)	84.5	91.1	91.2	90.5	89.3	90.6	89.3	90.7	86	87.1	88.9	86.7

Table 5.2. The major glucosinolates occurring in the *Brassicaceae* (nomenclature adopted from Bjerg and Sorensen, 1987).

Chemical grouping	Chemical name	Common name
Aliphatic glucosinolates	2-Propenyl or allyl glucosinolate	Sinigrin
	But-3-enyl glucosinolate	Gluconapin
	Pent-4-enyl glucosinolate	Glucobrassicanapin
	2-Hydroxybut-3-enyl glucosinolate	Progoitrin
	2-Hydroxypent-4-enyl glucosinolate	Gluconapoleiferin
	3-Methylsulfinylpropyl glucosinolate	Glucoiberin
	4-Methylsulfinylbutyl glucosinolate	Glucoraphanin
Aromatic glucosinolates	2-Phenethyl glucosinolate	Gluconasturtiin
Indole glucosinolates	Indole-3-ylmethyl glucosinolate	Glucobrassicin
	1-Methoxyindol-3-ylmethyl glucosinolate	Neoglucobrassicin
	4-Hydroxyindol-3-ylmethyl glucosinolate	4-Hydroxyglucobrassicin
	4-Methoxyindol-3-ylmethyl glucosinolate	4-Methoxyglucobrassicin

that is released from damaged plant cells (Mithen, 2001; Agerbirk *et al.*, 2008). Postharvest physical disruption of the plants (e.g. chewing, cooking, freezing/thawing and high temperature) leads to loss of cellular compartmentalization and subsequent mixing of glucosinolates and myrosinase (usually confined to specialized myrosin cells) to form the biologically active compounds (Rosa *et al.*, 1997). Hydrolysis products are governed by glucosinolate side-chain structure, the plant species, hydrolysis conditions, such as pH, and the presence of epithiospecifier protein (ESP) and thiocyanate-forming protein (TFP) (Bernardi *et al.*, 2000; Foo *et al.*, 2000; Burow and Wittstock, 2009).

The anticarcinogenic activity in broccoli is most likely due to activity of the isothiocyanates, iberin and sulforaphane, which are degradation products of, respectively, 3-methylsulfinylpropyl (glucoiberin) and 4-methylsulfinylbutyl (glucoraphanin) glucosinolates that accumulate in the florets of broccoli. Other members of *B. oleracea* contain differing concentrations of the cleavage products, 2-propenyl (sinigrin), 3-butenyl (gluconapin) and 2-hydroxy-3-butenyl (progoitrin) (Mithen *et al.*, 2003; Halkier and Gershenzon, 2006).

Sulforaphane has been shown to be a powerful inhibitor of phase 1 and inducer of phase 2 enzymes in human and animal cell lines (Maheo *et al.*, 1997; Juge *et al.*, 2007). Important phase 1 enzymes are the cytochrome P450s, and phase 2 enzymes include glutathione-*S*-transferases (GSTs) and UDP-glucuronyl transferases. These enzymes act to metabolize and excrete potential carcinogens and this activity could explain the potential anticarcinogenic effects of these molecules. Other potential activities of such products include initiating apoptosis and inducing cell cycle arrest (Bonnesen *et al.*, 1999; Sarikamis *et al.*, 2006). Another sulfur-containing compound, *S*-methyl cysteine sulfoxide, and its breakdown product, methyl methane thiosulfinate, have been found to inhibit chemically induced genotoxicity in mice (Stoewsand, 1995).

The anticarcinogenic activities of *Brassica* vegetables are more effective with certain human genotypes. Individuals who have homozygous null mutations in the GST genes, *GSTM1* and *GSTT1*, appear to gain less cancer protection from broccoli than those who can express the functional gene (Hayes and Strange, 2000; Joseph *et al.*, 2004). Between 39 and 63% of the population have the homozygous null *GSTM1* gene. While these people may gain less cancer protection from consuming broccoli, it is likely that they gain more cancer protection from eating other types of crucifers, such as cabbages and Chinese cabbage. The high content of different chemical compounds that may modulate carcinogenesis identify *Brassica* vegetables as very attractive subjects for chemopreventative studies. Most studies have been carried out with single, pure substances. By analysing the health-promoting effects of complex mixtures of compounds, as found in broccoli for example, the combined effects of various anticarcinogens may be elucidated (for example, Finley, 2003a, 2005; Jeffery and Araya, 2009). Combined effects are likely to be different from those observed using single substances and, in the case of cancer prevention, the combined effect is especially relevant since carcinogenesis is a multistage disease divided into several qualitative steps.

5.2.3 Flavour

The characteristic sulfurous and bitter taste of *Brassica* vegetables is assumed to be due mainly to the glucosinolate composition (Fenwick *et al.*, 1983; Hansen *et al.*, 1997; van Doorn *et al.*, 1998). The glucosinolates sinigrin and progoitrin have been correlated to bitterness in Brussels sprouts (van Doorn *et al.*, 1998), while a combination of neoglucobrassicin and sinigrin causes the bitter taste in cooked cauliflower (Engel *et al.*, 2002). The lengths and structures of the aliphatic side-chains of the glucosinolate moieties determine their biological activity including flavour. Following tissue disruption, myrosinase acts to release isothiocyanates, the major volatile flavour components of cruciferous crops. The flavour of the crops is determined partially by the total amount of glucosinolates, the side-chain structures and the activity of the myrosinase. The presence of certain isothiocyanates results in varying degrees of bitterness, which make particular cultivars unacceptable to the consumer (Schonhof *et al.*, 2004). Taste tests indicated that consumers preferred cultivars with low levels of bitter-tasting glucosinolates and higher sucrose content, and it was suggested that the bitterness of the beneficial glucosinolates could be masked by increasing sucrose contents (Schonhof *et al.*, 2004). However, Baik *et al.* (2003) compared flavour profiles among 19 different broccoli cultivars and did not find a link between glucosinolate content and flavour. Interestingly, the major isothiocyanate cleavage compounds in broccoli, sulforaphane and iberin, are non-volatile; therefore, they do not contribute to the flavour. To add to this complexity, another investigation of 113 varieties of turnip greens concluded that other phytochemicals as well as glucosinolates and their breakdown products were probably involved in flavour attributes (Padilla *et al.*, 2007).

5.2.4 Antinutrients

As well as having a potential beneficial health-promoting function, glucosinolates can be considered as having antinutrient activity in certain situations. This is clearly the case in oilseed rape varieties where the breakdown products may exhibit goitrogenic or antithyroid activity. The isothiocyanates give rise to the most actively goitrogenic compounds by being cyclized to form oxazolidone-2-thiones (Chubb, 1982). The most goitrogenic compound is 5-vinyl-oxazolidone-2-thione, commonly known as goitrin. The glucosinolate that gives rise to goitrin is 2-hydroxy-3-butenyl glucosinolate, or progoitrin (Chubb, 1982). This is the predominant glucosinolate in oilseed rape, representing between 50 and 70% of the total glucosinolate concentration (Zhao *et al.*, 1994). However, the glucosinolate composition of vegetable *Brassicas*, apart from contributing to an unacceptable taste, does not appear to have an antinutrient effect. Levels of progoitrin are low in broccoli, for example (Table 5.3), and there is no evidence for any goitrogenic effect on humans from *Brassica* consumption (Mithen, 2001). The toxicity of isothiocyanates against insect pests such as the black vine weevil (*Otiorhynchus sulcatus* F.) has been demonstrated (Borek *et al.*, 1988). These data suggest that *Brassica* spp. can be used in integrated pest management systems; see Hopkins *et al.* (2009).

Table 5.3. Glucosinolate content and standard deviation (mg/100 FM) of cauliflower and broccoli (n = 9). Redrawn from Schonhof et al. (2004).

Glucosinolate (mg/100 g FM)	Broccoli cultivars						Cauliflower cultivars			HSD
	Emperor	Shogun	Marathon	Viola	Chinese broccoli	Marine	Minarett	Alverda	Rosalind	
Glucoibeverin	ND	ND	ND	ND	0.01 ± 0.01	0.97 ± 0.25	ND	4.76 ± 3.17	0.03 ± 0.06	1.64
Glucoiberin	0.27 ± 0.16	1.97 ± 1.97	2.65 ± 1.57	3.78 ± 1.18	1.47 ± 1.12	2.90 ± 1.24	1.22 ± 0.79	27.72 ± 27.12	4.61 ± 1.84	13.56
Glucoerucin	ND	0.75 ± 0.24	ND	ND	1.33 ± 1.65	0.21 ± 0.14	ND	ND	0.58 ± 0.16	0.81
Glucoraphanin	11.58 ± 10.62	33.95 ± 11.77	21.10 ± 14.33	6.56 ± 1.56	39.67 ± 51.40	0.12 ± 0.11	3.41 ± 2.44	1.15 ± 1.01	11.60 ± 2.23	24.50
Sum alkyl glucosinolates	11.58 ± 10.76	36.67 ± 12.39	23.75 ± 15.80	10.34 ± 2.31	42.48 ± 54.08	4.21 ± 1.61	4.63 ± 2.96	33.63 ± 25.77	16.82 ± 3.65	26.29
Sinigrin	ND	0.88 ± 0.40	ND	ND	0.94 ± 1.01	3.39 ± 1.53	ND	ND	ND	0.94
Gluconapin	ND	1.01 ± 0.33	ND	ND	75.99 ± 88.83	ND	ND	0.42 ± 1.00	ND	42.90
Progoitrin	0.57 ± 0.23	10.28 ± 3.39	1.10 ± 0.47	1.03 ± 0.33	19.13 ± 36.45	0.60 ± 0.33	1.77 ± 1.45	0.80 ± 0.38	0.79 ± 0.15	17.64
Gluconapoleiferin	0.43 ± 0.18	0.53 ± 0.49	0.86 ± 0.28	0.65 ± 0.22	0.99 ± 1.39	0.32 ± 0.09	0.81 ± 0.66	0.42 ± 0.11	0.38 ± 0.21	0.80
Sum alkenyl glucosinolates	1.00 ± 0.36	12.70 ± 3.44	1.97 ± 0.66	1.68 ± 0.48	97.05 ± 127.87	4.30 ± 1.78	2.58 ± 2.96	1.64 ± 1.14	1.17 ± 0.34	60.06
Glucobrassicin	6.96 ± 6.14	7.63 ± 2.51	10.29 ± 4.79	12.09 ± 4.79	5.69 ± 7.72	4.77 ± 1.80	7.51 ± 3.14	9.09 ± 2.98	14.78 ± 0.29	5.31
Neoglucobrassicin	2.60 ± 3.69	6.35 ± 3.74	2.66 ± 1.63	0.86 ± 0.47	2.08 ± 2.59	0.45 ± 0.15	0.97 ± 0.56	1.00 ± 0.51	1.41 ± 0.29	2.76
4-Hydroxyglucobrassicin	0.06 ± 0.04	0.36 ± 0.48	0.29 ± 0.27	0.23 ± 0.30	1.53 ± 2.37	0.02 ± 0.03	1.32 ± 0.87	ND	0.33 ± 0.20	1.27
4-Methoxyglucobrassicin	0.43 ± 0.35	0.92 ± 0.38	1.32 ± 0.70	1.11 ± 0.58	0.63 ± 0.82	0.34 ± 0.12	0.62 ± 0.40	1.55 ± 0.46	1.15 ± 0.38	0.60
Sum indole glucosinolates	10.05 ± 10.02	15.26 ± 6.15	14.55 ± 6.70	14.29 ± 7.73	9.93 ± 13.53	5.57 ± 2.04	10.42 ± 4.79	11.64 ± 3.82	17.67 ± 3.10	8.15

Note: HSD = Tukey's honest significant difference; ND = not detectable.

5.3 Factors Affecting Composition

The chemical composition of *Brassica* vegetables is affected by a combination of genetics and both pre- and postharvest environmental conditions and treatments. For example, studies by Jeffery *et al.* (2003) found that the glucosinolate, vitamin and flavonoid contents of broccoli were affected by genotype and environment, and by processing (Fig. 5.2).

5.3.1 Genotype

Phytochemical profiles vary considerably between *Brassica* species and the levels of individual compounds vary between cultivars (Kurilich *et al.*, 1999; Kushad *et al.*, 1999; Jeffery *et al.*, 2003). An analysis of 50 different broccoli cultivars indicated that glucosinolate levels varied up to 20-fold, showing a considerable genotype-dependent effect. For example, the predominant glucosinolate in broccoli, glucoraphanin, showed levels ranging from 0.8 μmol/g DW in one broccoli cultivar to 21.7 μmol/g DW in another. Concentrations of the other glucosinolates in broccoli varied similarly over a wide range. In Brussels sprouts, cabbage, cauliflower and kale, the predominant glucosinolates measured were sinigrin (8.9, 7.8, 9.3 and 10.4 μmol/g DW, respectively) and glucobrassicin (3.2, 0.9, 1.3 and 1.2 μmol/g DW, respectively). Brussels sprouts also had significant

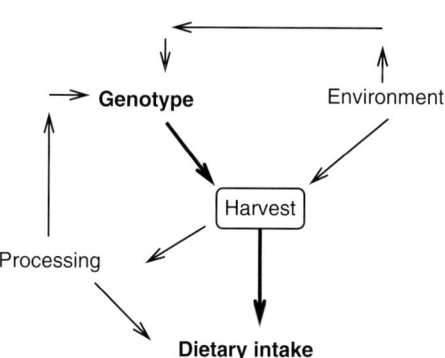

Fig. 5.2. Pre- and postharvest factors that affect chemical composition of *Brassica* vegetables. Redrawn from Jeffery *et al.* (2003).

amounts of gluconapin (6.9 μmol/g DW) (Kushad *et al.*, 1999). In another study, several cultivars of broccoli and cauliflower were assayed for glucosinolate composition and similar results were obtained, as shown in Table 5.3 (Schonhof *et al.*, 2004). These authors concluded that the composition of glucosinolates was determined genetically, but the growing conditions had an impact on the concentrations.

Genetic manipulation of glucosinolate levels has been suggested as a means to provide a better source for the consumption of anticarcinogenic compounds. A hybrid broccoli cultivar containing increased levels of glucoraphanin (the precursor of sulforaphane) has been generated by crossing a commercial broccoli line with a wild species of *Brassica* (*B. villosa*) (Sarikamis *et al.*, 2006). Following cooking, this line of broccoli contained around threefold higher levels of sulforaphane than commercial broccoli; also, when consumed as a soup, it has been shown to induce the expression of carcinogen-inactivating phase 2 enzymes (Gasper *et al.*, 2007). Further manipulation of the glucosinolate pathway in *Brassica* requires improved understanding of the mechanisms of biosynthesis. This is being addressed using *Arabidopsis* as a model system (Hall *et al.*, 2001; Kliebenstein, 2009).

Antioxidant levels (carotene, tocopherol and ascorbate) measured in different *Brassica* species and in 50 different broccoli cultivars indicated that there was considerable genotype-dependent variation (Kurilich *et al.*, 1999). Genotype influence on mineral composition (particularly Ca, Mg and S) was also identified in an analysis of 11 broccoli cultivars (Rosa *et al.*, 2002), and differences in levels of flavonoids and vitamin C were reported in an analysis of 14 commercial and experimental cultivars of broccoli (Vallejo *et al.*, 2002).

5.3.2 Agronomy

Growth conditions and time of harvesting are likely to have an effect on the chemical composition of *Brassicas*. Glucosinolate levels were assessed in a range of genotypes grown in thre different years, and there were significant environmental effects on levels of glucoraphanin (Farnham *et al.*, 2004). However, genotype effects were greater than environmental effects. A comparison of spring versus summer planting of 11 different broccoli lines showed that total glucosinolate levels were higher in the late (summer) crop (Rosa and Rodrigues, 2001). A study of sulforaphane levels in broccoli grown in three different seasons showed concentrations between 36.7 and 74.5 mg/100 g (Howard *et al.*, 1997); this difference was probably due to different average temperatures in the different seasons. Cooler and/or drier conditions might have resulted in higher glucosinolate production. This hypothesis was supported by another study in which glucosinolate content of cauliflower and broccoli was influenced strongly by daily temperature (Schonhof *et al.*, 2004). Low mean temperature in one season resulted in considerably higher levels of glucosinolates than those obtained in two subsequent warmer seasons. However, another study (Ciska *et al.*, 2000) reported that low rainfall and high temperature resulted in higher glucosinolate levels. The conclusions from all these reports is that more controlled experiments are needed to identify the environmental factors that have a role in controlling glucosinolate content; and a better understanding of the interaction between genotype and environment is also needed, since comparisons between different cultivars in different conditions are not appropriate.

Nutritional content is affected as soon as the crop is harvested. During storage and handling of these vegetables, many sequential changes take place, both in the appearance of the vegetable and also in the content of health-promoting components, many of which are likely to be degraded rapidly (reviewed in Jones *et al.*, 2006).

5.3.3 Recommended storage regimes for *Brassica* vegetables

Low temperature is extremely important in achieving adequate shelf life in broccoli. A temperature of 0°C is required to optimize broccoli storage life. Heads stored at 5°C can have a storage life of up to 14 days, yet storage at 10°C reduces shelf life by one-third. Brussels sprouts are moderately perishable and can be stored for 3–5 weeks at temperatures near 0°C, but they are often left in the field during early winter. Brussels sprouts are often hydrocooled, but can be air-cooled as well. Although they have considerable wax on their leaves, they become flaccid due to water loss if high relative humidity is not maintained. This clearly shows that the most effective way to maintain shelf life, and consequently the desired levels of bioactives in *Brassica* vegetables, is by cooling rapidly following harvest and maintaining the product in cool conditions at high humidity. Storage of broccoli at 20°C resulted in 82% loss of glucoraphanin content after 5 days; at 5°C this loss was only 31% (Rodrigues and Rosa, 1999). Similar results were obtained by Howard *et al.* (1997), who reported approximately 50% loss in sulforaphane after 21 days at 4°C. However, in another cultivar, indole glucosinolates were found to increase after 10 days' storage at 10°C, yet total glucosinolate levels did not change significantly (Hansen *et al.*, 1995). Table 5.4 shows the change in individual glucosinolates in four types of *Brassica* vegetables after 7 days' storage at 4°C (Song and Thornalley, 2007).

Postharvest storage in controlled atmospheres (CAs) aimed at reducing respiration is a well-tested methodology for many fruit and vegetable products, but is rarely now applied commercially to *Brassicas*. It appears that extending shelf life also has a positive effect on maintaining glucosinolate levels, but this has not been demonstrated clearly (Jones *et al.*, 2006). The shelf life of broccoli can be improved using an atmosphere of 1–2 kPa O_2 with 5–10 kPa CO_2 at a temperature range of 0–5°C. CA has been shown to reduce ethylene-induced yellowing (chlorophyll breakdown) and reduced butt-end discoloration. Similar results are observed with modified atmosphere packaging (MAP), which is a common format for some *Brassicas*. Broccoli quality was optimum when packaging was used to maintain 1–2 kPa O_2 and 5–10 kPa CO_2 (Jacobsson *et al.*, 2004), and levels of carotenoids and vitamin C were also retained by this treatment (Barth and Zhuang, 1996). Glucosinolate levels appeared to be more stable under MAP

Table 5.4. Baseline individual and total glucosinolate analyte contents of fresh vegetables (adopted from Song and Thornalley, 2007).

Produce type	Broccoli		Brussels sprout		Cauliflower		Green cabbage	
Day	0	7	0	7	0	7	0	7
Glucosinolate (µmol/100 g fresh weight)								
Glucoiberin	17.1	10.6	1.5	1.14	1.34	0.92	3.88	3.09
Glucoraphanin	29.4	21.3	0.55	0.25	0.31	0.25	0.35	0.41
Glucoalyssin	3.86	3.05	0.33	0.21	< LOD	< LOD	< LOD	< LOD
Sinigrin	1.4	1	8.56	7.08	5.28	4.79	5.09	4.57
Gluconapin	2.87	2.71	2.77	2.34	3.36	3.18	0.38	0.44
Progoitrin	3.33	3.14	2.41	1.98	0.45	0.4	0.62	0.39
Gluconasturtiin	4.44	3.71	1.06	0.76	2.79	2.47	< LOD	< LOD
Total	62.4	45.5	17.2	13.8	13.5	12	10.3	8.9
Loss in storage (%)	27		20		11		14	

Note: LOD = limit of detection.

treatment, especially at higher temperatures (Rangkadilok *et al.*, 2002).

5.4 Processing

5.4.1 Effects of cooking

Most traditional *Brassica* vegetables such as cabbage, broccoli, Brussels sprouts, etc., are consumed following cooking and this process can be very detrimental to the nutrient quality of the product that is actually consumed. This said, without cooking many *Brassicas* are near inedible. Table 5.1 shows the effects of cooking (boiling) on vitamin levels in several *Brassica* vegetables and it is clear that some vitamins, such as vitamin C, are reduced following cooking. Other components, such as vitamin A, appear to increase following cooking, possibly because they are easier to extract.

Measurement of total and individual flavonoid components in freshly harvested broccoli cooked in four different ways showed that there were large differences with the different cooking methods (Vallejo *et al.*, 2003). Microwaving was the most detrimental method,

with 97% loss of flavonoid content. Steaming was the optimum cooking method, showing a loss of only 11%. In another paper, the same authors compared the effects of the four cooking processes on the levels of vitamin C and glucosinolates in freshly harvested broccoli (Vallejo *et al.*, 2002). This study also showed that microwaving was very detrimental, with 40% loss of vitamin C and 74% loss of glucosinolates, and, again, steaming was found to have minimal effects on the levels of these important phytonutrients. The problem with these pieces of work is that, although they provide useful information, they do not represent reality as most consumers eat broccoli that has been stored either at retail or in the home. Another study investigated the effects of cooking (steaming and boiling) on glucosinolate profiles in broccoli (Gliszczynska-Swiglo *et al.*, 2006). Total glucosinolate levels were reduced around 50% by conventional boiling, but were increased following steaming (Table 5.5).

Gliszczynska-Swiglo *et al.* (2006) also showed that levels of total polyphenols and several vitamins actually appeared to increase following steaming (Table 5.6). Increased levels of carotene and lutein following both types of cooking are likely due to increased

Table 5.5. Glucosinolate content (µmol/g dry weight) in fresh, steamed and water-cooked broccoli. Redrawn from Gliszczynska-Swiglo *et al.* (2006) with permission from Taylor and Francis Ltd (http://www.informaworld.com).

	Compound	Fresh broccoli	Steamed broccoli	Water-cooked broccoli
Aliphatics	Glucoiberin	1.43	1.58	0.81
	Progoitrin	0.18	0.19	0.14
	Glucoraphanin	9.6	10.19	5.09
	Napoleiferin	0.31	0.26	0.14
	Glucoalyssin	0.07	0.15	0.1
	Gluconapin	Traces	Traces	Traces
	Glucoibervirin	Traces	0.05	Traces
	Glucoerucin	Traces	Traces	Traces
Aromatic indoles	Gluconasturtiin	0.1	0.14	0.05
	4-Hydroxyglucobrassicin	0.63	0.76	0.43
	Glucobrassicin	1.76	2.78	0.93
	4-Methoxyglucobrassicin	0.36	0.48	0.3
	Neoglucobrassicin	1.6	2.21	0.59
	Total	16.04	18.79	8.58

Table 5.6. Distribution of compounds analysed in fresh and domestically processed broccoli (mg/100 g). Redrawn from Gliszczynska-Swiglo *et al.* (2006) with permission from Taylor and Francis Ltd (http://www.informaworld.com).

Compound		Fresh	Steamed	Water-cooked
Total polyphenols	Dry weight	886.3	1409.1	775.8
	Fresh weight	109.9	167.3	89.2
Flavonoids	Dry weight	25.4	38.7	7.0
	Fresh weight	3.15	4.59	0.81
Phenolic acids	Dry weight	328.1	417.3	155.9
	Fresh weight	40.55	50.05	17.93
Vitamin C	Dry weight	681.2	652.6	524.8
	Fresh weight	84.5	77.7	60.3
β-Carotene	Dry weight	10.5	19.93	24.61
	Fresh weight	1.3	2.37	2.83
Lutein	Dry weight	6.47	26.6	38.8
	Fresh weight	0.8	3.16	4.46
Vitamin E_4	Dry weight	0.798	0.999	1.397
	Fresh weight	0.099	0.119	0.161

availability for extraction. The action of myros inase on the dominant broccoli glucosinolate, glucoraphanin, results in the production of the isothiocyanate derivative, sulforaphane, which has anticarcinogenic activity. However, an alternative degradation pathway involving the ESP group of enzymes results in an inactive nitrile derivative of this compound (Mithen *et al.*, 2003; Agerbirk *et al.*, 2008). This is not a problem when broccoli is cooked, since cooking destroys the myrosinase and the enzymes present in the gut microflora degrade the glucosinolates to active compounds. Heat treatments that inactivate the more labile ESP but not the myrosinase were found to increase the levels of active sulforaphane in the sample (Matusheski *et al.*, 2004).

5.4.2 Effects of freezing

Warmer weather induces faster crop growth in broccoli and consequently promotes greater horticultural maturity. Because broccoli has to be harvested at a physiologically immature stage, this can cause problems as higher temperatures also lead to reduced demand from consumers. Broccoli is usually frozen into florettes when product availability outstrips demand; however, freezing can have a negative affect on health-promoting compounds. Table 5.1 shows data on the levels of nutrients in cooked frozen broccoli and Brussels sprouts compared with cooked fresh vegetables. It is clear that some nutrients such as vitamin C are lost in the freezing process. However, there is no comparison with broccoli or Brussels sprout samples that have been stored for a few days before consumption.

Freezing broccoli is usually preceded by a blanching step to inactivate enzymes that cause product deterioration. Enzymes such as lipoxygenase, peroxidase and cystine lyase contribute to the development of 'off' aromas in stored vegetables. The blanching step inactivates myrosinase, which means that glucosinolate levels are stable (Rodrigues and Rosa, 1999). The blanching process also causes the thermal inactivation of ESPs. ESPs act to redirect the production of isothiocyanates (such as sulforaphane from glucoraphanin) towards nitrile production (in this example, sulforaphane nitrile) (Matusheski *et al.*, 2004; see Agerbirk *et al.*, 2008). In the absence of blanching, glucosinolates are degraded rapidly following thawing (Rosa *et al.*, 1997), the degree of breakdown governed primarily by the classes of glucosinolate present as a function of genotype used.

5.5 Conclusions

Increasing awareness of the beneficial phytochemicals present within the *Brassicaceae* has led to the promotion of these vegetable crops to 'super-food' status. It has become apparent that modern breeding techniques will enable the directed action required to capture the allelic variation necessary to improve public health further. In line with an increasing understanding of the interactions between genotype, preharvest environment and postharvest conditions, an understanding of nutrient × human allele interactions is required in order to maximize the benefits associated with the consumption of this crop type. The ever-growing field of nutritional genomics seeks to integrate these diverse areas of research in a systematic approach, to deliver directed advice that may be adopted for policy guidance.

References

Agerbirk, N., De Vos, M., Kim, J. H. and Jander, G. (2008) Indole glucosinolate breakdown and its biological effects. *Phytochemistry Reviews* 8, 101–120.

Al-Shehbaz, I.A., Beilstein, M.A. and Kellogg, E.A. (2006) Systematics and phylogeny of the Brassicaceae (Cruciferae): an overview. *Plant Systematics and Evolution* 259, 89–120.

Baik, H., Juvik, J., Jeffery, E., Wallig, M., Kushad, M. and Klein, B. (2003) Relating glucosinolate content and flavour of broccoli cultivars. *Journal of Food Science* 68, 1043–1050.

Bailey, L. and Gregory, J. (1999) Folate metabolism and requirements. *Journal of Nutrition* 129, 779–782.

Barth, M.M. and Zhuang, H. (1996) Packaging design affects antioxidant vitamin retention and quality of broccoli florets during postharvest storage. *Postharvest Biology and Technology* 9, 141–150.

Bernardi, R., Negri, A., Ronchi, S. and Palmieri, S. (2000) Isolation of the epithiospecifier protein from oil-rape (*Brassica napus* spp. *oleifera*) seed and its characterization. *FEBS Letters* 467, 296–298.

Bjerg, B. and Sørensen, H. (1987) Isolation of intact glucosinolates by column chromatography and determination of their purity. In: Wathelet, J.P. (ed.) *World Crops: Production, Utilization, Description. Volume 13: Glucosinolates in Rapeseed: Analytical Aspects*. Martinus Nijhoff Publishers, Dordrecht/Boston/Lancaster, pp. 59–75.

Block, G., Patterson, B. and Subar, A. (1992) Fruit, vegetables, and cancer prevention: a review of the epidemiological evidence. *Nutrition and Cancer* 18, 1–29.

Bones, A.M. and Rossiter, J.T. (2006) The enzymic and chemically induced decomposition of glucosinolates. *Phytochemistry* 67, 1053–1067.

Bonnesen, C., Stephensen, P., Andersen, O., Sørensen, H. and Vang, O. (1999) Modulation of cytochrome P-450 and glutathione S-transferase isoform expression *in vivo* by intact and degraded indolyl glucosinolates. *Nutrition and Cancer* 33, 178–187.

Borek, V., Leslie, R., Elberson, L.R., McCaffrey, J.P. and Morra, M.J. (1988) Toxicity of isothiocyanates produced by glucosinolates in Brassicaceae species to black vine weevil eggs. *Journal of Agricultural and Food Chemistry* 46, 5318–5323.

Broadley, M.R., Hammond, J.P., King, G.J., Astley, D., Bowen, H.C., Meacham, M.C., Mead, A., Pink, D.A.C., Teakle, G.R., Hayden, R.M., Spracklen, W.P. and White, P.J. (2008) Shoot calcium (Ca) and magnesium (Mg) concentrations differ between subtaxa, are highly heritable, and associate with potentially pleiotropic loci in *Brassica oleracea*. *Plant Physiology* 146, 1707–1720.

Burow, M. and Wittstock, U. (2009) Regulation and function of specifier proteins in plants. *Phytochemical Reviews* 8, 87–99.

Chu, Y., Sun, J., Wu, X. and Liu, R. (2002) Antioxidant and antiproliferative activities of common vegetables. *Journal of Agricultural and Food Chemistry* 50, 6910–6916.

Chubb, L.G. (1982) Anti-nutritive factors in animal foodstuffs. In: Haresign, W. (ed.) *Recent Advances in Animal Nutrition*. Butterworths, London, pp. 21–37.

Ciska, E., Martyniak-Przybyszewska, B. and Kozlowska, H. (2000) Content of glucosinolates in cruciferous vegetables grown at the same site for two years under different climatic conditions. *Journal of Agricultural and Food Chemistry* 48, 2862–2867.

De Pascale, S., Maggio, A. and Barbieri, G. (2005) Soil salinisation affects growth, yield, and mineral composition of cauliflower and broccoli. *European Journal of Agronomy* 23, 254–264.

De Rijke, E., Out, P., Niessen, W.M.A., Ariese, F., Gooijer, C. and Brinkman, U.A.T. (2006) Analytical separation and detection methods for flavonoids. *Journal of Chromatography A* 1112, 31–63.

Engel, E., Baty, C., Le Corre, D., Souchon, I. and Martin, N. (2002) Flavor-active compounds potentially implicated in cooked cauliflower acceptance. *Journal of Agricultural and Food Chemistry* 50, 6459–6467.

FAOSTAT (2008) Food and Agriculture Organization of the United Nations (http://faostat.fao.org/site/567/DesktopDefault.aspx?PageID=567#ancor, accessed 20 March 209).

Farnham, M.W., Grusak, M.A. and Wang, M. (2000) Calcium and magnesium concentration of inbred and hybrid broccoli heads. *Journal of the American Horticultural Society* 125, 344–349.

Farnham, M., Wilson, P., Stephenson, K. and Fahey, J. (2004) Genetic and environmental effects on glucosinolate content and chemoprotective potency of broccoli. *Plant Breeding* 123, 60–65.

Fenwick, G., Griffiths, N. and Heaney, R. (1983) Bitterness in Brussels sprouts (*Brassica oleracea* L. var. *gemmifera*): the role of glucosinolates and their breakdown products. *Journal of the Science of Food and Agriculture* 34, 73–80.

Finley, J., Davis, C. and Feng, Y. (2000) Selenium from high selenium broccoli protects rats from colon cancer. *Journal of Nutrition* 130, 2384–2389.

Finley, J.W. (2003a) Reduction of cancer risk by consumption of selenium-enriched plants, enrichment of broccoli with selenium increases the anticarcinogenic properties of broccoli. *Journal of Medicinal Food* 6, 19–26.

Finley, J.W. (2003b) The antioxidant responsive element (ARE) may explain the protective effects of cruciferous vegetables on cancer. *Nutrition Reviews* 61, 250–254.

Finley, J.W. (2005) Proposed criteria for assessing the efficacy of cancer reduction by plant foods enriched in carotenoids, glucosinolates, polyphenols and selenocompounds. *Annals of Botany* 95, 1075–1096.

Flannery, M.L., Mitchell, F.J.G., Coyne, S., Kavanagh, T.A., Burke, J.I., Salamin, N., Dowding, P. and Hodkinson, T.R. (2006) Plastid genome characterisation in *Brassica* and Brassicaceae using a new set of nine SSRs. *Theoretical and Applied Genetics* 113, 1221–1231.

Foo, H.L., Grønning, L.M., Goodenough, L., Bones, A.M., Danielsen, B., Whiting, D.A. and Rossiter, J.T. (2000) Purification and characterization of epithiospecifier protein from *Brassica napus*: enzymic intramolecular sulphur addition within alkenyl thiohydroximates derived from alkenyl glucosinolate hydrolysis. *FEBS Letters* 468, 243–246.

Gasper, A.V., Traka, M., Bacon, J.R., Smith, J.A., Taylor, M.A., Hawkey, C.J., Barrett, D.A. and Mithen, R.F. (2007) Consuming broccoli does not induce genes associated with xenobiotic metabolism and cell cycle control in human gastric mucosa. *Journal of Nutrition* 137, 1718–1724.

Gliszczynska-Swiglo, A., Ciska, E., Pawlak-Lemaska, K., Chmielewski, J., Borkowski, T. and Tyrakowska, B. (2006) Changes in the content of health-promoting compounds and antioxidant activity of broccoli after domestic processing. *Food Additives and Contaminants* 23, 1088–1098.

Halkier, B.A. and Gershenzon, J. (2006) Biology and biochemistry of glucosinolates. *Annual Review of Plant Biology* 57, 303–333.

Hall, C., McCallum, D., Prescott, A. and Mithen, R. (2001) Biochemical genetics of glucosinolate modification in Arabidopsis and Brassica. *Theoretical and Applied Genetics* 102, 369–374.

Hall, J.C., Sytsma, K.J. and Iltis, H.H. (2002) Phylogeny of Capparaceae and Brassicaceae based on chloroplast sequence data. *American Journal of Botany* 89(11), 1826–1842.

Hansen, M., Moller, P. and Sorensen, H. (1995) Glucosinolates in broccoli stored under controlled atmosphere. *Journal of the American Horticultural Science* 120, 1069–1074.

Hansen, M., Laustsen, A., Olsen, C., Poll, L. and Sorensen, H. (1997) Chemical and sensory quality of broccoli (*Brassica oleracea* L. var. *italica*). *Journal of Food Quality* 20, 441–459.

Hayes, J.D. and Strange, R.C. (2000) Glutathione S-transferase polymorphisms and their biological consequences. *Pharmacology* 61, 154–166.

Heaney, R.P., Weaver, C.M., Hinders, S.M., Martin, B. and Packard, P.T. (1993) Absorbability of calcium from Brassica vegetables, broccoli, bok choy, and kale. *Journal of Food Science* 58, 1378–1380.

Heber, D. (2004a) Phytochemicals beyond antioxidation. *Journal of Nutrition* 134, 3175S–3176S.

Heber, D. (2004b) Vegetables, fruits and phytoestrogens in the prevention of diseases. *Journal of Postgraduate Medicine* 50, 145–149.

Heimler, D., Vignolini, P., Giulia Dini M., Vincieri, F. and Romani, A. (2006) Antiradical activity and polyphenol composition of local Brassicaceae edible varieties. *Food Chemistry* 99, 464–469.

Hertog, M., Hollman, P., Katan, M. and Kromhout, D. (1993) Intake of potentially anticarcinogenic flavonoids and their determinants in adults in the Netherlands. *Nutrition and Cancer* 20, 21–29.

Hopkins, R.J., van Dam, N.M. and van Loon, J.J.A. (2009) Role of glucosinolates in insect–plant relationships and multitrophic interactions. *Annual Reviews of Entomology* 54, 57–83.

Howard, L., Jeffery, E., Wallig, M. and Klein, B. (1997) Retention of phytochemicals in fresh and processed broccoli. *Journal of Food Science* 62, 1098–1100.

Hu, F.B. (2003) Plant-based foods and prevention of cardiovascular disease: an overview. *American Journal Clinical Nutrition* 78, 544S–551S.

Jacobsson, A., Nielsen, T. and Sjoholm, I. (2004) Influence of temperature, modified atmosphere packaging, and heat treatment on aroma compounds in broccoli. *Journal of Agricultural and Food Chemistry* 52, 1607–1614.

Jeffery, E.H. and Araya, M. (2009) Physiological effects of broccoli consumption. *Phytochemical Reviews* 8, 283–298.

Jeffery, E.H., Brown, A.F., Kurilich, A.C., Keck, A.S., Matusheski, N., Klein, B.P. and Juvik, J.A. (2003) Variation in content of bioactive compounds in broccoli. *Journal of Food Composition and Analysis* 16, 323–330.

Jones, R., Faragher, J. and Winkler, S. (2006) A review of the influence of postharvest treatments on quality and glucosinolate content in broccoli (*Brassica oleracea* var. *italica*) heads. *Postharvest Biology and Technology* 41, 1–8.

Joseph, M., Moysich, K., Freudenheim, J., Shields, P., Bowman, E., Zhang, Y., Marshall, J. and Ambrosone, C. (2004) Cruciferous vegetables, genetic polymorphisms in glutathione S-transferases m1 and t1, and prostate cancer risk. *Nutrition and Cancer* 50, 206–213.

Juge, N., Mithen, R.F. and Traka, M. (2007) Molecular basis for chemoprevention by sulforaphane, a comprehensive review. *Cellular and Molecular Life Sciences* 64, 1105–1127.

Karpechenko, G.D. (1927) The production of polyploidy gametes in hybrids. *Hereditas* 9, 349–368.

King, G.J. (2006) Utilization of *Arabidopsis* and *Brassica* genomic resources to underpin genetic analysis and improvement of *Brassica* crops. In: Varshney, R.K. and Koebner, R.M.D. (eds) *Model Plants, Crop Improvement*. CRC Press, Boca Raton, Florida, pp. 33–70.

Kliebenstein, D.J. (2009) A quantitative genetics and ecological model system, understanding the aliphatic glucosinolate biosynthetic network via QTLs. *Phytochemical Reviews* 8, 243–254.

Knekt, P., Jarvinen, R., Reunanen, A. and Maatela, J. (1996) Flavonoid intake and coronary mortality in Finland: a cohort study. *British Medical Journal* 312, 478–481.

Koch, M.A., Dobeš, C., Kiefer, C., Schmickl, R., Klimeš, L. and Lysak, M. (2007) Supernetwork identifies multiple events of plastid *trn*F(GAA) pseudogene evolution in the Brassicaceae. *Molecular Biology and Evolution* 24(1), 63–73.

Kurilich, A.C., Tsau, G.J., Brown, A., Howard, L., Klein, B.P., Jeffery, E.H., Kushad, M., Wallig, M.A. and Juvik, J.A. (1999) Carotene, tocopherol and ascorbate contents in subspecies of *Brassica oleracea*. *Journal of Agricultural and Food Chemistry* 47, 1576–1581.

Kushad, M.M., Brown, A.F., Kurilich, A.C., Juvik, J.A., Klein, B.P., Wallig, M.A. and Jeffery, E.H. (1999) Variation of glucosinolates in vegetable crops of *Brassica oleracea*. *Journal of Agricultural and Food Chemistry* 47, 1541–1548.

Lampe, J.W. and Peterson, S. (2002) *Brassica*, biotransformation and cancer risk, genetic polymorphisms alter the preventive effects of cruciferous vegetables. *Journal of Nutrition* 132, 2991–2994.

Lange, W., Toxopeus, H., Lubberts, J.H., Dolstra, O. and Harrewijn, J.L. (1989) The development of Raparadish (× Brassicoraphanus, 2n = 38), a new crop in agriculture. *Euphytica* 40, 1–14.

Lysak, M.A., Kock, M.A., Pecinka, A. and Schubert, I. (2005) Chromosome triplication found across the tribe Brassiceae. *Genome Research* 15, 516–525.

McKillop, D.J., Pentieva, K., Daly, D., McPartlin, J.M., Hughes, J., Strain, J.J., Scott, J.M. and McNulty, H. (2002) The effect of different cooking methods on folate retention in various foods that are amongst the major contributors to folate intake in the UK diet. *British Journal of Nutrition* 88, 681–688.

Maheo, K., Morel, F., Langouet, S., Kramer, H., Le Ferrec, E., Ketterer, B. and Guillouzo, A. (1997) Inhibition of cytochromes P-450 and induction of glutathione S-transferases by sulforaphane in primary human and rat hepatocytes. *Cancer Research* 57, 3649–3652.

Matusheski, N., Juvik, J. and Jeffery, E. (2004) Heating decreases epithiospecifier protein activity and increases sulforaphane formation in broccoli. *Phytochemistry* 65, 1273–1281.

Metz, P.L.J. (1995) Hybridization of radish (*Raphanus sativus* L.) and oilseed rape (*Brassica napus* L.) through a flower-culture method. *Euphytica* 83(2), pp. 159–168.

Mithen, R. (2001) Glucosinolates – biochemistry, genetics and biological activity. *Plant Growth Regulation* 34, 91–103.

Mithen, R., Faulkner, K., Magrath, R., Rose, P., Williamson, G. and Marquez, J. (2003) Development of isothiocyanate-enriched broccoli and its enhanced ability to induce phase 2 detoxification enzymes in mammalian cells. *Theoretical and Applied Genetics* 106, 727–734.

Moreno, D.A., Carvajal, M., Lòpez-Berenguer, C. and Garcìa-Viguera, C. (2006) Chemical and biological characterisation of nutraceutical compounds of broccoli. *Journal of Pharmaceutical and Biomedical Analysis* 41, 1508–1522.

Myers, J.R. (2006) Outcrossing potential for *Brassica* species and implications for vegetable crucifer seed crops of growing oilseed *Brassicas* in the Willamette Valley. *Oregon State University Extension Service Special Report 1064*.

Padilla, G., Cartea, M., Velasco, P., de Haro, A. and Ordás, A. (2007) Variation of glucosinolates in vegetable crops of *Brassica rapa*. *Phytochemistry* 68, 4536–4545.

Paterson, A.H. (2006) Leafing through the genomes of our major crop plants, strategies for capturing unique information. *Nature Reviews Genetics* 7, 174–184.

Rangkadilok, N., Tomkins, B., Nicolas, M.E., Premier, R.R., Bennett, R.N., Eagling, D.R. and Taylor, P.W.J. (2002) The effect of post-harvest and packaging treatments on glucoraphanin concentration in broccoli (*Brassica oleracea* var. *italica*). *Journal of Agricultural and Food Chemistry* 50, 7386–7391.

Rodrigues, A.S. and Rosa, E. (1999) Effect of post-harvest treatments in the level of glucosinolates in broccoli. *Journal of the Science of Food and Agriculture* 79, 1028–1032.

Rosa, E.A.S. (1999) Chemical composition. In: Gòmez-Campo, C. (ed.) *Biology of Brassica Coenospecies*. Elsevier, Amsterdam, pp. 315–357.

Rosa, E. and Rodrigues, A. (2001) Total and individual glucosinolate content in 11 broccoli cultivars grown in early and late seasons. *Hortscience* 36(1), 56–59.

Rosa, E.A.S., Heaney, R.K., Fenwick, G.R. and Portas, C.A.M. (1997) Glucosinolates in crop plants. *Horticultural Reviews* 19, 99–215.

Rosa, E., Haneklaus, S. and Schnug, E. (2002) Mineral content of primary and secondary inflorescences of eleven broccoli cultivars grown in early and late seasons. *Journal of Plant Nutrition* 25, 1741–1751.

Sarikamis, G., Marquez, J., MacCormack, R., Bennett, R., Roberts, J. and Mithen, R. (2006) High glucosinolate broccoli, a delivery system for sulforaphane. *Molecular Breeding* 18, 219–228.

Schonhof, I., Krumbein, A. and Brückner, B. (2004) Genotypic effects on glucosinolates and sensory properties of broccoli and cauliflower. *Nahrung/Food* 48, 25–33.

Song, L. and Thornalley, P. (2007) Effect of storage, processing and cooking on glucosinolate content of *Brassica* vegetables. *Food and Chemical Toxicology* 45, 216–224.

Spooner, D.M., van den Berg, R.G., Hetterscheid, W.L.A. and Brandenburg, W.A. (2003) Plant nomenclature and taxonomy: an horticultural and agronomic perspective. *Horticultural Reviews* 28, 1–60.

Stoewsand, G.S. (1995) Bioactive organosulfur phytochemicals in *Brassica oleracea* vegetables – a review. *Food Chemistry and Toxicology* 33, 537–543.

Traka, M. and Mithen, R. (2009) Glucosinolates, isothiocyanates and human health. *Phytochemical Reviews* 8, 269–282.

U, N. (1935) Genome analysis in *Brassica* with special reference to the experimental formation of *B. napus* and peculiar mode of fertilization. *Journal of Japanese Botany* 7, 389–452.

Vallejo, F., Tomas-Barberan, F. and Garcia-Viguera, C. (2002) Potential bioactive compounds in health promotion from broccoli cultivars grown in Spain. *Journal of the Science of Food and Agriculture* 82(5), 1293–1297.

Vallejo, F., Tomas-Barberan, F. and Garcia-Viguera, C. (2003) Phenolic compound contents in edible parts of broccoli inflorescences after domestic cooking. *Journal of the Science of Food and Agriculture* 83, 1511–1516.

van Doorn, H., van der Kruk, G., van Holst, G., Raaijmakers-Ruijs, C., Postma, E., Groeneweg, B. and Jongen, W. (1998) The glucosinolates sinigrin and progoitrin are important determinants for taste preference and bitterness of Brussels sprouts. *Journal of the Science of Food and Agriculture* 78, 30–38.

van Poppel, G., Verhoeven, D.T., Verhagen, H. and Goldbohm, R.A. (1999) Brassica vegetables and cancer prevention. Epidemiology and mechanisms. *Advances in Experimental Medicine and Biology* 472, 159–168.

Verhoeven, D.T., Verhagen, H., Goldbohm, R.A., van den Brandt, P.A. and van Poppel, G. (1997) Review of mechanisms underlying anticarcinogenicity by *Brassica* vegetables. *Chemico-Biological Interactions* 103(2), 79–129.

Vinson, J., Hao, Y., Su, X. and Zubik, L. (1998) Phenol antioxidant quantity and quality in foods, vegetables. *Journal of Agricultural and Food Chemistry* 46, 3630–3634.

Warwick, S.I., Francis, A. and Al-Shehbaz, I.A. (2006) Species checklist and database on CD-Rom. *Plant Systematics and Evolution* 259, 249–258.

Zhao, F., McGrath, S.P. and Crosland, A.R. (1994) Comparison of three wet digestion methods for the determination of plant sulphur by inductively coupled plasma atomic emission spectrometry (ICP-AES). *Community in Soil Science and Plant Analysis* 25, 407–418.

6 Citrus
[Orange, Lemon, Mandarin, Grapefruit, Lime and Other Citrus Fruits]

Amarat H. Simonne and Mark A. Ritenour

6.1 Introduction

Citrus, a genus of the plant family *Rutaceae*, is one of the most important horticultural commodities in the world. First domesticated 4000 years ago in South–east Asia, some suggest that secondary diversification of *Citrus* occurred later in the Mediterranean and Caribbean (Mukhopadhyay, 2004). Others believe the orange, tangerine and pomelo (or pummelo) originated in China and South–east Asia, lemon and lime in northern India and grapefruit in the Caribbean (Saunt, 2000). As one of the ancient crops, *Citrus* has been subjected to years of natural and man-made genetic transformation, resulting in many new species and hybrids and leading to disagreements and inconsistencies among experts on the classification and taxonomy of the plants (Mukhopadhyay, 2004; Moore *et al.*, 2005; Khan, 2007; Laszlo, 2007; Ladaniya, 2008). To complicate the classification and taxonomy further, most citrus species are polyembryonic and apomictic, which means that seeds contain one zygotic embryo and other embryos made entirely of maternal, nucellar DNA and tissue (Mukhopadhyay, 2004; Khan, 2007; Ladaniya, 2008). To date, it is impossible to give an exact number of species for citrus. While Moore *et al.* (2005) stated that 150 genera and 900 species were attributed to the *Rutaceae* family, others stated 150 genera and 1500 species (Laszlo, 2007). Nevertheless, new genetic data derived from modern methods of analyses suggest that all citrus species are derived from three basic species: *C. grandis* (pomelo), *C. medica* (citron) and *C. reticulata* (mandarin) (Mukhopadhyay, 2004). Because of its long history and ability to grow in various regions of the world ranging from latitudes 40°N to 40°S (Mukhopadhyay, 2004), many scientific and contemporary reports, books and records have been published on citrus throughout the world, in many languages.

This chapter will focus on the major economically important true citrus plants that are now widely grown throughout the globe, including *Citrus*, *Fortunella* (kumquat) and *Poncirus* (trifoliate orange), which is used mainly as rootstock for citrus production because of its cold hardiness (Saunt, 2000). The *citrus* genera are divided further into the sweet orange (*C. sinensis*), mandarin (*C. reticulata*), grapefruit (*C. paradisi*), pummelo (*C. grandis*), lemon (*C. limon*), sour lime (*C. aurantifolia*), citron (*C. medica*) and sour orange (*C. aurantium*).

Total world production of citrus was estimated to be 116 million tonnes (Mt) in 2007 (FAOSTAT, 2009). Based on the world production data and availability of current research, similar crops will be grouped together. For example, mandarin, tangerine and clementine will be grouped into the same class, despite some slight differences in genetic background (Saunt, 2000; Mukhopadhyay, 2004; Ladaniya, 2008). The sweet orange groups are by far the most important citrus species for both fresh and juice consumption worldwide, with almost 64 Mt produced in 2007, followed by tangerine (27 Mt), lemon/lime (13 Mt) and grapefruit/pummelo (5 Mt) (FAOSTAT, 2009). The top orange producing countries are Brazil, the USA and China (Ladaniya, 2008).

Mukhopadhyay (2004) classified oranges from different producing countries around the world into four types, including common or round, navel, pigmented (blood-coloured) and acidless oranges. According to Ladaniya (2008), oranges for fresh market are divided into two large groups: blood (pigmented) or non-blood (non-pigmented) oranges.

Although citrus has been grown since ancient times, only in the past 100 years has its production and processing become fully commercialized. Citrus commercialization relied on advancements in its production and postharvest research, including postharvest handling, processing and nutritional research that developed hand in hand with other scientific advances in various fields during the 20th century (Nagy et al., 1977a,b). As a result, citrus is one of the world's major fruit commodities consumed, around the world, as fresh fruit or juice and other citrus-derived products (Boriss, 2009; UNCTAD, 2009).

6.2 Identity and Role of Bioactive Compounds

Since ancient times, citrus fruit and products have been recognized as important components of a healthy human diet due to various constituents such as vitamin C, folic acid, potassium, flavonoids, pectin, pigments and limonoids (Manners, 2007; Ladaniya, 2008). Citrus fruit contain numerous bioactive compounds that offer potential human health benefits. In addition to vitamin C and folic acid, which have been the traditional focus of citrus nutritional components, this chapter will focus on the potential health benefits

of other citrus components (bioactive compounds). The majority of studies attribute the health benefits of natural citrus to phytochemicals such as flavonoids, limonoids, furocoumarins and pectins, which are present in citrus fruit (Patil et al., 2006a,b).

6.2.1 Pigments (chlorophylls, carotenoids and anthocyanins)

Major pigments found in citrus fruit include chlorophylls, carotenoids and anthocyanins. Chlorophylls (a and b) impart green colours and predominate in the peel of citrus fruit during growth and maturation. Chlorophyll levels are high when the fruit are immature, and disappear in most citrus fruit after maturation and exposure to cool night temperatures. Most research on citrus fruit chlorophyll composition was conducted from the 1940s to 1980s. It was shown that changes in peel chlorophyll content were not a good indicator of maturity because the fruit often reached maturity while remaining green, especially in tropical or subtropical growing environments (Gross, 1987; Ladaniya, 2008). However, high chlorophyll content in the peel of most citrus fruit is often considered a negative quality characteristic because it masks the presence of carotenoids, which give the fruit its characteristic attractive orange and yellow colour. For lime, however, green colour is considered an important positive quality indicator, and thus chlorophyll maintenance is highly desirable during postharvest handling and marketing of the fruit (Win et al., 2006a,b).

Understanding the nature of chlorophyll degradation was important for developing effective degreening procedures for many fresh citrus varieties (Ladaniya, 2008). Optimum conditions for promoting chlorophyll degradation differ depending on variety and growing location and are related to chlorophyllase activity (Gross, 1987). Although the potential health benefits of chlorophyll derivatives have been explored (Egner et al., 2001; Ferruzzi et al., 2001, 2002; Fahey et al., 2005), limited documentation is available on the potential intake, bioactivity and specific health benefit of chlorophyll pigments from citrus.

Carotenoids are major pigments found in the peel, flesh, flavedo and juice of citrus fruit, giving rise to colours ranging from light pale yellow to deep orange. For non-blood-type orange, most of the pigments are carotenoids. The citrus carotenoids are by far one of the most complex and numerous carotenoid groups reported of any fruit (Gross, 1987). In addition to commonly known provitamin A carotenoids (e.g. α-carotene, β-carotene, β-cryptoxanthin), citrus also contains many other lesser-known provitamin A carotenoids, as well as a number of unique carotenoids such as C_{30}apocarotenoid (a genus-specific carotenoid found only in citrus), citrus apocarotenenal, trollizanthin, citraurin and α-cryptoxanthin, which is a derivative of α-carotene, to name just a few. Relatively few comprehensive reviews of citrus carotenoids have been conducted since Gross (1987), as summarized by Ladaniya (2008). Additional works in the past decade have focused on carotenoids of specific hybrids or new varieties formed from citrus mutations (Goodner et al., 2001; Lee, 2001; Lee and Castle, 2001; Dhuique-Mayer et al., 2005; Xu, C.J. et al., 2006; Xu, J. et al., 2006; Matsumoto et al., 2007; Meléndez-Martínez et al., 2007a; Wang et al., 2007, 2008 ; Xu et al., 2008). In summary, these studies show that the concentration and distribution of carotenoids in citrus are affected by variety, maturity, tissue type, climate and season. Most carotenoids are concentrated in the fruit peel and the distribution of carotenoids in juice and in flavedo are very similar, except for a few varieties of citrus (Matsumoto et al., 2007). A review by Meléndez-Martínez et al. (2007b) of carotenoid composition data in citrus fruit revealed that the carotenoid diversity in cultivated citrus was linked more to the global evolution process of the cultivated citrus rather than to the recent mutation or human selection process (Meléndez-Martínez, et al., 2007b). It was also observed that the citrus varieties grown in Mediterranean climates tended to accumulate more carotenoids than those grown in other geographic areas. This is because of the unique stress conditions found in the Mediterranean. Among ten citrus species and varieties (Salustiana, Hamlin, Maltaise, Shamouti, Sanguinelli, Valencia, Pera, Cara Cara, mandarin and clementine) used in the study, mandarin and clementine contained the most provitamin A activity, with activity ranging between 900 and 1000 retinol equivalents (Dhuique-Mayer et al., 2005). Additional new carotenoid works have focused on gene expression in different varieties (Shamouti orange, Sanginelli orange, Cara Cara navel orange and Huang pi Chen orange) as related to fresh fruit colour (Fanciullino et al., 2008), role of specific enzymes during maturation (Satsuma mandarin, Valencia orange and Lisbon lemon) (Kato et al., 2006), biosynthetic pathway of carotenoids in citrus (25 species) (Fanciullino et al., 2007) and regulation of colour break in mandarin fruit (Alós et al., 2006). These studies reveal that carotenoid distribution in citrus is highly complex and can be explained by many factors, such as diversity of the genes, mutation and changes in plant hormones, among other things.

Ladaniya (2008) summarized research on the distribution of various carotenoids in orange and other citrus fruit and reported that the carotenoids increased in both peel and pulp as the fruit ripened. The same author also reported specific content of some apocarotenoids in the peel of hybrid citrus, although these compounds were only a minor component of the peel of C. sinensis (Ladaniya, 2008). The presence of lycopene was also observed in some citrus species, including pink and red grapefruit and some mutants of sweet orange, such as red navel (Lee, 2001; Xu, C.J. et al., 2006; Xu, J. et al., 2006; Liu et al., 2007). Although citrus species are listed among more than 500 carotenoid sources (Babosa-Filho et al., 2008), very limited work has been done evaluating the activities of the carotenoids in citrus because the contribution of citrus carotenoids to antioxidative activity is negligible in the edible portion of orange juice (Gardner et al., 2000) and also because most of the citrus carotenoids (non-provitam A carotenoids) are located in the fruit peel, which typically is discarded. In addition, a recent review by Hooper and Cassidy (2006) did not include carotenoids in its list of bioactive compounds with health care potential. Therefore, relatively speaking, carotenoids in citrus may appear to have less prominent health-promoting roles than other classes of citrus bioactive compounds, yet it is impossible to disregard their importance because of other possible synergistic effects of citrus to human health as a whole.

Anthocyanins (members of the phenolic grouping of compounds) are another class of pigments found in pigmented or blood orange (sweet orange). These include glycosides of pelargonidin, peonidin, delphinidin and petunidin (Robards and Antolovich, 1997), and the concentration of these pigments increases when the fruit reach maturity. The Mediterranean growing conditions of hot days and cool nights facilitate the development of the anthocyanins (Davies and Albrigo, 1994). The blood orange is commercially important in Mediterranean countries. It has a deep colour in the flesh, and occasionally in the peel (Fig. 6.1).

6.2.2 Phenolics

As secondary metabolic products, phenolics are a class of organic compounds with one or more hydroxyl (OH) groups attached to an aromatic ring (benzene ring). Ladaniya (2008) has summarized the presence and changes of phenolic compounds which, to a certain extent,

affect the taste and colour of orange juice during processing. According to Kanes *et al.* (1993), four different classes of phenolic compounds absorbing at 285 nm (flavones/ols, flavanones, coumarins/cinnamic acid derivatives and psoralens) are found in *Rutaceae* species and cultivars; these include glycosides and the highly methoxylated flavones called polymethoxylated flavones (Manthey and Guthrie, 2002). Others (Shahidi and Ho, 2005; Ladaniya, 2008) classify phenolics into three groups based on the complexity of the chemical structures: (i) simple phenolics (monocyclic) such as catechol and hydroquinone; (ii) dicyclic phenolics with two benzene rings such as flavones; and (iii) polyphenolic compounds such as cyanins and anthocyanin pigments found in the blood-type oranges. Because there is no consistent way to group phenolic compounds, they often appear in the literature under different headings. It is well known that plant phenolic levels increase after infection by pathogens (Ladaniya, 2008). Many phenolics in citrus peels are often discarded as waste products after processing for juice. Recent research has attempted to recover some of these bioactive compounds from discarded citrus peel, as a source of natural antioxidants (Li *et al.*, 2006).

Fig. 6.1. From top left: Flame red grapefruit,[1] Marsh white grapefruit,[1] blood orange[2] and navel orange.[2] From bottom left: Pineapple orange,[2] Hamlin oranges,[2] pomelo fruit[3] and pomelo juice vesicles.[3] Photo credits: [1]Mark Ritenour, [2]Frederick S. Davies, [3]Amarat Simonne.

6.2.3 Flavonoids

Flavonoids are a subclass of polyphenols or polyphenolic compounds and they consist of two aromatic rings, each containing at least one hydroxyl (–OH) group, which are connected via a three-carbon 'bridge' and exist as a six-member heterocyclic ring (C_6-C_3-C_6). Citrus flavanones have a B-ring connected to the C-ring at position 2 (on the C-ring) without unsaturation on the C-ring and with another functional group on position 4-Oxo (Benavente-Garcia *et al.*, 1997; Robards and Antolovich, 1997; Mouly *et al.*, 1998; Beecher, 2003; Fig. 6.2). Furthermore, the citrus flavonoids (flavanones) can exist in conjugated forms with sugar (flavanone glycosides) or without sugar, but with one or more methyl groups (aglycones or polymethoxylated flavones), causing much complication in identification and analyses (Robards and Antolovich, 1997). Early works on the flavonoid content of citrus were somewhat fragmented, without a full representation of the citrus species, tissue types, or types of flavonoid, and this might have been due to the lack of suitable analytical methods (Miyake *et al.*, 1997; Mouly *et al.*, 1998; Nogota *et al.*, 2006). For example, Robards and Antolovich (1997) reported that major flavonoids found in grapefruit included naringin, narirutin, hesperidin and neohesperidin, as well as others such as tangeretin and polymetholylated flavones. Major flavonoids found in sweet orange include hesperidin, narirutin, eriocitrin and narirutin-4'-glucoside, as well as other flavones (sinensetin, nobiletin, tangeretin and isosinensitin) (Robards and Antolovich, 1997). Another study reported that orange (*C. sinensis* L.) contained mainly flavonoid glycoside, hesperidine, and its flavone analogue, diosmin, which have been shown to have anticarcinogenic activities (Manthey and Guthrie, 2002). In citrus plants, flavonoids have been attributed to the protection against some infectious plant diseases such as green mould caused by *Penicillium digitatum* (Ortuno *et al.*, 2006).

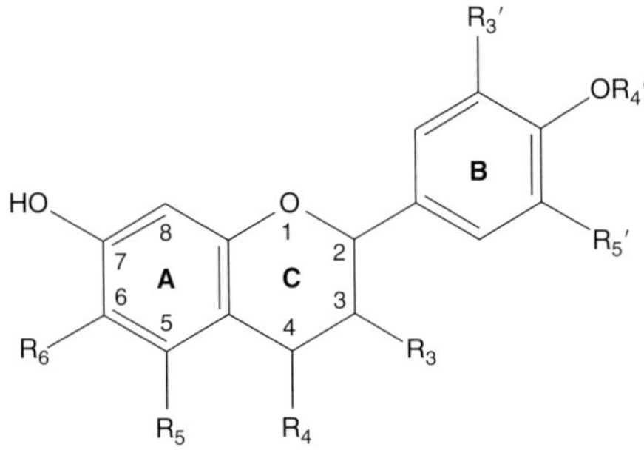

Nogata *et al.* (2006) evaluated concentrations of flavonoids in various tissues (whole fruit, peel, juice vesicles, flavedo, albedo and segment epidermis) of 45 citrus species

Fig. 6.2. General structure and numbering pattern of several classes of flavonoids contained in citrus fruit that are important for human health (based on Beecher, 2003).

classified by Tanaka's system (Tanaka, 1969). This is one of the most comprehensive studies on citrus flavonoids in various cross sections of citrus species. According to the Tanaka system, citrus is classified into the following sections or groups: I – *Papeda* (e.g. *C. macroptera* or Cabuyao); II – *Limonellus* (e.g. lime, Bergamot and Biroro); III – *Citrophorum* (e.g. citron, lemon and lumie); IV – *Cephacitrus* (e.g. Marsh grapefruit, pummelo); V – *Aurantium* (e.g. sour orange, Valencia); VI – *Osmocitrus* (e.g. Yuzu, Sudachi); VII – *Acrumen* (e.g. Ponkan, Satsuma, Clementine and Dancy tangerine); and VIII – *Pseudofortunella* (e.g. *C. madurensis*), *Fortunella-Eufortunella* (e.g. kumquat), and *Poncirus* (trifoliate orange) (Tanaka, 1969; Nogata *et al.*, 2006). Nogata *et al.* (2006) found that the flavonoid composition of citrus was in agreement in each section, with the exception of the *Aurantium* (V) section and others with a peculiar flavonoid composition. The profiles may be different if they are classified by a different system (Nogata *et al.*, 2006). Distribution of the predominent flavonoids in various tissues of selected common citrus species, as reported by Nogota *et al.* (2006), is provided in Table 6.1.

Peterson *et al.* (2006a,b) reviewed the available analytical data on flavonoids in many common citrus species including grapefruit and orange, and their relatives. They reported that, overall, grapefruit had a distinct flavanone profile, with naringin as a dominant compound, but was similar to the sour orange. These studies confirmed some of the findings of Nogata *et al.* (2006), with

Table 6.1. Distribution of flavonoids in various tissues from some selected citrus species based on the Tanaka grouping system.

Citrus group	Dominant flavonoids[a]					
	Whole fruit	Peels	Juice vesicles	Flavedo	Albedo	Segment epidermis
I – *Papeda*						
Cabuyao	PON, NHP	PON, NHP	PON	SNT	PON	PON
II – *Limonellus*						
Mexican lime	HSP	HSP	HSP	HSP	HSP	HSP
Tahiti lime	HSP	HSP	HSP, ERC	HSP, ERC	HSP	N/F
Bergamot	PON, NHP	PON, NRG, NHP	PON, NHP, NRG	NER, NRG	PON, NHP, NRG	PON, NER, NHP
Biroro	HSP	HSP	NDM, NRG	HSP	HSP	HSP
III – *Citrophorum*						
Citron	HSP	ERC, DSM	HSP, RTN, ERC	RTN	ERC, DSM	HSP, DSM
Eureka lemon	HSP	HSP	ERC, HSP	HSP	HSP	HSP
Sweet lemon	HSP	HSP	HSP	HSP	HSP	HSP
Lumie	HSP	HSP	HSP, ERC	ERC, HSP	HSP	HSP
IV – *Cephacitrus*						
Hirado buntan	NRG	NRG	NRG	NRG	NRG	NRG
Shaten yu	NRG	NRG	NRG, RTN	NRG, PON	NRG	NRG
Marsh grapefruit	NRG	NRG	NRG	NRG	NRG	NRG
Kinukawa	NHP, NRG	NHP, NRG	NRG, NHP	NHP	NHP, NRG	NRG, NHP
Hassaku	NRG	NRG	NRG, NRT	NHP	NRG	
V – *Aurantium*						
Natsudaidai	NRG	NRG	NRG	NHP, NRG	NRG	NRG, NHP
Sour orange	NRG, NHP	NRG, NHP	NRG, NHP	PON, NHP, NRG	NRG, NHP, PON	NRG, NHP, PON
Valencia	HSP	HSP	HSP	HSP	HSP	HSP
Morita navel	HSP	HSP	NRT, HSP	HSP	HSP	HSP
Shunkokan	NRT, HSP	HSP, NRT	NRT	HSP	NRT, HSP	NRT, HSP
VI – *Osmocitrus*						
Yuzu	NRT, HSP	HSP, NHP, NRG	NRT, HSP	NHP	HSP, NRT	NRT, HSP, NRG
Sudachi	NHP, NRT	NHP, RTN, NRT	NRT, HSP, NHP	NHP	NHP, RTN	NRT, NRG
Kabosu	HSP, NRT	HSP, NRT, RTN	NRT, HSP	NHP, HSP	HSP, RTN, NRT	NRT, HSP

(Continued)

Table 6.1. *Continued*

Citrus group	Whole fruit	Peels	Juice vesicles	Flavedo	Albedo	Segment epidermis
			Dominant flavonoids[a]			
VII – *Acrumen*						
King	HSP	HSP	NRT, HSP	HSP	HSP	NRT, HSP
Satsuma	HSP	HSP	NRT, HSP	HSP	HSP	HSP, NRT
Yatsushiro	HSP	HSP	HSP	HSP	HSP	HSP
Ponkan	HSP	HSP	HSP, NRT	HSP	HSP	HSP
Dancy tangerine	HSP	HSP	HSP	HSP	HSP	HSP
Clementine	HSP	HSP	HSP	HSP	HSP	HSP
Kishu	HSP	HSP	HSP	HSP	HSP	HSP
VIII – *Pseudofortunella*						
Shikikitsu	NRT	NRT	NRT	NRT	NRT	NRT
Fortunella-Eufortunella						
Oval kumquat	NRT	NRT	NRT	NRT	NRT	NRT
Meiwa kumquat	NRT	NRT	NRT	NRT	NRT	NRT
Poncirus						
Trifoliate orange	PON	PON	PON	NRG	PON	PON

Notes: [a]Adapted from Nogata *et al.* (2006). Within each cell of the table, the compounts are listed in order of magnitude. ERC = eriocitrin; NRT = narirutin; NRG = naringin; HSP = hesperidin; NHP = neohesperidin; PON = poncirin; SNT = sinensetin; RTN = rutin; DSM = diosmin; N/F = not found.

additional details on the flavonoid distribution in different types of grapefruit, orange and other related citrus species. As for lemon and lime, the flavanone profiles were dominated by hesperidin and eriocitrin, and were similar to sweet orange. Peterson *et al.* (2006a) reported total mean flavanone aglycone contents of 27 mg/100 g (white grapefruit) and 18 mg/100 g (red grapefruit). Naringin was the major (>60%) of the reported total flavanones in the different types of grapefruit studied. Pink and red grapefruit appeared to contain a lower amount of flavonoids than white grapefruit or grapefruit in general. Other detectable flavanones in grapefruit (in general) include didymin (<1%), eriocitrin (1.7%), hesperidin (10%), naritrutin (18%), neoeriocitrin (1.3%), neohesperidin (5%) and poncirin (<1%). For white grapefruit, after naringin, the distribution of flavonoids is didymin (<1%), eriocitrin (<1%), hesperidin (14%), naritrutin (20%), neoeriocitrin (<1%), neohesperidin (<1%) and poncirin (<1%). Flavonoids in red and pink grapefruit include hesperidin (1.5%), naritrutin (19%) and neohesperidin (2%). Overall, red and pink grapefruit contain lower flavonones than white grapefruit (Peterson *et al.*, 2006a). Total flavanones in lemon and lime were reported to be 26.58 and 17.29 mg aglycone/100 g fresh fruit or juice, respectively, and the major flavonone was hesperidin, with 59% and 90% in lemon and lime, respectively. Eriocitrin was the second major flavanone in lemon and lime, representing approximately 36% and 8% of the total flavonones, respectively. Other detectable flavanones in lemon were naringin and narirutin, and in lime were eriocitrin, narirutin and neoeriocitrin (Peterson *et al.*, 2006a). According to Peterson *et al.* (2006b), the major flavonoids in sweet (*C. sinensis*) and sour orange (*C. auranthum*) are flavanones. In general, sour orange has a distinctive flavonone profile, with naringin and neohesperidin as dominant flavones, and this finding is consistent with Nogota *et al.* (2006). Peterson *et al.* (2006b) found that, among all the oranges included in their study, sour oranges (*C. aurantium* varieties Bergamot, Chinotto, Daidai and Seville) had the highest flavanone content, with an average of 48 mg aglycones/100 g FW. Lowest levels of flavanone content

and similar compounds (19–25 mg aglycones/100 g ranges) were found in sweet oranges (blood, Hemlin, Navel and Valencia), tangors (a cross of *C. reticulata* and *C. sinensis*) and tangerines. Tangelos (a cross of *C. reticulata* and *C. paradisi*) have moderate levels of flavanones, with a mean content of 29 mg aglycones/100 g (Peterson *et al.*, 2006b). Overall, the flavanone content of the tangelo was different from the sweet and sour orange, tangerine and tangor. The Peterson *et al.* (2006a,b) data were incorporated into the USDA flavonoid database (Bhagwat *et al.*, 2006). Additional information on flavonoids in citrus fruit continue to be elucidated for various citrus species, including *C. bergamia* Risso (Gattuso *et al.*, 2007).

6.2.4 Limonoids

Limonoids are chemically related triterpene derivatives typically found in the *Rutaceae* and *Meliceae* families (Roy and Saraf, 2006; Manners, 2007). The significant amounts of highly oxygenated triterpenoid compounds (limonoids) to be found in citrus have not been used to their potential (Manners, 2007). Limonin was first identified as a citrus constituent in 1841, but all other limonoids have been isolated in the past 50–60 years. Limonoids have been well documented for the control of many species of insects (Patil *et al.*, 2006a,b). Although the nutritional roles of limonoids have not been defined (Manners *et al.*, 2003; Manners, 2007), much research has been conducted on this family of compound because limonin causes bitterness in many citrus juices. Manners *et al.* (2003) examined the bioavailability of pure limonin glucoside from citrus juice in healthy humans and found significant variations in plasma concentrations among the subjects, but they found that the average time to reach the highest concentrations (1.74–5.37 nmol/l) was 6 h. Manners (2007) indicated that the association of triterpenoid (limonoid) with bitter taste in orange juice was first established in the 19th century. It was not until 1938 that a specific compound, limonin, was isolated from orange juice. Early research focused on understanding the mechanism of bitterness formation and how to eliminate or prevent the bitterness in citrus juices (Nagy *et al.*, 1977a; Drewnoski and Gomez-Carneros, 2000). The first documented antioxidative properties of citrus limonoids (limonin, limonin glucoside and neoeriocitrin) were reported by Yu *et al.* (2005). A recent review (Ejaz *et al.*, 2006) indicated that citrus limonoids, in either natural fruit or purified form, might provide substantial anticancer effects. Limonoids have been screened and tested for anticancer properties in both laboratory animals and human cancer cells because of their structures. Citrus limonoids are absorbed and metabolized in humans, but the mechanisms of absorption and metabolism are still unclear (Breksa *et al.*, 2009). Furthermore, citrus limonoids are complex triterpenoid compounds and are found in different forms (aglycones, glucosides or A-ring lactone). Thus, isolation and identification of these compounds are important in relation to the evaluation of health benefit in human or animal studies (Jayaprakasha *et al.*, 2006; Patil *et al.*, 2006b; Breksa *et al.*, 2009).

6.2.5 Pectic substances

Although pectic substances can be found in many plant tissues, they are found in large quantities only in citrus fruit tissues. Pectic substances belong to a class of complex polysaccharides that serve as hydrating agents and cementing materials for the cellulosic network in plants (Nagy *et al.*, 1977a; Liu *et al.*, 2001). Commercial pectins are derived mostly from lime, lemon, grapefruit, orange and apple, for the production of jam and jellies. Citrus pectins have also been shown to have health benefits and have been used as wound treatment ingredients, homeostatic agents, immune complement activators, for the treatment of chronic diseases such as diabetes and high blood cholesterol, and for cancer inhibition (Liu *et al.*, 2001, 2002; Salman *et al.*, 2008). In recent years, additional health benefits associated with citrus pectins, used alone or in combination with other citrus bioactive compounds, continue to have evolved, including use in heavy metal detoxification (Eliaz *et al.*, 2007), treatment of advanced solid tumours (Azémar *et al.*, 2007) and lowering cholesterol in eggs (Lien *et al.*, 2008). As

pectic substances are predominently present in citrus peel and typically are discarded as waste, many research efforts have focused on recovering these pectic substances, as a source of dietary fibre (Figuerola *et al.*, 2005; Ubando-Rivera *et al.*, 2005; Marin *et al.*, 2007), and other bioactive compounds, as functional food ingredients or supplements (Schieber *et al.*, 2001; Ubando-Rivera *et al.*, 2005).

6.2.6 Other phytonutrients

In addition to the previously mentioned phytonutrients (carotenoids, phenolic compounds, flavonoids, limonoids, polysaccharides and pectic substances), citrus contain other compounds which may be beneficial or detrimental to human health that are as yet unknown. These include furocoumarins, which are found in grapefruit and other citrus species (Stanley and Jurd, 1971; Gattuso *et al.*, 2007). On the one hand, the furocoumarins (psoralens) have been found to have strong antioxidative properties and to protect against vascular injury caused by catheter-directed arterial intervention procedure (also known as balloon injury-related neointima formation) (Mollace *et al.*, 2008) but, on the other hand, they have been linked recently to increased cutaneous melanoma (Sayre and Dowdy, 2008).

6.3 Chemopreventive Activity and Bioavailability

Citrus is considered to be one of the largest suppliers of vitamin C, as well as other basic nutrients (sugars, folate, provitamin A, carotenoids, other vitamins and minerals) to the human diet (Nagy *et al.*, 1977a; Nagy, 1980; Manners, 2007; Ladaniya, 2008). In the 1980s and 1990s a plethora of epidemiological studies and reviews revealed the positive benefits of consuming fruit and vegetables for reducing the risk of some chronic diseases (Peterson *et al.*, 2006a,b; Table 6.2). Citrus fruit consumption was shown to have positive protective effects on chronic diseases (Economos and Clay, 1999) and to have anticancer (Silalahi, 2002; Cuthrell and Le Marchand,

2006), antiosteoporosis (Deyhim *et al.*, 2006) and antioxidative (Deyhim *et al.*, 2007) effects. Others did not find an association of citrus fruit consumption with the risks of breast cancer (Gaudet *et al.*, 2004) or prostate cancer (Bae *et al.*, 2008). Based on a comparative study of carotenoid intake, it was found that β-cryptoxanthin was obtained primarily from citrus in France, the UK, Republic of Ireland, Spain and the Netherlands (O'Neill *et al.*, 2001). New research also shows that serum β-cryptoxanthin concentration and circulating bone metabolic markers exist in healthy individuals following prolonged consumption of orange juice containing β-cryptoxanthin (Yamaguchi *et al.*, 2005). Other research revealed preferential uptake of β-carotene and free β-cryptoxanthin from the ester forms of β-cryptoxanthin by Caco-2 cells (Dhuique-Mayer *et al.*, 2006). More recently, citrus and citrus-derived bioactive compounds have been linked to many biological functions, including antioxidative, anti-inflammatory, antiallergic, antiviral, antiproliferative, antimutagenic and anticarcinogenic, to name just a few (Jayaprakasha *et al.*, 2006; Patil *et al.*, 2006a; Tripoli *et al.*, 2007; Benavente-Garcia and Castillo, 2008; Table 6.2). Furthermore, several citrus bioactive compounds (naringenin, hesperetin, hesparedin, eriocitrin, naringin, meoeriocitrin, natrituin, *p*-coumaric acid, cafeic acid and ferulic acid) have been identified as having a role in the prevention of cardiovascular disease (Joshipura *et al.*, 2001; Mennen *et al.*, 2004) and cancers (Benavente-Garcia *et al.*, 1997; Kris-Etherton *et al.*, 2002). After the positive associations between citrus bioactive compounds and human health were reported, many researchers examined the plasma kinetics of these compounds. Erlund *et al.* (2001) studied plasma concentrations or plasma kinetics of flavonones (naringenin and hesperetin) in humans after ingestion of orange and grapefruit juice, in order to see if flavonone concentrations in urine could be used as biomarkers of intakes. However, they found that the urine concentration of flavonones was not a good biomarker of dietary intake. Miyake *et al.* (2006) examined the fate of flavonoids in humans after ingestion of flavone glycosides and aglycone forms and found that the absorption of both forms was affected by the coexisting solutions, and the aglycone form was absorbed faster than the glycoside form. At this point, limited information is available on the absorption of these compounds from juice compared with whole fruit. Specific reports on prevention or curing of specific diseases are summarized in Table 6.2, respectively.

6.3.1 General health benefits as related to citrus

The pharmacological properties of citrus have been well documented in various regions of the world since medieval times. The early literature from the Mediterranean region revealed the use of citron and lemon as antidotes for 'poison and venom'. However, in the modern literature citron and bitter orange are documented to have anticancer activity, lime to have immunomodulatory effects in humans, pomelo to be of use in treating circulatory problems and lemon to be useful in easing hangover symptoms (Arias and Ramón-Laca, 2005). The protective benefits against chronic diseases (such as cardiovascular diseases, hypertension, cataract, diabetes, Alzeimers's disease, stroke and cancer) of diets rich in vegetables and fruit including citrus have been documented extensively in recent years (Joshipura *et al.*, 1999; Silalahi, 2002; Sun *et al.*, 2002; Bazzano *et al.*, 2003; Hung *et al.*, 2004; Mennen *et al.*, 2004; Norman *et al.*, 2004; Ladaniya, 2008). Citrus bioactive components, such as flavonoids, limonoids and dietary fibre, have been linked to the reduction of degenerative disease risk or to a protective effect. Despite the documentation of certain medicinal and nutritional values of citrus in ancient times, the specific basic nutritive values of citrus have only been well documented in the last 50 years (Ladaniya, 2008). For example, since the 1560s, citrus fruit has been known to prevent and cure scurvy (Nagy, 1980). However, it was not until the 1920s and 1930s that ascorbic acid (vitamin C) was identified as the compound that prevented scurvy. Gardner *et al.* (2000) examined the relative contributions of vitamin C, carotenoids and phenolics to the antioxidant potentials of fruit juices

Table 6.2. Health-promoting action of citrus fruit and fruit products including orange (C. sinensis), grapefruit, lemon and lime, as well as pure chemical extracts.

Activity	Disease conditions	Commodities	Population	Type of study	Reference
Protective effect	Cardiovascular, cancer, neural tube defects, anaemia, cataracts, bone metabolism and osteoporosis, kidney stone disease, cognitive function and asthma	Citrus fruit	N/A	Review (25 refs)	Economos and Clay (1999)
Protective effect	Association with reduced risk of ischaemic stroke	Citrus fruit and juice	Nurses and health professional (men and women)	Large population study	Joshipura et al. (1999)
Anticancer	Reduce risk of degenerative diseases	Citrus fruit	N/A	Review (35 refs)	Silalahi (2002)
No-anticancer	No association of citrus fruit consumption with breast cancer among pre- or postmenopausal women	Citrus fruit	Pre- and postmenopausal women	Analysis of large population case control study	Gaudet et al. (2004)
Antiosteoporosis	Positively affect serum antioxidant status and bone strength	Citrus juice	Senescent rat model of osteoporosis	Original research	Deyhim et al. (2006)
Anticancer	Protective effects against cancer	Grapefruit	N/A	Review of original research	Cuthrell and Le Marchand (2006)
No-anticancer	No association of citrus fruit consumption with prostate cancer risk	Citrus fruit	N/A	Review (54 refs)	Bae et al. (2008)

Pure chemical or extracts derived from citrus

Activity	Action	System	Dose	Extract type	Reference
Anticancer	Inhibit the proliferation of a number of cancer cell lines	Human cancer cell lines	IC50 below 10 µM	Pure citrus flavonoids (orange peel and other sources)	Manthey and Guthrie (2002)
Anticancer	Induce cancer cell death	Neuroblastoma cells (undifferentiated human SH-SY5Y)	0.1–10 mmol/l	Four citrus limonoid glucocides	Poulose et al. (2005)

Activity	Effect	Model/system	Dose	Extract/compound	Reference
Anticancer	Inhibit growth and apoptosis-inducing activity (cell death)	Human breast carcinoma cell lines (MCF-7)	Varied	Sweet orange peel extract containing hydroxylated polymethoxyflavones (PMFS) and non-PMFS	Sergeev et al. (2007)
Anticancer	Antineoplastic by reduction of tumour burden	Hamster cheek pouch model	2.0–2.5%	Pure citrus flavonoids	Miller et al. (2008)
Anti-inflammatory	Reduce the formation of intestine oedema in mice model	Mice model for induced colitis	Naringin 15.8 mg/kg/day, 5 days; naringenin 5 mg/kg/day	Pure citrus naringin and naringenin	Amaro et al. (2009)
Anti-inflammatory	Inhibition of pro-inflammatory cytokinin IL-1B secretion	Human peripheral blood mononuclear cells	0.25–1.0 mg/ml	Citrus pectin (30–90% esterified)	Salman et al. (2008)
Orange (C. sinensis)					
Antioxidative stress	Increase plasma antioxidant concentrations	Human (60 healthy females)	600 ml/day, 21 days	Blood orange juice	Riso et al. (2005)
Antioxidative stress	Improve antioxidant status and suppress peroxidation	Orchidectomized male rats	2.5–10% of orange pulp to the diet	Orange pulp	Deyhim et al. (2007)
Lime					
Antiplatelet activity	Inhibit platelet aggregation induced by ADP and epinephrine	Human plasma (platelet suspension)	IC50 = 0.40 and 0.32 mg/ml, respectively	Hydroalcoholic extract of lime (C. aurantifolia) leaves	Piccinelli et al. (2008)
Lemon					
Antitumour	Degradation of DNA in cancer cells	HL-60 cells	10–500 µM	Eriocitrin, 3,4-dihydroxyhydrocinnamic acid and phloroglucinal	Ogata et al. (2000)

(Continued)

Table 6.2. *Continued*

Activity	Action	System	Dose	Extract type	Reference
Anticancer	Inhibit human liver cancer cell proliferation	Human liver cancer cell (HepG2)	EC50 = 30.56 mg/ml	Fruit (acetone) extract	Sun et al. (2002)
Lowering blood lipids	Reducing total cholesterol, VLDL, LDL, triglyceride and phospholipid	Rats	0.35 and 0.70%	Eriocitrin (eriodictyol 7-O-beta-rutinoside) from lemon fruit	Miyake et al. (2006)
Grapefruit					
Anticancer	Inhibit human liver cancer cell proliferation	Human liver cancer cell (HepG2)	EC50 = 130.09 mg/ml	Fruit (acetone) extract	Sun et al. (2002)

Notes: VLDL = very low-density lipoprotein; LDL = low-density lipoprotein.

and found that vitamin C accounted for 65–100% of the antioxidant potential of beverages derived from citrus fruit; they also found that the contribution of carotenoids to antioxidant potential was negligible. As research from *in vitro* as well as population-based studies revealed the association of dietary flavonoid intake and human cancer risk, many researchers started examining the sources of flavonoid in human diets (Table 6.2). Lowe *et al.* (2003), based on a population survey, reported that low intake of citrus fruit possibly might play a role in low vitamin C level and a predisposition to low-grade inflammation and thrombosis, cardiovascular diseases and tooth loss. Mukhopadhyay (2004) summarized several reported therapeutic properties of citrus fruit constituents such as pectins and flavonoids. For example, tangeretin (a flavonoid) can prevent invasion of normal tissue by cancer cells. Hesperidin, naringin, tangeretin and nobiletin (also flavonoids) have anti-inflammatory and antiallergic activities. In addition, limonoids and flavonoids have been shown to have some synergistic effects in preventing certain types of cancers and cardiovascular diseases (Patil *et al.*, 2007). Somerset and Johannot (2008) examined sources of dietary flavonoids in Australian adults and found major contrition of flavonoids from various citrus sources, including orange (hesperetin and naringenin), lemon (eriodictyol), mandarin (hesperetin) and grapefruit (naringenin). Several types or classes of bioactive compounds have been identified in citrus fruit and they include non-nutritive fibre (cellulose, hemicelluloses, lignin and pectin), phenolic compounds, terpenoids and steroids, and pigments, which currently have no known vitamin activity (Ladaniya, 2008). Recent evidence has also attributed citrus fruit with anti-inflammatory properties (Ho and Lin, 2008; Amaro *et al.*, 2009; Cardile *et al.*, 2009; Pan *et al.*, 2009). Because citrus fruit include many classes of compounds including carbohydrates (mono, oligo and polysaccharide and their derivatives), organic acid, nitrogenous compounds (amino acid, amine, proteins, nucleotides and nucleic acids, enzymes), lipids (cutin, waxes, terpenoids and steroids), pigments, phenolics (phenols, flavonoids), vitamins,

volatile flavouring compounds and other inorganic constituents, it is fair to say that research has yet to link one single bioactive compound conclusively to any one single disease. Furthermore, distribution (quantity) and profile (chemical forms) of these bioactive compounds are different in the various citrus species, which gives rise to the unique appearance, flavour and other nutritional or medical properties of each citrus type (Nogota *et al.*, 2006; Ladaniya, 2008). In recent years, research efforts have focused more on identifying and quantifying new compounds and how they are affected by processing and postharvest treatments. As compositional and consumption data continue to accumulate along with advancements in other fields of science, it is expected that links between specific citrus components and human health will continue to be elucidated.

6.3.2 Cancer studies

Many researchers have attributed various components of citrus fruit to reducing the risk of certain types of cancers, or have suggested that they could be used to treat cancers (Del Rio *et al.*, 2004; Vanamala *et al.*, 2006; Huang *et al.*, 2007; Tripoli *et al.*, 2007; Cutler *et al.*, 2009; Lim *et al.*, 2009). However, a recent systematic review by Bae *et al.* (2008) did not show any association between citrus intake and incidence/severity of prostate cancer. Studies that have shown the protective effects of citrus against cancer have attributed them to flavonoids, limonoids and other phenolic compounds (e.g. coumarins). A recent comprehensive literature review (Tripoli *et al.*, 2007) revealed that citrus flavonoids exerted anticancer effects in various ways, including selective cytotoxicity, antiproliferative action and apoptosis; the conclusions were reached based on results of *in vitro* (using human and animal cell systems) and *in vivo* studies. Flavonoids can exert their effects at several stages of cancer formation, including induction (DNA damage), promotion (tumour development) and proliferation (invasion). Although the mechanisms of cancer prevention by flavonoids, or citrus flavonoids in particular, have been studied extensively and reviewed by many authors (Middleton *et al.*, 2000; Ogata *et al.*, 2000; Le Marchand, 2002; Huang *et al.*, 2007; Morley *et al.*, 2007; Rossi *et al.*, 2007; Tripoli *et al.*, 2007; Akao *et al.*, 2008; Linseisen and Rohrmann, 2008; Miller *et al.*, 2008; Miyata *et al.*, 2008; Park *et al.*, 2008), the results remain inconsistent and most reports recommend further research. These inconsistent results or outcomes may be due to other factors such as the forms, concentrations and interactions of the chemicals and the types of systems used in the studies, among others. After citrus limonoids were first linked to potential anti-cancer properties (Lam *et al.*, 1994), the subject of citrus limonoids and cancer has been studied extensively (Tanaka *et al.*, 2000; Manners *et al.*, 2003; Miller *et al.*, 2004; Vanamala *et al.*, 2005; Poulose *et al.*, 2005; Manners, 2007). Proposed mechanisms of cancer prevention of limonoids include antioxidative properties and the ability to detoxify carcinogens and harmful chemicals, stimulate the immune system, effect cell differentiation, block formation of nitrosamines, alter oestrogen metabolism, preserve the integrity of intracellular matrixes, maintain normal DNA repair, increase programme cell death (apoptosis) and decrease cell proliferations (Ejaz *et al.*, 2006). In 2007, a patent (US Patent 7201928) was awarded to a group of scientists for use of extracts of orange peel for the prevention and treatment of cancer (Huang *et al.*, 2007). Prince *et al.* (2009) compared different citrus coumarin effects on carcinogen-detoxifying enzymes in Nrf2 knockout mice and the results suggest different modes of actions for these compounds. In coming years, it is expected that much more research will further our understanding of the effects of citrus bioactive compounds on cancers. Additional information on cancer studies is provided in Table 6.2.

6.3.3 Cardiovascular diseases (CVD)

A body of research has revealed a clear association between increased consumption of vegetables and fruit, including citrus, and reduced risk of major chronic diseases such as cardiovascular disease (Middleton *et al.*, 2000; Josphipura *et al.*, 2001; Kris-Etherton *et al.*, 2002; Verhaar *et al.*, 2002; Vinson *et al.*, 2002;

Bazzano *et al.*, 2003; Lowe *et al.*, 2003; Hung *et al.*, 2004; Whitman *et al.*, 2005; Takachi *et al.*, 2007; Tripoli *et al.*, 2007; Jayprakasha *et al.*, 2008). Citrus components initially reported to have protective effects against CVD include antioxidative nutrients such as the carotenoids, vitamin C and folate as well as bioactive compounds which have no nutritional value (e.g. limonoid and phenolic compounds). Possible modes of action of folate in the prevention or protection against cardiovascular disease include homocysteine–folate interaction, antioxidative action, cofactor bioavailability or direct interactions of folate with enzyme endothelial NO synthase (Verhaar *et al.*, 2002). Among the bioactive compounds with protective effects on CVD, flavonoids have been researched substantially. The potential modes of action of citrus flavonoid include cholesterol-lowering potential (Kurowska and Manthey, 2004), antioxidative properties and anti-inflammatory antiatherosclerosis activities (Tripoli *et al.*, 2007). While citrus bioactive components such as phenylephrine from *C. aurantium*, or bitter orange, have been reported to reduce fat absorption, thereby preventing obesity and reducing the risk of CVD indirectly (Klein *et al.*, 2004), other compounds may help reduce damage in blood vessels during some intervention blockage procedures (Mollace *et al.*, 2008). Other research showed *Phellodendron* and citrus extracts to be beneficial on lipid levels, blood pressure and fasting glucose levels in osteoarthritis patients (Oben *et al.*, 2009). To date, it is impossible to isolate any one component of a particular fruit or vegetable as a 'magic bullet' for preventing CVD; it appears to be a combination of many factors.

6.3.4 Other beneficial effects

In addition to the previously mentioned benefits, citrus bioactive compounds have been reported to have many other health-promoting effects. For example, citrus flavonoids (tangeretin, naringenin) were reported to have a neuroprotective effect in rat models for Parkinson's disease (Datla *et al.*, 2001; Zbarsky *et al.*, 2005), Alzheimer's disease (Heo *et al.*, 2004), neuroblastoma cells

(Poulose *et al.*, 2005), neurodegenerative diseases (Cho, 2006) and H_2O_2-induced cytotoxicity in PC12 cells (Hwang and Yen, 2008). Additionally, citrus has also been documented to benefit weight loss, reducing the risk factor for metabolic syndrome (e.g. obesity, lowering cholesterols) (Patil *et al.*, 2006a), and to improve resistance to oxidative stress in streptozotocin-induced diabetic rat liver (Sugiura *et al.*, 2006). Furanocoumarins have also been tested for treating skin disorders such as vitiligo and psoriasis (Conforti *et al.*, 2009). Furthermore, citrus and citrus by-products (e.g. other phenolic compounds) have been associated with anti-inflammatory activities and antimicrobial activities (Patil *et al.*, 2006a; Malhotra *et al.*, 2008). Additional information on the health-promoting action of citrus fruit and fruit products is provided in Table 6.2.

6.3.5 Detrimental effects

The most reported detrimental effect associated with citrus is an interaction of grapefruit and grapefruit juice with the availability of specific medications (Patil *et al.*, 2006a,b). According to a recent review by Farkas and Greenblatt (2008), grapefruit juice is known to interact with more than 30 prescription drugs by increasing their bioavailability. Although such interaction was discovered two decades ago, the mechanisms behind it continue to be elucidated. Other citrus juices, such as orange, Seville orange, pomelo, lime and lemon, and tangerine, also seem to have some degree of influence on drug disposition (Farkas and Greenblatt, 2008). Specifically in grapefruit, it is the furocoumarins and their derivatives such as furanocoumarins that appear responsible for this drug interaction (Fukuda *et al.*, 1997). Furocoumarins and related compounds are formed by only four plant families, including *Rutaceae*, especially in grapefruit (Fukuda *et al.*, 1997; De Castro *et al.*, 2006; Peroutka *et al.*, 2007) and pommelo (Egashira *et al.*, 2004). The average furocoumarin content of white grapefruit juice was reported to be higher than that of red grapefruit juice (Fukuda *et al.*, 2000).

Furocoumarins specifically inhibit CYP3A4 and cytochrome P450 (Guo and Yamazoe, 2004), resulting in drug interactions. Furanocoumarins are toxic secondary metabolites, which may demonstrate antifungal activities and are phototoxic (Peroutka *et al.*, 2007; Larbat *et al.*, 2009). Although the content of furanocoumarins in citrus fruit is low (Peroutka *et al.*, 2007), many researchers have demonstrated that grapefruit and pommelo juice can interact with many types of prescription drugs (Egashira *et al.*, 2004; Dahan and Altman, 2004; De Castro *et al.*, 2006; Mertens-Talcott *et al.*, 2006; Farkas and Greenblatt, 2008; Boobis *et al.*, 2009; Pillai *et al.*, 2009). Another study by Yoo *et al.* (2007) revealed that a flavonoid glycoside, diosmin, might interact with P-gp-mediated efflux in Caco-2 cells, causing increased absorption of medications that are P-gp substrates. Although citrus limonoids have many potential health benefits, one study revealed that high levels of limonoid exposure in diets might cause a problem with weight gain, as seen in pregnant rats (Miller *et al.*, 2006).

6.4 Effect of Preharvest and Postharvest Continuum

It is clear that bioactive compounds in citrus are important to the human diet. However, there are many pre- and postharvest factors that influence these concentrations. The concentration of bioactive compounds can be influenced by cultivar (both rootstock and scion); climate and cultural practices during production; position of the fruit on the tree; fruit maturation; postharvest treatments; handling, storage and transportation of the fruit; juice processing and storage conditions; and the type of container used for juice packaging. For those involved in growing, packing and marketing citrus in its various forms, it is important to understand how these factors influence the nutritional value of the product ultimately delivered to the consumer, with the aim of maximizing the content of health-promoting compounds (Mukhopadhyay, 2004; Patil *et al.*, 2006a,b; Ladaniya, 2008).

6.4.1 Genotype

As might be expected, different citrus geno-types have been found to contain different levels of bioactive compounds. Nagy (1980) developed an excellent table listing the vitamin C content for a large number of orange, grapefruit, mandarin, lemon and lime cultivars reported in the literature. Overall, he found that orange generally had the greatest vitamin C content, followed by grapefruit, lemon, mandarin and finally lime. He found agreement from several sources that juice from early (cvs. Navel and Hamlin) and midseason (cv. Pineapple) orange varieties contained greater ascorbic acid concentrations than did late (i.e. Valencia) orange varieties. He also found that there were minor, usually insignificant, differences in the vitamin C content among grapefruit varieties, and that the vitamin C content among mandarin varieties was more variable but tended to be lower than either orange or grapefruit varieties. Differences in flavonoid composition and concentrations within citrus species were reported by Nogata *et al.* (2006) and discussed previously in this chapter (Table 6.1). The concentrations of flavonoids in whole fruit, peel, juice vesicles, flavedo, albedo and segment epidermis can vary among species by several orders of magnitude, but within citrus groups, member species tend to contain similar predominant flavonoids.

6.4.2 Fruit maturity and size

While total vitamin C content in citrus increases as the fruit matures, the actual concentration of vitamin C and flavonoid (hesperidin) decreases as the fruit size and volume increases (McDonald and Hildebrand, 1980; Nagy, 1980; Vandercook and Tisserat, 1989; Aparicio *et al.*, 1990). Vitamin C content also declines late in the season after the fruit begin to experience a decrease in water content (Harding *et al.*, 1940).

6.4.3 Production condition

Climatic conditions such as field temperature, light exposure, etc., have long been known to exert a profound effect on the external and internal quality attributes of fruit and vegetables, and have also been shown to affect the concentrations of bioactive compounds in citrus. For example, higher field temperatures have been shown to reduce citrus vitamin C content, so that fruit grown under cooler climates tend to have more vitamin C than those growing in warmer climates (Nagy, 1980). Furthermore, greater exposure to light during fruit growth and development results in greater vitamin C accumulation in the fruit. Izumi *et al.* (1992) found that when citrus trees received 80% or more shading for 3 months, the fruit developed significantly reduced ascorbic acid content, in both the juice and flavedo, compared with the unshaded control. Shaded fruit in the tree canopy also develop lower concentrations of vitamin C than do fruit located on the outside of the canopy (Sites and Reitz, 1951). Even the direction of sun exposure affects vitamin C content, with outer canopy fruit on the north side of the tree (in the northern hemisphere) accumulating lower levels of vitamin C than fruit on the south side. Citrus usually is grown on scions budded on to various rootstocks and these rootstocks often exert a strong influence on fruit quality. Nagy (1980) summarized the effects of rootstock on the vitamin C content of several citrus varieties, showing that the effects depended on the scion. For example, whereas Valencia orange, Marsh grapefruit and Orlando tangelo budded on rough lemon often resulted in fruit with the least vitamin C content compared with the results budding on sour orange, the results were reversed for Temple orange (Nagy, 1980). Rootstock was also found to influence total flavonoid content of lemon juice (Gil-Izquierdo *et al.*, 2004), the influence of rootstock on juice total flavonoid content being greater than that of the influence of the interstock.

In terms of tree nutrition, increased levels of nitrogen fertilization generally result in decreased ascorbic acid content in the juice of citrus fruit (Nagy, 1980). The same happens when phosphorus is added in excess of what is needed for optimum fruit yield (Sinclair, 1961). Conversely, higher rates of potassium fertilization generally increase ascorbic acid content (Sites, 1944). Deficiencies in minerals such as zinc, magnesium, manganese and copper will also result in reduced ascorbic acid content compared with soils without these deficiencies.

There is much interest in stimulating natural plant defence mechanisms to combat various plant diseases, and salicylic acid (SA) has been shown to act as an important signalling molecule in the process. Huang *et al.* (2008) found that 'Cara Cara' navel orange peel and pulp concentrations of lycopene, α-carotene, ascorbic acid, glutathione, total phenolics and total flavonoids were all increased during storage if the trees were sprayed with SA five times over a period of 71 days before harvest.

6.4.4 Postharvest treatments and storage

Postharvest treatments and storage affect the general quality of citrus fruit greatly, as well as their levels of basic nutrients and bioactive compounds. Acidic conditions help preserve ascorbic acid content in fruit and vegetables and the loss of acid is correlated with the loss of vitamin C, and these are both affected greatly by storage temperature (Nagy, 1980). For example, while storage of oranges at low temperatures (~ 3°C) results in little or no loss in total acids or vitamin C content, the loss of these compounds increases as storage temperatures increase (Nagy, 1980; Verma and Dashora, 2000; Al-Zubaidy and Khalil, 2007). The effect of temperature on the rate of vitamin C loss depends on the type of citrus fruit (Nagy, 1980). During storage, flavanone content tends to increase (Patil, 2004; Patil *et al.*, 2004).

In general, vitamin C losses can also be reduced by storing fruit and vegetables under controlled atmospheres with low O_2 or CO_2 levels up to 10 kPa (Lee and Kader, 2000). Ultra-low O_2 (0.05 kPa) treatments were evaluated as a quarantine treatment and found to promote higher concentrations of β-carotene, lycopene and vitamin C compared with control 'Rio Red' grapefruit (Patil and Shellie, 2004).

Low doses of irradiation (≤1 kGy) are believed generally to have no effect on the vitamin C content of fruit and vegetables (Lee and Kader, 2000). However, Mahrouz *et al.* (2002) found that exposure of Moroccan C. *clementina* (Nour) to 300 Gy resulted in significantly higher vitamin C content than found in untreated fruit during subsequent storage for 49 days at 3°C. Exposure of citrus fruit to ≤300 Gy also stimulated an increase in total phenolics (including flavonoids) and other health-promoting compounds during subsequent storage (Oufedjikh *et al.*, 2000; Mahrouz *et al.*, 2002; Vanamala *et al.*, 2003, 2005, 2007; Patil *et al.*, 2004). Higher irradiation doses, between 400 and 700 Gy, tended to result in decreased flavanone concentrations in grapefruit, especially in early season fruit (Patil *et al.*, 2004).

In terms of processed orange juice products, frozen concentrated orange juice contained 24–55% more ascorbic acid than did ready-to-drink juice (Johnston and Bowling, 2002). After reconstituting frozen concentrates or opening ready-to-drink orange juices, all lost about 2% of their reduced ascorbic acid content per day. However, Johnston and Hale (2005) found that the ascorbic acid content of reconstituted juice decreased 24% after 8 days of storage at 4°C, but did not decrease significantly in juice not-from-concentrate (NFC).

The use of high-pressure treatments on Valencia orange juice, instead of pasteurization, resulted in increased carotenoid content (including provitamin A), but decreased ascorbic acid content compared with that in freshly squeezed juice (Sanchez-Moreno *et al.*, 2003). After subsequent storage at 4°C for 10 days, both total ascorbic acid and carotenoid contents were lower in high-pressure treated juice compared with untreated juice. On the other hand, Butz *et al.* (2003) did not find reductions in vitamin C, carotenoids, or other antioxidative factors after high-pressure treatment of orange or lemon juices.

As in fresh fruit, acids help prevent the breakdown of vitamin C in juice products (Nagy, 1980). Phenolic compounds also help prevent the loss of vitamin C in citrus juices, by protecting against oxidative destruction (Miller and Rice-Evans, 1997). Conversely, higher concentrations of fructose in the juice result in accelerated vitamin C loss (Nagy, 1980).

The loss of vitamin C in citrus juices increases as storage duration and temperature increase (Robertson and Samaniego, 1986; Robertson and Samaniego-Esguerra, 1990).

Vitamin C is lost primarily in processed juice due to non-enzymatic aerobic and, to a much lesser extent, anaerobic reactions (Nagy, 1980). Thus, oxygen is excluded as much as possible from the juice during processing. Vitamin C loss in juices is slower when packed in plain tin cans, compared with enamel-lined cans or bottles, because tin reacts with some of the residual oxygen that would otherwise react with vitamin C (Moore *et al.*, 1944; Nagy, 1980). However, orange juice packed in glass still lost vitamin C content substantially slower than juice packed in hermetically sealed polyethylene containers, which in turn lost vitamin C content slower than juice packed in polystyrene bottles or waxed cartons (Bissett and Berry, 1975). About 5%, and usually no more than 10%, of the vitamin C content of citrus is lost during the manufacturing of juice products (Moore *et al.*, 1944; Nagy 1980). Knowledge of the stability and interaction of bioactive compounds in citrus and citrus products, such as flavonoids, in relation to pre- and postharvest treatment and storage, has begun to evolve in recent years thanks to the development of new analytical methods (Del Caro *et al.*, 2004; Dhuique-Mayer *et al.*, 2007; Miguel *et al.*, 2009).

6.5 Future Research Needs

In order to evaluate the health effects of citrus fruit and citrus products in humans, additional research is needed to address the analytical challenges with the bioactive compounds themselves (sensitivity, fine structural differentiation, detection limit) and then the effect in various human tissues. The isolation and characterization of citrus bioactive compounds must include the development of effective methods for the extraction, separation, identification and quantification of the compounds. Furthermore, accurate consumption data from various populations of the world need to be assessed so that more precise public health information can be obtained. Effects of pre- and postharvest treatments have been investigated for previously known or identified compounds. However, since citrus trees may have a long lifespan, a database on long-term changes in bioactive compounds, as affected by age and other environmental factors (i.e. disease pressure) and other cultural practices (i.e. pruning), would help in evaluating the health benefits of these compounds (individually and in combination).

6.6 Conclusion

The apparent health benefits of citrus bioactive compounds are promising yet inconsistent, due to many factors. Despite the wealth of information from previous research, much is still unknown. Therefore, it is important that collaborative research continues across disciplines ranging from horticulture, postharvest technology, food science, chemistry, pharmacy, plant and human physiology, epidemiology and human nutrition, to name just a few, in order to understand better and to optimize the potential health benefits from citrus bioactive compounds.

References

Akao, Y., Itoh, T., Ohguchi, K., Iinuma, M. and Nozawa, Y. (2008) Interactive effects of polymethoxy flavones from Citrus on cell growth inhibition in human neuroblastoma SH-SY5Y cells. *Bioorganic and Medical Chemistry* 16, 2803–2810.

Alós, E., Cercos, M., Rodrigo, M.-J., Zacarías, L. and Talon, M. (2006) Regulation of color breaks in citrus fruits. Changes in pigment profiling and gene expression induced by gibbrellins and nitrate, two ripening retardants. *Journal of Agricultural and Food Chemistry* 54, 4888–4895.

Al-Zubaidy, M.M.I. and Khalil, R.A. (2007) Kinetic and prediction studies of ascorbic acid degradation in normal and concentrate local lemon juice during storage. *Food Chemistry* 101, 254–259.

Amaro, M.I., Rocha, J., Vila-Real, H., Eduardo-Figuerira, M., Mota-filipe, H., Sepodes, B. and Ribeiro, M.H. (2009) Anti-imflammaotry activity of naringin and the biosynthesised naringenin by naringinase immobilized in microstructures material in a model of DSS-induced colitis in mice. *Food Research International* 42, 1010–1017.

Aparicio, J., Harding, P.L. and Soule, M.J. Jr (1990) A study of grapefruit (*Citrus paradisi* Macf.) cv. 'Marsh seedless,' physicochemical characterization according to circumference of the Murcia region. *Fruits* 45, 489–495.

Arias, B.A. and Ramón-Laca, L. (2005) Pharmacological properties of citrus and their ancient and medieval uses in the Mediterranean region. *Journal of Ethnopharmacology* 97, 89–95.

Azémar, M., Hildenbrand, B., Haering, B., Heim, M.E. and Unger, C. (2007) Clinical benefit in patients with advanced solid tumors treated with modified citrus pectin: a prospective pilot study. *Clinical Medicine: Oncology* 1, 73–80.

Babosa-Filho, J.M., Alencar, A.A., Nunes, X.P., de Andrade Tomaz, A.C., Sena-Filho, J.G., Athaylde-Fiho, P.F., Silva, M.S., Vanderlei de Souza, M.F. and Leitao da-Cunha, E.V. (2008) Sources of alpha-, beta-, gamma-, delta- and epsilon-carotenes: a twentieth century review. *Brazilian Journal of Pharmacognosy* 18, 135–154.

Bae, J.-M., Lee, E.J. and Guyatt, G. (2008) Citrus fruit intake and prostate cancer risk: a quantitative systematic review. *Journal of Preventive Medicine and Public Health* 41, 159–164.

Bazzano, L.A., Serdula, M.K. and Liu, S. (2003) Dietary intake of fruits and vegetables and risk of cardiovascular disease. *Current Atherosclerosis Reports* 5, 492–499.

Beecher, G.R. (2003) Overview of dietary flavonoids: nomanclature, occurrence and intake. Proceedings of the Third International Scientific Symposium on Tea and Human Health: Role of Flavonoids in the Diet. American Society for Nutritional Sciences. *Journal of Nutrition* 133, 3248S–3254S.

Benavente-Garcia, O. and Castillo, J. (2008) Update on uses and properties of citrus flavonoids: new findings in anticancer, cardiovascular, and anti-inflammatory activity. *Journal of Agricultural and Food Chemistry* 56, 6185–6205.

Benavente-Garcia, O., Castillo, J., Marin, F.R., Ortuno, A. and Del Rio, J.A. (1997) Uses and properties of *Citrus* flavonoids. *Journal of Agricultural and Food Chemistry* 45, 4505–4515.

Bhagwat, S., Gebhardt, S., Haytowitz, D., Holden, J. and Harnly, J. (2006) USDA database for the flavonoid content of selected foods. Release 2 (http://www.ars.usda.gov/nutrientdata, accessed 11 December 2009).

Bissett, O.W. and Berry, R.E. (1975) Ascorbic acid retention in orange juice as related to container type. *Journal of Food Science* 40, 178–180.

Boobis, A., Watelet, J.B., Whomsley, R., Benedetti, M.S., Demoly, P. and Tipton, K. (2009) Drug interactions. *Drug Metabolism Review* 41, 486–525.

Boriss, H. (2009) Citrus profile (http://www.agmrc.org/commodities__products/fruits/citrus/citrus_profile.cfm, accessed 25 August 2009).

Breksa, A.P. III, Hildlgo, M.B. and Yuen, M.L. (2009) Food chemistry. Liquid chromatography – electrospray ionization mass spectrometry method for the rapid identification of citrus limonoid glucosides in citrus juices and extracts. *Food Chemistry* 117, 739–744.

Butz, P., Fernandez Garcia, A., Lindauer, R., Dieterich, S., Bognar, A. and Tauscher, B. (2003) Influence of ultra high pressure processing on fruit and vegetable products. *Journal of Food Engineering* 56, 233–236.

Cardile, V., Grasca, G., Rizza, L., Rapisarda, P. and Bonia, F. (2009) Antiinflammatory effects of a red orange extract in human keratinocytes treated with interferon-gamma and histamine. *Phytotherapy Research* 24, 414–418.

Cho, J. (2006) Antioxidant and neuroprotective effects of hesperidin and its aglycone hesperetin. *Archives of Pharmacal Research* 29, 699–706.

Conforti, F., Marrelli, M., Menichini, F., Bonesi, M., Statti, G., Provenzano, E. and Menichini, F. (2009) Natural and synthetic furanocoumarins as treatment for vitiligo and psoriasis. *Current Drug Therapy* 4, 38–58.

Cuthrell, K. and Le Marchand, L. (2006) Grapefruit and cancer: a review. In: Patil, B.S., Turner, N.D., Miller, E.G. and Brodbelt, J.S. (eds) *Potential Health Benefits of Citrus*. American Chemical Society, Washington, DC, pp. 235–252.

Cutler, G.J., Nettleton, J.A., Ross, J.A., Harnack, L.J., Jacobs, D.R. Jr, Scrafford, C.G., Barraj, L.M., Mink, P.J. and Robien, K. (2009) Dietary flavonoid intake and risk of cancer in postmenopausal women: the Iowa women's health study. *International Journal of Cancer* 123, 664–671.

Dahan, A. and Altman, H. (2004) Food–drug interaction: grapefruit juice augments drug bioavailability – mechanism, extent and relevance. *European Journal of Clinical Nutrition* 58, 1–9.

Datla, K., Christidou, M., Widmer, W.W., Roopri, H.K. and Dexter, D.T. (2001) Tissue distribution and neuroprotective effects of citrus flavonoid tangeretin in rat model of Parkinson's disease. *Neuropharmacology and Neurotoxicology* 12, 3871–3875.

Davies, F.S. and Albrigo, L.G. (1994) *Citrus*. CAB International, Wallingford, UK.

De Castro, W.V., Mertens-Talcott, S., Rubner, A., Butterweck, V. and Derendorf, H. (2006) Variation of flavonoids and furanocoumarins in grapefruit juices: a potential source of variability in grapefruit juice–
drug interaction studies. *Journal of Agricultural and Food Chemistry* 54, 249–255.

Del Caro, A., Piga, A., Vacca, V. and Agabbio, M. (2004) Changes of flavonoids, vitamin C and antioxidant
capacity in minimally processed citrus segments and juices during storage. *Food Chemistry* 84, 99–105.

Del Rio, J.A., Fuster, M.D., Gomez, P., Porras, I., Garcia-Lidon, A. and Ortuno, A. (2004) *Citrus limon*: a
source of flavonoids of pharmaceutical interest. *Food Chemistry* 84, 457–461.

Deyhim, F., Garica, K., Lopez, E., Gonzalez, J., Ino, S., Garcia, M. and Patil, B.S. (2006) Citrus juice modulates bone strength in male senescent rat model of osteoporosis. *Nutrition* 22, 559–563.

Deyhim, F., Villarreal, A., Garcia, K., Rios, R., Garcia, C., Gonzales, C., Mandadi, K. and Patil, B. (2007)
Orange pulp improves antioxidant status and suppresses lipid peroxidation in orchidectomized male
rats. *Nutrition* 23, 617–621.

Dhuique-Mayer, C., Caris-Veyrat, C., Ollitrault, P., Curk, F. and Amiot, M.-J. (2005) Varietal and interspecific influence on micronutrient content in citrus from the Mediterranean area. *Journal of Agricultural
and Food Chemistry* 53, 2140–2145.

Dhuique-Mayer, C., Borel, P., Reboul, E., Capriccio, B., Besancon, P. and Amiot, M.J. (2006) β-Cryptoxanthin
from citrus juice: assessment of bioaccessibility using an *in vitro* disgestion/Cao-2 cell culture model.
British Journal of Nutrition 97, 883–890.

Dhuique-Mayer, C., Tbatou, M., Carail, M., Caris-Veyrat, C., Dornier, M. and Amiot, M.J. (2007) Thermal
degradation of antioxidant micronutrients in citrus juice: kinetics and newly formed compounds.
Journal of Agricultural and Food Chemistry 55, 4209–4216.

Drewnowski, A. and Gomez-Carneros, C. (2000) Bitter taste, phytonutrients, and the consumer: a review.
American Journal of Clinical Nutrition 72, 1424–1435.

Economos, C. and Clay, W.D. (1999) Nutritional and health benefits of citrus fruits. *Food, Nutrition and
Agriculture* 24, 11–18.

Egashira, K., Ohtani, H., Itoh, S., Koyabu, N., Tsujimoto, M., Murakami, H. and Sawada, Y. (2004) Inhibitory effects of pomelo on the metabolism of tacrolimus and the activities of CYP3A4 and P-glycoprotein.
Drug Metabolism and Disposition 32, 828–833.

Egner, P.A., Wang, J.B., Zhu, Y.R., Zhang, B.C., Wu, Y., Zhang, Q.N., Qian, G.S., Kuang, S.Y., Gange, S.J.,
Jacobson, L.P., Helzlsouer, K.J., Bailey, G.S., Groopman, J.D. and Kensler, T.W. (2001) Chlorophyllin
intervention reduces aflatoxin-DNA adducts in individuals at high risk for liver cancer. *Communication, The National Academy of Sciences* (http://www.pnas.org/content/98/25/14601.abstract, accessed
27 August 2009).

Ejaz, S., Ejaz, A., Matuda, K. and Lim, C.W. (2006) Limonoids as cancer chemopreventive agents. *Journal of
the Science of Food and Agriculture* 86, 339–345.

Eliaz, I., Weil, E. and Wilk, B. (2007) Integrative medicine and the role of modified citrus pectin/algenates
in heavy metal chelation and detoxification – five case reports. *Forsch Komplementmed* 14, 358–364.

Erlund, I., Meririnne, E., Alfthan, G. and Aro, A. (2001) Plasma kinetics and urinary excretion of the flavonones naringenin and hesperetin in human after ingestion of orange juice and grapefruit juice. *Journal
of Nutrition* 131, 235–241.

Fahey, J.W., Stephenson, K.K., Dinkova-Kostova, A., Egner, P.A., Kensler, T.W. and Talalay, P. (2005) Chlorophyll, chlorophyllin and related tetrapyrroles are significant inducers of mammalian phase 2 cytoprotective genes. *Carcinogenesis* 26, 1247–1255.

Fanciullino, A.-L., Dhuique-Mayer, C., Luro, F., Morillon, R. and Ollitrault, P. (2007) Carotenoid biosynthetic pathway in the Citrus genus: number of copies and phylogenetic diversity of seven genes.
Journal of Agricultural and Food Chemistry 55, 7405–7417.

Fanciullino, A.-L., Cercos, M., Dhuique-Myer, C., Froelicher, Y., Talon, M., Ollitrault, P. and Morllon, R.
(2008) Changes in carotenoid content and biosynthetic gene expression in juice sacs of four orange
varieties (*Citrus sinensis*) differing in flesh fruit color. *Journal of Agricultural and Food Chemistry* 56,
3628–3638.

FAOSTAT (2009) Food and Agriculture Statistics Division (http://faostat.fao.org/site/567/default.
aspx#ancor, accessed 4 June 2009).

Farkas, D. and Greenblatt, D.J. (2008) Influence of fruit juices on drug disposition: discrepancies between
in vitro and clinical studies. *Expert Opinion on Drug Metabolism and Toxicology* 4, 381–393.

Ferruzzi, M.G., Failla, M.L. and Schwartz, S.J. (2001) Assessment of degradation and intestinal cell uptake
of carotenoids and chlorophyll derivatives from spinach puree using *in vitro* digestion and Caco-2
human cell model. *Journal of Agricultural and Food Chemistry* 49, 2082–2089.

Ferruzzi, M.G., Böhm, V., Courtney, P.D. and Schwartz, S.J. (2002) Antioxidant and antimutagenic activity of dietary chlorophyll derivatives determined by radical scavenging and bacterial reverse mutagenesis assays. *Journal of Food Science* 67, 2589–2595.

Figuerola, F., Hurtado, M.L., Estevez, A.M., Chiffelle, I. and Asenjo, F. (2005) Fibre concentrates from apple pomace and citrus peel as potential fibre sources for food enrichment. *Food Chemistry* 91, 395–401.

Fukuda, K., Ohta, T., Oshima, Y., Ohashi, N., Yoshikawa, M. and Yamazoe, Y. (1997) Specific CYP3A4 inhibitors in grapefruit juice: furocoumarin dimmers as components of drug interaction. *Pharmacogenetics* 7, 391–396.

Fukada, K., Guo, L., Ohashi, N., Yoshikawa, M. and Yamazoe, Y. (2000) Amounts and variation in grapefruit juice of the main components causing grapefruit–drug interaction. *Journal of Chromatography* B 741, 195–203.

Gardner, P.T., White, T.A.C., McPhail, D.B. and Duthie, G.G. (2000) The relative contributions of vitamin C, carotenoids and phenolics to the antioxidant potential of fruits juices. *Food Chemistry* 68, 471–474.

Gattuso, G. Barreca, D., Caristi, C., Gargiulli, C. and Leuzzi, U. (2007) Distribution of flavonoids and furocoumarins in juices from cultivars of *Citrus bergamia Risso*. *Journal of Agricultural and Food Chemistry* 55, 9921–9927.

Gaudet, M.M., Britton, J.A., Kabat, G.C., Steck-Scott, S., Eng, S.M., Teitelbaum, S.L., Terry, M.B., Neugut, A.I. and Gammon, M.D. (2004) Fruit, vegetables, and micronutrients in relation to breast cancer modified by menopause and hormone receptor status. *Cancer Epidemiology, Biomarkers and Prevention* 13, 1485–1494.

Gil-Izquierdo, A., Riquelme, M.T., Porras, I. and Ferreres, F. (2004) Effect of the rootstock and interstock grafted in lemon tree (*Citrus limon* (L.) Burm.) on the flavonoid content of lemon juice. *Journal of Agricultural and Food Chemistry* 52, 324–331.

Goodner, K.L., Rouseff, R.L. and Hofsommer, H.J. (2001) Orange, mandarin, and hybrid classification using multivariate statistics based on carotenoid profiles. *Journal of Agricultural and Food Chemistry* 49, 1146–1150.

Gross, J. (1987) *Pigment in Fruits*. Alden Press, Oxford, UK.

Guo, L. and Yamazoe, Y. (2004) Inhibition of cytochrome P450 by furanocoumarins in grapefruit juice and herbal medicine. *Acta Pharmacol Sin* 25, 129–136.

Harding, P.L., Winston, J.R. and Fisher, D.F. (1940) Seasonal changes in Florida oranges. *United States Department of Agriculture Technical Bulletin* 753, 1–89.

Heo, H.J., Kim, D.O., Shin, S.C., Kim, M.J., Kim, B.G. II, and Shin, D.H. (2004) Effect of antioxidant flavanone, naringenin, from *Citrus junos* on neroprotection. *Journal of Agricultural and Food Chemistry* 52, 1520–1525.

Ho, S.C. and Lin, C.C. (2008) Investigation of heat treating conditions for enhancing the anti-inflammatory activity of citrus fruit (*Citrus reticulate*) peels. *Journal of Agricultural and Food Chemistry* 56, 7976–7982.

Hooper, L. and Cassidy, A. (2006) A review of the health care potential of bioactive compounds. *Journal of the Science of Food and Agriculture* 86, 1805–1813.

Huang, M.T., Ho, C.-T., Rosen, R.T., Ghai, G., Lipkin, M., Chen, K.Y., Telang, N., Boyd, C. and Ciszar, K. (2007) US Patent 7201928 – Extracts of orange peel for prevention and treatment of cancer (http://www.patentstorm.us/patents/7201928/fulltext.html, accessed 4 December 2009).

Huang, R., Xia, R., Lu, Y., Hu, L. and Xu, Y. (2008) Effect of pre-harvest salicylic acid spray treatment on post-harvest antioxidant in the pulp and peel of 'Cara cara' navel orange (*Citrus sinenisis* L. Osbeck). *Journal of the Science of Food and Agriculture* 88, 229–236.

Hung, H.C., Joshipura, K.J., Jiang, R., Hu, F.B., Hunter, D., Smith-Warner, S.A., Colditz, G.A., Rosner, B., Spiegelman, D. and Willett, W.C. (2004) Fruit and vegetable intake and risk of major chronic disease. *Journal of the National Cancer Institute* 96, 1577–1584.

Hwang, S.L. and Yen, G.C. (2008) Neuroprotective effects of the citrus flavonones against H_2O_2 induced cytotoxicity in PC12 cells. *Journal of Agricultural and Food Chemistry* 56, 859–864.

Izumi, H., Ito, T. and Yoshida, Y. (1992) Effect of light intensity during the growing period on ascorbic acid content and its histochemical distribution on the leaves and peel and fruit quality of Satsuma mandarin. *Journal of the Japanese Society for Horticultural Science* 61, 7–15.

Jayaprakasha, G.K., Brodbelt, J.S., Bhat, N.G. and Patil, B.S. (2006) Methods for the separation of limonoids from citrus. In: Patil, B.S., Turner, N.D., Miller, E.G. and Broadbelt, J.S. (eds) *Potential Health Benefits of Citrus*. American Chemical Society, Washington, DC, pp. 34–51.

Jayaprakasha, G.K., Girennavar, B. and Patil, B.S. (2008) Antioxidant capacity of pummelo and navel oranges: extraction efficiency of solvents in sequence. *LWT – Food Science and Technology* 41, 376–384.

Johnston, C.S. and Bowling, D.L. (2002) Stability of ascorbic acid in commercially available orange juices. *Journal of the American Dietetic Association* 102, 525–529.

Johnston, C.S. and Hale, J.C. (2005) Oxidation of ascorbic acid in stored orange juice is associated with re-duced plasma vitamin C concentrations and elevated lipid peroxides. *Journal of the American Dietetic Association* 105, 106–109.

Joshipura, K.J., Ascherio, A., Mason, J.E., Stampfer, M.J., Rimm, E.B., Speizer, F.E., Hennekens, C.H., Spi-gelman, D. and Willett, W.C. (1999) Fruit and vegetable intake in relation to risk of ischemic stroke. *Journal of the American Medical Association* 282, 1233–1239.

Joshipura, K.J., Hu, F.B., Manson, J.E., Stampfer, M.J., Rimm, E.B., Speizer, F.E., Colditz, G., Ascherio, A., Rosner, B., Spiegelman, D. and Willett, W.C. (2001) The effect of fruit and vegetable intake on risk for coronary heart disease. *Annals of Internal Medicine* 134, 1106–1114.

Kanes, K., Tisserat, B., Bershow, M. and Vandercook, C. (1993) Phenolic composition of various tissues of *Rutaceae* species. *Phytochemistry* 32, 967–974.

Kato, M., Matsumoto, H., Ikoma, Y., Okuda, H. and Yano, M. (2006) The role of carotenoid cleavage dioxy-genases in the regulation of carotenoid profiles during maturation in citrus fruit. *Journal of Experimental Botany* 57, 2153–2164.

Khan, I.A. (2007) *Citrus, Genetics, Breeding and Biotechnology*. CAB International, Wallingford, UK.

Klein, S., Burke, L.E., Bray, G.A., Blair, S., Allison, D.B., Pi-Sunyer, X., Hong, Y. and Eckel, R.H. (2004) Clinical implications of obesity with specific focus on cardiovascular disease: a statement for profes-sionals from the American Heart Association Council on Nutrition, Physical Activity, and Metabo-lism: endorsed by the American College of Cardiology Foundation. *Circulation* 110, 2952–2967.

Kris-Etherton, P.M., Hecker, K.D., Bonanome, A., Coval, S.M., Binkoski, A.E., Hilpert, K.F., Griel, A.E. and Etherton, T.D. (2002) Bioactive compounds in foods: their role in the prevention of cardiovascular disease and cancer. *American Journal of Medicine* 113, 71–88.

Kurowska, E.M. and Manthey, J.A. (2004) Hypolipidemic effects and absorption of citrus polymethoxyl-ated flavones in hamsters with diet-induced hypercholesterolemia. *Journal of Agricultural and Food Chemistry* 52, 2879–2886.

Ladaniya, M.S. (2008) *Citrus Fruit: Biology, Technology and Evaluation*. Academic Press, New York.

Lam, L.K.T., Zhang, J., Hasegawa, S. and Schut, H.A.J. (1994) Inhibition of chemically induced carcino-genesis by citrus limonoids. In: Huang, M., Osawa, T., Ho., C. and Rosen, R.T. (eds) *Food Phyto-chemicals for Cancer Prevention. I. Fruit and Vegetables*. American Chemical Society, Washington, DC, pp. 209–219.

Larbat, R., Hehn, A., Hans, J., Schneider, S., Jugde, H., Schneider, B., Matern, U. and Bourgaud, F. (2009) Isolation and functional characterization of CYP71AJ4 encoding for the first P450 monooxygenase of angular furanocoumarin biosynthesis. *Journal of Biological Chemistry* 284, 4776–4785.

Laszlo, P. (2007) *Citrus: a History*. The University of Chicago Press, Chicago, Illinois.

Le Marchand, L. (2002) Cancer preventive effects of flavonoids – a review. *Biomedicine and Pharmacotherapy* 56, 296–301.

Lee, H. (2001) Characterization of carotenoids in juice of red navel orange (Cara Cara). *Journal of Agricul-tural and Food Chemistry* 49, 2563–2568.

Lee, H. and Castle, W. (2001) Seasonal changes of carotenoid pigments and color in Hamlin, Earlygold, and Budd blood orange juices. *Journal of Agricultural and Food Chemistry* 49, 877–882.

Lee, K.S. and Kader, A.A. (2000) Preharvest and postharvest factors influencing vitamin C content in hor-ticultural crops. *Postharvest Biology and Technology* 20, 207–220.

Li, B.B., Smith, B. and Hossain, M.M. (2006) Extraction of phenolics from citrus peels I. Solvent extraction methods. *Separation and Purification Technology* 48, 182–188.

Lien, T.F., Shuang, H. and Su, W.T. (2008) Effect of adding extracted hesperitin, naringenin and pectin on egg cholesterol, serum traits and antioxidant activity in laying hens. *Archives of Animal Nutrition* 62, 33–43.

Lim, H.K., Moon, J.Y., Kim, H., Cho, M. and Cho, S.K. (2009) Induction of apoptosis in U937 human leu-kaemia cells by the hexane fraction of extract of immature *Citrus grandis* Osbeck fruits. *Food Chemistry* 114, 1245–1250.

Linseisen, J. and Rohrmann, S. (2008) Biomarkers of dietary intake of flavonoids and phenolic acids for studying diet–cancer relationship in humans. *European Journal of Nutrition* 47(Supplement 2), 60–68.

Liu, Q., Xu, J., Liu, Y., Zhao, X., Deng, X., Guo, L. and Gu, J. (2007) A novel bud mutation that confers ab-normal patterns of lycopene accumulation in sweet orange fruit (*Citrus sinensis* L. Osbeck). *Journal of Experimental Botany* 58, 4161–4171.

Liu, Y., Ahmad, H., Luo, Y., Gardiner, D.T., Gunasekera, R.S., McKeehn, W.L. and Patil, B.S. (2001) Citrus pectin: characterization and inhibitory effect on fibroblast growth factor–receptor interaction. *Journal of Agricultural and Food Chemistry* 49, 3051–3050.

Liu, Y., Ahmad, H., Luo, Y., Gardiner, D.T., Gunasekera, R.S., McKeehan, W.L. and Patil, B.S. (2002) Influence of harvest time on citrus pectin and its *in vitro* inhibition of fibroblast growth factor signal transduction. *Journal of the Science of Food and Agriculture* 82, 469–477.

Lowe, G., Woodward, M., Rumley, A., Morrison, C., Tunstall-Pedoe, H. and Stephen, K. (2003) Total tooth loss and prevalent cardiovascular disease in men and women. Possible roles of citrus fruit consumption, vitamin C, and inflammatory and thrombotic variables. *Journal of Clinical Epidemiology* 56, 694–700.

McDonald, R.E. and Hildebrand, B.M. (1980) Physical chemical characteristics of lemons from several countries. *Journal of the American Society for Horticultural Science* 105, 135–141.

Mahrouz, M., Lacroix, M., D'Aprano, G., Oufedjikh, H., Boubekri, C. and Gagnon, M. (2002) Effect of γ-irradiation combined with washing and waxing treatment on physicochemical properties, vitamin C, and organoleptic quality of *Citrus clementina* Hort. Ex. Tanaka. *Journal of Agricultural and Food Chemistry* 50, 7271–7276.

Malhotra, S., Shakya, G., Kumar, A., Vanhoecke, B.W., Cholli, A.L., Raj, H.G., Saso, L., Ghosh, B., Bracke, M.E., Prasad, A.K., Biswal, S. and Parmar, V.S. (2008) Antioxidant, anti-inflammatory and antiinvasive activities of biopolyphenolics. *ARKIVOC* (vi), 119–139.

Manners, G.D. (2007) Citrus limonoids: analysis, bioactivity, and biomedical prospects. *Journal of Agricultural and Food Chemistry* 55, 8285–8294.

Manners, G.D., Jacob, R.A., Breksa, A.P. III, Schoch, T.K. and Hasegawa, S. (2003) Bioavailability of citrus limonoids in humans. *Journal of Agricultural and Food Chemistry* 51, 4156–4161.

Manthey, J.A. and Guthrie, N. (2002) Antiproliferative activities of citrus flavonoids against six human cancer cell lines. *Journal of Agricultural and Food Chemistry* 50, 5837–5843.

Marin, F.R., Soler-Rivas, C., Benavente-Garcia, O., Castillo, J. and Perez-Alvarez, J.A. (2007) By-products from different citrus processes as a source of customized functional fibres. *Food Chemistry* 100, 736–741.

Matsumoto, H., Ikoma, Y., Kato, M., Kuniga, T., Nakajima, N. and Yoshida, T. (2007) Quantification of carotenoids in citrus fruits by LC-MS and comparison of pattern of seasonal changes for carotenoids among citrus varieties. *Journal of Agricultural and Food Chemistry* 55, 2356–2368.

Meléndez-Martínez, A.J., Vicario, I.M. and Heredia, F.J. (2007a) Provitamin A carotenoids and ascorbic acid contents of different types of orange juices marketed in Spain. *Food Chemistry* 101, 177–184.

Meléndez-Martínez, A.J., Vicario, I.M. and Heredia, F.J. (2007b) Review: Analysis of carotenoids in orange juice. *Journal of Food Composition and Analysis* 20, 608–649.

Mennen, L.I., Sapinho, D., de Bree, A., Arnalt, N., Bertrais, S., Galan, P. and Hercberg, S. (2004) Nutritional epidemiology – research communication. *Journal of Nutrition* 134, 923–926.

Mertens-Talcott, S.U., Zadezensky, I., De Castro, W.V., Derendorf, H. and Butterweck, V. (2006) Grapefruit–drug interactions: can interactions with drugs be avoided? *Journal of Clinical Pharmacology* 46, 1390–1416.

Middleton, E., Kandaswami, C. and Theoharides, T. (2000) The effects of plant flavonoids on mammalian cells: implications for inflammation, heart disease, and cancer. *Pharmacological Review* 52, 673–751.

Miguel, M.G., Duarte, A., Nunes, S., Sustelo, V., Martins, D. and Dandlen, S.A. (2009) Ascorbic acid and flavanone glycosides in citrus: relationship with antioxidant activity. *Journal of Food, Agriculture and Environment* 7, 222–227.

Miller, E.G., Porter, J.L., Binnie, W.H., Guo, I.Y. and Hasegawa, S. (2004) Further studies on the anticancer activity of citrus limonoids. *Journal of Agricultural and Food Chemistry* 52, 4908–4912.

Miller, E.G., Gibbins, R.P., Taylor, S.E., McIntosh, J.E. and Patil, B.S. (2006) Long-term screening study on the potential toxicity of limonoids. In: *Potential Benefits of Citrus*. ACS Symposium Series 936, American Chemical Society, Washington, DC, pp. 82–94.

Miller, E.G., Peacock, J.J., Bourlan, C., Taylor, S.E., Wright, J.M. and Patil, B.S. (2008) Inhibition of oral carcinogenesis by citrus flavonoids. *Nutrition and Cancer* 60, 69–74.

Miller, N.J. and Rice-Evans, C.A. (1997) The relative contributions of ascorbic acid and phenolic antioxidants to the total antioxidant activity of orange and apple fruit juices and blackcurrant drink. *Food Chemistry* 60, 331–337.

Miyake, Y., Yamamoto, K., Morimitsu, Y. and Osawa, T. (1997) Isolation of C-glucosylflavone from lemon peel and antioxidative activity of flavonoid compounds in lemon fruit. *Journal of Agricultural and Food Chemistry* 45, 4619–4623.

Miyake, Y., Suzuki, E., Ohya, S., Fukumoto, S., Hiramitsu, M., Sakaida, K., Osawa, T. and Furuichi, Y. (2006) Lipid-lowering effect of eriocitrin, the main flavonoid in lemon fruits, in rats on a high fat and high cholesterol diet. *Journal of Food Science* 71, S633–S637.

Miyata, Y., Sato, T., Imada, K., Dobashi, A., Yano, M., and Ito, A. (2008) A citrus poly methoxyflavonoid, nobilitin is a novel MEK inhibitor that exhibits antitumor metastasis in human fibrosarcoma HT-1080 cells. *Biochemical and Biophysical Research Communications* 366, 168–173.

Mollace, V., Ragusa, S., Sacco, I., Muscoli, C., Sculco, F., Visalli, V., Palma, E., Muscoli, S., Mondello, L., Dugo, P., Rotiroti, D. and Rotroti, D. (2008) The protective effect of Bergamot oil extract on lecitine-like oxy LDL receptor-1 expression in ballon injury-related neointima formation. *Journal of Cardiovascular Pharmacology* 13, 120–129.

Moore, E.L., Wiederhold, E., Atkins, C.D. and MacDowell, L.G. (1944) Ascorbic acid retention in Florida grapefruit juices. *Canner* 98, 24–26.

Moore, G.A., Grosser, J.W. and Gmitter, F.G. Jr (2005) Rutaceae: citrus grapefruits, lemon, lime, orange, etc. In: Litz, R.E. (ed.) *Biotechnology of Fruit and Nut Crops*. CAB International, Wallingford, UK, pp. 583–625.

Morley, K.L., Ferguson, P.J. and Koropatnick, J. (2007) Tangeretin and nobiletin induce G1 cell cycle arrest but not apoptosis in human breast and colon cancer cells. *Cancer Letters* 251, 168–178.

Mouly, P., Gaydou, E.M. and Auffray, A. (1998) Simultaneous separation of flavavone glycosides and polymethoxylated flavones in citrus juices using liquid chromatography. *Journal of Chromatography* 800, 171–179.

Mukhopadhyay, S. (2004) *Citrus: Production, Post Harvest, Disease and Pest Management*. Science Publisher, Inc., Enfield, New Hampshire.

Nagy, S. (1980) Vitamin C contents of citrus fruit and their products: a review. *Journal of Agricultural and Food Chemistry* 28, 8–18.

Nagy, S., Shaw, P.E. and Veldhuis, M.K. (1977a) *Citrus Science and Technology*, Volume 1. The AVI Publishing Company, Inc, Wesport, Connecticut.

Nagy, S., Shaw, P.E. and Veldhuis, M.K. (1977b) *Citrus Science and Technology*, Volume 2. The AVI Publishing Company, Inc, Wesport, Connecticut.

Nogata, Y., Sakamoto, K., Shiratsuchi, H., Ishii, T., Yano, M. and Ohta, H. (2006) Flavonoid composition of fruit tissues of citrus species. *Bioscience, Biotechnology, and Biochemistry* 70, 178–192.

Norman, H.G., Go, V.L. and Butrum, R.R. (2004) Potential synergy of phytochemicals in cancer prevention: mechanism of action. *Journal of Nutrition* 134, 3479S–3485S.

Oben, J., Enonchong, E., Kothari, S., Chambliss, W., Garrison, R. and Dolnick, D. (2009) Phellodendron and citrus extracts benefit cardiovascular health in osteoarthritis patients: a double-blind, placebo-controlled pilot study. *Nutrition Journal* 8, 38 (http://www.nutritionj.com/content/pdf/1475-2891-8-38.pdf).

Ogata, S., Miyake, Y., Yamamoto, K., Okumura, K. and Taguchi, H. (2000) Apoptosis induced by the flavonoid from lemon fruit (*Citrus limon* Burm.f.) and its metabolites in HL-60 cells. *Bioscience, Biotechnology, and Biochemistry* 64, 1075–1078.

O'Neill, M.E., Carroll, Y., Corridan, B., Olmedilla, B., Granado, F., Blanco, I., Van den Berg, H., Hininger, I., Rousell, A.M., Chopra, M., Southon, S. and Thurnham, D.I. (2001) A European carotenoid database to assess carotenoid intakes and its use in a five-country comparative study. *British Journal of Nutrition* 85, 499–507.

Ortuno, A., Baidez, A., Gomez, P., Arcas, M.C., Porras, I., Garcia-Lidon, A. and Del Rio, J.A. (2006) *Citrus paradise* and *Citrus sinensis* flavonoids: their influence in the defence mechanism against *Penicillium digitatum*. *Food Chemistry* 98, 351–358.

Oufedjikh, H., Mahrouz, M., Amiot, M.J. and Lacroix, M. (2000) Effect of γ-irradiation on phenolic compounds and phenylalanine ammonia-lyase activity during storage in relation to peel injury from peel of *Citrus clementina* Hort. Ex. Tanaka. *Journal of Agricultural and Food Chemistry* 48, 559–565.

Pan, M.H., Lai, C.S., Dushenkov, S. and Ho, C.T. (2009) Modulation of inflammatory genes by natural dietary bioactive compounds. *Journal of Agricultural and Food Chemistry* 57, 4467–4477.

Park, H.J., Kim, M.J., Ha, E. and Chung, J.H. (2008) Apoptotic effect of hesperidin through caspase 3 activation in human colon cancer cells, SNU-C4. *Phytomedicine* 15, 147–151.

Patil, B.S. (2004) Irradiation applications to improve functional components of fruits and vegetables. In: Komolprasert, V. and Morehouse, K.M. (eds) *Irradiation of Food and Packaging: Recent Developments*. American Chemical Society, Washington, DC, pp. 117–137.

Patil, B.S. and Shellie, K.C. (2004) Carotenoids and vitamin C changes by semi-commercial ultra-low oxygen storage in grapefruit. *Acta Horticulturae* 632, 321–328.

Patil, B.S., Vanamala, J. and Hallman, G. (2004) Irradiation and storage influence on bioactive components and quality of early and late season 'Rio Red' grapefruit (*Citrus paradisi* Macf.). *Postharvest Biology and Technology* 34, 53–64.

Patil, B.S., Brobelt, J.S., Miller, E.G. and Turner, N.D. (2006a) Potential health benefits of citrus; an overview. In: Patil, B.S., Turner, N.D., Miller, E.G. and Brodbelt, J.S. (eds) *Potential Health Benefits of Citrus.* American Chemical Society, Washington, DC, pp. 1–16.

Patil, B.S., Yu, J., Dandekar, D.V., Toledo, R.T., Singh, R.K. and Pike, L.M. (2006b) Citrus bioactive limonoids and flavonoids extraction by supercritical fluids. In: Patil, B.S., Turner, N.D., Miller, E.G. and Brodbelt, J.S. (eds) *Potential Health Benefits of Citrus.* American Chemical Society, Washington, DC, pp. 18–33.

Patil, B.S., Jayaprakasha, G.K. and Harris, E.D. (2007) Impact of citrus limonoids on human health. *Acta Horticulturae* 744, 127–134.

Peroutka, R., Schulzova, V., Botek, P. and Hajslova, J. (2007) Analyses of furanocoumarins in vegetables (Apiaceae) and citrus fruits (Rutaceae). *Journal of the Science of Food and Agriculture* 87, 2152–2163.

Peterson, J.J., Beecher, G.R., Bhagwat, S.A., Dwyer, J.T., Gebhardt, S.E., Haytowitz, D.B. and Holden, J.M. (2006a) Flavonones in grapefruit, lemons, and limes: a compilation and review of the data from the analytical literature. *Journal of Food Composition and Analyses* 19, S74–S80.

Peterson, J.J., Dwyer, J.T., Beecher, G.R., Bhagwat, S.A., Gebhardt, S.E., Haytowitz, D.B. and Holden, J.M. (2006b) Flavonones in oranges, tangerines (mandarins), tangors, and tangelos: a compilation and review of the data from the analytical literature. *Journal of Food Composition and Analyses* 19, S66–S73.

Piccinelli, A.L., Mesa, M.G., Armenteros, D.M., Alfonso, M.A., Arevalo, A.C., Campone, L. and Rastrelli, L. (2008) HPLC-PDA-MS and NMR characterization of C-glycosyl flavones in a hydroalcoholic extract of *Citrus aurantifilia* leaves with antiplatelet activity. *Journal of Agricultural and Food Chemistry* 56, 1574–1581.

Pillai, U., Muzaffar, J., Sen, S. and Yencey, A. (2009) Grapefruit juice and varapamil: a toxic cocktail. *Southern Medical Journal* 102, 308–309.

Poulose, S.M., Harris, E.D. and Patil, B.S. (2005) Citrus limonoids induce apoptosis in human neuroblastoma cells and have radical scavenging activity. *Journal of Nutrition* 135, 870–877.

Prince, M., Li, Y., Childers, A., Itoh, K., Yamamoto, M. and Kleiner, H.E. (2009) Comparison of citrus coumarins on carcinogen-detoxifying enzumes in Nrf2 knockout mice. *Toxocology Letters* 185, 180–186.

Riso, P., Vistoli, F., Gardana, C., Grande, S., Brusamolino, A., Galvano, F., Galvano, G. and Perrini, M. (2005) Effects of blood orange juice intake on antioxidant bioavailability and on different markers related to oxidative stress. *Journal of Agricultural and Food Chemistry* 53, 941–947.

Robards, K. and Antolovich, M.A. (1997) Critical review: analytical chemistry of fruit bioflavonoids. *Analyst* 122, 11R–34R.

Robertson, G.L. and Samaniego, C.M.L. (1986) Effect of initial dissolved oxygen levels on the degradation of ascorbic acid and the browning of lemon juice during storage. *Journal of Food Science* 51, 184–187, 192.

Robertson, G.L. and Samaniego-Esguerra, C.M. (1990) Effect of soluble solids and temperature on ascorbic acid degradation in lemon juice stored in glass bottles. *Journal of Food Quality* 13, 361–374.

Rossi, M., Garvello, W., Talamini, R., La Vecchia, C., Franceschi, S., Lagiou, P., Zambon, P., Del Maso, L., Bosetti, C. and Negri, E. (2007) Flavonoids and risk of squamous cell esophageal cancer. *International Journal of Cancer* 120, 1560–1564.

Roy, A. and Saraf, S. (2006) Limonoids: overview of significant bioactive triterpenes distributed inplant kingdom. *Biological and Pharmaceutical Bulletin* 29, 191–201.

Salman, H., Bergman, M., Djaldetti, M., Orlin, J. and Bessler, H. (2008) Citrus pectin affects cytokine production by human perpherial blood mononuclear cells. *Biomedicine and Pharmacotherapy* 62, 579–582.

Sanchez-Moreno, C., Plaza, L., de Ancos, V. and Cano, M.P. (2003) Vitamin C, provitamin A carotenoids, and other carotenoids in high-pressurized orange juice during refrigerated storage. *Journal of Agricultural and Food Chemistry* 51, 647–653.

Saunt, J. (2000) *Citrus Varieties of the World.* Sinclair International Ltd, Norwich, UK.

Sayre, R.M. and Dowdy, J.C. (2008) The increase in melanoma: are dietary furocumarins responsible? *Medical Hypotheses* 70, 855–859.

Schieber, A., Stintzing, F.C. and Carle, R. (2001) By-products of plant food processing as a source of functional compounds – recent developments. *Trends in Food Science and Technology* 12, 401–413.

Sergeev, I.N., Ho, C.-T., Li, S., Colby, J. and Dushenkov, S. (2007) Apoptosis-inducing activity of hydroxylated polymethoxyflavones and polymethoxyflavones from orange peel in human breast cancer cells. *Molecular Nutrition and Food Research* 51, 1478–1484.

7   Cucurbits
[Cucumber, Melon, Pumpkin and Squash]

D. Mark Hodges and Gene E. Lester

7.1 Introduction

The focus of this chapter is on the edible members of the *Cucurbitaceae* family, namely cucumber, sweet melon, pumpkin, squash and watermelon. The *Cucurbitaceae* family is comprised of a taxonomic group of closely related genera with diverse origins, with the term cucurbit (or cucurbits) denoting all species within the *Cucurbitaceae*. The three important food-grade cucurbit genera, *Citrullus*, *Cucumis* and *Cucurbita*, include the species *Citrullus lanatus* (watermelons), *Cucumis melo* (cantaloupes and other sweet melons), *Cucum sativa* (cucumbers and pickles) and *Cucurbita maxima*, *Cucur mixta*, *Cucur moshata* and *Cucur pepo* (squashes, gourds and pumpkins); all have edible fruit, with the exception of gourds (Bailey and Bailey, 1976).

All species of *Citrullus*, *Cucumis* and *Cucurbita* have many similar morphological characteristics. Cucurbit plants are mostly monoecious, the flowers are insect pollinated and the fruit are many-seeded berries (pepos) of various exotic shapes, sizes and beautiful colours. A complete morphological and taxonomic description along with nomenclature, evolution and genetics of cucurbits can be found in Robinson and Decker-Walters (1997) and Janick *et al.* (2007). Two exquisite photography books are available; one on melons (Goldman, 2002) and the other on gourds,

pumpkins and squash (Goldman, 2004), which capture beautifully the artistic morphology of these fruit, as well as their food attributes.

According to Maynard and Maynard (2000), cucurbits have centres of origin throughout the tropics and subtropics of Africa, South–east Asia and the Americas. Most cucurbit species are adapted to moist, warm conditions, with some adapted to warm, arid climates. All are frost sensitive, requiring protection in temperate areas or production during annual warm cycles. Currently, China is the leading producer of nearly all cucurbits (Table 7.1), although other major producing countries include Brazil, Cameroon, Cuba, Egypt, India, Iran, Russia, Spain, Turkey and the USA (USDA-ERS, 2007). The significance of cucurbits in human commerce is demonstrated by an abundance of references (Whitaker and Davis, 1962; Robinson and Decker-Walters, 1997; Maynard and Maynard, 2000; Janick *et al.*, 2007; USDA-ERS, 2007), highlighting their major economic value.

In the USA alone the annual (2007) retail value of *Citrullus*, *Cucumis* and *Cucurbita* species is greater than US$2.8 billion (Table 7.2). In addition to cucurbits providing much needed economic livelihood worldwide, humans have come to depend on marketable food-grade cucurbit fruit for critical daily nutrition.

Table 7.1. Leading countries and overall world cucurbit production, 2007 data.[a]

Cucurbit commodity	Country (hectares × 1000)	World (hectares × 1000)
Cantaloupe and other melons	China (608)	1289
Cucumber and pickles	China (1604)	2531
Pumpkins and squash	India (378)	1546
Watermelon	China (2314)	3804

Note: [a]USDA-ERS (2007).

Table 7.2. Retail values of US cucurbit production and imports, 2007 data.[a]

Cucurbit commodity	US production (US$ million)	US imports (US$ million)
Cantaloupe melon	313	156
Cucumbers	228	471
Cucumbers (pickling)	167	77
Honeydew and other melons	82	35
Pumpkins	117	61
Squash	227	213
Watermelon	476	158

Note: [a]USDA-ERS (2007).

The intent of this chapter is to provide a brief description of the botany and physiology of edible cucurbit species and to discuss their phytochemical (human health interactive compound) composition and the bioavailability of some major nutrients. Pre- and postharvest factors that may influence the availability of specific major phytonutrients of fresh cucumber, cantaloupe and honeydew melon, squash, pumpkin and watermelon will also be discussed.

7.2 Botany/Physiology of *Cucurbitaceae*

There are many botanical similarities among the cucurbit species (Robinson and Decker-Walters, 1997). Cucurbits have a strong taproot penetrating the soil to a depth of 1 m or more, with secondary roots occurring near the soil surface extending beyond the spread of aboveground stems. Stems are angled in cross section, centrally hollow, sap-filled and branched, herbaceous and often suffrutescent (softly woody), prostrate, trailing or climbing. Palmately veined leaves are usually simple with no leaflets, are three- to seven-lobed, have one leaf per node and are arranged along the stem in a helical fashion. Most cucurbits have solitary, unbranched tendrils, originating from the leaf axis, which are coiled, allowing stems to cling to supports.

Flowers are large and showy, usually orange-yellow, yellow or yellowish-white, and have symmetrically fused calyx and corolla which are cup- to bell-shaped and predominately five-lobed. Staminate and pistillate flowers are originally bisexual. During ontogeny, the pistil is retarded in staminate flowers, and undeveloped stamens can be seen in mature pistillate flowers. Stamen number is five and pistillate flowers have inferior ovaries with a fused style and lobed stigma. Floral nectaries are borne inside and at the base of both staminate and pistillate flowers.

Cucurbit fruit are extremely diverse and have signature characteristics in size, shape, colour and ornamentation familiar to all as a cucumber, melon, pumpkin, squash or watermelon. Fruit growth is sigmodial (Sinnott, 1945), with expansion occurring continuously, but with a growth rate greater at night than during the day (Lester, 1998). Cucurbit fruit are indehiscent 'pepos' (pepo is a fleshy fruit with a leathery, non-septate rind derived from an inferior ovary), with one or three ovary sections or locules. Although the fruit have hard, lignified rinds, some squash cultivars have been bred to have a tender, edible rind (i.e. summer squash). Fruit contain tens to hundreds of seeds, rarely winged, usually flat with an oily embryo, rich in tocopherols, especially the larger seeds typically found in pumpkin and squash (DellaPenna and Pogson, 2006). Seed maturation, optimal for human consumption and germination, occurs after fruit senescence; mature seed germination occurs under low light at 25–30°C, with adequate non-soaking moisture, within 2–3 days.

7.2.1 Cucumber

Fruit generally are oblong or narrowly cylindrical, with small warts (tubercles) and spines, and enlarge (mostly elongate) with maturation. Spine colour is associated with maturation (Robinson and Decker-Walters, 1997). Fruit having white spines are light green to yellow at maturity and not netted. Black-spined fruit become orange or brown and often develop netting at maturity. Immature fruit have dark green internal flesh (yet are edible), whereas at maturity cucumbers have white or colourless flesh. Cucumbers have the highest moisture content (95%) of all cucurbits and contain 15 kcal, 1.0 g protein, 0.1 g fat, 3.4 g carbohydrates and 0.6 g fibre/100 g fresh weight (FW) (USDA-ARS, 2007). The biochemistry of the cucumber fruit is being studied in the USA to decipher the identity of *compound Q*, an extract used in China and credited with remedial and relief properties in AIDS sufferers (Hoareau and DaSilva, 1999).

7.2.2 Cantaloupe and honeydew melon

Fruit are generally round or oval, although some of the less common cultigens may be oblong (Robinson and Decker-Walters, 1997). Some cultigens have a reticulated (netted) rind (reticulatus types, e.g. cantaloupes), with various levels of aroma, while others are odourless, smooth-skinned (inodorous types, e.g. honeydew). Some less common inodorous types have a furrowed rind (e.g. Casaba); concave vein tracks (sutures) define other cultigens (e.g. Charentais). Mature rind colour varies from green, with or without whitish stripes, to yellow, tan or white. Flesh colour may be orange (β-carotene), green or white. An abscission zone at attachment of the peduncle and fruit occurs in some cultigens at physiological maturity. Melons have a moisture content of 91% and contain 32 kcal, 0.7 g protein, 0.2 g fat, 7.6 g carbohydrates and 0.5 g fibre/100 g FW (USDA-ARS, 2007). Melon seed extract, in some 'traditional' cultures, is used as an antidiabetic and for treating chronic or acute eczema (Yanty *et al.*, 2007).

7.2.3 Pumpkins and squash

Plants usually have large, orange-coloured, edible flowers and the fruit come in a vast variety of sizes, shapes, colours and surface characteristics. Pumpkin is a corruption of the Old English word 'pompion', which stems from the Latin 'pepo' meaning large, ripe, round melon, but it has no botanical meaning (Robinson and Decker-Walters, 1997). Thus, all pumpkins are squashes, but are differentiated from 'squash' in use. Almost all pumpkins are used for pies, jack-o'-lanterns or stock feed.

Squash is a corruption of a Native American word 'askutasquash', meaning to be eaten raw or cooked, depending on whether the fruit is immature (summer squash) or mature (winter squash). Squash species are distinguished on the basis of peduncle, stem, leaf and seed characteristics (Robinson and Decker-Walters, 1997). Peduncles can be hard, soft, smooth and/or angular. Stems can be hard, soft, round, smooth and/or angular, whereas leaves can be lobed, deeply cut, round, palmate, soft and/or prickly. Seeds can be white, black, brown, white to tan, white to brown, smooth, wrinkled, pitted and with or without smooth or ragged margins. Pumpkins have a moisture content of 92% and contain 26 kcal, 1.0 g protein, 0.1 g fat, 6.5 g carbohydrates and 1.1 g fibre/100 g FW (USDA-ARS, 2007). Immature (summer) squash have a moisture content of 94% and contain 17–20 kcal, 0.9–1.2 g protein, 0.1 g fat, 3.6–5.1 g carbohydrates and 0.6 g fibre/100 g fresh weight (USDA-ARS, 2007). Mature (winter) squash have a moisture content of 84–88% and contain 39–54 kcal, 1.5 g protein, 0.2 g fat, 9.4–14.0 g carbohydrates and 1.4 g fibre/100 g FW (USDA-ARS, 2007). Pumpkin and squash traditionally have been useful, in terms of human health, for blood cleansing, an aid to constipation and digestion and a source of energy (Rahman *et al.*, 2008).

7.2.4 Watermelon

Watermelon is differentiated from its family members by having pinnatified leaves, stems that are hairy, thin, angular and grooved, and branched tendrils (Robinson and Decker-Walters, 1997). Flowers are less showy than other cucurbits. Fruit are large, round to oblong or cylindrical, measuring as long as 60 cm. The rind is light to dark green, either solid in coloration, striped or marbled. The flesh can be bland to extremely sweet and is usually red (*cis*-isomers of lycopene), but there are also orange (β-carotene), salmon, yellow, green and white cultigens. The salmon and yellow cultigens of watermelon result from differing concentrations of a variety of carotenoids. Watermelon seeds can be black, brown, red, green or white; size and shape, along with colour, can be an aid in varietal identification. Seedless watermelons were developed for disease resistance, earliness of maturity, high yield, improved flesh characteristics (e.g. greater sugar content) and a durable rind. Watermelon fruit have a moisture content of 93% and contain 26 kcal, 0.5 g protein, 0.2 g fat, 6.4 g carbohydrates and 0.3 g fibre/100 g FW (USDA-ARS, 2007). In some 'traditional' cultures, watermelon fruit is used as a cooling agent to allay thirst, a source of energy, an aphrodisiac or blood purifier, and is considered good for sore eyes, scabies and itches (Rahman *et al.*, 2008).

7.3 Phytochemicals

An inclusive listing of phytochemicals and their health bioactive attributes, important to humans, in cucurbits has been provided recently by Kader *et al.* (2004). Bioactive components/nutrients appearing in appreciable concentrations in cucurbits include carotenoids, ascorbate, folate (B_9), potassium, citrulline and phenolics.

7.3.1 Ascorbate

The ascorbate dietary reference intake (DRI) for adult males is 90 mg/day (USDA-ARS, 2009). Cucurbit fruit range in total ascorbic acid (ascorbate) from as little as 9 mg/100 g FW in pumpkin to as much as 37 mg/100 mg FW in cantaloupe (Table 7.3). Less than one-half of a cantaloupe is an excellent source

(supplying more than 100%) of the DRI of ascorbate. The water-soluble compound ascorbate, commonly known as vitamin C, is related structurally to C_6 sugars ($C_6H_8O_6$), being an aldono-1,4-lactone of a hexonic acid (Davey *et al.*, 2000). Ascorbate has three primary functions in plants and animals, namely: (i) as an enzyme cofactor, (ii) as an antioxidant and (iii) as donor/acceptor in plasma membrane or chloroplastic electron transport. As such, in plants it can function in a number of physiological processes such as cell division and growth, re-reduction of α-tocopherol, photosynthesis, transmembrane electron transport and stress tolerance (Hodges, 2001). In humans, ascorbate can promote health in its role as an antioxidant (and thus defend against oxidative stress-related diseases such as cancer, various neurological disorders and cardiovascular disease), and as an enzyme cofactor it can participate in collagen hydroxylation (collagen being important in the synthesis/maintenance of cartilage, skin, teeth, bones and gums) and carnitine synthesis, among others. It has been noted that ascorbate can exhibit pro-oxidant activity in the presence of metal ions; however, there is no evidence that this is a significant problem *in vivo*.

7.3.2 Carotenoids

The β-carotene (provitamin A) DRI for adult males is 1800 µg/day (USDA-ARS, 2009). Cucurbit fruit range in β-carotene (Fig. 7.1) from as little as 30 µg/100 g FW in green-flesh honeydew melon to as much as 3100 µg/100 mg FW in pumpkin (Table 7.3). Thus, cantaloupe, orange-fleshed honeydew melon and pumpkin are an excellent source of carotenoids (i.e. provitamin A) as they provide more than 100% DRI of vitamin A. The lipophilic carotenoids are essentially long chains of conjugated double bonds and are differentiated by cyclization of the end group or by addition of oxygen (Rao and Rao, 2007). Carotenoids often provide the cucurbits with their distinctive colours; for example the orange-yellow pigment of cantaloupe and orange-fleshed honeydew melon is attributable to β-carotene, while the red hue of certain

Table 7.3. Bioactive/nutrient content per 100 g fresh weight of various cucurbits. Unless otherwise noted, data are from the USDA Nutrient Database (http://www.nal.usda.gov/fnic/foodcomp/search/).

Phytonutrient	Cantaloupe	Green-fleshed honeydew melons (orange-fleshed)	Cucumber	Pumpkin	Summer squash	Winter squash	Watermelon
Total ascorbate (mg)	36.7	18 (20.7)[a]	2.8	9	17	12.3	8.1
ß-Carotene (µg)	2020	30 (3000)[a]	45	3100	120	820	303
α-Carotene (µg)	16	0	11	515	0	0	0
Lycopene (µg)	1	0	26	0	0	0	4532
ß-Cryptoxanthin (µg)	0	0	0	2145	0	0	78
Lutein/zeaxanthin (µg)	26	27	23	1500	2125	38	8
ι-citrulline (mg)	ND[b]	ND	ND	ND	ND	ND	240[c]
Total folate (µg)	21	19	7	16	29	24	3
Total phenolics (mg)	24[a]	25 (30)[a]	20[d]	7[e]	83.3[f]	83.3[f]	6.4[g]
Potassium (mg)	267	228	147	340	262	350	112

Notes: [a]Hodges and Lester (2006). [b]ND = not determined. [c]Rimando and Perkins-Veazie (2005). [d]Chu et al. (2002). [e]Kwon et al. (2007). [f]Turkmen et al. (2005). Data reported as units/DM. 90% water content estimated in order to calculate as units/FW. [g]Gil et al. (2006).

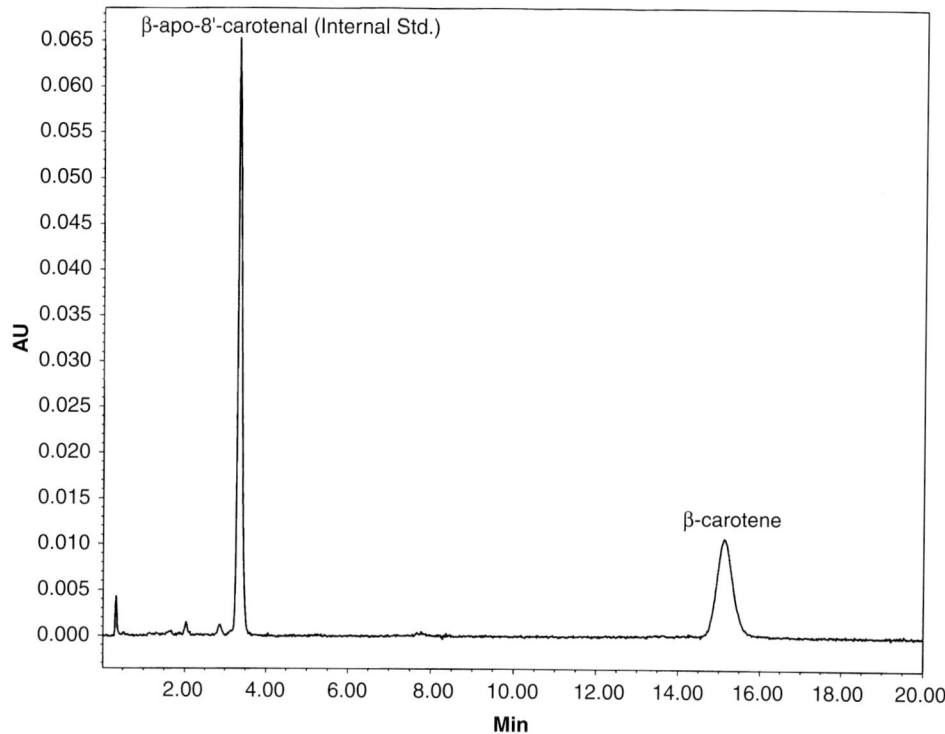

Fig. 7.1. Carotenoids were separated by HPLC-DAD on a C-18 column by the method of Okuno *et al.* (1998). β-Carotene was extracted from *Cucumis melo* (cv. Cruiser) mesocarp tissue with heptane and monitored at 454 nm. β-Apo-8'-carotenal (internal standard) elutes at ~ 3.5 min and β-carotene elutes at ~15 min. β-Carotene levels determined for this melon sample (per g fresh weight) were 26.3 µg/g.

cultivars of watermelon is due to lycopene (Fig. 7.2). In plants and humans, carotenoids are essential as accessory pigments in light harvesting, function as photoprotectants (e.g. skin cancer; Heinrich *et al.*, 2003) and, as active oxygen scavengers (particularly singlet oxygen), serve as precursors for such compounds as abscisic acid and serve to quench the excited triplet state of chlorophyll. Approximately 600 carotenoid compounds have been characterized, and of these about 50 are consumed in the human diet, though only 12 account for the majority of the intake (Voutilainen *et al.*, 2006). The most commonly consumed primary carotenoids from cucurbits include β-carotene and lycopene, with additional secondary carotenoids including α-carotene, β-cryptoxanthin and lutein/zeaxanthin.

Consumption of carotenoids has been associated with reduced incidences of various forms of cancer, AMD (advanced macular degeneration), cardiovascular diseases and osteoporosis; linkages have also been made between carotenoids and diseases such as hypertension, various neurodegenerative diseases (e.g. Alzheimer's, Parkinson's, amyotrophic lateral sclerosis) and emphysema (Voutilainen *et al.*, 2006; Rao and Rao, 2007).

Cucurbits can be an abundant source of various carotenoids (Table 7.3). The health functionality of carotenoids has been attributed primarily to their antioxidant activities (e.g. protection of low-density lipoproteins, scavenging of active oxygen). In addition, they have also been shown to play roles in such processes as immune response, modulation of particular drug-metabolizing enzymes, regulation of cell growth and gap junction communications (Rao and Rao, 2007). Moreover, α- and β-carotene, as well as β-cryptoxanthin, function as precursors to vitamin A. However,

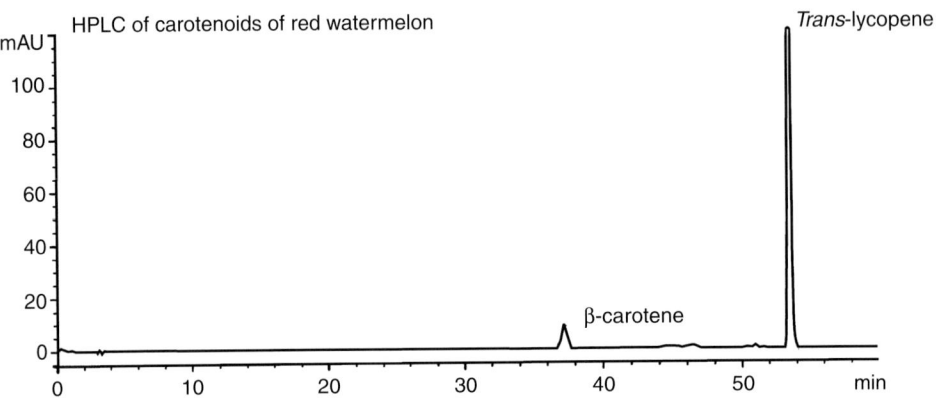

Fig. 7.2. Carotenoids were separated on a C-30 column by the method of Craft (2001). The carotenoids were extracted from a ripe watermelon (cv. Sangria) with hexane before injection on to the column. Eluting carotenoids were monitored at 503 nm. *Cis*-isomers of lycopene elute at ~ 45, 46 and 51 min. The carotenoid levels determined for this particular watermelon (per g fresh weight) were 31.7 µg/g *trans*-lycopene, 1.2 µg/g *cis*-isomers of lycopene and 4.9 µg/g ß-carotene. (Chromatogram courtesy of W. Fish, USDA-ARS, Lane, Oklahoma).

as the intervention study with smokers famously demonstrated (Alpha-Tocopherol, Beta-Carotene Cancer Prevention Group, 1994), pharmacological levels (as opposed to normal dietary levels) of supplementation with β-carotene can lead to pro-oxidant activities, thus promoting active oxygen species-related dysfunctions such as cancer and heart disease. The results of this and other trials have led to the postulation that β-carotene promotes health when consumed at normal dietary levels, but may have detrimental affects when taken at higher levels. There is now a moratorium on similar intervention studies with β-carotene and the focus has switched to lycopene.

7.3.3 Folate

The folate DRI for adult males is 400 µg/day (USDA-ARS, 2009). Cucurbit fruit range in folate from as little as 9 µg/100 g FW in cucumber to as much as 29 µg/100 mg FW in pumpkin (Table 7.3), making curcurbit fruit a good, but not an excellent, source of the DRI for folate. Tetrahydrofolate (composed of a pterin, a *p*-aminobenzioic acid and a glutamate chain with a number varying from 1 to 14 moieties) and its derivatives (which differ in the oxidation state of the pterin ring, the number of glutamate moieties and in the oxidation state of

the transfer C1 unit) are commonly grouped under the name of folates (Rébeillé *et al.*, 2006). These compounds are involved in 'carbon one' (C1) transfer reactions, which occur in two important plant and animal cycles: (i) the DNA biosynthesis cycle where C1 is donated during synthesis of purines and pyrimidines; and (ii) the methylation cycle whereby the folates participate in methylation reactions via *S*-adenosyl methionione (Scott *et al.*, 2000). They are also involved in the biosynthesis of such compounds as methionine, pantothenate, glycine and serine. Additionally, in plants, folates are involved in lignin formation and photorespiration (Basset *et al.*, 2005).

Humans cannot synthesize folate, thus plant foods are the most important sources of folate (Scott *et al.*, 2000). Folate deficiency in humans has been associated with neural tube defects, higher risk of cardiovascular diseases, neurodegenerative diseases such as Alzheimer's and increased incidences of colorectal, breast, pancreative, bronchial and cervical cancers, as well as leukaemia (Lucock *et al.*, 2003, as referenced in Rébeillé *et al.*, 2006).

7.3.4 Potassium

The potassium (K) DRI for adult males is 3500 mg/day (USDA-ARS, 2009). Cucurbit fruit range in K from as little as 147 mg/100 g FW in cucumber to as much as 350 mg/100g FW in winter squash (Table 7.3), making cantaloupe, pumpkin, and squash good sources of potassium DRI. There are many reviews that describe the role of potassium (K) in plant metabolism. Suffice to say for this chapter that, in plants, K plays important roles in enzyme activation, nutrient/assimilate transport into sink organs, photosynthesis and other energetic processes, cell-wall elasticity and osmotic potential (Jordan-Meille and Pellerin, 2008). A potassium deficiency can alter plant metabolism; examples include reduced photosynthetic CO_2 fixation and impairment in partioning/utilization of photosynthates (Cakmak, 2005). With regards to human health, potassium is a highly significant mineral. Potassium, along with sodium, can regulate the water and the acid–base balance in blood and tissues. It helps with normal kidney function and, through its participation in electrical potential gradient generation, plays a role in nerve impulses and cardiac, smooth and skeletal muscle contraction. Potassium is required in protein synthesis and in carbohydrate metabolism. Potassium deficiencies have been associated with high blood pressure, stroke, inflammatory bowel disease and asthma. Low K diets lead to increased calcium losses through excretion (thereby influencing osteoporosis) (Lanham-New, 2008).

7.3.5 L-Citrulline

Few natural foods are enriched in L-citrulline but this amino acid is found in abundance in the edible tissues of watermelon (*Ci. lanatus*) (Rimando and Perkins-Veazie, 2005) and at lower levels in most members of the cucurbit family. In adults, L-citrulline is converted to arginine, a substrate for endothelial production of nitric oxide (NO; a vasodilator), an agent considered to be important in the regulation of blood pressure and cardiovascular health. Studies with L-citrulline have also shown that this amino acid can increase the capacity of red blood cells to transport oxygen, and thus can act as a treatment in sickle cell anaemia (Waugh *et al.*, 2001). A recent study indicated that rats dosed with synthetic

L-citrulline/arginine reduced glucose uptake and weight gain, and improved aortic flexibility (Wu *et al.*, 2007). Currently, human intervention trials, feeding watermelon, are under way to determine the impact of pre-operation ingestion of L-citrulline/arginine on aiding patients' recovery from general anaesthesia (Mark Arney, National Watermelon Promotion Board, 2009, personal communication).

7.3.6 Phenolics

The phenolic structure is characterized by at least one aromatic ring bearing one or more hydroxyl groups; various subgroups are distinguished by their number of carbons and the basic structure of the carbon skeleton (Hodges and Kalt, 2003). The structure of the phenolic compound influences its bioactivity with regards to human health, and a myriad of health-promoting activities has been ascribed to phenolics and their derivatives. Hydroxyl groups and conjugated double bonds allow the phenolic to assume an antioxidant function through electron donation, metal ion (e.g. Cu^{3+}, Fe^{3+}) chelation and active oxygen quenching.

Induction of antifungal control using cucumber phenolics (ferulic acid and *p*-hydroxybenzoic acid) has been effective in controlling microbial populations; but little is known of cucurbit phenolic antifungal properties with humans (Yu, 2001). However, as a last general example of the claimed bioactive effects of phenolics, high phenolic content has been associated with hypocholesterolaemic and hypotriglyceridaemic activities in hamsters (Lin *et al.*, 2008).

7.4 Cucurbits and Human Health

The majority of studies on the direct health-promoting effects of cucurbit consumption have focused on carotenoid bioavailability comparisons with other horticultural products. For example, a study compared bioavailability of lycopene and β-carotene between watermelon and tomato in non-smoking humans (Edwards *et al.*, 2003).

Four treatments were used: (i) control (base diet); (ii) base diet plus 20.1 mg/day lycopene and 2.5 mg/day β-carotene from fresh-frozen watermelon juice; (iii) base diet plus 40.2 mg/day lycopene and 5.0 mg/day β-carotene from fresh-frozen watermelon juice; and (iv) base diet plus 18.4 mg/day lycopene and 0.6 mg/day β-carotene from canned tomato juice. Increases in plasma concentrations of lycopene (100–200%) were similar between the ~ 20 mg/day treatments of watermelon and tomato juices; increases in lycopene bio-absorption were not dependent on heat treatment of the fresh-frozen watermelon juice. A dose–response effect did not occur in either lycopene or β-carotene plasma levels following the doubling of the watermelon dosage, though there was a large degree of variation; the background diet included < 0.4 mg/day lycopene but 0.84 mg/day β-carotene, and the statistical power was weakened as only half the subjects consumed the double dose.

In another carotenoid bioavailability study, spinach (*Spinacia oleracea*) serving portions containing 3.0 mg/day and pumpkin squash (*Cucur moschata*) serving portions containing 1.4 mg/day β-carotene were fed to 7- to 12-year-old children for 3 weeks following a 3-week run-in period of 0.8 mg/day β-Carotene from long yard beans (*Vigna unguiculata*) (Van Lieshout *et al.*, 2003). β-Carotene bioavailability from pumpkin was found to be 1.7-fold greater than that from spinach when measured in the serum, emphasizing the importance of the food matrix/composition with regards to bioavailability. In comparing the bioavailability of β-carotene between spinach leaves and pumpkin, the ease of freeing this compound from pumpkin chromoplasts (where it is dissolved in oil droplets) relative to extracting it from leaf chloroplasts, along with the possible presence of dietary fibre in the spinach which may bind to the β-carotene molecules, may account for the greater bioavailability from pumpkin (de Pee *et al.*, 1998).

7.5 Preharvest and Postharvest Effects on Bioactive Content

Market quality characteristics, maturity indices, grades, sizes, packaging and postharvest storage protocols and disorders have been summarized for cucumber (Saltveit, 2004), melon (Lester and Shellie, 2004; Shellie and Lester, 2004), squash and pumpkin (Brecht, 2004; McCollum, 2004) and watermelon (Rushing, 2004). These important parameters, having been well documented recently, along with genetic diversity, physiology and biotechnology features of melon (Nunez-Palenius *et al.*, 2008), will not be addressed here. Instead, consideration will be given to pre- and postharvest environmental and some genetic factors influencing production and retention of human health bioactive compounds.

7.5.1 Cucumber fruit

A 240 ml serving of cucumber puree (peel included, weighing 150 g FW) provides 3.8% of the DRI for provitamin A (as β-carotene) and 4.7% of the DRI for ascorbate. Although there is a considerable amount of published literature on pre- and postharvest factors that influence the bioactive content of cucumbers, the vast majority of this work focuses on seedlings, roots or leaves versus fruit. In a postharvest study, cucumber fruit were stored for 15 days in 5, 21 or 100% kPa O_2 at 5, 10 or 20°C (Srilaong and Tatsumi, 2003). Among the parameters measured were antioxidant enzyme (superoxide dismutase (SOD), catalase) activities and compound (ascorbate, glutathione) concentrations. At 5°C catalase activities were higher in fruit stored in 21% kPa O_2; however, the typical ascorbate decline in postharvest products was not affected by storage atmosphere.

7.5.2 Cantaloupe and honeydew melon

Orange-fleshed varieties provide more than 100% (per 236 g FW) of the DRI of β-carotene and more than 33% of the DRI for ascorbic acid. Green-fleshed varieties provide slightly less than one-third of the daily requirement of ascorbic acid; but both melon types provide an extensive list of bioactive nutrients and minerals (Lester, 1997). Melons are a good source of two additional important human health bioactive components, potassium (K) and 5-methyl-teterahydrofolic acid (folate or B$_9$), even though concentrations may not satisfy the recommended daily allotment (Lester and Crosby, 2002). Concentrations of vitamins (ascorbic acid, β-carotene, folate) and K in melon fruit are affected by environmental factors such as soil texture (sand versus clay), production season (autumn versus spring) and production year (Table 7.4). Compared with sandy soils, clay produces fruit with generally significantly higher concentrations of ascorbic acid, β-carotene, folate and K. High mineral soils (clay) are naturally higher in the cation exchange capacities necessary for the biosynthesis of many biochemical processes, especially vitamins, found in melon fruit (Marschner, 2002). In the USA, autumn-grown melon fruit, when compared with spring-produced, have higher levels of vitamins and minerals due to cooler autumn temperatures, allowing for a greater retention of phytonutrients. Comparison of production years also had an impact, likely due to a combination of photosynthetic flux (light) and temperature variations from year to year. Although environmental factors impact vitamin and K concentrations in melon fruit, genetics appear to have an even greater influence. The ranges in ascorbic acid, β-carotene, folate and K concentrations are greatest when mature fruit are harvested based on marketable size classifications. The range in phytonutrient concentrations due to fruit size is even greater when comparing cultivars. A strong genetic effect is a positive finding because it is easier to control for cultivar and fruit size in any given production site than it is to maintain consistent natural light and temperature variations on a seasonal and annual basis.

The edible tissue of *Cucum melo* cultivars is known to be sweeter the closer it is to the seed cavity (Lester and Dunlap, 1985). Human health-promoting compounds, ascorbic acid (vitamin C), β-carotene, (provitamin A) and folic acid, also are more concentrated the closer the edible tissue is to the seed cavity, as are enzymatic antioxidants (ascorbate peroxidase, catalase and SOD) and the health-promoting mineral K (Lester, 2008). Enzymatic antioxidants like the aforementioned usually are destroyed in the human gut on ingestion. However, the biopolymeric matrix of muskmelon (*Cucum melo*) fruit, i.e. the edible tissue, protects active molecules like SOD against the digestive process by interacting with the intestinal epithelial barrier, allowing for bio-absorptivity (Vouldoukis *et al.*, 2004).

Postharvest decay of melon fruit is reduced when melons are stored between 4 and 10°C (Lester and Shellie, 2004; Shellie and Lester, 2004). Various melon cultivars stored as whole fruit at 5 or 10°C for 17 or 24 days and then sampled for ascorbic acid, β-carotene, folate and K content showed the greatest retention of vitamins at the coldest temperature (5°C) over the shortest period (17 days) (Table 7.5). Storage duration and temperature have no effect on K concentrations. However,

Table 7.4. Preharvest environmental/genetic influence on melon fruit vitamin and K concentrations. Data from Lester and Eischen (1996), Lester and Crosby (2002) and Lester and Hodges (2008).

Environmental/ genetic factors	Ascorbic acid (µg/g FW)	β-Carotene (µg/g FW)	Folate (µg/g FW)	Potassium (mg/g FW)
Clay soil	160–250[a]	21–26	0.38–0.68	1.6–2.0
Sandy soil	140–220[a]	17–21	0.29–0.57	1.3–1.7
Autumn	160–220[b]	12–19	0.36–0.45	2.1–3.8
Spring	140–180[b]	10–16	0.57–0.69	2.1–2.9
Year	150–260[c]	15–21	0.28–0.68	1.6–2.0
Cultivar	140–260[d]	17–21	0.28–0.69	1.6–2.0
Fruit size	70–250[e]	12–21	0.20–0.68	1.2–2.0

Notes: [a]Range of all cultivars at the same size across seasons, and years; [b]range of all cultivars at the same size across soil types, seasons, and years; [c]range of all cultivars at the same size across seasons; [d]range of all cultivars across sizes, soil types, seasons and years; [e]range of all sizes, across cultivars, soil types, seasons and years.

Table 7.5. Postharvest storage temperature and duration influence on melon fruit vitamin and K concentrations. Data from Lester and Hodges (2008).

Storage		Ascorbic acid	β-Carotene	Folate	Potassium
days	°C	(μg/g FW)	(μg/g FW)	(μg/g FW)	(mg/g FW)
17[a]	5	124–181[c]	9–16	0.48–0.59	2.1–3.8
17[a]	10	122 173[c]	7–15	0.28–0.35	2.0–3.0
24[b]	5	123–179[c]	9–16	0.30–0.53	2.0–3.3
24[b]	10	90–142[c]	7–12	0.26–0.29	2.1–3.0

Notes: [a]Fruit were stored 14 days at 5 or 10°C, followed by 3 days at 20°C; [b]fruit were stored 21 days at 5 or 10°C, followed by 3 days at 20°C; [c]range across all cultivars.

some minerals such as calcium do decrease in melon with storage duration (Lester and Grusak, 1999, 2001). Of the vitamins (ascorbic acid, β-carotene, folate) that decrease in content during storage, folate demonstrates the greatest rate of decline, followed by ascorbic acid then β-carotene.

Volatile or aroma compounds generally are not considered as having direct health-promoting properties. However, 3-methylthio-propionic acid ethyl ester (MTPE), found in certain *Cucum melo* varieties, has anticancer properties (Nakamura *et al.*, 2008). Japanese pickling melon (*Cucum melo* var. conomon) at the overly ripe fruit stage and orange-fleshed muskmelon (*Cucum melo* var. reticulatus) are the only *Cucumis* species to have significant concentrations of MTPE. As an anticancer compound, MTPE targets human colon cancer cells. The only down side is that the average MTPE concentrations found in orange-fleshed melon and overly ripe Japanese pickling melon are 2.5 and 3.8 μg/100 g FW, respectively; thus ~ 1000-fold too dilute. These melon varieties, therefore, would have to be processed and the volatile concentrated to have a pharmacological effect.

7.5.3 Pumpkin and squash

A 240 ml serving of pumpkin squash puree (~ 140 g FW) offers about 250% of the DRI of β-carotene and about 10% of the required K. Pumpkin also represents an appreciable source of α-carotene, β-cryptoxanthin and lutein/zeaxanthin. A 140g FW portion of winter squash (e.g. acorn squash, buttercup squash) generally contains 65 and 10.4% of

the recommended intakes of β-carotene and K, respectively. Summer squash (e.g. zucchini/courgette) provides 8% of the daily requirements of β-carotene, 22.7% of ascorbate and 6.6% of K in a 120 g (FW) cup.

Not only can production/techniques, growth environment (elevated temperature and light favour carotenoid synthesis) and maturity/ripening stage affect carotenoid levels drastically (Rodriguez-Amaya *et al.*, 2008), but carotenoid content can also be highly dependent on the often overlooked cultivars within a genotype. For example, α-carotene levels ranged from 0.42 to 7.5, β-carotene from 1.4 to 7.2 and lutein/zeaxanthin from 0.8 to 17 mg/100 g FW in 12 varieties of *Cucur maxima* grown under similar conditions (albeit only for 1 year) (Murkovic *et al.*, 2002). However, a detailed listing of specific cultivars and their individual carotenoids was not included.

Processing and storage can also have dramatic effects on the bioactive content of postharvest commodities (Hodges and Toivonen, 2008). Concentrations of β-carotene were found to increase dramatically (0.065–0.733 mg/100 g FW) in whole pumpkins stored for 3 months, concomitant with darkening flesh colour, although the storage conditions were not specified in this study (Chavasit *et al.*, 2002). With respect to processing/food preparation, total phenolics and total antioxidant activity (as measured by the 2,2-diphenyl-1-picryl-hydrazyl (DPPH) method) were assessed in squash (type not reported) prior to cooking and following microwaving (100 g squash in 6 ml water microwaved for 1 min), steaming (7.5 min) or boiling (100 g squash in 150 ml water boiled for 5 min) (Turkmen *et al.*, 2005). All three processing protocols resulted in decreases in total phenolic contents of about 34%, with no significant difference in losses noted between cooking procedures. Although these cooking protocols induced an apparent increase in total antioxidant capacities (by approximately 132%), post-cooking values were not significantly different from the control data.

Pumpkin and gourd seeds are an excellent source of health-promoting oils (Stevenson *et al.*, 2007). Nearly 100 different pumpkin and gourd varieties were assayed for seed oil content, fatty acid composition and tocopherol content. Oil content ranged from 11 to 31% of the seed weight, and the oil total unsaturated fatty acids content ranged from 73 to 81%. α-Tocopherol, the predominant tocopherol in plant tissues and the molecular form with relatively all the vitamin E activity (DellaPenna and Pogson, 2006), ranged from 75 to 500 µg/g oil. This study extracted oil from fresh-frozen seeds and concluded pumpkins seeds were a good commercial source of highly unsaturated oil and tocopherol, with the potential to improve the nutrition of human diets. The authors did not assay seeds as a roasted product, or study how seed harvesting conditions, storage period, storage environment and processing procedures influenced oil characteristics.

7.5.4 Watermelon

Watermelon fruit is an excellent source of antioxidant vitamin C and provitamin A. A 240 ml serving of watermelon puree (156 g FW) provides 13.5% of the DRI for vitamin C, 25.2% of vitamin A (from β-carotene), 11% of vitamin B_6 (pyridoxine), 8% of vitamin B_1 (thiamine), 3.6% of potassium and 4% of magnesium. Red and pink coloured watermelon cultivars are excellent sources of lycopene. Red-fruited watermelon have been grouped as low (< 50 mg/kg FW), medium (50–70 mg/kg FW), high (70–90 mg/kg FW) or very high (> 90 mg/kg FW) lycopene-containing cultivars (Perkins-Veazie *et al.*, 2006). Watermelon fruit are also exceptionally high in citrulline (Rimando and Perkins-Veazie, 2005). Ingestion of watermelon

increases plasma arginine concentration in adults (Collins *et al.*, 2007).

Minimal processing (fruit cutting, packaging and chilling) of watermelon pieces does not affect their nutritional content significantly even after 6, and up to 9, days' storage at 5°C; no loss in carotenoids occurred and vitamin C declined only 5% (Gil *et al.*, 2006). A study of carotenoid changes in three types of watermelon (open-pollinated seeded, hybrid seeded and seedless) stored at 5, 13 or 21°C for 14 days found carotenoid levels increased in fruit stored at 21°C versus fresh-harvested fruit (Perkins-Veazie and Collins, 2006). Watermelons stored at 21°C gained between 11 and 40% in lycopene, and β-carotene content increased by between 50 and 139%; fruit stored at 5°C and 13°C, however, showed very small changes in carotenoid content.

Lycopene, the carotenoid that gives the characteristic red colour in watermelon, degrades even at freezing (–20°C) temperatures (Fish and Davis, 2003), whereas β-carotene, the carotenoid that gives the characteristic orange colour in muskmelon, squash and orange-fleshed watermelon, is more heat stable and is increased with cooking time and temperature (Park, 1987). It is hypothesized that the enzyme, lypoxygenase, which is associated with carotene degradation, is degraded with heating, thus preserving β-carotene (Reeve, 1943).

7.6 Future Research Needs

Cucurbits, a family of diverse species, are in fact limited in germplasm (i.e. parental lines) used in breeding current commercial cultivars (Kevin Crosby, cucurbit breeder, Texas A&M, 2008, personal communication). Selection of potential commercial cultivars is based almost entirely on disease and insect resistance, suitable phenotype and sugar content (e.g. sweet melon), whereas postharvest keeping quality or human health-promoting properties are largely ignored. Exotic cucurbit germplasm, likely available from national/international seed bank repositories and commercial seed companies' heirloom selections, represent a potentially untapped phytonutrient source and should be investigated for variability in the critical human health compounds such as ascorbic acid, carotenes, folates and other B vitamins, tocopherols, phylloquinone and potassium. Other, lesser known health compounds such as essential amino acids (e.g. citrulline/arginine), phenolics and enzymatic antioxidants (e.g. SOD; the mesocarp matrix unique to the *Cucum melo* species provides a protective layer keeping SOD viable during the digestive process, and, therefore, is able to elicit *in vivo* pharmacological effects (Vouldoukis *et al.*, 2004)), also require further investigation in current and future cucurbit germplasm. Besides conventional breeding and molecular engineering approaches (e.g. to alter regulatory factors that influence bioactive content), numerous preharvest production factors as well as postharvest protocols can influence bioactive content (for reviews, see Lee and Kader, 2000; Hodges *et al.*, 2004; Hodges and DeLong, 2007). Once cucurbit cultivars inherently enriched in bioactives are identified/ developed, production and postharvest technologies designed to enhance optimally bioactive content can be characterized.

7.7 Conclusions

Although cucurbits are an excellent source of a variety of bioactive compounds, it is the outstanding carotenoid content of certain cucurbits that places them in the top tier of fruit and vegetables with health-promotion properties. Pumpkin, cantaloupe melon, orange-fleshed honeydew melon and watermelon, in particular, are commonly consumed members of the *Cucurbitaceae* containing good to excellent carotenoid concentrations (β-carotene for pumpkin, cantaloupe and orange-fleshed honeydew melon and lycopene for watermelon); in all cases, a 240 ml serving contains enough carotenoids to meet the recommended DRI. The netted rind of cantaloupe can harbour illness-related enteric bacteria such as *Salmonella* Lignieres, *Shigella* Chatellani and Dawson and *Escherichia coli* O157:H7; however, there are a number of cultivars or an alternative cucurbit, orange-fleshed honeydew melon, that contain equal or greater amounts of carotenoids and other bioactives (Lester and Hodges, 2008). The concept of growing crops for health rather than just for food or fibre is slowly diversifying the focus of plant biotechnology and medicine and has been well reviewed by Raskin *et al.* (2002). There are a number of other avenues to enhancing bioactive content besides conventional/molecular breeding and cultivar/species selection. For example, growing practices, as demonstrated by the relatively few well-conducted studies comparing organic versus conventional and sustainable cropping systems on plant phytonutrients, have demonstrated a production system effect, and this should be investigated further in cucurbits. Additionally, now that the US Food and Drug Administration is clearing more and more fruit and vegetables for food safety irradiation, updated investigations into the effect of ionizing irradiation on cucurbits must occur, given 'irradiation treatments can stimulate the biosynthesis of bioactive compounds' (Lacroix and Vigneault, 2007). These and other methods of enhancing the already notable bioactive content of cucurbits will result in consumer access to cucurbits such as melon, squash and pumpkin enriched in health-promoting compounds. As bioabsorption of bioactive compounds within a natural food matrix has often been considered to be more effective than with supplements/nutraceuticals (e.g. Kader *et al.*, 2004), the enrichment of cucurbits already high in bioactive complements will almost certainly improve consumer health and well-being.

References

Alpha-Tocopherol, Beta-Carotene Cancer Prevention Group, The (1994) The effect of vitamin E and beta carotene on the incidence of lung cancer and other cancers in male smokers. *New England Journal of Medicine* 330, 1029–1035.

Bailey, L.H. and Bailey, E.Z. (1976) *Hortus Third, a Concise Dictionary of Plants Cultivated in the United States and Canada*. MacMillan Publishing Company, New York.

Basset, G.J.C., Quinlivan, E.P., Gregory, J.F. III and Hanson, A.D. (2005) Folate synthesis and metabolism in plants and prospects for biofortification. *Crop Science* 45, 449–453.

Brecht, J.K. (2004) Pumpkin and winter squash. In: Gross, K.C., Wang, C.Y. and Saltveit, M.E. (eds) *The Commercial Storage of Fruits, Vegetables, and Florist and Nursery Stocks* (http://www.ba.ars.usda/hb66/contents.html, accessed 6 July 2008).

Cakmak, I. (2005) The role of potassium in alleviating detrimental effects of abiotic stresses in plants. *Journal of Plant Nutrition and Soil Science* 168, 521–530.

Chavasit, V., Pisaphab, R., Sungpuag, P., Jittinandana, S. and Wasantiwisut, E. (2002) Changes in β-carotene and vitamin A contents of vitamin A-rich foods in Thailand during preservation and storage. *Journal of Food Science* 67, 375–379.

Chu, Y.-F., Sun, J., Wu, X. and Liu, R.H. (2002) Antioxidant and antiproliferative activities of common vegetables. *Journal of Agricultural and Food Chemistry* 59, 6910–6916.

Collins, J.K., Wu, G., Perkins-Veazie, P., Spears, K., Claypool, P.L., Baker, R.A. and Clevidence, B.A. (2007) Watermelon consumption increases plasma arginine concentrations in adults. *Nutrition* 23, 261–266.

Craft, N.E. (2001) Chromatographic techniques for carotenoid separation. In: Wrolstad, R.E., Acree, T.E., Decker, E.A., Penner, M.H., Reid, D.S., Schwartz, S.J., Shoemaker, C.F. and Sporns, P. (eds) *Current Protocols in Food Analytical Chemistry*. John Wiley and Sons, New York, pp. F2.3.1–2.3.15.

Davey, M.W., Van Montagu, M., Inzé, D., Sanmartin, M., Kanellis, A., Smirnoff, N., Benzie, I.J.J., Strain, J.J., Favell, D. and Fletcher, J. (2000) Plant L-ascorbic acid, chemistry, function, metabolism, bioavailability and effects of processing. *Journal of the Science of Food and Agriculture* 80, 825–860.

de Pee, S., West, C.E., Permaesih, D., Martuti, S. and Hautvast, J.G.A.J. (1998) Orange fruit is more effective than are dark-green, leafy vegetables in increasing serum concentrations of retinol and β-carotene in schoolchildren in Indonesia. *American Journal of Clinical Nutrition* 68, 1058–1067.

DellaPenna, D. and Pogson, B.J. (2006) Vitamin synthesis in plants: tocopherols and carotenoids. *Annual Review Plant Biology* 57, 711–738.

Edwards, A.J., Vinyard, B.T., Wiley, E.R., Brown, E.D., Collins, J.K., Perkins-Veazie, P., Baker, R.A. and Clevidence, B.A. (2003) Consumption of watermelon juice increases plasma concentrations of lycopene and β-carotene in humans. *Journal of Nutrition* 123, 1043–1050.

Fish, W.W. and Davis, A.R. (2003) The effect of frozen storage conditions on lycopene stability in watermelon tissue. *Journal of Agriculture and Food Chemistry* 51, 3582–3585.

Gil, M.I., Aguayo, E. and Kader, A. (2006) Quality changes and nutrient retention in fresh-cut versus whole fruits during storage. *Journal of Agricultural and Food Chemistry* 54, 4284–4296.

Goldman, A. (2002) *Melon for the Passionate Grower, with Practical Advice on Growing, Pollinating, Picking, and Preparing an Extraordinary Harvest.* Artisan Books, Workman Publishing Inc, New York.

Goldman, A. (2004) *The Complete Squash, a Passionate Grower's Guide to Pumpkins, Squash, and Gourds.* Artisan Books, Workman Publishing Inc, New York.

Heinrich, U., Gartner, C., Wiebusch, M., Eicher, O., Sies, H., Tronnier, H. and Stohl, W. (2003) Supplementation with β-carotene or a similar amount of mixed carotenoids protects humans from uv-induced erythema. *Journal of Nutrition* 133, 98–101.

Hoareau, L. and DaSilva, E.J. (1999) Medical plants: a re-emerging health aid. *Journal of Biotechnology* 2, 56–70.

Hodges, D.M. (2001) Chilling effects on active oxygen species and their scavenging systems in plants. In: Basra, A. (ed.) *Crop Responses and Adaptations to Temperature Stress*. Food Products Press, Binghamton, New York, pp. 53–78.

Hodges, D.M. and DeLong, J.M. (2007) The relationship between antioxidants and postharvest quality of fruits and vegetables. *Stewart Postharvest Review* 3, 1–9.

Hodges, D.M. and Kalt, W. (2003) Health functionality of small fruit. *Acta Horticulturae* 626, 17–23.

Hodges, D.M. and Lester, G.E. (2006) Comparisons between orange- and green-fleshed non-netted and orange-fleshed netted muskmelons, antioxidant changes following different harvest and storage periods. *Journal of the American Society for Horticultural Science* 131, 110–117.

Hodges, D.M. and Toivonen, P.M.A. (2008) Quality of fresh-cut fruits and vegetables as affected by exposure to abiotic stress. *Postharvest Biology and Technology* 48, 155–162.

Hodges, D.M., Lester, G.E., Munro, K.M. and Toivonen, P.M.A. (2004) Oxidative stress. Importance for postharvest quality. *HortScience* 39, 924–929.

Janick, J., Paris, H.S. and Parrish, D.D. (2007) The cucurbits of Mediterranean antiquity, identification of taxa from ancient images and descriptions. *Annals of Botany* 100, 1441–1457.

Jordan-Meille, L. and Pellerin, S. (2008) Shoot and root growth of hydroponic maize (*Zea mays* L.) as influenced by K deficiency. *Plant Soil* 304, 157–168.

Kader, A.A., Perkins-Veazie, P. and Lester, G.E. (2004) Nutritional quality: importance to human health. In: Gross, K.C., Wang, C.Y., and Saltveit, M.E. (eds) *The Commercial Storage of Fruits, Vegetables, and Florist and Nursery Stocks* (http://www.ba.ars.usda/hb66/contents.html, accessed 6 July 2008).

Kwon, Y.-I., Apostolidis, E., Kim, Y.-C. and Shetty, K. (2007) Health benefits of traditional corn, beans, and pumpkin: *in vitro* studies for hyperglycemia and hypertension management. *Journal of Medicinal Food* 10, 266–275.

Lacroix, M. and Vigneault, C. (2007) Irradiation treatment for increasing fruit and vegetable quality. *Stewart Postharvest Review* 3, 1–8.

Lanham-New, S.A. (2008) The balance of bone health: tipping the scales in favor of potassium-rich, bicarbonate-rich foods. *Journal of Nutrition* 138, 172S–177S.

Lee, S.K. and Kader, A.A. (2000) Preharvest and postharvest factors influencing vitamin C content of horticultural crops. *Postharvest Biology and Technology* 20, 207–220.

Lester, G. (1997) Melon (*Cucumis melo* L.) fruit nutritional quality and health functionality. *HortTechnology* 7, 222–227.

Lester, G. (1998) Diurnal growth measurements of honeydew and muskmelon fruits. *HortScience* 33, 156.

Lester, G.E. (2008) Antioxidant, sugar, mineral, and phytonutrient concentrations across edible fruit tissue of orange-fleshed honey dew melon (*Cucumis melo* L.). *Journal of Agriculture and Food Chemistry* 56, 3694–3698.

Lester, G.E. and Crosby, K.M. (2002) Ascorbic acid, folic acid, and potassium content in postharvest green-fleshed honeydew muskmelons: influence of cultivar, fruit size, soil type and year. *Journal of the American Society for Horticultural Science* 127, 843–847.

Lester, G.E. and Dunlap, J.R. (1985) Physiological changes during development and ripening of 'Perlita' muskmelon fruit. *Sciencia Horticulturae* 26, 323–331.

Lester, G.E. and Eischen, F. (1996) Beta-carotene content of postharvest orange-fleshed muskmelon fruit: effect of cultivar, growing location and fruit size. *Plant Foods and Human Nutrition* 49, 191–197.

Lester, G.E. and Grusak, M.A. (1999) Postharvest application of calcium and magnesium to honey dew and netted muskmelons: effects on tissue ion concentrations, quality and senescence. *Journal of the American Society for Horticultural Science* 124, 545–552.

Lester, G.E. and Grusak, M.A. (2001) Postharvest application of chelated and nonchelated calcium dip treatments to commercially grown honey dew melons: effects on peel attributes, tissue calcium concentration, quality, and consumer preference following storage. *HortTechnology* 11, 561–566.

Lester, G.E. and Hodges, D.M. (2008) Antioxidants associated with fruit senescence and human health: novel orange-fleshed non-netted honey dew genotype comparisons following different seasonal productions and cold storage durations. *Postharvest Biology and Technology* 48, 347–354.

Lester, G. and Shellie, K. (2004) Honey dew melon. In: Gross, K.C., Wang, C.Y. and Saltveit, M.E. (eds) *The Commercial Storage of Fruits, Vegetables, and Florist and Nursery Stocks* (http://www.ba.ars.usda/hb66/contents.html, accessed 6 July 2008).

Lin, L.-Y., Peng, C.-C., Yang, Y.-L. and Peng, R.Y. (2008) Optimization of bioactive compounds in buckwheat sprouts and their effects on blood cholesterol in hamsters. *Journal of Agricultural and Food Chemistry* 56, 1216–1223.

Lucock, M., Yates, Z., Glanville, T., Leeming, R., Simpson, N. and Daskalakis, I. (2003) A critical role for B-vitamin nutrition in human developmental and evolutionary biology. *Nutrition Research* 23, 1463–1475.

McCollum, T.G. (2004) Squash (summer). In: Gross, K.C., Wang, C.Y. and Saltveit, M.E. (eds) *The Commercial Storage of Fruits, Vegetables, and Florist and Nursery Stocks* (http://www.ba.ars.usda/hb66/contents.html, accessed 6 July 2008).

Marschner, H. (2002) *Mineral Nutrition of Higher Plants*, 2nd edn. Academic Press, New York.

Maynard, D. and Maynard, D.N. (2000) Cucumber, melon and watermelon. In: Kniple, K.F. and Orneles, K.C. (eds) *The Cambridge World History of Food*. Cambridge University Press, Cambridge, UK, pp. 298–313.

Murkovic, M., Mülleder, U. and Neunteufl, H. (2002) Carotenoid content in different varieties of pumpkins. *Journal of Food Composition and Analysis* 15, 633–638.

Nakamura, Y., Nakayama, Y., Ando, H., Tanaka, A., Matuso, T., Okamoto, S., Upham, B.L., Chang, C.-C., Trosko, J.E., Park, E.Y. and Satao, K. (2008) 3-methylthiopropionic acid ethyl ester, isolated from

Katsura-uri (Japanese pickling melon, *Cucumis melo* var. conomon), enhanced differentiation in human colon cancer cells. *Journal of Agriculture and Food Chemistry* 56, 2977–2984.

Nunez-Palenius, H.G., Gomez-Lim, M., Ochoa-Alejo, N., Gumet, R., Lester, G. and. Cantliffe, D.J. (2008) Melon fruits; genetic diversity, physiology and biotechnology. *Critical Reviews in Biotechnology* 19, 13–55.

Okuno, S., Yoshimoto, M., Kumagai, T. and Yamakawa, O. (1998) Contents of β-carotene and α-tocopherol of sweetpotato cultivars newly developed for processing purposes. *Tropical Agriculture (Trinidad)* 75, 174–176.

Park, Y.W. (1987) Effect of freezing, thawing, drying, and cooking on carotene retention in carrots, broccoli and spinach. *Journal of Food Science* 52, 1022–1025.

Perkins-Veazie, P. and Collins, J.K. (2006) Carotenoid changes of intact watermelons after storage. *Journal of Agricultural and Food Chemistry* 54, 5868–5874.

Perkins-Veazie, P., Collins, J.K., Davis, A.R. and Roberts, W. (2006) Carotenoid content of 50 watermelon cultivars. *Journal of Agricultural and Food Chemistry* 54, 2593–2597.

Rahman, A.H.M.M., Anisuzzaman, M., Ahmed, F., Rafiul Islam, A.K.M. and Naderussaman, A.T.M. (2008) Study of nutritive value and medicinal uses of cultivated cucurbits. *Journal of Applied Sciences Research* 4, 555–558.

Rao, A.V. and Rao, L.G. (2007) Carotenoids and human health. *Pharmacological Research* 55, 207–216.

Raskin, I., Ribnicky, D.M., Komarnytsk, S., Nebojsa, L., Poulev, A., Borisjuk, N., Brinker, A., Moreno, D.A., Ripoll, C., Yakoby, N., O'Neal, J.M., Cornwell, T., Pastor, I. and Fridlender, B. (2002) Plants and human health in the twenty-first century. *Trends in Biotechnology* 20, 522–531.

Rébeillé, F., Ravanel, S., Jabrin, S., Douce, R., Storozhenko, S. and Van Der Straeten, D. (2006) Folates in plants: biosynthesis, distribution, and enhancement. *Physiologia Plantarum* 126, 330–342.

Reeve, R.M. (1943) Microscopy of the oils and carotene bodies in dehydrated carrot. *Food Research* 8, 137–141.

Rimando, A.M. and Perkins-Veazie, P.M. (2005) Determination of citrulline in watermelon rind. *Journal of Chromatography* 1078, 196–200.

Robinson, R.W. and Decker-Walters, D.S. (1997) *Cucurbits*. CAB International, New York.

Rodrigeuz-Amaya, D.B., Kimura, M., Godoy, H.T. and Amaya-Farfan, J. (2008) Updated Brazilian database on food carotenoids: factors affecting carotenoid composition. *Journal of Food Composition and Analysis* 21, 445–463.

Rushing, J.W. (2004) Watermelon. In: Gross, K.C., Wang, C.Y. and Saltveit, M.E. (eds) *The Commercial Storage of Fruits, Vegetables, and Florist and Nursery Stocks* (http://www.ba.ars.usda/hb66/contents.html, accessed 6 July 2008).

Saltveit, M.E. (2004) Cucumber. In: Gross, K.C., Wang, C.Y. and Saltveit, M.E. (eds) *The Commercial Storage of Fruits, Vegetables, and Florist and Nursery Stocks* (http://www.ba.ars.usda/hb66/contents.html, accessed 6 July 2008).

Scott, J., Rébeillé, F. and Fletcher, J. (2000) Folic acid and folates: the feasibility for nutritional enhancement in plant foods. *Journal of the Science of Food and Agriculture* 80, 795–824.

Shellie, K.C. and Lester, G. (2004) Netted melon. In: Gross, K.C., Wang, C.Y. and Saltveit, M.E. (eds) *The Commercial Storage of Fruits, Vegetables, and Florist and Nursery Stocks* (http://www.ba.ars.usda/hb66/contents.html, accessed 6 July 2008).

Sinnott, E.W. (1945) The relation of growth to size in cucurbit fruits. *American Journal of Botany* 32, 439–446.

Srilaong, V. and Tatsumi, Y. (2003) Changes in respiratory and antioxidative parameters in cucumber fruit (*Cucumis sativa* L.) under high and low oxygen concentrations. *Journal of the Japanese Society for Horticulture* 72, 525–532.

Stevenson, D.G., Eller, F.J., Wang, L., Jane, J.-L., Wang, T. and Inglett, G.E. (2007) Oil and tocopherol content and composition of pumpkin seed oil in 12 cultivars. *Journal of Agriculture and Food Chemistry* 55, 4005–4013.

Turkmen, N., Sari, F. and Velioglu, Y.S. (2005) The effect of cooking methods on total phenolics and antioxidant activity of selected green vegetables. *Food Chemistry* 93, 713–718.

USDA-ARS (US Department of Agriculture-Agricultural Research Service) (2007) USDA National Nutrient Database for Standard Reference, Release 20 (http://www.ars.usda.gov/ba/bhnrc/ndl, accessed 6 July 2008).

USDA-ARS (US Department of Agriculture-Agricultural Research Service) (2009) Dietary Reference Intakes Recommended for Individuals (http://fnic.nal.usda.gov/nal_display/index, accessed 12 August 2009).

USDA-ERS (US Department of Agriculture-Economic Reporting Service) (2007) Vegetables and melons outlook/VGS-319 (http://www.ers.usda.gov/publications/agoutlook, accessed 22 February 2008).

Van Lieshout, M., West, C.E., Van de Bovenkamp, P., Wang, Y., Sun, Y., Van Breemen, R.B., Muhilal, D.P., Verhoeven, M.A., Creemers, A.F.L. and Lugtenburg, J. (2003) Extraction of carotenoids from feces, enabling the bioavailability of β-carotene to be studied in Indonesian children. *Journal of Agricultural and Food Chemistry* 51, 5123–5130.

Vouldoukis, I., Lacan, D., Kamate, C., Coste, P., Calenda, A., Mazier, D., Conti, M. and Dugas, B. (2004) Antioxidant and anti-inflammatory properties of *Cucumis melo* LC. Extract rich in superoxide dismutase activity. *Journal of Ethnopharmacology* 94, 67–75.

Voutilainen, S., Nurmi, T., Mursu, J. and Rissanen, T.H. (2006) Carotenoids and cardiovascular health. *American Journal of Clinical Nutrition* 83, 1265–1271.

Waugh, W.H., Daeschner, C.W., Files, B.A., McConnell, M.E. and Strandjord, S.E. (2001) Oral citrulline as arginine precursor may be beneficial in sickle cell disease: early phase two results. *Journal of the National Medical Association* 93, 363–371.

Whitaker, T.W. and Davis, G.N. (1962) *Cucurbits. Botany, Cultivation, and Utilization.* Interscience Publishers Inc, New York.

Wu, G., Collins, J.K., Perkins-Veazie, P.M., Siddiq, M., Dolan, K.D., Kelley, K.A., Heaps, C.L. and Meininger, C.J. (2007) Dietary supplementation with watermelon pomace juice enhances arginine availability and ameliorates the metabolic syndrome in Zucker diabetic fatty rats. *Journal of Nutrition* 137, 2680–2685.

Yanty, N.A.M., Lai, O.M., Osman, A., Long, K. and Ghazali, H.M. (2007) Physiochemical properties of *Cucumis melo* var. inodorus (honeydew melon) seed and seed oil. *Journal of Food Lipids* 15, 42–55.

Yu, J.Q. (2001) Autotoxic potential protection of cucurbit crops. *Journal of Crop Protection* 4, 335–348.

8 Exotics
[Litchi, Longan, Rambutan, Pomegranate, Mangosteen, Kiwifruit, Passion Fruit, Persimmon, Carambola]

Nettra Somboonkaew and Leon A. Terry

8.1 Introduction

Exotic fruit are usually subtropical to tropical in origin. They are becoming more important fresh commodities in both producer and overseas countries. Exotic fruit are a significant source of potential functional dietary agents, for example, phenolic compounds, ascorbic acid and provitamin A (Wall, 2006). However, detailed information about the potential health benefits of fresh exotic fruit is limited. As a consequence, and for the purpose of this chapter, the nutritional and medicinal benefits of nine exotic fruit, namely litchi, rambutan, longan, mangosteen, kiwifruit, persimmon, passion fruit and carambola, are discussed.

8.2 Identity of Bioactive Compounds

Fruit are an important source of bioactive substances, e.g. provitamin, plant pigments and polyphenolic compounds (Zhang *et al.*, 2001). Increasingly, edible plant products rich in phytochemicals are recognized worldwide for their potential to reduce the risk of several diseases, including cancer, diabetes and cardiovascular disease (Charoensiri *et al.*, 2009), and to promote human health. The extraction of fruit and the isolation and purification of

active substances are dependent on the type of plant material, extraction solvent and isolation methodology. The extraction solvents are often acetone, methanol, ethanol and water, while acetonitrile, methanol, ethanol, acid and base and water are common chemicals used in isolation methods (Table 8.1).

8.3 Litchi, Longan and Rambutan

Litchi (*Litchi chinensis* Sonn.; Fig. 8.1) and longan (*Dimocarpus longan* Lour. or *Euphoria longana* Lam.; Fig. 8.2) are subtropical to tropical fruit, while rambutan (*Nephelium lappaceum* Linn.; Fig. 8.3) is a warm-tropical fruit. They belong to the family *Sapindaceae*. Litchi and longan originated in China, whereas rambutan was derived from Malaysia and the Sumatra region. However, the fruit are now grown widely in North Australia, South–east Asia, South Asia, Israel, South and equatorial Africa, Madagascar, Central America and the USA (O'Hare, 1997).

The edible portion of litchi and longan is sweet and very fragrant. Rambutan pulp is similar in flavour to the litchi but less aromatic, while the texture is firmer and not as juicy. The pericarp colour of litchi and rambutan varies between a yellow and crimson, whereas the pericarp of longan is light brown.

Table 8.1. Analytical methods for health-related compounds in exotic fruit.

	Analysis			
	Method: column	Extracting	Mobile phase	Reference
Vitamins				
Vitamin A, provitamin A and other carotenoids	HPLC: C-18	THF	CH$_3$CN: THF: H$_2$O	Wall (2006)
	HPLC: C-18	KOH	CH$_3$CN: THF: CH$_3$OH: TEA + ammonia acetate	Charoensiri et al. (2009)
Vitamin C	HPLC: C-18	H$_3$PO$_4$	NaH$_2$PO$_4$	Wall (2006)
	HPLC: RP-18	H$_3$PO$_4$	Bu$_4$NOH: CH$_3$OH: KH$_2$PO$_4$	Nishiyama et al. (2004)
	HPLC: LC-NH2	Oxalic+DCPIP	–	Al-Maiman and Ahmad (2002)
	HPLC: supelcosil LC-18	H$_2$O	75% Formic acid: 25% Acetonitrile (pH 5.0)	Vinci et al. (1995)
	Titrimetric method	CH$_3$COOH		AOAC (2000)
Vitamin E	HPLC: C-18	KOH	CH$_3$OH	Charoensiri et al. (2009)
Organic acids	HPLC: C-18	H$_2$O	0.2% H$_3$PO$_4$	Somboonkaew and Terry (2010)
	Capillary electrophoresis	–	CTAB (pH 2.8)	Tezcan et al. (2009)
	HPLC: C-18	H$_3$PO$_4$	H$_2$SO$_4$ (A), CH$_3$OH (B)	Ozgen et al. (2008)
	Capillary electrophoresis	–	C$_4$H$_8$N$_2$O$_3$ (pH 12.85)	Tezcan et al. (2009)
Sugars	HPLC: C-18	CH$_3$COOH: H$_2$O	H$_2$O	Somboonkaew and Terry (2010)
	HPLC: Exsil amino	H$_2$O	CH$_3$CN: H$_2$O	Ozgen et al. (2008)
	HPLC: LC-NH2	–	–	Al-Maiman and Ahmad (2002)
Fibre	AOAC method	–		Yapo and Koffi (2008)
Fatty acids	FAMEs	–	–	Al-Maiman and Ahmad (2002)
Phenolics	HPLC: C-18	50% C$_2$H$_6$O: 50% H$_2$O	2% Acetic acid (A), 10% Acetonitrile: 15% Methanol (B)	Prasad et al. (2009)
	HPLC: Nucleosil	CH$_3$OH: HCl	H$_2$SO$_4$ (A), CH$_3$CN: H$_2$O: H$_2$SO$_4$ (B)	Sarni-Manchaco et al. (2000)
	HPLC: RP-18	CH$_3$OH: H$_2$O	H$_2$SO$_4$: CH$_3$OH	Rangkadilok et al. (2005)
	GC-MS: capillary column	–	–	Zadernowski et al. (2009)
	HPLC: RP Amide C16	CH$_3$OH		Mahattanatawee et al. (2006)
Anthocyanins	HPLC: LC-18	CH$_3$CN: C$_2$H$_6$O	H$_2$O: CH$_3$CN: H$_2$SO$_4$	Talcott (2003)
	–	CH$_3$OH		Chen et al. (2008)
	HPLC: C-18	CH$_3$OH: H$_2$O: HCl	H$_2$O: CH$_3$OH: H$_3$PO$_4$	Somboonkaew and Terry (2010)
Xanthones	HPLC: silica gel	C$_2$H$_6$O	H$_3$PO$_4$: CH$_3$COOH (A), CH$_3$CN (B)	Tewtrakul et al. (2009)
	HPLC: RP-18	C$_2$H$_6$O	CH$_3$CN: H$_2$O	Destandau et al. (2009)

Note: FAMES = fatty acid methyl esters.

Fig. 8.1. Litchi (*Litchi chinensis* Sonn.).

Fig. 8.2. Longan (*Dimocarpus longan* Lour. or *Euphoria longana* Lam.).

Fig. 8.3. Rambutan (*Nephelium lappaceum* Linn.).

The skin of litchi and longan is thin, while the skin of rambutan is thicker and is covered with hair-like spinterns.

8.3.1 Health-promoting compounds in *Sapindaceae* fruit

Sapindaceae fruit are used traditionally as a tonic to the heart, brain and liver, and they also allay thirst (Souza *et al.*, 2007). Modern medicine has studied the health-promoting compounds in *Sapindaceae* fruit. Ascorbic acid, or vitamin C (L-ascorbic acid and L-dehydroascorbic acid form), is only a minor constituent in fresh fruit and vegetables but is of major importance in human nutrition for the prevention of several diseases such as scurvy. Litchi, longan and rambutan pulp are good sources of ascorbic acid. Longan flesh had the highest ascorbic content (63.3–88.8 mg/100 g fresh weight (FW)) among these three fruit, while litchi pulp and rambutan flesh had 17.9–27.6 mg/100 g and 36.4–56.0 mg/100 g FW, respectively (Jiang and Fu, 1999; Srilaong *et al.*, 2002; Song *et al.*, 2006; Wall, 2006). Litchi pulp extract (in water or alcohol and analysed by HPLC) was reported to show promising hepatoprotective activity, which was influenced by the antioxidant activity of ascorbic acid (Souza *et al.*, 2007).

Carotenoids are water-insoluble and isoprenoid lipid molecules which normally are found in membrane structures and result in the yellow, orange or red colour of the fruit (Britton and Hornero-Méndez, 1997). The carotenoid backbone is either linear or contains end cyclic end groups. The most abundant end is the β-ionone ring of β-Carotene (β,β-carotene) and its derivatives. β,β-carotene is a major provitamin A due to the fact that it can provide two molecules of retinol, which is a precursor for vitamin A. However, other carotenoids such as ε carotene (β,ε-carotene), lycopene, lutein and zeaxantin also supply vitamin A from fruit but they can contribute only one molecule of vitamin A. A lack of vitamin A causes impaired iron mobilisation, growth retardation and blindness, to depressed immune response and increased susceptibility to infectious disease (Sommer and Davidson, 2002). Vitamin A may help to prevent many types of cancer, cardiovascular disease and diabetes (Maughan, 2005). However, the edible part of *Sapindaceae* fruit is not a good source for carotenoid i.e. β-carotene and lycopene with 0–2 µg/100 g FW concentration (Holden *et al.*, 1999; Setiawan *et al.*, 2001; Charoensiri *et al.*, 2009).

Polyphenols are a large group of compounds which contribute to the organoleptic and nutritional qualities of fruit and vegetables (Macheix *et al.*, 1990) found in varied ranges, for instance, flavonoids, cinnamic acid and coumarins. They have been reported to prevent degenerative diseases, and are antiallergenic, antiatherogenic, anti-inflammatory, antimicrobial, antioxidant, antithrombotic, cardioprotective and vasodilatory agents (Parada and Aguilera, 2007). However, very small concentrations of polyphenols have been reported in *Sapindaceae* fruit pulp. Only 3.19–3.51 and 1.57–1.77 mg/100 g FW of total phenolic compounds were detected by HPLC in litchi and rambutan, respectively (Gorinstein *et al.*, 1999; Somboonkaew and Terry, 2010). Several studies have reported high concentrations of total polyphenols from the seed and pericarp of Sapindaceae fruit. The extracts from seed and pericarp, therefore, have been applied to cosmetic, nutraceutical and pharmaceutical applications. For instance, ethanolic rambutan pericarp extract consisted of high phenolic concentration (> 700 mg GAE/g extract), low pro-oxidant capacity and effective antioxidant activities (strong scavenging activities against 2,2-diphenyl-1-picrylhydrazyl (DPPH), with 0.26 1/IC50 (µg/ml), galvinoxyl and 2,2'-azino-*bis*(3-ethylbenzothiazoline-6-sulfonic acid) (ABTS) with 1.7 IC50 (µg/ml) and superoxide anions and inhibition of induced lipid autooxidation) (Palanisamy *et al.*, 2008). These results showed higher antioxidant activities than vitamin C and commercial grape seed supplements. Ethanolic extracts of rambutan peel inhibited extremely high value of IC_{50} to scavenged DPPH free radical (>100 µg/ml) and showed non-toxic property to the Caco-2 cells (derived from human colon adenoma and used for drug absorption screening) and peripheral blood mononuclear cells (PBMC), they could be a potential source of natural antioxidants for food or drugs (Oko-nogi *et al.*, 2007). However, Thitilertdecha *et al.* (2008) documented a higher total phenolic content in a methoanolic fraction of rambutan peel (542.2 mg catechin/g dry extract) than in ether and aqueous extractions, with corresponding antioxidation activity (50% DPPH inhibition concentration (4.94 µg/ml)) and antibacterial activity against five pathogenic bacteria, namely *Pseudomonas aeruginosa*, *Vibrio cholera*, *Enterococcus faecalis*, *Staphylococcus aureus* and *S. epidermidis*, measured using a disc diffusion assay.

Furthermore, the three main polyphenols in pulp, pericarp and seed of 70% (v/v) methanolic longan extracts are gallic acid, ellagic acid and corilagin, with the highest concentrations found in the seed: gallic acid 0.8–2.3 mg/g dry weight (DW), ellagic acid 1.4–4.5 mg/g DW and corilagin 3.7–8.6 mg/g DW (Rangkadilok *et al.*, 2005). The high content of ellagic acid in longan seed was found to inhibit *Plasmodium falciparum* (*in vitro*) (Banzouzi *et al.*, 2002), while corilagin from seed extracts reduced blood pressure of spontaneously hypertensive rats by blocking noradrenaline and/or direct vasorelaxation (Cheng *et al.*, 1995). The antioxidant activities and polyphenolic contents in dry litchi pericarp have also been reported (Hu *et al.*, 2010); the levels of phenolics, total flavonoids and proanthocyanidins among eight cultivars (cvs. Nuomici, Feizixiao, Edanli, Yuhebao, Wuye, Guiwei, Dingxiang and Lanzhu) varied between 10–24, 15–38 and 16–44 mg/g pericarp DW, respectively, while the antioxidant activities (free-radical scavenging activities) were 18–72% in a DPPH decoloration test and 22–57% using a ferric-reducing antioxidant assay.

Dietary fibre (DF), for example cellulose, pectin substance, lignin and gum, is partially digestible in the human colon. DF intake appears to be negatively correlated with the incidence of colon cancer, ischaemic heart disease and diabetes mellitus (Burkitt *et al.*, 1972). Jenkins *et al.* (1978) described how water-soluble DF decelerated sugar digestion and absorption, resulting in insulin and hormone reduction (Monnier *et al.*, 1982; Mann *et al.*, 2004). The contents of total DF in litchi fruit and rambutan were 2.20 and 1.64 g/100 g aril FW, respectively (Gorinstein *et al.*, 1999)

and only 0.19 g/100 g FW was found in longan aril (Nititham *et al.*, 2004).

Phytate (phytic acid or inositol hexakisphosphate; IP6) was investigated in DF. The amounts of IP6 found in litchi, rambutan and longan pulp were 1.84–4.30, 2.04 and 0.04–0.116 mg/100 g FW, respectively (Nititham *et al.*, 2004; Charoensiri *et al.*, 2009). IP6 provides the antioxidant effect and starch digestibility (Jenkins *et al.*, 1978), prevents colon cancer (*in vitro*) by increasing cell apoptosis and differentiation, affects colon morphology favourably, and protects against a fatty liver (*in vitro*) resulting from elevated hepatic lipogenesis (Onomi *et al.*, 2004).

8.3.2 Effect of preharvest and postharvest treatments on bioactive compounds in *Sapindaceae* fruit

Although the effects of preharvest factors, for example climate, growing area, agricultural practices and biofortification (genetic and breeding), on the quantity and quality of litchi, longan and rambutan fruit have been widely studied, few of the reports have been linked to the potential effect on health-promoting compounds.

The requirement of potassium (K) in litchi fruit development is much higher compared with nitrogen (N) and phosphorus (P). K enhanced photosynthesis activity, CO_2 uptake (Debnath *et al.*, 2006), stomatal conductance and water use efficiency (Pathak *et al.*, 2007), which led to better quantity and quality of litchi fruit. However, the rate of K application varies with fruit cultivar, environment and agricultural practices. Pathak and Mitra (2008) stated that the application of 600 g K_2O/ plant/year, split into two treatments (15 days after fruit set and 30 days before flowering) increased leaf K content and resulted in higher total soluble solids (TSS):acid ratio (64:1) and ascorbic acid concentration (49.67 mg/100 g fresh aril) in litchi fruit. Baomei *et al.* (2008) also reported that the use of NPK plus K fertilizer increased yield (38.8%), fruit size (by 0.4 g) and soluble solid contents (0.9%) in litchi but gave no details on the health-promoting compounds in the fruit.

Potassium chlorate ($KClO_3$) has been found to induce flowering in longan and litchi and to accelerate cytokinin (CK) synthesis (Bangerth, 2008; Hegele *et al.*, 2008) in buds. CKs play an important role in fruit development. Stern *et al.* (2006) described how applying cytokinin N-(2-chloro-4-pyridyl)-N'-phenylurea (CPPU) at 5–10 mg/l to litchi tree extended litchi harvesting time by 2–3 weeks. CPPU treatment also produced more than 50% higher soluble solid contents:titrable acidity (SSC:TA) ratios than seen with the control fruit and gave a red colour (anthocyanins) comparable with that seen in the untreated group. A high ratio of SSC:TA resulted in less browning, lower decay development and less aril discoloration. Sprays of ethrel (0.25 ml/l) to litchi fruit cv. Bombai at aril development stage also increased the ratio of total soluble solids (TSS) to acid, while enhancing the anthocyanin content in pericarp at harvest stage (Dhua *et al.*, 2005).

Hu *et al.* (2010) suggested that mulching the tree basins with fallen litchi leaves, combined with sprinkle irrigating the orchard at 40% pan coefficient during fruit development, with shade nets 30 days before harvesting, minimized pericarp cracking and sunburn disorder in litchi cvs. Bombai, Bedana and Elaichi. Control of fruit disorders reduces the loss of health-promoting compounds. Fruit bagging is another preharvest practice that influences health-promoting compounds in several fruit due to the effect of microclimate changes around the fruit. However, there was no significant difference in β-carotene, anthocyanin and phenolic concentrations between bagged and unbagged longan fruit, but significant differences in all three compounds between maturation stages were reported in the same study (Jaroenkit *et al.*, 2008).

Appropriate postharvest handling can prolong the postharvest life of fruit and preserve the amount and quality of health-promoting compounds. The effects of postharvest treatment on phytochemical compounds in litchi, longan and rambutan fruit have been widely studied over the past decades, but little emphasis has been placed on monitoring the temporal changes in bioactives. Rather, most postharvest work has concentrated on the role of phenolics in the postharvest browning of pericarp tissue and rarely has considered phenolic compounds in the edible aerial tissue.

8.4 Pomegranate

Pomegranate (*Punica granatum* L.; Fig. 8.4) belongs to the family *Punicaceae*, which originates from the Near to Middle East (Roy and Waskar, 1997). However, the fruit is now grown extensively in Iran, Afghanistan, India, Turkey, Egypt, Tunisia, Morocco and Spain (Saxena *et al.*, 1987; Vardin and Fenercioglu, 2003). The fruit can be grown in a wide range of climates (temperate to desert conditions) and growing areas (Patil and Karale, 1985). The edible portion of pomegranate is juicy, sweetly acidic yet slightly astringent, with small angular hard seeds. The pomegranate aril varies between crimson and yellowish-white in colour, depending on the cultivar, whereas the fruit skin ranges from a glossy reddish-yellow to red colour.

8.4.1 Health-promoting compounds in pomegranates

The fruit arils hold high concentrations of sugars, acids, polyphenols, vitamins and minerals (Al-Maiman and Ahmed, 2002), while the skin is an important source of tannin and other phenols (Roy and Waskar, 1997). Vitamin contents in pomegranate vary with cultivar, season and growing area. Vitamin C, or ascorbic acid (10.2–69.0 mg/100 g fresh aril), is especially high in concentration in pomegranate aril tissue (USDA, 2008). High contents of vitamin B, especially pantothenic acid (vitamin B_5), tocopherol (vitamin E) and phylloquinone (vitamin K) have also been reported in pomegranate aril tissue (Tezcan *et al.*, 2009).

The main sugars in pomegranate are fructose and glucose (5.80–7.06 and 5.80–7.62 g/100 g aril FW, respectively), with low content of sucrose (0.02–0.04 g/100 g fresh aril) (Tezcan *et al.*, 2009). Total sugar in fruit aril is 11.60–14.3 g/100 g aril FW. However, organic acids are relatively low in pomegran-

Fig. 8.4. Pomegranate (*Punica granatum* L.).

ate. Total organic acids are *c*.0.36–3.34 g/100 g aril FW. Citric acid is the most important acid in aril, with small concentrations of malic and ascorbic acid (0.2–3.2, 0.09–0.15 and 0.07–0.14 g/100 g aril FW, respectively).

Pomegranate aril contains high concentrations of phenolic compounds, particularly anthocyanins (328–815 mg/l). Delphinidins (3-glucoside and 3,5-diglucoside) are predominant, with lower levels of cyanidins (3-glucoside and 3,5-diglucoside) and pelargonidins (3-glucoside and 3,5-diglucoside). Ellagic acid and tannins are other polyphenols that have been detected in aril tissue. Commercial pomegranate juice, prepared from aril, peel and seed, contains high levels of gallotannin, quercetin, kaempferol, lutrolin glycoside, catechin and epicatechin; however, punicalagin, a member of the tannin family, and ellagic acid are the most abundant phenols in pomegranate juice.

The high content of bioactive compounds in pomegranate results in strong antioxidant properties (total monomeric anthocyanin: 6.1–219.0 mg cyanidin 3-glucoside/l; Trolox equivalent antioxidant capacity (TEAC): 4.38–7.70 mmol Trolox equivalents (TE)/l; and ferric ion reducing antioxidant power (FRAP): 4.6–10.9 mmol TE/l)). Recent studies have shown potential antioxidant effects of pomegranate in reducing heart disease and cancer risk, eliciting antimicrobial, antidiarrhoeal and antiulcer activities, increasing epididymal sperm concentration and reducing abnormal sperm rate. Cuccioloni *et al.* (2009) documented the potential role of bioactive metabolites in pomegranate components in

the regulation of physiopathological processes involving thrombin. Pomegranate extract may also be of therapeutic use for the treatment of inflammatory diseases, by suppressing human mast cells/basophils (Rasheed *et al.*, 2009). Such an extract can reverse proatherogenic effects induced by shear stress perturbation (*in vivo* and *in vitro*) (de Nigris *et al.*, 2007).

Daily consumption of pomegranate juice improved stress-induced myocardial ischaemia in patients who had coronary heart disease (Sumner *et al.*, 2005). Syed *et al.* (2009) summarized data involving the chemoprotective and chemotherapeutic potentials of pomegranate against various cancers, including skin cancers. Ellagitannin from pomegranate juice was reported to inhibit androgen-dependent and androgen-independent prostate cancer cell growth in humans (Hong *et al.*, 2008). Therefore, pomegranate juice has attracted increasing interest recently for its health benefits.

8.5 Mangosteen

Mangosteen (*Garcinia mangostana* Linn.; Fig. 8.5) is a tropical fruit belonging to the family Guttiferae. Although mangosteen is found mainly in South–east Asia, specifically Thailand, Indonesia, the Philippines, Malaysia, Myanmar, Vietnam, Cambodia and Papua New Guinea, the fruit has spread to other warm and humid tropical areas, namely Madagascar, Sri Lanka, India, Honduras, Brazil and Australia (Osman and Milan, 2006).

Fig. 8.5. Mangosteen (*Garcinia mangostana* Linn.).

The edible part of the mangosteen is juicy, with a slightly acidic sweet taste, soft, fragrant and has a cream- to white-coloured flesh, whereas the peel is dark purple.

8.5.1 Health-promoting compounds in mangosteen

Mangosteen fruit have been used for over a century as a traditional medicine in the South-east Asian region, namely in the treatment of chronic ulcer, dysentery, diarrhoea, cystitis, thrush, eczema, gonorrhoea and infected wounds (Osman and Milan, 2006; Pedraza-Chaverri *et al.*, 2008). Current research has been centred on understanding these effects and identifying the causal bioactives.

In general, vitamin concentrations in mangosteen are relatively low compared with other exotic fruit, yet both folate and ascorbic acid levels are comparatively high (USDA, 2008). Dangcham and Siriphanich (2001) reported that citric, succinic and malic acids (approximately 2 mg/g FW) were the main organic acids in the edible part of mangosteen. Total sugar is 0.175 g/g FW aril (Charoensiri *et al.*, 2009). The major sugars in the fruit aril are fructose, glucose and sucrose (Morton, 1987; Osman and Milan, 2006; Charoensiri *et al.*, 2009) and a small content of dextrose and kerrelose (Jayaweera, 1981), while D-galacturonic acid and a small amount of neutral sugars (L-arabinose as the major one and L-rhamnose and D-galactose) were

reported as carbohydrates in mangosteen peel (Chanarat *et al.*, 1997). Total dietary fibre in mangosteen is 0.02 g/g FW aril.

Similarly to most exotic fruit, pericarp phenolic compounds are significantly more abundant than in the aril. Zadernowski *et al.* (2009) documented that the total phenolic content of the peel and rind was 288.3 g/kg DW, whereas it was only 6.4 g/kg in aril. The main phenolic acid in aril is *p*-hydroxybenzoic. The pericarp is a rich source of phenolic compounds such as xanthones, phenolic acids, tannin and anthocyanins (Jung *et al.*, 2006; Fu *et al.*, 2007). However, xanthone is the only phenolic compound in mangosteen that has been widely studied. Fifty xanthones have been found recently in mangosteen (Pedraza-Chaverri *et al.*, 2008). Walker (2007) reported a rapid analysis method for xanthones in mangosteen rind. Dry rind powder was extracted with 80:20 (v/v) acetone: water and six xanthones quantified, namely α-mangostins, β-mangostins, gartanin, 9-hydroxycalabaxanthone, 3-isomangostin and 8-desoxygartanin, using standard HPLC-photodiode array detector. Xanthones have been documented as having antioxidant (Jung *et al.*, 2006), antimalarial (Likhitwitayawuid *et al.*, 1998), antimicrobial (Sundaram *et al.*, 1983; Mahabusarakam *et al.*, 1986; Rukachaisirikul *et al.*, 2003) and anti-inflammatory activities (Chen *et al.*, 2008), antimyocardial toxicity and oxidative stress (Sampath and Vijayaragavan, 2008), ROS scavenging capacity (Pedraza-Chaverri *et al.*, 2008), antileukemic activity (Chiang *et al.*, 2004) and antiacne properties (Chomnawang *et al.*, 2007).

8.6 Kiwifruit

Kiwifruit, or Chinese gooseberry, belongs to the *Actinidaceae* family. There are more than 50 species in the genus *Actinidia* but the commercial fruit is *A. deliciosa* (Ferguson, 1999). Although kiwifruit originated from China, it has been developed commercially in New Zealand, and subsequently elsewhere. The fruit is egg-shaped, with pale-brown skin and covered with downy hairs. The fruit flesh is a translucent green in colour with a massive amount of little black seeds around the fruit core.

8.6.1 Health-promoting compounds in kiwifruit

Kiwifruit is a rich source of ascorbic acid (68.24–96.00 mg/100 g FW), vitamins A, E and K and folate (Kvesitadze *et al.*, 2001; Guldas, 2003). Apart from ascorbic acid, citric, quinic and malic acids are the major organic acids in kiwifruit pulp (Boyes *et al.*, 1996; Marsh *et al.*, 2003). Major sugars are fructose and glucose (4.35 and 4.11 g/100 g FW pulp, respectively), with minor contents of maltose, galactose and sucrose (USDA, 2008). Amino acids (e.g. aspartic and glutamic acids) were shown to be present in the fruit aril, whereas α-linolenic and linoleic acids (fatty acids) have been reported in fruit seed (Donaldson and Quirin, 2008). Fresh fruit contains 2.6–3.0 g total dietary fibre in 100 g FW (USDA, 2008). Phenolic compounds in kiwifruit are low in comparison with other exotic fruit. Total phenols in fresh kiwifruit were 1–7 mg/l juice (Dawes and Keene, 1999). Cinnamic acids were the only phenolic compounds detected in kiwifruit, and caffeic acid was the main cinnamic acid present (Wijngaard *et al.*, 2009). However, chlorogenic acid, protocatechuic acid, hydroxybenzoic acid, epicatechin, catechin, procyanidins and kaempferol have also been reported in kiwifruit (Dawes and Keene, 1999; Mattila *et al.*, 2006). Du *et al.* (2009) stated that the concentrations of phenols and vitamin C in eight *Actinidia* genotypes were highly correlated with total antioxidant capacity (as measured in DPPH, ABTS, ORAC, FRAP, SASR and MCC tests; see chapters 18 and 19 of this

volume). Kiwi seed oil contains more than 65% omega-3 fatty acid in the α-linolenic acid (ALA) form, providing a vegetable alternative to fish oil (Donaldson and Quirin, 2008).

The health benefits of kiwifruit have been widely discussed. The fruit may increase the antioxidant status of plasma and lymphocytes (Collins *et al.*, 2001, 2003). Kiwifruit may also have a beneficial effect in reducing the levels of carcinogens absorbed from the diet, by promoting laxation (Kestell *et al.*, 1999). Rush *et al.* (2006) documented that the fruit might provide a sustainable population intervention that could decrease risk factors associated with cancer. Although the health benefits of kiwifruit have been widely reported, people can be allergic to the fruit due to the protein-dissolving enzyme, actinidin, and calcium oxalate crystals. These two chemicals can cause itching and soreness of the mouth.

8.7 Passion Fruit

Passion fruit originated from Brazil and Ecuador and is classified in the *Passifloraceae* family. *Passiflora edulis* Sims. is the most well-known species, which contains two forms, purple and yellow. Yellow passion fruit has a thick and hard yellow rind, brown seed and aromatic and acidic pulp, while purple fruit is smaller, with purple peel, black seed and a sweeter taste (Bora and Narain, 1997; Talcott *et al.*, 2003).

8.7.1 Health-promoting compounds in passion fruit

High levels of vitamins have been detected in passion fruit, including vitamin C and vitamin A (40–65 mg and 1272 IU/100 g fresh fruit, respectively) (Vinci *et al.*, 1995; Suntornsuk *et al.*, 2002; USDA, 2008). Mercadente *et al.* (1998) documented 13 provitamin A in passion fruit, including β-carotene, α-carotene, β-cryptoxanthin, lycopene and xanthophyll. Total sugar in 100 g fresh passion fruit is about 7.21–14.45 g, which includes glucose, fructose and sucrose (USDA, 2008; Vera *et al.*, 2009). The fruit contains high content of organic acids, resulting in an acidic flavour (c.3.2–4.2 pH; Godoy and Rodriguez-Amaya, 1994; USDA, 2008). Vera *et al.* (2009) reported that the predominant organic acids in passion fruit were citric (5.41 g/100 g FW) and malic (0.33 g/100 g FW). Six different sugar residues, xylose, glucose, galactose, galactosamine, unknown and fucose, were found in extracts of passion fruit peel with a relative ratio of 1:0:0.2:0.06:0.05:trace (Tommonaro *et al.*, 2007). Passion fruit (100 g) contains about 9.14–28.33 g total dietary fibre (DF) (USDA, 2008). Passion fruit seed oil consists of up to 89.43% unsaturated fatty acid which is 72% linoleic acid (Liu *et al.*, 2009). Seed oil also contains passifin, which is a novel distinctive dimeric antifungal protein (Lam and Ng, 2009). HPLC analysis indicated that passion fruit contained a number of phenolic compounds including syringic acid, tryptophan, flavonoids glycoside, quercetin, luteolin, cyanidins, catechin and epicatechin (Talcott *et al.*, 2003; Foo *et al.*, 2010). Total phenols in passion fruit are about 43.5–61 mg GAE/100 g FW, with antioxidant capacity FRAP: 17.5–50.0 μmol Trolox/100 g FW and DPPH: 94.0% (Talcott *et al.*, 2003; Vasco *et al.*, 2008) and TEAC: 0.32 μm/mg FW (Tommonaro *et al.*, 2007).

The health benefits of passion fruit have been documented widely. Barbosa *et al.* (2008) reported that an extract of passion fruit (cvs. Alata and Edulis) induced anxiolytic-like effects in rats without disruption to the memory process. Administration of oral *P. incarnata* as a premedication (in outpatient surgery) reduced anxiety without inducing sedation (Movafegh *et al.*, 2008). Brown *et al.* (2007) also described the anxiolytic properties of chrysin (passion fruit extract) in rats. Rebello *et al.* (2008) documented that passion fruit extract affected the biodistribution of sodium ^{99m}Tc in male rats but did not influence the shape of red blood cells. The fruit peel is a rich source of health-promoting compounds. Rebello *et al.* (2007) illustrated that the pectin in fruit peel extract decreased cholesterol levels and increased glucose tolerance in rats and humans. Extracts from purple fruit peel decreased systolic blood pressure and serum nitric oxide levels significantly in rats and systolic and diastolic blood pressure in humans (Zibadi *et al.*, 2007).

8.8 Persimmon

Persimmon (Fig. 8.6) is a subtropical fruit and is grown commercially in China, Japan, Korea and the USA. *Diospyros kaki*, oriental or Japanese persimmon, is the most well-known worldwide, while *D. virginiana* is widely grown in the USA. The fruit is pumpkin-shaped, a bright to dark orange colour and varies in size from 1.5–9 cm.

The major vitamins in persimmon fruit include vitamins A and C. Total vitamin A

Fig. 8.6. Persimmon (*Diospyros kaki*).

(1627 IU; USDA, 2008) includes β-carotene, β-cryptoxanthin, zeaxanthin, lutein and lycopene. Vitamin C contents in persimmon fruit ranged from 35 to 66 mg/100 g FW according to cultivar and fruit maturity (Homnava *et al.*, 1990; USDA, 2008). Organic acids in the fruit include succinic, malic, citric and quinic, with total acids of 0.88–1.36 g/100 g DW (Senter *et al.*, 1991). Persimmon contains about 12.53 g sugar/100 g FW, with a majority of fructose, glucose and sucrose and a minor concentration of arabinose, galactose, sorbitol and inositol (Senter *et al.*, 1991). Total DF in persimmon ranges between 1.5 and 3.6 g/100 g fresh fruit (Gorinstein *et al.*, 2001; USDA, 2008).

Persimmon fruit can be divided into two varieties, astringent and non-astringent. The non-astringent type contains fewer tannin cells and less water-soluble tannins, resulting in a sweeter flavour. However, both varieties are rich in phenolic compounds, namely phenolic acid (e.g. *p*-coumaric, ferulic and protocatechuic acids) and proanthocyanidins (Haslam *et al.*, 1988; Gorinstein *et al.*, 1994; Jung *et al.*, 2005). Chen *et al.* (2009) reported that persimmon tannins consisted mainly of (–)-catechin, (–)-epicatechin, (–)-epigallocatechin, chlorogenic acid and caffeic acid, with 32.31 mg total phenols in 100 g fresh fruit. Kato (1984) found a high correlation between astringency level and tannin concentration, with fruit containing about 0.25% tannin being slightly astringent, while those having less than 0.1% were non-astringent. Astringent varieties contain phenolic compounds at concentrations 4–6 times higher than those in non-astringent fruit (Suzuki *et al.*, 2005). Concentrations of volatile compounds such as acetaldehyde and ethanol, produced by seeds, also affect the degree of astringency in persimmon fruit (Taira *et al.*, 1986). Sugiura *et al.* (1979) reported that non-astringent fruit accumulate less volatile compounds.

Persimmon pulp phenols, in methanolic extracts, contain high molecular weight tannins (epigallocatechin, epigallocatechin-3-*O*-gallate, epicatechin-3-*O*-gallate and unknown polymers) with powerful antioxidant activities (Gu *et al.*, 2008). Phenols in fruit peel reduced glucose-induced cytotoxity, intracellular reactive oxygen species, nitric oxide,

superoxide and peroxunitrile concentrations and the overexpression of inducible nitric oxide synthase (iNOS) and cyclooxygenase-2 (COX-2), indicating the benefit of persimmon peel phenols to antioxidant activity in the diabetic condition (Yokozawa *et al.*, 2007). Several recent studies have documented health-promoting compounds in persimmon leaves, root and bark extracts, including antimicrobial activity in the leaves (Sakanaka *et al.*, 2005; Lee *et al.*, 2007), antiprotozoal activity in the root (Ganapaty *et al.*, 2006), antioxidant and lipoxygenase inhibitory activity in the root and bark (Maiga *et al.*, 2006) and antiplasmodial properties (Kantamreddi and Wright, 2008). Furthermore, persimmon has been associated with an inhibitory effect on human lymphoid leukaemia cells (Achiwa *et al.*, 1996, 1997), mutagenicity of *c*-nitro and *c*-nitroso compounds, incidence of stroke and extension of the lifespan in stroke-prone spontananeously hypertensive rats (Uchida *et al.*, 1995) and the treatment of streptozotocin-induced diabetic rats and, possibly, morphological changes in the livers, kidneys and hearts of such rodents (Azadbakhta *et al.*, 2010).

8.9 Carambola

Carambola, or starfruit (*Averrhoa carambola*; Fig. 8.7), belongs to the *Oxalidaceae* family and originates from South-east Asia (Watson *et al.*, 1988). The fruit has been grown recently in South–east Asia, India, China, the USA, Central America and Brazil. It is oblong-shaped, with three to six longitudinal ribs, resulting in a star-shaped cross section. The fruit size ranges from 80 to 250 mm in length and 50 to 100 mm in width, depending on the cultivar (Watson *et al.*, 1988). Starfruit pulp is tart in taste when immature and slightly acidic sweet when ripe, with a smooth to fibrous texture. The skin of the fruit is waxy, thin and firm, and is a pale greenish-yellow to orange colour, and thus a similar colour to the fruit pulp.

Vitamins A and C are the major vitamins in carambola fruit. Gross *et al.* (1983) reported that ζ-carotene was the highest provitamin A (25% of total carotenoids in carambola),

Fig. 8.7. Carambola or starfruit (*Averrhoa carambola*).

while trace levels of β-cryptoxanthin and β-carotene were present. USDA (2008) found 25 μg β-carotene, 24 μg α-carotene and 61 IU vitamin A in 100 g fresh carambola. Vitamin C contents ranged between 14.4 and 53.1 mg/100 g of fresh edible portion, depending on cultivar and maturity stage (Morton, 1987; Cooper *et al.*, 1995; Luximon-Ramma *et al.*, 2003). Fresh green fruit (immature fruit) contained 12.51 mg/g organic acid, consisting of 5 mg oxalic, 4.37 tartaric, 1.32 citric, 1.21 malic, 0.39 α-ketoglutaric, 0.22 succinic and a trace of fumaric, while 13 mg of total acids, made up of 9.58 mg oxalic, 0.91 tartaric, 2.20 α-ketoglutaric and 0.31 fumaric acid, were found in yellow (mature) carambola (Morton, 1987; Maharaj and Badrie, 2006). Total sugar in 100 g fresh carambola is about 4 mg (USDA, 2008); the predominant sugars are glucose and fructose (51.0 and 35.9%, respectively), with a small amount of sucrose (13.1%) (Chan and Heu, 1975). The fruit possess a high level of total DF: 2 g/100 g fresh fruit (USDA, 2008) or 46.0–58.2 g/100 g pomace (Chau *et al.*, 2004a). Amino acids and fatty acids are found in carambola, but only in minor quantities (Morton, 1987; USDA, 2008).

Carambola fruit could offer an inexpensive source of antioxidants. Fresh carambola (100 g) contained total phenolics 142.9–209.9 mg gallic acid, flavonoids 10.3–14.8 mg quercetin, and proanthocyanidins 89.6–132.1 mg

cyanidin chloride (Luximon-Ramma *et al.*, 2003). Luximon-Ramma *et al.* (2003) also reported strong correlations between antioxidant activities and total phenol and proanthocyanidin concentrations. The antioxidant activities of 1 g fresh carambola, measured as TEAC and FRAP, were between 11 and 17 μmol Trolox and 9–22 μmol Fe(II), respectively. The antioxidant activities in carambola were mainly attributed to singly linked proanthocyanidins that existed as dimers, trimers, tetramers and pentamers of catechin or epicatechin (Shui and Leong, 2004).

Starfruit was used traditionally for treating restlessness, headache, nausea and coughs (Burkhill, 1935). High DF contents resulted in good swelling properties, water and oil capacities and cation-exchange capacity (Chau *et al.*, 2004a,b). Carolino *et al.* (2005) reported a neurotoxic fraction from carambola that changed animal behaviour, including tonic-clonic seizures that evolved into status epilepticus, accompanied by cortical epileptiform activity. These effects indicate that carambola may be considered as a new tool for neurochemical and neuroethological research. However, there have been reports of hiccups, confusion and death after fruit ingestion in uraemic patients (Neto *et al.*, 1998, 2003). A neurotoxin is present in carambola that can cross the blood–brain barrier, being excreted by the kidneys (Tse *et al.*, 2003).

References

AOAC International (2000) *Official Methods of AOAC International*, 17th edn. AOAC International, Gaithersburg, Maryland, 967.21.

Achiwa, Y., Hibasami, H., Katsuzaki, H. and Komiya, T. (1996) Inhibitory effect of persimmon extract on skin tumor promotion. *Food Science and Technology International* 2, 183–186.

Achiwa, Y., Hibasami, H., Katsuzaki, H., ImaI, K. and Komiya, T. (1997) Inhibitory effects of persimmon (*Diospyros kaki*) extract and related polyphenol compounds on growth of human lymphoid leukemia cells. *Bioscience, Biotechnology and Biochemistry* 61, 1099–1101.

Al-Maiman, S.A. and Ahmad, D. (2002) Changes in physical and chemical properties during pomegranate (*Punica granatum* L.) fruit maturation. *Food Chemistry* 76, 437–441.

Azadbakhta, M., Safapour, S., Ahmadi, A., Ghasemi, M. and Shokrzadeh, M. (2010) Anti-diabetic effects of aqueous fruits extract of *Diospyros lotus* L. on streptozotocin-induced diabetic rats and the possible morphologic changes in the liver, kidney and heart. *Journal of Pharmacognosy and Phytotherapy* 2, 10–16.

Bangerth, K.F. (2008) Nature and significance of correlative hormonal signals in growth and development of annual and perennial plants. *Acta Horticulturae* 774, 379–390.

Banzouzi, J.T., Prado, R., Menan, H., Valentin, A., Roumestan, C., Mallie, M., Pelissier, Y. and Blache, Y. (2002) *In vitro* antiplasmodial activity of extracts of *Alchornea cordifolia* and identification of an active constituent: ellagic acid. *Journal of Ethnopharmacology* 81, 399–401.

Baomei, Y., Lixian, Y., Guoliang, L., Zhaohuan, H., Xiuchong, Z. and Shihua, T. (2008) Investigating on the soil nutrient properties of litchi plantation in Guangdong. *Guangdong Agricultural Science* 5, 51–53.

Barbosa, P.R., Valvassori, S.S., Bordignon, C.L.J., Kappel, V.D., Martins, M.R., Gavioli, E.C., Quevedo, J. and Reginatto, F.H. (2008) The aqueous extracts of *Passiflora alata* and *Passiflora edulis* reduce anxiety-related behaviours without affecting memory process in rats. *Journal of Medicinal Food* 11, 282–288.

Bora, P.S. and Narain, S. (1997) Passion fruit. In: Mitra, S.K. (ed.) *Postharvest Physiology and Storage of Tropical and Sub-tropical Fruits*. CAB International, Wallingford, UK.

Boyes, S., Strübi, P. and Marsh, H. (1996) Sugar and organic acid analysis of *Actinidia arguta* and rootstock-scion combinations of *Actinidia arguta*. *LWT – Food Science and Technology* 30, 390–397.

Britton, G. and Hornero-Méndez, D. (1997) Carotenoids and colour in fruits and vegetables. In: Tomás-Barberán, F.A. and Robins, R.J. (eds) *Photochemistry of Fruits and Vegetables*. Clarendon Press, Oxford, UK, pp. 11–28.

Brown, E., Hurd, N.S., McCall, S. and Ceremuga, T.E. (2007) Evaluation of the anxiolytic effects of chrysin, a *Passiflora incarnata* extract, in the laboratory rat. *Journal of the American Association of Nurse Anesthetists* 75, 333–337.

Burkill, I.H. (1935) *A Dictionary of the Economic Products of the Malay Peninsula*. Published on behalf of the Government of the Straits Settlements and Federated Malay States by the Crown Agents for the Colonies, London, UK.

Burkitt, D.P. and Trowell, H. (1975) *Refined Carbohydrate Foods and Diseases*. Academic Press, New York.

Burkitt, D.P., Walker, A.R.P. and Painter, N.S. (1972) Effect of dietary fiber on stools and transit times and its relation to disease. *Lancet* 2, 1408–1411.

Carolino, R.O.G., Beleboni, R.O., Pizzo, A.B., Vecchio, F.D., Garcia-Cairasco, N., Moyses-Neto, M., Santos, W.F.D. and Coutinho-Netto, J. (2005) Convulsant activity and neurochemical alterations induced by a fraction obtained from fruit Averrhoa carambola (Oxalidaceae: Geraniales). *Neurochemistry International* 46, 523–531.

Chan, H. T. and Heu, R.A. (1975) Identification and determination of sugars in starfruit, sweetsop, green sapote, jack fruit and wi apple. *Journal of Food Science* 40, 1329–1330.

Chanarat, P., Chanarat, N., Fujihara, M. and Nagumo, T. (1997) Immunopharmacological activity of polysaccharide from the pericarp of mangosteen garcinia: phagocytic intracellular killing activities. *Journal of the Medical Association of Thailand* 1, S149–154.

Charoensiri, R., Kongkachuichai, R., Suknicom, S. and Sungpuag, P. (2009) Beta-carotene, lycopene, and alpha-tocopherol contents of selected Thai fruits. *Food Chemistry* 113, 202–207.

Chau, C.F., Chen, C.H. and Lee, M.H. (2004a) Characterization and physicochemical properties of some potential fibres derived from *Averrhoa carambola*. *Nahrung* 48, 43–46.

Chau, C.F., Chen, C.H. and Lin, C.Y. (2004b) Insoluble fiber-rich fractions derived from *Averrhoa carambola*: hypoglycemic effects determined by *in vitro* methods. *LWT – Food Science and Technology* 37, 331–335.

Chen, G., Xue, J. and Feng, X. (2009) Inhibitory effect of water extract and its main contents of persimmon leaves on stimulus-induced superoxide generation in human neutrophils. *Journal of Food Biochemistry* 33, 113–121.

Chen, L.G., Yang, L.L. and Wang, C.C. (2008) Anti-inflammatory activity of mangostins from *Garcinia mangostana*. *Food and Chemical Toxicology* 46, 688–693.

Cheng, J.T., Lin, T.C. and Hsu, F.L. (1995) Antihypertensive effect of corilagin in the rat. *Canadian Journal of Physiology and Pharmacology* 73, 1425–1429.

Chiang, L.C., Cheng, H.W., Liu, M.C., Chiang, W. and Chun-Ching, L.C.C. (2004) *In vitro* evaluation of antileukemic activity of 17 commonly used fruits and vegetables in Taiwan. *LWT – Food Science and Technology* 37, 539–544.

Chomnawang, M.T., Surassmo, S., Nukoolkarn, V.S. and Gritsanapan, W. (2007) Effect of *Garcinia mangostana* on inflammation caused by *Propionibacterium acnes*. *Fitoterapia* 78, 401–408.

Collins, A.R., Dusinska, M., Horvathova, E., Munro, E., Savio, M. and Stetina, R. (2001) Inter-individual differences in DNA base excision repair activity measured *in vitro* with the comet assay. *Mutagenesis* 16, 297–301.

Collins, A.R., Harrington, V., Drew, J. and Melvin, R. (2003) Nutritional modulation of DNA repair in a human intervention study. *Carcinogenesis* 24, 511–515.

Cooper, A., Poirier, S., Murphy, M. and Oswald, M.J. (1995) *South Florida Tropicals: Carambola*. Florida Cooperative Extension Service, University of Florida, Florida.

Cuccioloni, M., Mozzicafreddo, M., Sparapani, L., Spina, M., Eleuteri, A.M., Fioretti, E. and Angeletti, M. (2009) Pomegranate fruit components modulate human thrombin. *Fitoterapia* 80, 301–305.

Dangcham, S. and Siriphanich, J. (2001) Mechanism of flesh translucent disorder development of mangosteen fruits (*Garcinia mangostana* L.). In: *Proceeding of 39th Kasetsart University Conference*, 5–7 February 2001, Bangkok, pp. 483–489.

Dawes, H.M. and Keene, J.B. (1999) Phenolic composition of kiwifruit juice. *Journal of Agricultural and Food Chemistry* 47, 2398–2403.

Debnath, S., Duttaray, S.K. and Mitra, S.K. (2006) Relationship of leaf position and CO_2 assimilation with fruit growth in litchi. *Indian Journal of Plant Physiology* 11, 195–201.

de Nigris, F., Balestrieri, M.L., Williams-Ignarro, S., D'Armiento, F.P., Fiorito, C., Ignarro, L.J. and Napoli, C. (2007) The influence of pomegranate fruit extract in comparison to regular pomegranate juice and seed oil on nitric oxide and arterial function in obese Zucker rats. *Nitric Oxide* 17, 50–54.

Destandau, E., Toribio, A., Lafosse, M., Pecher, V., Lamy, C. and André, P. (2009) Centrifugal partition chromatography directly interfaced with mass spectrometry for the fast screening and fractionation of major xanthones in *Garcina mangostana*. *Journal of Chromatography A* 1216, 1390–1394.

Dhua, R.S., Roychoudhury, R., Ray, S.K.D. and Kabir, J. (2005) Staggering the lychee fruit harvest. *Acta Horticulturae* 665, 347–354.

Donaldson, B.W. and Quirin, K.W. (2008) *Kiwifruit Oil Extraction Method and Product*. International patent classification C11B1/10; A01H5/10.

Du, G., Li, M., Ma, F. and Liang, D. (2009) Antioxidant capacity and the relationship with polyphenol and vitamin C in *Actinidia* fruits. *Food Chemistry* 113, 557–562.

Ferguson, A.R. (1999) New temperate fruits: *Actinidia chinensis* and *Actinidia deliciosa*. In: Janick, J. (ed.) *Proceedings of the Fourth National Symposium New Crops and New Uses: Biodiversity and Agricultural Sustainability*. ASHS Press, Alexandria, Virginia.

Foo, L.Y., Lu, Y. and Watson, R.R. (2010) *Method of Treating Inflammation Disorders Using Extracts of Passion Fruit*. United States patent application 12/178390.

Fu, C., Loo, A.E., Chia, F.P. and Huang, D. (2007) Oligomeric proanthocyanidins from mangosteen pericarps. *Journal of Agricultural and Food Chemistry* 55, 7689–7694.

Ganapaty, S., Thomas, P., Karagianis, G., Waterman, P.G. and Brun, R. (2006) Antiprotozoal and cytotoxic naphthalene derivatives from *Diospyros assimilis*. *Phytochemistry* 67, 1950–1956.

Godoy, H.T. and Rodriguez-Amaya, D.B. (1994) Occurrence of *cis*-isomers of provitamin A in Brazilian fruits. *Journal of Agricultural and Food Chemistry* 42, 1306–1313.

Gorinstein, S., Zemser, M., Weitz, M., Halevy, S., Deutsch, J., Tilis, K., Feintuch, D., Guerra, N., Fishman, M. and Bartnikowska, E. (1994) Fluorometric analysis of phenolics in persimmons. *Bioscience, Biotechnology, and Biochemistry* 58, 1087–1092.

Gorinstein, S., Zemser, M., Haruenkit, R., Chuthakorn, R., Grauer, F., Belloso, O. and Trakhtenberg, S. (1999) Comparative content of total polyphenols and dietary fiber in tropical fruits and persimmon. *Journal of Nutritional Biochemistry* 10, 367–371.

Gorinstein, S., Zachwieja, Z., Folta, M., Barton, H., Piotrowicz, J., Zemser, M., Weisz, M., Trakhtenberg, S. and Martın-Belloso, O. (2001) Comparative contents of dietary fiber, total phenolics, and minerals in persimmons and apples. *Journal of Agricultural and Food Chemistry* 49, 952–957.

Gross, J., Ikan, R. and Eckhardt, G. (1983) Carotenoids of the fruit of *Averrhoa carambola*. *Phytochemistry* 22, 1479–1481.

Gu, H.F., Li, C.M., Xu, Y.J., Hu, W.F., Chen, M.H. and Wan, Q.H. (2008) Structural features and antioxidant activity of tannin from persimmon pulp. *Food Research International* 41, 208–217.

Guldas, M. (2003) Peeling and the physical and chemical properties of kiwi fruit. *Journal of Food Processing and Preservation* 27, 271–284.

Haslam, E., Lilleyb, T.H. and Butler, L.G. (1988) Natural astringency in foodstuffs – a molecular interpretation. *Critical Reviews in Food Science and Nutrition* 27, 1–40.

Hegele, M., Manochai, P., Naphrom, D., Sruamsiri, P. and Wunsche, J. (2008) Flowering in longan (*Dimocarpus longan* L.) induced by hormonal changes following $KClO_3$ applications. *European Journal of Horticultural Science* 73, 49–54.

Holden, J.M., Eldridge, A.L., Beecher, G.R., Marilyn, B.I., Bhagwat, S., Davis, C.S., Douglass, L.W., Gebhardt, S., Haytowitz, D. and Schakel, S. (1999) Carotenoid content of US foods: an update of database. *Journal of Food Composition Analysis* 12, 169–196.

Homnava. A., Payne. J., Koehler. P. and Eitenmiller. R. (1990) Provitamin A (alpha-carotene, beta-carotene and beta-cryptoxanthin) and ascorbic acid content of Japanese and American persimmons. *Journal of Food Quality* 13, 85–95.

Hong, M.Y., Seeram, N.P. and Heber, D. (2008) Pomegranate polyphenols downregulate expression of androgen synthesizing genes in human prostate cancer cells overexpressing the androgen receptor. *Journal of Nutritional Biochemistry* 19, 848–855.

Hu, Z.Q., Huang, X.M., Chen, H.B. and Wang, H.C. (2010) Antioxidant capacity and phenolic compounds in litchi (*Litchi chinensis* Sonn.) pericarp. *Acta Horticulturae* 863, 567–574.

Jaroenkit, T., Ussahatanonta, S. and Phimphimol, J. (2008) Postharvest methods to reduce sulfer dioxide residues in fresh longan. *Acta Horticulturea* 804, 183–189.

Jayaweera, D.M.A. (1981) *Medicinal Plants used in Ceylon Part 3*. National Science Council of Sri Lanka, Colombo, Sri Lanka.

Jenkins, D.J., Wolever, T.M. and Leeds, A.R. (1978) Dietary fibres, fibre analogues, and glucose tolerance: importance of viscosity. *British Medical Journal* 1, 1392–1394.

Jiang, Y.M. and Fu, J.R. (1999) Postharvest browning of litchi fruit by water loss and its prevention by controlled atmosphere storage at high relative humidity. *LWT – Food Science and Technology* 32, 278–283.

Jung, H.A., Su, B.N., Keller, W.J., Mehta, R.G. and Kinghorn, A.D. (2006). Antioxidant xanthones from the pericarp of *Garcinia mangostana* (Mangosteen). *Journal of Agricultural and Food Chemistry* 54, 2077–2082.

Jung, S.T., Park, Y.S., Zachwieja, Z., Folta, M., Barton, H., Piotrowicz, J., Katrich, E., Trakhtenberg, S. and Gorinstein, S. (2005) Some essential phytochemicals and the antioxidant potential in fresh and dried persimmon. *International Journal of Food Sciences and Nutrition* 56, 105–113.

Kantamreddi, V.S.S. and Wright, C.W. (2008) Investigation of Indian *Diospyros* species for antiplasmodial properties. *Evidence-based Complementary and Alternative Medicine* 5, 187–190.

Kato, K. (1984) The condition of tannin and sugar extraction, the relation of tannin concentration to astringency and the behaviour of ethanol during the de-astringency by ethanol in persimmon fruit. *Journal of the Japanese Society for Horticultural Science* 53, 127–134.

Kestell, P., Zhao, L., Zhu, S., Harris, P.J. and Ferguson, L.R. (1999) Studies on the mechanism of cancer protection by wheat bran: effects on the absorption, metabolism and excretion of the food carcinogen 2-amino-3-methylimidazo[4,5-f] quinoline (IQ). *Carcinogenesis* 20, 2253–2260.

Kvesitadze, G.I., Kalandiya, A.G., Papunidze, S.G. and Vanidze, M.R. (2001) Identification and quantification of ascorbic acid in kiwi fruit by high-performance liquid chromatography. *Applied Biochemistry and Microbiology* 37, 215–218.

Lam, S.K. and Ng, T.B. (2009) Passiflin, a novel dimeric antifungal protein from seeds of the passion fruit. *Phytomedicine* 16, 172–180.

Lee, H.H., Seob, S.W., Jang, E.S., Ryu, M.J., Yoo, J.S., Park, Y. and Lee, S.Y. (2007) Cytotoxic and antioxidant effects of the methanol extract of persimmon (*Diospyros kaki*) leaf. Available at: http://210.101.116.28/W_kiss3/08412478_pv.pdf (accessed 6 January 2010).

Likhitwitayawuid, K., Phadungcharoen, T. and Krungkrai, J. (1998) Antimalarial xanthones from *Garcinia cowa*. *Planta Medica* 64, 70–72.

Liu, S., Yang, F., Zhang, C., Ji, H., Hong, P. and Deng, C. (2009) Optimization of process parameters for supercritical carbon dioxide extraction of *Passiflora* seed oil by response surface methodology. *The Journal of Supercritical Fluids* 48, 9–14.

Luximon-Ramma, A., Bahorun, T. and Crozier, A. (2003) Antioxidant actions and phenolic and vitamin C contents of common Mauritian exotic fruits. *Journal of the Science of Food and Agriculture* 83, 496–502.

Macheix, J.J., Fleuriet, A. and Billot, J. (1990) *Fruit Phenolics*. CRC Press, Boca Raton, Florida.

Mahabusarakam, W., Wiriyachitra, P. and Phongpaichit, S. (1986) Antimicrobial activities of chemical constituents from *Garcinia mangostana* Linn. *Journal of the Science Society of Thailand* 12, 239–243.

Maharaj, L.K. and Badrie, N. (2006) Consumer acceptance and physicochemical quality of osmodehydrated carambola (*Averrhoa carambola* L.) slices. *International Journal of Consumer Studies* 30, 16–24.

Mahattanatawee, K., Manthey, J.A., Luzio, G., Talcott, S.T., Goodner, K. and Baldwin, E.A. (2006) Total antioxidant activity and fiber content of select Florida-grown tropical fruits. *Journal of Agricultural and Food Chemistry* 54, 7355–7363.

Maiga, A., Malterud, K.E., Diallo, D. and Paulsen, B.S. (2006) Antioxidant and 15-lioxygenase inhibitory activities of the Malian medicinal plants *Diospyros abyssinica* (Hiern) F. White (Ebenaceae), *Lannea velutina* A. Rich (Anacardiaceae) and *Crossopteryx febrifuga* (Afzel) Benth (Rubiaceae). *Journal of Ethnopharmacology* 104, 132–137.

Mann, J.I., De Leeuw, I., Hermansen, K., Karamanos, B., Karlstrom, B., Katsilambros, N., Riccardi, G., Rivellese, A.A., Rizkalla, S., Slama, G., Toeller, M., Uusitupa, M. and Vessby, B. (2004) Evidence-based nutritional approaches to the treatment and prevention of diabetes mellitus. *Nutrition, Metabolism and Cardiovascular Diseases* 14, 373–394.

Marsh, K., Rossiter, K., Lau, K., Walker, S., Gunson, A. and Macrae, E.A. (2003) The use of fruit pulps to explore flavour in kiwifruit. *Acta Horticulturea* 610, 229–237.

Mattila, P., Hellström, J. and Törrönen, R. (2006) Phenolic acids in berries, fruits, and beverages. *Journal of Agricultural and Food Chemistry* 54, 7193–7199.

Mercadente, A.Z., Britton, G. and Rodriguez, A.D.B. (1998) Carotenoids from yellow passion fruit (*Passiflora*). *Journal of Agricultural and Food Chemistry* 46, 4102–4106.

Monnier, L.H., Colette, C. and Aquirre, L. (1982) Restored synergistic enterohormonal response after addition of dietary fibre to patients which impaired glucose tolerance and reactive hypoglycemia. *Diabetes Metabolism* 8, 217–222.

Morton, J.F. (1987) Carambola. In: Morton, J.F. (ed.) *Fruits of Warm Climates*. Self-published, Miami, Florida.

Movafegh, A., Alizadeh, R., Hajimohamadi, F., Esfehani, F. and Nejatfar, M. (2008) Preoperative oral *Passiflora incarnata* reduces anxiety in ambulatory surgery patients: a double-blind, placebo-controlled study. *Anesthesia and Analgesia* 106, 1728–1732.

Neto, M.M., Robl, F. and Netto, J.C. (1998) Intoxication by star fruit (*Averrhoa carambola*) in six dialysis patients? (Preliminary report). *Nephrology Dialysis Transplantation* 13, 570–572.

Neto, M.M., Costa, J.A.C., Garcia-Cairasco, N., Netto, J.C., Nakagawa, B. and Dantas, M. (2003) Intoxication by star fruit (*Averrhoa carambola*) in 32 uraemic patients: treatment and outcome. *Nephrology Dialysis Transplantation* 18, 120–125.

Nishiyama, I., Yamashita, Y., Yamanaka, M., Shimohashi, A., Fukuda, T. and Oota, T. (2004) Varietal difference in vitamin C content in the fruit of kiwifruit and other *Actinidia* species. *Journal of Agricultural and Food Chemistry* 52, 5472–5475.

Nititham, S., Kommindr, S. and Nichachotsalid, A. (2004) Phylate and fiber content in Thai fruits commonly consumed by diabetic patients. *Journal of the Medical Association of Thailand* 87, 1444–1446.

O'Hare, T.J. (1997) Rambutan. In: Mitra, S.K. (ed.) *Postharvest Physiology and Storage of Tropical and Subtropical Fruits*. CAB International, New York, pp. 309–334.

Okonogi, S., Duangrat, C., Anuchpreeda, S., Tachakittirungrod, S. and Chowwanapoonpohn, S. (2007) Comparison of antioxidant capacities and cytotoxicities of certain fruit peels. *Food Chemistry* 103, 839–846.

Onomi, S., Okazaki, Y. and Katayama, T. (2004) Effect of dietary level of phytic acid on hepatic and serum lipid status in rats fed a high-sucrose diet. *Bioscience, Biotechnology and Biochemistry* 68, 1379–1381.

Osman, M.B. and Milan, A.R. (2006) *Fruits for the Future 9 – Mangosteen (Garcinia mangostana)*. Southampton Centre for Underutilised Crops, Chichester, UK.

Ozgen, M., Durgaç, C., Serçe, S. and Kaya, C. (2008) Chemical and antioxidant properties of pomegranate cultivars grown in the Mediterranean region of Turkey. *Food Chemistry* 111, 703–706.

Palanisamy, U., Cheng, H.M., Masilamani, T., Subramaniam, T., Ling, L.T. and Radhakrishnan, A.K. (2008) Rind of the rambutan, *Nephelium lappaceum*, a potential source of natural antioxidants. *Food Chemistry* 109, 54–63.

Parada, J. and Aguilera, J.M. (2007) Food microstructure affects the bioavailability of several nutrients. *Journal of Food Science* 72, R21–R32.

Pathak, P.K. and Mitra, S.K. (2008) Effect of phosphorus, potassium, sulphur and boron on litchi. *Indian Journal of Horticulture* 65, 137–140.

Pathak, P.K., Majumdar, K. and Mitra, S.K. (2007) Leaf potassium content influences photosynthesis activity, yield, and fruit quality of litchi. *Better-Crops-India* 1, 12–14.

Patil, A.V. and Karale, A.R. (1985) Pomegranate. In: Bose, T.K. (ed.) *Fruits of India Tropical and Subtropical Calcutta*. Naya Prakash, Calcutta, India.

Pedraza-Chaverri, J., Cárdenas-Rodríguez, N., Orozco-Ibarra, M. and Pérez-Rojas, J.M. (2008) Medicinal properties of mangosteen (*Garcinia mangostana*). *Food and Chemical Toxicology* 46, 3227–3239.

Prasad, K.N.,Yang, B.,Yang, S., Chen, Y., Zhao, M., Ashraf, M. and Jiang, Y.M. (2009) Identification of phenolic compounds and appraisal of antioxidant and antityrosinase activities from litchi (*Litchi sinensis* Sonn.) seeds. *Food Chemistry* 116, 1–7.

Rangkadilok, N., Worasuttayangkurn, L., Bennett, R.N. and Satayavivad, J. (2005) Identification and quantification of polyphenolic compounds in longan (*Euphoria longana* Lam.) fruit. *Journal of Agricultural and Food Chemistry* 53, 1387–1392.

Rasheed, Z., Akhtar, N., Anbazhagan, A., Ramamurthy, S., Shukla, M. and Haqqi, T. (2009) Polyphenol-rich pomegranate fruit extract (POMx) suppresses PMACI-induced expression of pro-inflammatory cytokines by inhibiting the activation of MAP kinases and NF-kappaB in human KU812 cells. *Journal of Inflammation* 6. Available at: http://www.journal-inflammation.com/content/pdf/1476-9255-6-1.pdf (accessed 18 September 2009).

Rebello, B.M., Moreno, S.R.F., Ribeiro, C.G., Neves, R.F., Fonseca, A.S., Caldas, L.Q.A., Bernardo-Filho, M. and Medeiros, A.C. (2007) Effect of a peel passion fruit flour (*Passiflora edulis* f. flavicarpa) extract on the labeling of blood constituents with technetium-99m and on the morphology of red blood cells. *Brazilian Archives of Biology and Technology* 50, 153–159.

Rebello, B.M., Moreno, S.R.F., Godinho, C.R., Neves, R.F., Fonseca, A.S., Bernardo-Filho, M. and Medeiros, A.C. (2008) Effects of *Passiflora edulis* flavicarpa on the radiolabeling of blood constituents, morphology of red blood cells and on the biodistribution of sodium pertechnetate in rats. *Applied Radiation and Isotopes* 66, 1788–1792.

Roy, S.K. and Waskar, D.P. (1997) Pomegranate. In: Mitra, S.K. (ed.) *Postharvest Physiology and Storage of Tropical and Subtropical Fruits*. CAB International, Wallingford, UK.

Rukachaisirikul, V., Pailee, P., Hiranrat, A., Tuchinda, P., Yoosook, C., Kasisit, J., Taylor, W.C. and Reutrakul, V. (2003) Anti-HIV-1 protostane triterpenes and digeranylbenzophenone from trunk, bark and stems of *Garcinia speciosa*. *Planta Medica* 69, 1141–1146.

Rush, E., Ferguson, L.R., Cumin, M., Thakur, V., Karunasinghe, N. and Plank, L. (2006) Kiwifruit consumption reduces DNA fragility: a randomized controlled pilot study in volunteers. *Nutrition Research* 26, 197–201.

Sakanaka, S., Tachibana, Y. and Okada, Y. (2005) Preparation and antioxidant properties of extract of Japanese persimmon leaf (kakinoha-cha). *Food Chemistry* 89, 569–575.

Sampath, P.D. and Vijayaragavan, K. (2008) Ameliorative prospective of alpha-mangostin, a xanthone derivative from *Garcinia mangostana* against beta-adrenergic cathecolamine-induced myocardial toxicity and anomalous cardiac TNF-alpha and COX-2 expressions in rats. *Experimental and Toxicologic Pathology* 60, 357–364.

Sarni-Manchado, P.L., Le Roux, E., Le Guerneve, C., Lozano, Y. and Cheynier, V. (2000) Phenolic composition of litchi fruit pericarp. *Journal of Agricultural and Food Chemistry* 48, 5995–6002.

Saxena, A.K., Mevah, J.K. and Berry, S.K. (1987) Pomegranate – post harvest technology, chemistry and processing. *Indian Food Packer* 41, 43–60.

Senter, S.D., Chapman, G.W., Forbus, W.R. and Payne, J.A. (1991) Sugar and nonvolatile acid composition of persimmons during maturation. *Journal of Food Science* 56, 989–991.

Setiawan, B., Sulaeman, A., Giraud, D.W. and Driskell, J.A. (2001) Carotenoid content of selected Indonesian fruits. *Journal of Food Composition Analysis* 14, 169–196.

Shui, G. and Leong, L.P. (2004) Analysis of polyphenolic antioxidants in star fruit using liquid chromatography and mass spectrometry. *Journal of Chromatography A* 1022, 67–75.

Somboonkaew, N. and Terry, L.A. (2010) Physiological and biochemical profiles of imported litchi fruit under modified atmosphere packaging. *Postharvest Biology and Technology* 56, 246–253.

Sommer, A. and Davidson, F.R. (2002) Assessment and control of vitamin A deficiency: the annecy accords. *Journal of Nutrition* 132, 2845S–2850S.

Song, L.L., Jiang, Y.M., Gao, H.Y. and Li, C.T. (2006) Effects of adenosine triphosphate on browning and quality of harvested litchi fruit. *American Journal of Food Technology* 1, 173–178.

Souza, M.G., Singh, R., Reddy, P.P., Hukkeri, V.I. and Byahatti, V.V. (2007) Hepatoprotective activity of fruit pulp extract of *litchi chinensis* Sonner on carbon tetrachloride induced hepatotoxicity in albino rats. *Internet Journal of Alternative Medicine* 4. Available at: http://www.ispub.com/ostia/index.php?xmlFilePath=journals/ijam/vol4n1/litchi.xml (accessed 20 November 2009).

Srilaong, V., Kanlayanarat, S. and Tatsumi, Y. (2002) Changes in commercial quality of 'Rong-rien' rambutan in modified atmosphere packaging. *Food Science and Technology Research* 8, 2845S–2852S.

Stern, R.A., Nerya, O. and Ben Arie, R. (2006) The cytokinin CPPU delays maturity in litchi cv. 'Mauritius' and extends storage-life. *Journal of Horticultural Science and Biotechnology* 81, 158–162.

Sugiura, A., Yonemori, K., Harada, H. and Tomana, T. (1979) Changes in ethanol and acetaldehyde contents in Japanese persimmon fruits and their relation to natural deastringency. *Studies Institute of Horticulture, Kyoto University* 9, 41–47.

Sumner, M.D., Elliott-Eller, M., Weidner, G., Daubenmier, J.J., Chew, M.H., Marlin, R., Raisin, C.J. and Ornish, D. (2005) Effects of pomegranate juice consumption on myocardial perfusion in patients with coronary heart disease. *American Journal of Cardiology* 96, 810–814.

Sundaram, B.M., Gopalakrishnan, C., Subramanian, S., Shankaranarayanan, D. and Kameswaran, L. (1983) Antimicrobial activities of *Garcinia mangostana* L.. *Planta Medica* 48, 59–60.

Suntornsuk, L., Gritsanapun, W., Nilkamhank, S. and Aochom, A. (2002) Quantitation of vitamin C content in herbal juice using direct titration. *Journal of Pharmaceutical and Biomedical Analysis* 28, 849–855.

Suzuki, T., Someya, S., Hu, F. and Tanokura, M. (2005) Comparative study of catechin compositions in five Japanese persimmons (*Diospyros kaki*). *Food Chemistry* 93, 149–152.

Syed, D., Schmitt, C., Hadi, N., Mukhtar, H., Afaq, F. and Malik, A. (2009) Molecular mechanisms of chemoprevention of cancer by pomegranate. In: Seeram, N.P., Schulman, R.N. and Heber, D. (eds) *Pomegranates: Ancient Roots to Modern Medicine – Industrial Profiles*. CRC Press, Boca Raton, Florida.

Taira, S., Itamura, H. and Watanabe, T. (1986) Effect of alcohol concentration for removal of astringency and fruit maturity on the remained astringency and quality of treated fruits in persimmon, cv. Hiratanenashi. *Japanese Society for Horticultural Science*, Spring meet, pp. 444–445.

Talcott, S.T., Percival, S.S., Pittet-Moore, J. and Celoria, C. (2003) Phytochemical composition and antioxidant stability of fortified yellow passion fruit (*Passiflora edulis*). *Journal of Agricultural and Food Chemistry* 51, 935–941.

Tewtrakul, S., Tansakul, P. and Panichayupakaranant, P. (2009) Anti-allergic principles of *Rhinacanthus nasutus* leaves. *Phytomedicine* 16, 929–934.

Tezcan, F., Gültekin-Özgüven, M., Diken, T., Özçelik, B. and Erim, F.B. (2009) Antioxidant activity and total phenolic, organic acid and sugar content in commercial pomegranate juices. *Food Chemistry* 115, 873–877.

Thitilertdecha, N., Teerawutgulrag, A. and Rakariyatham, N. (2008) Antioxidant and antibacterial activities of *Nephelium lappaceum* L. extracts. *LWT – Food Science and Technology* 41, 2029–2035.

Tommonaro, G., Segura Rodríguez, C.S., Santillana, M., Immirzi, B., De Prisco, R., Nicolaus, B. and Poli, A. (2007) Chemical composition and biotechnological properties of a polysaccharide from the peels and antioxidative content from the pulp of *Passiflora liguralis* fruits. *Journal of Agricultural and Food Chemistry* 55, 7427–7433.

Tse, K.C., Yip, P.S., Lam, M.F., Choy, B.Y., Li, F.K., Lui, S.L., Lo, W.K. and Chan, T.M. (2003) Star fruit intoxication in uraemic patients: case series and review of the literature. *Internal Medicine Journal* 33, 314–316.

Uchida, S., Ozaki, M., Akashi, T., Yamashita, K., Niwa, M. and Taniyama, K. (1995) Effects of (−)-epigalloc atechin-3-O-gallate (green tea tannin) on the life span of stroke-prone spontaneously hypertensive rats. *Clinical and Experimental Pharmacology and Physiology (Supplement)* 22, S302–303.

USDA (United States Department of Agriculture) (2008) The USDA national nutrient database for standard reference; Release 22. United States Department of Agriculture.

Vardin, H. and Fenercioglu, H. (2003) Study on the development of pomegranate juice processing technology: clarification of pomegranate juice. *Nahrung* 42, 300–303.

Vasco, C., Ruales, J. and Kamal-Eldin, A. (2008) Total phenolic compounds and antioxidant capacities of major fruits from Ecuador. *Food Chemistry* 111, 816–823.

Vera, E., Sandeaux, J., Persin, F., Pourcelly, G., Dornier, M. and Ruales, J. (2009) Deacidification of passion fruit juice by electrodialysis with bipolar membrane after different pretreatments. *Journal of Food Engineering* 90, 67–73.

Vinci, G., Botrè, F., Mele, G. and Ruggieri, G. (1995) Ascorbic acid in exotic fruits: a liquid chromatographic investigation. *Food Chemistry* 53, 211–214.

Walker, E.B. (2007) HPLC analysis of selected xanthones in mangosteen fruit. *Journal of Separation Science* 30, 1229–1234.

Wall, M.M. (2006) Ascorbic acid and mineral composition of longan (*Dimocarpus longan*), lychee (*Litchi chinensis*), and rambutan (*Nephelium lappaceum*) cultivars grown in Hawaii. *Journal of Food Composition and Analysis* 19, 655–663.

Watson, B.J., George, A.P., Nissen, R.J. and Brown, B.I. (1988) Carambola: a star on the horizon. *Queensland Agricultural Journal* 114, 45–51.

Wijngaard, H.H., Rößle, C. and Brunton, N. (2009) A survey of Irish fruit and vegetable waste and by-products as a source of polyphenolic antioxidants. *Food Chemistry* 116, 202–207.

Yapo, B.M. and Koffi, K.L. (2008) Dietary fiber components in yellow passion fruit rind – a potential fiber source. *Journal of Agricultural and Food Chemistry* 56, 5880–5883.

Yokozawa, T., Kim, Y.A., Kim, H.Y., Lee, Y.A. and Nonaka, G.I. (2007) Protective effect of persimmon peel polyphenol against high glucose-induced oxidative stress in LLC-PK1 cells. *Food and Chemical Toxicology* 45, 1979–1987.

Zadernowski, R., Czaplicki, S. and Naczk, M. (2009) Phenolic acid profiles of mangosteen fruits (*Garcinia mangostana*). *Food Chemistry* 112, 685–689.

Zhang, Y.J., Abe, T., Tanaka, T., Yang, C.R. and Kouno, I. (2001) Phyllanemblinins A–F, new ellagitannins from *Phyllanthus emblica*. *Journal of Natural Products* 64, 1527–1532.

Zibadi, S., Faridc, R., Moriguchi, S., Lue, Y., Fooe, L.Y., Tehrani, P.M., Ulreichf, J.B. and Watson, R.R. (2007) Oral administration of purple passion fruit peel extract attenuates blood pressure in female spontaneously hypertensive rats and humans. *Nutrition Research* 27, 408–416.

9 Grape

Pierre-Louis Teissedre and Christian Chervin

9.1 Introduction

The grape is the fruit of the grapevine (*Vitis* species). This is the second most widely grown fruit in the world. Indeed, according to a report by the Food and Agriculture Organization (FAO) about the world market for fruit (FAO, 2000), grapes represent 14.6% of the global production of fruit, just after oranges; equivalent to nearly 68 million tonnes (Mt) per annum. Grapes are a valuable source of numerous phytonutrients and many studies have suggested cardiovascular benefits, while some work has indicated cancer chemopreventive activity (Pezzuto, 2008).

Grape berries are classified as white (green yellow or golden yellow) or red grapes (pink, purple or black). Grapes are used primarily for making wine from fermented juice (one speaks in this case of wine grapes), but they can also be consumed as fruit, either fresh as table grapes (namely, cvs. Chasselas, Thompson Seedless, Cardinal, Lavalée, Hamburg Muscat, Danlas, Prima, Italia, etc.) or as dried fruit, such as raisins, which are used primarily in baking or cooking. Grapes are also used to produce plain juices. Most research to date has concentrated on establishing the health-promoting properties of wine grapes rather than table grapes, and thus this inevitably will form the basis of this chapter.

More than 5000 grape varieties are listed and today about 250 of these are cultivated commercially. The varieties are distinguished by their different shapes of leaf, berries and colours, and have different fragrances and taste profiles. The study of grapes is called ampelography, which has the etymology of the Greek words, ampelos: 'grapevine' and graphein: 'write'. The most important species of grapes are: *Vitis vinifera* (from Europe, and from which are derived all major varieties for wine and table grapes) and *V. labrusca*, *V. riparia* or *V. rupestris* (from North America, used mainly as table grapes and relatively few for wines). During the attack of European vineyards by phylloxera in the 19th century, the cultivation of European grape varieties (Cabernet-sauvignon, Merlot, Cabernet Franc, Pinot Noir, Syrah (Shiraz), Grenache, Cinsault, Carignan, Sauvignon Blanc, etc.) was allowed to continue thanks to grafting on to North American *Vitis* rootstocks. Grapes are a major source of polyphenols (Waterhouse *et al.*, 1996), which constitute a family of organic molecules characterized by the presence of several phenol groups, leading to molecules of high molecular weight.

9.2 Identity and Role of Bioactives

9.2.1 Polyphenols

Polyphenols are products of secondary metabolism in plants and are becoming increasingly important, particularly because of their reported beneficial effects on health (Stanley et al., 2003). Indeed, their role as natural antioxidants is attracting increasing interest in the prevention and treatment of cancer (Chen et al., 2004) and inflammatory (Laughton et al., 1991), cardiovascular (Frankel et al., 1993) and neurodegenerative (Orgogozo et al., 1997) diseases. They are also used as additives in the food, pharmaceutical and cosmetic industries (Anon., 2006). The term 'polyphenol' was introduced in 1980 (Anon., 1980), replacing the former term 'tannin', and was defined as follows: 'water-soluble phenolic compounds, molecular weight of between 500 and 3000 Da, which, in addition to the properties of usual phenols, have the ability to precipitate alkaloids, gelatine and other proteins'. In addition to this definition, polyphenols possess high antioxidant properties (Teissedre et al., 1996). The natural polyphenols include a wide range of chemicals, each including at least one aromatic nucleus and one or more hydroxyl groups, in addition to other constituents (Bamforth, 1999). Polyphenols can range from simple molecules such as phenolic acids to compounds that are highly polymerized (more than 30,000 Da), as for tannins. The latter are molecules containing at least one cycle benzene and hydroxy groups. Because some hydroxyl groups bind to salivary proteins, some polyphenols, such as, tannins are defined as astringent, giving a sensation of dryness in the mouth. Some fruit have high concentrations

of phenolic compounds, such as plums and persimmons, in which they can reach 1–2 % of fresh weight (Macheix et al., 1990).

Grape berries also contain large amounts of phenolic compounds, including catechins, concentrated mainly in the seeds and skins (Table 9.1). Up to maturity, some quantitative and qualitative changes occur during grape ripening. The harvesting date of grapes (directly correlated with the degree of horticultural maturity) influences their chromatic characteristics. In general, higher intensities of blue or violet tones are detected in black grapes, collected at a later harvest maturity date, in which the ratios of anthocyanins/proanthocyanidins are the lowest. Polyphenolic maturity or skin maturity is achieved on a chemical level when the potential for anthocyanin synthesis starts to diminish after reaching a plateau. With increased fruit maturity there is a change in the skin phenols that corresponds to an increase in phenol maturity or polymerization. Skin tannin polymerization parallels a sensory transformation, an evolution from hard and astringent to dusty to soft and supple. Polymerized skin tannins have a large molecular weight and are smoother on the palate than smaller, low molecular weight tannins, which are considered to be hard and astringent. The pips change from green to a brown or uniform yellow colour; once all traces of green have disappeared, the tannins become less dry, less astringent; the pips become less

Table 9.1. Catechin (mg/kg) content in grapes.

Cultivar	Catechins	Per cent in stalk	Per cent in seeds	Per cent in skins	Per cent in pulp
Alicante-bouschet	551	15	64	10	11
Cabernet Sauvignon	344	10	83	7	T
Carignan	94	27	54	19	T
Cinsault	154	47	37	9	T
Grenache Blanc	144	17	51	32	T
Grenache Noir	173	25	64	11	T
Merlot	601	9	81	11	T
Mourvedre	171	25	58	17	T
Pinot Noir	1165	4	94	2	T
Colobel (hybrid)	862	8	79	7	6
Mean	377	20	65	14	–

Note: T = trace.
Source: Bourzeix et al. (1986).

hard, less brittle, and lose their grassy aromas, which are replaced first by toasted notes and then by roasted, once polyphenolic maturity has been reached (Kennedy et al., 2000). This colour change represents oxidative reactions and corresponds to the degree of extractable tannins. Finally, a limited degree of polymerization, and perhaps depolymerization, occurs in the fruit during maturation. In wines, tannin polymerization continues (although there is some depolymerization) until an anthocyanin molecule binds the terminal end; polymerization is then believed to stop. The ratio of free anthocyanins to tannins is important in impacting polymerization (Teissedre, 2008). The main phenolic molecules (non-flavonoids or flavonoids) are given in Fig. 9.1.

Phenolic compounds are synthesized primarily from carbohydrates via the shikimic acid or acetate pathways; that of shikimic acid leading, via transamination and deamination, to cinnamic acids and their derivatives, and that of acetate leading to polycetoesters or polyacetate (malonate). The structure of phenolic compounds ranges from a simple aromatic nucleus in low molecular weight tannins to complex high molecular weight polyphenols. Such compounds can be ordered by the nature of their carbon skeleton and the length of the aliphatic chain linked to a benzene ring (Cheynier et al., 1997). Phenolic compounds can be divided into two major groups: flavonoids and non-flavonoids.

Phenolic compounds of 14 pomace samples originating from red and white winemaking were identified and quantified by HPLC-MS-DAD (13 anthocyanins, 11 hydroxybenzoic and hydroxycinnamic acids, 13 catechins and flavonols, as well as two stilbenes) in the skins and seeds. Large variability in all of the individual phenolic compounds was observed, and was dependent on cultivar and vintage. Grape skins proved to be a rich source of anthocyanins, hydroxycinnamic acids, flavanols and flavonol glycosides, whereas flavanols were present mainly in the seeds. Both skins and seeds of most grape cultivars constitute a promising source of polyphenols (Kammerer et al., 2004).

Flavonoids

Flavonoids are polyphenolic compounds containing 15 carbon atoms forming a C6-C3-C6 structure (Fig. 9.2): two aromatic cores connected by a bridge of three carbons. These compounds are the most abundant among all phenolic compounds. They have a variety of roles in plants as secondary metabolites, being involved in UV protection, pigmentation and resistance to diseases. The C6-C3-C6 structure is the product of two synthesis pathways, the B ring and the three-carbon bridge constituting a phenylpropanoid unit synthesized from phenylalanine, while the A ring comes from the acetic–malonic acid pathway. The merger of these two parties involves condensation of a phenylpropanoid, 4-coumaryl, with three molecules of malonyl CoA, each giving two carbon atoms. The reaction is catalysed by chalcone synthase, giving tetrahydroxychalcone, which in turn will lead to all flavonoids (Crozier, 2003). There are several groups of flavonoids: flavonols, flavan-3-ols, flavones, isoflavones, flavanones and anthocyanidins (Fig. 9.3). It should be noted that isoflavones are not present in grapes. The basic structure of flavonoids may undergo numerous substitutions; the hydroxyl groups generally are in positions 4, 5 and 7. Most flavonoids exist in the form of glycosides; the nature of sugar varies greatly, depending on the species. The substitutions change the solubility of flavonoids; hydroxylations and glycosylations make compounds generally more hydrophilic, while other substitutions, such as methylation, make them more lipophilic.

Flavonols

Flavonols are the most widespread flavonoids among fruit. Flavonols such as myricetin, quercetin, isorhamnetin and kaempferol are most often present in the form of O-glycosides. The combination is most often in position 3 of the aromatic ring C (Fig. 9.1), although substitutions in positions 5, 7, 4′, 3′ and 5′ are possible. The number of aglycones is quite low, but there are a very large number of conjugated forms; kaempferol alone can be conjugated in 200 different glycosidic forms. There

Phenolic acids

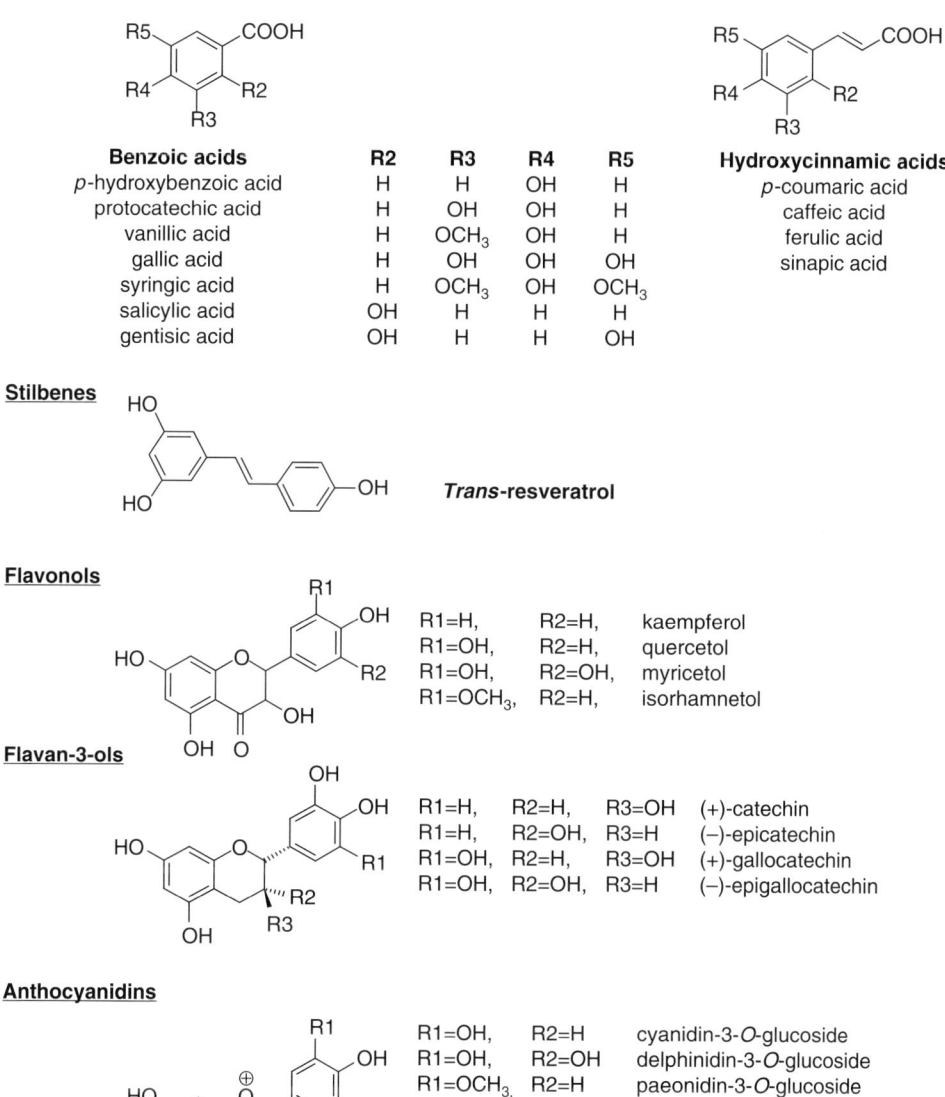

Benzoic acids	R2	R3	R4	R5	Hydroxycinnamic acids
p-hydroxybenzoic acid	H	H	OH	H	p-coumaric acid
protocatechic acid	H	OH	OH	H	caffeic acid
vanillic acid	H	OCH₃	OH	H	ferulic acid
gallic acid	H	OH	OH	OH	sinapic acid
syringic acid	H	OCH₃	OH	OCH₃	
salicylic acid	OH	H	H	H	
gentisic acid	OH	H	H	OH	

Stilbenes

Trans-resveratrol

Flavonols

R1=H,	R2=H,	kaempferol
R1=OH,	R2=H,	quercetol
R1=OH,	R2=OH,	myricetol
R1=OCH₃,	R2=H,	isorhamnetol

Flavan-3-ols

R1=H,	R2=H,	R3=OH	(+)-catechin
R1=H,	R2=OH,	R3=H	(−)-epicatechin
R1=OH,	R2=H,	R3=OH	(+)-gallocatechin
R1=OH,	R2=OH,	R3=H	(−)-epigallocatechin

Anthocyanidins

R1=OH,	R2=H	cyanidin-3-O-glucoside
R1=OH,	R2=OH	delphinidin-3-O-glucoside
R1=OCH₃,	R2=H	paeonidin-3-O-glucoside
R1=OCH₃,	R2=OH	petunidin-3-O-glucoside
R1=OCH₃,	R2=OCH₃	malvidin-3-O-glucoside

Fig. 9.1. Examples of phenolic compounds: grape non-flavonoids and flavonoids.

is a high variability in their concentration, depending on season and cultivars (Crozier, 2003). The flavonol structure is flat. Four flavonols are mostly found in grapes: kaempferol, quercetin (5–10 mg/kg), myricetin and isorhamnetin. The derivatives of quercetin are predominant. The average flavanol content in grapes is about 50 mg/kg, but can vary between 10 and 285 mg/kg. The presence of flavonols was investigated in the berry skins of 91 grape varieties (*V. vinifera*), in order to produce a classification based on the flavonol profile (Mattivi *et al.*, 2006). In red grapes, the main flavonol was quercetin (mean = 43.99%), followed by myricetin (36.81%), kaempferol (6.43%), laricitrin (5.65%), isorhamnetin (3.89%) and syringetin (3.22%). In white grapes, the main flavonol was quercetin (mean = 81.35%), followed by kaempferol (16.91%) and isorhamnetin (1.74%).

Flavanones

Flavanones are the first products of the flavonoid biosynthesis pathway. They are characterized by the absence of the double bond between C2 and C3 and by the presence of a chirality centre in C2. Most flavanones encountered in nature have the B cycle attached to the C cycle in C2 (Fig. 9.1). The structure of flavanones is very reactive and gives rise to reactions of hydroxylation, *O*-methylation and glycosylation. Flavanones are present at concentrations of a few mg/kg in grapes.

FLAVAN-3-OLS. Flavan-3-ols are the most complex category of flavonoids. These compounds range from simple monomers such as (+)-catechin and its isomer (−)-epicatechin, to oligomers and polymers of proanthocyanidins. Proanthocyanidins are formed of catechin and epicatechin, with oxidative coupling

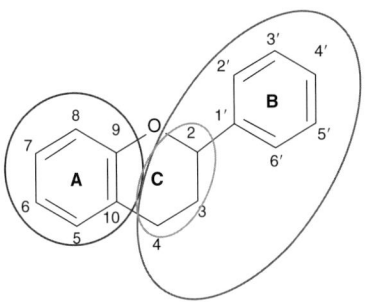

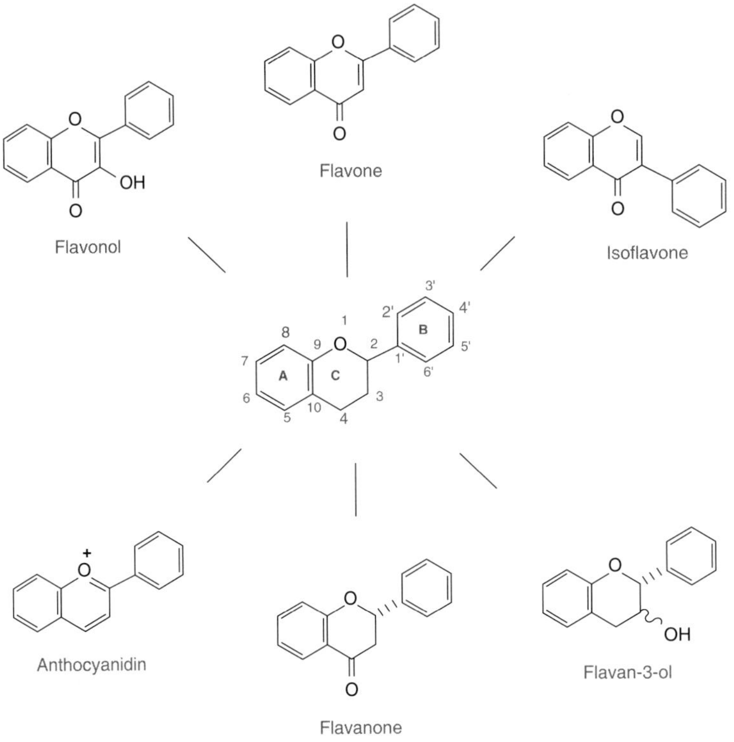

Fig. 9.2. Flavonoid skeleton. In light grey, the three-carbon bridge; in dark grey, the moiety arising from the shikimate pathway and in black the moiety arising from the acetate pathway.

Flavonol

Flavone

Isoflavone

Anthocyanidin

Flavanone

Flavan-3-ol

Fig. 9.3. Structure of the main flavonoids.

between positions C4 and C6 or C8 of the adjacent monomer. The oligomers of procyanidins are formed by two to five units of catechin or epicatechin, the polymers being formed by six or more units. In addition, flavan-3-ols may be esterified with gallic acid or hydroxylated to form gallocatechins (epicatechin gallate, epigallocatechin, epigallocatechin gallate) and gallotannins. The flavan-3-ols present in grapes are mostly in the form of polymers. The seed tannins are made up of procyanidins (polymers of catechin and epicatechin), partially galloylated, while those of the skins also contain prodelphinidins (polymers of gallocatechin and epigallocatechin). The average number of monomeric units, defined as the mean degree of polymerization (mDP), may go up to 18 in a fraction from seeds and to around 30 in a skin extract. It has been found that mDP can change between vintages. For grapes of cv. Cabernet Sauvignon from the Bordeaux area, the mDP in seeds was 4.7 while that in skins was 25.7, in 2006 (Chira *et al.*, 2009), and these levels were found to be double in 2007. On the other hand, the same authors found, in cv. Merlot grapes from 2006, an mDP of 3.6 in seeds and one of 35.4 in skins. However, the mDP in skins was found to be only 24.2 in 2007. Soil and climate can create variability in tannin mDP levels in grape skins and seeds. The GPC (gel permeation chromatography) molecular weight (MW) distribution was used by Weber *et al.* (2007) and indicated components ranging from 1180 to 5000 MW in grape seed. In this work, MALDI-TOF mass spectrometry (MS) analyses showed that grape seed contained oligomers with both odd and even numbers of gallate. Reflectron MALDI-TOF MS identified oligomers up to a pentamer and heptamer, and linear MALDI-TOF MS showed a mass range nearly double that of the reflectron analyses. Recent studies have speculated that, as well as their antioxidant role, flavonoids can act by modulating cell signalling pathways and/or gene expression. In streptozotocin-induced diabetic rats (used as an oxidative stress model), grape seed procyanidin extract (GSPE) was used to study regulation of copper/zinc-superoxide dismutase (Cu/Zn-SOD), an enzyme that defends against oxidative stress (Puiggros *et al.*, 2009). The results indicated that the

Table 9.2. Average content of phenolic compounds in different parts of the red grape berry (mg/kg).

	Pulp	Skins	Seeds
Tannins	Traces	100–500	1000–6000
Anthocyanins	–	500–3000	–
Phenolic acids	20–170	50–200	–

expression profile of Cu/Zn-SOD in diabetic rats was similar to the profile in non-diabetic rats. Nevertheless, the administration of GSPE increased Cu/Zn-SOD activity in both diabetic and non-diabetic rats, and a direct interaction between some small or medium-sized GSPE components and the enzyme was found to be responsible for the increase in Cu/Zn-SOD activity. The levels of catechins and procyanidins (oligomers of catechins) vary depending on the type of grape or wine. Levels of these compounds are between 243 and 1108 mg/kg, of which over 89% generally is located in the seeds (Revilla *et al.*, 1997). Table 9.2 provides a breakdown of the families of polyphenols based on the different parts of the berry of red grapes.

Flavones

Flavones are structurally very close to flavonols; the difference is the absence of hydroxyl in C3. There are also numerous substitutions on the flavone skeleton, such as hydroxylation, methylation, O- and C-alkylation and glycosylation. In plants, flavones are present mainly in the form of glycosides (Bohm *et al.*, 1998). The flavone content in grape is very low.

Anthocyanidins

Anthocyanidins are widely present in the plant kingdom, mainly in the form of glycosides, and are found in black/red grape skins, where they are responsible for the colours red, blue and purple, depending on the pH of the cell compartment. These compounds are involved in protecting plants against excessive sunlight. The most common anthocyanidins are pelargonidin, cyanidin, delphinidin, peonidin and malvidin, but these compounds are present in glycosylated forms only and are referred to as anthocyanins. Anthocyanidins are also capable of conjugating with hydroxy-cinnamic acids and other organic acids (e.g. malic and acetic acids). Unlike other species (North American *Vitis* species) that have significant levels of diglycosylated anthocyanins in positions C3' and C5', *V. vinifera* contains only traces of these and is characterized by the presence of a majority of anthocyanin monoglucosides, particularly the malvidin 3-*O*-glucoside and its acylated derivatives. Anthocyanins are present in red grapes at about 500–3000 mg/kg, but can reach higher values in cultivars called 'dyers', particularly cv. Alicante Bouschet (5000 mg/kg), in which the anthocyanin concentration in pulp is also high. The delphinidin-like flavonols, myricetin, laricitrin and syringetin, are missing from all white varieties, indicating that the enzyme flavonoid 3',5'-hydroxylase is not expressed in white grape varieties (Mattivi *et al.*, 2006).

Non-flavonoids

The main non-flavonoids important in fruit are phenolic acids, hydroxycinnamic acids and stilbenes. In grapes, although not coloured themselves, the non-flavonoid constituents are known to enhance and stabilize the colour of red wines by intra- and intermolecular reactions. Furthermore, they contribute to wine flavour (volatile phenolic acids), and some of them exhibit potent biological activities.

Simple phenols

Simple phenols are derivatives of the C6 benzene ring, are rare in their natural state and are obtained by decarboxylation of shikimic acid. They include hydroquinol, pyrocatechol and phloroglucinol.

Phenolic acids

An enzyme-assisted release of phenolic antioxidants from grape pomace from wine production was examined (Meyer *et al.*, 1998). The enzymes used were grindamyl pectinase and celluclast. Total phenols released ranged from 820 to 6055 mg/l gallic acid equivalents (GAE) and varied in response to enzyme type, time of enzyme treatment, particle size of the pomace and type of extraction solvent employed. The yield of total phenols was correlated to the degree of plant cell wall breakdown within the pomace. Grindamyl pectinase catalysed degradation of grape pomace polysaccharides, whereas celluclast did not. Reduction of the particle size of grape pomace to 125–250 µm increased enzymatic polysaccharide hydrolysis and the recovery of phenols. The grape pomace extracts retarded human low-density lipoprotein oxidation significantly *in vitro*. When evaluated at 3.0 µM GAE, phenolic extracts of grindamyl pectinase-treated pomace of small particle size (125–250 µm) appeared to release more active antioxidant phenols than seen with the other types of enzyme treatments tested.

Hydroxybenzoic acids have a C6-C1 structure, composed of a benzene ring on which a one-carbon aliphatic chain is bound. They include vanillic acid, syringic acid, gentisic acid and gallic acid. The main compound in grapes is gallic acid, the content of which is between 100 and 230 mg/kg. The differences in levels of phenolic acids in seeds and skins from grapes of *V. vinifera* cvs. Merlot and Chardonnay and in seeds from grapes of *V. rotundifolia* cv. Muscadine were determined, and the antioxidant activities of these components assessed (Yilmaz and Toledo, 2004). The contribution of phenolic acid to the total antioxidant capacity of grape seeds and skins was also determined. Gallic acid concentrations were 99 mg/100 g of dry matter (DM) in cv. Muscadine seeds, 15 mg/100 g of DM in cv. Chardonnay seeds and 10 mg/100 g of DM in cv. Merlot seeds, respectively. This phenolic constituent of grape contributed to the antioxidant capacity, measured as oxygen radical absorbance capacity (ORAC), on the basis of the corrected concentrations of gallic acid.

Cinnamic acid is a C6-C3 compound produced by a deamination of phenylalanine catalysed by phenylalanine ammonia-lyase; para-coumaric acid (*p*-coumaric) is then produced by hydroxylation of cinnamic acid. Cinnamic acid and hydroxycinnamic acids are also called phenylpropanoids. Their basic skeleton is a benzene ring with a three-carbon aliphatic chain, with one or more hydroxyl groups often esterified as esters of aliphatic alcohols. Common hydroxycinnamic acids are caffeic, *p*-coumaric, ferulic and sinapic acids. They are produced by a series of hydroxylations and methylations and they often accumulate in the form of tartaric acid esters: coutaric, caftaric and fertaric acids, which are esters of *p*-coumaric, ferulic and sinapic acids, respectively. These constituents are present mainly in the flesh of grape berries. The main hydroxycinnamic acid in grapes is caftaric acid (cafeoyl-tartrate ester), which may reach about 200 mg/kg.

Stilbenes

Stilbenes are polyphenolic compounds that have a C6-C2-C6 structure, with two benzenes linked by a methylene bridge. They are produced by plants in response to fungal, bacterial and viral attacks; this has been particularly demonstrated for *trans*-resveratrol. Resveratrol is synthesized by condensation of 4-coumaryl CoA with three molecules of malonyl CoA, each giving two carbon atoms. The reaction is catalysed by stilbene synthase; the products involved are the same as for the synthesis of flavonoids, the only difference being the enzyme catalysing the reaction. Resveratrol is present in the *cis* and *trans* forms, and is present in plant tissue mainly in the form of *trans*-resveratrol-3-*O*-glucosides (*trans*-piceid and *trans*-astringine). There are oligomers of stilbenes, identified in grapes, such as pallidol and viniferins (Ribeiro *et al.*, 1999; Landrault *et al.*, 2002; Vitrac *et al.*, 2005) and, more recently, a tetramer of resveratrol: hopeaphenol (Guebailia *et al.*, 2006).

In grapes, stilbene synthesis occurs in the skin and is induced by biotic and abiotic stresses. Stilbene biosynthesis has been investigated in healthy grapes, at both biochemical and molecular levels, by measuring the concentration of resveratrols (*trans*-resveratriol, *trans*-piceid and *cis*-piceid) in the ripe berries of 78 *V. vinifera* varieties for 3 years (Gatto *et al.*, 2008). Significant differences appeared among genotypes, providing the first tentative varietal classification based on resveratrol content. Furthermore, an increasing stilbene accumulation from veraison to ripening phase was also observed. The highest resveratrol-producing varieties found are cvs. Pinot Noir (22 mg/l), Pinot Tete de Negre (18 mg/l), Tarrango (16 mg/l), Franconia

(16 mg/l), Marsanne, Roussanne and Malvasia (*c.*12 mg/l). The lowest resveratrol producers include cvs. Xarello, Refosco and Primitivo (between 3 and 5 mg/l), Petit Manseng and Nebiollo (< 3 mg/l). Moreover, macroarray data analysis revealed that high resveratrol levels were also accompanied by the upregulation of genes involved in plant defence, and the concomitant underexpression of genes related to the ripening process and to indole alkaloid synthesis. Results obtained for the cardiovascular benefits and cancer chemopreventive activity of resveratrol might be relevant to grape consumption, especially responses that could be mediated by low concentrations of the substance (Pezzuto, 2008).

9.2.2 Sugars

Grape is one of the sweetest fruit, with sugar levels reaching 20% of fresh weight at full maturity. In raisins, the sugar content can reach 60%. Obviously, this is important for a nutritionist given the consequences of a lack or excess of sugars in a diet.

Glucose, fructose, galactose, sucrose, maltose, melibiose, raffinose and stachyose were identified in the berries of grape *V. vinifera* L. cv. Thompson Seedless (Kliewer, 1966). The grapes accumulated mainly glucose and fructose, at equal levels. On average, a grape contains 15–18 g of sugars in 100 g of fresh weight at full maturity. Total soluble solids can reach 20, 22 or even 25% (when in many fresh fruit it does not exceed 12%).

9.2.3 Organic acids, minerals and aromas

Organic acids found in grape berries are: malic, tartaric, citric, isocitric, ascorbic, *cis*-aconitic, oxalic, glycolic, glyoxylic, succinic, lactic, glutaric, fumaric, pyrrolidone-carboxylic, α-ketoglutaric, pyruvic, oxaloacetic, galacturonic, glucuronic, shikimic, quinic, chlorogenic and caffeic (Kliewer, 1966). The organic acids reach 1–1.5 g/100 g of FW.

In the water reserve of grape berries, which represents over 80% of the fruit fresh weight, many inorganic compounds (K^+, Ca^{2+}, Mg^{2+}, Fe^{2+}, SO_4^{2-}, PO_4^{3-}) are also present.

It should be noted that among the minerals present, potassium (especially in the form of bitartrate) dominates, with a level of 250 mg/100 g FW. The level of sodium remains very low (2 mg/100 g FW). There are also trace elements (reducing or oxidizing), essential for carrying out the chemical reactions required for cell multiplication (Se, Ni, Cr, I, Zn, Cu, Mn, F, V, Co,…), which give the grape some original nutritional qualities. The aroma compounds in grapes are either free, such as terpenes, or bound, mainly as glycosides. They are detailed in a recent review on grape berry biochemistry (Conde *et al.*, 2007).

9.2.4 Vitamins

One interesting point with grapes is that they accumulate very low levels of vitamin C, ascorbic acid, in comparison with other fruit. This may be due to the fact that grape berry tissues transform ascorbic acid to tartaric acid, as shown in a recent study (De Bolt *et al.*, 2006). The group B vitamins are all well represented, particularly B_1, B_3 and B_5. The level varies with the degree of maturity of the grapes. Vitamin C (more abundant in the outer grain) varies between 4 and 10 mg/100 g. The mean nutritional profile of a grape, including the percentage of the RDA (recommended daily allowance) of minerals and vitamins represented by a 100g serving, is given in Table 9.3.

9.3 Chemopreventive Activity and Bioavailability

The grape offers some potential health benefits via its phenolic composition, flavonoids and non-flavonoids. It also has sugars, acids and vitamins that may have interesting properties which can be modulated by an adapted nutrition.

9.3.1 Cancer studies

Many grape polyphenols have been shown to have anticancerous effects. The polyphenol,

3,5,4′-trihydroxystilbene, commonly called resveratrol, was shown to be an effective candidate for chemoprevention of lung cancer due to its ability to induce apoptosis (Weng *et al.*, 2009). It is worth noting that polydatin, a glycoside of resveratrol, which is present in grape juice and wine, has been shown to be metabolized rapidly to resveratrol in the small intestine and liver of rats, then metabolized further to glucuronidated resveratrol (Zhou *et al.*, 2009).

Other studies have shown that grape polyphenols play several roles in limiting cancer development. For example, proanthocyanidins have been shown to prevent the metastatic cascade by mediating the inhibitory signals for cancer cell migration, an essential step in invasion and metastasis (Punathil and Katiyar, 2009).

9.3.2 Cardiovascular diseases

The moderate consumption of grapes or grape products (containing polyphenols) may lead to a decrease of platelet aggregation and vasodilatory effects in blood vessels. In the case of atherosclerosis, polyphenols reduce the formation of the plaque, cholesterol, and increase rates of certain enzymatic antioxidant defences against free radicals. Recently, Décordé *et al.* (2008) were able to demonstrate that the phenolic compounds of red grapes prevented the development of atherosclerosis, induced earlier by an atherogenic diet, in Syrian golden hamsters. In this work, consumption of grapes (cv. Hambourg Muscat) or derived juice by the hamsters, at doses equivalent to 600 g/day or 500 ml/day, respectively, for a man weighing 70 kg, led to a reduction in the surface fat deposition in the aortic arch by 78 and 93%, respectively, compared with that seen in the control animals (atherogenic food only). Likewise, with the consumption of grapes or grape juice, total cholesterol was decreased by 30.4 and 34.6%, and non-high-density lipoprotein cholesterol by 64 and 58.9%, respectively. In parallel, levels of antioxidant capacity of blood plasma in the animals receiving the grapes or grape juice increased by 41 and 61%, respectively. In

Table 9.3. Nutritional profile of grape (for 100 g).

Class of nutrients	Specific nutrients	Quantity	Adult EU RDA (%)[a]
Calories		67.0 kcal	
Water		81.3 g	
Carbohydrates		17 g	
	Monosaccharides	13.5	
	Disaccharides	2.9	
	Soluble fibre	0.26	
	Insoluble fibre	0.76	
	Other saccharides	0.15	
Proteins		0.63 g	
	Arginine	0.04	
	Aspartic acid	0.09	
	Glutamic acid	0.13	
	Glycine	0.02	
	Histidine	0.02	
	Leucine	0.02	
	Lysine	0.02	
	Methionine	0.02	
	Phenylalanine	0.02	
	Proline	0.02	
	Serine	0.02	
	Threonine	0.02	
	Tyrosine	0.02	
	Valine	0.02	
Fats		0.32 g	
	Saturated fatty acids	0.11	
	Oleic acid	0.02	
	Linoleic acid	0.09	
	Linolenic acid	0.02	
Vitamins			
	β-Carotene	59.1 µg	7.4
	Vitamin B_1	0.09 mg	6.4
	Vitamin B_2	0.07 mg	4.3
	Niacine	0.30 mg	1.7
	Vitamin B_6	0.11 mg	5.5
	Biotin	0.30 µg	0.2
	Vitamin C	4.00 mg	6.7
	Vitamin E	0.35 mg	3.5
	Folic acid	3.89 µg	1.9
	Pantotenic acid	0.02 mg	0.3
Minerals			
	Potassium	191.7 mg	5.5
	Calcium	14.0 mg	1.4
	Phosphorus	10.0 mg	1.0
	Magnesium	5.0 mg	1.3
	Sodium	2.0 mg	–
	Manganese	0.7 mg	35.0
	Iron	0.28 mg	1.6
	Copper	0.04 mg	2.0
	Zinc	0.04 mg	0.3
	Iodine	1.0 µg	0.7
	Selenium	0.2 µg	0.3
Organic acids		563.0 mg	
	Malic acid	540.0 mg	
	Citric acid	23.0 mg	

Note: [a]Adult European Union Recommended Daily Allowance (%).

agreement with other work (Vinson *et al.*, 2001), a hypolipidaemic effect was also found. Indeed, the antioxidant potential of grape juice polyphenols has been shown to limit atherosclerosis linked to the oxidation of human low-density lipoproteins. A normalization of the systolic pressure, plus a significant reduction of cardiac hypertrophy and of free radical generation in the thoracic aorta, have also been observed.

A study compared the platelet activity *ex vivo* of human volunteers before and after they had drunk black grape juice, orange juice or grapefruit juice for 7–10 days (Keevil *et al.*, 2000). Drinking red grape juice for a week reduced blood platelet aggregation by 77%, while the consumption of orange juice or grapefruit juice had no effect on this parameter. The red grape juice contained approximately three times more polyphenols than the total citrus juice and had platelet inhibitor potential in the healthy human subjects. The platelet inhibitory effect of grape juice may decrease the risk of coronary thrombosis and myocardial infarction. The bioavailability of other polyphenols, such as resveratrol and quercetin, has also been shown (Meng *et al.*, 2004).

Despite its relatively low level, the vitamin C in grapes is important. Indeed, it reinforces the roles of anthocyanins and other polyphenols at the level of the arterial wall, as it acts as a cofactor. It is therefore worth noting that 4 mg of vitamin C/100 g fresh grapes may have an impact on polyphenol actions.

9.3.3 Antibiotic effects

Extracts from grape seeds have been tested for antibacterial activity against *Bacillus cereus* strains, *B. coagulans*, *B. subtilis*, *Staphylococcus aureus*, *Escherichia coli* and *Pseudomonas aeruginosa*. Jayaprakasha *et al.* (2003) demonstrated that Gram-positive bacteria could be inhibited completely with 850–1000 µl/l of phenolic extract of grape seed, while Gram-negative bacteria were inhibited by levels of 1250–1500 µl/l extracts of grape seed.

The antibacterial action of organic acids against oral streptococci, responsible for cavity development, and against *Streptococcus pyogenes*, responsible for pharyngitis, have been studied (Daglia *et al.*, 2007). The compounds found responsible for such activities were succinic, malic, lactic, tartaric, citric and acetic acids. Findings show that organic acids found in grapes are active against oral streptococci and *S. pyogenes* and suggest that it enhances oral health. There are, however, concerns over the effect of fruit acids on enamel integrity.

Some antimicrobial activities were observed with water-soluble extracts prepared from cv. Muscadine seeds (Kim *et al.*, 2008). The extracts had strong antimicrobial activity against a cocktail of three strains of *E. coli* O157:H7. Extracts had high acidity (pH 3.39–3.43), total phenolics (2.21–3.49 mg/ml), tartaric acid (5.6–10.7 mg/ml), tannic acid (5.7–8.1 mg/ml) and gallic acid (0.33–0.59 mg/ml). Heat treatment of the extracts increased antimicrobial activity, possibly because of increased acidity, tartaric acid, total phenolics and individual phenolics.

With regard to the relationship between the structure and activity of triterpene acids (ursolic, oleanoic, gypsogenic and sumaresinolic acids) and derivatives, it was found that both hydroxy and carboxy groups present in the triterpenes were important for their antibacterial activity against several oral pathogens (*S. mutans*, *S. mitis*, *S. sanguinis*, *S. salivarius*, *S. sobrinus* and *Enterococcus faecalis*), potentially responsible for the formation of dental caries in humans (Scalon Cunha *et al.*, 2007).

Flavonoids and related polyphenols also possess promising anti-HIV effects. A study showed that grape seed extracts (GSE) downregulated the expression of HIV-1 entry coreceptors significantly and that GSE-treated cultures showed a significantly lower number of HIV-positive cells (Nair *et al.*, 2002).

9.3.4 Metabolic diseases

In order to improve defences against oxidative stress in diabetes and stimulate oxygen secretion, it is necessary to maintaina high antioxidant capacity of the plasma. The high levels of fructose and glucose in grapes make them a useful supply of rapidly available sugars to counteract hypoglycaemia.

The grape, with its wealth of organic acids and minerals, can affect the acid–alkaline balance of the cellular environment. The grape facilitates urinary excretion because of its high water and potassium contents, which promote diuretic action. Grape suspensions in water or ethanol have been shown to prevent obstructive bladder dysfunction, which often affects ageing men (Agartan *et al.*, 2005). The stimulation of intestinal transit is also facilitated by the presence of effective fibres in grapes. This dual activity on elimination (urine and faeces) has sometimes resulted in the prescription of grapes in detoxification diets. The potassium/sodium ratio is high (around 125) in grapes and this may also contribute to the diuretic properties of this fruit (Szentmihályi *et al.*, 1998).

9.3.5 Other beneficial effects

Fresh and dried grapes provide phenolic antioxidants, which are believed to contribute to potential health effects. In a recent study, Parker *et al.* (2007) checked the antioxidant capacity of grape polyphenols from (fresh, dried) cv. Thompson Seedless and observed the effects of their consumption for 4 weeks in 15 healthy human subjects. The ORAC (oxygen radical absorbacne capacity) was increased after 2 weeks of grape consumption (250 g/day) and after 3 weeks of raisin consumption (50 g/day). Even the oxidation of serum was significantly limited by the consumption of sultanas (golden raisins) after 4 weeks (time latency).

The main sugars accumulating in grape berries are glucose and fructose, as mentioned previously. Even if they are not regarded generally as bioactives, they provide readily available energy, thanks to their high bioavailability, which is increased by the fact that these sugars are accompanied by natural organic acids and group B vitamins that ensure their good assimilation at the cellular level. The grape is therefore a recommended natural food for high-energy needs or when the body is subjected to intense muscle activity. The grape provides an average of 72 kcal (301 kJ)/100 g FW,

with extremes ranging between 60 and 80 kcal, depending on the sugar content.

Organic acids give grapes a refreshing flavour, which compensates for their pleasant but high sweetness. They also have a slightly stimulating action on digestive secretions, which facilitates good assimilation. The presence of some vitamins in grape berries plays a beneficial role, especially in the functioning of the nervous system, thanks to the presence of group B vitamins (vitamin B deficiency has been shown to be linked with mouse cognitive dysfunction (Troen *et al.*, 2008)), and for the protection of capillary blood vessels, due to 'vitamin P' action.

9.4 Effects of Postharvest Continuum

The literature is scarce on postharvest treatments affecting grape bioactives and nutrition. It is likely that there may be more studies in the future, now that bioactive compounds have attracted more attention.

The main physiological change that occurs during the postharvest life of grapes is loss of water. During grape dehydration, Moreno *et al.* (2008) observed that the soluble solid content increased. The same study showed that the amount of anthocyanin per berry remained unchanged over the dehydration process (weight loss of 15% in 4 days) and that the terpene contents increased in wines made from dried grapes, but the study did not check whether this was due to concentration changes and/or modulation of the glycosylated compound content. Glycosylation may indeed modify compound availability in the digestive process, thus modulating the nutritional value of grapes. However, Bellincontro *et al.* (2006) showed that, in a different grape cultivar, cv. Aleatico, a longer postharvest period (13 days at a relatively high humidity) induced a decrease in anthocyanins, but that a postharvest treatment with ethylene increased the global polyphenol content. Postharvest ethylene treatment was also shown to increase the concentrations of some alcohols. In another study, the amount of sodium metabisulfite used to preserve table grapes (as an antioxidant and antifungal agent) was shown to preserve the vitamin C content (Sharayei *et al.*, 2004).

Several studies have demonstrated that postharvest UV treatment and wilting can induce some increase of the stilbene content in various cultivars of grapes (Versari *et al.*, 2001; Cantos *et al.*, 2003). Sanchez-Ballesta *et al.* (2006) showed that a high CO_2 pretreatment could delay stilbene accumulation over the first part of storage; this compound increasing again during the shelf-life period. Treatment of table grapes, cv. Autumn Seedless, with ozone increased total phenolics, while modified atmosphere packaging maintained their concentration (Artes-Hernandez *et al.*, 2007). These postharvest treatments were tested as an alternative to the ubiquitous industrial SO_2 application, which can cause bleaching, berry drop and flavour taint.

The main postharvest problem affecting table grapes is the development of fungi. Disease is manifested principally as a result of infection by *Botrytis cinerea*, the causal agent of grey mould. Rots increase waste and reduce grape quality by consuming sugars and producing some acids, and render the clusters unattractive so they loose their commercial value. Although *B. cinerea* generally is not considered a hazard to human health, some fungi, like *Aspergilus* spp., may produce ochratoxin A (Hocking *et al.*, 2007), and this is a mycotoxin that has attracted recent attention. Hocking *et al.* (2007) reviewed the pre- and postharvest factors affecting *Aspergilus* development and the solutions available to avoid ochratoxin accumulation. One strategy that may be used to counteract these fungi in the coming years is the use of antagonistic yeasts (Bleve *et al.*, 2006) isolated from the associated microflora found on the surface of the berries.

9.5 Conclusions

Grape berries, and the derived products thereof, could play a preventive role in many diseases when consumed regularly and moderately. The phenolic compounds of grapes and wine undeniably have therapeutic properties (Auger *et al.*, 2002; Landrault *et al.*,

2003; Al-Awwadi *et al.*, 2004a,b, 2005), particularly for certain chronic diseases such as atherosclerosis, diabetes, hypertension and some cancers. Among the mechanisms of action of phenolic compounds involved in the prevention of chronic diseases, the following should be recognized:

- a direct effect by trapping free radicals.
- preserving endogenous antioxidants (vitamin E, vitamin C, β-carotene…).
- preserving antioxidant enzymes (SOD: superoxide dismutase, catalase, SeGSH-Px: glutathione peroxidase).
- reducing the re-equilibration of cholesterol and blood lipids (HDL/LDL).
- chelation of cofactors of fatty acid oxidation such as some metals (Fe^{2+}, Cu^{2+}).
- inhibitory effect on oxidative enzymes such as cyclooxygenases and lipooxygenases.
- effects on the synthesis of endothelial nitric oxide: at the cellular level of the arterial wall leading to vasorelaxation and a hyperpolarization of the membrane by human extracellular potassium.
- inhibiting effects on the genesis of producing the NAPH oxidase-level cells of the vascular wall (thoracic aorta and heart), thus reducing the production of free radicals.

Polyphenols have received a lot of attention in the past two decades regarding their chemopreventive role, but other compounds such as sugars are present at very high concentration in grapes and present interesting or negative nutritional values, depending on the consumer and the quantities ingested; sugars (derived from both natural and synthetic sources) are now under the scrutiny of many nutritionists. Whether future specific health claims will be sought from or allowed by regulatory authorities is not known but, based on existing data, it is clear that grapes should be considered an integral component of the fruit- and vegetable-enriched diets that are widely recommended by health authorities. Further research on the effects and mechanisms of action of compounds in grape and its derivatives on chronic diseases needs to be pursued.

References

Agartan, C.A., Whitbeck, C., Chichester, P. and Levin, R.M. (2005) Effect of ethanol on protection of urinary bladder function by grape suspensions. *Urology* 66, 213–217.

Al-Awwadi, N.A., Azay, J., Poucheret, P., Cassanas, G., Krosniak, M., Auger, C., Gasc, F., Rouanet, J.M., Cros, G. and Teissedre, P.L. (2004a) Antidiabetic activity of red wine polyphenols, ethanol or both, in streptozotocin-treated rats. *Journal of Agricultural and Food Chemistry* 52, 1008–1016.

Al-Awwadi, N.A., Bornet, A., Azay, J., Araiz, C., Delbosc, S., Cristol, J.P., Linck, N., Cros, G. and Teissedre, P.L. (2004b) Red wine polyphenols alone or in association with ethanol prevent hypertension, cardiac hypertrophy, and production of reactive oxygen species in the insulin-resistant fructose-fed rat. *Journal of Agricultural and Food Chemistry* 52, 5593–5597.

Al-Awwadi, N.A., Araiz, C., Bornet, A., Delbosc, S., Cristol, J.P., Linck, N., Azay, J., Teissedre, P.L. and Cros, G. (2005) Extracts enriched in different polyphenolic families normalize increased cardiac NADPH oxidase expression while having differential effects on insulin resistance, hypertension and cardiac hypertrophy in high fructose-fed rats. *Journal of Agricultural and Food Chemistry* 53, 151–157.

Anon. (1980) Polyphenols [Substance Name]; use the precise structure header, most commonly in the Flavonoids group; this term only refers vaguely to phenolic (aromatic) hydroxyls. Date introduced: 18 August 1980 in MeSH database.

Anon. (2006) 3rd International Conference on Polyphenols Applications. The International Society for Antioxidants in Nutrition and Health (ISANH), 26–27 October, 2006, St Julian, Malta.

Artes-Hernandez, F., Aguayo, E., Artes, F. and Tomas-Barberan, F.A. (2007) Enriched ozone atmosphere enhances bioactive phenolics in seedless table grapes after prolonged shelf life. *Journal of the Science of Food and Agriculture* 87, 824–831.

Auger, C., Caporiccio, B., Landrault, N., Teissedre, P.L., Laurent, C., Cros, G., Besançon, P. and Rouanet, J.M. (2002) Red wine phenolic compounds reduce plasma lipids and apolipoprotein B, and prevent early aortic atherosclerosis in hypercholesterolemic golden Syrian hamsters (*Mesocricetus auratus*). *Journal of Nutrition* 132, 1207–1213.

Bamforth C.W. (1999) Beer haze. *Journal of the American Society of Brewing Chemists* 57, 81–90.

Bellincontro, A., Fardelli, A, Santis, D. de, Botondi, R. and Mencarelli, F. (2006) Postharvest ethylene and 1-MCP treatments both affect phenols, anthocyanins and aromatic quality of Aleatico grape and wine. *Australian Journal of Grape and Wine Research* 12, 141–149.

Bleve, G., Grieco, F., Cozzi, G., Logrieco, A. and Visconti, A. (2006) Isolation of epiphytic yeasts with potential for biocontrol of *Aspergillus carbonarius* and *A. niger* on grape. *International Journal of Food Microbiology* 108, 204–209.

Bohm, H., Boeing, H., Hempel, J., Raab, B. and Kroke, A. (1998) Flavonols, flavone and anthocyanins as natural antioxidants of food and their possible role in the prevention of chronic diseases. *Zeitschrift für Ernährungswissenschaft* 37, 147–163.

Bourzeix, M., Weyland, D., Heredia, N. and Desfieux, C. (1986) Study of catechins and procyanidins in grape clusters, wine, and other vine products. *Bulletin de l'Office International de la Vigne et du Vin* 59, 1171–1253.

Cantos, E., Tomas-Barberan, F.A., Martinez, A. and Espin, J.C. (2003) Differential stilbene induction susceptibility of seven red wine grape varieties upon UV-C irradiation. *European Food Research and Technology* 217, 253–258.

Chen, D., Daniel, K.G., Kuhn, D.J., Kazi, A., Bhuiyan, M., Li, L., Wang, Z., Wan, S.B., Lam, W.H., Chan, T.H. and Dou, Q.P. (2004) Green tea and tea polyphenols in cancer prevention. *Frontiers in Bioscience* 9, 2618–2631.

Cheynier, V., Fulcrand, H., Sarni, P. and Moutounet, M. (1997) Application des techniques analytiques à l'étude des composés phénoliques et leurs réactions au cours de leur vinification. *Analusis Magazine* 25, 14–44.

Chira, K., Schmauch, G., Saucier, C., Fabre, S. and Teissedre, P.L. (2009) Grape variety effect on proanthocyanidin composition and sensory perception of skin and seed tannin extracts from bordeaux wine grapes (Cabernet Sauvignon and Merlot) for two consecutive vintages (2006 and 2007). *Journal of Agricultural and Food Chemistry* 57, 545–553.

Conde, C., Silva, P., Fontes, N., Dias, A.C.P., Tavares, R.M., Sousa, M.J., Agasse, A., Delrot, S. and Geros, H. (2007) Biochemical changes throughout grape berry development and fruit and wine quality. *Food* 1, 1–22.

Crozier, A. (2003) Classification and biosynthesis of secondary plant products: an overview. In: Goldberg, G. (ed.) *Plants: Diet and Health*. British Nutrition Foundation, Chapman Hall, London, pp. 27–48.

Daglia, M., Papetti, A., Grisoli, P., Aceti, C., Dacarro, C. and Gazzani, G. (2007) Antibacterial activity of red and white wine against oral streptococci. *Journal of Agricultural and Food Chemistry* 55, 5038–5042.

De Bolt, S., Cook, D.R. and Ford, C.M. (2006) L-Tartaric acid synthesis from vitamin C in higher plants. *Proceedings of the National Academy of Sciences of the United States of America* 103, 5608–5613.

Décordé, K., Teissèdre, P.-L., Auger, C., Cristol, J.-P. and Rouanet, J.-M. (2008) Phenolics from purple grape, apple, purple grape juice and apple juice prevent early atherosclerosis induced by an atherogenic diet in hamsters. *Molecular Nutrition and Food Research* 53(5), 659–666.

FAO (2000) Fruit production and consumption, data from World Fruit Program.

Frankel, E.N., Kanner, J., German, J.B., Parks, E. and Kinsella, J.E. (1993) Inhibition of *in vitro* oxidation of human low-density lipoprotein with phenolic substances in red wine. *Lancet* 341, 454–457.

Gatto, P., Vrhovsek, U., Muth, J., Segala, C., Romualdi, C., Fontana, P., Pruefer, D., Stefanini, M., Moser, C., Mattivi, F. and Velasco, R. (2008) Ripening and genotype control stilbene accumulation in healthy grapes. *Journal of Agricultural and Food Chemistry* 56, 11773–11785.

Guebailia, H.A., Chira, K., Richard, T., Mabrouk, T., Furiga, A., Vitrac, X., Monti, J.P., Delaunay, J.C. and Mérillon, J.M. (2006) Hopeaphenol: the first resveratrol tetramer in wines from North Africa. *Journal of Agricultural and Food Chemistry* 54, 9559–9564.

Hocking, A.D., Leong, S.L.L., Kazi, B.A., Emmett, R.W. and Scott, E.S. (2007) Fungi and mycotoxins in vineyards and grape products. *International Journal of Food Microbiology* 119, 84–88.

Jayaprakasha, G.K., Selvi, T. and Sakariah, K.K. (2003) Antibacterial and antioxidant activities of grape (*Vitis vinifera*) seed extracts. *Food Research International* 36, 117–122.

Kammerer, D., Claus, A., Carle, R. and Schieber, A. (2004) Polyphenol screening of pomace from red and white grape varieties (*Vitis vinifera* L.) by HPLC-DAD-MS/MS. *Journal of Agricultural and Food Chemistry* 52, 4360–4367.

Keevil, J.G., Osman, H.E., Reed, J.D. and Folts, J.D. (2000) Grape juice, but not orange juice or grapefruit juice, inhibits human platelet aggregation. *Journal of Nutrition* 130, 53–56.

Kennedy, J.A., Troup, G.J., Pilbrow, J.R., Hutton, D.R., Hewitt, D., Hunter, C.R., Ristic, R., Iland, P.G. and Jones, G.P. (2000) Development of seed polyphenols in berries from *Vitis vinifera* L. cv. Shiraz. *Australian Journal of Grape and Wine Research* 6, 244–254.

Kim, T.J., Weng, W.L., Stojanovic, J., Lu, Y., Jung, Y.S. and Silva, J.L. (2008) Antimicrobial effect of water-soluble muscadine seed extracts on *Escherichia coli* O157:H7. *Journal of Food Protection* 71, 1465–1468.

Kliewer, W.M. (1966) Sugars and organic acids of *Vitis vinifera*. *Plant Physiology* 41, 923–931.

Landrault, N., Larronde, F., Delaunay, J.C., Castagnino, C., Vercauteren, J., Merillon, J.M., Gasc, F., Cros G. and Teissedre, P.L. (2002) Levels of stilbene oligomers and astilbin in French varietal wines and in grapes during noble rot development. *Journal of Agricultural and Food Chemistry* 50, 2046–2052.

Landrault, N., Poucheret, P., Azay, J., Krosniak, M., Gasc, F., Jenin, C., Cros, G. and Teissedre, P.L. (2003) Effect of a polyphenols-enriched chardonnay white wine in diabetic rats. *Journal of Agricultural and Food Chemistry* 51, 311–318.

Laughton, M.J., Evans, P.J., Moroney, M.A., Hoult, J.R. and Halliwell, B. (1991) Inhibition of mammalian 5-lipoxygenase and cyclo-oxygenase by flavonoids and phenolic dietary additives. Relationship to antioxidant activity and to iron ion-reducing ability. *Biochemical Pharmacology* 42, 1673–1681.

Macheix, J.J., Fleuriet, A. and Billot, J. (1990) *Fruit Phenolics*. CRC Press, Boca Raton, Florida.

Mattivi, F., Guzzon, R., Vrhovsek, U., Stefanini, M. and Velasco, R. (2006) Metabolite profiling of grape: flavonols and anthocyanins. *Journal of Agricultural and Food Chemistry* 54, 7692–7702.

Meng, X., Maliakal, P., Lu, L., Lee, M.J. and Yang, C.S. (2004) Urinary and plasma levels of resveratrol and quercetin in humans, mice, and rats after ingestion of pure compounds and grape juice. *Journal of Agricultural and Food Chemistry* 52, 935–942.

Meyer, A.S., Jepsen, S.M. and Sørensen, N.S. (1998) Enzymatic release of antioxidants for human low-density lipoprotein from grape pomace. *Journal of Agricultural and Food Chemistry* 46, 2439–2446.

Moreno, J.J., Cerpa-Calderon, F., Cohen, S.D., Fang, Y., Qian, M. and Kennedy, J.A. (2008) Effect of postharvest dehydration on the composition of pinot noir grapes (*Vitis vinifera* L.) and wine. *Food Chemistry* 109, 755–762.

Nair, M.P., Kandaswami, C., Mahajan, S., Nair, H.N., Chawda, R., Shanahan, T. and Schwartz, S.A. (2002) Grape seed extract proanthocyanidins downregulate HIV-1 entry coreceptors, CCR2b, CCR3 and CCR5 gene expression by normal peripheral blood mononuclear cells. *Biological Research* 35, 421–431.

Orgogozo, J.M., Dartigues, J.F., Lafont, S., Letenneur, L., Commenges, D., Salamon, R., Renaud, S. and
    Breteler, M.B. (1997) Wine consumption and dementia in the elderly: a prospective community study
    in the Bordeaux area. *Revue Neurologie (Paris)* 153, 185–192.

Parker, T.L., Wang, X.H., Pazmino, J. and Engeseth, N.J. (2007) Antioxidant capacity and phenolic content
    of grapes, sun-dried raisins, and golden raisins and their effect on *ex vivo* serum antioxidant capacity.
    Journal of Agricultural and Food Chemistry 55, 8472–8477.

Pezzuto, J.M. (2008) Grapes and human health: a perspective. *Journal of Agricultural and Food Chemistry* 56,
    6777–6784.

Puiggros, F., Sala, E., Vaqué, M., Ardévol, A., Blay, M., Fernandez-Larrea, J., Arola, L., Bladé, C., Pujadas, G.
    and Salvado, M.J. (2009) *In vivo, in vitro,* and *in silico* studies of Cu/Zn-superoxide dismutase regula-
    tion by molecules in grape seed procyanidin extract. *Journal of Agricultural and Food Chemistry* 57,
    3934–3942.

Punathil, T. and Katiyar, S.K. (2009) Inhibition of non-small cell lung cancer cell migration by grape seed
    proanthocyanidins is mediated through the inhibition of nitric oxide, guanylate cyclase, and ERK1/2.
    Molecular Carcinogenesis 48, 232–242.

Revilla, E., Alonso, E. and Kovac, V. (1997) The content of catechins and procyanidins in grapes and wines
    as affected by agroecological and technological practices. In: Watkins, T.R. (ed.) *Wine: Nutritional and
    Therapeutic Benefits*. American Chemistry Society, Washington, DC, pp. 69–80.

Ribeiro de Lima, M.T., Waffo-Teguo, P., Teissedre, P.-L., Pujolas, A., Vercauteren, J., Cabanis, J.C. and
    Merillon, J.M. (1999) Determination of stilbenes (*trans*-astringin, cis and *trans* piceid, and *cis* and *trans*
    resveratrol) in Portuguese wines. *Journal of Agricultural and Food Chemistry* 47, 2666–2670.

Sanchez-Ballesta, M.T., Jimenez, J.B., Romero, I., Orea, J.M., Maldonado, R., Urena, A.G., Escribano, M.I.
    and Merodio, C. (2006) Effect of high CO_2 pretreatment on quality, fungal decay and molecular regu-
    lation of stilbene phytoalexin biosynthesis in stored table grapes. *Postharvest Biology and Technology* 42,
    209–216.

Scalon Cunha, L.C., Andrade e Silva, M.L., Cardoso Furtado, N.A., Vinhólis, A.H., Gomes Martins, C.H.,
    da Silva Filho, A.A. and Cunha, W.R. (2007) Antibacterial activity of triterpene acids and semi-
    synthetic derivatives against oral pathogens. *Zeitschrift für Naturforschung C* 62, 668–672.

Sharayei, P., Shahbake, M.A. and Mokhtarian, A. (2004) Investigation of the effects of Grapeguard® on the
    quality and incidence of fungal contamination in grape during cold storage. *Journal of Agricultural
    Engineering Research* 5, 1–16.

Stanley, F., Wainapel, M.D., Avital, M.P.H. and Fast, M.D. (2003) Antioxidants and the free radical theory of
    degenerative disease. In: Hoffman, R.L. (ed.) *Nutritional Therapy in Rehabilitation. Alternative Medicine
    and Rehabilitation*. Demos Medical Publishing, New York.

Szentmihályi, K., Kéry, Á., Then, M., Lakatos, B., Sándor, Z. and Vinkler, P. (1998) Potassium–sodium ratio
    for the characterization of medicinal plant extracts with diuretic activity. *Phytotherapy Research* 12,
    163–166.

Teissedre, P.-L. (2008) Oxygen and phenolic compounds: from the barrel to the chips. *Biofutur* 294,
    40–43.

Teissedre, P.-L., Frankel, E.N., Waterhouse, A.L., Peleg, H. and German, J.G. (1996) Inhibition of *in vitro*
    human LDL oxidation by phenolic antioxidants from grape and wine. *Journal of the Science of Food and
    Agriculture* 50, 55–61.

Troen, A.M., Shea-Budgell, M., Shukitt-Hale, B., Smith, D.E., Selhub, J. and Rosenberg, I. (2008) B-vitamin
    deficiency causes hyperhomocysteinemia and vascular cognitive impairment in mice. *Proceedings of
    the National Academy of Sciences of the United States of America* 105, 12474–12479.

Versari, A., Parpinello, G.P., Tornielli, G.B., Ferrarini, R. and Giulivo, C. (2001) Stilbene compounds and
    stilbene synthase expression during ripening, wilting and UV treatment in grape cv. Corvina. *Journal
    of Agricultural and Food Chemistry* 9, 5531–5536.

Vinson, J.A., Teufel, K. and Wu, N. (2001) Red wine, dealcoholized red wine, and especially grape juice,
    inhibit atherosclerosis in a hamster model. *Atherosclerosis* 156, 67–72.

Vitrac, X., Bornet, A., Vanderlinde, R., Valls, J., Richard, T., Delaunay, J.C., Mérillon, J.M. and Teissedre, P.L.
    (2005) Determination of stilbenes (delta-viniferin, trans-astringin, trans-piceid, cis- and trans-resvera-
    trol, epsilon-viniferin) in Brazilian wines. *Journal of Agricultural and Food Chemistry* 53, 5664–5669.

Waterhouse, A.L., German, J.B., Frankel, E.N., Walzem, R.L., Teissedre, P.L. and Folts, J. (1996) The pheno-
    lic phytochemicals in wine fruit and tea: potential nutritional effects and dietary levels. In: Finley,
    J.W., Armstrong, D.J., Nagy, S. and Robinson, S. (eds) *Hypernutrition Foods*. Agscience, Auburndale,
    Florida, pp. 219–238.

Weber, H.A., Hodges, A.E., Guthrie, J.R., O'Brien, B.M., Robaugh, D., Clark, A.P., Harris, R.K., Algaier, J.W. and Smith, C.S. (2007) Comparison of proanthocyanidins in commercial antioxidants: grape seed and pine bark extracts. *Journal of Agricultural and Food Chemistry* 55, 148–156.

Weng, C.J., Yang, Y.T., Ho, C.T. and Yen, G.C. (2009) Mechanisms of apoptotic effects induced by resveratrol, dibenzoylmethane, and their analogues on human lung carcinoma cells. *Journal of Agricultural and Food Chemistry* 57, 5235–5243.

Yilmaz, Y. and Toledo, R.T. (2004) Major flavonoids in grape seeds and skins: antioxidant capacity of catechin, epicatechin, and gallic acid. *Journal of Agricultural and Food Chemistry* 52, 255–260.

Zhou, S., Yang, R., Teng, Z., Zhang, B., Hu, Y., Yang, Z., Huan, M., Zhang, X. and Mei, O. (2009) Dose-dependent absorption and metabolism of trans-polydatin in rats. *Journal of Agricultural and Food Chemistry* 57, 4572–4579.

10 Leafy Vegetables and Salads

Peter M.A. Toivonen and D. Mark Hodges

10.1 Introduction

The diversity in consumption of leafy vegetable types is significant worldwide (Table 10.1). Most leafy vegetables are consumed in a cooked format. However, consumption in the raw state is becoming a more common practice, partly attributable to expanded production of salad vegetables in regions of the world that historically have not consumed the products in a raw format. The need for cooking in many leafy vegetables may arise from the need to detoxify components present in the raw product (Kuti and Konuru, 2006; Orech *et al.*, 2006). Also, some very bitter indigenous leafy vegetables are very high in polyphenols, but will show significant losses of these bitter compounds with cooking, which thus renders them more palatable (Kuti and Konuru, 2004). Hence, it is important to keep this in mind when evaluating the consumption format (cooked versus raw) for adaptation of indigenous leafy vegetables to conventional Western diets.

In this chapter we will deal with lettuces and other salad greens (e.g. spinach). Lettuces of different types, such as romaine and leafy lettuces, increasingly are cultivated globally, depending on the type and preferences of the population (de Vries, 1997). There are other leafy vegetables, such as spinach and the leafy Asian crucifers, which make up a large dietary distribution in Asia as well as North America and Europe (FAO, 2008). Less ubiquitous are indigenous green leafy vegetables, which can form a significant local requirement for nutrition; some, like tree spinach, are also eaten more widely as part of ethnic diets in the developed world (Kuti and Konuru, 2006). Leafy green vegetables make up a significant component of healthful diets around the world (FAO, 2008).

The production of leafy vegetables intended for salads has expanded significantly and forms a key agrifood sector in North America, Europe, Australia and New Zealand (Hedges and Lister, 2005). The industry is so well developed that now a significant portion (over 30% in the USA) of the fresh vegetable industry is involved in the ready-to-eat format as packaged salads (Cardwell, 2005). Consumption of leafy vegetables such as iceberg and romaine lettuces has shown a large increase over the past several decades, as demonstrated by US data (Putnam and Allshouse, 1999), and this has been attributed partly to the convenience factor of the ready-to-eat (fresh-cut) format in an increasingly busy world (Rocha and Morais, 2007). Consumer awareness of the healthfulness of leafy vegetables in their diet has also contributed to increased consumption of leafy vegetables and related products (Goldman, 2003; Rai *et al.*, 2006; Rocha and Morais,

Table 10.1. Listing of leafy green vegetables, distribution and their usage format.

Common name	Scientific name	Distribution	Usage format
Iceberg lettuce or crisphead lettuce	*Lactuca sativa* L. var. capitata	USA, Canada, Europe	Fresh salad
Romaine (cos) lettuce	*Lactuca sativa* L. var. longifolia	Europe, North America, Australia	Raw or cooked
Loose-leaf lettuce, green leaf lettuce, red leaf lettuce or oak leaf lettuce	*Lactuca sativa* L. var. crispa	Worldwide	Fresh salad
Butterhead lettuce, Boston lettuce or Bibb lettuce	*Lactuca sativa* L. var. capitata	Europe, North America	Fresh salad
Celtuce, stem lettuce, celery lettuce, asparagus lettuce or Chinese lettuce	*Lactuca sativa* L. var. asparagina, augustana or angustata	China	Stir-fried
Escarole, chicory, witloof, Belgian endive, French endive or frisée	*Cichorium endivia* L.	Europe, North America	Raw or cooked
Radicchio, leaf chicory or Italian chicory	*Cichorium intybus* L.	Europe, North America	Raw or cooked
Spinach	*Spinacia oleracea* L.	Worldwide	Raw or cooked
New Zealand spinach	*Tetragonia tetragonioides* (Pallas) O. Ktze	Worldwide	Raw or cooked
Chard, Swiss chard, silverbeet, perpetual spinach and mangold	*Beta vulgaris* L. var. cicla	Worldwide	Raw or cooked
Pak choi	*Brassica campestris* L. ssp. *chinensis* or *Brassica rapa* ssp. *chinensis*	Worldwide	Raw or cooked
Chinese leaf mustard, gai choi	*Brassica juncea* Coss	China, Vietnam	Stir-fried
Yun tai	*Brassica rapa* ssp. *nipposinica* var. *chinoleifera*	China	Stir-fried
Garden cress	*Lepidium sativum* L.	Europe, North America, Asia	Raw or cooked
Watercress	*Nasturtium officinale* W.T. Aiton, *Nasturtium microphyllum* (Boenn.) Rchb.	Europe, Asia	Raw or cooked
Sorrel, dock, sour dock	*Rumex acetosa*	Worldwide	Raw or cooked
Tree spinach or chaya	*Cnidoscolus chayamansa* McVaugh or *Cnidoscolus aconitifolius* (P. Mill.) I.M. Johnston	Guatemala, Belize, Mexico, Honduras, Cuba, USA	Raw or cooked
Drumstick tree, horseradish tree	*Moringa oleifera* Lam.	Tropics and subtropics of Asia and Africa	Cooked
Amaranth, bayam, kulitis, yin choi, lenga lenga, biteku teku, callaloo, efo tete, arowo jeja and vleeta	*Amaranthus* sp.	Indonesia, Malaysia, India, China, Congo, Carribean, Guatamala, East Africa, Nigeria, Greece	Cooked

Coriander, cilantro, dhania, kindza, Chinese parsley and Mexican parsley	*Coriandrum sativum* L.	Worldwide	Raw or cooked
Bell tree dahlia or txoloj	*Dahlia imperialis* Roezl.	Guatamala	Cooked
Cowpea or rijol	*Vigna sesquipedalis* Fruwirth	Guatamala	Cooked
Tziton	*Tinantia erecta* Jacq.	Guatamala	Cooked
Sweet protato	*Ipomoea batatas*		Cooked
Garland chrysanthemum	*Chrysanthemum coronarium*	China	Cooked
Water convolvulus leaves, Chinese water spinach, kangkong, swamp cabbage	*Ipomoea aquatica*	Asia	Cooked
Arugula, rocket, eruca	*Eruca sativa*	Mediterranean, Europe, North America	Raw

2007). As a consequence, it is quite likely that consumption of leafy vegetables will continue to grow. Moreover, as the developing world begins to tackle health-based issues relating to diet, there will be the need to adapt indigenous as well as mainstream leafy vegetables to support health-promoting initiatives (Nakahara *et al.*, 2002; Fahey, 2005; Islam, 2006; Smith and Eyzaguirre, 2007).

10.2 Common Commodities

There is great variety in the leafy vegetables that are consumed in different areas of the world, including those that are consumed in indigenous diets (Table 10.1). While the word 'indigenous' is used to describe some leafy vegetables, globalization of world cuisine and emigration have broadened the geographic consumption of many of these vegetables. In many cases there is a movement to enhance the consumption of very healthful indigenous leafy vegetables as part of a strategy to ensure good nutrition and health in developing countries (AVRDC, 2004). There is a wide range of plant genera that provide leafy vegetables that can be consumed by humans. Their use and distribution will be discussed briefly, to gain a perspective on consumption rates and impact on human diets. While an attempt has been made to review as many leafy vegetables as possible, the discussions will focus necessarily on the

vegetables for which there is a significant amount of bioactive constituent data in the literature.

10.2.1 Lettuce and specialty salad vegetables

Three main families of plants provide the bulk of the commonly consumed lettuce and specialty salad vegetables. The largest variety of lettuce and lettuce-type vegetables consumed is included under the *Compositae* family of vegetables. These vegetables are generally labelled as lettuces or chicories. Many members of the *Brassicaceae* family of vegetables have become popular in salad mixes. Finally, *Amaranthaceae* family leafy vegetables are used extensively, either cooked or as salad vegetables or garnishes.

The largest representation from the *Compositae* family comes from the leafy vegetables classified as lettuces, and they are all classified as the species *Lactuca sativa* L. and are generally divided into six edible genotypes: butterhead, cos, latin, crisphead, cutting and stalk lettuce (de Vries, 1997). The common names associated with lettuces include iceberg lettuce or crisphead lettuce, romaine or cos lettuce, loose-leaf lettuce, green leaf lettuce, red leaf lettuce, oak leaf lettuce, butterhead lettuce, Boston or Bibb lettuce, celtuce or stem lettuce, celery lettuce and asparagus or Chinese lettuce (Table 10.1). While most are eaten in a raw format as a salad vegetable, some are eaten cooked, depending on the food customs of the area in question. Celtuce or stem lettuce, also known as asparagus or Chinese lettuce, for which the stem is the primary edible component, generally is eaten in the cooked form. Lettuce production and consumption are worldwide in scope (de Vries, 1997).

The chicory leafy vegetables are considered specialty salad vegetables and are also members of the *Compositae* family. There are many common names used within the species *Cichorium endivia* L., including escarole, chicory, witloof, Belgian endive, French endive or frisée (Lucchin *et al.*, 2008). There are also several common names associated with the leafy portions of a related species, *C. intybus* L.: radicchio, leaf chicory or Italian chicory (Lucchin *et al.*, 2008). It is consumed in fresh and cooked forms, mostly in Europe, but has gained popularity in other parts of the world.

The *Amaranthaceae* are represented by vegetables commonly known as chard, Swiss chard, silverbeet, perpetual spinach and mangold (*Beta vulgaris* L. var. *cicla*) or beet leaves (*B. vulgaris* subsp. *vulgaris*). These two types of leafy vegetables have been cultivated in Europe since antiquity and are now grown worldwide, including parts of Asia (Pyo *et al.*, 2004; Goldman and Navazio, 2008).

The *Brassicaceae* are represented by the vegetables commonly called cresses. Watercress (*Nasturtium officinale* W.T. Aiton, *N. microphyllum* (Boenn.) Rchb.) and garden cress (*Lepidium sativum* L.) are widely consumed in cooked or salad forms throughout Europe, North America and Australasia (Nuez and Hernández Bermejo, 1994; Fennell, 2006).

Coriander (also known as cilantro, dhania, kindza, Chinese parsley, Mexican parsley) is classified under the *Apiaceae* family. Coriander is consumed in diets around the world (Alberta Agriculture, Food and Rural Development, 1998).

10.2.2 Spinach

Spinach (*Spinacia oleracea* L.) belongs to the *Amaranthaceae* family and New Zealand spinach (*Tetragonia tetragonioides* (Pallas) O. Ktze)

belongs to the family *Aizoaceae*. While these two leafy vegetables are unrelated, they are consumed in similar formats, as leafy salad vegetables or cooked, on a worldwide basis (FAO, 2008).

10.2.3 Asian leafy brassicas

These vegetables are species of the genus *Brassica* and include leafy vegetables commonly termed pak choi, Chinese leaf mustard or gai choi and yun tai. While the consumption of these leafy vegetables is primarily in China, it is increasing in transplanted populations of Chinese and Asians in other parts of the world (Nöthlings *et al.*, 2006; Chen *et al.*, 2008). They are all used in cooked formats, but most especially as part of stir-fried mixtures.

10.2.4 Indigenous greens

The indigenous leafy vegetable grouping includes a broad range of species generally eaten in a cooked form. While some are widely consumed, their consumption is in traditional diets and hence these vegetables are classified in this chapter as indigenous.

Tree spinach or chaya (*Cnidoscolus chayamansa* McVaugh or *C. aconitifolius* (P. Mill.) I.M. Johnston) is in the family *Euphorbiaceae* and is always eaten in the cooked form since thorough cooking destroys most of the toxic cyanogenic glycosides present in the raw leaves. Tree spinach is consumed primarily in Central America, Mexico and southern USA (Kuti and Konuru, 2006).

Several different species of amaranth (*Amaranthus* sp.) are considered indigenous vegetables, even though they are consumed globally in a cooked format. Amaranth species are cultivated and consumed as a leafy vegetable in such diverse countries as Indonesia, Malaysia, India, China, Congo, Carribean, Guatamala, East Africa, Nigeria and Greece (Enama, 1994; Costea, 2003).

Young cowpea (*Vigna* sp.) shoots have been used as a green leafy staple in many tropical and subtropical countries of the world (Imungi and Potter, 1983; Booth *et al.*, 1992; Mosha *et al.*, 1997). Cowpea shoots are now consumed worldwide in the developed and developing world (Fatokun *et al.*, 2002).

Sweet potato leaves (*Ipomoea batatas* (L.) Lam) are consumed predominantly in African and Asian countries (Nwinyi, 1992; Almazan *et al.*, 1997). The shoots of sweet potato are rich in bioactive content, the tuber of the sweet potato is an important crop worldwide and the crop is tolerant of a range of climatic conditions (including monsoon season) (Islam, 2006). Therefore, it is felt that there is good potential for this leafy vegetable to become an important part of a healthy diet in most of the developing world (Islam, 2006).

Water convolvulus leaves (*I. aquatica*) are widely consumed in Asian countries (Prasad *et al.*, 2005). The vegetable has many common names, including Chinese water spinach, kangkong and swamp cabbage.

10.3 Major Phytochemicals

Leafy vegetables are highly regarded as carotenoid-enriched foods and many studies have shown that consumption of green leafy vegetables, such as spinach and collard greens, rich in the carotenoids lutein and zeaxanthin is associated with a substantially reduced risk of cataracts and advanced macular degeneration, one of the leading causes of blindness among the elderly in North America (Seddon *et al.*, 1994). In another carotenoid study, lutein intake from dietary sources that included spinach, lettuce and greens was associated inversely (after adjustment for fibre and folate) with colon cancer in both men and women, with the greatest inverse association observed in patients who were diagnosed when they were young (Slattery *et al.*, 2000). The food matrix itself can have a dramatic effect on bioavailability of leafy vegetable bioactive carotenoids. For example, either a 300 g dose (containing 20.8 µmol *trans*-β-carotene equivalents) of pureed spinach or a 100 g dose (containing 19.2 µmol *trans*-β-carotene equivalents) of carrots that had been deuterated intrinsically was fed to each of a group of men and women, with a standard liquid diet containing 13.5 g fat, for a 21-day period (Tang *et al.*, 2005). The subjects fed the spinach exhibited a blood serum

retinol yield of 21 ± 11 nmol/dose, whereas those who received the pureed carrot showed a yield of 32 ± 16 nmol/dose. The authors explained the yield difference between spinach and carrot as due to the β-carotene in spinach being associated with pigment proteins in the chloroplasts, whereas it was in the form of more easily digestible carotene crystals in carrot chromoplasts.

Bioactive compounds other than carotenoids present in leafy vegetables have also been studied for their efficacy in health promotion. For example, in its role of promoting bone health, a study of phylloquinone (vitamin K) and incidences of hip fractures demonstrated that women who consumed one or more daily servings of lettuce (the study food that contributed most to the dietary phylloquinone intakes) exhibited significantly reduced risk of hip fracture (Feskanich *et al.*, 1999). Out of more than 72,000 studied patients, 65 cases of hip fracture occurred with less than one lettuce serving/week, 52 cases occurred with two to four servings/week and 46 cases with five or six servings/week. A serving in that study was defined as one cup (~ 227 g) of lettuce.

In a comparison between glycolipid extracts from spinach, parsley, green onion, chive, sweet pepper, green tea, carrot and garlic, spinach extracts demonstrated the strongest inhibition of DNA polymerase α and human cancer cell proliferation (in cell cultures of gastric cancer and promyelocytic leukaemia cell lines), which the authors attributed to spinach having the highest levels of sulfoglycolipids (Kuriyama *et al.*, 2005).

As a final example of the effects of consumption of leafy vegetable products on human health, several polyphenolics isolated from sweet potato leaves were applied to three human cancer cell lines (stomach cancer, colon cancer and promyelocytic leukaemia cell lines) (Kurata *et al.*, 2007). Results indicated that the polyphenolics, especially 3,4,5-tri-*O*-caffeoylquinic acid, inhibited both the mutation of normal cells as well as the growth of cancer cells by apoptosis induction. Moreover, 3,5-di-*O*-caffeoylquinic acid, 4,5-di-*O*-caffeoylquinic acid, 3,4-di-*O*-caffeoylquinic acid and 3,4,5-tri-*O*-caffeoylquinic acid exhibited highest sensitivity towards the cancer cell lines in the order of the promyelocytic leukaemia, stomach cancer, then colon cancer cell lines.

10.3.1 Carotenoids

Most green leafy tissue has significant levels of lipophilic carotenoids and, as such, all leafy vegetables surveyed in this chapter are known to contain quantities of various carotenoids (Table 10.2). In this chapter the word 'carotenoid' is used as a general term to refer to both pure hydrocarbon carotenoids and oxygen-containing xanthophylls. The carotenoids are essentially long chains of conjugated double bonds and are differentiated by cyclization of the end group or by addition of oxygen (Rao and Rao, 2007). Xanthophylls have identical structures to carotenoids, other than specific sites where hydrogen atoms are substituted with hydroxyl groups or oxygen. The compounds considered as carotenoids are β-carotene, α-carotene, lutein, lycopene, neoxanthin, zeaxanthin and violaxanthin. All of these species have significant antioxidant capacities and it is thought this is one of their primary roles with regards to human health benefits. In addition, they have also been shown to play roles in such processes as immune response, modulation of particular drug-metabolizing enzymes, regulation of cell growth and gap junction communications (as referenced in Rao and Rao, 2007). Moreover, α-carotene and β-carotene, as well as β-cryptoxanthin, can be converted to vitamin A, and lutein and zeaxanthin have a specific function in protecting against high-energy blue light in the eye (Johnson, 2002). In general, β-carotene tends to be the predominant species in most leafy vegetables, with lutein plus zeaxanthin making up the majority of the remainder of the carotenoids (USDA, 2008).

The variability in carotenoid content between the leafy vegetables listed in Table 10.2 is much less than for most of the other bioactives that are reported. This may be due to the fact that carotenoids are associated with photosynthesis and protection against high-light injury in the chloroplast membranes (Young, 1991). All plant leafy tissues would have a similar requirement for such a protection mechanism and, as such, the content in leafy tissues from different plants would be expected to be relatively stable.

10.3.2 Tocopherols/tocotrienols

In most cases the data in Table 10.2 have derived from data values representing vitamin E content, which includes various forms of both tocopherols (e.g. α, β and γ) and tocotrienols. As it is well established that tocopherols are found in the vegetative or green portions of a plant, while tocotrienols are localized in the seeds (Munné-Bosch and Alegre, 2002), it can be assumed that all the data referring to vitamin E can be considered to be the same, as referring specifically to tocopherols. The tocopherol levels in leafy vegetables are relatively low, but stable in amounts over all of the vegetables that are listed in Table 10.2.

As discussed for carotenoids, tocopherol contents of the leafy vegetables listed in Table 10.2 are very similar for all. This uniformity can be explained by the role that this lipophilic bioactive plays in leaf cells. Tocopherol is a component of the plant membrane and acts as an important free radical protection system in both the chloroplast and mitochrondria of leaf tissues (Munné-Bosch, 2005). Since all plant leaves require a similar protection capability, the contents of this bioactive would be expected to be extremely stable across all species. The consumption of one leafy green vegetable versus another should not result in great differences in tocopherol uptake.

Tocopherol consumption has been associated with a number of mammalian health benefits. For example, α-tocopherol acts as a potent chain-breaking antioxidant and can inhibit vascular smooth muscle proliferation (thus reducing incidences of atherosclerosis and hypertension). γ-Tocopherol reduces prostaglandin E_2 synthesis and cyclooxygenase-2 activity (i.e. possesses anti-inflammatory activity) and may reduce the risk of diabetes and Alzheimer's disease (via protection against reactive nitrogen species) (for review, see Saldeen and Saldeen,

Table 10.2. The reported bioactive contents of various leafy vegetables.

Vegetable common name	Carotenoids (mg/100 g FW)	Tocopherols/tocotrienols (mg/100 g FW)	Ascorbate (mg/100 g FW)	Phenolics (mg/100 g)	Phylloquinones (µg/100 g)	Folate (µg/100 g)	Sulfur compounds (mg/100 g FW)
Lettuce	4–13	0.2–0.3	3–24	105–453	16–173	30–136	ND
Spinach	4–18	2	5–28	2	380–498	172–302	ND
Chicory	14	2	24	320–537	298	110	ND
Endive	5–16	0.5	7–41	432–1093	231	48–142	ND
New Zealand spinach	16–18	1	30–36	123	~	15	~
Swiss chard	15	2	30	145–1320	830	14	~
Asian leafy brassicas	4–12	1–7	20–139	93–234	43–497	333–425	4–282
Garden cress	17	1	69	~	542	80–186	120–390
Watercress	8	1	43	263	250	9	17–145
Coriander	3–11	3	27–72	580	310	62–196	~
Bell tree dahlia	3	~	~	~	630	~	~
Cowpea (leafy tips)	6	~	9–36	~	~	101–154	~
Tziton	5	~	~	~	250	~	~
Tree spinach	0.1–3	~	165–172	122–291	250	~	~
Amaranth	6–8	1–2	43–59	247	72–130	85–332	ND
Drumstick tree leaves	7	9	36–245	691–1300	~	40	ND
Sweet potato leaves	3	13	11–35	684–1111	185–427	80	~
Garland chrysanthemum	3–5	~	45	257–281	230–350	177	~
Sorrel	1	0.4–0.6	26–48	1456	~	13	~

(Continued)

Table 10.2. *Continued*

Vegetable common name	Carotenoids (mg/100 g FW)	Tocopherols/ tocotrienols (mg/100 g FW)	Ascorbate (mg/100 g FW)	Phenolics (mg/100 g)	Phylloquinones (µg/100 g)	Folate (µg/100 g)	Sulfur compounds (mg/100 g FW)
Water convolvulus leaves	10	~	16–45	726	~	35–225	ND
Arugula	5	0.4–2	15–254	132–235	109	97–198	95–139
Beet greens	5	1.5	30	128	400	15	~

Notes: ND = measured but not detected; ~ = no reported measurement found in the literature.
Source: Data are extracted and/or calculated from: Imungi and Potter (1983); Makkar and Becker (1997); Mosha et al. (1997); Ahenkora et al. (1998); Booth and Suttie (1998); Jiao et al. (1998); Kuti and Kuti (1999); Chu et al. (2000); Alzoreky and Nakahara (2001); Ching and Mohamed (2001); Islam et al. (2003); Iwatani et al. (2003); McNaughton and Marks (2003); Ninfali and Bacchiocca (2003); Rao (2003); Seshadra and Nambiar (2003); AVRDC (2004); Kuti and Konuru (2004); Pyo et al. (2004); Thu et al. (2004); Damon et al. (2005); de Azevedo-Meleiro and Rodriguez-Amaya (2005); Higdon (2005); Innocenti et al. (2005); Krumbein et al. (2005); Alfawaz (2006); Kidmose et al. (2006); Kim et al. (2006); Bergquist et al. (2007); Johansson et al. (2007); Kamao et al. (2007); Liu et al. (2007); Chen et al. (2008); Lavelli (2008); Martínez-Sánchez et al. (2008); USDA (2008); van der Walt et al. (2009).

2005). The positive health benefits associated with tocopherol (vitamin E) have led to recommendations to use vitamin E supplements; however, recent clinical research has raised concern that this high-dose supplementation can actually increase mortality in humans (Miller *et al.*, 2005). As a consequence, current medical advice is to rely on dietary tocopherol from fruit and vegetables while avoiding high-dosage supplements until more is known.

10.3.3 Ascorbate

Ascorbic acid is a ubiquitous antioxidant, providing protection against most, if not all, free radical species (Smirnoff, 1996). Unlike carotenoids and tocopherol, ascorbic acid is water soluble and shows a high degree of variation in content between leafy vegetable types, and even within a particular vegetable species (Table 10.2). Since it is a major metabolite in plant cells and has multiple functions, it is found in significant quantities in plant tissues, particularly photosynthetic tissues (Table 10.2; Smirnoff, 1996). Ascorbic acid is quite labile; hence, a significant ingestion is required to prevent deficiency symptoms such as scurvy (Smirnoff, 1996). In humans, ascorbate can promote health in its role as an antioxidant (i.e. defend against oxidative stress-related diseases such as cancer, various neurological disorders and cardiovascular disease) and, as an enzyme cofactor, it can participate in such reactions as collagen hydroxylation (collagen being important in the synthesis/maintenance of cartilage, skin, teeth, bones and gums) and carnitine synthesis.

As indicated above, there is a high degree of variation in ascorbate content in leafy vegetables. Some of this variation is due to differences in metabolism in tissues, but some variation may be related to the fact that ascorbate is so labile and is often the first bioactive compound to show decline under even optimal postharvest handling conditions (Gil *et al.*, 1998). This effect of postharvest handling conditions on bioactives content will be discussed later in this chapter.

10.3.4 Phenolics

The term 'phenolic', for the purposes of this discussion, refers to flavonoids and phenolic acids. Flavonoid components probably have been the most widely studied of all of the phenolics in regards to human health effects (Chu *et al.*, 2000; Ekman and Patterson, 2005). The structure of the phenolic compound influences its bioactivity with regards to human health and a myriad of health-promoting activities has been ascribed to phenolics and their derivatives. Hydroxyl groups and conjugated double bonds allow the phenolic to assume an antioxidant function through electron donation, metal ion (e.g. Cu^{3+}, Fe^{3+}) chelation and active oxygen quenching. The presence of a conjugated sugar appears to enhance specific anticancer activities of anthocyanins (Katsube *et al.*, 2003). Anthocyanins also possess anti-inflammatory activities due to their abilities to reduce chemoattractants, cyclooxygenase activities, chemokine content, platelet aggregation and cell wall adhesion factors in endothelial cells (Youdim *et al.*, 2002; Sreenivasan and Gaffar, 2008). As a last example of the claimed bioactive effects of phenolics, high phenolic content has been associated with hypocholesterolaemic and hypotriglyceridaemic activities in hamsters (Lin *et al.*, 2008). Leafy vegetables generally are very good sources of phenolics (Table 10.2; Ekman and Patterson, 2005).

Plant phenolics in the leaf are usually localized in the cell vacuole, a storage area for cell excreta (Toivonen and Brummell, 2008). In plants they can behave as anti- or pro-oxidants, depending on the environment in which they are present (Sakihama *et al.*, 2002). Their role has been considered previously to be as preformed antifungals and antifeedants, due to their reactivity and bitterness (Sakihama *et al.*, 2002; Ekman and Patterson, 2005). There is now consideration that they may also have protective effects for leaves against light damage, due to their ability to absorb high-energy light strongly (Close and McArthur, 2002). Also, because of their reactivity, phenolics become an important component of browning (healing) reactions after leaf cells have been damaged (Toivonen and Brummell, 2008).

Phenolic concentrations show a wide range of variation between and within the leafy vegetable types listed in Table 10.2. The lettuces generally have a moderate content of phenolics and the levels can range fourfold between types (Table 10.2; Liu *et al.*, 2007); iceberg lettuce has the lowest concentrations; butterhead and Batavia intermediate concentrations; romaine and leaf lettuce have the highest concentrations. Other leafy vegetables that are commonly used in salads (i.e. chicory, endive, chard, coriander and sorrel) can have much higher levels of phenolics than lettuce (Table 10.2), and hence these vegetables add a bitter character to salads, which is associated with their high phenolic levels (Ekman and Patterson, 2005). Some of the indigenous vegetables, such as drumstick tree leaves, sweet potato leaves and water convolvulus leaves, have extremely high levels of phenolics, but this is likely moderated at the consumer level because all of these vegetables generally are cooked in water before consumption and these compounds are leached out of the leaves during cooking (Table 10.1; Kuti and Konuru, 2004).

10.3.5 Phylloquinones

Phylloquinones are very abundant in dark green vegetables (Table 10.2; Booth *et al.*, 1993). There are two major forms of phylloquinones from natural sources and they are phylloquinone (vitamin K_1) and menaquinone (vitamin K_2); only phylloquinone is found in plant tissues (Booth and Suttie, 1998). In lettuces, phylloquinone content has a wide range, with greatest content in green, red and Boston leaf lettuces, followed by romaine and then iceberg lettuce (Damon *et al.*, 2005). However, phylloquinone uptake is considered to be greater from iceberg lettuce in North America since that type of lettuce is consumed most (Damon *et al.*, 2005). While much is known about phylloquinone content in some leafy vegetables, little is known for those not so commonly consumed (Table 10.2). In relation to human health, phylloquinone plays a role in blood coagulation (through action as a cofactor in conversion of specific glutamyl residues to γ-carboxyl

glutamyl residues) and in increased bone-mineral density and reduced bone resorption (Shearer *et al.*, 1996; Kamao *et al.*, 2007).

10.3.6 Folate

While folate is considered generally to be a significant nutrient from leafy green vegetables, it is clear from Table 10.2 that there is a wide range of folate concentrations within and between each type of leafy green vegetable. In terms of lettuce, the variation in folate concentration is fourfold, which is similar to the range found for phenolic concentrations in lettuces (Table 10.2). Romaine lettuce has the highest folate content, butterhead has about half that concentration and iceberg and leaf lettuces have a quarter of the concentration in romaine (USDA, 2008). Asian leafy brassicas and spinach have the highest contents of all of the most widely consumed leafy vegetables (Table 10.2). There are also several leafy vegetables that have very low folate concentrations, including Swiss chard, watercress and New Zealand spinach (Table 10.2).

Higher folate levels are known to exist in the leaves of plants than in other plant tissues (Gambonnet *et al.*, 2001) and hypotheses for this higher concentration have been suggested. Gambonnet *et al.* (2001) found that folate was involved with nucleic acid synthesis in all parts of the plant. However, there are two potentially significant roles for folate (present as methyltetrahydrofolate) in the leaves: (i) as a substrate to enable *S*-adenosyl homocysteine recycling in mitochondrial photorespiration, a process that requires methylation reactions; and (ii) as a source for methyl groups for the assembly of the highly methylated chloroplast photosynthetic apparatus. Hence, it is expected that folate levels generally would correlate with chloroplast concentrations in a particular leafy vegetable, i.e. darker green vegetables generally should have relatively high folate levels.

Animals cannot synthesize folate; thus, plant foods are the most important sources of folate (Scott *et al.*, 2000). Folate supplementation is most often recommended for pregnant women to reduce the risks of spina bifida and neural tube defects in developing embryos (Daly *et al.*, 1995). Folate deficiency in humans has been associated with higher risk of cardiovascular diseases, neurodegenerative diseases such as Alzheimer's and increased incidences of colorectal, breast, pancreatic, bronchial and cervical cancers, as well as leukaemia (Lucock *et al.*, 2003, as referenced in Rébeillé *et al.*, 2006).

10.3.7 Sulfur compounds

Although there are several forms of sulfur compound found in plants, the discussion in this chapter will focus on the glucosinolate levels found in leafy vegetables. This group of sulfur compounds can be grouped into three chemical classes, aliphatic, indolyl and aromatic glucosinolates, based on whether their amino acid precursor is methionine, tryptophan or an aromatic amino acid (tyrosine or phenylalanine) (Rosa, 1999). While there are many families of plants that contain glucosinolates (Fahey *et al.*, 2001), most of those listed in Table 10.2 are from the family *Brassicaceae* (see Chapter 5 of this volume). The only exception may be the drumstick tree. While the leaves of that tree have no measurable levels of glucosinolate (Makkar and Becker, 1997), the alternative common name, horseradish tree (Table 10.1), suggests that there must be some sulfur compounds that confer a pungent character in that leafy vegetable.

The main role of glucosinolate in plant leaves appears to be for defence against herbivory (Shroff *et al.*, 2008). Physical injury to cells causes the native glucosinolates to be hydrolysed by myrosinase, a process which requires that the two become desegregated (Fahey *et al.*, 2001; Shroff *et al.*, 2008). The hydrolysis of glucosinolates results in the formation of acrid isothiocyanates, which are the active defence compounds (Fahey *et al.*, 2001; Shroff *et al.*, 2008). The isothiocyanates are the compounds that provide the sharp mustard- or horseradish-type character to vegetables containing glucosinolates, since mastication will initiate hydrolysis of the glucosinolates (Fahey *et al.*, 2001).

Although glucosinolate breakdown products have been noted to exert a plethora of toxic and antinutritional effects in animals (e.g. detrimental effects on thyroid metabolism, embryotoxicity, growth impairment), epidemiological evidence also indicates that glucosinolate consumption reduces colon, rectum and thyroid cancers, and, when part of a diet enriched with other fruit and vegetables, protects against other cancers as well (Mithen *et al.*, 2000). The anticancer activity of glucosinolates has been attributed primarily to their ability to induce phase 1 (e.g. cytochrome P450s) and 2 (e.g. glutathione-*S*-transferase) detoxification enzymes that modify and conjugate carcinogens prior to excretion.

10.4 Preharvest and Postharvest Effects on Bioactive Content

There are some reports on the effects of pre- and postharvest factors on the bioactive content of some leafy vegetables. However, this information is limited to only a few representative crops within this category. Hence, the following discussion is very limited for many of the leafy vegetables listed in Table 10.1.

10.4.1 Preharvest effects

In the past, many cultivars have been chosen for their shelf-life qualities (Toivonen and DeEll, 2002), but in some cases this selection has also resulted in opting for cultivars having greater bioactive content. For example, Hodges and Forney (2003) associated higher basal ascorbate levels with improved shelf-life potential when comparing two cultivars of spinach. This finding is supported by another study finding that ascorbic acid concentration at harvest could be correlated with post-storage visual quality of spinach, thus indicating that the antioxidative capacity of ascorbic acid protects plant tissue against oxidative stress and ensuing quality loss (Bergquist, 2006). As databases develop for leafy vegetable cultivar differences in bioactive content, the commercial selection of cultivars that provide improved healthfulness to the consumer can be implemented. An additional benefit may also be improved postharvest handling characteristics.

There are a few reports identifying cultivar differences in bioactive content in the current literature. Differences in carotenoid, ascorbate, phenolic and phylloquinone content of lettuce types and cultivars have been reported by various groups (Degl'Innocenti *et al.*, 2005; Johansson *et al.*, 2007; Liu *et al.*, 2007; USDA, 2008). The differences in the total phenolic content among leaf lettuce cultivars can be more than fourfold, from the lowest content cv. Two Star, to the highest content cv. Galactica (Liu *et al.*, 2007). A study comparing the bioactive content of two cultivars of leaf lettuce showed that the cultivar having the highest ascorbate content also had the lowest levels of phenolics (Degl'Innocenti *et al.*, 2005), suggesting that not all bioactive compounds might be optimized simultaneously in single cultivar selection. In regards to phenolics, it must be noted that cultivars having higher phenolic content may be less desirable for fresh-cut processing, since higher phenolic content is associated with greater severity in cut-surface browning (Toivonen and Brummell, 2008). Therefore, there must be a level of compromise between improving the functionality of the vegetable and maintaining its suitability for the market place.

Spinach also shows significant cultivar-dependent variation in bioactive content. Howard *et al.* (2002) and Pandjaitan *et al.* (2005) have demonstrated significant differences in phenolic and flavonoid content in numerous commercial spinach cultivars and advanced selections. Hodges and Forney (2003) found large differences in ascorbate content when they compared two commercial cultivars of spinach.

Cultivar evaluations in Swiss chard have focused on the phenolic components in the tissues. Pyo *et al.* (2004) have reported significantly greater levels of phenolic content in red- versus green-type Swiss chard, with the differences between the Swiss chards due to differences in both the phenolic acid as well as the flavonoid content/profiles. Gil *et al.* (1998) found that yellow-type Swiss chard had significantly higher flavonoid content than the green type. Differences in phenolic content have also been reported in cultivar comparisons of a vegetable related closely to Swiss chard, beet green (Ninfali and Bacchiocca, 2003).

Chicory cultivars show extremely wide variations in total phenolic content (Innocenti *et al.*, 2005). The reported differences in phenolic content of chicory cultivars were associated largely with chicoric acid content, which accounted for over 50% of the phenolics extracted. Rocket, a leafy vegetable of similar usage as chicory, displays significant variation in both glucosinolate and flavonoid content among the different commercial cultivars (Bennett *et al.*, 2006).

In regard to indigenous leafy vegetables, there has been a reported twofold difference in total flavonoid content between the red and green type of sweet potato leaf (Chu *et al.*, 2000). In another study where ten sweet potato leaf cultivars were evaluated, there was a twofold range in total phenolic content between the lowest and highest content cultivars (Yashimoto *et al.*, 2002). In a later study comparing 1389 sweet potato genotypes, a two orders of magnitude (i.e. 100-fold) difference was found in total phenolic content (Islam, 2006).

Asian leafy vegetables generally have been highly regarded for their health-promoting properties (Jiao *et al.*, 1998; McNaughton and Marks, 2003), but very little is known about cultivar differences for this class of leafy vegetable. The glucosinolate content of Chinese cabbage cultivars can show up to threefold difference between the cultivar with the lowest and the one with the highest content (Lewis and Fenwick, 1988). Even greater differences in glucosinolate content can be found when comparing types of Chinese leafy vegetables (He *et al.*, 2003).

There is limited information on production practices and their effects on the bioactive contents in leafy vegetables. Studies on the effects of nitrate levels in hydroponic feeding systems have shown that reducing nitrogen feeding levels by half induces a greater than twofold increase in ascorbate levels of butterhead lettuce (Chiesa *et al.*, 2006). Not only is the nitrogen rate in the nutrient medium important with regards to bioactive content in leafy vegetables, but also important is the ratio of ammonium to nitrate. Kim *et al.* (2006) found that as the per cent molar ratio of ammonium to nitrate increased to 100 in the nutrient media, the glucosinolate content of rocket leaves declined. In a comparison between organic and conventional production systems, there was no difference in phenolic content for lettuce, but organically grown pak choi had higher phenolic content than conventionally grown pak choi (Young *et al.*, 2005). However, the authors attributed the significant result in pak choi simply to the fact that the organically grown product suffered from flea beetle attack and therefore the leaves would have produced more phenolics in response to the wounding caused by the flea beetle feeding.

The growing environment can also influence the bioactives that accumulate in some leafy vegetables. Butterhead lettuce grown in the open field had a much higher content of flavonols and caffeic acid derivatives than those grown in polycarbon-covered greenhouses (Romani *et al.*, 2002). In a study on production practices in spinach, it was found that shading decreased ascorbic acid content, while in contrast it increased carotenoid concentration (Bergquist, 2006). However, in that same study, shading did not have a consistent effect on flavonoid content of spinach. Some of these responses may be explained by the effect of UV-B light exposure. Caldwell and Britz (2006) found that supplemental UV-B light led to higher carotenoid content in green leaf lettuce cultivars, whereas it also led to decline in carotenoid content in red leaf cultivars.

The geographic location where the leafy vegetable is grown may also affect the bioactive content of the vegetable. A study comparing the concentrations of ascorbate, phenolics and flavonoids in drumstick tree leaves demonstrated that their contents varied depending on whether the tree was grown in Niger, India or Nicaragua (Siddhuraju and Becker, 2003). Ascorbate content did not parallel the total phenolic or flavonoid content among the regions. A wide range in ascorbate, total phenolics and flavonoids was also determined in drumstick tree leaves in different growing regions within one country, Pakistan (Iqbal and Bhanger, 2006). Similarly, folate content has been found to vary in different lettuce types grown in various European countries (Johansson *et al.*, 2007).

The age of the leafy vegetable at harvest will have a significant effect on the functional component levels in the leaves. Bergquist (2006) found that ascorbate and flavonoid content in spinach declined with the chronological age of the leaves and so suggested harvest of young leaves to ensure the best bioactive levels. However, carotenoid content did not change between leaves of different maturities. In apparent contrast, Pandjaitan *et al.* (2005) concluded that mid-maturity leaves of spinach had higher levels of flavonoids than younger or older leaves. These apparent differences in conclusions may be related to the fact that Bergquist (2006) harvested leaves from plants of differing chronological age, whereas Pandjaitan *et al.* (2005) harvested leaves of differing chronological age on a single, more mature plant. In endive and Boston lettuce, the carotenoid concentrations of mature leaves were two- to fourfold greater than those of young leaves (de Azevedo-Meleiro and Rodriguez-Amaya, 2005). In contrast, the younger leaves of New Zealand spinach had slightly higher carotenoid levels than the mature leaves (de Azevedo-Meleiro and Rodriguez-Amaya, 2005).

Another aspect of production is the time in a growing season when the leafy vegetable is harvested (i.e. harvest date). There are several of reports showing that the harvest season is extremely important with regards to bioactive content in a particular leafy vegetable. For example, the carotenoid contents of minimally processed endive and New Zealand spinach grown in Brazil were significantly higher in the summer than in the winter growing season (de Azevedo-Meleiro and Rodriguez-Amaya, 2005; Rodriguez-Amaya *et al.*, 2007). Howard *et al.* (2002) found that spring-planted spinach had almost twofold greater phenolic concentrations than autumn-planted spinach in Arkansas. They suggested that the reason spinach planted during the spring developed higher phenolic content and antioxidant capacity than autumn-grown spinach was due to the higher temperatures and greater light intensity typical of the earlier growing season. Similar patterns were seen in Sweden, where spring-sown spinach had higher phenolic content than plants sown in August (Bergquist, 2006). Finally, phylloquinone content also appears to show the same pattern in spring-grown versus autumn-grown iceberg lettuce in Finland (Koivu et al., 1997).

10.4.2 Postharvest effects

Washing is ideally a process that most, if not all, fresh produce undergoes before use or consumption. There is some evidence to support the contention that washing protocols affect the functional value of leafy vegetables and their products. For example, Baur et al. (2004) showed that washing trimmed heads or shredded iceberg lettuce with chlorinated water (100–200 mg/l free chlorine) reduced phenylalanine ammonia lyase (PAL) activity significantly and led to concomitant rise in 3,5-di-O-caffeoylquinic acid (isochlorogenic acid) concentration in the tissues during storage, compared with washes using tap or ozonated water. However, there was little effect of washing treatment on O-caffeoyltartaric (caftaric acid), di-O-caffeoyltartaric (chicoric acid), 5-O-caffeoylquinic (chlorogenic acid) and O-caffeoylmalic acid contents. Esparza Rivera et al. (2006) found that immersion in hydro-cooling water containing 1% ascorbic acid would increase ascorbate content in green leaf lettuce threefold, whereas a hydro-cool spray with the same concentration would not have a significant effect on ascorbate residues in the tissues. They also determined that there was no impact of either immersion or spray on the phenolic or antioxidant contents in the lettuce. While the immersion-treated lettuce had significantly more ascorbate, there was a rapid decline in this compound in the tissue, so that the differences between treatments had disappeared by 21 days (Esparza Rivera et al., 2006), well past the expiry date of such a product.

Heat has been widely studied as a postharvest or processing tool to reduce browning in lettuce (Hodges and Toivonen, 2007). The effect of heat on lettuce is to inhibit PAL activity, thus preventing the wound-induced accumulation of phenolics in fresh-cut product (Saltveit, 2000). While the original work was performed on lettuce rib tissue, subsequent work has also shown that there is a similar pattern of response in both midrib and lamella tissue of iceberg lettuce (Fukumoto et al., 2002). The greatest and most consistent response to heat treatment was the reduction in the accumulation of 3,5-di-O-caffeoylquinic acid (isochlorogenic acid). In addition, Moreira et al. (2006) demonstrated that heat treatments accelerated the loss in ascorbate in romaine lettuce. These examples suggest that the use of heat treatments to preserve visual quality in lettuce can lead to loss of bioactive content and functional quality.

Cutting, a common process during handling fresh leafy vegetables, is known to induce either increases or accelerated losses of bioactive components. Wound-induced increase in phenolics, a commonly known response to cutting in lettuce (Kang and Saltveit, 2002; Choi et al., 2005; Reyes et al., 2007), is achieved through upregulation of PAL activity (Campos-Vargas et al., 2005). In contrast, the cutting process leads to significant reduction in ascorbic acid content in iceberg lettuce, the actual level of loss being determined by cutting method and blade sharpness (Barry-Ryan and O'Beirne, 1999). In the latter study, the best retention of ascorbate was obtained by tearing, as opposed to cutting, the lettuce leaves.

Modified atmosphere packaging (MAP) is a widely used technology for whole and minimally processed leafy vegetables. Nitrogen-flushed MAP (called active MAP) retains ascorbate content of minimally processed iceberg lettuce better than either passive MAP (i.e. sealed without gas flushing the package) or non-sealed plastic wrapping of the processed lettuce (Barry-Ryan and O'Beirne, 1999). Schreiner et al. (2003) found that MAP in lettuce did not affect retention of carotenoids when compared with non-packaged lettuce. MAP has been shown to enhance flavonoid contents in Swiss chard stored at 6°C, while at the same time, ascorbate content loss was accelerated by MAP (Gil et al., 1998). Storage of Chinese leafy vegetables at 4 and 20°C for 6 days resulted in an increase of phenolic acid content, with the greatest response occurring at 20°C (Harbaum et al., 2008). The authors based their explanation for this increase during storage on the belief that the storage environment imposed stress on the vegetables, which resulted in stress-induced upregulation of PAL activity, with a consequent rise in phenolics. However, other constituents may not be affected by MAP conditions. In contrast to these many studies, Serafini *et al.* (2002) found that MAP packaging of lettuce results in complete loss of bioavailability of phenolics. However, these authors did not provide accurate information on the MAP treatment, to allow interpretation as to whether the results reflected those found for commercially packaged products. Certainly, more work is required to understand better the impact of MAP packaging and storage on bioavailability of bioactives in leafy salad vegetables.

Recently, the effects of superatmospheric oxygen levels (i.e. levels about 21 kPa) have been reviewed and, while there are many reports of the effects of such apparently detrimental atmospheres on bioactive contents on many fruit, only one report deals with the effects on the bioactive content of leafy vegetables (Zheng and Wang, 2007). In that one study, Heimdal *et al.* (1995) reported that superatmospheric oxygen in MAP delays the degradation of ascorbic acid in shredded iceberg lettuce; however, it is unclear how superatmospheric oxygen atmospheres can be used practically and safely.

Storage temperature is also a well-documented factor regulating bioactive content in fresh leafy vegetables (Watada, 1987). Fresh-cut iceberg lettuce stored at 3°C retains 20% more ascorbic acid content than does the same product stored at 8°C (Barry-Ryan and O'Beirne, 1999). Total phenolic content increased in storage for minimally processed butterhead and romaine lettuces, but the increase was greater at 13°C storage temperature than at 5°C (Castañer *et al.*, 1999). The increase was associated with an upregulation in PAL activity, as would be expected with *de novo* synthesis of phenolics (Degl'Innocenti *et al.*, 2005). Flavonoid losses in sweet potato leaves were very much accelerated by storage at 25°C as compared with 4°C, with losses at day 1 for the higher temperature equating to the losses that occurred at day 9 if the leaves were stored at 4°C (Chu *et al.*, 2000). In another

example, carotenoid content declined by an average of 14% in minimally processed endive over 5 days' storage at 7–9°C, while losses in New Zealand spinach were on average 29% under the same conditions (de Azevedo-Meleiro and Rodriguez-Amaya, 2005). Pandrangi and Laborde (2004) found that folate and carotenoids in commercially packaged spinach declined much more rapidly at 10 and 20°C than at 4°C, and that packaging had no effect on the decline. In that study, the commercial standard microperforated film was used to package the spinach.

Duration of fresh storage can be a significant factor in determining the content of some bioactive compounds at the consumer level. There were considerable losses in ascorbic acid during storage of spinach, whereas carotenoids and flavonoids were more stable and sometimes increased in concentration (Bergquist, 2006). Pandrangi and Laborde (2004) found declines in carotenoids in the range of 39% over 6 days at the same storage temperature as used by Bergquist (2006). The differences in the two studies may reflect differences in production systems, cultivar selection and/or the chronological age of the spinach used in the research. Declines in carotenoids similar to those shown by Pandrangi and Laborde (2004) have been shown, with time in storage at 7–9°C, for endive and New Zealand spinach (de Azevedo-Meleiro and Rodriguez-Amaya, 2005). Similarly, ascorbate decline increases with storage time in minimally processed, packaged iceberg lettuce (Barry-Ryan and O'Beirne, 1999). From these examples it becomes clear that ascorbate is the most labile of the bioactive compounds in leafy vegetables and that it may be the most limiting constituent when considering storage recommendations. This is a new concept, since research and commercial definitions of shelf life have been focused previously on the appearance of disorders or decay on the vegetable tissues. The nutritional value of fresh vegetables, whether minimally processed or whole, should be included in the consideration for determining the end of useful shelf life.

The use of irradiation to sanitize fresh leafy vegetables is increasing with the increased concern for food safety, particularly for packaged salads that will not be cooked prior to consumption (Goularte *et al.*, 2004). Although the impact of this treatment has not been studied to a large degree, there are some indications that doses effective in controlling microorganism growth on the cut leafy product also lead to increased phenolic content in lettuce (Fan *et al.*, 2003) and arugula (Nunes *et al.*, 2008). However, in contrast, irradiation treatment reduces ascorbate content significantly in iceberg and romaine lettuce (Fan and Sokorai, 2008). Irradiation can lead to quality defects in lettuce, but warm-water pretreatments have been found to improve quality retention and prevent losses of ascorbate (Fan *et al.*, 2003). However, enhanced phenolic content in response to irradiation is prevented with warm-water treatment.

10.5 Future Research Needs

There are large gaps in understanding the factors that affect the content of most bioactive compounds in leafy vegetables. If the nutritional and functional values of these commodities are considered to be central to human health, then how healthfulness can be modulated through intentional preharvest and/or postharvest practices needs to be clearly understood. For instance, there is a need to develop an accepted threshold based on the minimum acceptable loss of ascorbic acid levels when modelling shelf life for whole and fresh-cut packaged salads.

Another emerging concern relates to the indigenous leafy vegetables that are central to the diet and health of those in the developing world. Can these indigenous leafy vegetables play a role in economic development concomitant with maintaining population health in developing countries? If such new products are introduced to the Western diet, then knowledge of the effects of transport, handling and storage on bioactive content will also require further work. Another aspect of indigenous vegetables is that a few, such as tree spinach, have been shown to have specific therapeutic benefits for managing some prevalent diseases in the Western world, such as diabetes (Kuti and Torres, 1996). It could be that inclusion of some of these exotic leafy

vegetables may provide alternatives or adjuncts to medical interventions. However, such use must be tempered with clinical research to set appropriate consumption dosages and limits.

While there are measures of bioactive contents available for many of the leafy vegetables, the biological heath-promoting significance of some of these compounds has yet to be confirmed. While health benefits attributed to ascorbate, folate, phylloquinones, carotenoids and tocopherols have been relatively well documented in the literature, there are questions as to the specific significance of phenolics as antioxidants and preventers of diseases such as cancer (Seifried *et al.*, 2003). One reason for raising this question is that, while there are many reports showing generally good relationships between antioxidant capacity and phenolic content, there are other reports showing significant discrepancies. One study showed that there was a lack of a good correlation between ORAC and phenolic contents of beet greens and spinach, with the authors stating that the ORAC value was a measure of the quality of antioxidants and their interactions in the tissue matrix (Ninfali and Bacchiocca, 2003). Another study showed a poor relationship between the DPPH (2,2-diphenyl-1-picrylhydrazyl; another total antioxidant assay) assay and phenolic content of different types of lettuce (Liu *et al.*, 2007). In a last example, researchers found that many antioxidant capacity assays (DPPH radical-scavenging activity, superoxide radical-scavenging activity in riboflavin/light/NBT system, hydroxyl radical-scavenging activity and inhibition of lipid peroxidation induced by $FeSO_4$ in egg yolk) did not exhibit a strong relationship with the phenolic content of leafy vegetables (Dasgupta and De, 2007). They found that *Asteracantha longifolia* Nees and *I. reptans* (Linn.) Poir. had high phenolic content (as measured by the Folin–Ciocalteu reaction), but the measured antioxidant capacity of the phenolic fraction was relatively low. In both cases, there was a much higher level of antioxidant capacity associated with ascorbate, which was known to be reactive in the Folin–Ciocalteu reaction (Huang *et al.*, 2005). Therefore, the use of the Folin–Ciocalteu reaction to estimate total phenolic content must be evaluated carefully, to determine if there are significant levels of interfering compounds in the tissue matrix. These discrepancies point to the need for future research to characterize the functionality of complete bioactive extracts, as well as identification and quantification of the bioactive molecules present in leafy vegetable extracts. It may be that there is significant interaction between bioactive constituents in a complete extract that modulates the functionality of the extract. However, research into the identification and quantification of bioactive constituents needs to be continued, to provide the necessary information to bolster the health-based argument for increased consumption of leafy vegetables (Goldman, 2003; van Dokkum *et al.*, 2008).

Another area that needs further exploration is the bioavailability of the pure, isolated bioactive compounds themselves. More than one research report has shown that the food matrix is an important determinant of bioavailability of a bioactive molecule (e.g. β-carotene and lutein in spinach; Castenmiller *et al.*, 1999). The relative bioavailability of specific molecules in differing leafy matrices is poorly understood. For example, the bioavailability of carotenoids is not uniform among different compounds – lutein is generally much more bioavailable than β-carotene in the leafy vegetable matrix (Erdman, 1999). Folate bioavailability in vegetable matrices is also poorly understood. While leafy vegetables are particularly rich in folates, the biological importance of these high levels in a specific leaf-tissue matrix is not truly understood with regards to human health, since the relative bioavailability varies according to the vegetable in question (Gregory, 2001). There is some evidence that microstructural and biochemical interactions occurring during digestion of leafy vegetable matrices are responsible for the consequent bioavailability of specific bioactive components (Parada and Aguilera, 2007). As a good example, recent work by Su and Arab (2006) has demonstrated that regular consumption of leafy green vegetables in a salad format, with dressing, led to above-median blood serum levels of folic acid, vitamins C and E, lycopene and α- and β-carotene in test sub-ject populations. The authors found that the improved bioavailability of the bioactives was due to the addition of the salad dressing, which modified the matrix in the gut during digestion. Therefore, it appears that tissue content is only one determinant of the bioavailability of a bioactive molecule, which is also affected by many other factors, such as the tissue matrix itself and also the preparation format of the vegetable at the time of consumption.

Recently, the concept of high botanical diversity in human diets has emerged. There are clinical research findings suggesting that botanical diversity is important in expanding the bioactivity of diets high in fruit and vegetables, and that smaller amounts of many bioactive compounds have greater beneficial effects than larger amounts of relatively few bioactive compounds (Thompson *et al.*, 2006). The authors of that work cite that this may be why the concept of antioxidant supplements for managing cancer has had mixed results, as concluded by Seifried *et al.* (2003). Leafy vegetables have a significant diversity in types of bioactive compounds (Table 10.2), and often within the same botanical family there are differences in the specific molecules that are found between leafy vegetable species. For example, among Brassicaceous leafy vegetables, the forms of flavonoids found in watercress are completely different from those found in rocket (Martínez-Sánchez *et al.*, 2008). In another example, the relative quantities and content of specific glucosinolates vary significantly among different Asian leafy Brassicaceous vegetables (He *et al.*, 2003). Consequently, leafy vegetables considered in this chapter could provide at least a partial source for a high botanical diversity diet if this becomes an approach that is supported in the long-term by further research results.

10.6 Conclusions

All leafy vegetables are generally rich sources of a wide range of bioactive health-promoting compounds, including carotenoids, tocopherols, phenolics, phylloquinones, folate and ascorbate. Some leafy vegetables are also rich sources of glucosinolates, which also are considered to offer health benefits. The wide range of leafy vegetables is enhanced by the deployment of indigenous examples into mainstream consumption as populations migrate and cuisines follow. It is arguable to state that leafy vegetables have the most potential to provide health benefits, particularly in developing world economies.

While much is understood about the bioactive content of mainstream leafy vegetables, little is known about the health-promoting properties of most of the indigenous leafy vegetables that are consumed around the world. This is an area that requires more research, as the potential for inclusion of indigenous leafy vegetables into common diets increases concomitant with globalization.

In regards to increasing the consumption of mainstream leafy vegetables, there is some good preliminary information demonstrating the clinical benefits of such consumption, but still more work is required to underpin fully the recommendations for increased consumption. Also, bioavailability studies should focus on the availability and absorption of bioactives of leafy vegetables in the complete matrix in which they are consumed. For example, if the vegetable is eaten as part of a salad mix with salad dressing, then the analysis of benefits should be based on that complete product. One reason for this approach is that, in reference to fat-soluble bioactive compounds (e.g. carotenoids and tocopherols), the addition of a salad dressing containing oil can influence the bioavailability of those bioactives dramatically. Hence, future analysis must take into account the format for consumption.

It is difficult to conclude whether work should be conducted to create (via classical and/or molecular breeding) leafy vegetables that are extremely high in a particular bioactive. There are two reasons for this reticence: (i) some evidence from the literature indicates that it is not the quantity of a particular bioactive from a single vegetable that is important to health; rather, it is the diversity of bioactives in the diet that is important; and (ii) if many leafy vegetables are consumed in a diet, then there may be sufficient uptake of bioactives, rendering an enhancement in content irrelevant to further health improvement. Some have also raised the concern of potential toxicity and/or bitter taste if some classes of bioactives are increased excessively.

Bioactive content can be affected and/or manipulated by many factors. Future research should be conducted to optimize production (including cultivar selection) and postharvest management of leafy vegetables better, such that bioactive values are maintained at desirable levels. Some treatments/technologies potentially may enhance concentrations of some bioactive compounds.

Finally, leafy green vegetables are a significant source of most of the important bioactive compounds found in fruit and vegetables. The nature of the function of many compounds, particularly phylloquinones and folate, in the leaf provides a mechanistic explanation for the richness of these compounds in leaves. They potentially can provide a diversity of tastes and flavours that also enrich the pleasure of a healthy and healthful diet. In a world that will require options to ensure health in developing economies, the inclusion of indigenous leafy vegetables can provide a sustainable mechanism to deliver such a diet to less affluent consumers.

References

Ahenkora, K., Adu Dapaah, H.K. and Agyemang, A. (1998) Selected nutritional components and sensory attributes of cowpea (*Vigna unguiculata* [L.]Walp) leaves. *Plant Foods for Human Nutrition* 52, 221–229.

Alberta Agriculture, Food and Rural Development (1998) Coriander. *Alberta Agriculture, Food and Rural Development Agri-Facts*, Agdex 147/20-2, 4 pp.

Alfawaz, M.A. (2006) Chemical composition of hummayd (*Rumex vesicarius*) grown in Saudi Arabia. *Journal of Food Composition and Analysis* 19, 552–555.

Almazan, A.M., Begum, F. and Johnson, C. (1997) Nutritional quality of sweetpotato greens from green-house plants. *Journal of Food Composition and Analysis* 10, 246–253.

Alzoreky, N. and Nakahara, K. (2001) Antioxidant activity of some edible Yemeni plants evaluated by fer-rylmyoglobin/ABTS⁺ assay. *Food Science and Technology Research* 7, 141–144.

AVRDC (2004) *AVRDC Report 2003.* AVRDC Publication Number 04-599. AVRDC – The World Vegetable Center, Shanhua, Taiwan, 194 pp.

Barry-Ryan, C. and O'Beirne, D. (1999) Ascorbic acid retention in shredded iceberg lettuce as affected by minimal processing. *Journal of Food Science* 64, 498–500.

Baur, S., Klaiber, R.G., Koblo, A. and Carle, R. (2004) Effect of different washing procedures on phenolic metabolism of shredded, packaged iceberg lettuce during storage. *Journal of Agricultural and Food Chemistry* 52, 7017–7025.

Bennett, R.N., Rosa, E.A.S., Mellon, F.A. and Kroon, P.A. (2006) Ontogenic profiling of glucosinolates, fla-vonoids, and other secondary metabolites in *Eruca sativa* (salad rocket), *Diplotaxis erucoides* (wall rocket), *Diplotaxis tenuifolia* (wild rocket), and *Bunias orientalis* (Turkish rocket). *Journal of Agricultural and Food Chemistry* 54, 4005–4015.

Bergquist, S. (2006) Bioactive compounds in baby spinach (*Spinacia oleracea* L.) effects of pre- and posthar-vest factors. Doctoral thesis, Swedish University of Agricultural Sciences, Alnarp, Sweden.

Bergquist, S.Å.M., Gertsson, U.E. and Olsson, M.E. (2007) Bioactive compounds and visual quality of baby spinach – changes during plant growth and storage. *Acta Horticulturae* 744, 343–348.

Booth, F.E.M. and Wickens, G.E. (1988) *Non-Timber Uses of Selected Arid Zone Trees and Shrubs in Africa,* FAO Conservation Guide 19. Food and Agriculture Organization, Rome, 176 pp.

Booth, S.L. and Suttie, J.W. (1998) Dietary intake and adequacy of Vitamin K¹. *Journal of Nutrition* 128, 785–788.

Booth, S., Bressani, R. and Johns, T. (1992) Nutrient content of selected indigenous leafy vegetables con-sumed by Kekchi people of Alta Verapaz, Guatemala. *Journal of Food Composition and Analysis* 5, 25–34.

Booth, S.L., Sadowski, J.A., Weihrauch, J.L. and Ferland, G. (1993) Vitamin K_1 [phylloquinone] content of foods: a provisional table. *Journal of Food Composition and Analysis* 6, 109–120.

Caldwell, C.R. and Britz, S.J. (2006) Effect of supplemental ultraviolet radiation on the carotenoid and chlorophyll composition of green house-grown leaf lettuce (*Lactuca sativa* L.) cultivars. *Journal of Food Composition and Analysis* 19, 637–644.

Campos-Vargas, R., Nonogakia, H., Suslow, T. and Saltveit, M.E. (2005) Heat shock treatments delay the increase in wound-induced phenylalanine ammonia-lyase activity by altering its expression, not its induction in Romaine lettuce (*Lactuca sativa*) tissue. *Physiologia Plantarum* 123, 82–91.

Cardwell, M. (2005) Fresh-cut sales shoot up like weeds. *Food in Canada* July/August, 24–30.

Castañer, M., Gil, M.I., Ruíz, M.V. and Artés, F. (1999) Browning susceptibility of minimally processed baby and romaine lettuces. *European Food Research and Technology* 209, 52–56.

Castenmiller, J.J.M., West, C.E., Linssen, J.PH., van het Hof, K.H. and Voragen, A.G.J. (1999) The food ma-trix of spinach is a limiting factor in determining the bioavailability of β-carotene and to a lesser ex-tent of lutein in humans. *Journal of Nutrition* 129, 349–355.

Chen, X., Zhu, Z., Gerendás, J. and Zimmermann, N. (2008) Glucosinolates in Chinese *Brassica campestris* vegetables: Chinese cabbage, purple cai-tai, choysum, pakchoi, and turnip. *HortScience* 43, 571–574.

Chiesa, A., Frezza, D., León, A., Mayorga, I. and Logegaray, V. (2006) Technological strategies for assuring and maintaining the quality of minimally processed lettuce (*Lactuca sativa* L.). *Acta Horticulturae* 712, 483–490.

Ching, L.S. and Mohamed, S. (2001) Alpha-tocopherol content in 62 edible tropical plants. *Journal of Agri-cultural and Food Chemistry* 49, 3101–3105.

Choi, Y.J., Tomás-Barberán, F.A. and Saltveit, M.E. (2005) Wound-induced phenolic accumulation and browning in lettuce (*Lactuca sativa* L.) leaf tissue is reduced by exposure to n-alcohols. *Postharvest Biol-ogy and Technology* 37, 47–55.

Chu, Y.-H., Chang, C.-L. and Hsu, H.-F. (2000) Flavonoid content of several vegetables and their antioxi-dant activity. *Journal of the Science of Food and Agriculture* 80, 561–566.

Close, D.C. and McArthur, C. (2002) Rethinking the role of many plant phenolics – protection from photodamage not herbivores? *Oikos* 99, 166–172.

Costea, M. (2003) Notes on economic plants. *Economic Botany* 57, 646–649.

Daly, L.E., Kirke, P.M., Molloy, A., Weir, D.G. and Scott, J.M. (1995) Folate levels and neural tube defects. Implications for prevention. *Journal of the American Medical Association* 274, 1698–1702.

Damon, M., Zhang, N.Z., Haytowitz, D.B. and Booth, S.L. (2005) Phylloquinone (vitamin K1) content of vegetables. *Journal of Food Composition and Analysis* 18, 751–758.

Dasgupta, N. and De, B. (2007) Antioxidant activity of some leafy vegetables of India: a comparative study. *Food Chemistry* 101, 471–474.

de Azevedo-Meleiro, C.H. and Rodriguez-Amaya, D.B. (2005) Carotenoids of endive and New Zealand spinach as affected by maturity, season and minimal processing. *Journal of Food Composition and Analysis* 18, 845–855.

de Vries, I.M. (1997) Origin and domestication of *Lactuca sativa* L. *Genetic Resources and Crop Evolution* 44, 165–174.

Degl'Innocenti, E., Guidi, L., Pardossi, A. and Tognoni, F. (2005) Biochemical study of leaf browning in minimally processed leaves of lettuce (*Lactuca sativa* L. var. *Acephala*). *Journal of Agricultural and Food Chemistry* 53, 9980–9984.

Ekman, J.H. and Patterson, B.D. (2005) Why fruits and vegetables are good for health. In: Ben-Yehoshua, S. (ed.) *Environmentally Friendly Technologies for Agricultural Produce Quality*. Taylor and Francis, Boca Raton, Florida, pp. 333–396.

Enama, M. (1994) Culture: the missing nexus in ecological economics perspective. *Ecological Economics* 10, 93–95.

Erdman, J.W. Jr (1999) Variable bioavailability of carotenoids from vegetables. *American Journal of Clinical Nutrition* 70, 179–180.

Esparza Rivera, J.R., Stone, M.B., Stushnoff, C., Pilon-Smits, E. and Kendall, P.A. (2006) Effects of ascorbic acid applied by two hydrocooling methods on physical and chemical properties of green leaf lettuce stored at 5°C. *Journal of Food Science* 71, S270–S276.

Fahey, J.W. (2005) *Moringa oleifera*: a review of the medical evidence for its nutritional, therapeutic, and prophylactic properties. Part 1. *Trees for Life Journal* 1, 5–20.

Fahey, J.W., Zalcmann, A.T. and Talalay, P. (2001) The chemical diversity and distribution of glucosinolates and isothiocyanates among plants. *Phytochemistry* 56, 5–51.

Fan, X. and Sokorai, K.J.B. (2008) Retention of quality and nutritional value of 13 fresh-cut vegetables treated with low-dose radiation. *Journal of Food Science* 73, S367–S372.

Fan, X., Toivonen, P.M.A., Rajkowski, K.T. and Sokorai, K.B. (2003) Warm water treatment in combination with modified atmosphere packaging reduces undesirable effects of irradiation on the quality of fresh-cut iceberg lettuce. *Journal of Agricultural and Food Chemistry* 51, 1231–1236.

FAO (2008) Food and Agricultural Organization of the United Nations Statistical Databases and Data-sets (http://faostat.fao.org/site/567/DesktopDefault.aspx?PageID=567, accessed 6 October 2008).

Fatokun, C.A., Tarawali, S.A., Singh, B.B., Kormawa, P.M. and Tamò, M. (eds) (2002) *Challenges and Opportunities for Enhancing Sustainable Cowpea Production*. Proceedings of the World Cowpea Conference III held at the International Institute of Tropical Agriculture (IITA), Ibadan, Nigeria, 4–8 September 2000. IITA, Ibadan, Nigeria, 396 pp.

Fennell, J.F.M. (2006) *Potential for Watercress Production in Australia: Scoping Study*. Rural Industries Research and Development Corporation Publication No 06/105, Lenswood, South Australia, 73 pp.

Feskanich, D., Weber, P., Willett, W.C., Rockett, H., Booth, S.L. and Colditz, g.A. (1999) Vitamin K intake and hip fractures in women: a prospective study. *American Journal of Clinical Nutrition* 69, 74–79.

Fukumoto, L.R., Toivonen, P.M.A. and Delaquis, P.J. (2002) Effect of wash water temperature and chlorination on phenolic metabolism and browning of stored iceberg lettuce photosynthetic and vascular tissues. *Journal of Agricultural and Food Chemistry* 50, 4503–4511.

Gambonnet, B., Jabrin, A., Ravanel, S., Karan, M., Douce, R. and Rébeillé, F. (2001) Folate distribution during higher plant development. *Journal of the Science of Food and Agriculture* 81, 835–841.

Gil, M.I., Ferreres, F. and Tomás-Barberán, F.A. (1998) Effect of modified atmosphere packaging on the flavonoids and vitamin C content of minimally processed Swiss chard (*Beta vulgaris* subspecies *cycla*). *Journal of Agricultural and Food Chemistry* 46, 2007–2012.

Goldman, I.L. (2003) Recognition of fruit and vegetables as healthful: vitamins and phytonutrients. *HortTechnology* 13, 252–258.

Goldman, I.L. and Navazio, J.P. (2008) Table beet. In: Prohens, J. and Nuez, F. (eds) *Handbook of Plant Breeding. Vegetables I (Asteraceae, Brassicaceae, Chenopodicaceae, and Cucurbitaceae)*. Springer, New York, pp. 219–238.

Goularte, L., Martins, C.G., Morales-Aizpurúa, I.C., Destro, M.T., Franco, B.D.G.M., Vizeu, D.M., Hutzler, B. and Landgraf, M. (2004) Combination of minimal processing and irradiation to improve the microbiological safety of lettuce (*Lactuca sativa* L.). *Radiation Physics and Chemistry* 71, 155–159.

Gregory, J.F. III (2001) Case study: folate bioavailability. *Journal of Nutrition* 131, 1376S–1382S.

Harbaum, B., Hubbermann, E.M., Zhu, Z. and Schwarz, K. (2008) Free and bound phenolic compounds in leaves of pak choi (*Brassica campestris* L. ssp. *chinensis* var. *communis*) and Chinese leaf mustard (*Brassica juncea* Coss). *Food Chemistry* 110, 838–846.

He, H., Liu, L., Song, S., Tang, X. and Wang, Y. (2003) Evaluation of glucosinolate composition and contents in Chinese brassica vegetables. *Acta Horticulturae* 620, 85–92.

Hedges, L.J. and Lister, C.E. (2005) *Nutritional Attributes of Salad Vegetables*. Crop and Food Research Confidential Report No. 1473. New Zealand Institute for Crop and Food Research Limited, Christchurch, New Zealand.

Heimdal, H., Kuhn, B.F., Poll, L. and Larsen, L.M. (1995) Biochemical changes and sensory quality of shredded and MA-packaged iceberg lettuce. *Journal of Food Science* 60, 1265–1268.

Higdon, J. (2005) Isothiocyanates. Micronutrient Information Center, Linus Pauling Institute, Oregon State University, Corvallis, Oregon (http://lpi.oregonstate.edu/infocenter/phytochemicals/isothio/, accessed 25 September 2008).

Hodges, D.M. and Forney, C.F. (2003) Postharvest ascorbate metabolism in two cultivars of spinach differing in their senescence rates. *Journal of the American Society for Horticultural Science* 128, 930–935.

Hodges, D.M. and Toivonen, P.M.A. (2007) Quality of fresh-cut fruits and vegetables as affected by exposure to abiotic stress. *Postharvest Biology and Technology* 48, 155–162.

Howard, L.R., Pandjaitan, N., Morelock, T. and Gil, M.I. (2002) Antioxidant capacity and phenolic content of spinach as affected by genetics and growing season. *Journal of Agricultural and Food Chemistry* 50, 5891–5896.

Huang, D., Ou, B. and Prior, R.L. (2005) The chemistry behind antioxidant capacity assays. *Journal of Agricultural and Food Chemistry* 53, 1841–1856.

Imungi, J.K. and Potter, N.N. (1983) Nutrient contents of raw and cooked cowpea leaves. *Journal of Food Science* 48, 1252–1254.

Innocenti, M., Gallori, S., Giaccherini, C., Ieri, F., Vincieri, F.F. and Mulinacci, N. (2005) Evaluation of the phenolic content in the aerial parts of different varieties of *Cichorium intybus* L. *Journal of Agricultural and Food Chemistry* 53, 6497–6502.

Iqbal, S. and Bhanger, M.I. (2006) Effect of season and production location on antioxidant activity of *Moringa oleifera* leaves grown in Pakistan. *Journal of Food Composition and Analysis* 19, 544–551.

Islam, S. (2006) Sweetpotato (*Ipomoea batatas* L.) leaf: its potential effect on human health and nutrition. *Journal of Food Science* 71, R13–R21.

Islam, S., Yoshimoto, M., Ishiguro, K. and Yamakawa, O. (2003) Bioactive compounds in *Ipomoea batatas* leaves. *Acta Horticulturae* 628, 693–699.

Iwatani, Y., Arcot, J. and Shrestha, A.K. (2003) Determination of folate contents in some Australian vegetables. *Journal of Food Composition and Analysis* 16, 37–48.

Jiao, D., Yu, M.C., Hankin, J.H., Low, S.-H. and Chung, F.-L. (1998) Total isothiocyanate contents in cooked vegetables frequently consumed in Singapore. *Journal of Agricultural and Food Chemistry* 46, 1055–1058.

Johansson, M., Jägerstad, M. and Frølich, W. (2007) Folates in lettuce: a pilot study. *Scandinavian Journal of Food Science and Nutrition* 51, 22–30.

Johnson, E.J. (2002) The role of carotenoids in human health. *Nutrition in Clinical Care* 5, 56–65.

Kamao, M., Suhara, Y., Tsugawa, N., Uwano, M., Yamaguchi, N., Uenishi, K., Ishida, H., Sasaki, S. and Okano, T. (2007) Vitamin K content of foods and dietary vitamin K intake in Japanese young women. *Journal of Nutritional Science and Vitaminology* 53, 464–470.

Kang, H.-M. and Saltveit, M.E. (2002) Antioxidant capacity of lettuce leaf tissue increases after wounding. *Journal of Agricultural and Food Chemistry* 50, 7536–7541.

Katsube, N., Iwashita, K., Tsushida, T., Yamaki, K. and Kobori, M. (2003) Induction of apoptosis in cancer cells by bilberry (*Vaccinium myrtillus*) and the anthocyanins. *Journal of Agricultural and Food Chemistry* 51, 68–75.

Kidmose, U., Yang, R.-Y., Thilsted, S.H., Christensen, L.P. and Brandt, K. (2006) Content of carotenoids in commonly consumed Asian vegetables and stability and extractability during frying. *Journal of Food Composition and Analysis* 19, 562–571.

Kim, S.-J., Kawaharada, C. and Ishii, G. (2006) Effect of ammonium: nitrate nutrient ratio on nitrate and glucosinolate contents of hydroponically-grown rocket salad (*Eruca sativa* Mill.) *Soil Science and Plant Nutrition* 52, 387–393.

Koivu, T.J., Piironen, V.I., Henttonen, S.K. and Mattila, P.H. (1997) Determination of phylloquinone in vegetables, fruits, and berries by high-performance liquid chromatography with electrochemical detection. *Journal of Agricultural and Food Chemistry* 45, 4644–4649.

Krumbein, A., Schonhof, I. and Schreiner, M. (2005) Composition and contents of phytochemicals (gluco-sinolates, carotenoids and chlorophylls) and ascorbic acid in selected *Brassica* species (*B. juncea, B. rapa* subsp. *nipposinica* var. *chinoleifera, B. rapa* subsp. *chinensis* and *B. rapa* subsp. *rapa*). *Journal of Applied Botany and Food Quality* 79, 168–174.

Kurata, R., Adachi, M., Yamakawa, O. and Yoshimoto, M. (2007) Growth suppression of human cancer cells by polyphenolics from sweetpotato (*Ipomoea batatas* L.) leaves. *Journal of Agricultural and Food Chemistry* 55, 185–190.

Kuriyama, I., Musumi, K, Yonezawa, Y., Takemura, M., Maeda, N., Iijima, H., Hada, Y., Yoshida, H. and Mizushina, Y. (2005) Inhibitory effects of glycolipids fraction from spinach on mammalian DNA polymerase activity and human cancer cell proliferation. *Journal of Nutritional Biochemistry* 16, 594–601.

Kuti, J.O. and Konuru, H.B. (2004) Antioxidant capacity and phenolic content in leaf extracts of tree spinach (*Cnidoscolus* spp.). *Journal of Agriculture and Food Chemistry* 52, 117–121.

Kuti, J.O. and Konuru, H.B. (2006) Cyanogenic glycosides content in two edible leaves of tree spinach (*Cnidoscolus* spp.). *Journal of Food Composition and Analysis* 19, 556–561.

Kuti, J.O. and Kuti, H.O. (1999) Proximate composition and mineral content of two edible species of *Cnidoscolus* (tree spinach). *Plant Foods for Human Nutrition* 53, 275–283.

Kuti, J.O. and Torres, E.S. (1996) Potential nutritional and health benefits of tree spinach. In: Janick, J. (ed.) *Progress in New Crops*. ASHS Press, Arlington, Virginia, pp. 516–520.

Lavelli, V. (2008) Antioxidant activity of minimally processed red chicory (*Cichorium intybus* L.) evaluated in xanthine oxidase-, myeloperoxidase-, and diaphorase-catalyzed reactions. *Journal of Agricultural and Food Chemistry* 56, 7194–7200.

Lewis, J. and Fenwick, G.R. (1988) glucosinolate content of brassica vegetables – Chinese cabbages pe-tsai (*Brassica pekinensis*) and pak-choi (*Brassica chinensis*). *Journal of the Science of Food and Agriculture* 45, 379–386.

Lin, L.-Y., Peng, C.-C., Yang, Y.-L. and Peng, R.Y. (2008) Optimization of bioactive compounds in buckwheat sprouts and their effects on blood cholesterol in hamsters. *Journal of Agricultural and Food Chemistry* 56, 1216–1223.

Liu, X., Ardo, S., Bunning, M., Parry, J., Zhou, K., Stushnoff, C., Stoniker, F., Yu, L. and Kendall, P. (2007) Total phenolic content and DPPH· radical scavenging activity of lettuce (*Lactuca sativa* L.) grown in Colorado. *LWT – Food Science and Technology* 40, 552–557.

Lucchin, M., Varotto, S., Barcaccia, G, and Paolo Parrini, P. (2008) Chicory and endive. In: Prohens, J. and Nuez, F. (eds) *Handbook of Plant Breeding. Vegetables I (Asteraceae, Brassicaceae, Chenopodicaceae, and Cucurbitaceae)*. Springer, New York, pp. 3–8.

Lucock, M., Yates, Z., Glanville, T., Leeming, R., Simpson, N. and Daskalakis, I. (2003) A critical role for B-vitamin nutrition in human developmental and evolutionary biology. *Nutrition Research* 23, 1463–1475.

McNaughton, S.A. and Marks, G.C. (2003) Development of a food composition database for the estimation of dietary intakes of glucosinolates, the biologically active constituents of cruciferous vegetables. *British Journal of Nutrition* 90, 687–697.

Makkar, H.P.S. and Becker, K. (1997) Nutrients and antiquality factors in different morphological parts of the *Moringa oleifera* tree. *Journal of Agricultural Science* 128, 311–322.

Martínez-Sánchez, A., Gil-Izquierdo, A., Gil, M.I. and Ferreres, F. (2008) A comparative study of flavonoid compounds, vitamin C, and antioxidant properties of baby leaf *Brassicaceae* species. *Journal of Agricultural and Food Chemistry* 56, 2330–2340.

Miller, E.R. III, Pastor-Burriuso, R., Dalal, D., Riemersma, R.A., Appel, L.J. and Guallar, E. (2005) Meta-analysis: high-dosage vitamin E supplementation may increase all-cause mortality. *Annals of Internal Medicine* 142, 37–46.

Mithen, R.F., Dekker, M., Verkerk, R. and Rabot, S. (2000) The nutritional significance, biosynthesis and bioavailability of glucosinolates in human foods. *Journal of the Science of Food and Agriculture* 80, 967–984.

Moreira, M.R., Ponce, A.G., del Valle, C.E. and Roura, S.I. (2006) Ascorbic acid retention, microbial growth, and sensory acceptability of lettuce leaves subjected to mild heat shocks. *Journal of Food Science* 71, S188–S192.

Mosha, T.C., Pace, R.D., Adeyeye, S., Laswai, H.S. and Mtebe, K. (1997) Effect of traditional processing practices on the content of total carotenoid, β-carotene, α-carotene and vitamin A activity of selected Tanzanian vegetables. *Plant Foods for Human Nutrition* 50, 189–201.

Munné-Bosch, S. (2005) The role of α-tocopherol in plant stress tolerance. *Journal of Plant Physiology* 162, 743–748.

Munné-Bosch, S. and Alegre, L. (2002) The function of tocopherols and tocotrienols in plants. *Critical Reviews in Plant Sciences* 21, 31–57.

Nakahara, K., Roy, M.K., Alzoreky, N.S., Na Thalang, V. and Trakoontivakorn, G. (2002) Inventory of indigenous plants and minor crops in Thailand based on bioactivities. In: Mori, Y., Hayashi, T. and Highley, E. (eds) *9th Japan International Research Center for Agricultural Sciences International Symposium 2002 – 'Value-Addition to Agricultural Products'*. Japan International Research Centre for Agricultural Sciences, Ibaraki, Japan, pp. 135–139.

Ninfali, P. and Bacchiocca, M. (2003) Polyphenols and antioxidant capacity of vegetables under fresh and frozen conditions. *Journal of Agricultural and Food Chemistry* 51, 2222–2226.

Nöthlings, U., Wilkens, L.R., Murphy, S.P., Hankin, J.H., Henderson, B.E. and Kolonel, L.N. (2006) Vegetable intake and pancreatic cancer risk: the multiethnic cohort study. *American Journal of Epidemiology* 165, 138–147.

Nuez, F. and Hernández Bermejo, J.E. (1994) Neglected crops: 1492 from a different perspective. In: Hernándo Bermejo, J.E. and León, J. (eds) *Plant Production and Protection Series No. 26*. FAO, Rome, pp. 303–332.

Nunes, T.P., Martins, C.G., Behrens, J.H., Souza, K.L.O., Genovese, M.I., Destro, M.T. and Landgraf, M. (2008) Radioresistance of *Salmonella* species and *Listeria monocytogenes* on minimally processed arugula (*Eruca sativa* Mill.): effect of irradiation on flavonoid content and acceptability of irradiated produce. *Journal of Agricultural and Food Chemistry* 56, 1264–1268.

Nwinyi, S.C.O. (1992) Effect of age at shoot removal on tuber and shoot yields at harvest of five sweet potato (*Ipomoea batatas* (L.) Lam) cultivars. *Field Crops Research* 29, 47–54.

Orech, F.O., Akenga1, T., Ochora, J., Friis, H. and Aagaard-Hansen, J. (2006) Potential toxicity of some traditional leafy vegetables consumed in Nyang'oma Division, Western Kenya. In: Midiwo, J.O., Yenesew, A. and Derese, S. (eds) *11th NAPRECA Symposium Book of Proceedings*. Natural Product Research Network for Eastern and Central Africa, Nairobi, pp. 78–87.

Pandjaitan, N., Howard, L.R., Morelock, T. and Gil, M.I. (2005) Antioxidant capacity and phenolic content of spinach as affected by genetics and maturation. *Journal of Agricultural and Food Chemistry* 53, 8618–8623.

Pandrangi, S. and Laborde, L.F. (2004) Retention of folate, carotenoids, and other quality characteristics in commercially packaged fresh spinach. *Journal of Food Science* 69, C702–C707.

Parada, J. and Aguilera, J.M. (2007) Food microstructure affects the bioavailability of several nutrients. *Journal of Food Science* 72, R21–R32.

Prasad, K.N., Divakar, S., Shivamurthy, G.R. and Aradhya, S.M. (2005) Isolation of a free radical-scavenging antioxidant from water spinach (*Ipomoea aquatica* Forsk). *Journal of the Science of Food and Agriculture* 85, 1461–1468.

Putnam, J.J. and Allshouse, J.E. (1999) Food, consumptions, prices, and expenditures, 1970–97. USDA Economic Research Service, Statistical Bulletin Number 965. 196 pp (http://www.ers.usda.gov/publications/sb965/, accessed 13 July 2008).

Pyo, Y.-H., Lee, T.-C., Logendra, L. and Rosen, R.T. (2004) Antioxidant activity and phenolic compounds of Swiss chard (*Beta vulgaris* subspecies *cycla*) extracts. *Food Chemistry* 85, 19–26.

Rai, A., Mohapatra, S.C. and Shukla, H.S. (2006) Correlates between vegetable consumption and gallbladder cancer. *European Journal of Cancer Prevention* 15, 134–137.

Rao, A.V. and Rao, L.G. (2007) Carotenoids and human health. *Pharmacological Research* 55, 207–216.

Rao, B.N. (2003) Bioactive phytochemicals in Indian foods and their potential in health promotion and disease prevention. *Asia Pacific Journal of Clinical Nutrition* 12, 9–22.

Rébeillé, F., Ravanel, S., Jabrin, S., Douce, R., Storozhenko, S. and Van Der Straeten, D. (2006) Folates in plants: biosynthesis, distribution, and enhancement. *Physiologia Plantarum* 126, 330–342.

Reyes, L.F., Villarreal, J.E. and Cisneros-Zevallos, L. (2007) The increase in antioxidant capacity after wounding depends on the type of fruit or vegetable tissue. *Food Chemistry* 101, 1254–1262.

Rocha, A. and Morais, A.M.M.B. (2007) Role of minimally processed fruit and vegetables on the diet of the consumers in the XXI century. *Acta Horticulturae* 746, 265–271.

Rodriguez-Amaya, D.B., Porcu, O.M. and Azevedo-Meleiro, C.H. (2007) Variation in the carotenoid composition of fruits and vegetables along the food chain. *Acta Horticulturae* 744, 387–394.

Romani, A., Pinelli, P., Galardi, C., Sani, G., Cimato, A. and Heimler, D. (2002) Polyphenols in greenhouse and open-air-grown lettuce. *Food Chemistry* 79, 337–342.

Rosa, E.A.S. (1999) Chemical composition. In: Gómez-Campo, C. (ed.) *Biology of Brassica Coenospecies*. Elsevier Science BV, Amsterdam, pp. 315–357.

Sakihama, Y., Cohen, M.F., Grace, S.C. and Yamasaki, H. (2002) Plant phenolic antioxidant and prooxidant activities: phenolics-induced oxidative damage mediated by metals in plants. *Toxicology* 177, 67–80.

Saldeen, K. and Saldeen, T. (2005) Importance of tocopherols beyond α-tocopherol: evidence from animal and human studies. *Nutrition Research* 25, 877–889.

Saltveit, M.E. (2000) Wound-induced changes in phenolic metabolism and tissue browning are altered by heat shock. *Postharvest Biology and Technology* 21, 61–69.

Schreiner, M., Huyskens-Keil, S., Krumbein, A., Prono-Widayat, H. and Lüdders, P. (2003) Effect of film packaging and surface coating on primary and secondary plant compounds in fruit and vegetable products. *Journal of Food Engineering* 56, 237–240.

Scott, J., Rébeillé, F. and Fletcher, J. (2000) Folic acid and folates: the feasibility for nutritional enhancement in plant foods. *Journal of the Science of Food and Agriculture* 80, 795–824.

Seddon, J.M., Ajani, U.A., Sperduto, R.D., Hiller, R., Blair, N., Burton, T.C., Farber, M.D., Gragoudas, E.S., Haller, J., Miller, D.T., Yannuzzi, L.A. and Willett, W. (1994) Dietary carotenoids, vitamins A, C and E, and advanced age-related macular degeneration. Eye disease case–control study group. *Journal of the American Medical Association* 272, 1413–1420.

Seifried, H.E., McDonald, S.S., Anderson, D.E., Greenwald, P. and Milner, J.A. (2003) The antioxidant conundrum in cancer. *Cancer Research* 63, 4295–4298.

Serafini, M., Bugianesi, R., Salucci, M., Azzini, E., Raguzzini, A. and Maiani, G. (2002) Effect of acute ingestion of fresh and stored lettuce (*Lactuca sativa*) on plasma total antioxidant capacity and antioxidant levels in human subjects. *British Journal of Nutrition* 88, 615–623.

Seshadra, S. and Nambiar, V.S. (2003) Kanjero (*Digera arvensis*) and drumstick leaves (*Moringa oleifera*): nutrient profile and potential for human consumption. *World Review of Nutrition and Dietetics* 91, 41–59.

Shearer, M.J., Bach, A. and Kohlmeier, M. (1996) Chemistry, nutritional sources, tissue distribution and metabolism of vitamin K with special reference to bone health. *Journal of Nutrition* 126, 1181S–1186S.

Shroff, R., Vergara, F., Muck, A., Svatoš, A. and Gershenzon, J. (2008) Non-uniform distribution of glucosinolates in *Arabidopsis thaliana* leaves has important consequences for plant defense. *Proceedings of the National Academy of Sciences of the United States of America* 105, 6196–6201.

Siddhuraju, P. and Becker, K. (2003) Antioxidant properties of various solvent extracts of total phenolic constituents from three different agroclimatic origins of drumstick tree (*Moringa oleifera* Lam.) leaves. *Journal of Agricultural and Food Chemistry* 51, 2144–2155.

Slattery, M.L., Benson, J., Curtin, K., Ma, K.-N., Schaeffer, D. and Potter J.D. (2000) Carotenoids and colon cancer. *American Journal of Clinical Nutrition* 71, 575–582.

Smirnoff, N. (1996) The function and metabolism of ascorbic acid in plants. *Annals of Botany* 78, 661–669.

Smith, F. and Eyzaguirre, P. (2007) African leafy vegetables: their role in the world health organization's global fruit and vegetables initiative. *African Journal of Food Agriculture and Development* 7, 1–17.

Sreenivasan, P.K. and Gaffar, A. (2008) Antibacterials as anti-inflammatory agents: dual action agents for oral health. *Antonie van Leeuwenhoek* 93, 227–239.

Su, L.J. and Arab, L. (2006) Salad and raw vegetable consumption and nutritional status in the adult US population: results from the third national health and nutrition examination survey. *Journal of the American Dietetic Association* 106, 1394–1404.

Tang, G., Qin, J., Dolnikowski, G.G., Russell, R.M. and Grusak, M.A. (2005) Spinach or carrots can supply significant amounts of vitamin A as assessed by feeding with intrinsically deuterated vegetables. *American Journal of Clinical Nutrition* 82, 821–828.

Thompson, H.J., Heimendinger, J., Diker, A., O'Neill, C., Haegele, A., Meinecke, B., Wolfe, P., Sedlacek, S., Zhu, Z. and Jiang, W. (2006) Dietary botanical diversity affects the reduction of oxidative biomarkers in women due to high vegetable and fruit intake. *Journal of Nutrition* 136, 2207–2212.

Thu, N.N., Sakurai, C., Uto, H., van Chuyen, N., Yamamoto, S., Ohmori, R. and Kondo, K. (2004) The polyphenol content and antioxidant activities of the main edible vegetables in northern Vietnam. *Journal of Nutritional Science and Vitaminology (Tokyo)* 50, 203–210.

Toivonen, P.M.A. and Brummell, D. (2008) Biochemical bases of appearance and texture changes in fresh-cut vegetables and fruits. *Postharvest Biology and Technology* 48, 1–14.

Toivonen, P.M. and DeEll, J.R. (2002) Physiology of fresh-cut fruits and vegetables. In: Lamikanra, O. (ed.) *Fresh-Cut Fruits and Vegetables: Science, Technology, and Market.* CRC Press, Boca Raton, Florida, pp. 91–123.

USDA (2008) *Composition of Foods Raw, Processed, Prepared. USDA National Nutrient Database for Standard Reference, Release 21*. US Department of Agriculture, Agricultural Research Service (http://www.ars.usda.gov/nutrientdata, accessed 24 September 2008).

Van der Walt, A.M., Ibrahim, M.I.M., Bezuidenhout, C.C. and Loots, D.T. (2009) Linolenic acid and folate in wild-growing African dark leafy vegetables (morogo). *Public Health Nutrition* 12, 525–530.

van Dokkum, W., Frølich, W., Saltmarsh, M. and Gee, J. (2008) The health effects of bioactive plant components in food: results and opinions of the EU COST 926 action. *Nutrition Bulletin* 33, 133–139.

Watada, A.E. (1987) Vitamins. In: Weichmann, J. (ed.) *Postharvest Physiology of Vegetables*. Marcel Dekker, Inc, New York, pp. 455–468.

Yashimoto, M., Yahara, S., Okuno, S., Islam, S., Ishiguro, K. and Yamakawa, O. (2002) Antimutagenicity of mono-, di-, and tricaffeoylquinic acid derivatives isolated from sweetpotato (*Ipomoea batatas* L.) leaf. *Bioscience, Biotechnology and Biochemistry* 11, 2336–2341.

Youdim, K.A., McDonald, J., Kalt, W. and Joseph, J.A. (2002) Potential role of dietary flavonoids in reducing microvascular endothelium vulnerability to oxidative and inflammatory insults. *Journal of Nutritional Biochemistry* 13, 282–288.

Young, A.J. (1991) The photoprotective role of carotenoids in higher plants. *Physiologia Plantarum* 83, 702–708.

Young, J.E., Zhao, X., Carey, E.E., Welti, R., Yang, S.-S. and Wang, W. (2005) Phytochemical phenolics in organically grown vegetables. *Molecular Nutrition and Food Research* 49, 1136–1142.

Zheng, Y. and Wang, C.Y. (2007) Effect of high oxygen atmospheres on quality of fruits and vegetables. *Stewart Postharvest Review* 2, 1–8.

11 Pome Fruit

Chris B. Watkins and Rui Hai Liu

11.1 Introduction

A pome is an accessory fruit composed of five or more carpels in which the exocarp forms an inconspicuous layer. The mesocarp is usually fleshy and the endocarp forms a leathery case around the seed. Outside of the endocarp is the most edible part of this fruit, derived from the floral tube (torus) and other parts, which corresponds to what is commonly called the core. The best-known example of pome fruit is the apple, but other plants that produce fruit classified as a pome include cotoneaster, hawthorn, medlar, pear, pyracantha, toyon, quince, rowan and whitebeam. Of these species, a large literature exists about the apple and thus this chapter also has an unavoidable bias towards only one of the many pome fruit. However, examples of the health benefits of other pome fruit are included in this chapter wherever possible.

The health benefits of the apple fruit have long been recognized; in the Middle Ages the English said 'To eat an apple before going to bed will make the doctor beg his bread', which is commonly known in modern terms as the proverb 'An apple a day keeps the doctor away'. Apples are one of the most popular and healthy fruit that are commonly enjoyed by people all over the world. In the East, the Oriental pear has long been associated with antitussive, anti-inflammatory and diuretic properties (Cui *et al.*, 2005).

Apple nutrients are listed in Table 11.1. Apples are a natural source of fibre, minerals and vitamin C. Besides an excellent source of fibre, apples are a very significant source of phenolics in people's diet. For example, of the top 25 fruit consumed in the USA, apples are the number one source of phenolics in the American diet and provide Americans with 33% of the phenolics they consume (Boyer and Liu, 2004; Wolfe *et al.*, 2008). In contrast, the pear contributes only 3% (Wolfe *et al.*, 2008).

11.2 Phytochemicals

11.2.1 Phenolics

Phenolics are compounds possessing one or more aromatic rings with one or more hydroxyl groups and generally are categorized as phenolic acids, flavonoids, stilbenes, coumarins and tannins (Liu, 2004; Fig. 11.1). They are the products of secondary metabolism in plants, providing essential functions in the reproduction and growth of the plant, and acting as defence mechanisms against pathogens, parasites and predators, as well as contributing to plant colour. In addition to

Table 11.1. Apple nutrient composition.[a]

	Concentration (100 g fresh weight)
Water (g)	85.6
Energy (kcal)	52
Energy (kJ)	218
Protein (g)	0.26
Total lipid (fat) (g)	0.17
Total carbohydrate (g)	13.81
Total sugars (g)	10.39
Sucrose	2.07
Glucose (dextro se)	2.43
Fructose	5.90
Total dietary fibre (g)	2.4
Pectin (g)	0.5
Ash, total minerals (g)	0.19
Potassium, K (mg)	107
Calcium, Ca (mg)	6
Magnesium, Mg (mg)	5
Phosphorus, P (mg)	11
Iron, Fe (mg)	0.12
Vitamin C, total ascorbic acid (mg)	4.6
Vitamin A (IU)	54
Vitamin E (α-tocopherol) (mg)	0.18
β-Carotene (μg)	27
Lutein + zeaxanthin (μg)	29

Source: [a]USDA (2008).

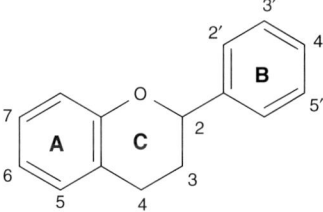

Fig. 11.1. The generic structure of flavonoids.

their roles in plants, phenolic compounds in diets may provide health benefits associated with reduced risk of chronic diseases.

Apples are a good source of phenolic compounds (Eberhardt *et al.*, 2001; Sun *et al.*, 2002; Boyer and Liu, 2004). The total extractable phenolic content has been investigated and ranges from 110 to 357 mg/100 g fresh apple (Podsedek *et al.*, 2000; Liu *et al.*, 2001). Phenolic concentrations in pear fruit are generally lower than those in apples, being in the range of

63–98 mg/100 g fresh weight (FW) (Vinson *et al.*, 2001; Kevers *et al.*, 2007; Wolfe *et al.*, 2008). A 100 g serving of pear fruit (with peel) provides between 27.2 and 40.7 mg of phenolics, depending on the cultivar, compared with between 5.5 and 8.4 mg of L-ascorbic acid (L-AA) (Galvis-Sanchez *et al.*, 2003).

11.2.2 Flavonoids and other phytochemicals

Flavonoids are a group of phenolic compounds with antioxidant and biological activity that have been identified in fruit such as apples, as well as in other fruit, vegetables and other plant foods. Flavonoids have been linked to reducing the risk of major chronic diseases. More than 4000 distinct flavonoids have been identified (Liu, 2004). They commonly have a generic structure consisting of two aromatic rings (A and B rings) linked by three carbons that are usually in an oxygenated heterocycle ring, or C ring (Fig. 11.1).

Differences in the generic structure of the heterocycle C ring classify them as flavonols, flavones, flavanols (catechins), flavanones, anthocyanidins, or isoflavonoids (isoflavones) (Fig. 11.2). Flavonols (quercetin, kaempferol and myricetin), flavones (luteolin and apigenin), flavanols (catechin, epicatechin, epigallocatechin (EGC), epicatechin gallate (ECG) and epigallocatechin gallate (EGCG)), flavanones (naringenin), anthocyanidins (cyanidin and malvidin), or isoflavonoids (genistein and daidzein) are common flavonoids in the diet. Flavonoids are found most frequently in nature as conjugates in glycosylated or esterified forms, but can occur as aglycones, especially as a result of the effects of food processing. Many different glycosides can be found in nature as more than 80 different sugars have been discovered bound to flavonoids.

Five major groups of phenolic compounds were detected in eight apple cultivars by HPLC: flavan-3-ols, flavonols, dihydrochalcones, anthocyanins and phenolic acids (hydroxybenzoic acid and hydroxycinnamic acids) (Tsao et al., 2003). The major group was the flavan-3-ols, or catechins, which accounted for 56% of polyphenols in the flesh (0–583.0 µg/g FW) and 60% of pholyphenols

in the peel (151–1655 µg/g FW). The dominant catechins in apples were catechin and epicatechin, and the dimers, procyanidin B1 (epicatechin plus catechin) and procyanidin B2 (two epicatechin molecules) (Tsao et al., 2003). The second major group of polyphenols in apples was the flavonols, such as quercetin, kaempferol and myricetin. Quercetin and its glycosides are by far the most abundant flavonols, can contribute up to 18% of total phenolics and occur exclusively in the peel. Concentrations were found to be 220–350 µg/g FW (Tsao et al., 2003).

The other three minor polyphenolic groups found in these eight apple cultivars were dihydrochalcones, anthocyanins and phenolic acids (Tsao et al., 2003). The main dihydrochalcones were phloretin 2′-glycoside (phlorizdin) and phloretin 2′-xyloglucoside, which occurred mostly in the peel at an average concentration of 124 µg/g FW. Anthocyanins, which occurred as cyanidin glycosides in red apples, ranged from 43 to 208 µg/g FW.

The two main groups of phenolic acids found in apples were hydroxybenzoic acids and hydroxycinnamic acids. Hydroxybenzoic acids accounted for less than 5% of phenolics in apples. In the peel, total benzoic acids

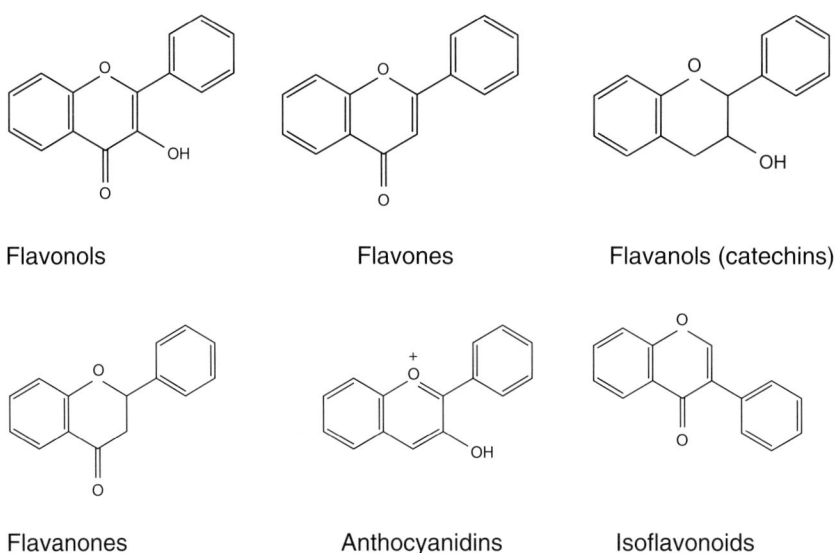

Flavonols　　　　　　　　　Flavones　　　　　　　　Flavanols (catechins)

Flavanones　　　　　　　　Anthocyanidins　　　　　　Isoflavonoids

Fig. 11.2. Structures of main classes of dietary flavonoids.

might be between 40 and 80 µg/g FW. The major hydroxycinnamic acid found in the apples was chlorogenic acid, 136 µg/g FW in the peel and 177 µg/g FW in the flesh. Two other hydroxycinnamic acids, *p*-coumaric acid and caffeic acid, occur in apples in lower concentrations (Tsao *et al.*, 2003).

Vrhovsek *et al.* (2004) found that flavanols (catechin and proanthocyanidins) were the major class of apple polyphenols (71–90%), followed by hydroxycinnamates (4–18%), flavonols (1–11%), dihydrochalcones (2–6%) and, in red apples, anthocyanins (1–3%). Wojdylo *et al.* (2008) reported similar results in a study of old and new cultivars, flavanols (catechin and oligomeric procyanidins) being the major class of apple polyphenols, representing more than 80%, followed by hydroxycinnamic acids (1–31%), flavonols (2–10%), dihydrochalcones (0.5–5%) and, in red apples, anthocyanins (1%).

He and Liu (2008) isolated 29 compounds from cv. Red Delicious apple peels, including triterpenoids, flavonoids, organic acids and plant sterols. Quercetin-3-*O*-β-D-glucopyranoside comprised 83% of the total flavonoids in the peel, while quercetin-3-*O*-β-D-galactopyranoside comprised 17%. Trace amounts of quercetin (0.2%) were detected along with (–)-catechin, (–)-epicatechin and quercetin-3-*O*-α-L-arabinofuranoside. Two new compounds were isolated from apple peel and identified for the first time: 3β-*trans*-*p*-coumaroyloxy-2α,3β,13β-trihydroxy-urs-11-en-28-oic acid and its *cis* counterpart, 3β-*cis*-*p*-coumaroyloxy-2α,3β,13β-trihydroxy-urs-11-en-28-oic acid (He and Liu, 2008).

Apple peels contain high amounts of triterpenoids. Thirteen triterpenoids were isolated from the peels of cv. Red Delicious apples (He and Liu, 2008). Ursolic acid was the most abundant triterpenoid in the apple peels. The cv. Annura apple, which is reddened after harvest, has antioxidant activity attributed to a ursane triterpene, annurcoic acid (D'Abrosca *et al.*, 2006), as well as other secondary metabolites (Cefarelli *et al.*, 2006).

Phenolics in pear fruit include 5'-caffeoylquinic, *p*-coumaroylquinic, *p*-coumaroylmalic and dicaffeoylquinic esters, quercetin 3-*O*-glycosides identified as rutinoside, glucoside and malonylglucoside, and the isorhamnetin 3-*O*-glycosides, rutinoside, galactorhamnoside, glucoside, malonyl galactoside and malonyl glucoside (Oleszek *et al.*, 1994). In pear fruit, only chlorogenic acid was detected in the flesh, whereas higher concentrations of the acid as well as flavonols and arbutin were found in the peel (Galvis-Sanchez *et al.*, 2003). The major phenolic compounds in Oriental pears are arbutin and chlorogenic acid (Cui *et al.*, 2005).

11.2.3 Ascorbic acid

Vitamin C (L-AA) is an essential nutrient, required from the diet by humans because of the absence of the enzyme L-gulonolactone oxidase needed in its synthetic pathway. Vitamin C is widely distributed in foods, with the highest levels found in fruit and vegetables. Apples contain an average of 46 µg vitamin C/g FW of fruit. Once absorbed, it is present in the body in two forms in equilibrium: ascorbic acid and the reduced form, dehydroascorbic acid (DHA). Vitamin C is essential in the human body and participates in important physiological functions, including: as an enzyme cosubstrate in collagen synthesis, neurotransmitter metabolism and carnitine synthesis; as a redox agent responsible for maintaining enzyme-bound metals in the correct oxidation state for functionality; as a regulator of antihistamine reactions; and as a stimulator of immunoglobulins (Davey *et al.*, 2000). In addition to the metabolic roles, vitamin C also functions as an antioxidant. Ascorbic acid can react with free radicals in the body to form a relatively unreactive ascorbyl radical, which then disproportionates to ascorbate and dehyroascorbate. Besides scavenging harmful free radicals, such as peroxyl radicals, hydroxyl radicals and superoxide, ascorbic acid may reduce the chromanoxyl radical to regenerate vitamin E (Davey *et al.*, 2000).

11.2.4 Antioxidant activity

Although the L-AA content of raw cv. Red Delicious apples with skin was 5.7 mg/100 g, the antioxidant activites of 1 g of apple with skin and without skin were 83 and 46 µmol vitamin C equivalents/g, respectively (Eberhardt *et al.*, 2001). The calculated antioxidant activity of L-AA in 1 g of cv. Red Delicious apple with skin was only 0.32 µmol vitamin C equivalents/g. The L-AA in apple with skin of this cultivar accounts for as little as 0.4% of total antioxidant activity (Eberhardt *et al.*, 2001). Other research has shown a larger contribution, of 10% of the total antioxidant capacity (Lee *et al.*, 2003; Vanzani *et al.*, 2005), but nevertheless the majority of antioxidant activity of apples is not from L-AA but from other phytochemicals in the fruit. The combinations of different phytochemicals in apples may function additively or synergistically to be responsible for this potent antioxidant activity. This is likely to be true also for pears and hawthorn, in which L-AA is a small component of total antioxidant activity (Galvis-Sanchez *et al.*, 2003; Guo *et al.*, 2003; Kevers *et al.*, 2007).

Pear fruit have lower total antioxidant capacity than apple fruit, as measured with a number of assay types (Wang *et al.*, 1996), although the differences are small compared with fruit with relatively high capacity, e.g. berry fruit (Vinson *et al.*, 2001; Leontowicz *et al.*, 2002, 2003; Wolfe *et al.*, 2008). Hawthorn fruit extracts contained chlorogenic acid, epicatechin, hyperoside, isoquercitrin, quercetin, rutin and protocatechuic acid, which were shown to inhibit oxidation of human low-density lipoprotein and α-tocopherol (Zhang *et al.*, 2001) and linoleic acid (Sokol-Letowska, 2007). No comparisons of hawthorn fruit with either pear or apple appear to be available, although antioxidant capacity (FRAP assay) of hawthorn was 27 times that of duck pear (Guo *et al.*, 2003). Antiradical activity of quince fruit is correlated with the concentrations of caffeoylquinic acid, phenolic, ascorbic acid and citric acid concentrations (Silva *et al.*, 2004), and Chinese quince and quince phenolics gave greater inhibition of gastric ulcers in rats than did apples (Hamauzu *et al.*, 2006). Compared with apple, quince and Chinese quince phenolic extracts were less effective in a linoleic acid peroxidation system, but phenolics from Chinese quince had the strongest anti-influenza vial activity in a haemagglutination inhibition test (Hamauzu *et al.*, 2005).

11.2.5 Bioavailability of phytochemicals

Phytochemicals cannot exert any biological effects unless they are absorbed, metabolized and distributed in the body. The extent to which this happens depends on many factors, including the nature of the compound, the food matrix, the presence of microorganisms and subject conditions (age and gender).

Human intake of all flavonoids is estimated at a few hundred mg to 650 mg/day. The total average intake of flavonols (quercetin, myricetin and kaempferol) and flavones (luteolin and apigenin) was estimated as 23 mg/day, of which quercetin contributed ~ 70%, kaempferol 17%, myricetin 6%, luteolin 4% and apigenin 3%.

Several studies have been performed to investigate the bioavailability of quercetin. Quercetin administered orally in an ethanol solution was found to be 36.4–53.0% absorbed (Walle *et al.*, 2001). To determine if different forms of quercetin can be absorbed, Hollman *et al.* (1995) fed nine ileostomy patients, who lacked colons, a single dose of quercetin in onions, which contained mostly quercetin 3-glucoside, pure quercetin 3-rutinoside (rutin), the predominant form of quercetin in tea, and pure quercetin aglycone. The average absorption of quercetin was 52% from onions, 17% for rutin and 24% for quercetin aglycone.

The same research group later fed quercetin to nine subjects as a single large dose using onions (see Chapter 2 of this volume), apples (which contained a variety of quercetin glycosides) and pure rutin, and monitored plasma quercetin levels over 36 h (Hollman *et al.*, 1997). Quercetin from onions was absorbed the most rapidly and rutin the least rapidly, indicating that quercetin glucoside was likely absorbed from the stomach or small intestine, while quercetin from rutin was probably absorbed from the colon after microbial cleavage of the sugar. Peak plasma levels were 224 ng/ml after the onion meal, 92 ng/ml after the apples and 90 ng/ml after the rutin, showing quercetin from apples and rutin to have only 30% of the bioavailability of quercetin from onions. The half-life of quercetin in plasma was found to be about 24 h, suggesting that accumulation of the compound was possible.

A study involving examination of querce tin, quercetin 4'-glucoside and quercetin 3,4'-diglucoside transport through Caco-2 cells showed quercetin aglycone was absorbed more rapidly across the intestinal epithelial cells than either glucoside (Walgren *et al.*, 1998). It also refuted the earlier theory of an active transport mechanism for quercetin. McAnlis *et al.* (1999) fed each of five healthy volunteers a single dose of onions, containing about 50 mg of quercetin, and measured plasma quercetin levels. Quercetin in the plasma increased from 28.4 ± 1.9 ng/ml at baseline to 248.4 ± 103.9 ng/ml after 2 h before returning to baseline in 24 h, an observation not noted by other experimenters. In another study, six subjects were fed a meal containing fried onions and fresh cherry tomatoes (Boyle *et al.*, 2000). Plasma quercetin levels increased from 16.5 ± 2.7 ng/ml to 104.9 ± 10.42 ng/ml 4 h after ingestion and remained elevated at 8 h. After 24 h, plasma quercetin levels were still higher than baseline. Quercetin was shown to be absorbed from the stomach of rats, but the glycosides, rutin and isoquercitrin, were not, indicating that the aglycones might be partly absorbed in the stomach of humans (Crespy *et al.*, 2002). It is obvious quercetin can be absorbed, but the glycosylation of quercetin largely determines to what extent.

11.2.6 Antimicrobial activity

Though little work is available, Fattouch *et al.* (2008) have found antimicrobial activity of apple, pear and quince extracts against a number of microorganisms associated with foodborne diseases and/or spoilage of contaminated product. Antimicrobial activity is not highly correlated with phenolic contents and further studies are warranted.

11.3 Health Benefits

11.3.1 Cancer

Consumption of apples has been linked to the prevention of chronic disease. Several epidemiological studies have linked apple consumption

specifically with a reduced risk of cancer (Knekt *et al.*, 1997; Feskanich *et al.*, 2000; Le Marchand *et al.*, 2000; Gallus *et al.*, 2005).

In a study in Finland involving 10,000 men and women and a 24-year follow-up, a strong inverse association was seen between flavonoid intake and lung cancer development (Knekt *et al.*, 1997). The mean flavonoid intake was 4.0 mg/day and 95% of the total flavonoid intake was quercetin. Apples and onions together provided 64% of all flavonoid intakes. The reduced risk of lung cancer associated with increased flavonoid consumption was especially strong in younger people and in non-smokers. Apples were the only specific foods that were related inversely to lung cancer risk in Finland. Men and women with the highest quartile of apple intake had a relative risk of 0.42 for lung cancer compared with those in the lowest quartile of apple intake (Knekt *et al.*, 1997). Since apples were the main source of flavonoids in the Finnish population, it was concluded that the flavonoids from apples were most likely responsible for the decreased risk of lung cancer.

A Hawaiian case–control study found that apple and onion intake was associated with a reduced risk of lung cancer in both males and females (Le Marchand *et al.*, 2000). Smoking history and food intake were assessed for 582 patients with lung cancer and 582 control subjects without lung cancer. The relative risk for lung cancer was decreased by 40% in individuals with high apple and onion intake when compared with those who consumed the lowest amount of these fruit and vegetables. No associations were seen with red wine, black tea or green tea. Both onions and apples are high in flavonoids, especially quercetin and quercetin conjugates. Le Marchand *et al.* (2000) found an inverse association between lung cancer and quercetin intake, although the trend was not statistically significant. Interestingly, the inverse association seen between apple and onion intake and lung cancer was stronger for squamous cell carcinomas than for adenocarcinomas.

In women from the Nurses' Health Study, significantly lower risks of lung cancer were found with an increase of one serving/ day of apples and pears (relative risk = 0.63)

(Feskanich *et al.*, 2000). In the Nurses' Health Study and the Health Professionals' Follow-up Study, involving over 77,000 women and 47,000 men, fruit and vegetable intake was associated with a 21% reduced risk in lung cancer in women, but this association was not seen in men (Feskanich *et al.*, 2000). Very few of the individual fruit and vegetables examined had a significant effect on lung cancer risk in women; however, apples were one of the individual fruit associated with a decreased risk in lung cancer. Women who consumed at least one serving per day of apples and pears had a reduced risk of lung cancer. Of the men involved, there was no association seen between any individual fruit or vegetable and lung cancer risk.

Another epidemiological study including 2569 breast cancer patients showed that apple consumption was linked to a lower risk of breast cancer (Gallus *et al.*, 2005). When compared with subjects consuming less than one apple/day, the multivariate odds ratio (OR) for at least one apple/day was 0.82 (95% CI 0.73–0.92) for breast cancer. After further allowance for consumption of vegetables and other fruit, the association with apples became even stronger for breast cancer (OR 0.76, CI 0.67–0.85) (Gallus *et al.*, 2005).

The relationship between dietary catechins and epithelial cancer was examined in 728 men (aged 65–84), as part of the Zutphen Elderly Study (Arts *et al.*, 2001). Tea, a naturally high source of catechins, contributed 87% of the total catechin intake in this study, while apples contributed 8.0% of catechin consumption. It was found that total catechin and tea consumption did not have an effect on lung cancer, but apple consumption was associated with decreased epithelial lung cancer incidence (Arts *et al.*, 2001). This supported the findings of the previous studies discussed above, where apples were associated significantly inversely with lung cancer, and might suggest that catechins alone do not have an effect against lung cancers. Other data from the Zutphen Elderly study showed an inverse association between fruit and vegetable flavonoids and total cancer incidence and tumours of the alimentary and respiratory tract (Hertog *et al.*, 1994). Again, tea flavonoids were not associated with a decrease in cancer risk.

Apples have been shown to have strong antioxidant activity (Wang *et al.*, 1996; Sun *et al.*, 2002). In previous studies, apple extracts have been shown to have potent antiproliferative activity against colon, liver and breast cancer cells *in vitro*, in a dose-dependent manner (Eberhardt *et al.*, 2001; Sun *et al.*, 2002; Wolfe *et al.*, 2003). Apples with peel inhibit the growth of human liver cancer cells to a greater extent than those without peel (Eberhardt *et al.*, 2001; Wolfe *et al.*, 2003). Apple extracts inhibited 7,12-dimethylbenz(a) anthracene (DMBA)-induced mammary cancer in rats, in a dose-dependent manner, in doses equivalent to human consumption of one, three or six apples/day (Liu *et al.*, 2005). The apple extracts downregulated proliferating cell nuclear antigen (PCNA), Cyclin D1 and Bcl-2, and upregulated Bax and nuclear fragments, reduced tumour size and tumour burden and delayed tumour onset in a clear dose-dependent manner (Liu *et al.*, 2009). It was reported that apple extracts had activity inhibiting NFkB activation in human breast cancer MCF-7 cells (Yoon and Liu, 2007, 2008).

11.3.2 Cardiovascular diseases

Apple consumption has also been related to the reduced incidence of cardiovascular disease. Coronary mortality was reduced by 43% in Finnish women in the highest quartile of apple intake compared with the lowest quartile. Total flavonoid intake was found to be associated significantly and inversely with coronary mortality in women, but no significant reduction in risk of coronary mortality in men was found (Knekt *et al.*, 1996).

The risk of thrombotic stroke was reduced in women in the highest quartile of apple intake when compared with those who consumed the lowest amounts of apples in a Finnish cohort study (Knekt *et al.*, 2000). Onion intake and quercetin intake were not associated with thrombotic stroke or other cerebrovascular diseases.

The Women's Health Study found apple consumption was associated with a reduced risk of cardiovascular disease, in a study examining the association of flavonoid intake and cardiovascular disease (Sesso *et al.*, 2003). Women ingesting the highest amounts of flavonoids had a 35% reduction in risk of cardiovascular events. Flavonoid intake was not associated with risk of stroke, myocardial infarction or cardiovascular disease death. Quercetin did not have any association with cardiovascular disease, cardiovascular events, myocardial infarction or stroke. However, both apple intake and broccoli intake were associated with reductions in the risk of both cardiovascular disease and cardiovascular events. Women ingesting apples had a 13–22% decrease in cardiovascular disease risk.

Apple and wine consumption was also associated inversely with death from coronary heart disease in postmenopausal women, in a study of nearly 35,000 women in Iowa (Arts *et al.*, 2001). The intakes of catechin and epicatechin, both constituents of apples, were strongly inversely associated with coronary heart disease death. Although total catechin intake was inversely associated with coronary heart disease mortality, Arts *et al.* (2001) found that tea catechins were not associated with coronary heart disease mortality in postmenopausal women. Apple catechins may be more bioavailable than the catechin and epicatechin gallates commonly found in teas.

The Zutphen Elderly Study examined the relationship between flavonoid intake and risk of coronary heart disease and found that flavonoid intake was negatively correlated with mortality from heart disease in elderly men and also with myocardial infarction (Hertog *et al.*, 1993). Tea was the main source of flavonoids in this study and was also negatively correlated with coronary heart disease. Apple intake contributed to approximately 10% of the total ingested flavonoids and was also associated with a reduced risk of death from coronary heart disease in men, although the relationship was not statistically significant (Hertog *et al.*, 1993).

11.3.3 Pulmonary function and asthma

Apple consumption has been linked inversely with asthma and has also been associated positively with general pulmonary health. Shaheen

et al. (2001) conducted a study about diet and lifestyle involving nearly 600 individuals with asthma and 900 individuals without asthma. Total fruit and vegetable intake was associated weakly with asthma, and apple intake showed a stronger inverse relationship with asthma. The association was very strong in subjects who consumed at least two apples/week. Onion, tea and red wine consumptions were not related to asthma incidence, suggesting the specific beneficial effect of apple flavonoids. Vitamin C and vitamin E were not correlated with asthma incidence, whereas carotene intake was associated weakly, but positively, with asthma.

In the Dutch MORGEN study of over 13,000 adults, it was found that apple and pear intake was positively associated with pulmonary function and negatively associated with chronic obstructive pulmonary disease (Tabak *et al.*, 2001). Catechin intake was also positively associated with pulmonary function and negatively associated with chronic obstructive pulmonary disease, but there was no association between tea, the main source of catechins, and chronic obstructive pulmonary disease (Tabak *et al.*, 2001).

In another study involving 1600 adults in Australia, apple and pear intake was associated with a decreased risk of asthma and a decrease in bronchial hypersensitivity, but total fruit and vegetable intake was not associated with asthma risk or severity (Woods *et al.*, 2003). Specific antioxidants, such as vitamin E, vitamin C, retinol and β-carotene, were not associated with asthma or bronchial hypersensitivity.

11.4 Effect of Cultivar

Comparative studies on the health-promoting components of different apple and pear cultivars have tended to focus either on total and individual phenolic concentrations, or on L-AA and associated antioxidant systems. An extensive literature shows that cultivars can vary greatly in phenolic composition (Amiot *et al.*, 1995; Lister *et al.*, 1996a; van de Sluis *et al.*, 2001; Imeh and Khokhar, 2002; Galvis-Sanchez *et al.*, 2003; Lee *et al.*, 2003; Leja *et al.*, 2003; Wolfe *et al.*, 2003; Napolitano *et al.*, 2004;

Vrhovsek et al., 2004; Khanizadeh et al., 2007; Lata and Tomala, 2007). The concentration ranges for all individual phenolic compounds are available in Treutter (2001). L-AA concentrations also vary greatly among cultivars (Galvis-Sanchez et al., 2003; Davey and Keulemans, 2004; Vrhovsek et al., 2004; Lata et al., 2005; Davey et al., 2007). Also, it has been observed that older cultivars tend to have higher concentrations than newer cultivars (Planchon et al., 2004), as do later harvested cultivars compared with early harvested cultivars (Davey et al., 2007).

Total phenolic concentrations of ten different apple cultivars, with and without skin, were reported by Liu et al. (2001). The cv. Fuji apples with skin had the highest total phenolic content (230.49 ± 4.4 mg/100 g apple), followed by cvs. Red Delicious, Gala, Liberty, Northern Spy, Golden Delicious, Fortune, Jonagold, Empire and NY674. The total phenolic content in apples without skin was highest for cv. Red Delicious (167.82 ± 1.7 mg/100 g apple) followed by cvs. Northern Spy, Fortune, Gala, Fuji, Liberty, Golden Delicious, NY674, Jonagold and Empire. Total phenolic content was higher in all cultivars for apples with skin when compared with apples without skin, with the exception of NY674 (Liu et al., 2001).

The distribution of phenolics among the flesh, flesh + peel and peel of four apple cultivars have been reported (Wolfe et al., 2003). In cvs. Idared, Rome Beauty, Golden Delicious and Cortland apples, total phenolic concentrations of the flesh were 75.7 ± 4.0, 93.0 ± 4.1, 97.7 ± 8.9 and 103.2 ± 12.3 mg/100 g fruit, respectively. In the same order, total phenolic concentrations of the flesh with the peel were 120.1 ± 15.0, 159.0 ± 15.1, 129.7 ± 9.7 and 119.0 ± 14.9 mg/100 g fruit. Total phenolic concentrations of the peel were 588.9 ± 83.2, 500.2 ± 13.7, 309.1 ± 32.1 and 388.5 ± 82.4 mg/100 g fruit. Most notable is the fact that the phenolics are highly concentrated in the peel. The total phenolic concentrations of the peels were higher than the flesh and flesh + peel values in all cultivars ($P < 0.05$) (Wolfe et al., 2003).

Total antioxidant capacity and antiproliferation activity also vary among cultivars. Total antioxidant capacity (TEAC assay) of five apple genotypes (whole fruit) varied by two- to threefold in the hydrophilic phase, but much less in the lipophilic phase (Scalzo et al., 2005). Wolfe et al. (2003) found that concentrations of total flavonoids, anthocyanins and antioxidant activity (TOSC assay) varied between cvs. Rome Beauty, Idared, Cortland and Golden Delicious, especially in the peel. Interestingly, antiproliferative activity of the flesh + peel of cv. Rome Beauty was greater than expected from the values for peel or flesh alone, suggesting synergistic effects between phytochemicals from the two tissue types. In addition, Wolfe et al. (2003) found that peels had the highest antioxidant capacity and antiproliferation activity against HepG$_2$ human liver cancer cells. Total phenolic concentrations of 11 apple and six pear genotypes were associated closely with antioxidant capacities (FRAP assay (Khanizadeh et al., 2007) and DPPH assay (Galvis-Sanchez et al., 2003), respectively) although associations could be poor (Kahkonen et al., 1999; Mareczek et al., 2000).

These studies, and others, also show that total phenolic concentrations, flavonoids and antioxidant activity (Burda et al., 1990; Ju et al., 1996; Escarpa and Gonzalez, 1998; Galvis-Sanchez et al., 2003; Wolfe et al., 2003; McGhie et al., 2005; Lata, 2007) and L-AA (Galvis-Sanchez et al., 2003, 2006; Davey et al., 2004; Li et al., 2008) are higher in the peel than in the flesh of apples and pears. However, the available studies differ greatly in the tissue type (peel, flesh or combined) and components that are analysed, and therefore comparisons between studies are often difficult. For both phenolics and L-AA, the variation between fruit of the same cultivar and between years can be very high (Planchon et al., 2004; Lata et al., 2005). Geographical growing region may also affect the phenolic composition of apple cultivars (McGhie et al., 2005), although the absence of replication of orchards within regions and a study carried out in only 1 year makes this conclusion tentative. Seasonal variations of L-AA, flavonols, anthocyanins and phenolics, as well as other bioactive compounds, were greater in peel than whole fruit (Lata and Tomala, 2007).

11.5 Effect of Maturation

Phenolic concentrations typically decrease during apple and pear fruit development, though sometimes after an initial increase (Harel et al., 1966; Mosel and Herrman, 1974; Coseteng and Lee, 1987; Burda et al., 1990; Awad et al., 2001b). After the decline, however, concentrations remain relatively unchanged during the maturation period. The main phenolic compounds, epicatechin and procyanidin B2, and phloretin glycosides in the peel and flesh remain constant during maturation (Burda et al., 1990). Decreased phenol concentrations during development are associated with lower antioxidant capacities of flesh extracts (Hamauzu et al., 1999). However, while concentrations of individual phenolic components decline during development, the total amount per apple increases (Awad et al., 2001a).

L-AA concentrations in cv. Conference pears fluctuate in young fruit, remain stable during fruit maturation and start to decline 1 week before commercial harvest (Franck et al., 2003). Amiot et al. (1995) found that differences in phenolic concentrations and browning susceptibility of pear purees were associated more with cultivar than with maturity effects.

11.6 Effect of Preharvest and Postharvest Continuum

11.6.1 Orchard management

Modern orchard systems emphasize the high yields of large fruit and requirements of red coloration appropriate to the cultivar, whether it is bicoloured or full red. The major flavonoid, anthocyanin, is therefore the most visible phytochemical for red apple cultivars. Anthocyanin concentrations are affected by environmental and cultural practices (Saure, 1990; Lancaster, 1992), but generally increase markedly during ripening (Knee, 1972). Anthocyanin synthesis is reported to be associated closely with the activity of phenylalanine ammonia lyase (PAL) (Tan, 1979; Wang et al., 2000), which causes the non-oxidative

deamination of L-phenylalanine to form trans-cinnamic acid and a free ammonium ion. However, associations between PAL activity and anthocyanin accumulation are not always apparent (Lister et al., 1996b); PAL activity is only the first step in the biosynthesis of a large range of phenylpropanoid-derived secondary products in plants, such as flavonoids and isoflavonoids, coumarins, lignins, wound-protective hydroxycinnamic acid esters and other phenolic compounds (Tomas-Barberan and Espin, 2001). PAL activity can be activated by ethylene in other storage and plant tissues (Ritenour et al., 1995; Tomas-Barberan and Espin, 2001) and its activity is associated with increasing ethylene concentrations in apple fruit (Faragher and Brohier, 1984; Blankenship and Unrath, 1988).

Concentrations and amounts of anthocyanin and quercetin glycosides were higher in the skin of cvs. Elstar and Jonagold apples from the outer canopy compared with the inner canopy, whereas concentrations of catechins, phloridzin and chlorogenic acid were independent of canopy position (Awad et al., 2001a,c). Exposure of fruit on the tree also has a major effect on ascorbic acid concentrations which are much greater on the red (sunny) side of individual fruit than on the green (shady) side and much greater in exposed fruit than in shaded fruit in the canopy (Davey et al., 2004; Planchon et al., 2004; Hagen et al., 2007).

Crop load did not affect chlorogenic acid or total and individual flavonoid concentrations in cvs. Jonagold and Red Elstar apples (Awad et al., 2001a), but lower crop loads resulted in higher concentrations of both total and individual phenolic compounds (Stopar et al., 2002). Nitrogen is applied to soils to increase productivity and fruit size, but may decrease red coloration (Johnson and Samuelson, 1990; Raese and Drake, 1997). Calcium is applied to trees to decrease the incidence of certain storage disorders, such as bitter pit and senescent breakdown. Awad and de Jager (2002b) found that chlorogenic acid concentrations were related negatively to the concentrations of N, K and Mg and the N/Ca ratio, but related positively to P and Ca in cv. Elstar flesh tissue. Anthocyanins, catechins and total flavonoids were associated with lower red coloration in the skin of nitrogen-treated fruit. Calcium sprays were associated with higher anthocyanins, total flavonoids, epicatechin, chlorogenic and total phenols in apples (Sannomuru *et al.*, 1998; Awad and de Jager, 2002b). Preharvest sprays of ethephon and phosphorus–calcium mixtures (Seniphos) stimulated red colour and concentrations of anthocyanins, proanthocyanidins and flavonols in the skin of cv. Fuji apples (Li *et al.*, 2002).

The effects of photosynthesis on phenolics have been illustrated in a study of branch orientation in pear fruit (Colaric *et al.*, 2006). Lower catechin, epicatechin, sinapic acid, syringic acid and total phenolic concentrations, as well as individual sugars, were found in fruit from branches bent down in summer compared with no branch manipulations. However, fruit from spring-bent branches had constituent concentrations closer to those of the control.

No studies that relate these preharvest management practices to total antioxidant capacity and antiproliferation activity appear to be available for pome fruit. However, the effects of organic versus non-organic production on phytochemical content of apples have been the focus of numerous studies. Many studies show no differences, between production systems, for phenolics, antioxidant bioavailability or protection against oxidative DNA damage in flesh samples, with or without peel (Tarozzi *et al.*, 2004; Briviba *et al.*, 2007; Roth *et al.*, 2007). However, Weibel *et al.* (2000) found that organically grown cv. Golden Delicious (flesh and peel), and Hecke *et al.* (2006) that fruit from several organically grown cultivars (juice from pressed fruit), had higher total phenolic concentrations than fruit from integrated fruit production systems. Organically grown fruit sometimes had the highest sugar concentrations, although acid levels were not consistently different between the management systems (Reganold *et al.*, 2001; DeEll and Prange, 1992; Hecke *et al.*, 2006). In studies conducted in Washington State on cv. Gala apples, Peck *et al.* (2006) did not find any effects of cultivation system on soluble solids concentration (SSC), but they did find higher total antioxidant activity in organically grown fruit than in those from

conventional or integrated fruit production systems. However, in similar studies carried out in New York on cv. Liberty apples, no difference for total phenolic concentrations or antioxidant capacity was found between organic and integrated fruit production systems (Peck *et al.*, 2009). Cultivation system had little effect on ascorbic acid or thiobarbituric acid reactive substances in pears, but total phenolics and polyphenol oxidase (PPO) activity were higher in organic than in conventional systems (Carbonaro *et al.*, 2002). With numerous differences between how and where the farming systems were implemented, results for apples and other fruit crops have been variable (Zhao *et al.*, 2006), and no universal conclusions can be drawn. Long-term rigorous and comprehensive comparisons of these growing systems under a variety of climatic regions are needed to determine the consistency of fruit responses.

11.6.2 Effects of postharvest treatment

Apples and pears are horticultural products that are stored for long periods using a range of technologies such as refrigerated and controlled atmosphere (CA) and, more recently, the ethylene inhibitor, 1-methylcyclopropene (1-MCP) (Watkins, 2003, 2008). The majority of available research has focused on physical quality attributes, and an emphasis on characterization of the health-related components of apple after harvest has been relatively recent. Phenolics, for example, have been of interest as substrates for browning enzymes in both apples and pears (Ranadive and Haard, 1971; Coseteng and Lee, 1987; Richard-Forget *et al.*, 1992). Browning of fruit during storage as a result of senescence and/or CA-induced disorders has been of increasing interest (Volz *et al.*, 1998; Fernandez-Trujilo *et al.*, 2001; de Castro *et al.*, 2008). An important disorder of apples known as superficial scald, which results in browning of the apple skin, has also been the subject of considerable interest in terms of phenolics (Ju *et al.*, 1996; Piretti *et al.*, 1996; Golding *et al.*, 2001). Phenolics may also be involved in scab resistance (Mayr *et al.*, 1997), and changes in concentrations of benzoic acid, *p*-coumaryl-quinic acids and chlorogenic acid may also be important in resistance of apple fruit against disease (Brown and Swinburn, 1973; Ndubizu, 1976; Noble and Drysdale, 1983; Lattanzio et al., 2001; Michalek et al., 2005). Finally, phenolics, in addition to L-AA, are important antioxidants involved in normal cellular function (Rice-Evans et al., 1996). As described earlier, however, total antioxidant capacity is associated more closely with total phenolic concentrations than those of L-AA (Eberhardt et al., 2001; Lee et al., 2003; Hagen et al., 2007). L-AA contributes as little as 0.4% (Sun et al., 2002) to 10% of the total antioxidant capacity of apples (Lee et al., 2003; Vanzani et al., 2005), but it may contribute to antioxidant activity of phenolics (Saucier and Waterhouse, 1999). Moreover, the contribution of individual antioxidant components is much smaller than total antioxidant capacity, as determined by assays such as DPPH (Chinnici et al., 2004) and peroxyl radical trapping efficiency (Wolfe et al., 2003; Vanzani et al., 2005). Polymeric anthocyanidins may be a factor (Vanzani et al., 2005), but the available data do not exclude the possibility that synergistic interactions among antioxidant components explain the discrepancy (Eberhardt et al., 2001; Chinnici et al., 2004).

Little information is available for pear fruit. L-AA concentrations have been investigated in relation to core breakdown of pears stored in CA. Development of this disorder is associated with decreasing L-AA concentrations in the core and may occur only when concentrations fall below a threshold value (Veltman et al., 1999, 2000; Zerbini et al., 2002).

Postharvest changes of phenolics in apples appear modest compared with cultivar and preharvest effects such as exposure to light. Although phenolics undergo constant turnover and degradation (Barz and Hoesel, 1979; Stafford, 1990), little is known about the biochemical processes underlying these changes. The major exception is PAL activity, described earlier.

Although total antioxidant capacity of apples is tested routinely and antiproliferation activity of apples has been well described (Liu et al., 2001; Wolfe et al., 2003), there is an absence of information about antiproliferation

activity against cancer cell systems in any postharvest system.

Air storage

Early research with nine apple cultivars found that phenolic concentrations remained relatively constant over 4 months of cold storage (Coseteng and Lee, 1987). Individual phenolic compounds in peel and flesh tissues of cvs. Rhode Island Greening, Empire and Golden Delicious also changed little over 6 months of storage, although epicatechin concentrations appeared to increase in peel and flesh tissues of cv. Rhode Island Greening after harvest (Burda et al., 1990). However, no statistical information for changes over time was provided by the authors. In cvs. Red Delicious and Ralls, chlorogenic acid, flavonoid and anthocyanin concentrations in the peel remained stable for 4–5 months of storage (Ju et al., 1996). Golding et al. (2001) found that concentrations of total phenolics, total benzoic acid derivatives, total procyanidins, phloridzin, total cinnamic derivatives and chlorogenic acid in peels differed greatly among cvs. Lady Williams, Crofton and Granny Smith. These concentrations increased in the early stages of storage, but then generally declined over a 9-month storage period. Piretti et al. (1994) also found that epicatechin, quercetin glycosides and procyanidins in peel of cv. Granny Smith apples decreased from day 100 to day 205 of storage. Lattanzio et al. (2001) found an increase of chlorogenic acid, phlorodzin and phloretin glycoside concentrations in the peel after 60 days of air storage, which then either levelled off or declined. In a detailed study of harvest date effects, MacLean et al. (2006) found that chlorogenic acid concentrations in cv. Red Delicious peel tissues increased during air storage, while those of anthocyanins decreased. Total antioxidant activity of peel tissues increased during storage of cv. Empire, but decreased in cv. Red Delicious (MacLean et al., 2003). Total phenolic concentrations increased in the skin of cv. Jonagold but not cv. S'ampion, while anthocyanin concentrations decreased in both cultivars (Leja et al., 2003). However, total antioxidant activity (inhibition of linoleic acid peroxidation and

DPPH assays) increased during storage of both cultivars (Leja *et al.*, 2003). Total quercetin glycoside, phloridzin, cyanidin galactoside and chlorogenic acid concentrations were unaffected by storage in whole fruit of four cultivars, whereas catechin decreased over storage in each cultivar, depending on harvest year (van de Sluis *et al.*, 2001). Napolitano *et al.* (2004) found that epicatechin concentrations increased in flesh tissues of cvs. Golden Delicious, Red Delicious and Empire, while a major increase in chlorogenic acid and catechin was found in cv. Annurca, a cultivar that was harvested at an immature stage and then treated to exposure to sunlight for a month before storage. The total antioxidant activity (ABTS) increased during storage of all cultivars, but especially cv. Annura; activity was correlated with catechin and phloridzin concentrations and not that of chlorogenic acid (Napolitano *et al.*, 2004). Air storage did not affect antioxidant activity (inhibition of lipid peroxidation) in whole fruit of four cultivars (van de Sluis *et al.*, 2001). However, Tarozzi *et al.* (2004) found that antioxidant bioactivity *in vitro*, measured in terms of intracellular antioxidant, cryoprotective and antiproliferation activity in human colon carcinoma (Caco-2) cells, decreased over time.

During the shelf-life period at warmer temperatures after cold storage, phenolics may be stable (Awad and de Jager, 2000), but decreases in various phenolic components, including chlorogenic acid, flavonoids and anthocyanins, in cvs. Granny Smith, Red Delicious and Ralls, have been reported (Piretti *et al.*, 1994; Ju *et al.*, 1996). Phenolic compounds in the peel, but not the flesh, increased during a shelf-life period of 21 days (Perez-Ilzarbe *et al.*, 1997).

Overall, the literature, while contradictory, suggests that phenolics are relatively stable and therefore that the health benefits of phenolics are likely to be maintained during storage. Where increases in total phenolics and individual components have been reported, these occur in the early stages of storage. It is possible that these changes during early storage are associated with ethylene stimulation of PAL activity. What is unclear, however, is the bioactivity of these phenolics, with poor understanding of the interactions

among individual compounds and ascorbic acid, or even the effects of changing acid and sugar levels in the fruit.

In contrast, total L-AA concentrations generally decrease during air storage (Vilaplana *et al.*, 2006; de Castro *et al.*, 2008; Fawbush *et al.*, 2009), although not always (Tarozzi *et al.*, 2004). Cultivars vary in the rates of L-AA loss over time (Davey and Keulemans, 2004). Davey *et al.* (2004) reported that ascorbic acid (L-AA) could not be synthesized *de novo* by fruit. Although Li *et al.* (2008) found that both peel and flesh tissues were able to synthesize ascorbic acid, and L-AA concentrations increased in fruit exposed to visible and UV-B light after harvest (Hagen *et al.*, 2007). Losses of L-AA might be associated with declining cellular function and might be especially important during defence against physiological disorders.

Controlled atmosphere (CA) storage

Compared with air storage, relatively few studies on the effects of CA on antioxidative compounds of apple are available. CA storage conditions had little effect on total quercetin glycoside, phloridzin and cyanidin galactoside concentrations in whole fruit of four cultivars (van de Sluis *et al.*, 2001). Chlorogenic acid and catechin concentrations decreased in cv. Jonagold, whereas only catechin decreased in cv. Golden Delicious, and storage did not affect antioxidant activity (inhibition of lipid peroxidation) in any cultivar (van de Sluis *et al.*, 2001). Catechin, epicatechin and quercetin glycoside concentrations in the peel decreased in CA storage of cvs. Granny Smith, Jonagold and Elstar apples in a similar fashion to that in air storage (Piretti *et al.*, 1994; Awad and de Jager, 2000, 2003). In contrast, total phenolic concentrations and total antioxidant activity in the peel of cvs. Jonagold and S'ampion increased, while anthocyanin concentrations during CA storage were maintained at levels similar to those at harvest (Leja *et al.*, 2003). Total phenolics in peel of cv. Golden Reiders, but not cv. Gala Must, increased during 7 months of CA storage (Mareczek *et al.*, 2000); these concentrations decreased only in cv. Gala Must during a subsequent shelf life. In pear fruit, hydroxycinnamic derivatives and flavonols were stable in air and CA (0, 0.5 and 5% kPa CO_2 in 2% kPa O_2) for 4 months, but flava-3-ol concentrations decreased (Galvis-Sanchez *et al.*, 2006). Accumulation of phenolic compounds that was observed in air-stored cv. Williams pears was inhibited by CA storage (Amiot *et al.*, 1995).

L-AA levels generally decline in apples during CA storage (de Castro *et al.*, 2008; Fawbush *et al.*, 2009), although Lata (2008) found increased L-AA concentrations (in addition to thiols and phenolic compounds) in apple peel during CA storage. Loss of L-AA in elevated CO_2 concentrations was reduced with lower O_2 concentrations and might be associated with susceptibility of fruit to flesh browning (de Castro *et al.*, 2008).

In pears, L-AA concentrations decreased more rapidly in CA than in air storage for 22 days (Larrigaudiere *et al.*, 2001a) but regenerated during subsequent storage, depending on gas concentration (Larrigaudiere *et al.*, 2001b). L-AA concentrations also increased when the CO_2 concentration in the storage atmosphere was increased from 0 to 10% kPa (Veltman *et al.*, 2000). However, Galvis-Sanchez *et al.* (2006) found that L-AA concentrations decreased, while those of DHA decreased during storage in air and CA. Franck *et al.* (2003) found that losses of L-AA from pears was similar in fruit cooled slowly or quickly in conjunction with CA storage, but specific CA regimes had significant differences.

1-Methylcyclopropene (1-MCP)

An inhibitor of ethylene perception, 1-MCP, has been registered recently for use on food crops around the world, and it is used extensively to maintain quality of apple fruit during air and CA storage (Watkins, 2006, 2008). To date, studies on the effects of 1-MCP on antioxidant components are limited. MacLean *et al.* (2006) found that 1-MCP treatment inhibited an increase in chlorogenic acid in peel tissues of cv. Red Delicious apples during storage, and resulted in higher flavonoid concentrations. The effects of 1-MCP on chlorogenic acid decreased as fruit harvest date and internal ethylene increased; this effect was consistent with a role of ethylene in

modulating PAL activity, as later harvest date presumably would have a smaller inhibition of ethylene production and thus less effect on PAL activity. MacLean *et al.* (2003) also showed that total antioxidant activity (TOSC) was higher in peels of 1-MCP-treated cvs. Empire and Red Delicious apples than in untreated fruit. The total antioxidant activity (DPPH) of cv. Golden Smoothee flesh was unaffected by 1-MCP treatment, but total L-AA concentrations were slightly lower after 30 and 90 days of air storage (Vilaplana *et al.*, 2006).

In pears, 1-MCP treatment inhibited an increase of chlorogenic acid observed in untreated fruit, and overall the results suggested that 1-MCP inhibited the transcription of the key flavonoid biosynthetic enzymes, PAL and chalcone synthase (MacLean *et al.*, 2007).

Postharvest chemical treatments

Postharvest treatment of fruit with diphenylamine (DPA), an antioxidant used to prevent superficial scald development, resulted in higher total phenolics during air storage, but the overall effects were minor (Golding *et al.*, 2001). In contrast, Duvenage and DeSward (1973) concluded that DPA inhibited both the synthesis and oxidation of flavonols during storage. DPA did not affect the loss of L-AA in fruit during the first 2 months of storage, but concentrations were higher, in both air- and CA-stored fruit, after 4 months (de Castro *et al.*, 2008).

Calcium is often applied to apple fruit as a postharvest drench or dip at the same time as DPA treatment. Little is known about the effects of such applications on nutritional components, but Bangerth (1976) found that L-AA concentrations in fruit were increased by calcium dips or infiltration.

UV irradiation

Postharvest irradiation of fruit, usually as combined visible and UV-B radiation, induces anthocyanin accumulation in both non-red and red apples (Arakawa *et al.*, 1985; Reay, 1999; Lancaster *et al.*, 2000; Reay and Lancaster, 2001; Ubi *et al.*, 2006). Hagen *et al.*

(2007) extended these studies to show that exposure of fruit from the inner (shaded) and outer (exposed) part of the tree to visible light and UV-B radiation could be used to increase health-related components; radiation caused higher accumulations of anthocyanin, quercetin glycosides, chlorogenic acid, ascorbic acid, total phenols and antioxidant (ORAC) capacity in the peel, but not flesh, compared with untreated fruit. Total antioxidant capacity was correlated more strongly with the sum of phenols than with ascorbic acid concentrations (Hagen *et al.*, 2007). Methyl jasmonate (MJ) interacts synergistically with ethylene in UV/white light-treated apple fruit to influence peel pigment synthesis pathways, the effects being specific to key pigment components (Rudell *et al.*, 2002; Rudell and Mattheis, 2008).

11.7 Conclusion

The apple has been well studied compared with other pome fruit. The evidence from epidemiological studies strongly suggests that consumption of apples is associated with a decreased risk of developing chronic diseases such as cardiovascular disease, cancer and asthma. *In vitro* studies have demonstrated that apples have high antioxidant activity and antiproliferative activity against colon, liver and breast cancer cells, and can decrease lipid oxidation and lower cholesterol. Apple extracts inhibit NFκB activation, induce G1 arrest and decrease expression of Cyclin D1 and Cdk4 in human breast cancer MCF-7 cells. Most recently, several studies have found that apple extracts prevent DMBA-induced mammary cancer in rats, downregulate PCNA, Cyclin D1 and Bcl-2, upregulate Bax and induce apoptosis *in vivo*. The health benefits of apples, and probably other pome fruit, are attributed to the complex mixture of phytochemicals, plant bioactive compounds with a variety of functions. The interaction of the phytochemicals, especially the additive and synergistic effects, warrants more study to explain further the mechanism behind the ability of apple and other fruit to reduce risk of chronic disease. More work is needed to understand better the bioavailability of pome fruit phytochemicals. While more research is warranted with apple fruit, this review illustrates clearly the paucity of information for pear and other pome fruit.

Understanding of the processes involved in intake and bioavailability of phytochemicals in pome fruit may lead to much better knowledge of exactly how important pre- and postharvest management and handling practices are in maintaining nutritional quality. Selection processes by breeders inevitably are focused on appearance and yield, resistance to disease and physiological disorders. These factors may or may not be associated with nutritional quality.

References

Amiot, M.J., Tacchini, M., Aubert, S.Y. and Oleszek, W. (1995) Influence of cultivar, maturity stage, and storage-conditions on phenolic composition and enzymatic browning of pear fruits. *Journal of Agricultural and Food Chemistry* 43, 1132–1137.

Arakawa, O., Hori, Y. and Ogata, R. (1985) Relative effectiveness and interaction of ultraviolet-B, red and blue light in anthocyanin synthesis of apple fruit. *Physiologia Plantarum* 64, 323–327.

Arts, I.C.W., Hollman, P.C.H., De Mesquita, H.B.B., Feskens, E.J.M. and Kromhout, D. (2001) Dietary catechins and epithelial cancer incidence: The Zutphen elderly study. *International Journal of Cancer* 92, 298–302.

Awad, M.A. and de Jager, A. (2000) Flavonoid and chlorogenic acid concentrations in skin of 'Jonagold' and 'Elstar' apples during and after regular and ultra low oxygen storage. *Postharvest Biology and Technology* 20, 15–24.

Awad, M.A. and de Jager, A. (2002a) Formation of flavonoids, especially anthocyanin and chlorogenic acid in 'Jonagold' apple skin: influences of growth regulators and fruit maturity. *Scientia Horticulturae* 93, 257–266.

Awad, M.A. and de Jager, A. (2002b) Relationships between fruit nutrients and concentrations of flavonoids and chlorogenic acid in 'Elstar' apple skin. *Scientia Horticulturae* 92, 265–276.

Awad, M.A. and de Jager, A. (2003) Influences of air and controlled atmosphere storage on the concentration of potentially healthful phenolics in apples and other fruits. *Postharvest Biology and Technology* 27, 53–58.

Awad, M.A., de Jager, A., Dekker, M. and Jongen, W.M.F. (2001a) Formation of flavonoids and chlorogenic acid in apples as affected by crop load. *Scientia Horticulturae* 91, 227–237.

Awad, M.A., de Jager, A., van der Plas, L.H.W. and van der Krol, A.R. (2001b) Flavonoid and chlorogenic acid changes in skin of 'Elstar' and 'Jonagold' apples during development and ripening. *Scientia Horticulturae* 90, 69–83.

Awad, M.A., Wagenmakers, P.S. and de Jager, A. (2001c) Effects of light on flavonoid and chlorogenic acid levels in the skin of 'Jonagold' apples. *Scientia Horticulturae* 88, 289–298.

Bangerth, F. (1976) Relationships between Ca content respectively Ca treatments and ascorbic acid content of apple, pear and tomato fruits. *Qualitas Plantarum – Plant Foods for Human Nutrition* 26, 341–348.

Barz, W. and Hoesel, W. (1979) Metabolism and degradation of phenolic compounds in plants. In: Swain, T., Harborne, J.B. and Sumere, C.F.V. (eds) *Biochemistry of Plant Phenolics*. Plenum Press, New York, pp. 339–369.

Blankenship, S.M. and Unrath, C.R. (1988) PAL and ethylene content during maturation of Red and Golden Delicious apples. *Phytochemistry* 27, 1001–1003.

Boyer, J. and Liu, R.H. (2004) Apple phytochemicals and their health benefits. *Nutrition Journal* 3, 5 (http://www.nutritionj.com/content/3/1/5, accessed 13 April 2011).

Boyer, J., Brown, D. and Liu, R.H. (2004) Uptake of quercetin and quercetin 3-glucoside from whole onion and apple peel extracts by Caco-2 cell monolayers. *Journal of Agricultural and Food Chemistry* 52, 7172–7179.

Boyle, S.P., Dobson, V.L., Duthie, S.J., Kyle, J.A.M. and Collins, A.R. (2000) Absorption and DNA protective effects of flavonoid glycosides from an onion meal. *European Journal of Nutrition* 39, 213–223.

Briviba, K., Stracke, B.A., Rufer, C.E., Watzl, B., Weibel, F.P. and Bub, A. (2007) Effect of consumption of organically and conventionally produced apples on antioxidant activity and DNA damage in humans. *Journal of Agricultural and Food Chemistry* 55, 7716–7721.

Brown, A.E. and Swinburn, T.R. (1973) Factors affecting accumulation of benzoic acid in Bramleys Seedling apples infected with *Nectria galligena*. *Physiological Plant Pathology* 3, 91–99.

Burda, S., Oleszek, W. and Lee, C.Y. (1990) Phenolic-compounds and their changes in apples during maturation and cold-storage. *Journal of Agricultural and Food Chemistry* 38, 945–948.

Carbonaro, M., Mattera, M., Nicoli, S., Bergamo, P. and Cappelloni, M. (2002) Modulation of antioxidant compounds in organic vs conventional fruit (peach, *Prunus persica* L., and pear, *Pyrus communis* L.). *Journal of Agricultural and Food Chemistry* 50, 5458–5462.

Cefarelli, G., D'Abrosca, B., Fiorentino, A., Izzo, A., Mastellone, C., Pacifico, S. and Piscopo, V. (2006) Free-radical-scavenging activities of secondary metabolites from reddened cv. Annurca apple fruits. *Journal of Agricultural and Food Chemistry* 54, 803–809.

Chinnici, F., Bendini, A., Gaiani, A. and Riponi, C. (2004) Radical scavenging activities of peels and pulps from cv. Golden Delicious apples as related to their phenolic composition. *Journal of Agricultural and Food Chemistry* 52, 4684–4689.

Colaric, M., Stampar, F., Solar, A. and Hudina, M. (2006) Influence of branch bending on sugar, organic acid and phenolic content in fruits of 'Williams' pears (*Pyrus communis* L.). *Journal of the Science of Food and Agriculture* 86, 2463–2467.

Coseteng, M.Y. and Lee, C.Y. (1987) Changes in apple polyphenoloxidase and polyphenol concentrations in relation to degree of browning. *Journal of Food Science* 52, 985–989.

Crespy, V., Morand, C., Besson, C., Manach, C., Demigne, C. and Remesy, C. (2002) Quercetin, but not its glycosides, is absorbed from the rat stomach. *Journal of Agricultural and Food Chemistry* 50, 618–621.

Cui, T., Nakamura, K., Ma, L., Li, J.Z. and Kayahara, H. (2005) Analyses of arbutin and chlorogenic acid, the major phenolic constituents in oriental pear. *Journal of Agricultural and Food Chemistry* 53, 3882–3887.

D'Abrosca, B., Fiorentino, A., Monaco, P., Oriano, P. and Pacifico, S. (2006) Annurcoic acid: a new antioxidant ursane triterpene from fruits of cv. Annurca apple. *Food Chemistry* 98, 285–290.

Davey, M.W. and Keulemans, J. (2004) Determining the potential to breed for enhanced antioxidant status in *Malus*: mean inter- and intravarietal fruit vitamin C and glutathione contents at harvest and their evolution during storage. *Journal of Agricultural and Food Chemistry* 52, 8031–8038.

Davey, M.W., Van Montagu, M., Inze, D., Sanmartin, M., Kanellis, A., Smirnoff, N., Benzie, I.J.J., Strain, J.J., Favell, D. and Fletcher, J. (2000) Plant L-ascorbic acid: chemistry, function, metabolism, bioavailability and effects of processing. *Journal of the Science of Food and Agriculture* 80, 825–860.

Davey, M.W., Franck, C. and Keulemans, J. (2004) Distribution, developmental and stress responses of antioxidant metabolism in *Malus*. *Plant Cell and Environment* 27, 1309–1320.

Davey, M.W., Auwerkerken, A. and Keulemans, J. (2007) Relationship of apple vitamin C and antioxidant contents to harvest date and postharvest pathogen infection. *Journal of the Science of Food and Agriculture* 87, 802–813.

de Castro, E., Barrett, D.M., Jobling, J. and Mitcham, E.J. (2008) Biochemical factors associated with a CO_2-induced flesh browning disorder of Pink Lady apples. *Postharvest Biology and Technology* 48, 182–191.

DeEll, J.R. and Prange, R.K. (1992) Postharvest quality and sensory attributes of organically and conventionally grown apples. *HortScience* 27, 1096–1099.

Duvenage, A.J. and DeSwardt, G.H. (1973) Superficial scald on apples: the effect of maturity and diphenylamine on the flavonoid content in the skin of two cultivars. *Zeitschrift Fur Pflanzenphysiologie* 70, 222–234.

Eberhardt, M.V., Liu, R.H., Smith, N.L. and Lee, C.V. (2001) Antioxidant activity of various apple cultivars. *Abstracts of papers of The American Chemical Society*. American Chemical Society, Washington, D.C., 221(1), 118-AGFD.

Escarpa, A. and Gonzalez, M.C. (1998) High-performance liquid chromatography with diode-array detection for the determination of phenolic compounds in peel and pulp from different apple varieties. *Journal of Chromatography A* 823, 331–337.

Faragher, J.D. and Brohier, R.L. (1984) Anthocyanin accumulation in apple skin during ripening: regulation by ethylene and phenylalanine ammonia-lyase. *Scientia Horticulturae* 22, 89–96.

Fattouch, S., Caboni, P., Coroneo, V., Tuberoso, C., Angioni, A., Dessi, S., Marzouki, N. and Cabras, P. (2008) Comparative analysis of polyphenolic profiles and antioxidant and antimicrobial activities of Tunisian pome fruit pulp and peel aqueous acetone extracts. *Journal of Agricultural and Food Chemistry* 56, 1084–1090.

Fawbush, F., Nock, J.F. and Watkins, C.B. (2009) Antioxidant contents and activity of 1-methylcyclopropene (1-MCP)-treated 'Empire' apples in air and controlled atmosphere storage. *Postharvest Biology and Technology* 52, 30–37.

Fernandez-Trujillo, J.P., Nock, J.F. and Watkins, C.B. (2001) Superficial scald, carbon dioxide injury, and changes of fermentation products and organic acids in 'Cortland' and 'Law Rome' apples after high carbon dioxide stress treatment. *Journal of the American Society for Horticultural Science* 126, 235–241.

Feskanich, D., Ziegler, R.G., Michaud, D.S., Giovannucci, E.L., Speizer, F.E., Willett, W.C. and Colditz, G.A. (2000) Prospective study of fruit and vegetable consumption and risk of lung cancer among men and women. *Journal of the National Cancer Institute* 92, 1812–1823.

Franck, C., Baetens, M., Lammertyn, J., Verboven, P., Davey, M.W. and Nicolai, B.M. (2003) Ascorbic acid concentration in cv. Conference pears during fruit development and postharvest storage. *Journal of Agricultural and Food Chemistry* 51, 4757–4763.

Gallus, S., Talamini, R., Giacosa, A., Montella, M., Ramazzotti, V., Franceschi, S., Negri, E. and La Vecchia, C. (2005) Does an apple a day keep the oncologist away? *Annals of Oncology* 16, 1841–1844.

Galvis-Sanchez, A.C., Gil-Izquierdo, A. and Gil, M.I. (2003) Comparative study of six pear cultivars in terms of their phenolic and vitamin C contents and antioxidant capacity. *Journal of the Science of Food and Agriculture* 83, 995–1003.

Galvis-Sanchez, A.C., Fonseca, S.C., Gil-Izquierdo, A., Gil, M.I. and Malcata, F.X. (2006) Effect of different levels of CO_2 on the antioxidant content and the polyphenol oxidase activity of 'Rocha' pears during cold storage. *Journal of the Science of Food and Agriculture* 86, 509–517.

Golding, J.B., McGlasson, W.B., Wyllie, S.G. and Leach, D.N. (2001) Fate of apple peel phenolics during cool storage. *Journal of Agricultural and Food Chemistry* 49, 2283–2289.

Guo, C., Yang, J., Wei, J., Li, Y., Xu, J. and Jiang, Y. (2003) Antioxidant activities of peel, pulp and seed fractions of common fruits as determined by FRAP assay. *Nutrition Research* 23, 1719–1726.

Hagen, S.F., Borge, G.I.A., Bengtsson, G.B., Bilger, W., Berge, A., Haffner, K. and Solhaug, K.A. (2007) Phenolic contents and other health and sensory related properties of apple fruit (*Malus domestica* Borkh., cv. Aroma): effect of postharvest UV-B irradiation. *Postharvest Biology and Technology* 45, 1–10.

Hamauzu, Y., Ueda, Y. and Banno, K. (1999) Relationship between the antioxidant capacity and polyphenols in 'Tsugaru' apple fruit. *Journal of the Japanese Society for Horticultural Science* 68, 675–682.

Hamauzu, Y., Yasui, H., Inno, T., Kume, C. and Omanyuda, M. (2005) Phenolic profile, antioxidant property, and anti-influenza viral activity of Chinese quince (*Pseudocydonia sinensis* Schneid.), quince (*Cydonia oblonga* Mill.), and apple (*Malus domestica* Mill.) fruits. *Journal of Agricultural and Food Chemistry* 53, 928–934.

Hamauzu, Y., Inno, T., Kume, C., Irie, M. and Hiramatsu, K. (2006) Antioxidant and antiulcerative properties of phenolics from Chinese quince, quince, and apple fruits. *Journal of Agricultural and Food Chemistry* 54, 765–772.

Harel, E., Mayer, A.M. and Shain, Y. (1966) Catechol oxidases, endogenous substrates and browning in developing apples. *Journal of the Science of Food and Agriculture* 17, 389–392.

He, X.J. and Liu, R.H. (2008) Triterpenoids isolated from apple peels have potent antiproliferative activity and may be partially responsible for apple's anticancer activity. *Journal of Agricultural and Food Chemistry* 55, 4366–4370.

Hecke, K., Herbinger, K., Veberic, R., Trobec, H., Toplak, H., Stampar, F., Keppel, H. and Grill, D. (2006) Sugar-, acid- and phenol contents in apple cultivars from organic and integrated fruit cultivation. *European Journal of Clinical Nutrition* 60, 1136–1140.

Hertog, M.G., Feskens, E.J., Hollman, P.C., Katan, M.B. and Kromhout, D. (1993) Dietary antioxidant flavonoids and risk of coronary heart disease: the Zutphen Elderly Study. *Lancet* 342, 1007–1011.

Hertog, M.G., Feskens, E.J., Hollman, P.C., Katan, M.B. and Kromhout, D. (1994) Dietary flavonoids and cancer risk in the Zutphen Elderly Study. *Nutrition and Cancer* 22, 175–184.

Hollman, P.C.H., Devries, J.H.M., Vanleeuwen, S.D., Mengelers, M.J.B. and Katan, M.B. (1995) Absorption of dietary quercetin glycosides and quercetin in healthy ileostomy volunteers. *American Journal of Clinical Nutrition* 62, 1276–1282.

Hollman, P.C.H., vanTrijp, J.M.P., Buysman, M., VanderGaag, M.S., Mengelers, M.J.B., deVries, J.H.M. and Katan, M.B. (1997) Relative bioavailability of the antioxidant flavonoid quercetin from various foods in man. *FEBS Letters* 418, 152–156.

Imeh, U. and Khokhar, S. (2002) Distribution of conjugated and free phenols in fruits: antioxidant activity and cultivar variations. *Journal of Agricultural and Food Chemistry* 50, 6301–6306.

Johnson, D.S. and Samuelson, T.J. (1990) Short-term effects of changes in soil management and nitrogen fertilizer application on 'Bramley's Seedling' apple trees. 2. Effects on mineral composition and storage quality of fruit. *Journal of Horticultural Science* 65, 495–502.

Ju, Z.G., Yuan, Y.B., Liu, C.L., Zhan, S.M. and Wang, M.X. (1996) Relationships among simple phenol, flavonoid and anthocyanin in apple fruit peel at harvest and scald susceptibility. *Postharvest Biology and Technology* 8, 83–93.

Kahkonen, M.P., Hopia, A.I., Vuorela, H.J., Rauha, J.P., Pihlaja, K., Kujala, T.S. and Heinonen, M. (1999) Antioxidant activity of plant extracts containing phenolic compounds. *Journal of Agricultural and Food Chemistry* 47, 3954–3962.

Kevers, C., Falkowski, M., Tabart, J., Defraigne, J.O., Dommes, J. and Pincemail, J. (2007) Evolution of antioxidant capacity during storage of selected fruits and vegetables. *Journal of Agricultural and Food Chemistry* 55, 8596–8603.

Khanizadeh, S., Tsao, R., Rekika, D., Yang, R. and DeEll, J. (2007) Phenolic composition and antioxidant activity of selected apple genotypes. *Journal of Food Agriculture and Environment* 5, 61–66.

Knee, M. (1972) Anthocyanin, carotenoid, and chlorophyll changes in the peel of Cox's Orange Pippin apples during ripening on and off the tree. *Journal of Experimental Botany* 23, 184–196.

Knekt, P., Jarvinen, R., Reunanen, A. and Maatela, J. (1996) Flavonoid intake and coronary mortality in Finland: a cohort study. *British Medical Journal* 312, 478–481.

Knekt, P., Jarvinen, R., Seppanen, R., Heliovaara, M., Teppo, L., Pukkala, E. and Aromaa, A. (1997) Dietary flavonoids and the risk of lung cancer and other malignant neoplasms. *American Journal of Epidemiology* 146, 223–230.

Knekt, P., Isotupa, S., Rissanen, H., Heliovaara, M., Jarvinen, R., Hakkinen, S., Aromaa, A. and Reunanen, A. (2000) Quercetin intake and the incidence of cerebrovascular disease. *European Journal of Clinical Nutrition* 54, 415–417.

Lancaster, J.E. (1992) Regulation of skin color in apples. *Critical Reviews in Plant Sciences* 10, 487–502.

Lancaster, J.E., Reay, P.F., Norris, J. and Butler, R.C. (2000) Induction of flavonoids and phenolic acids in apple by UV-B and temperature. *Journal of Horticultural Science and Biotechnology* 75, 142–148.

Larrigaudiere, C., Lentheric, I., Pinto, E. and Vendrell, M. (2001a) Short-term effects of air and controlled atmosphere storage on antioxidant metabolism in conference pears. *Journal of Plant Physiology* 158, 1015–1022.

Larrigaudiere, C., Pinto, E., Lentheric, I. and Vendrell, M. (2001b) Involvement of oxidative processes in the development of core browning in controlled-atmosphere stored pears. *Journal of Horticultural Science and Biotechnology* 76, 157–162.

Lata, B. (2007) Relationship between apple peel and the whole fruit antioxidant content: year and cultivar variation. *Journal of Agricultural and Food Chemistry* 55, 663–671.

Lata, B. (2008) Apple peel antioxidant status in relation to genotype, storage type and time. *Scientia Horticulturae* 117, 45–52.

Lata, B. and Tomala, K. (2007) Apple peel as a contributor to whole fruit quantity of potentially healthful bioactive compounds. Cultivar and year implication. *Journal of Agricultural and Food Chemistry* 55, 10,795–10,802.

Lata, B., Przeradzka, M. and Binkowska, M. (2005) Great differences in antioxidant properties exist between 56 apple cultivars and vegetation seasons. *Journal of Agricultural and Food Chemistry* 53, 8970–8978.

Lattanzio, V., Di Venere, D., Linsalata, V., Bertolini, P., Ippolito, A. and Salermo, M. (2001) Low temperature metabolism of apple phenolics and quiescence of *Phlyctaena Vagabunda*. *Journal of Agricultural and Food Chemistry* 49, 5817–5821.

Le Marchand, L., Murphy, S.P., Hankin, J.H., Wilkens, L.R. and Kolonel, L.N. (2000) Intake of flavonoids and lung cancer. *Journal of the National Cancer Institute* 92, 154–160.

Lee, K.W., Kim, Y.J., Kim, D.O., Lee, H.J. and Lee, C.Y. (2003) Major phenolics in apple and their contribution to the total antioxidant capacity. *Journal of Agricultural and Food Chemistry* 51, 6516–6520.

Leja, M., Mareczek, A. and Ben, J. (2003) Antioxidant properties of two apple cultivars during long-term storage. *Food Chemistry* 80, 303–307.

Leontowicz, H., Gorinstein, S., Lojek, A., Leontowicz, M., Ciz, M., Soliva-Fortuny, R., Park, Y.S., Jung, S.T., Trakhtenberg, S. and Martin-Belloso, O. (2002) Comparative content of some bioactive compounds in apples, peaches and pears and their influence on lipids and antioxidant capacity in rats. *Journal of Nutritional Biochemistry* 13, 603–610.

Leontowicz, M., Gorinstein, S., Leontowicz, H., Krzeminski, R., Lojek, A., Katrich, E., Ciz, M., Martin-Belloso, O., Soliva-Fortuny, R., Haruenkit, R. and Trakhtenberg, S. (2003) Apple and pear peel and pulp and their influence on plasma lipids and antioxidant potentials in rats fed cholesterol-containing diets. *Journal of Agricultural and Food Chemistry* 51, 5780–5785.

Li, M.J., Ma, F.W., Zhang, M. and Pu, F. (2008) Distribution and metabolism of ascorbic acid in apple fruits (*Malus domestica* Borkh cv. Gala). *Plant Science* 174, 606–612.

Li, Z.H., Gemma, H. and Iwahori, S. (2002) Stimulation of 'Fuji' apple skin color by ethephon and phosphorus-calcium mixed compounds in relation to flavonoid synthesis. *Scientia Horticulturae* 94, 193–199.

Lister, C.E., Lancaster, J.E. and Walker, J.R.L. (1996a) Developmental changes in enzymes of flavonoid biosynthesis in the skins of red and green apple cultivars. *Journal of the Science of Food and Agriculture* 71, 313–320.

Lister, C.E., Lancaster, J.E. and Walker, J.R.L. (1996b) Phenylalanine ammonia-lyase (PAL) activity and its relationship to anthocyanin and flavonoid levels in New Zealand-grown apple cultivars. *Journal of the American Society for Horticultural Science* 121, 281–285.

Liu, J.R., Dong, H.W., Chen, B.Q., Zhao, P. and Liu, R.H. (2009) Fresh apples suppress mammary carcinogenesis and proliferative activity and induce apoptosis in mammary tumors of the Sprague–Dawley rat. *Journal of Agricultural and Food Chemistry* 57, 297–304.

Liu, R.H. (2004) Potential synergy of phytochemicals in cancer prevention: mechanism of action. *Journal of Nutrition* 134, 3479S–3485S.

Liu, R.H., Eberhardt, M.V. and Lee, C.Y. (2001) Antioxidant and antiproliferative activities of selected New York apple cultivars. *New York Fruit Quarterly* 9, 15–17.

Liu, R.H., Liu, J.R. and Chen, B.Q. (2005) Apples prevent mammary tumors in rats. *Journal of Agricultural and Food Chemistry* 53, 2341–2343.

McAnlis, G.T., McEneny, J., Pearce, J. and Young, I.S. (1999) Absorption and antioxidant effects of quercetin from onions, in man. *European Journal of Clinical Nutrition* 53, 92–96.

McGhie, T.K., Hunt, M. and Barnett, L.E. (2005) Cultivar and growing region determine the antioxidant polyphenolic concentration and composition of apples grown in New Zealand. *Journal of Agricultural and Food Chemistry* 53, 3065–3070.

MacLean, D.D., Murr, D.P. and DeEll, J.R. (2003) A modified total oxyradical scavenging capacity assay for antioxidants in plant tissues. *Postharvest Biology and Technology* 29, 183–194.

MacLean, D.D., Murr, D.P., DeEll, J.R. and Horvath, C.R. (2006) Postharvest variation in apple (*Malus* × *domestica* Borkh.) flavonoids following harvest, storage, and 1-MCP treatment. *Journal of Agricultural and Food Chemistry* 54, 870–878.

MacLean, D.D., Murr, D.P., DeEll, J.R., Mackay, A.B. and Kupferman, E.M. (2007) Inhibition of PAL, CHS, and ERS1 in 'Red d'Anjou' pear (*Pyrus communis* L.) by 1-MCP. *Postharvest Biology and Technology* 45, 46–55.

Mayr, U., Michalek, S., Treutter, D. and Feucht, W. (1997) Phenolic compounds of apple and their relationship to scab resistance. *Journal of Phytopathology – Phytopathologische Zeitschrift* 145, 69–75.

Mareczek, A., Leja, M. and Ben, J. (2000) Total phenolics, anthocyanins and antioxidant activity in the peel of the stored apples. *Journal of Fruit and Ornamental Plant Research* 8, 59–64.

Michalek, S., Klebel, C. and Treutter, D. (2005) Stimulation of phenylpropanoid biosynthesis in apple (*Malus domestica* Borkh.) by abiotic elicitors. *European Journal of Horticultural Science* 70, 116–120.

Mosel, H.D. and Herrmann, K. (1974) Changes in catechins and hydroxycinnamic acid derivatives during development of apples and pears. *Journal of the Science of Food and Agriculture* 25, 251–256.

Napolitano, A., Cascone, A., Graziani, G., Ferracane, R., Scalfi, L., Di Vaio, C., Ritieni, A. and Fogliano, V. (2004) Influence of variety and storage on the polyphenol composition of apple flesh. *Journal of Agricultural and Food Chemistry* 52, 6526–6531.

Ndubizu, T.O.C. (1976) Relations of phenolic inhibitors to resistance of immature apple fruits to rot. *Journal of Horticultural Science* 51, 311–319.

Noble, J.P. and Drysdale, R.B. (1983) The role of benzoic acid and phenolic compounds in latency in fruits of 2 apple cultivars infected with *Pezicula malicorticis* or *Nectria galligena*. *Physiological Plant Pathology* 23, 207–216.

Oleszek, W., Amiot, M.J. and Aubert, S.Y. (1994) Identification of some phenolics in pear fruit. *Journal of Agricultural and Food Chemistry* 42, 1261–1265.

Peck, G.M., Andrews, P.K., Reganold, J.P. and Fellman, J.K. (2006) Apple orchard productivity and fruit quality under organic, conventional, and integrated management. *HortScience* 41, 99–107.

Peck, G.M., Merwin, I.A., Watkins, C.B., Chapman, K.W. and Padilla-Zakour, O. (2009) Maturity and quality of Liberty apple fruit under integrated and organic fruit production systems are similar in a New York orchard. *HortScience* 44, 1382–1389.

Perez-Ilzarbe, J., Hernandez, T., Estrella, I., Vendrell, M. (1997) Cold storage of apples (cv. Granny Smith) and changes in phenolic compounds. *Zeitschrift für Lebensmittel-Untersuchung und -Forschung* 204, 52–55.

Piretti, M.V., Gallerani, G., Pratella, G.C. (1994) Polyphenol fate and superficial scald in apple. *Postharvest Biology and Technology* 4, 213–224.

Piretti, M.V., Gallerani, G. and Brodnik, U. (1996) Polyphenol polymerisation involvement in apple superficial scald. *Postharvest Biology and Technology* 8, 11–18.

Planchon, V., Lateur, M., Dupont, P. and Lognay, G. (2004) Ascorbic acid level of Belgian apple genetic resources. *Scientia Horticulturae* 100, 51–61.

Podsedek, A., Wilska-Jeszka, J., Anders, B. and Markowski, J. (2000) Compositional characterisation of some apple varieties. *European Food Research and Technology* 210, 268–272.

Raese, J.T. and Drake, S.R. (1997) Nitrogen fertilization and elemental composition affects fruit quality of 'Fuji' apples. *Journal of Plant Nutrition* 20, 1797–1809.

Ranadive, A.S. and Haard, N.F. (1971) Changes in polyphenolics on ripening of selected pear varieties. *Journal of the Science of Food and Agriculture* 22, 86–89.

Reay, P.F. (1999) The role of low temperatures in the development of the red blush on apple fruit ('Granny Smith'). *Scientia Horticulturae* 79, 113–119.

Reay, P.F. and Lancaster, J.E. (2001). Accumulation of anthocyanins and quercetin glycosides in 'Gala' and 'Royal Gala' apple fruit skin with UV-B-visible irradiation: modifying effects of fruit maturity, fruit side, and temperature. *Scientia Horticulturae* 90, 57–68.

Reganold, J.P., Glover, J.D., Andrews, P.K. and Hinman, H.R. (2001) Sustainability of three apple production systems. *Nature* 410, 926–930.

Rice-Evans, C.A., Miller, N.J. and Paganga, G. (1996) Structure-antioxidant activity relationships of flavonoids and phenolic acids. *Free Radical Biology and Medicine* 20, 933–956.

Richard-Forget, F.C., Rouetmayer, M.A., Goupy, P.M., Philippon, J. and Nicolas, J.J. (1992) Oxidation of chlorogenic acid, catechins, and 4-methylcatechol in model solutions by apple polyphenol oxidase. *Journal of Agricultural and Food Chemistry* 40, 2114–2122.

Ritenour, M.A., Ahrens, M.J. and Saltveit, M.E. (1995) Effects of temperature on ethylene-induced phenylalanine ammonia-lyase activity and russet spotting in harvested iceberg lettuce. *Journal of the American Society for Horticultural Science* 120, 84–87.

Roth, E., Berna, A., Beullens, K., Yarramraju, S., Lammertyn, J., Schenk, A. and Nicolai, B. (2007) Postharvest quality of integrated and organically produced apple fruit. *Postharvest Biology and Technology* 45, 11–19.

Rudell, D.R. and Mattheis, J.P. (2008) Synergism exists between ethylene and methyl jasmonate in artificial light-induced pigment enhancement of 'Fuji' apple fruit peel. *Postharvest Biology and Technology* 47, 136–140.

Rudell, D.R., Mattheis, J.P., Fan, X. and Fellman, J.K. (2002) Methyl jasmonate enhances anthocyanin accumulation and modifies production of phenolics and pigments in 'Fuji' apples. *Journal of the American Society for Horticultural Science* 127, 435–441.

Sannomaru, Y., Katayama, O., Kashimura, Y. and Kaneko, K. (1998) Changes in polyphenol content and polyphenoloxidase activity of apple fruits during ripening process. *Journal of the Japanese Society for Food Science and Technology-Nippon Shokuhin Kagaku Kogaku Kaishi* 45, 37–43.

Saucier, C.T. and Waterhouse, A.L. (1999) Synergetic activity of catechin and other antioxidants. *Journal of Agricultural and Food Chemistry* 47, 4491–4494.

Saure, M.C. (1990) External control of anthocyanin formation in apple. *Scientia Horticulturae* 42, 181–218.

Scalzo, J., Politi, A., Pellegrini, N., Mezzetti, B. and Battino, M. (2005) Plant genotype affects total antioxidant capacity and phenolic contents in fruit. *Nutrition* 21, 207–213.

Sesso, H.D., Gaziano, J.M., Liu, S. and Buring, J.E. (2003) Flavonoid intake and the risk of cardiovascular disease in women. *American Journal of Clinical Nutrition* 77, 1400–1408.

Shaheen, S.O., Sterne, J.A., Thompson, R.L., Songhurst, C.E., Margetts, B.M. and Burney, P.G. (2001) Dietary antioxidants and asthma in adults: population-based case–control study. *American Journal of Respiratory and Critical Care Medicine* 164, 1823–1828.

Silva, B.M., Andrade, P.B., Valentao, P., Ferreres, F., Seabra, R.M. and Ferreira, M.A. (2004) Quince (*Cydonia oblonga* Miller) fruit (pulp, peel, and seed) and jam: antioxidant activity. *Journal of Agricultural and Food Chemistry* 52, 4705–4712.

Sokol-Letowska, A., Oszmianski, J. and Wojdylo, A. (2007) Antioxidant activity of the phenolic compounds of hawthorn, pine and skullcap. *Food Chemistry* 103, 853–859.

Stafford, H.A. (1990) *Flavonoid Metabolism.* CRC Press, Boca Raton, Florida.

Stopar, M., Bolcina, U., Vanzo, A. and Vrhovsek, U. (2002) Lower crop load for cv. Jonagold apples (*Malus × domestica* Borkh.) increases polyphenol content and fruit quality. *Journal of Agricultural and Food Chemistry* 50, 1643–1646.

Sun, J., Chu, Y.F., Wu, X.Z. and Liu, R.H. (2002) Antioxidant and anti proliferative activities of common fruits. *Journal of Agricultural and Food Chemistry* 50, 7449–7454.

Tabak, C., Arts, I.C., Smit, H.A., Heederik, D. and Kromhout, D. (2001) Chronic obstructive pulmonary disease and intake of catechins, flavonols, and flavones: the MORGEN Study. *American Journal of Respiratory and Critical Care Medicine* 164, 61–64.

Tan, S.C. (1979) Relationships and interactions between phenylalanine ammonia-lyase, phenylalanine ammonia-lyase inactivating system, and anthocyanin in apples. *Journal of the American Society for Horticultural Science* 104, 581–586.

Tarozzi, A., Marchesi, A., Cantelli-Forti, G. and Hrelia, P. (2004) Cold-storage affects antioxidant properties of apples in caco-2 cells. *Journal of Nutrition* 134, 1105–1109.

Tomas-Barberan, F. and Espin, J.C. (2001) Phenolic compounds and related enzymes as determinants of quality in fruits and vegetables. *Journal of the Science of Food and Agriculture* 81, 853–876.

Treutter, D. (2001) Biosynthesis of phenolic compounds and its regulation in apple. *Plant Growth Regulation* 34, 71–89.

Tsao, R., Yang, R., Christopher, J., Zhu, Y. and Zhu, H.H. (2003) Polyphenolic profiles in eight apple cultivars using high-performance liquid chromatography (HPLC). *Journal of Agricultural and Food Chemistry* 51, 6347–6353.

Ubi, B.E., Honda, C., Bessho, H., Kondo, S., Wada, M., Kobayashi, S. and Moriguchi, T. (2006) Expression analysis of anthocyanin biosynthetic genes in apple skin: effect of UV-B and temperature. *Plant Science* 170, 571–578.

USDA (United States Department of Agriculture) (2008) Agriculture Research Service. Available at: http://www.ars.usda.gov/nutrientdata:2008.

van der Sluis, A.A., Dekker, M., de Jager, A. and Jongen, W.M.F. (2001) Activity and concentration of polyphenolic antioxidants in apple: effect of cultivar, harvest year, and storage conditions. *Journal of Agricultural and Food Chemistry* 49, 3606–3613.

Vanzani, P., Rossetto, M., Rigo, A., Vrhovsek, U., Mattivi, F., D'Amato, E. and Scarpa, M. (2005) Major phytochemicals in apple cultivars: contribution to peroxyl radical trapping efficiency. *Journal of Agricultural and Food Chemistry* 53, 3377–3382.

Veltman, R.H., Larrigaudiere, C., Wichers, H.J., Van Schaik, A.C.R., Van der Plas, L.H.W. and Oosterhaven, J. (1999) PPO activity and polyphenol content are not limiting factors during brown cove development in pears (*Pyrus communis* L. cv. Conference). *Journal of Plant Physiology* 154, 697–702.

Veltman, R.H., Kho, R.M., van Schaik, A.C.R., Sanders, M.G. and Oosterhaven, J. (2000) Ascorbic acid and tissue browning in pears (*Pyrus communis* L. cvs Rocha and Conference) under controlled atmosphere conditions. *Postharvest Biology and Technology* 19, 129–137.

Vilaplana, R., Valentines, M.C., Toivonen, P. and Larrigaudiere, C. (2006) Antioxidant potential and peroxidative state of 'Golden Smoothee' apples treated with 1-methylcyclopropene. *Journal of the American Society for Horticultural Science* 131, 104–109.

Vinson, J.A., Su, X.H., Zubik, L. and Bose, P. (2001) Phenol antioxidant quantity and quality in foods: fruits. *Journal of Agricultural and Food Chemistry* 49, 5315–5321.

Volz, R.K., Biasi, W.V. and Mitcham, E.J. (1998) Fermentative volatile production in relation to carbon dioxide-induced flesh browning in 'Fuji' apple. *HortScience* 33, 1231–1234.

Vrhovsek, U., Rigo, A., Tonon, D. and Mattivi, F. (2004) Quantitation of polyphenols in different apple varieties. *Journal of Agricultural and Food Chemistry* 52, 6532–6538.

Walgren, R.A., Walle, U.K. and Walle, T. (1998) Transport of quercetin and its glucosides across human intestinal epithelial Caco-2 cells. *Biochemical Pharmacology* 55, 1721–1727.

Walle, T., Walle, U.K. and Halushka, P.V. (2001) Carbon dioxide is the major metabolite of quercetin in humans. *Journal of Nutrition* 131, 2648–2652.

Wang, H., Cao, G.H. and Prior, R.L. (1996) Total antioxidant capacity of fruits. *Journal of Agricultural and Food Chemistry* 44, 701–705.

Wang, H.Q., Arakawa, O. and Motomura, Y. (2000) Influence of maturity and bagging on the relationship between anthocyanin accumulation and phenylalanine ammonia-lyase (PAL) activity in 'Jonathan' apples. *Postharvest Biology and Technology* 19, 123–128.

Watkins, C.B. (2003) Principles and practices of postharvest handling and stress. In: Feree, D. and Warrington, I.J. (eds) *Apples: Crop Physiology, Production and Uses*. CAB International, Wallingford, UK, pp. 585–614.

Watkins, C.B. (2006) The use of 1-methylcyclopropene (1-MCP) on fruits and vegetables. *Biotechnology Advances* 24, 389–409.

Watkins, C.B. (2008) Overview of 1-methylcyclopropene trials and uses for edible horticultural crops. *HortScience* 43, 86–94.

Weibel, F.P., Bickel, R., Leuthold, S. and Alfoldi, T. (2000) Are organically grown apples tastier and healthier? A comparative field study using conventional and alternative methods to measure fruit quality. *Acta Horticulturae* 517, 417–426.

Wojdylo, A., Oszmianski, J. and Laskowski, P. (2008) Polyphenolic compounds and antioxidant activity of new and old apple varieties. *Journal of Agricultural and Food Chemistry* 56, 6520–6530.

Wolfe, K., Wu, X.Z. and Liu, R.H. (2003) Antioxidant activity of apple peels. *Journal of Agricultural and Food Chemistry* 51, 609–614.

Wolfe, K.L., Kang, X.M., He, X.J., Dong, M., Zhang, Q.Y. and Liu, R.H. (2008) Cellular antioxidant activity of common fruits. *Journal of Agricultural and Food Chemistry* 56, 8418–8426.

Woods, R.K., Walters, E.H., Raven, J.M., Wolfe, R., Ireland, P.D., Thien, F.C.K. and Abramson, M.J. (2003) Food and nutrient intakes and asthma risk in young adults. *American Journal of Clinical Nutrition* 78, 414–421.

Yoon, H. and Liu, R.H. (2007) Effect of selected phytochemicals and apple extracts on NF-kappa B activation in human breast cancer MCF-7 cells. *Journal of Agricultural and Food Chemistry* 55, 3167–3173.

Yoon, H. and Liu, R.H. (2008) Effect of 2 alpha-hydroxyursolic acid on NF-kappa B activation induced by TNF-alpha in human breast cancer MCF-7 cells. *Journal of Agricultural and Food Chemistry* 56, 8412–8417.

Zerbini, P.E., Rizzolo, A., Brambilla, A., Cambiaghi, P. and Grassi, M. (2002) Loss of ascorbic acid during storage of conference pears in relation to the appearance of brown heart. *Journal of the Science of Food and Agriculture* 82, 1007–1013.

Zhang, Z., Chang, Q., Zhu, M., Ho, W.K.K. and Chen, Z.Y. (2001) Characterization of antioxidants present in hawthorn fruits. *Journal of Nutritional Biochemistry* 12, 144–152.

Zhao, X., Carey, E.E., Wang, W.Q. and Rajashekar, C.B. (2006) Does organic production enhance phytochemical content of fruit and vegetables? Current knowledge and prospects for research. *HortTechnology* 16, 449–456.

12   Potato and Other Root Crops

Anne Pihlanto

12.1 Introduction

Diet is known to play an important role in many major diseases of our society, such as cardiovascular diseases, cancer, obesity and diabetes. Certain bioactive compounds in food plants have been known for a long time for their beneficial effects, whereas others have been recognized more recently. Diets rich in polyphenolics and carotenoids have been associated with a lower incidence of atherosclerotic heart disease and certain cancers. Reduction of atherosclerotic heart disease in association with antioxidant-rich diets is hypothesized to be related to a reduction in the oxidative polymerization of low-density lipoproteins and consequent lesion formation and plaque build-up in key coronary arteries. Cancer reduction is further hypothesized to be due to protection of DNA from destruction by reactive oxidative species. Consumption of diets high in fruit and vegetables increases the antioxidant levels in blood serum in human subjects. In recent years, some studies have focused on the role of polyphenols and potato proteins in the prevention and treatment of obesity.

Potatoes (*Solanum tuberosum* L.) and other root crops are perhaps the most important vegetables consumed in the Western diet. This chapter reviews the current knowledge about the major beneficial compounds derived from potatoes and carrots, as well as sweet potato, red beet, red radish, cassava and yam. Special emphasis is given to their potential health effects (Table 12.1). Suggestions for future research are also mentioned.

12.2 Identity and Role of Bioactivities

12.2.1 Potato

The potato is one of the world's most widely grown crops. Potato production is about 320 million tonnes (Mt) globally, of which about 66% is used as food, 12% as feed and 10% as seed (FAO, 2010). The rest is used mainly in the starch industry, since increasing amounts of starch are now extracted from potato tubers and modified for further uses in processed foods and in non-foods, including the paper industry. Potatoes represent a staple source of nutrients and energy in many different countries. They are a proven source of proteins, carbohydrates and minerals such as calcium, potassium and phosphorus, and their value in the human diet is often underestimated or overlooked. Several small molecular weight compounds, many of which have reported beneficial effects on health, have been found in potatoes (Friedman, 1997). These compounds include secondary metabolites,

Table 12.1. Bioactive compounds and potential health effects (*in vitro*, cell, animal and human studies).

Food	Component	Potential benefit	Reference
Proven effect on animal and human studies			
Potato	Tuber protease inhibitors	Prevents experimental protease-induced dermatitis (*in vivo*) (humans)	Ruseler-van Embden *et al.* (2004)
		Prevents protease-induced perianal dermatitis (faeces of subjects suffering perianal dermatitis): Suppresses proteolytic activity in faeces	
	Peel extract	Protection of liver injury (in rats)	Singh *et al.* (2008)
		Increases the activities of hepatic antioxidant enzymes	
		Protects acute liver damage induced by CCl$_4$	
	Protein hydrolysate	Protective effect against ethanol-induced mucosal damage in rats	Kudoh *et al.* (2003)
Carrot		Beneficial effects on cholesterol metabolism (mice and rats):	Nicolle *et al.* (2003, 2004)
		Increases antioxidant status	
		Inhibits cholesterol absorption	
		Decreases serum cholesterol concentration (5 human subjects)	Robertson *et al.* (1979)
	Insoluble fibre and cellulose	Lowers serum triglyceride, total cholesterol and liver lipids (in hamster)	Chou *et al.* (2008)
	Falcarinol	Chemopreventive impact on lymphoblastic leukaemia cell line CEM-C7H2	Christensen and Brandt (2006)
		Reduces development of tumours in rats	
In vitro evidence on positive effect on cardiovascular health and prevention of certain cancer cells			
Potato	Patatin	Radical, hydroxyradical scavenging activity, prevention of LDL peroxidation	Liu *et al.* (2003)
	Protein hydrolysate	Antihypertensive and antioxidative activity *in vitro*	Pihlanto *et al.* (2008)
			Mäkinen *et al.* (2008)
	Early potato extracts	Free radical, oxyradical scavenging activity	Leo *et al.* (2008)
		Inhibition of breast cancer MCF-7 cell proliferation	
	Methanol extract	Free radical, oxyradical, hydroxyradical scavenging activity, inhibition of lipid peroxidation	Chu *et al.* (2000)
	Glycoalkaloids	Inhibition of cholinesterases, complexes with cholesterols and other phytosterols	Friedman (2006)
Sweet potato	Phenolic	Antioxidant activity as Trolox equivalent, ranging from 1.3 to 4.6 mg/g DW	Padda and Picha (2008)
	Trypsin inhibitor	Induced NB4 promyelocytic leukaemia cell apoptosis through the inhibition of cell growth	Huang *et al.* (2007)
Satiety in vitro and in vivo			
Potato	Protein hydrolysate	Induce CCK release from enteroendocrine cells	Foltz *et al.* (2008)
	Proteinase inhibitor II	Reduced energy intake in lean subjects	Hill *et al.* (1990)

such as polyphenols, protease inhibitors and glycoalkaloids. Storage and heat treatments, during commercial processing or domestic cooking, may affect the total content of these secondary metabolites, which has to be taken into account when evaluating the intake of these compounds. Outside the centre of origin of cultivated potato in the Andes of South America, it is rare to find varieties with anthocyanidin pigments conferring red or purple flesh. However, much of the world's production is occupied by the yellow-flesh potatoes, which have higher total carotenoid than the white-fleshed varieties of North America and Great Britain.

Polyphenolic compounds

Potato tubers are one of the richest sources of polyphenols. Content of polyphenols is affected mainly by variety, year of cultivation, stress factors (mechanical damage of tubers, attack of pathogens, action of light on tubers or irrigation) and by storage and cooking treatment. To a lesser extent, the effect of geographical location, soil type, potassium fertilization, storage temperature, γ-irradiation and other factors could be involved, but there is only a little demonstrable empirical evidence for this in the literature (Friedman, 1997).

Phenolic acids are distributed mostly between the cortex and skin (peel) tissues of the potato (Fig. 12.1). About 50% of the phenolic compounds are located in the potato peel and adjoining tissues, which are often wasted, while the remainder decrease in concentration from outside toward the centre of the potato tubers (Friedman, 1997). Lewis *et al.* (1998a)

found that cultivated potato tuber skin contained 2–5 mg/g fresh weight (FW) phenolic acid and 0.2–0.3 mg flavonoids. In wild *Solanum* species, phenolic acids ranged from 6 to 27 µg/g FW in skin and from 1 to 6 µg/g FW in the flesh (Lewis *et al.*, 1998b). Purple- and red-skinned tubers contained twice the concentration of phenolic acids as white-skinned tubers. Tuber flesh contained lower concentrations, ranging from 0.1 to 0.6 mg of phenolic acids and from 0 to 0.03 mg/g FW of flavonoids. Furthermore, purple- and red-fleshed cultivars had twice the flavonoid concentration of white-fleshed cultivars and three to four times the concentration of phenolic acids. Mattila and Hellström (2007) found the highest content of phenolic acids in raw and boiled potato peels, varying from 0.23 to 0.45 mg/g FW. When peeled potatoes were boiled for 18–25 min with or without NaCl, the phenolic acid content was clearly lower.

Studies on the phenolic composition of potatoes have reported that the predominant phenolic acid is chlorogenic acid, which constitutes about 80% of the total phenolic acids. Chlorogenic acid is formed between caffeic and quinic acid and is hydrolysed rapidly to caffeic acid under alkaline conditions. Traces of other phenolic acids, such as protocatechic, ferulic, vanillic and *p*-coumaric acids, can be found. Chlorogenic acid content varied from 0.1 to 0.2 mg/g in different potato varieties (Friedman, 1997). In skin extracts of wild *Solanum* species, chlorogenic acid accounted for 40–50% of the total phenolic acid content, with caffeic acid also present at high levels (10–30%). Flesh phenolic acids comprised 30–40% protocatechic acid, 20–30%

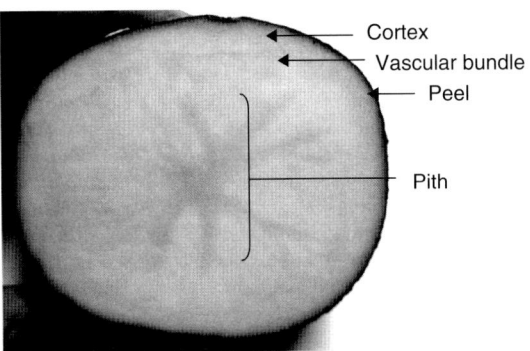

Fig. 12.1. Cross section of a potato tuber.

Cortex
Vascular bundle
Peel

Pith

chlorogenic acid and 20–30% p-coumaric acid (Lewis et al., 1998b). In early potatoes, the cultivars had high concentrations of free chlorogenic acid (0.47–0.92 mg/g dry weight (DW)) and moderate amounts of caffeic acid (0.05–0.12 mg/g DW), lower concentrations of ferulic acid (0.006–0.039 mg/g DW) and traces of p-coumaric acid and some unidentified compounds. Nearly all the phenolic acids in potatoes are in a soluble form (Mattila and Hellström, 2007). In the wild species (white or light purple tubers), concentrations of total flavonoids ranged between 20 and 170 µg/g FW in skin and 0 and 25 µg/g FW in flesh. The major flavonoids were catechin, epicatechin, eriodictyol and naringenin (Lewis et al., 1998b). Chu et al. (2000) reported that in potatoes purchased from local markets, the total amount of flavonols and flavones was 0.13 mg/kg and the major compound was kaempferol. The major flavonoid in early potatoes was catechin (0.43–1.57 mg/g DW) and this was the major compound present in a bound form (Leo et al., 2008).

Anthocyanins are the major group of pigments in potatoes and tubers which exhibit a range of colours, with heterogeneous flushes of red, and have been studied by several workers (Fig. 12.2). The pigments have been determined to be various types of acylated anthocyanidin glucosides. Rodriguez-Saona et al. (1998) reported anthocyanidin contents of partially and solidly red-fleshed potatoes ranging from 0.03 to 0.40 mg/g FW. The major pigments were identified to be acylated glucosides of pelargonidin. Lewis et al. (1998a) found much higher concentration of anthocyanins in certain cultivars, extending up to 3.68 mg/g FW in the purple-fleshed cv. Urenika and up to 0.22 mg/g FW in red-fleshed types. Concentrations were considerably higher in skin, approaching 9 mg/g FW in purple-fleshed and 5 mg/g FW (of the skin alone) in red-fleshed types. p-Coumaryl conjugates of anthocyanins were present in peel (up to 7 mg/g) and flesh (up to 2 mg/g) (Lewis et al., 1998a). Pelargonidin and peonidin were in nearly equal amounts in the red flesh, while petunidin and malvidin were predominant in the purple flesh. Wild species had no anthocyanins in the flesh but up to 0.27 mg/g (FW) in the skin (Lewis et al., 1998b). Jansen and

Flamme (2006) found, on average, higher amounts of anthocyanins in skin (0.65 mg/g FW) in 31 analysed potato cultivars/breeding clones. The corresponding values of whole tubers (0.31 mg/g FW) and flesh (0.22 mg/g FW) were significantly lower. The cv. Peru Purple was shown to have the highest anthocyanin content in the skin: 2.96 mg/g FW. Fossen and Andersen (2000) determined the anthocyanins of the purple-fleshed cv. Congo to consist of ferulyl gluco- and rhamnopyranosides of malvidin and petudin. Fossen et al. (2003) further reported the finding of acylation with caffeic acid in extracts from an unnamed, purple-fleshed, Norwegian-derived cultivar.

Several results support the theory that one of the reasons why potatoes produce phenolic acids and flavonoids is to aid in plant defence. In diseased tubers, increased output via the cinnamic acid, benzoic acid and flavonoid, especially flavonol, pathways has been found. Also supporting this hypothesis is the observation that much higher concentrations of phenolic acids and flavonoids (plus anthocyanins) have been found in tuber skin than in the flesh, and this may be related to the skin being the first line of defence against pathogens and pests (Friedman, 1997; Brown, 2005).

There are only a few publications describing the effect of cooking on the polyphenolic compounds in potatoes, a situation that is bizarre considering that potatoes are not eaten raw. Phenolic acids seemed to remain quite stable in potatoes during boiling. Mattila and Hellström (2007) found that 71–92% and 65–100% of soluble and total phenolic acids on a dry matter basis, respectively, were left in potato peels after boiling (18–25 min), as compared with unboiled peels. Boiling destroyed 65%, microwave baking 45% and oven baking 100% of the original amount of chlorogenic acid. Anthocyanins in potatoes survived to a large degree during various cooking methods, including frying in oil (Brown, 2005).

Potato antioxidant compounds are associated closely with a strong antioxidant capacity and their antioxidant effect is due mainly to their redox properties and is the result of various possible mechanisms: free

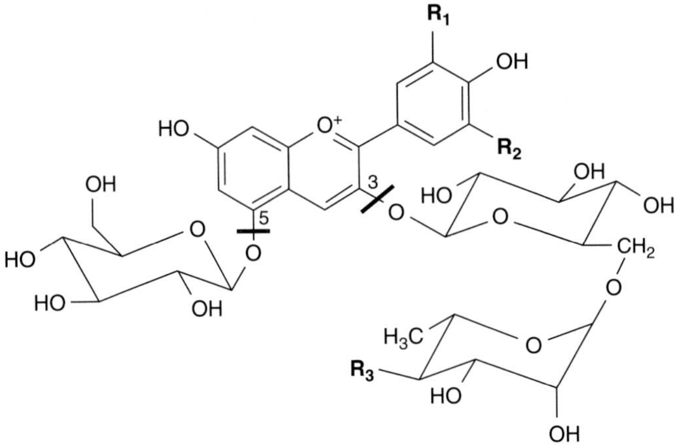

R₃ = OH for non-acylated compounds

R₃ = coumaric acid for acylated anthocyanins = HO —

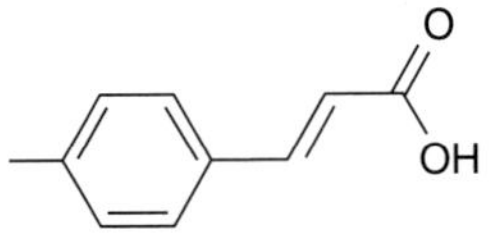

—— = cleavage position during LC-ESI-MS^2

aglycon	R$_1$	R$_2$
petunidin	OCH$_3$	OH
pelargonidin	H	H
peonidin	OCH$_3$	H
malvidin	OCH$_3$	OCH$_3$

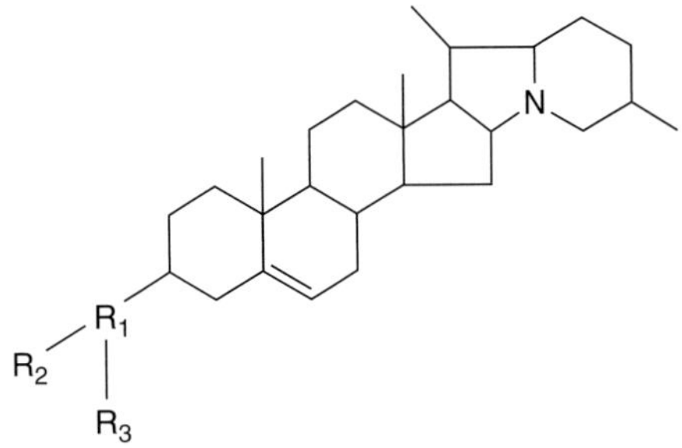

Fig. 12.2. Structure of potato anthocyanins (Eichhorn and Winterhalter, 2005).

radical scavenging activity, transition metal chelating activity and/or singlet-oxygen-quenching capacity. Phenolic content and antioxidative activity of potato cultivars have been shown to be genotype-dependent and not related to flesh colour. Polyphenolic compounds in potatoes show antioxidative activity in several food systems. For example, freeze-dried extracts from peels of potato varieties prevented soybean or sunflower oil oxidation. These results suggest the possible value of potato peel in the prevention of oxidative rancidity of food oils (Brown, 2005).

Glycoalkaloids

Glycoalkaloids (GAs) are secondary metabolites that are produced following the general steroid biosynthesis pathway, starting from acetyl-coenzyme A and followed by the intermediates, mevalonic acid, squalene, cycloartenol and cholesterol. α-Chaconine and α-solanine are the main GAs of the cultivated potato (Fig. 12.3), whereas many other GAs are known in the wild potato species. GAs may be toxic to bacteria, fungi, viruses, insects, animals, and even humans. Their natural function is probably to serve as stress metabolites for the protection of potato when attacked by insects, fungi, etc. The potential

human toxicity of GAs has led to the establishment of strict guidelines limiting the GA content of new cultivars before they can be released for commercial use. In Finland, as in many countries, the official recommended upper limit for total GA concentration in cultivated *S. tuberosum* tubers is 0.20 mg/g FW.

At least 90 structurally different steroidal alkaloids have been isolated and characterized in over 300 *Solanum* species. The GAs identified to date are composed of C-27-steroidal alkaloid along with various sugar

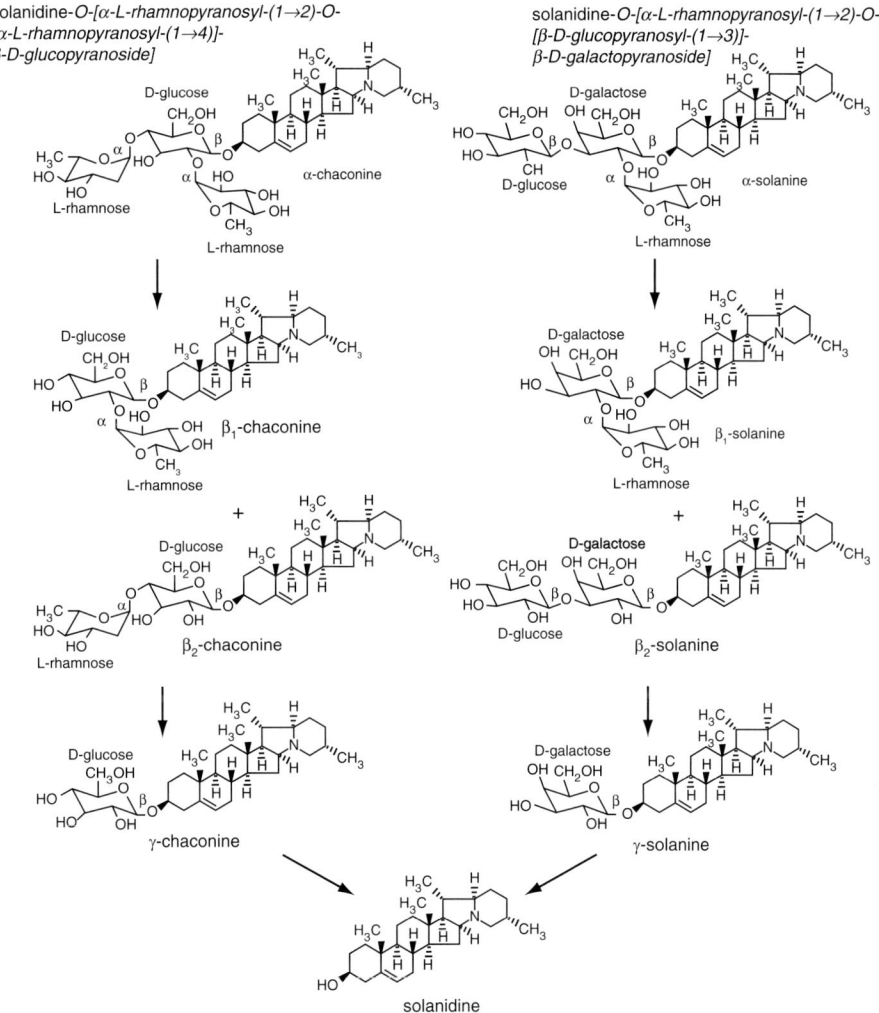

Fig. 12.3. Intermediates in the hydrolysis of the trisaccharide side chains of α-chaconine and α-solanine to the aglycon solanidine (Friedman, 2006).

moieties, usually composed of di-, tri- or tet-rasaccharides. In commercially cultivated potato, the major GAs, α-chaconine and α-solanine, are triglycosylated derivatives of the same aglycone, solanidine, but differ in their carbohydrate moiety. In α-solanine, the carbohydrate moiety is composed of galactose, glucose and rhamnose (β-solatriose), while in α-chaconine glucose, rhamnose and rhamnose β-chacotriose are present. The trisaccharide chains can be cleaved by acid or enzymatic hydrolysis to form aglycon solanidine. Other GAs that are present, but in much lower concentration, are α- and β-solanines and chaconines, α- and β-solamarines, aglycones demissidine and 5-α-solanidan-3-β-ol. In wild potatoes such as *S. chacoense*, numerous other GAs and aglycones have been identified. Two other structural classes of potato GAs, the leptines and leptidines, are present in the leaves of *S. chacoense* but not in the leaves of *S. tuberosum*, and are not present in potato tubers (Friedman, 2006; Nema *et al.*, 2008).

GAs are found throughout the potato plant, with levels varying considerably between different tissues and between the same parts in different plants and varieties. The highest GA concentration is found in the outer layer of the tuber, especially next to eyes, wounds and in the sprout, yet they decrease towards the centre of the tuber. The total GA concentration in the peels was reported to vary from 0.25 to 0.95 mg/g FW. For peeled potatoes, the average content can vary from 1 µg/g to 0.20 mg/g FW in the whole tuber. When potato tubers are exposed to light, the solanine content in the peel may increase by as much as tenfold. The ratio of α-solanine to α-chaconine concentration differs, depending on the part of the potato plant and the variety. Most studies report the value of this ratio to be between 1:2 and 1:7. Both rates and patterns of accumulation, as well as the ratio of α-chaconine to α-solanine during tuber growth and development, are influenced strongly by genotype (Friedman, 2006; Nema *et al.*, 2008).

Proteins and peptides

Potato tuber proteins are classified into three groups: patatins, protease inhibitors and other proteins. The patatin and protease inhibitor classes, which form the bulk of the potato tuber protein, are considered to be mostly storage proteins. The majority of the isoforms have defined enzymatic and inhibitory activities that might be of physiological relevance.

Patatin, a glycoprotein of molecular weight *c*.45 kDa, accounted for about 40% of the total soluble protein in potato. Patatin exists in a number of charge forms, which differ between potato cultivars. A high degree of homology between the isoforms of patatin was also indicated by NH_2-terminal amino acid sequence analysis (Shewry, 2003). Pots *et al.* (1999) separated the patatins of cv. Bintje into four pools with different chromatographic and electrophoretic characteristics, but similar biophysical properties. Little information on other proteins present in the tuber is available and, until recently, no systematic gene discovery or protein sequencing had been undertaken. Jørgensen *et al.* (2006) determined 43 different potato tuber proteins from the starchy potato cv. Kuras. Potato tuber protease inhibitors are a diverse group of proteins that inhibit a variety of proteases and some other enzymes, for example invertase. They differ in their amino acid sequence, chain length and subunit composition (monomer to pentamer).

The role of patatin in potato tubers remains unclear. It has been speculated that, besides being the main storage protein of potato tubers, patatin might also be involved in the resistance reaction induced by pathogen attack. The lipid acyl hydrolase activity of patatin could be important for the rapid degradation of cell membranes, and thus rapid degradation of certain metabolites. In addition to enzymatic and inhibitory activities, patatin was shown to have antioxidant or antiradical activity. Liu *et al.* (2003) found that purified patatin showed antioxidant or antiradical activity in a series of *in vitro* tests, including 1,1-diphenyl-2-picrylhydrazyl (DPPH) radical scavenging activity assays, anti-human low-density lipoprotein peroxidation tests, and protections against hydroxyl radical-mediated DNA damages and peroxynitrite-mediated dihydrorhodamine 123 oxidations. Gaillard and Matthew (2006) measured two lipid-degrading enzymes, namely a hydrolytic enzyme (lipolytic acyl hydrolase) and an oxidizing enzyme (lipoxygenase), in tubers from 23 varieties of potato at harvest. All varieties contained high levels of lipoxygenase activity, and 22 varieties had very high levels of the hydrolytic enzyme, ranging from 5 to 50 µmol of substrate hydrolysed/min/g of FW of tuber. One variety, Desiree, contained a much lower level of this enzyme (0.06–0.2 units/g FW). From these results it may be concluded that the major soluble protein may be beneficial when it is consumed, owing to its antioxidant activity. In addition, recent findings indicate that potato isolates and by-products from the potato industry could be useful sources of bioactive compounds (Pihlanto et al., 2008).

Peptides with specific biological activity may be located in the amino acid sequence of a given protein. Enzymatic degradation of foodstuffs in the gut and in food processing releases short-chain peptide sequences from intact proteins, glycoproteins and lipoproteins. In some cases, peptides may act as regulatory compounds with a hormone-like activity, based on their amino acid composition and sequence. The beneficial health effects may be attributed to numerous known peptide sequences exhibiting, for example, antimicrobial, antioxidative, antithrombotic, antihypertensive and immunomodulatory activities. Such activities are based on the peptides' amino acid composition and sequence. Bioactive peptides usually contain between three and 20 amino acid residues per molecule. In order to elicit a biological response, peptides must both cross the intestinal epithelium and enter the blood circulation, or bind directly to specific epithelial cell-surface receptor sites (Pihlanto and Korhonen, 2003).

There are a few reports on the bioactivities of potato hydrolysates; however, no peptides have been identified so far. Based on the amino acid sequences of potato proteins, there are several potential precursor proteins, from which peptides with various activities could be released. Recently, it has been found that hydrolysis of protein isolates from potato tubers results in enhanced inhibition of angiotensin I converting enzyme (ACE) (Pihlanto et al., 2008), which plays a major role in the control of blood pressure. Further studies showed that autolysis of protein isolates from vascular bundle and inner tuber tissues of potato enhanced the inhibition of the ACE. In addition, the physiological age of the tuber affected the strength of ACE inhibition, the rate of its increase during autolysis and the tuber tissue where ACE inhibition was most pronounced. Many tuber proteins, including two aspartic protease inhibitors, were observed to degrade during autolysis (Mäkinen et al., 2008). Foltz et al. (2008) found that potato protein hydrolysate stimulated cholesystokinin (CCK) release from a murine cell line of enteroendocrine origin (ST-1 cell line) and directly stimulated CCK_1R-expressing cells. As CCK_1R is expressed in the gastrointintestinal tract, direct interaction of CCK_1R with dietary peptides may contribute to their satiety effects.

Other bioactivities

White- and yellow-fleshed potatoes are very familiar to people around the world. Iwanzik et al. (1983), in one of the most complete studies, compared potatoes with various degrees of yellow intensity, finding a range of total carotenoids from 0.27 to 3.29 µg/g FW. They listed lutein, neoxanthin, violaxanthin and lutein-5,6-epoxide as components and found a strong correlation between carotenoid concentration and colorimetric measurements of yellowness. The summer and spring potatoes studied were found to contain 0.13 and 0.60 µg/g FW, respectively, of xanthophyll identified as lutein. Furthermore, a range of 0.97–5.36 µg/g FW in a series of cultivars and breeding lines has been obtained, although, in this study, carotenoid content did not appear to be related to colour of flesh (Brown, 2005).

The potato pulp resulting from industrial starch processing is a potential source of dietary fibre. Potato pulp has a high content of soluble fibre. Meyer et al. (2009) reported that in potato pulp about 22% by weight of the non-starch polysaccharide material was made up of cellulose, leaving about 38–40% by weight of the dry matter as pectin and hemicellulose polymers containing different monosaccharides, with a remarkably high proportion of galactose and relatively high uronic acid levels – the latter indicating galacturonic acids and hence presence of pectin.

12.2.2 Carrots

Carrot (*Daucus carota*) is one of the major vegetables consumed in the Western world, whatever the season. Its consumption has been increasing in Western countries. Carrot is one of the richest vegetables in fibres and carotenoids. It also contains other antioxidants such as vitamin E (0.52 mg/g), vitamin C (0.07 mg/g) and polyphenols such as *p*-coumaric, chlorogenic and caffeic acids.

Carotenoids

Carotenoids are a group of natural pigments responsible for the yellow, orange or red colour of many foods. Carotenoids are defined by their chemical structure. The majority of carotenoids are derived from a 40-carbon polyene chain, which could be considered the backbone of the molecule. This chain may be terminated by cyclic end-groups and may be complemented with oxygen-containing functional groups. Besides the well-known provitamin A activity of some of these compounds, they have also been associated with lowered risk of developing degenerative diseases such as cancer, cardiovascular diseases, cataracts and macular degeneration.

Carrots are especially rich in carotenoids. α-Carotene and β-carotene are the principal carotenoids in carrots, while lutein is a minor component. Total carotenoid content and distribution of carotenoids at harvest are influenced by genetic and environmental factors. The data for α-carotene in carrot tissue ranged from 5.3 to 51.6 µg/g, for β-carotene they ranged from 33.0 to 130 µg/g and for minor component, lutein, they ranged from 3.6 to 5.6 µg/g (Müller, 1997; Niizu and Rodriguez-Amaya, 2005).

Thermal treatments reduce the total carotenoid content in carrots. During home processing, 14.4–39.9% of carotenoids are lost. Water cooking, without pressure, carried out under controlled time and temperature (6 min, 96°C), was the best method, as the losses in the amounts of α- and β-carotenoids were approximately 24% in relation to total (Pinheiro-Santana *et al.*, 1998). Nguyen *et al.* (2007) found that pressure-assisted thermal processing (temperatures up to 105°C) resulted in higher carotenoid content than thermal processing alone.

Polyacetylenes

Carrot storage roots normally contain three major polyacetylenes, namely (Z)-heptadeca-1,9-diene-4,6-diyn-3-ol (falcarinol), (Z)-heptadeca-1,9-diene-4,6-diyne-3,8-diol (falcarindiol) and (Z)-3-acetoxyheptadeca-1,9-diene-4,6-diyn-8-ol (falcarindiol 3-acetate); however, most investigations have concentrated on falcarinol and falcarindiol. Polyacetylenes have been identified as the dominant cause of the commonly experienced bitter taste in carrot. The correlation between their concentration and individual bitter detection threshold clearly indicated falcarindiol as the main compound responsible for the bitter off-taste of fresh and stored carrots. Falcarindiol is present mainly in the peel and outer layer of the carrot root. In peeled carrots, the content varied significantly between 0.006 and 0.030 mg/g FW for falcarindiol, between 0.010 and 0.022 mg/g FW for falcarindiol 3-acetate and between 0.035 and 0.067 mg/g FW for falcarinol. Carrot peel contained over a tenfold higher concentration of falcarindiol than the corresponding peeled carrots.

Polyacetylenes are potent antifungal and antibacterial compounds. They are also known to be inhibitors of a number of enzymes such as diacylglycerol acyltransferase, inducible nitric oxide synthase and cholesteryl ester transfer protein, as well as microsomal and mitochondrial enzymes. *In vitro* experiments indicate that some polyacetylenes might exhibit antiallergenic and anti-inflammatory activities. In addition, polyacetylenes have proven to be cytotoxic against a number of solid and leukaemic cancer cell lines and to potentiate cytotoxicity of other anticancer drugs (Christensen and Brandt, 2006).

Flavonoids and other bioactivities

Gębczyński (2006) found that fresh carrot (cv. Koral) contained polyphenols at 20.9 mg/100 g, with a corresponding antioxidative activity of 19.4% (expressed as radical scavenging activity). After 12 months stor age, carrot retained 73–78% of polyphenols, while its antioxidative activity was reduced to 30–39% of the initial level. Mattila and Hellström (2007) reported that carrots contained 0.081–0.17 mg/g FW chlorogenic acid, 0.086–0.20 mg/g FW unknown phenolic compounds, and traces of other phenolic acids. The major anthocyanins in black carrot (*D. carota* ssp. *sativus* var. Atrorubens Alef) plant tissues and cell cultures possess cyanidin as an aglycon. In addition, traces of anthocyanins possessing a peonidin- or pelargonidin-type aglycon have been described (Kammerer *et al.*, 2003; Schwarz *et al.*, 2004). Also, carrot peels are a good source of antioxidant dietary fibre (Chantaro *et al.*, 2008).

12.2.3 Other root crops

Yang *et al.* (2008) analysed the flavonoid contents as aglycones (for quercetin, kaempferol, isorhamnetin, luteolin and apigenin) for 115 edible plants (91 species), among them sweet potato (*Ipomoea batatas*). The latter's total flavonoid content varied from 0 to 0.446 µg/g DW. The purple sweet potato variety contained the highest amount of flavonoids and was especially rich in quercetin. Padda and Picha (2008) found significant differences in total phenolics, in different genotypes of sweet potato, ranging from 1.4 to 4.7 µg/g DW. The highest total phenolic content and antioxidant activity was observed in a purple-fleshed genotype. Chlorogenic acid and 3,5-dicaffeoylquinic acid were the predominant phenolic acids, while caffeic acid was the least abundant in most genotypes. The highest content of chlorogenic acid (422.4 µg/g DW) was present in a white-fleshed cv. Quarter Million imported from Jamaica. However, a purple-fleshed genotype had the highest amounts of 3,5-dicaffeoylquinic (485.6 µg/g DW), 3,4-dicaffeoylquinic (125.6 µg/g DW), 4,5-dicaffeoylquinic (284.4 µg/g DW) and caffeic (20.5 µg/g DW) acids.

Several studies have shown that red beet/beetroot (*Beta vulgaris* L.) is a good source of natural antioxidants. The red beet colour results from the betalains, which are used extensively as natural food colorants.

In vitro studies have demonstrated that the betalains from red beets possess high antiradical and antioxidant activity. Consequently, it has become a popular belief that betalains provide protection against oxidative stress-related disorders by acting as antioxidants *in vivo* (Stintzing and Carle, 2004). If this belief is correct, consumers may benefit from regular consumption of products rich in betalains (e.g. red beet juice). Besides these hydrophilic pigments, the presence of further health-promoting constituents of red beets, such as phenolics (phenolic acids, phenolic acid esters and flavonoids) and folic acid, has been reported.

Schreiner *et al.* (2002) found bioactive indolyl glucosinolates in red radish (*Raphanus sativus* L. var. *sativus*) and production process effects on the content of these compounds. Furthermore, there are several other root and tuber crops that are grown worldwide but usually have low commercial value for direct consumption.

The yams (*Dioscorea villosa*) are some the most important tuber crops in West Africa. Yams have a complex phytochemical profile; the most predominant compounds are dioscorine alkaloid and diosgenin saponin. These compounds traditionally are considered toxic, but such toxicity can be removed by washing, boiling and cooking. Okwu and Ndu (2006) reported that different varieties of yam expressed different levels of alkaloids and flavonoids. The highest (0.195 mg/g) saponin content was found in the *D. rotundata* hybrid and the lowest (0.03 mg/g) in the *D. alata* hybrid (TDa 117). Alkaloids apart from saponins were found in concentrations from 0.01 to 0.02 mg/g. Good quantities of flavonoids were found in tubers (0.06–1.0 mg/g), whereas tannins and phenolic compounds were in smaller quantities (Okwu and Ndu, 2006). Apart from food, yams are used for medicinal purposes; the sapogenins, aglycons of yam saponins, are medically important, mainly because of their steroid structure.

Cassava (*Manihot esculenta*) is an important tropical root crop, ranking fourth on the list of major food crops in developing countries. An item of concern for public health authorities, plant breeders, producers, processors and, principally, the consumers is its toxic potential. The potential toxicity is due to the presence of two cyanogenic glycosides, linamarin and lotaustralin, in a ratio of approx 20:1, which, after hydrolysis into cyanohydrins, yield the toxic compound, hydrogen cyanide. Cassava root contains linamarin, equivalent to 0.002–1.0 mg HCN/g of cassava flesh. Traditional methods of processing cassava, such as sun drying, soaking, boiling and fermentation, eliminate most of the cyanide. The cyanogenic potential of insufficiently processed cassava has been identified as a factor in health problems like acute toxic effects, iodine deficiency disorders, tropical ataxic neuropathy and the paralytic disease, konzo (Essers, 1995; Egan *et al.*, 1998).

12.3 Potential Health Benefits

12.3.1 Bioavailability

Any suggested health-promoting agent in foods would, after consumption, probably be required to appear in the blood to a significant extent to have an effect on target sites other than those of the gastrointestinal tract. The notion of bioavailability integrates several variables, such as intestinal absorption, metabolism by the microflora, intestinal and hepatic metabolism, plasma kinetics, the nature of circulating metabolites, cellular uptake and intracellular metabolism. The difficulty lies in integrating all the information and relating the variables to health effects at the organ level.

The health effects of polyphenols depend on the amount consumed and on their bioavailability. The structural diversity of polyphenols makes the estimation of their content in food difficult. Furthermore, they are not evenly distributed and processing may result in a loss or enrichment of some phenolic compounds. In potato, phenolic acids and flavones are found in the peel. It should be pointed out that peel polyphenols have little dietary significance if the polyphenols are destroyed during processing (baking, cooking, frying) or they are discarded during cooking preparation (e.g. peeling). Peels of baked or fried potatoes are the principal source of potato polyphenols in the human diet.

Direct evidence on the bioavailability of a few polyphenolic compounds has been obtained by measuring their concentrations in plasma and urine after the ingestion of either pure compounds or foodstuffs with known content of the compounds. Although very abundant in our diet, proanthocyanidins and anthocyanins are absorbed either very poorly or not at all, or their metabolites have not yet been identified. Thus, their action is restricted to the intestine. Intakes of monomeric flavonols, flavones and flavanols are relatively low, and plasma concentrations rarely exceed 1 µmol/l because of limited absorption and rapid elimination. Flavanones and isoflavones are the flavonoids with the best bioavailability profiles, and plasma concentrations may reach 5 µmol/l. However, the distribution of these substances is restricted to citrus fruit and soybean. Finally, hydroxycinnamic acids are found in a wide variety of foods, often at high concentrations, but esterification decreases their intestinal absorption. As a general rule, the metabolites of polyphenols are eliminated rapidly from plasma, which indicates that consumption of plant products on a daily basis is necessary to maintain high concentrations of metabolites in the blood (Scalbert and Williamson, 2000; Manach *et al.*, 2004).

The consumption of carotenoid-rich foods such as carrots has been associated with a decrease in the risk of developing certain types of degenerative and chronic diseases. Processing of food and the interaction of carotenoids with lipophilic food components or ingredients may modify the amount of the pigments released from the food matrix, and therefore potentially increase or decrease their bioavailability. Carotenoids derived from food have been detected in human and rat plasma; however, carotenoids in foods are much less available than are purified sources of β-carotene. The relative bioavailability of β-carotene from vegetables compared with purified β-carotene ranged between 19 and 34% for carrots (Micozzi *et al.*, 1992; Nicolle *et al.*, 2003). Hornero-Méndez and Mínguez-Mosquera (2007) found that food processing and mainly lipid content improved carotenoid bioaccessibility from carrots significantly, and therefore might increase bioavailability in humans. The amount of dietary fat required to ensure carotenoid absorption seems to be low (3–5 g/meal), although it depends on the physicochemical characteristics of the carotenoids ingested (Van Hof *et al.*, 2000). The presence of dietary fibre in vegetables and fruit may explain in part the lower bioavailability of carotenoids from plant foods. It has been suggested that fibre interferes with micelle formation by partitioning bile salts and fat in the gel phase of dietary fibre. Some studies have shown that carotenes from cooked carrots are absorbed more efficiently than are those from raw carrots, because the plant cells are disrupted during preparation and because the mechanical homogenization of carrots allows greater efficiency of absorption (Van Hof *et al.*, 2000).

Although the potency of the exogenous peptides is lower than that of endogenous peptides or peptide-based drugs, they may well have physiological effects, because food proteins are usually ingested in fairly large amounts. To exert physiological effects *in vivo*, bioactive peptides must be released during intestinal digestion or/and be resistant to digestive enzymes, and then reach their target sites at the luminal side of the intestinal tract or, after absorption, in the peripheral organs. The gastrointestinal tract of humans contains a number of enzymes involved in the hydrolysis of proteins and peptides and they are located in a number of sites. It is important to recognize that peptidase enzymes never occur alone. Throughout the gastrointestinal tract there is always a mixture of peptidases working synergistically. The main event in the intraluminal digestion of proteins consists of cleavage of polypeptides by pancreatic proteases, such as trypsin, chymotrypsin, elastase and carboxypeptidase. Furthermore, the microorganisms of the colon produce large numbers of peptidase enzymes, in considerable quantities, that participate in protein digestion. The small intestine is the principal site of absorption. Di- and tripeptides, such as immunopeptides and several ACE inhibitors, may pass across the intestine, in quantitatively significant amounts, to reach peripheral target sites.

After absorption in the intestinal tract, serum peptidases can hydrolyse the peptide bonds further. Resistance to peptidase degradation may, in fact, be a prerequisite for a physiological effect following oral ingestion and/or the intravenous infusion of biologically active peptides/hydrolysates (Pihlanto and Korhonen, 2003).

12.3.2 Cancer studies

Carrots are one of the plant products that have shown the strongest protective anticancer effects. Carrots contain β-carotene, dietary fibres and polyphenols, which have been shown in experimental studies to be potentially anticarcinogenic. Antioxidant properties of carotenoids are thought to be at least partly responsible for protecting against colon cancer. There are large amounts of *in vitro* data supporting this hypothesis but there is little known about the antioxidant effects of carotenoid-rich food *in vivo*, particularly in the gastrointestinal tract. Briviba *et al.* (2004) found that consumption of carotenoid-rich juices for 2 weeks increased the carotenoid level in plasma and faeces; the antioxidant capacity of low-density lipoprotein tended to be increased by only approximately 4.5%, and lipid peroxidation in men's plasma and faeces was not affected. Thus, processes other than lipid peroxidation could be responsible for the preventive effects of tomatoes and carrots against colon cancer. A human intervention study with carotenoid-rich juices led to only minor changes in luminal biomarkers relevant to colon carcinogenesis (Schnäbele *et al.*, 2008). The results by Nyberg *et al.* (1998) supported evidence linking a diet rich in vegetables and non-citrus fruit with decreased lung cancer risk and suggested that, among vegetables, carrot consumption was the most important component or marker for this effect in Sweden.

A recent *in vitro* study aiming to screen for potential health-promoting compounds from vegetables showed that falcarinol, but not β-carotene, could stimulate differentiation of primary mammalian cells in concentrations between 0.004 and 0.4 µM falcarinol. Therefore, falcarinol appears to be one of the bioactive components in carrots and related vegetables that could explain their health-promoting properties, rather than carotenoids or other types of primary and/or secondary metabolites. This hypothesis is further supported by recent studies on the bioavailability of falcarinol in humans. The carrot and falcarinol treatments showed a significant tendency to reduce numbers of (pre)cancerous lesions with increasing size of lesion in rat models (Christensen and Brandt, 2006). Traditionally, polyacetylenes in food are generally considered undesirable toxicants but this belief may need to be revised, and perhaps these compounds may instead be regarded as important nutraceuticals. The major challenge with regard to bioactive polyacetylenes is to test their health-promoting effects *in vivo* in clinical as well as in further preclinical studies.

12.3.3 Cardiovascular diseases

The biological activities of polyphenols have often been evaluated *in vitro* on pure enzymes, cultured cells or isolated tissues, by using polyphenol aglycones or some glycosides that are present in food. Very little is known about the biological properties of the conjugated derivatives present in plasma or tissues because of the lack of precise identification and commercial standards. Polyphenols exhibit strong *in vitro* antioxidant activity with heart disease-related lipoprotein. Since *in vivo* oxidation of low density lipoproteins (LDL) appears to be a major cause of heart disease, it is possible that chlorogenic acid and other polyphenols may also lessen such disease. Lazarov and Werman (1996) found that the consumption of potato peel induced a lowering of cholesterol in rats. They ascribed this result to the fibre content in the peel. However, it is likely that the polyphenols and other antioxidants as well as the glycoalkaloids in the peel contributed to the observed hypocholesterolaemia. Han *et al.* (2006) investigated the antioxidant effects of purple anthocyanin-rich potato flakes *in vivo* and found that the flakes enhanced the serum Trolox equivalent antioxidant capacity values and suppressed hepatic thiobarbituric acid

reactive substances (TBARS) levels in rats. Shindo *et al.* (2007) showed that administering anthocyanin-rich colours from purple sweet potato and red radish decreased the blood pressure of spontaneously hypertensive rats (SHR). Robert *et al.* (2006, 2008) suggested that consumption of cooked potatoes (consumed with skin) might enhance antioxidant defence and improve lipid metabolism. These effects limit oxidative stress and reduce the risk of developing the associated degenerative diseases, including cardiovascular disease, and could have potential in cardiovascular disease prevention.

The effect of carrot consumption on cholesterol metabolism has been studied. Robertson *et al.* (1979) found that 200 g of raw carrot eaten each day for 3 weeks reduced serum cholesterol significantly by 11%, increased faecal bile acid and fat excretion by 50% and increased stool weight modestly by 25%. The latter effect was attributed to carrot fibre. Nicolle *et al.* (2003, 2004) found that carrot consumption modified cholesterol absorption and bile acid excretion and increased antioxidant status. It is likely that these effects could be due to the synergistic effect of fibre and associated antioxidants.

12.3.4 Other beneficial effects

Gastrointestinal hormones such as cholecystokinin (CCK) are important in the regulation of food intake and in maintaining energy homeostasis. The role of CCK as a major endocrine determinant in food intake regulation is well documented. Studies in animals and humans using different protein sources suggest that, in order to stimulate CCK secretion efficiently, the protein needs to be hydrolysed to short-chain peptides and amino acids (Schwartz *et al.*, 2000). Discovering either specific protein hydrolysates or other food-grade bioactive compounds with optimized CCK-releasing properties is an interesting target in the development of functional food products for weight management purposes. Foltz *et al.* (2008) found that potato hydrolysates induced CCK-releasing properties and stimulated CCK_1R expression in enteroendocrine cells.

Potato tubers are an extraordinarily rich source of protease inhibitors, which represent 25–30% of potato juice protein. As well as containing serine, cysteine and aspartage proteinases, they also contain a metal-containing carboxypeptidase and pancreatic proteases. In 11 lean subjects, the addition of proteinase inhibitors extracted from potatoes was studied. Reduced energy intake and increased CCK release was found, suggesting proteinase inhibition might have therapeutic potential for reducing food intake (Hill *et al.*, 1990). Potato protein fractions were capable of inhibiting a large part of the high proteolytic activity in faeces from patients with gastrointestinal resections and infants (*in vitro*) and prevented experimental protease-induced dermatitis (*in vivo*) (Ruseler-van Embden *et al.*, 2004).

Free radicals and reactive oxygen species play a central role in liver disease pathology and progression. Accordingly, dietary antioxidants have been proposed as therapeutic agents to counteract liver damage. Studies by Singh *et al.* (2008) indicated that potato peel extract treatment had a potent protective effect against oxidative stress and acute liver damage induced by CCl_4 in rats, as revealed by the remarkable decrease in hepatic malondialdehyde (MDA) content accompanied with enhanced reduced glutathione (GSH) and superoxide dismutase (SOD) activities.

12.4 Preharvest and Postharvest Continuum

Climatic conditions, especially temperature and light intensity, have a strong effect on nutritional quality. Soil type, irrigation, fertilization and other cultural practices influence the water and nutrient supply to the plant, which affect the composition and quality attributes (appearance, texture, taste and aroma) of the harvested plant parts. Maturity at harvest, harvesting method and extent of physical injuries have an influence on the content of several components in the tubers. In addition, many biochemical and physical changes, including variations in the concentration of bioactive compounds, occur during storage. A better understanding of how changes in the health-promoting compounds vary with preharvest factors and postharvest storage conditions will allow optimization of the bioactive components in these products at the time of their ingestion.

12.4.1 Potato

Preharvest considerations must include cultivar selection, as individual cultivars have significantly different proportions of bioactive components. The differences in specific antioxidant activity observed in several studies suggest that potato genotype and harvest location influence the accumulation of polyphenols in different quantities and/or types of polyphenols. Pigmented potato cultivars are an especially rich source of anthocyanins. Jansen and Flamme (2006) analysed 27 potato cultivars and four breeding clones grown in field plots for tuber production. The analysis revealed considerable differences in the amounts of anthocyanins between the cultivars/breeding clones. Interestingly, two different rates of nitrogen fertilization, at 100 and 200 kg/ha, had no significant effect on the pigment content of the potatoes. In dry matter, starch and protein contents, the coloured potato cultivars/breeding clones were comparable with traditional cultivars. Reyes *et al.* (2005) found that the anthocyanin content of potatoes ranged from 0.11 to 1.74 mg cyanidin-3-glucoside/g FW in purple-fleshed tubers and from 0.21 to 0.55 mg cyanidin-3-glucoside/g FW in red-fleshed tubers, and showed large variation between genotypes and locations. The antioxidant capacity of purple- and red-fleshed potatoes ranged from 0.513 to 1.426 mg Trolox equivalents/g FW, and phenolic compounds, especially anthocyanins, were responsible for the antioxidant capacity. Lachman *et al.* (2009) analysed total anthocyanins and individual anthocyanidins in 15 red- and purple-fleshed potato cultivars produced in five different locations in the Czech Republic. A significantly different abundance of anthocyanins was found in the individual cultivars. Different site conditions affected polyphenol content in tubers significantly. Increased height above sea level, higher annual sum of precipitation and lower annual temperatures caused a higher level of antioxidant activity and total anthocyanins (Hamouz *et al.*, 2007; Lachman *et al.*, 2009).

Content of polyphenols is dissimilar at different stages of tuber maturity. Lewis *et al.* (1999) observed that, as tubers reached maturity, the concentration of anthocyanins at the stem end was always higher than at the bud end. In addition, the observed differences were more marked in smaller (50–70 g) tubers. The polyphenol content decreased with tuber growth and maturity (Reyes *et al.*, 2004).

Effective control of sprouting is a fundamental requirement of potato storage. This can be accomplished by very low temperature storage and the use of sprout suppressants (namely, maleic hydrazide (MH) and chlorpropham (CIPC)), which include the natural plant-growth regulator, ethylene. Potatoes are low producers of ethylene (< 0.1 µl/kg/h at 20°C) but, since the 2003 publication of a commodity approval, exogenous application of ethylene has been approved for use as a postharvest sprout suppressant (Terry, 2008). It is apparent that ethylene treatment can extend the storage life of potatoes; however, the biological mechanism(s) by which this occurs is currently unknown. To date, there has been a scarcity of research carried out that has profiled the detailed biochemical and rheological changes induced by ethylene treatment during storage and compared these against physiological effects such as sprout inhibition. Storage of potato tubers at different temperatures (2, 10 and 20°C) and fluctuating temperature (from 2 to 20°C) did not affect the total accumulation of polyhenols. Similarly, ethylene treatments of air-stored samples led to no significant differences in either anthocyanin or phenolic content. Only tubers with low initial anthocyanin levels treated with methyl jasmonate showed 60% anthocyanin accumulation, and wounding induced the accumulation of phenolic compounds (Reyes and Cisneros-Zevallos, 2003). Jansen and Flamme (2006) found that anthocyanins were preserved in stored tubers without the risk of any degradation over a longer period of time.

Total amounts of GAs are influenced by environmental factors such as soil and climate. The relative concentration of GA falls with increasing tuber size and maturation. Immature tubers contain 28–38 mg α-solanine and 66–84 mg α-chaconine per kg, whereas mature tubers contain 4–10 mg α-solanine and 11–30 mg α-chaconine per kg. Storage conditions, especially light and temperature, are responsible mainly for increases in solanine during marketing. Although the GA content can increase in the dark, the rate of formation is only about 20% of that in the light (Friedman, 2006; Nema *et al.*, 2008).

Studies have shown consistently that the developmental stage and physiological age of potato tubers are associated with significant alterations in the protein composition of tubers. Lehesranta *et al.* (2006) observed significant changes in the potato tuber proteome during tuber development and physiological aging, but different tissues or fractions of the tuber were not compared. Qualitative and quantitative changes have been detected in protein content during the development of tubers (Borgmann *et al.*, 1994) and dormancy (Espen *et al.*, 1999) and after dormancy (Desires *et al.*, 1995a,b). Differences in enzymatic activities were confined to different parts of the potato tuber at different physiological stages. These changes were also associated with differences in the production of ACE-inhibitory activity (Mäkinen *et al.*, 2008).

12.4.2 Carrots

β-Carotene content in carrots is affected by various environmental factors such as growing conditions, processing and storage. Purple and orange carrot have been found to contain the highest amount of total carotenoids; only trace amounts of these compounds were detected in yellow carrot and none was found in white carrot (Alasalvar *et al.*, 2001; Surles *et al.*, 2004). Lee (1986) found that the provitamin A content increased during maturation and then decreased. Moreover, during storage at 2°C and 90% relative humidity, the total carotene content increased during the first 100-day storage but decreased thereafter. Sanitation procedures affect carotenoid content during storage. Washing carrots with acidified sodium chlorite retained higher carotenoid content than seen in carrots washed with water or other sanitizers (Ruiz-Gruz *et al.*, 2007).

During handling and distribution of fresh carrots from field to consumer, a less desirable taste can often develop. Bitterness and harshness are two usual sensory characteristics mentioned in connection with less desirable taste and flavour of carrots. Part of the reduction in taste quality could be due to exposure of carrots to ethylene during transport or storage. The production of the bitter compound, 3-methyl-6-methoxy-8-hydroxy-3,4-dihydroisocoumarin (6-methoxymellein), after ethylene exposure is well documented. Different kinds of mechanical stress (e.g. dropping on hard surfaces, harsh handling during harvest, packing or slicing) are known to enhance ethylene production as well as respiration, and cause splitting of the carrot roots.

12.5 Future Research Needed

The occurrence of many natural biomolecules in plants, such as those found in potatoes and carrots, has been shown in several studies during the last decades. Several bioactive food components are now well established; their structure and concentration, as well as *in vitro* activities are known. However, the primary interest of bioactive components is their physiological functions in the human body after consumption. For example, the total antioxidant activity of food, including the bioactive components, could provide a potentially useful measure of the physiological function of that food. The most important future research needs related to bioactive components are summarized hereunder:

- Screening for potential bioactivity among proteins in potato.
- To use by-products from starch manufacture as a starting material to produce novel products, including health-promoting ingredients such as fibre, polyphenols, proteins and peptides.
- Study of the effects of conventional and novel processing technologies on bioactivities.

- Study of the interactions of polyphenols, peptides and fibres with other food components during processing and the effects of these interactions on bioactivity.
- Basic research on allergenicity and toxicity of the compounds.
- Evaluation of the efficacy of polyphenols, proteins and peptides in animal model and human clinical studies per se and in food systems.

12.6 Conclusions

An increasing number of studies have revealed that potatoes and other root crops such as carrots and sweet potatoes are good sources of various bioactive components. These components are mainly the secondary metabolite products, such as polyphenols and glycoalkaloids in potato and carotenoids, polyacetylenes and polyphenols in carrot. The influence of genetic and environmental factors on the concentration of polyphenols, glycoalkaloids and carotenoids needs greater attention. The activity of these compounds has been shown with various *in vitro* methods. Potatoes are also a good source of bioactive proteins, since they have shown antioxidant as well as enzymatic and inhibitory activities. Furthermore, the digestion of potato proteins produces hydrolysates with satiety and antihypertensive activities *in vitro*. Potato and carrot are promising sources of components that have beneficial effects on cardiovascular health. The data available suggest that carrot polyacetylenes may inhibit the growth of cancer cells, and that potato proteins and hydrolysates can have satiety-promoting effects. Further studies are needed to establish the *in vivo* efficacy of these components, as well as their long-term effects when administered orally on a regular basis. Based on present knowledge, potato could and perhaps should be included as one of the major important sources of bioactive compounds. Furthermore, potato and carrot components provide a highly interesting base to be used as active ingredients for the formulation of functional health-promoting foods.

References

Alasalvar, C., Grigor, H.M., Zhang, D., Quantick, P.C. and Shahidi, F. (2001) Comparison of volatiles, phenolics, sugars, antioxidant vitamins, and sensory quality of different colored carrot varieties. *Journal of Agricultural and Food Chemistry* 49, 1410–1416.

Borgmann, K., Sinha, P. and Frommer, W.B. (1994) Changes in the two-dimensional protein pattern and in gene expression during the sink-to-source transition of potato tubers. *Plant Sciences* 99, 97–108.

Briviba, K., Schnäbele, K., Rechkemmer, G. and Bub, A. (2004) Supplementation of a diet low in carotenoids with tomato or carrot juice does not affect lipid peroxidation in plasma and feces of healthy men. *Journal of Nutrition* 134, 1081–1083.

Brown, C.R. (2005) Antioxidants in potato. *American Journal of Potato Research* 82,163–172.

Chantaro, P., Devahastin, S. and Chiewchan, N. (2008) Production of antioxidant high dietary fiber powder from carrot peels. *LWT – Food Science and Technology* 41, 1987–1994.

Chou, S.-T., Chien, P.J. and Chau, C.-F. (2008) Particle size reduction effectively enhances the cholesterol-lowering activities of carrot insoluble fiber and cellulose *Journal of Agricultural and Food Chemistry* 56, 10,994–10,998.

Christensen, L.P. and Brandt, K. (2006) Bioactive polyacetylenes in food plants of the Apiaceae family: occurrence, bioactivity and analysis. *Journal of Pharmaceutical and Biomedical Analysis* 41, 683–693.

Chu, Y.-H., Chang, C.-L. and Hsu, H.-F. (2000) Flavonoid content of several vegetables and their antioxidant activity. *Journal of the Science of Food and Agriculture* 80, 561–566.

Desires, S., Couillerot, J., Hilbert, J. and Vasseur, J. (1995a) Protein changes in *Solanum tuberosum* during *in vitro* tuberization of nodal cuttings. *Plant Physiology and Biochemistry* 33, 303–310.

Desires, S., Couillerot, J., Hilbert, J. and Vasseur, J. (1995b) Protein changes in *Solanum tuberosum* during storage and dormancy breaking of *in vitro* microtubers. *Plant Physiology and Biochemistry* 33, 479–487.

Egan, S.V., Yeoh, H.H. and Bradbury, J.H. (1998) Simple picrate paper kit for determination of the cyanogenic potential of cassava flour. *Journal of Agricultural and Food Chemistry* 76, 39–48.

Eichhorn, S. and Winterhalter, P. (2005) Anthocyanins from pigmented potato (*Solanum tuberusum* L.) varieties. *Food Research International* 38, 943–948.

Espen, L., Morgutti, S. and Cocucci, S.M. (1999) Changes in the potato (*Solanum tuberosum* L.) tuber at the onset of dormancy and during storage at 23°C and 3°C. II. Evaluation of protein patterns. *Potato Research* 42, 203–214.

Essers, A.J.A. (1995) Removal of cyanogens from cassava roots: studies on domestic sun-drying and solid-substrate fermentation in rural Africa. Thesis, Landbouwuniversiteit, Wageningen, The Netherlands, ISBN 90-5485-378-6.

FAO (2010) Strengthening potato value chains: technical and policy options for developing countries. Coordinated by Cromme, N., Prakash, A.B., Lutaladio, N. and Ezeta, F. (http://www.fao.org/docrep/013/i1710e/i1710e00.pdf, accessed 13 April 2011.)

Foltz, M., Ansems, P., Schwarz, J., Tasker, M.C., Lourbakos, A. and Gerhardt, C.C. (2008) Protein hydrolysates induce CCK release from enteroendocrine cells and act as partial agonists of the CCK1 receptor. *Journal of Agricultural and Food Chemistry* 56, 837–843.

Fossen, T. and Andersen, Ø.M. (2000) Anthocyanins from tubers and shoots of the purple potato, *Solanum tuberosum*. *Journal of Horticultural Science and Biotechnology* 75, 360–363.

Fossen, T., Øvstedal, D.O., Slimestad, R. and Andersen, Ø.M. (2003) Anthocyanins from a Norwegian potato cultivar. *Food Chemistry* 81, 433–437.

Friedman, M. (1997) Chemistry, biochemistry, and dietary role of potato polyphenols. A review. *Journal of Agricultural and Food Chemistry* 45, 1523–1540.

Friedman, M. (2006) Potato glycoalkaloids and metabolites: roles in the plant and in the diet. *Journal of Agricultural and Food Chemistry* 54, 8655–8681.

Gaillard, T. and Matthew, J.A. (2006) Lipids of potato tubers. II. Lipid-degrading enzymes in different varieties of potato tuber. *Journal of Agricultural and Food Chemistry* 24, 623–627.

Gębczyński, P. (2006) Content of selected antioxidative compounds in raw carrot and in frozen product prepared for consumption. *Electronic Journal of Polish Agricultural Universities* 9, 3. (http://www.ejpau.media.pl/volume9/issue3/art-03.html. Accessed 13 April 2011).

Hamouz, K., Lachman, J., Čepl, J., Dvořák, P., Pivec V. and Prášilová, M. (2007) Site conditions and genotype influence polyphenol content in potatoes. *Horticultural Science (Prague)* 34, 132–137.

Han, K.-H., Sekikawa, M., Shimada, K., Hashimoto, M., Hashimoto, N., Noda, T., Tanaka, H. and Fukushima, M. (2006) Anthocyanin-rich purple potato flake extract has antioxidant capacity and improves antioxidant potential in rats. *British Journal of Nutrition* 96, 1125–1133.

Hill, A.J., Peikin, S.R., Ryan, C.A. and Blundell, J.E. (1990) Oral administration of proteinase inhibitor II from potatoes reduces energy intake in man. *Physiology and Behavior* 48, 241–246.

Hornero-Méndez, D. and Mínguez-Mosquera, M.I. (2007) Bioaccessibility of carotenes from carrots: effect of cooking and addition of oil. *Innovative Food Science and Emerging Technologies* 8, 407–412.

Huang, G.-J., Sheu, M.-J., Chen, H.-J., Chang, Y.-S. and Lin, Y.-H. (2007) Growth inhibition and induction of apoptosis in NB4 promyelocytic leukemia cells by trypsin inhibitor from sweet potato storage roots. *Journal of Agricultural and Food Chemistry* 55, 2548–2553.

Iwanzik, W., Tevini, M., Stute, R. and Hilbert, R. (1983) Carotinoidgehalt und Zusammensetzung verschiedener deutscher Kartoffelsorten und deren Bedeutung fur die Fleischfarbe der Knolle. *Potato Research* 26, 149–162.

Jansen, G. and Flamme, W. (2006) Coloured potatoes (*Solanum tuberosum* L.) – anthocyanin content and tuber quality. *Genetic Resources and Crop Evolution* 53, 1321–1331.

Jørgensen, M., Bauw, G. and Welinder, K.G. (2006) Molecular properties and activities of tuber proteins from starch potato cv. Kuras. *Journal of Agricultural and Food Chemistry* 54, 9389–9397.

Kammerer, D., Carle, D. and Schieber, A. (2003) Detection of peonidin and pelargonidin glycosides in black carrots (*Daucus carota* ssp. *sativus* var. *atrorubens* Alef.) by high-performance liquid chromatography/electrospray ionization mass spectrometry. *Rapid Communications in Mass Spectrometry* 17, 2407–2412.

Kudoh, K., Matsumoto, M., Onodera, S., Takeda, Y., Ando, K. and Hiomi, N. (2003) Antioxidative activity and protective effect against ethanol-induced gastric mucosal damage of a potato protein hydrolysate. *Journal of Nutritional Science and Vitaminology* 49, 451–455.

Lachman, J., Hamouz, K., Šulc, M., Orsák, M., Pivec, M., Hejtmánková, A., Dvorák, P. and Cepl, J. (2009) Cultivar differences of total anthocyanins and anthocyanidins in red and purple-fleshed potatoes and their relation to antioxidant activity. *Food Chemistry* 114, 836–843.

Lazarov, K. and Werman, M. (1996) Hypocholesterolaemic effect of potato peels as a dietary fibre source. *Medical Science Research* 9, 581–584.

Lee, C.Y. (1986) Changes in carotenoid content of carrots during growth and post-harvest storage. *Food Chemistry* 20, 285–293.

Lehesranta, S.J., Davies, H.V., Shepherd, L.V.T., Koistinen, K.M., Massat, N., Nunan, N., McNicol, J.W. and Kärenlampi, S.O. (2006) Proteomic analysis of the potato tuber life cycle. *Proteomics* 6, 6042–6052.

Leo, L., Leone, A., Longo, C., Lombardi, D.A., Raimo, F. and Zacheo, G. (2008) Antioxidant compounds and antioxidant activity in 'early potatoes'. *Journal of Agricultural and Food Chemistry* 56, 4154–4163.

Lewis, C.E., Walker J.R.L., Lancaster, J.E. and Sutton K.H. (1998a) Determination of anthocyanins, flavonoids and phenolic acids in potatoes. I: Coloured cultivars of *Solanum tuberosum* L. *Journal of the Science of Food and Agriculture* 77, 45–57.

Lewis, C.E., Walker, J.R.L., Lancaster, J.E. and Sutton, K.H. (1998b) Determination of anthocyanins, flavonoids and phenolic acids in potatoes. II: Wild, tuberous *Solanum* species. *Journal of the Science of Food and Agriculture* 77, 58–63.

Lewis, C.E., Walker, J.R.L. and Lancaster, J.E. (1999) Changes in anthocyanin, flavonoid and phenolic acid concentrations during development and storage of coloured potato (*Solanum tuberosum* L) tubers. *Journal of the Science of Food and Agriculture* 79, 311–316.

Liu, T.-W., Han, C.-H., Lee, M.-H., Hsu, F.-L. and Hou, W.-C. (2003) Patatin, the tuber storage protein of potato (*Solanum tuberosum* L.), exhibits antioxidant activity *in vitro*. *Journal of Agricultural and Food Chemistry* 51, 4389–4393.

Mäkinen, S., Kelloniemi, J., Pihlanto, A., Mäkinen, K., Korhonen, M., Hopia, A. and Valkonen, J.P.T. (2008) Inhibition of angiotensin converting enzyme I caused by autolysis of potato proteins by enzymatic activities confined to different parts of the potato tuber. *Journal of Agricultural and Food Chemistry* 56, 9875–9883.

Manach, C., Scalbert, A., Morand, C., Rémésy C. and Jiménez, L. (2004) Polyphenols: food sources and bioavailability. *American Journal of Clinical Nutrition* 79, 727–747.

Mattila, P. and Hellström, J. (2007) Phenolic acids in potatoes, vegetables, and some of their products. *Journal of Food Composition, Analysis* 20, 152–160.

Meyer, A.S., Dam, B.P. and Lærke, H.N. (2009) Enzymatic solubilization of a pectinaceous dietary fiber fraction from potato pulp: optimization of the fiber extraction process. *Biochemical Engineering Journal* 43, 106–112.

Micozzi, M.S., Brown, E.D., Edwards, B.K., Bieri, J.G., Taylor, P.R., Khachik, F., Beecher, F.G. and Smith, J.C. (1992) Plasma carotenoid response to chronic intake of selected foods and β-carotene supplements in men. *American Journal of Clinical Nutrition* 55, 1120–1125.

Müller, H. (1997) Determination of the carotenoid content in selected vegetables, fruit by HPLC, photodiode array detection. *Zeitschrift für Lebensmittel-Untersuchung und -Forschung* 204, 88–94.

Nema, P.K., Ramayya, N., Duncan, E. and Niranjan, K. (2008) Potato glycoalkaloids: formation and strategies for mitigation. *Journal of the Science of Food and Agriculture* 88, 1869–1881.

Nguyen, L.T., Rastogi, N.K. and Balasubramaniam, V.M. (2007) Evaluation of the instrumental quality of pressure-assisted thermally processed carrots. *Journal of Food Science* 72, E264–E270.

Nicolle, C., Cardinault, N., Aprikian, O., Busserolles, J., Grolier, P., Rock, E., Demigné, C., Mazur, A., Scalbert, A., Amouroux, P. and Rémésy, C. (2003) Effects of carrot intake on cholesterol metabolism and on antioxidant status in cholesterol-fed rat. *European Journal of Nutrition* 42, 254–261.

Nicolle, C., Gueux, E., Lab, C., Jaffrelo, L., Rock, E., Mazur, A., Amouroux, P. and Rémésy, C. (2004) Lyophilized carrot ingestion lowers lipemia and beneficially affects cholesterol metabolism in cholesterol-fed C57BL/6J mice. *European Journal of Nutrition* 43, 237–245.

Niizu, P.Y. and Rodriguez-Amaya, D.B. (2005) New data on the carotenoid composition of raw salad vegetables. *Journal of Food Composition and Analysis* 18, 739–749.

Nyberg, F., Agrenius, V., Svartengren, K., Svensson, C. and Pershagen, G. (1998) Dietary factors and risk of lung cancer in never-smokers. *International Journal of Cancer* 78, 430–436.

Okwu, D.E and Ndu, C.U. (2006) Evaluation of the phytonutrients, mineral and vitamin contents of some varieties of yam (*Dioscorea* sp.) *International Journal of Molecular Medicine and Advance Sciences* 2, 199–203.

Padda, M.S. and Picha, D.H. (2008) Quantification of phenolic acids and antioxidant activity in sweetpotato genotypes. *Scientia Horticulturae* 119, 17–20.

Pihlanto, A. and Korhonen, H.J.T. (2003) Bioactive peptides and proteins. *Advances in Food and Nutrition Research* 47, 175–276.

Pihlanto, A., Akkanen, S. and Korhonen, H.J. (2008) ACE-inhibitory and antioxidant properties of potato (*Solanum tuberosum*). *Food Chemistry* 109, 104–112.

Pinheiro-Santana, H.M., Stringheta, P.C., Brandao, S.C.C., Paez, H.H. and de Queiroz, V.M.V. (1998) Evaluation of total carotenoids, α- and β-carotene in carrots (*Daucus carota* L.) during home processing. *Ciencia e Tecnologia de Alimentos* 18, 39–44.

Pots, A.M., Gruppen, H., Hessing, M., van Boekel, M.A.J.S. and Voragen, A.G.J. (1999) Isolation and characterization of patatin isoforms. *Journal of Agricultural and Food Chemistry* 47, 4587–4592.

Reyes, L.F. and Cisneros-Zevallos, L. (2003) Wounding stress increases the phenolic content and antioxidant capacity of purple-flesh potatoes (*Solanum tuberosum* L.). *Journal of Agricultural and Food Chemistry* 51, 5296–5300.

Reyes, L.F., Miller, J.C. Jr and Cisneros-Zevallos, L. (2004) Environmental conditions influence the content and yield of anthocyanins and total phenolics in purple- and red-flesh potatoes during tuber dvelopment. *American Journal of Potato Research* 81, 187–193.

Reyes, L.F., Miller, J.C. and Cisneros-Zevallos, L. (2005) Antioxidant capacity, anthocyanins and total phenolics in purple and red-fleshed potato *(Solanum tuberosum* L.) genotypes. *American Journal of Potato Research* 82, 271–277.

Robert, L., Narcy, A., Rock, E., Demigne, C., Mazur, A. and Rémésy, C. (2006) Entire potato consumption improves lipid metabolism and antioxidant status in cholesterol-fed rat. *European Journal of Nutrition* 45, 267–274.

Robert, L., Narcy, A., Rayssiguier, Y., Mazur, A. and Rémésy, C. (2008) Lipid metabolism and antioxidant status in sucrose vs. potato-fed rats. *Journal of the American College of Nutrition* 27, 109–116.

Robertson, J., Brydon, W.G., Tadesse, K., Wenham, P., Walls, A. and Eastwood, M.A. (1979) The effect of raw carrot on serum lipids and colon function. *American Journal of Clinical Nutrition* 32, 1889–1892.

Rodriguez-Saona, L.E., Giusti, M.M. and Wrolstad, R.E. (1998) Anthocyanin pigment composition of red-fleshed potatoes. *Journal of Food Science* 63, 458–465.

Ruiz-Gruz, S., Islas-Osuna, M.A., Sotelo-Mundo, R.R., Vazquez-Ortiz, F. and Gonzales-Aguilar, G.A. (2007) Sanitation procedure affects biochemical and nutritional changes of shredded carrots. *Journal of Food Science* 72, S146–S152.

Ruseler-van Embden, J.G.H., van Lieshout, L.M.C., Smits, S.A., van Kessel, I. and Laman, J.D. (2004) Potato tuber proteins efficiently inhibit human faecal proteolytic activity: implications for treatment of perianal dermatitis. *European Journal of Clinical Investigation* 34, 303–311.

Scalbert, A. and Williamson, G. (2000) Dietary intake and bioavailability of polyphenols. *Journal of Nutrition* 130, 2073S–2085S.

Schnäbele, K., Briviba, K., Bub, A., Roser, S., Pool-Zobel, B.L. and Rechkemmer, G. (2008) Effects of carrot and tomato juice consumption on faecal markers relevant to colon carcinogenesis in humans. *British Journal of Nutrition* 99, 606–613.

Schreiner, M., Huyskens-Keil, S., Peters, P., Schonhof, I., Krumbein, A. and Widell, S. (2002) Seasonal climate effects on root colour and compounds of red radish. *Journal of the Science of Food and Agriculture* 82, 1325–1333.

Schwartz, M.W., Woods, S.C., Porte, D. Jr, Seeley, R.J. and Baskin, D.G. (2000) Central nervous system control of food intake. *Nature* 404, 661–671.

Schwarz, M., Wray, V. and Winterhalter, P. (2004) Isolation and identification of novel pyranoanthocyanins from black carrot (*Daucus carota* L.) juice. *Journal of Agricultural and Food Chemistry* 52, 5095–5101.

Shewry, P.R. (2003) Tuber storage proteins. *Annals of Botany* 91, 755–769.

Shindo, M., Kasai, T., Abe, A. and Kondo, Y. (2007) Effects of dietary administration of plant-derived anthocyanin-rich colors to spontaneously hypertensive rats. *Journal of Nutritional Science and Vitaminology* 53, 90–93.

Singh, N., Kamath, V., Narasimhamurthy, K. and Rajini, P.S.P. (2008) Protective effect of potato peel extract against carbon tetrachloride-induced liver injury in rats. *Environmental Toxicology and Pharmacology* 26, 241–246.

Stintzing, F.C. and Carle, R. (2004) Functional properties of anthocyanins and betalains in plants, food, and in human nutrition. *Trends in Food Science and Technology* 15, 19–38.

Surles, R.L., Weng, N., Simon, P.W. and Tanumihardjo, S.A. (2004) Carotenoid profiles and consumer sensory evaluation of specialty carrots (*Daucus carota*, L.) of various colors. *Journal of Agricultural and Food Chemistry* 52, 3417–3421.

Terry, L.A. (2008) Highlighting the value of potatoes. *Edibles Post-harvest Technology Horticulture Week* 31. (http://www.hortweek.com/news/864752/Highlighting-value-potatoes. Accessed 13 April 2011)

van Hof, K.H., West, C.E., Weststrate, J.A. and Hautvast, J.G.A.J. (2000) Dietary factors that affect the bioavailability of carotenoids. *Journal of Nutrition* 130, 503–506.

Yang, R.-Y., Lin, S. and Kuo, G. (2008) Content and distribution of flavonoids among 91 edible plant species. *Asia Pacific Journal of Clinical Nutrition* 17, 275–279.

13 *Prunus*

Ariel R. Vicente, George A. Manganaris, Luis Cisneros-Zevallos and
Carlos H. Crisosto

13.1 Introduction

The genus *Prunus* includes about 430 species
of deciduous or evergreen trees and shrubs
naturally widespread throughout temperate
regions. It belongs to the subfamily *Prunoi-
deae*, within the *Rosaceae* family. While some
species do not yield edible fruit and are used
for decoration, others are grown commercially for fruit and 'nut' production. Most of
these species are originally from Asia or
Southern Europe, such as peach (*Prunus persica*), nectarine (*P. persica* var. nectarina),
European plum (*P. domestica*), Japanese plum
(*P. salicina*), apricot (*P. armeniaca*), mume or
Japanese apricot (*P. mume*), sweet cherry
(*P. avium*), sour cherry (*P. cerasus*) and almond
(*P. amygdalus*).

The fruit of these species is defined
botanically as a 'drupe' (Brady, 1993). This
name is derived from the latin word *druppa*,
which means overripe olive. Drupes mostly
develop from flowers with superior ovaries
having a single carpel. The fruit usually have
a clear ventral suture, do not retain floral residues next to the pedicel and are characterized
by a membranous exocarp, with an outer
fleshy mesocarp consisting mainly of paren-
chyma cells (Romani and Jennings, 1971). The
mesocarp surrounds a shell (the pit or stone)
of hardened endocarp with a seed inside and,
due to this, *Prunus* species are also referred to
as 'stone fruit'. Unlike almonds, in which the
consumed portion is the seed within the pit,
the edible part in most stone fruits includes
the mesocarp, and eventually the exocarp.
Growth of stone fruit usually shows a double
sigmoid pattern, with an initial stage after
fertilization characterized by active cell division, followed by a phase in which all the
parts of the ovary besides the embryo and
endosperm grow. Later on, whole fruit
growth is decelerated, while seed development and endocarp lignification occur and,
lastly, mesocarp expansion resumes (Romani
and Jennings, 1971).

Prunus species cultivation is common in
temperate regions throughout the globe. The
most popular fruit within this group are
by far peaches/nectarines and plums, with
annual productions of around 17.5 and 9.7
million tonnes (Mt), respectively, in 2007
(Table 13.1). Apricot production is around 3
Mt/year, while almonds and cherries account
for 2 Mt annually (Table 13.1). The main
world producing countries of stone fruit
include China, the USA, Italy, Spain and Turkey. Serbia and Syria are also important producers of plums and almonds, respectively.

Probably in association with the differences in the economical importance of *Prunus*
species, much more research has been
devoted historically to peaches and nectarines (the biochemistry and physiology of

Table 13.1. Production of *Prunus* species in 2007 (FAOSTAT, 2009).

	Production (t)	Main producing countries
Almond	2,065,489	USA, Spain, Italy, Syria
Apricot	3,067,952	Turkey, Iran, Italy
Cherry	1,995,751	Turkey, Iran, USA
Peach and nectarine	17,502,245	China, Italy, Spain, USA
Plum	9,719,451	China, Serbia, USA
Total	34,350,888	

peaches and nectarines are similar, with the main difference between them being the appearance of the skin, with more variation within each of the groups). Cherries and plums have been studied to a lesser extent, and some other species, such as apricot and sour cherry, have received little attention. The biochemical and physiological changes associated with ripening of these fruits have been reviewed (Romani and Jennings, 1971; Brady, 1993). Some areas in which there have been significant advances include: (i) evaluation of the influence of orchard practices on fruit yield and quality (La Rue and Johnson, 1989; Crisosto *et al.*, 1997); (ii) development of proper harvest indices and definition of optimal harvest operations (Crisosto *et al.*, 1995; Crisosto and Mitchell, 2002); (iii) determination of optimal cooling strategies, temperatures for storage and transportation and modified atmospheres (Crisosto *et al.*, 2009); and (iv) the characterization of the physiological basis of some disorders, such as internal breakdown, mealiness and flesh reddening (Brummell *et al.*, 2004a,b; Lurie and Crisosto, 2005; Manganaris *et al.*, 2008).

Stone fruit are highly appreciated due to their unique aesthetic and organoleptic characteristics. Traditionally, attributes evaluated in relation to fruit quality include mostly appearance, sugars and acids. However, the fruit also contain a myriad of phytochemicals that, though present in relatively low concentrations, have a key role in overall quality. Some of these chemicals might be the main determinants of colour and flavour. In addition, many of these compounds have been found to play protective roles against some human diseases (Ames *et al.*, 1993; Rice-Evans and Miller, 1996; Olsson *et al.*, 2004). When

ingested regularly and in significant amounts as part of the diet, these metabolites may have noticeable long-term physiological effects (Espín *et al.*, 2007). Increasing the consumption of fruit and vegetables results in a significant increase of plasma antioxidants (Cao *et al.*, 1998). The metabolites that have received more attention include ascorbic acid, tocopherols, tocotrienols, carotenoids and phenolics (Robards, 2003). It is currently recommended to have five to ten servings of fruit and vegetables daily (http://www.mypyramid.gov), and in the past years there have been efforts to develop educational and promotional programmes related to fruit and vegetable consumption (http://www.5aday.nhs.uk/topTips/default.html).

Studying plant phytochemicals is not an easy endeavour, especially for carotenoids and phenolics, which include thousands of different compounds, commonly present at relatively low concentrations. In the past 20 years, the improvement in analytical techniques (based mostly on HPLC coupled with mass spectrometry detection) has allowed rapid progress in the identification, quantification and overall study of fruit phytochemicals, with special emphasis on those showing bioactivity.

Fresh *Prunus* species are significant contributors of bioactive compounds to the diet during spring and summer, although the increase in year-round supply in the developed world has lessened these seasonal eating habits. Most *Prunus* fruit and seeds are commonly used for processing, including jam production, canning, drying or roasting, and regularly are consumed year-round. Black plum varieties are among fruit with the highest antioxidant capacity and are very rich in phenolic compounds, while apricots are high in carotenoids. Cherries and almonds are also rich in phenolics and have an oxygen radical absorbing capacity (ORAC) similar to that found in strawberry or raspberry (which are commonly recognized as good sources of antioxidants; see Chapters 14 and 15 of this volume). Although peaches and nectarines do not rank top among fruit in terms of antioxidant capacity or content of bioactive compounds, they still do have moderate levels of carotenoids and ascorbic acid. Due to their popularity, either in fresh or processed form, they may contribute significantly to global intake of bioactives. Vinson *et al.* (2001) estimated that peach contribution of phenolics to the diet, despite the modest level of phenolic compounds relative to plums and cherries, is highest among stone fruit. Despite these general features, both the qualitative and quantitative profiles of these compounds in stone fruit vary markedly depending on the variety (Tomás Barberán *et al.*, 2001; Dalla Valle *et al.*, 2007; Vizzotto *et al.*, 2007; Díaz-Mula *et al.*, 2008; Ruiz *et al.*, 2008). Orchard practices, growing location, season, environmental conditions, postharvest management and processing operations may also determine great changes in the levels of phytochemicals. In this chapter, we describe the antioxidants and bioactive compounds in *Prunus* species and we analyse the main factors affecting their levels.

13.2 Antioxidants

The ORAC assay has been applied routinely to determine antioxidant capacity and currently there are databases generated for different foods (Wu *et al.*, 2004a; USDA, 2007a). The ORAC values of some common fruit are shown in Fig. 13.1. Blackcurrants and cranberries rank at the top, with ORAC values of around 10,000 μmol Tropax equivalents (TE)/100 g. Blueberries, which have been recognized repeatedly for their high content of antioxidants, have ORAC comparable to that observed in plums (Fig 13.1). Cherries also show relatively high ORAC, not far from that found in strawberries.

An original limitation of the ORAC method was the inability to determine both hydrophilic and lipophilic antioxidants. A modified ORAC method was developed for that purpose. Hydrophilic and lipophilic antioxidants were extracted in polar solvent and hexane, respectively. The use of randomly methylated β-cyclodextrin as a solubility enhancer allowed the use of the same peroxyl-free radical source (Wu *et al.*, 2004b). This method has the advantage that similar assay conditions and standards are used for both the hydrophilic and lipophilic assays and could be used to determine a total ORAC value for a given sample.

Table 13.2 shows the hydrophilic and lipophilic ORAC values for the main cultivated *Prunus* species. As mentioned, there is a broad range of antioxidant capacities among stone fruit. Plums are by far the stone fruit that rank at the top, with ORAC values of 6239 μmol TE/100 g. Black plum varieties have even higher ORAC than several fruit commonly recognized as antioxidant-rich fruit (e.g. blueberries). The values in Table 13.2 have to be interpreted cautiously since broad variation for a given fruit has been reported, depending on the variety evaluated. In a survey of eight plum cultivars, Blackamber showed highest antioxidant capacity, followed relatively closely by cvs. Golden Globe and Larry Ann and, with much lower levels, by cvs. TC Sun, Sonogold, Angeleno, Golden Japan and Black Diamond (Díaz-Mula *et al.*, 2008). Almonds and cherries also have high ORAC (3361 and 4454 μmol TE/100 g, respectively). The ORAC reported for peach, nectarine and apricot are usually lower than those found in other *Prunus* species (Wu *et al.*, 2004a).

In all cases, it is seen clearly that the lipophilic contribution to total ORAC is low. This suggests that the relative relevance of phenolic compounds, ascorbic acid and possibly xantophylls in the antioxidant capacity of stone fruit is higher than that of carotenes and tocopherol (however, whether or not the conditions for measurement of lipophilic antioxidant are distinct from those occurring *in vivo*, and that the real antioxidant contribution of these metabolites is underestimated, is not known). Stone fruit total antioxidant capacity

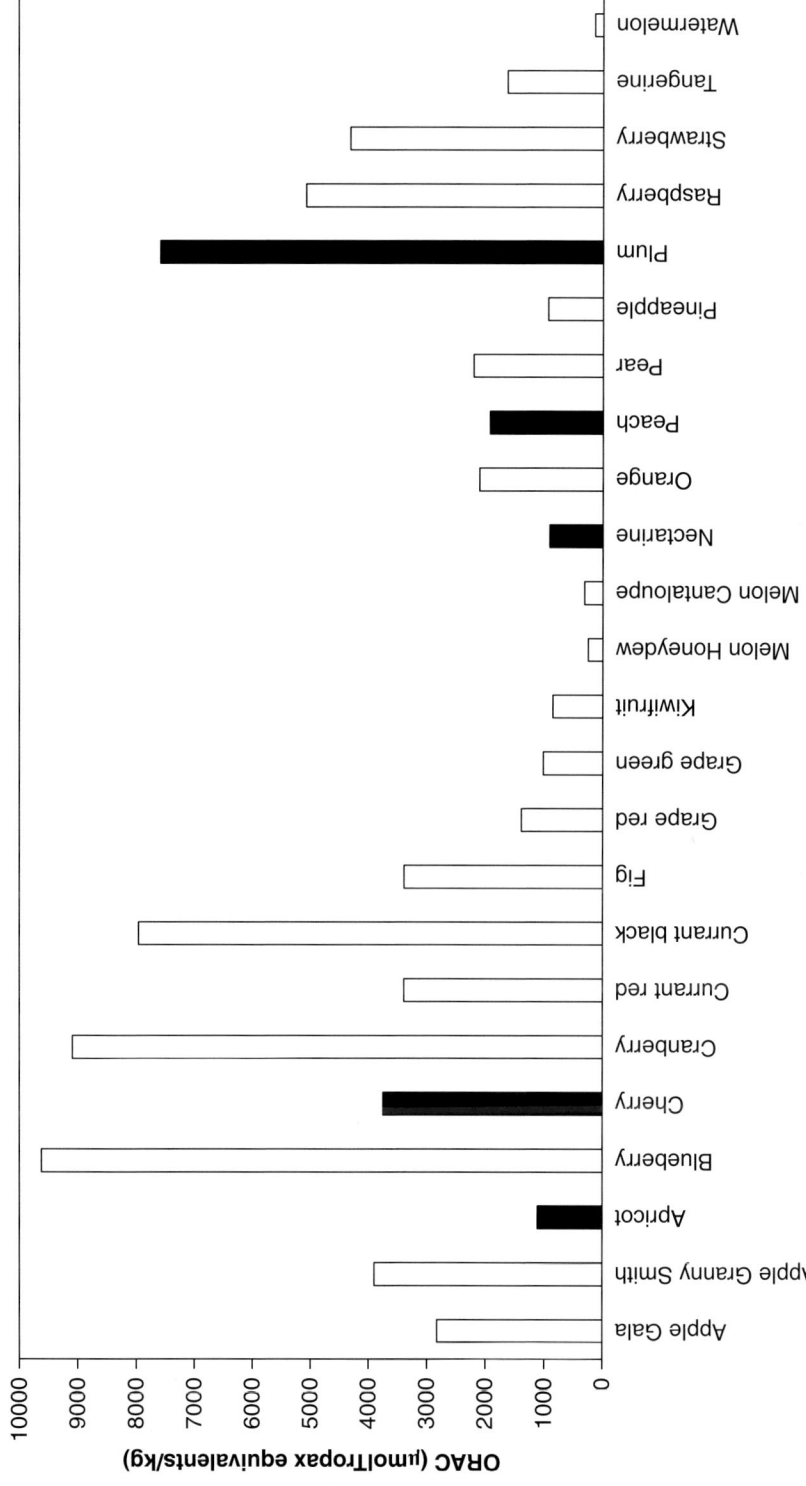

Fig. 13.1. Total oxygen radical absorbance capacity (ORAC) of different fruit. The black columns highlight *Prunus* species (USDA, 2007a).

Table 13.2. Hydrophilic, lipophilic and total oxygen radical absorbance capacity (ORAC) of different *Prunus* species (adapted from USDA, 2007a, and Wu *et al.*, 2004a).

	Hydrophilic ORAC (µmol TE/100 g)	Lipophilic ORAC (µmol TE/100 g)	Total ORAC (µmol TE/100 g)
Almond	4282	172	4454
Apricot	1309	32	1341
Cherry, sweet	3344	17	3361
Nectarine	720	29	749
Peach	1813	50	1863
Plum	6222	17	6239
Plum, black	7301	38	7339

Note: TE = Trolox equivalents.

(TAC) have shown high positive correlations with total phenolics and antioxidants, suggesting that this group of compounds is the predominant contributor to TAC (Serrano *et al.*, 2005; Drogoudi *et al.*, 2008).

The distribution of antioxidants also varies within the fruit. In plum and cherry, TAC is always higher in the skin than in the flesh (Serrano *et al.*, 2005; Díaz-Mula *et al.*, 2008). In peach, removal of the peel also results in a significant loss of total antioxidant capacity (Remorini *et al.*, 2008).

13.3 Main Bioactive Compounds in *Prunus* Species

13.3.1 Carotenoids

Carotenoids comprise a group of over 600 secondary plant metabolites with several functions in plants (Gross, 1991). They are essential structural components of the photosynthetic apparatus and contribute to light absorption in regions of the spectrum where chlorophyll absorption is low (Cuttriss and Pogson, 2004). They also provide protection against photo-oxidation. Carotenoids are responsible for the distinctive yellow and orange colours of the pulp and peel of most *Prunus* species. They are liposoluble terpenoids due to their relatively long (40 C) hydrocarbon structure (Sandmann, 2001). Carotenoids are synthesized *de novo* from geranyl-geranyl diphosphate by all photosynthetic organisms and accumulated in plastids (Rodríguez-Amaya, 2001). Then the enzyme phytoene synthase catalyses the condensation of two molecules of geranylgeranyl pyrophosphate to form phytoene, a colourless carotenoid (Gross, 1991). After that, double bonds are added sequentially to generate compounds with higher levels of conjugated double bonds (3 in phytoene, 5 in phytofluene, 7 in ζ-carotene, 9 in neurosporene and 11 in lycopene). Interestingly, the addition of double bonds is related to carotenoid spectral properties. Terpenoid polyenes absorb light in the UV and visible regions. Most show three absorbance maxima and the position of these peaks usually is related to the number of conjugated double bonds (as they increase, carotenoid colour shifts from yellow to orange and red) (Gross, 1991). Carotenoids can be subdivided into two groups, carotenes containing only carbon and hydrogen such as α-carotene, β-carotene and lycopene, and the oxygenated derivatives or xanthophylls such as lutein, cryptoxanthin, zeaxanthin and violaxanthin (Rodríguez-Amaya, 2001).

One of the main important features of carotenoids is being precursors of vitamin A (Sandmann, 2001). Humans depend on dietary carotenoids for making their retinoids, which are crucial for normal vision (Kopsell and Kopsell, 2006). Structurally, vitamin A (retinol) is basically a half molecule of β-carotene with an extra molecule of water, and retinoids are generated on cleavage by dioxygenases. Other carotenoids with at least one unsubstituted β-ionone ring (such as γ-carotene, α-carotene and cryptoxanthin) are precursors of vitamin A. In contrast, those carotenoids that are acyclic, or with β rings with hydroxy, epoxy and carbonyl substituents, do not have provitamin A activity (Rodríguez-Amaya, 2001). Retinol equivalents (μg RE) from fruit carotenoids are calculated as μg of β-carotene/12 or as μg of other provitamin A carotenoids/24, to account for differences in the absorption and biotransformation of carotenoids into retinol.

Besides being precursors of vitamin A, some benefits of carotenoids have been associated with their antioxidant properties (Bendich, 1993). The conjugated double bonds are not only responsible for the colour of carotenoids but also are main determinants of their antioxidant properties. Carotenoids have been shown to react with singlet oxygen and peroxyl radicals (Stahl and Sies, 1997). The different group members can vary on their antioxidant capacity in humans. Larger conjugated systems, such as astaxanthin, are known to have a higher antioxidant activity (Miki, 1991). Lycopene also has high antioxidant capacity, surpassing that of β-carotene (Di Mascio *et al.*, 1989). The fate of most carotenoids on consumption, besides lycopene and β-carotene, is not completely known (Rao and Rao, 2007). The bioavailability of carotenoids from plant foods is variable and depends on the type of compounds present in the food, release from the food matrix, absorption in the intestinal tract, and the nutritional status of the ingesting host, as well as the presence of other components in the food matrix (Rao and Rao, 2007). Recent studies have shown that absorption of carotenoids increases when they are ingested with dietary lipids. Processing activities usually increase bioavailability through increased release of bound carotenoids from the food matrix; after that, carotenoids are assimilated and oriented into lipid micelles, incorporated into chylomicrons, and eventually delivered to the liver (Kopsell and Kopsell, 2006).

From the whole group of more than 600 known carotenoids only 40 are present in a typical human diet, and just five of them (β-carotene, α-carotene, lutein, cryptoxanthin and lycopene) represent close to 90% (Gerster, 1997). The bicyclic β-carotene is the most widespread of all carotenoids in foods (Rodríguez-Amaya, 2001). In *Prunus* species, β-carotene is also by far the most abundant carotenoid; α-carotene is also found in apricots but at lower concentration. The most abundant xantophylls present in most *Prunus* fruit is lutein, but β-cryptoxanthin is also present in orange-fleshed commodities such as apricot, peach and nectarine, and to a lower extent in plum (Tourjee *et al.*, 1998; Rodríguez-Amaya, 2001; USDA, 2009). In this group, apricots show the highest levels of carotenoids and provitamin A, followed by cherry (Table 13.3). Peaches, nectarines and plums have lower concentrations of carotenoids. However, the variation among varieties is large. For instance, when studying eight plum varieties, Díaz-Mula *et al.* (2008) found fourfold variation in total carotenoid content. In apricot, Ruiz *et al.* (2005) found that total carotenoid content ranged from 1500 to 16,500 μg/100 g, depending on the variety being considered. Carotenoids were not distributed evenly in the food itself and various investigators found that they were usually more concentrated in the peel than in the pulp. In plums and peaches, the ratio of carotenoids in the peel to the pulp was around five (Díaz Mula *et al.*, 2008; Remorini *et al.*, 2008).

Though it has been suggested that breeding programmes might consider the selection of lines based on the content of bioactive compounds, using traditional analytical techniques for screening might be impractical. Yellow-flesh peaches have a higher concentration of carotenoids than light-coloured ones. Ruiz *et al.* (2005) reported that in apricot, carotenoid content showed high and positive correlation with the colour (hue) of both flesh and pulp. In addition, the authors developed a model to predict carotenoid content from single non-destructive colour measurements (Ruiz *et al.*, 2008).

13.3.2 Ascorbic acid

Vitamin C is an essential nutrient for humans and a small number of other mammalian species (Hancock and Viola, 2005). Ascorbic acid plays an important role in hydroxylation reactions (e.g. in the synthesis of collagen,

Table 13.3. Carotenoid content in *Prunus* species (USDA, 2009).

	Carotenoid (µg RE/100 g)	β-CAR (µg/100 g)	α-CAR (µg/100 g)	LUT-ZEA (µg/100 g)	β-CRY (µg/100 g)
Almond	0	1	0	1	0
Apricot	96	1094	19	89	104
Cherry, sweet	64	770	0	85	0
Nectarine	17	150	0	130	98
Peach	16	162	0	91	67
Plum	17	190	0	73	35

Notes: β-CAR = β-carotene; α-CAR = α-carotene; LUT-ZEA = lutein + zeaxanthin; β-CRY = β cryptoxanthin; RE = retinol equivalents 1 µg RE = 12 µg β-carotene or 24 µg of other carotenoids having provitamin A activity (in this case, α-car and β-cry).

and it is highly important for the *de novo* synthesis of bone, cartilage and wound healing). The disorder associated with vitamin C deficiency is known as scurvy. Scurvy leads to the formation of liver spots on the skin, spongy gums and bleeding from all mucous membranes. The compounds accounting for vitamin C activity present in foods include ascorbic acid and its initial oxidation product, dehydroascorbic acid (which can be reduced back in the human body) (Lee and Kader, 2000). Besides its role in some enzymatic reactions, ascorbic acid is also a potent antioxidant, protecting the cell against reactive oxygen species (Halliwell and Gutteridge, 1995; Noctor and Foyer, 1998). Currently, a dietary daily intake of 90 mg of vitamin C is recommended. Ascorbic acid is water soluble and highly available, but is also one of the bioactive compounds more susceptible to degradation (Rickman *et al.*, 2007). Plants are able to synthesize vitamin C through sequence steps that convert L-galactose or galacturonic acid to ascorbic acid (Smirnoff and Wheeler, 2000; Valpuesta and Botella, 2004). Its concentration in fruit, vegetables and nuts varies from 1 to 150 mg/100 g (Lee and Kader, 2000).

Prunus species are not particularly rich in ascorbic acid, with concentrations ranging from negligible levels in almond to around 10 mg/100 g in most stone fruit (Table 13.4). In some products such as broccoli and pepper, ascorbic acid seems to be one of the compounds contributing most to antioxidant capacity, but in peach it accounts for only 0.8% (Sun *et al.*, 2002).

Table 13.4. Ascorbic acid content in *Prunus* species (USDA, 2009).

	Ascorbic acid (mg/100 g)
Almond	0.0
Apricot	10.0
Cherry, sweet	10.0
Nectarine	5.4
Peach	6.6
Plum	9.5

13.3.3 Vitamin E

Vitamin E is liposoluble and exists in eight different forms (four tocopherols and four tocotrienols). All the isomers have aromatic rings and are named alpha (α), beta (β), gamma (γ) and delta (δ). The designation of the isomers is related to the number and position of methyl groups in the molecule ring. From all the possible forms, α-tocopherol is the most active. The main function of these compounds seems to be as antioxidants. α-Tocopherol has also been implicated in other biological processes, such as inhibition of some protein kinases, prevention of platelet aggregation, enhancement of vasodilation and modulation of the activities of enzymes associated with the immune system (Food and Nutrition Board, Institute of Medicine, 2000; Traber, 2001). Foods rich in vitamin E include vegetable oils, nuts and avocado. Broccoli and leafy vegetables have lower tocopherol content than fat-rich products, but they are good sources compared with other fruit and vegetables.

α-Tocopherol content in *Prunus* fruit varies from 0.07 to 26 mg/100 g (Table 13.5). Almonds are extremely rich in vitamin E, with 60 g covering the daily recommended intake (22 IU or 15 mg α-tocopherol).

13.3.4 Phenolic compounds

Plants produce thousands of phenolic compounds as secondary metabolites. They are synthesized via the shikimic acid pathway. This is the biosynthetic route to the aromatic amino acids and is restricted to microorganisms and plants (Robards, 2003). Phenolics contribute greatly to the sensory qualities (colour, flavour, taste) of fresh fruit and their products (Kim *et al.*, 2003a,b). Astringency generally is recognized as a loss of lubrication, a feeling of extreme dryness in the palate believed to be associated with the interaction of polyphenols with praline-rich proteins present in the saliva (Haslam, 2007). Phenolic compounds have been associated with ecological functions. Some benefits associated with phenolic compound consumption have been related to their antioxidant properties (Tall *et al.*, 2004). Phenolic compounds may exert their antioxidant activity by different mechanisms. They may act by direct-scavenging free radicals (Kris-Etherton *et al.*, 2002). Some phenolics can prevent the formation of free radicals by chelating copper and iron. Finally, they can regenerate other antioxidants such as tocopherol (McAnlis *et al.*, 1999). Fruit, vegetables and beverages are the major sources of phenolic compounds in the human diet (Balasundram *et al.*, 2006). Some *Prunus* fruit, such as red and black plum varieties, cherries and almonds, are very good sources of phenolic compounds (Vinson *et al.*, 2001; Wu *et al.*, 2004a) (Table 13.6). Sour cherries were reported to have a higher level of total phenolics than sweet cherries (Kim *et al.*, 2005). Peach, nectarine and apricot have lower total phenolics, as measured with the Folin–Ciocalteu reagent. However, wide intervarietal differences exist (Tomás-Barberán *et al.*, 2001). Pinelo *et al.* (2004) suggested that relatively high levels of phenolic antioxidants might be recovered from almond hulls for further use. The peel of *Prunus* fruit has also been found to contain higher amounts of phenolics than other fleshy parts (Balasundram *et al.*, 2006).

Various studies have shown that total antioxidant capacity shows better correlation with total phenolic compound content than with ascorbic acid, tocopherol or carotenoids (Serrano *et al.*, 2005; Drogoudi *et al.*, 2008). However, it is currently unknown which would be an appropriate polyphenol daily uptake that would exert a beneficial effect (Espín *et al.*, 2007). Regarding phenolic compound metabolism, the bowel microflora can hydrolyse phenolic compound glycosides, releasing aglycon, which could be transformed further to produce derivatives of acetic and phenyl propionic acids (Formica and Regelson, 1995).

Phenolic compounds might be divided in groups represented differently, depending on the species considered. Here, we summarize the main characteristics of phenolic acids,

Table 13.5. Vitamin E content in *Prunus* species (USDA, 2009).

	Vitamin E (mg α-TOC/100 g)
Almond	26.22
Apricot	0.89
Cherry, sweet	0.07
Nectarine	0.77
Peach	0.73
Plum	0.26

Note: α-TOC = α-tocopherol.

Table 13.6. Total phenolics (TP) in *Prunus* species (adapted from USDA, 2007a, and Wu *et al.*, 2004a).

Fruit	TP (mg GAE/100 g)
Almond	418
Apricot	79
Cherry, sweet	339
Nectarine	107
Peach	148
Plum	367
Plum, black	478

Note: GAE = gallic acid equivalents.

flavonoids, proanthocyanidins and tannins, describing the different metabolites found within these groups in *Prunus* species.

Phenolic acids

Phenolic acids occur in two classes: derivatives of benzoic acid or cinnamic acid. Hydroxycinnamic acids exhibit higher antioxidant activity than the corresponding hydroxybenzoic acids. These phenolic acid derivatives in turn combine with sugars to become glycosylated. Caffeic acid is generally the most abundant phenolic acid and accounts for between 75 and 100% of the total hydroxycinnamic acid content of most fruit, with the highest concentrations typically in the outer parts of ripe fruit. Caffeic and quinic acids may combine to form chlorogenic acid (5-*O*-caffeoylquinic acid, 5-CQA; Manach *et al.*, 2004). Apples and pears contain mainly 5-CQA (Möller and Herrmann, 1983). In cherries, another isomeric form is the most abundant (3-CQA) (Herrmann, 1989). Relatively high amounts of 4-CQA are characteristic of prunes. Interestingly, the concentrations of chlorogenic and caffeic acid, two major phenolic acids in the epidermis and subtending cell layers of peach, are especially high in peach genotypes with a high level of resistance to the brown rot fungus, *Monilinia fructicola* (Bostock *et al.*, 1999).

Flavonoids

Flavonoids are a large group of structurally related compounds with a chromane-type skeleton, and a phenyl substituent in the C2 or C3 position. They are present in all terrestrial vascular plants and have been used historically in chemotaxonomy. To date, more than 6000 flavonoids have been identified. They are classified into at least ten chemical groups. Among them, anthocyanins, flavones, flavonols, flavanones, flavan-3-ols and isoflavones are particularly common in the diet. Members of the first five groups can be found in *Prunus* species.

ANTHOCYANINS. The name anthocyanin derives from *anthos*, which means flower, and *kyanos*, which means blue (Kong *et al.*, 2003). Anthocyanins are the most widespread of the flavonoid pigments. As they confer red, blue and purple colours to plant tissues, they contribute largely to the visual quality of fruit (Mazza, 1995). Anthocyanin is actually a name used to designate glycosides of flavonoid molecules known as anthocyanidins. Six anthocyanidins occur most frequently in plants: pelargonidin, cyanidin, peonidin, delphinidin, petunidin and malvidin. Anthocyanins are either 3- or 3,5-glycosylated, with the most prevalent sugars being glucose, rhamnose, xylose, galactose, arabinose and fructose. Different patterns of hydroxylation and glycosylation in anthocyanins appear to modulate their antioxidant properties. Comparing the antioxidant capacity of different anthocyanins, cyanidin-3-glucoside (the most common anthocyanin found in *Prunus* species) had the highest ORAC activity, which was 3.5 times stronger than Trolox, while pelargonidin had the lowest antioxidant activity, but was still as potent as Trolox (Wang *et al.*, 1997). Cherries are by far the stone fruit with the highest content of cyanidin (75.18 mg/100 g), followed by Black Diamond® plums (40 mg/100 g) and red-flesh plums (13.7). Peonidin is found in cherry and plum in lower but still significant proportions (Table 13.7). Anthocyanin content in other *Prunus* species is low. In peach, anythocyanins are associated mainly with the peel. In some varieties, anthocyanin accumulation produces endocarp staining, a trait that can give an attractive appearance to fresh fruit, but is detrimental in the canning process, as the redness becomes brown and unsightly.

FLAVONES, FLAVONOLS, FLAVANONES AND FLAVAN-3-OLS. The main flavones in the diet are apigenin and luteolin. Among *Prunus* species, they are found at low concentrations in red plum varieties. Among flavonols, quercetin is the compound found most commonly in *Prunus* species. It is present in relatively high concentration, especially in cherry and red plum, but at levels lower than those found in flavonol-rich products such as onion. Flavonols are located almost exclusively in the peel

Table 13.7. Flavonoids in *Prunus* species (USDA, 2007b).

Fruit	Group	Compound	mg/100 g
Almond	Anthocyanidins	Cyanidin	2.46
	Flavan-3-ols	(–)-Epicatechin	0.60
		(–)-Epigallocatechin	2.59
		(+)-Catechin	1.28
	Flavanones	Eriodictyol	0.25
		Naringerin	0.13
	Flavonols	Isorhamnetin	7.05
		Kaempferol	0.52
		Quercetin	0.36
Apricot	Flavan-3-ols	(–)-Epicatechin	5.47
		(–)-Epigallocatechin-3-galate	4.79
Cherry, sweet	Anthocyanidins	Cyanidin	75.18
		Pelargonidin	0.54
		Peonidin	4.47
	Flavan-3-ols	(–)-Epicatechin	6.97
		(–)-Epigallocatechin-3-galate	0.05
		(–)-Epigallocatechin	0.34
		(+)-Catechin	1.31
	Flavonols	Quercetin	2.64
Nectarine	Anthocyanidins	Cyanidin	1.81
	Flavan-3-ols	(–)-Epicatechin	2.54
		(+)-Catechin	2.98
	Flavonols	Quercetin	0.69
Peach	Anthocyanidins	Cyanidin	1.61
	Flavan-3-ols	(–)-Epicatechin	2.34
		(–)-Epigallocatechin	1.04
		(–)-Epigallocatechin-3-galate	0.30
		(+)-Catechin	4.92
	Flavonols	Quercetin	0.68
Plum, red	Anthocyanidins	Cyanidin	4.73
		Delphinidin	0.02
		Pelargonidin	0.02
		Peonidin	2.21
	Flavones	Apigenin	0.01
		Luteolin	0.01
	Flavonols	Kaempferol	0.01
		Myricetin	0.02
		Quercetin	1.85

(Tomás-Barberán *et al.*, 2001). Flavanones are typical of citrus, but eriodictyol and naringenin are also present in almond nuts. Finally, flavan-3-ols, including (+)-catechin and (-)-epicatechin and their gallic acid esters, show six times higher levels in black plums than in most other *Prunus* fruit that have similar total contents.

Proanthocyanidins (PAs) and tannins

Proanthocyanidins are the second most abundant natural phenolics after lignin (Espín *et al.*, 2007). Proanthocyanidins, also known as condensed tannins, are oligomers. Despite their role as potent antioxidants, they have been associated with whole plant ecological functions (Rawat *et al.*, 1998). Proanthocyani-dins produced by *P. armeniaca* roots were sug-gested to limit the germination and growth of surrounding species (Rawat *et al.*, 1999). Poly-mers of catechin and epicatechin are the most common PAs in food. Colonic flora could metabolize these compounds, producing phe-nolic acids (Depréz *et al.*, 2000). PAs have received increasing attention due to evidence associating their action with some health ben-efits (Lazarus *et al.*, 1999). Plum and almonds are rich in PAs, with values even higher than those found in grape (USDA, 2004). PAs with a higher degree of polymerization are normal in these fruit as opposed to PAs from other *Prunus* species which show lower mean molecular mass (Table 13.8).

Ellagitannins (ETs) are a subgroup of tan-nins (Hümmer and Schreier, 2008). Structur-ally, they contain at least two galloyl units C–C coupled to each other, but lack glycosidi-cally linked catechin units (Khanbabaee and van Ree, 2001). With more than 500 natural products characterized so far, they are the largest group of tannins. They are found in pomegranates, black raspberries, raspberries, strawberries, walnuts and almonds.

13.4 Phytonutrients

Nutraceutical compounds are defined as extracts of foods that exert a medicinal effect on human health by preventing or limiting the progression of chronic diseases. Extracts from fruit and vegetables contain mixtures of phyto-chemicals, and recent studies on nutraceutical properties have been conducted isolating chemical compounds, defining fractions and using mixtures to determine the specific bio-active compounds present. Many of the phyto-chemical categories present in *Prunus* species, including carotenoids, vitamins, phenolics and others, may play a role as nutraceuticals through the prevention of chronic diseases such as cancer, cardiovascular disease, Alzheimer's, metabolic syndrome and others (Sun *et al.*, 2002; Kottová *et al.*, 2004; Heo and Lee, 2005; Chen *et al.*, 2007; Chen and Blumberg, 2008). The following is a short summary of nutraceu-tical properties of selected *Prunus* species.

Almond extracts have been character-ized for their effects against cancer and car-diovascular disease. For example, whole almonds and almond fractions have been shown to reduce aberrant crypt foci in a rat model of colon carcinogenesis (Davis and Iwahashi, 2001). Clinical studies have shown that inclusion of almonds in the daily diet may elevate the blood levels of high-density lipoproteins (HDL) while lowering levels of low-density lipoproteins (LDL), and these lipid-altering effects have been associated with the interactive or additive effects of the numerous bioactive constituents found in almonds (Spiller *et al.*, 1998). A dose–response study on hyperlipidaemic subjects confirmed these results, showing that 73 g of almonds in the daily diet reduced LDL by ~ 9.4%, increased HDL ~ 4.6% and reduced the LDL/HDL ratio ~ 12% (Jenkins *et al.*, 2002). Several constituents of almond have been associated with anti-inflammatory properties, antihepa-totoxicity effects and immunity boosting properties (Puri, 2003).

Studies with peach and plum have been conducted to determine their effects against cancer, cognitive deficits and hepatotoxicity. On the other hand, nectarines and apricots still await characterization for their nutraceu-tical properties.

Peach and plum phenolic compounds have been shown to inhibit growth and induc-tion of differentiation in colon cancer cells (Lea *et al.*, 2008). Furthermore, plum extracts (dried plums) have been shown to alter several risk factors related to colon carcinogenesis in rats, including a reduction of faecal total and sec-ondary bile acid concentrations, decrease of colonic β-glucoronidase and 7α-dehydroxylase activities and an increase in antioxidant activ-ity. The effect was associated with the extracts' phytochemical content (Yang and Gallaher, 2005). More recently, chlorogenic acid and neo-chlorogenic acid in plum and peach fruit have been identified as bioactive compounds with potential chemopreventive properties against an oestrogen-independent breast cancer cell line, while having little effect on normal cells (Noratto *et al.*, 2009). On the other hand, plum (*P. domestica* L.) juice was effective in mitigat-ing cognitive deficits in aged rats, and this effect was associated with the amount of

Table 13.8. Proanthocyanidins in *Prunus* species (USDA, 2004).

Fruit	Proanthocyanidin	mg/100 g
Almond	Monomers	7.77
	Dimers	9.52
	Trimers	8.82
	4–6 mers	39.97
	7–10 mers	37.68
	Polymers	80.26
	Total	184.02
Apricot	Monomers	1.32
	Dimers	11.33
	Trimers	0.70
	4–6 mers	4.90
	7–10 mers	2.20
	Polymers	0.60
	Total	21.05
Cherry, sweet	Monomers	5.11
	Dimers	3.25
	Trimers	2.39
	4–6 mers	6.51
	7–10 mers	1.87
	Polymers	0.00
	Total	19.13
Nectarine	Monomers	5.57
	Dimers	5.00
	Trimers	1.75
	4–6 mers	5.98
	7–10 mers	3.57
	Polymers	7.31
	Total	29.18
Peach	Monomers	4.48
	Dimers	12.24
	Trimers	4.41
	4–6 mers	17.66
	7–10 mers	10.94
	Polymers	22.02
	Total	71.75
Plum, red	Monomers	10.88
	Dimers	38.54
	Trimers	22.25
	4–6 mers	58.04
	7–10 mers	33.79
	Polymers	57.28
	Total	220.78

phenolics supplemented (Shukitt-Hale *et al.*, 2009). In addition, *P. persica* pericarp extracts were tested against cisplatin-induced acute toxicity in mice. Results showed a significant protection against the acute nephrotoxicity and hepatotoxicity, probably the result of a reduction in cisplatin-induced oxidative stress (Lee *et al.*, 2008). Similar results were observed with *P. persica* flesh (Lee *et al.*, 2007). Likewise, studies with methanolic extracts of immature plum fruit (*P. salicina* L. cv. Soldam, 20–40 days before final harvest) showed inhibitory effects against benzo(α)pyrene (B(α)P) induced toxicity in mice, by inhibiting the induction of

CYP1A1 expression, CYP1A1 being the primary cytochrome P450 involved in the metabolism and bioactivation of B(α)P (Kim *et al.*, 2008). Recent studies with peach, nectarine and apricot extracts have shown *in vitro* binding of a mixture of bile acids (secreted in human bile at a duodenal physiological pH of ~6.3). Using cholestyramine (a bile acid-binding, cholesterol-lowering drug) as a reference indicated that binding potential followed the order peach > apricot > nectarine. The effects have been associated with the mixture of phytochemicals present in the extracts (Kahlon and Smith, 2007).

Research with tart and sweet cherry fruit extracts and fractions has been reported in the areas of metabolic syndrome, cancer, inflammation and effects on neuronal cells. For example, studies have shown that cherry-enriched diets reduce metabolic syndrome and oxidative stress in rats (Seymour *et al.*, 2008), while regular tart cherry intake by obesity-prone rats fed with a high-fat diet altered the abdominal adiposity and affected adipose gene transcription and inflammation (Seymour *et al.*, 2009). Furthermore, intake of tart cherry in diets reduced several phenotypic factors that were associated with risk for metabolic syndrome and Type 2 diabetes in rats, altering hyperlipidaemia, hepatic steatosis and hepatic peroxisome proliferator-activated receptors (Seymour *et al.*, 2008). In clinical studies, it was shown that consumption of cherries lowered plasma urate in healthy women (Jacob *et al.*, 2003). This has important implications, since high levels of plasma urate have been associated with cardiovascular disease, diabetes and the metabolic syndrome.

Studies with anthocyanin-rich tart cherry extracts have shown inhibition of intestinal tumorogenesis in mice (Bobe *et al.*, 2006) and antiproliferation activity against human colon cancer cells (Kang *et al.*, 2003). Tart, sweet and sour cherries have shown cyclooxygenase inhibitory activities (Wang *et al.*, 1999; Seeram *et al.*, 2001; Mulabagal *et al.*, 2009), and suppressed inflammation pain behaviour has been associated with tart cherry anthocyanin intake in rats (Tall *et al.*, 2004). In fact, clinical trials have confirmed that consumption of cherries (Bing sweet cherries) lowers

circulating concentrations of inflammation markers in healthy men and women (Kelley *et al.*, 2006). Finally, studies using sweet and sour cherry phenolics have shown a protective effect on neuronal cells (Kim *et al.*, 2005).

There is a need to continue further research on the nutraceutical properties of *Prunus* species, as well as the different genotypes involved in each type of fruit, since the phytochemical make-up may vary, as mentioned earlier in this chapter. In addition, novel approaches may be used as tools for bioactive compound discovery and elucidating the mechanisms of action, such as exploring the human gene expression associated with chronic diseases. If a pattern of gene expression linked to a desired bioactive compound effect in the targeted cell is established, then post-treatment gene expression may be used to screen bioactive compounds from different species for their ability to induce the target phenotype (Evans and Guy, 2004).

13.5 Factors Affecting Bioactive Compounds in *Prunus* Species

The concentration of bioactive compounds can be affected by genetic factors, cultural practices, environmental conditions and storage processing treatments.

13.5.1 Genetic factors

As previously shown, large differences in total antioxidants, carotenoids and phenolics exist between *Prunus* species (Tables 13.2, 13.3 and 13.6). In addition, for any given species, the concentration of bioactives among varieties can also show dramatic differences. Analysis of 37 apricot varieties showed tenfold difference in the accumulation of carotenoids (Ruiz *et al.*, 2005). In another study, also in apricot, total phenolics varied from 30.3 to 742 mg of gallic acid equivalents per 100 g (Drogoudi *et al.*, 2008). In plum, varieties with dark purple-coloured skin showed 200% higher total phenolics than others (Rupasinghe *et al.*, 2006). The plum cvs. Black Beauty and Angeleno were especially rich in phenolics (Tomás-Barberán *et al.*, 2001). In peach, fruit showing more intense endocarp staining presented higher antioxidant capacity.

The effects induced by the rootstocks on controlling plant development and production are well known (Caruso *et al.*, 1996, 1997). In contrast, the influence on fruit bioactive compounds is quite variable. In some cases, no pronounced differences were observed (Drogoudi and Tsipouridis, 2007). On the other hand, some works reported that the levels of phytochemicals might be influenced significantly by rootstock (Giorgi *et al.*, 2005; Remorini *et al.*, 2008).

13.5.2 Cultural practices and environmental conditions

The accumulation of bioactive compounds is affected greatly by the fruit-ripening process. In most *Prunus* species, the concentration of ascorbic acid increases at advanced ripening stages (Table 13.9). In sweet cherry (cv. 4-70), high content of ascorbic acid was detected at early phases of development. As ripening proceded, ascorbic acid levels dropped and then increased progressively to reach maximum content at full maturity (Serrano *et al.*, 2005).

Carotenoid concentration also increases steadily during ripening in most stone fruit (Katayama *et al.*, 2006; Díaz-Mula *et al.*, 2008). Besides its contribution to full colour development, enzymatic carotenoid degradation of nectarines was shown to play a role in C-13 norisoprenoid aroma compound formation (Balderman *et al.*, 2005). The modifications in total phenolics showed no clear trend during ripening of peach, nectarine and plum cultivars (Tomás-Barberán *et al.*, 2001). In cherry, total phenolic compounds decrease during the early stages of development, but later on

skin colour development is associated with the accumulation of anthocyanins (Díaz-Mula *et al.*, 2008; Usenik *et al.*, 2008b). Plum colour development is also related directly to increased anthocyanidin content (Díaz-Mula *et al.*, 2008; Usenik *et al.*, 2008a), and delaying harvesting might lead to significant increases in these compounds and total antioxidant activity (by 10–20%) (Díaz-Mula *et al.*, 2008).

Sunlight has been shown to be associated with increased content of several bioactive compounds, including ascorbic acid (Lee and Kader, 2000) and anthocyanins. Shading of plum fruit resulted in poor colour development (Murray *et al.*, 2005). UV radiation is known to be associated with anthocyanin accumulation (Arakawa, 1993). Phenylalanine ammonia lyase, chalcone synthase and dihydroflavinol reductase (enzymes involved in phenylpropanoid metabolism) are induced by UV radiation (Tomás-Barberán and Espín, 2001). The differences between day and night temperatures have been shown to affect anthocyanin accumulation in plums (Tsuji *et al.*, 1983). Regulated water deficit resulted in reduced content of vitamin C and carotenoids in the fruit peel, while increasing anthocyanins and procyanidins (Buendía *et al.*, 2008).

Results comparing the effect of conventional or organic production on bioactive compounds in fruit and vegetables are variable. Organic peaches showed higher content of phenolic and ascorbic acid than conventionally produced fruit (Carbonaro and Mattera, 2001; Carbonaro *et al.*, 2002). Ascorbic acid, vitamin E and β-carotene were higher in organic plums grown on soil covered with natural meadow, while total phenolic content was higher in conventional plums (Lombardi-Boccia *et al.*, 2004). Bourn and Prescott (2002) concluded that further studies are required to understand fully the effect of

Table 13.9. Change in ascorbic acid content (mg/100 g) in some *Prunus* species during ripening (adapted from Lee and Kader, 2000).

Fruit (cultivar)	Green	Half-ripe	Ripe
Apricot (Tilton)	11.7	12.9	14.3
Peach (Elberta)	7.8	10.2	12.2

inorganic and organic fertilizers on the nutritional value of crops. Winter and Davis (2006) affirmed that it was not possible to ensure that organically grown products were nutritionally superior to those obtained by conventional agricultural techniques.

13.5.3 Storage and processing

In general, ascorbic acid is the compound showing the highest losses during storage and processing. In apricots a loss of around 26% ascorbic acid was found after 4 days at 20°C. However, the contribution of this compound to total antioxidant capacity in stone fruit is minimal compared with that of phenolic compounds. Several reports have shown that the content of total phenolics does not change or even increases during postharvest storage. Asami et al. (2002) reported no significant loss of total phenolics in clingstone peaches during cold storage. In dark purple and yellow plum varieties, increases in total phenolics were found even during storage (Díaz-Mula et al., 2008). Carotenoid content also increased in apricots during storage (Egea et al., 2007).

Although it seems clear that phenolic compound synthesis can proceed after harvest, the changes in phenolic compounds depend on the balance of de novo synthesis and degradation. Peaches are an excellent example of fruit susceptible to enzymatic browning, and this is a major problem encountered during handling and processing (Brandelli and Lopes, 2005). Improper storage conditions can lead in many cases to tissue browning due to phenolic compound degradation by the action of several enzymes, such as polyphenol oxidases and peroxidases. Refrigeration can reduce the activity of phenol oxidizing enzymes (Tomás-Barberán and Espín, 2001). In addition, fast cooling and low temperature storage reduce losses of other antioxidants such as ascorbic acid. Finally, gamma irradiation of almond skins increased the yield of total phenolic content, as well as enhanced antioxidant activity. Ultraviolet treatments have been shown to increase the accumulation of phenolic antioxidants.

A substantial amount of research literature has been published reporting the effects of processing on the nutritional quality of fruit and vegetables (Rickman et al., 2007). Various processing operations have profound effects on the level of bioactive compounds. Since bioactive compounds are generally present at higher concentration at the fruit surface, peel removal during processing results in significant losses. In addition, it increases leaching of soluble metabolites. Slicing favours ascorbic acid oxidation, the enzymatic degradation of phenolic compounds and tissue browning (McCarthy and Mattheus, 1994). Vitamin E is highly susceptible to oxidation during storage and processing. Blanching and other heat treatments might be useful to inactivate enzymes involved in phenolic compound oxidation, but might decrease the content of heat-sensitive compounds such as ascorbic acid. Carotenoids are quite heat stable (Nicoli et al., 1999) and their absorption could even be improved by heat treatments. Heating can lead to the formation of cis isomers of carotenoids (Lessin et al., 1997), which have lower relative provitamin A activity than the corresponding trans forms (Minguez-Mosquera et al., 2002).

Clingstone peaches are commonly preserved by thermal processing methods such as canning (Hong et al., 2004). Since simple phenolic compounds are water soluble, they are susceptible to leaching. Peach canning resulted in a 21% loss in total phenolics. The reduction was associated with migration of procyanidins into the canning syrup (Hong et al., 2004). The effects of freezing on bioactive compounds are less marked than those of other processing operations (Rickman et al., 2007).

Preservation methods usually are believed to reduce the level of antioxidants in food. While this might be true in some cases, the level of bioactive compounds in processed foods might still be higher than that found in many other food groups. In some cases, processing can even increase total antioxidant capacity. For instance, industrial drying (oven drying) of almond and roasting almond skins increased (twofold) the contents of phenolic compounds. In prunes, although drying results in marked losses of ascorbic acid and anthocyanins, total antioxidant capacity may be increased by formation of new antioxidants (Ryley and Kayda, 1993).

13.6 Future Prospects and Directions of Research and Development

Several years of active research have led to great advances in the area of fruit bioactive compounds, and the exploitation of these metabolites is increasing rapidly. Accumulating evidence suggests that having five to ten servings of fruit daily may play an important role in preventing some health disorders. However, it is important to interpret results cautiously, to avoid making premature and unjustified claims on health benefits when direct data are lacking.

Fresh *Prunus* species are significant contributors of bioactive compounds to the diet during the spring and summer. All the main groups of bioactive compounds (ascorbic acid, vitamin E, carotenoids, phenolic compounds) are represented to different degrees in most *Prunus* species.

In all cases, total antioxidant capacity seems to correlate best with the level of phenolics, suggesting that this is the predominant group. Plums rank top among *Prunus* species in antioxidant capacity, and black varieties have ORAC values close to those observed in fruit richest in antioxidants, such as blackcurrants and cranberries (Fig. 13.1). Almonds and cherries are rich in total phenolics and also show high antioxidant capacity. Peaches and nectarines show moderate levels of carotenoids and phenolics, but due to their high consumption, either fresh or processed, ultimately contribute significantly to dietary intake of bioactive compounds. Despite the general features of the main *Prunus* species, several works have shown that there is a wide variation among varieties for any given species. The accumulation of antioxidants is also variable within a single fruit, with the peel usually showing two- to 40-fold higher content of these substances than the other tissues. This may have implications where the peel is removed before consumption, a common practice in some parts of the Mediterranean.

Many *Prunus* species are also commonly used for processing. This might result in some losses of some antioxidants, such as ascorbic acid due to oxidation or leakage during heat treatments or long-term storage, or anthocyanins during drying. However, the contribution of bioactive compounds in frozen, canned or dried products still supersedes that found in many other food groups. In some cases (plum drying, almond roasting), processed foods could have more antioxidant capacity than fresh commodities due to the formation of new metabolites and a concentration effect. Consequently, intake of all forms of fruit and vegetables should be encouraged, as long as added ingredients, such as sugar, fat and salt, are not significant (Rickman *et al.*, 2007).

Complete understanding of many aspects of the bioactive compounds in fruit is lacking, and multiple aspects are to be learned. While the identification of these compounds and the study of differences among species have flourished, further combined research efforts (in chemical, biochemical, medical and agronomic areas) are required in order to:

- characterize the mechanisms that account for specific bioactive compound protection.
- study bioactive compound bioavailability, metabolism, dose–response and toxicity.
- start evaluating the interactions among bioactive compounds, since they are consumed almost exclusively in combinations.
- generate guidelines and protocols to bring some order related to the analytical determination of antioxidant capacity on food (Huang *et al.*, 2005).
- identify the molecular determinants of the distinct accumulation of bioactive compounds in external and internal fruit tissues or in different varieties.
- incorporate into breeding programmes traits related to higher production of bioactive compounds.
- determine more completely the influence of orchard variables on fruit bioactive compounds in order to design practices oriented to maximize accumulation of desirable metabolites.

In all, the positive attributes associated with bioactive compounds in fruit in general, and in *Prunus* species in particular, provide an opportunity that might be capitalized on.

References

Ames, B.N., Shigenaga, M.K. and Hagen, T.M. (1993) Oxidants, antioxidants, and the degenerative diseases of aging. *Proceedings of the National Academy of Sciences of the United States of Ameria* 90, 7915–7922.

Arakawa, O. (1993) Effect of ultraviolet light on anthocyanin synthesis in light colored sweet cherry cv. Nato Nishiki. *Journal of the Japanese Society for Horticultural Science* 62, 543–546.

Asami, D.K., Hong, Y.J., Barrett, D.M. and Mitchell, A.E. (2002) Processing-induced changes in total phenolics and procyanidins in clingstone peaches. *Journal of the Science of Food and Agriculture* 83, 56–63.

Balasundram, N., Sundram, K. and Samman, S. (2006) Fruits, vegetables and beverages are the major sources of phenolic compounds in the human diet. *Food Chemistry* 99, 191–203.

Balderman, S., Naim, M. and Fleischmann, P. (2005) Enzymatic carotenoid degradation and aroma formation in nectarines (*Prunus persica*). *Food Research International* 38, 833–836.

Bendich, A. (1993) Biological functions of carotenoids. In: Canfield, L.M., Krinsky, N.I. and Olson, J.A. (eds) *Carotenoids in Human Health*, Academy of Sciences, New York, pp. 61–67.

Bobe, G., Wang, B., Seeram, N.P., Nair, M.G. and Bourquin, L.D. (2006) Dietary anthocyanin-rich tart cherry extract inhibits intestinal tumorigenesis in APC(Min) mice fed suboptimal levels of sulindac. *Journal of Agricultural and Food Chemistry* 54, 9322–9328.

Bostock, R.M., Wilcox, S.M., Wang, G. and Adaskaveg, J.E. (1999) Suppression of *Monilinia fructicola* cutinase production by peach fruit surface phenolic acids. *Physiological and Molecular Plant Pathology* 54, 37–50.

Bourn, D. and Prescott, J. (2002) A comparison of the nutritional value, sensory qualities, and food safety of organically and conventionally produced foods. *Critical Reviews in Food Science and Nutrition* 42, 1–34.

Brady, C.J. (1993) Stone fruit. In: Seymour, G.B., Taylor, J.E. and Tucker, G.A. (eds) *Biochemistry of Fruit Ripening*. Chapman and Hall, London, pp. 379–404.

Brandelli, A. and Lopes, C.H.G.L. (2005) Polyphenoloxidase activity, browning potential and phenolic content of peaches during postharvest ripening. *Journal of Food Biochemistry* 29, 624–637.

Brummell, D.A., Dal Cin, V., Crisosto, C.H. and Labavitch, J.M. (2004a) Cell wall metabolism during maturation, ripening and senescence of peach fruit. *Journal of Experimental Botany* 55, 2029–2039.

Brummell, D.A., Dal Cin, V., Lurie, S., Crisosto, C.H. and Labavitch, J.M. (2004b) Cell wall metabolism during the development of chilling injury in cold-stored peach fruit: association of mealiness with arrested disassembly of cell wall pectins. *Journal of Experimental Botany* 55, 2041–2052.

Buendía, B., Allende, A., Nicolás, E., Alarcón, J.J. and Gil, M.I. (2008) Effect of regulated deficit irrigation and crop load on the antioxidant compounds of peaches. *Journal of Agricultural and Food Chemistry* 56, 3601–3608.

Cao, G., Booth, S.L., Sadowski, J.A. and Prior, R.L. (1998) Increases in human plasma antioxidant capacity following consumption of controlled diets high in fruits and vegetables. *American Journal of Clinical Nutrition* 68, 1081–1087.

Carbonaro, M. and Mattera, M. (2001) Polyphenoloxidase activity and polyphenol levels in organically and conventionally grown peach (*Prunus persica* L., cv. Regina bianca) and pear (*Pyrus communis* L., cv. Williams). *Food Chemistry* 72, 419–424.

Carbonaro, M., Mattera, M., Nicoli, S., Bergamo, P. and Cappelloni, M. (2002) Modulation of antioxidant compounds in organic vs. conventional fruit (peach, *Prunus persica* L., and pear, *Pyrus communis* L.). *Journal of Agricultural and Food Chemistry* 50, 5458–5462.

Caruso, T., Giovannini, D. and Liverani, A. (1996) Rootstock influences the fruit mineral, sugar and organic acid content of a very early ripening peach cultivar. *Journal of Horticultural Science and Biotechnology* 71, 931–937.

Caruso, T., Inglese, P., Sidari, M. and Sottile, F. (1997) Rootstock influences seasonal dry matter and carbohydrate content and partitioning in aboveground components of 'Flordaprince' peach trees. *Journal of the American Society for Horticultural Science* 122, 673–679.

Chen, C.Y. and Blumberg, J. (2008) *In vitro* activity of almond skin polyphenols for scavenging free radicals and inducing quinone reductase. *Journal of Agricultural and Food Chemistry* 56, 4427–4434.

Chen, C.Y., Milbury, P., Chung, S. and Blumberg, J. (2007) Effect of almond skin polyphenolics and quercetin on human LDL and apolipoprotein B-100 oxidation and conformation. *Journal of Nutritional Biochemistry* 18, 785–794.

Crisosto, C.H. and Mitchell, J.P. (2002) Preharvest factors affecting fruit and vegetable quality. In: Kader, A.A. (ed.) *Postharvest Technology of Horticultural Crops*. Publication 3311. Agriculture and Natural Resources, University of California, California, pp. 49–54.

Crisosto, C.H., Mitchell, F.G. and Johnson, R.S. (1995) Factors in fresh market stone fruit quality. *Postharvest News Information* 6, 17N–21N.

Crisosto, C.H., Johnson, R.S., DeJong, T. and Day, K.R. (1997) Orchard factors affecting postharvest stone fruit quality. *HortScience* 32, 820–823.

Crisosto, C.H., Mitcham, E.J. and Kader, A.A. (2009) Peach and nectarine. Recommendations for maintaining postharvest quality. (http://postharvest.ucdavis.edu/produce/producefacts/fruit/necpch.shtml, accessed 10 November 2009).

Cuttriss, A. and Pogson, B. (2004) Carotenoids. In: Davies, K. (ed.) *Plant Pigments and Their Manipulation*. Annual Plant Reviews, Vol 14. CRC Press, New Zealand, 352 pp. 57–91.

Dalla Valle, A.Z., Mignani, I., Spinardi, A., Galvano, F. and Ciappellano, S. (2007) The antioxidant profile of three different peaches cultivars (*Prunus persica*) and their short-term effect on antioxidant status in human. *European Food Research and Technology* 225, 167–172.

Davis, P.A. and Iwahashi, C.K. (2001) Whole almonds and almond fractions reduce aberrant crypt foci in a rat model of colon carcinogenesis. *Cancer Letters* 165, 27–33.

Depréz, S., Brezillon, C., Rabot, S., Philippe, C., Mila, I., Lapierre, C. and Scalbert, A. (2000) Polymeric proanthocyanidins are catabolized by human colonic microflora into low-molecular-weight phenolic acids. *Journal of Nutrition* 130, 2733–2738.

Di Mascio, P., Kaiser, S. and Sies, H. (1989) Lycopene as the most efficient biological carotenoid singlet oxygen quencher. *Archives of Biochemistry and Biophysics* 274, 532–538.

Díaz-Mula, H.M., Zapata, P.J., Guillén, F., Castillo, S., Martinez-Romero, D., Valero, D. and Serrano, M. (2008) Changes in physicochemical and nutritive parameters and bioactive compounds during development and on-tree ripening of eight plum cultivars: a comparative study. *Journal of the Science of Food and Agriculture* 88, 2499–2507.

Drogoudi, P.D. and Tsipouridis, C.G. (2007) Effects of cultivar and rootstock on the antioxidant content and physical characters of clingstone peaches. *Scientia Horticulturae* 115, 34–39.

Drogoudi, P.D., Vemmos, S., Pantelidis, G., Petri, E., Tzoutzoukou, C. and Karayiannis, I. (2008) Physical characters and antioxidant, sugar, and mineral nutrient contents in fruit from 29 apricot (*Prunus armeniaca* L.) cultivars and hybrids. *Journal of Agricultural and Food Chemistry* 56, 10754–10760.

Egea, M.I., Sanchez-Bel, P., Martinez-Madrid, M.C., Flores, F.B. and Romojaro, F. (2007) The effect of beta ionization on the antioxidant potential of 'Bulida' apricot and its relationship with quality. *Postharvest Biology and Technology* 46, 63–70.

Espín, J.C., García-Conesa, M.T. and Tomás-Barberán, F.A. (2007) Nutraceuticals: facts and fiction. *Phytochemistry* 68, 2986–3008.

Evans, W. and Guy, K. (2004) Gene expression as drug discovery tool. *Nature Genetics* 36(3), 214–215.

FAOSTAT (2009) (http://faostat.org, accessed 1 November 2009).

Food and Nutrition Board, Institute of Medicine (2000) Vitamin E. Dietary reference intakes for vitamin C, vitamin E, selenium, and carotenoids. National Academy Press, Washington, DC, pp. 186–283.

Formica, J.V. and Regelson, W. (1995) Review of the biology of quercetin and related bioflavonoids. *Food and Chemical Toxicology* 33, 1061–1080.

Gerster, H. (1997) The potential role of lycopene for human health. *Journal of the American College of Nutrition* 16, 109–126.

Giorgi, M., Capocasa, F., Scalzo, J., Murri, G., Battino, M. and Mezzetti, B. (2005) The rootstock effects on plant adaptability, production, fruit quality, and nutrition in the peach (cv. 'Suncrest'). *Scientia Horticulturae* 107, 36–42.

Gross, J. (1991) *Pigments in Vegetables: Chlorophylls and Carotenoids*. Van Nostrand Reinhold, New York, 351 pp.

Halliwell, B. and Gutteridge, J.M.C. (1995) The definition and measurement of antioxidants in biological systems. *Free Radical Biology and Medicine* 18, 125–126.

Hancock, R.D. and Viola, R. (2005) Improving the nutritional value of crops through enhancement of L-ascorbic acid (vitamin C) content: rationale and biotechnological opportunities. *Journal of Agricultural and Food Chemistry* 53, 5248–5257.

Haslam, E. (2007) Vegetable tannins – lessons of a phytochemical lifetime. *Phytochemistry* 68, 2713–2721.

Heo, H.J. and Lee, C.Y. (2005) Phenolic phytochemicals in cabbage inhibit amyloid β protein-induced neurotoxicity. *LWT – Food Science and Technology* 39, 331–337.

Herrmann, K. (1989) Occurrence and content of hydoxycinnamic and hydroxybenzoic acid compounds in foods. *Critical Reviews in Food Science and Nutrition* 28, 315–347.

Hong, Y.J., Barrett, D.M. and Mitchell, A.E. (2004) Liquid chromatography/mass spectrometry investigation of the impact of thermal processing and storage on peach procyanidins. *Journal of Agricultural and Food Chemistry* 52, 2366–2371.

Huang, D., Ou, B. and Prior, R.L. (2005) The chemistry behind antioxidant capacity assays. *Journal of Agricultural and Food Chemistry* 53, 1841–1856.

Hümmer, W. and Schreier, P. (2008) Analysis of proanthocyanidins. *Molecular Nutrition and Food Research* 52, 1381–1398.

Jacob, R.A., Spinozzi, G.M., Simon, V.A., Kelley, D.S., Prior, R.L., Hess-Pierce, B. and Kader, A.A. (2003) Consumption of cherries lowers plasma urate in healthy women. *Journal of Nutrition* 133, 1826–1829.

Jenkins, D.J., Kendall, C.W., Marchie, A., Parker, T.L., Connelly, P.W., Qian, W., Haight, J.S., Faulkner, D., Vidgen, E., Lapsley, K.G. and Spiller, G.A. (2002) Dose response of almonds on coronary heart disease risk factors: blood lipids, oxidized low-density lipoproteins, lipoprotein(a), homocysteine, and pulmonary nitric oxide: a randomized, controlled, crossover trial. *Circulation* 106, 1327–1332.

Kahlon, T.S. and Smith, G.E. (2007) *In vitro* binding of bile acids by bananas, peaches, pineapple, grapes, pears, apricots and nectarines. *Food Chemistry* 101, 1046–1051.

Kang, S.Y., Seeram, N.P., Nair, M.G. and Bourquin, L.D. (2003) Tart cherry anthocyanins inhibit tumor development in Apc(Min) mice and reduce proliferation of human colon cancer cells. *Cancer Letters* 194, 13–19.

Katayama, T., Nakayama, T.O.M., Lee, T.H. and Chichester, C.O. (2006) Carotenoid transformations in ripening apricots and peaches. *Journal of Food Science* 36, 804–806.

Kelley, D.S., Rasooly, R., Jacob, R.A., Kader, A.A. and Mackey, B.E. (2006) Consumption of Bing sweet cherries lowers circulating concentrations of inflammation markers in healthy men and women. *Journal of Nutrition* 136, 981–986.

Khanbabaee, K. and van Ree, T. (2001) Tannins: classification and definition. *Natural Product Reports* 18, 641–649.

Kim, D.O., Chun, K., Kim, Y.J., Moon, H. and Lee, C.Y. (2003a) Quantification of polyphenolics and their antioxidant capacity in fresh plums. *Journal of Agricultural and Food Chemistry* 51, 6509–6515.

Kim, D.O., Jeong, S.W. and Lee C.Y. (2003b) Antioxidant capacity of phenolic phytochemicals from various cultivars of plums. *Food Chemistry* 81, 321–326.

Kim, D.O., Heo, H.J., Kim, Y.J., Yang, H.S. and Lee, C.Y. (2005) Sweet and sour cherry phenolics and their protective effects on neuronal cells. *Journal of Agricultural and Food Chemistry* 53, 9921–9927.

Kim, H.J., Yu, M.-H. and Lee, I.-S. (2008) Inhibitory effects of methanol extract of plum (*Prunus salicina* L., cv. 'Soldam') fruits against benzo(α)pyrene-induced toxicity in mice. *Food and Chemical Toxicology* 46, 3407–3413.

Kong, J.M., Chia, L.S., Goh, N.K., Chia, T.F. and Brouillard, R. (2003) Analysis and biological activities of anthocyanins. *Phytochemistry* 64, 923–933.

Kopsell, D.A. and Kopsell, D.E. (2006) Accumulation and bioavailability of dietary carotenoids in vegetable crops. *Trends in Plant Science* 11, 499–507.

Kottová, N., Kazdová, L., Oliyarnyk, O., Veceta, R., Sobolova, L. and Ulrichova, J. (2004) Phenolics-rich extracts from *Silybum marianum* and *Prunella vulgaris* reduce a high-sucrose diet induced oxidative stress in hereditary hypertriglyceridemic rats. *Pharmacological Research* 50, 123–130.

Kris-Etherton, P.M., Hecker, K.D., Bonanome, A., Coval, S.M., Binkoski, A.E., Hilpert, K.F., Griel, A.E. and Etherton, T.D. (2002) Bioactive compounds in foods: their role in the prevention of cardiovascular disease and cancer. *American Journal of Medicine* 113, 71S–88S.

La Rue, J.H. and Johnson, R.S. (1989) *Peaches, Plums, and Nectarines – Growing and Handling for Fresh Market.* Postharvest Technology Research and Information Center, University of California, Davis, California, 252 pp.

Lazarus, S.A., Adamson, G.E., Hammerstone, J.F. and Schmitz, H.H. (1999) High-performance liquid chromatography/mass spectrometry analysis of proanthocyanidins in foods and beverages. *Journal of Agricultural and Food Chemistry* 47, 3693–3701.

Lea, M., Ibeh, C., desBordes, C., Vizzotto, M., Cisneros-Zevallos, L., Byrne, D., Okie, W. and Moyer, M. (2008) Inhibition of growth and induction of differentiation of colon cancer cells by peach and plum phenolic compounds. *Anticancer Research* 28, 2067–2076.

Lee, C.K., Park, K.K., Hwang, J.K., Lee, S.K. and Chung, W.Y. (2007) The extract of *Prunus persica* flesh (PPFE) attenuates chemotherapy-induced hepatotoxicity in mice. *Phytotherapy Research* 22, 223–227.

Lee, C.K., Park, K.K., Hwang, J.K., Lee, S.K. and Chung, W.Y. (2008) The pericarp extract of *Prunus persica* attenuates chemotherapy-induced acute nephrotoxicity and hepatotoxicity in mice. *Journal of Medicinal Food* 11, 302–306.

Lee, S.K. and Kader, A.A. (2000) Preharvest and postharvest factors influencing vitamin C content of horticultural crops. *Postharvest Biology and Technology* 20, 207–220.

Lessin, W.J., Catigani, G.L. and Schwartz, S.J. (1997) Quantification of *cis-trans* isomers of provitamin a carotenoids in fresh and processed fruits and vegetables. *Journal of Agricultural and Food Chemistry* 45, 3728–3732.

Lombardi-Boccia, G., Lucarini, M., Lanzi, S., Aguzzi, A. and Cappelloni, M. (2004) Nutrients and antioxidant molecules in yellow plums (*Prunus domestica* L.) from conventional and organic productions: a comparative study. *Journal of Agricultural and Food Chemistry* 52, 90–94.

Lurie, S. and Crisosto, C.H. (2005) Chilling injury in peach and nectarine. *Postharvest Biology and Technology* 37, 195–208.

McAnlis, G.T., McEneny, J., Pearce, J. and Young, I.S. (1999) Absorption and antioxidant effects of quercetin from onions, in man. *European Journal of Clinical Nutrition* 53, 92–96.

McCarthy, M.A. and Mattheus, R.H. (1994) Nutritional quality of fruits and vegetables subjected to minimal processes. In: Wiley, R.C. (ed.) *Minimally Processed Refrigerated Fruits and Vegetables*. Chapman and Hall, New York, pp. 313–326.

Manach, C., Scalbert, A., Morand, C., Remesy, C. and Jimenez, L. (2004) Polyphenols: food sources and bioavailability. *American Journal of Clinical Nutrition* 79, 727–747.

Manganaris, G.A., Vicente, A.R., Crisosto, C.H. and Labavitch, J.M. (2008) Cell wall modifications in chilling-injured plum fruit (*Prunus salicina*). *Postharvest Biology and Technology* 48, 77–83.

Mazza, G. (1995) Anthocyanins in grapes and grape products. *Critical Reviews in Food Science and Nutrition* 35, 341–371.

Miki, W. (1991) Biological functions and activities of animal carotenoids. *Pure and Applied Chemistry* 63, 141–146.

Minguez-Mosquera, M.I., Hornero-Mendez, D. and Perez-Galvez, A. (2002) Carotenoids and provitamin A in functional foods. In: Hurst, W.J. (ed.) *Methods of Analysis for Functional Foods and Nutraceuticals*. CRC Press, Boca Raton, London, New York, Washington, DC, pp. 101–157.

Möller, B. and Herrmann, K. (1983) Quinic acid esters of hydroxycinnamic acids in stone and pome fruit. *Phytochemistry* 22, 477–481.

Mulabagal, V., Lang, G.A., DeWitt, D.L., Dalavoy, S.S. and Nair, M.G. (2009) Anthocyanin content, lipid peroxidation and cyclooxygenase enzyme inhibitory activities of sweet and sour cherries. *Journal of Agricultural and Food Chemistry* 57, 1239–1246.

Murray, X.J., Holcroft, D.M., Cook, N.C. and Wand, S.J.E. (2005) Postharvest quality of 'Laetitia' and 'Songold' (*Prunus salicina* Lindell) plums as affected by preharvest shading treatments. *Postharvest Biology and Technology* 37, 81–92.

Nicoli, M.C., Anese, M. and Parpine, M. (1999) Influence of processing on the antioxidant properties of fruit and vegetables. *Trends in Food Science and Technology* 10, 94–100.

Noctor, G. and Foyer, C. (1998) Ascorbate and glutathione: keeping active oxygen under control. *Annual Review of Plant Physiology and Plant Molecular Biology* 49, 249–279.

Noratto, G., Porter, W., Byrne, D. and Cisneros-Zevallos, L. (2009) Identifying peach and plum polyphenols with chemopreventive potential against estrogen independent breast cancer cells. *Journal of Agricultural and Food Chemistry* 57, 5219–5226.

Olsson, M.E., Gustavsson, K.E., Andersson, S., Nilsson, A. and Duan, R.D. (2004) Inhibition of cancer cell proliferation *in vitro* by fruit and berry extracts and correlations with antioxidant levels. *Journal of Agricultural and Food Chemistry* 52, 7264–7271.

Pinelo, M., Rubilar, M., Sineiro, J. and Núñez, M.J. (2004) Extraction of antioxidant phenolics from almond hulls (*Prunus amygdalus*) and pine sawdust (*Pinus pinaster*). *Food Chemistry* 85, 267–273.

Puri, H.S. (2003) *Rosayana Ayurvedic Herbs for Longevity and Rejuvenation*. Taylor and Francis, London, pp. 71–73.

Rao, A.V. and Rao, L.G. (2007) Carotenoids and human health. *Pharmacological Research* 55, 207–216.

Rawat, M.S.M., Pant, G., Prasad, D., Joshi, R.K. and Pande, C.B. (1998) Plant growth inhibitors from *Prunus armeniaca*. *Biochemical Systematics and Ecology* 26, 13–23.

Rawat, M.S.M., Prasad, D., Joshi, R.K. and Pant, G. (1999) Proanthocyanidins from *Prunus armeniaca* roots. *Phytochemistry* 50, 321–324.

Remorini, D., Tavarini, S., Degl'Innocenti, E., Loreti, F., Massai, M. and Guidi, L. (2008) Effect of rootstocks and harvesting time on the nutritional quality of peel and flesh of peach fruits. *Food Chemistry* 110, 361–367.

Rice-Evans, C. and Miller, N.G.P. (1996) Structure–antioxidant activity relationships of flavonoids and phenolic acids. *Free Radical Biology and Medicine* 20, 933–956.

Rickman, J.C., Barrett, D.M. and Bruhn, C.M. (2007) Nutritional comparison of fresh, frozen and canned fruits and vegetables. Part 1. Vitamins C and B and phenolic compounds. *Journal of the Science of Food and Agriculture* 87, 930–944.

Robards, K. (2003) Strategies for the determination of bioactive phenols in plants, fruit and vegetables. *Journal of Chromatography* 1000, 657–691.

Rodríguez-Amaya, D.B. (2001) *A Guide to Carotenoid Analysis in Foods*. ILSI Press, International Life Sciences Institute, One Thomas Circle, NW Washington, DC.

Romani, R.J. and Jennings, W.G. (1971) Stone fruits. In: Hulme, A.C. (ed.) *The Biochemistry of Fruits and Their Products*, 2. Academic Press, New York, pp. 411–436.

Ruiz, D., Egea, J., Tomás Barberán, F.A. and Gil, M.I. (2005) Carotenoids from new apricot (*Prunus armeniaca* L.) varieties and their relationship with flesh and skin color. *Journal of Agricultural and Food Chemistry* 53, 6368–6374.

Ruiz, D., Reich, M., Bureau, S., Renard, C.M.G. and Audergon, J.M. (2008) Application of reflectance colorimeter measurements and infrared spectroscopy methods to rapid and non-destructive evaluation of carotenoids content in apricot (*Prunus armeniaca* L.). *Journal of Agricultural and Food Chemistry* 56, 4916–4922.

Rupasinghe, H.P.V., Jayasankar, S. and Lay, W. (2006) Variation in total phenolics and antioxidant capacity among European plum genotypes. *Scientia Horticulturae* 108, 243–246.

Ryley, J. and Kayda, P. (1993) Vitamins in thermal processing. *Food Chemistry* 49, 119–129.

Sandmann, G. (2001) Genetic manipulation of carotenoid biosynthesis, strategies, problems and achievements. *Trends in Plant Science* 6, 14–17.

Seeram, N.P., Momin, R.A., Nair, M.G. and Bourquin, L.D. (2001) Cyclooxygenase inhibitory and antioxidant cyanidin glycosides in cherries and berries. *Phytomedicine* 8, 362–369.

Serrano, M., Guillen, F., Martinez-Romero, D., Castillo, S. and Valero, D. (2005) Chemical constituents and antioxidant activity of sweet cherry at different ripening stages. *Journal of Agricultural and Food Chemistry* 53, 2741–2745.

Seymour, E.M., Singer, A.A.M., Kirakosyan, A., Urcuyo-Llanes, D.E., Kaufman, P.B. and Bolling, S.F. (2008) Altered hyperlipidemia, hepatic steatosis and hepatic PPARs in rats with intake of tart cherry. *Journal of Medicinal Food* 11, 252–259.

Seymour, E.M., Lewis, S.K., Urcuyo-Llanes, D.E., Kirakosyan, A., Kaufman, P.B. and Bolling, S.F. (2009) Regular tart cherry intake alters abdominal adiposity, adipose gene transcription and inflammation in obesity-prone rats fed a high fat diet. *Journal of Medicinal Food* 12, 935–942.

Shukitt-Hale, B., Kalt, W., Carey, A., Vinqvist-Tymchuk, M., McDonald, J. and Joseph, J. (2009) Plum juice, but not dried plum powder, is effective in mitigating cognitive deficits in aged rats. *Nutrition* 25, 567–573.

Smirnoff, N. and Wheeler, G.L. (2000) Ascorbic acid in plants, biosynthesis and function. *Critical Reviews in Biochemistry and Molecular Biology* 35, 291–314.

Spiller, G.A., Jenkins, D.A., Bosello, O., Gates, J.E., Cragen, L.N. and Bruce, B. (1998) Nuts and plasma lipids: an almond-based diet lowers LDL-C while preserving HDL-C. *Journal of the American College of Nutrition* 17, 285–290.

Stahl, W. and Sies, H. (1997) Antioxidant defense: vitamin E and C and carotenoids. *Diabetes* 46, S14–S18.

Sun, J., Chu, Y.F., Wu, X. and Liu, R. (2002) Antioxidant and antiproliferative activities of common fruits. *Journal of Agricultural and Food Chemistry* 50, 7449–7454.

Tall, J.M., Seeram, N.P., Zhao, C., Nair, M.G., Meyer, R.A. and Raja, S.N. (2004) Tart cherry anthocyanins suppress inflammation-induced pain behavior in rat. *Behavioural Brain Research* 153, 181–188.

Tomás-Barberán, F. and Espín, J.C. (2001) Phenolic compounds and related enzymes as determinants of quality in fruits and vegetables. *Journal of the Science of Food and Agriculture* 81, 853–876.

Tomás-Barberán, F.A., Gil, M.I., Cremin, P., Waterhouse, A.L., Hess-Pierce, B. and Kader, A.A. (2001) HPLC-DAD-ESIMS analysis of phenolic compounds in nectarines, peaches, and plums. *Journal of Agricultural and Food Chemistry* 49, 4748–4760.

Tourjee, K.R., Barrett, D.M., Romero, M.V. and Gradziel, T.M. (1998) Measuring flesh color variability among processing clingstone peach genotypes differing in carotenoid composition. *Journal of the American Society for Horticultural Science* 123, 433–437.

Traber, M.G. (2001) Does vitamin E decrease heart attack risk? Summary and implications with respect to dietary recommendations. *Journal of Nutrition* 131, 395S–397S.

Tsuji, M., Harakawa, M. and Komiyama, Y. (1983) Inhibition of increase of pulp colour and phenylalanine ammonia-lyase activity in plum fruit at high temperature (30°C). *Journal of the Japanese Society for Food Science and Technology* 30, 688–692.

USDA (2004) USDA database for the proanthocyanidin content of selected foods prepared by nutrient data laboratory. Beltsville Human Nutrition Research Center, Agricultural Research Service, US Department of Agriculture.

USDA (2007a) *Oxygen Radical Absorbance Capacity (ORAC) of Selected Foods*. US Department of Agriculture, Agricultural Research Service, Beltsville Human Nutrition Research Center Nutrient Data Laboratory, Maryland, 36 pp.

USDA (2007b) *USDA Database for the Flavonoid Content*. US Department of Agriculture, Agricultural Research Service, Beltsville Human Nutrition Research Center, Nutrient Data Laboratory, Maryland, 128 pp.

USDA (2009) *Composition of Foods, Raw, Processed, Prepared. USDA National Nutrient Database for Standard Reference, Release 20*. USDA-ARS, Beltsville Human Nutrition Research Center, Nutrient Data Laboratory, Beltsville, Maryland (http://www.ars.usda.gov/nutrientdata, accessed March 2009).

Usenik, V., Fabčič, J. and Štampar, F. (2008a) Sugars, organic acids, phenolic composition and antioxidant activity of sweet cherry (*Prunus avium* L.). *Food Chemistry* 107, 185–192.

Usenik, V., Kastelec, D., Veberič, R. and Štampar, F. (2008b) Quality changes during ripening of plums (*Prunus domestica* L.). *Food Chemistry* 111, 830–836.

Valpuesta, V. and Botella, M.A. (2004) Biosynthesis of L-ascorbic acid in plants, new pathways for an old antioxidant. *Trends in Plant Science* 9, 573–577.

Vinson, J.A., Su, X., Zubik, L. and Bose, P. (2001) Phenol antioxidant quantity and quality in foods: fruits. *Journal of Agricultural and Food Chemistry* 49, 5315–5321.

Vizzotto, M., Cisneros-Zevallos, L., Byrne, D., Ramming, D. and Okie, W. (2007) Large variation found in the phytochemical and antioxidant activity of peach and plum germplasm. *Journal of the American Society for Horticultural Science* 132, 334–340.

Wang, H., Cao, G. and Prior, R.L. (1997) Oxygen radical absorbing capacity of anthocyanins. *Journal of Agricultural and Food Chemistry* 45, 304–309.

Wang, H., Nair, M.G., Strasburg, G.M., Chang, Y.C., Booren, A.M., Gray, J.I. and DeWitt, D.L. (1999) Antioxidant and antiinflammatory activities of anthocyanins and their aglycon, cyanidin, from tart cherries. *Journal of Natural Products* 62, 294–296.

Winter, C. and Davis, S.F. (2006) Organic foods. *Journal of Food Science* 71, R117–R124.

Wu, X., Beecher, G.R., Holden, J.M., Haytowitz, D.B., Gebhardt, S.E. and Prior, R.L. (2004a) Lipophilic and hydrophilic antioxidant capacities of common foods in the United States. *Journal of Agricultural and Food Chemistry* 52, 4026–4037.

Wu, X., Gu, L., Golden, J., Haytowitz, D.B, Gebhardt, S.E., Beecher, G. and Prior, R.L. (2004b) Development of a database for total antioxidant capacity in foods: a preliminary study. *Journal of Food Composition and Analysis* 14, 407–422.

Yang, Y. and Gallaher, D. (2005) Effect of dried plums on colon cancer risk factors in rats. *Nutrition and Cancer* 53, 117–125.

14   *Ribes* and *Rubus* [Blackberry, Currants and Raspberry, etc.]

Jordi Giné Bordonaba and Leon A. Terry

14.1 Introduction

A great deal of berry fruit which are commercially available in fresh or processed form belong to the *Ribes* and *Rubus* genera, which encompass several species including blackberry (*Rubus* spp.), black raspberry (*Ru. occidentalis* L.; *Ru. leucodermis* Torr. & A. Grey), red raspberry (*Ru. idaeus* L.), blackcurrant (*Ribes nigrum* L.), redcurrant (*Ri. rubrum* L.), white currant (*Ri. glandulosum* Grauer), arctic bramble (*Ru. arcticus* L.), boysenberries (*Ru. ursinus* × *idaeus*), cloudberries (*Ru. chamaemorus* L.), gooseberry (*Ri. uva-crispa* L.), loganberry (*Ru. loganobaccus* L.H. Bailey), etc. Plants from both *Ribes* and *Rubus* are generally shrubs of small to medium size and usually are characterized by giving attractive small fruit rich in potential health-related compounds. As with many other fruit and vegetables, they represent an important source of micro- and macronutrients, including fibre, sugars, vitamins, minerals, etc.; however, most of their health-promoting properties have been associated largely with their high levels of bioactive compounds (namely, ascorbic acid, phenolic acids and flavonoids including anthocyanins) with known antioxidant capacity (Table 14.1). Nowadays, scientific evidence suggests that increased production or ineffective scavenging of reactive oxygen species (ROS) may play a crucial role in the development of certain pathologic conditions, especially cancer or chronic diseases (Wolfe *et al.*, 2008). Consumption of fruit and vegetables is likely to be responsible for decreasing the severity or incidence of these diseases, by reducing oxidative stress and modulating signal transduction pathways involved in cell proliferation and survival (Wolfe *et al.*, 2008). In this context, several studies to date have shown higher antioxidant activity, in cell-free systems, within *Ribes* and *Rubus* fruit than in many other food sources. Accordingly, numerous health-promoting properties have been attributed to fruit from either the *Ribes* or *Rubus* genera during the past few decades. Rather than health-related benefits due to single compounds, it is believed that most of their benefits come from the additive or synergistic effect from several bioactives (Seeram, 2008) present in these berries.

The present chapter describes the different key bioactive compounds of fruit from *Ribes* and *Rubus* species and discusses the latest scientific evidence on the health-promoting properties of these fruit.

14.1.1 *Ribes*

The genus *Ribes* embraces the shrubs of both currants and gooseberries and belongs to the family *Grossulariaceae*. It includes more than 150 described species of bushes that are

Table 14.1. Nutrient and mineral composition of main *Ribes* and *Rubus* berries.

	Blackcurrant	Blackberries	Raspberries	Redcurrants
Dry matter (%)	20	12	13	16
Sugars (g)	15.4	9.6	12	14
Organic acids	–	–	–	–
Proteins (g)	1.4	1.4	0.9	1.4
Fibre (g)	–	5	7	4
Vitamin C (mg)	85–500	11–28	25	22–53
Anthocyanins (mg)	152–400	160	–	1.1–136
Antioxidant activity (AU)	–	56.6–71.8	–	40–63

native throughout Northern Europe, North America and Asia, and in mountainous areas of north–west Africa and South America (Brennan, 2005). Five main *Ribes* subgenera are grown for their fruit and these include blackcurrants, redcurrants, white currants, gooseberries and jostaberries (Brennan, 2005). Historically, blackcurrants, for instance, were imported from Holland to England in 1611 by Tradescant (Brennan, 2005). Later, during the 18th century, blackcurrants were domesticated in Eastern Europe and spread over Russia, and in the particular case of the UK, blackcurrant cultivation was especially encouraged by the British government during the Second World War, due, in part, to the fruit's suitability to the UK weather and the berries' high content of vitamin C, as no other sources of this vitamin were really available. In those days, the major part of the production was made into blackcurrant syrup. Nowadays blackcurrants are the leading *Ribes* crop worldwide and are still mainly processed rather than used fresh, due to their strong flavour (Brennan *et al.*, 1997; Barney and Hummer, 2005). Redcurrants are grown to be eaten fresh or to be processed into juice and conserves, while white currants provide the greatest yields and are freshly consumed or used for baby food processing (Barney and Hummer, 2005). Similarly, gooseberries are cultivated mainly for the fresh market and for inclusion in jams and pies (Brennan, 2005).

Ribes fruit have been appreciated for centuries as a nutritious food. Berries, including blackcurrants, redcurrants, etc., may be considered as an ancient food in northern Europe. The use of blackcurrant fruit as a herbal medicine emerged in the Middle Ages. In the 16th century, European herbalists started to recommend *Ribes* berries or their syrups for the treatment of several illnesses, including bladder stones and liver disorders, coughs and lung ailments. However, it was not until the 18th century that the use of *Ribes* fruit became widespread among European herbalists and physicians. Several berry-derived products were employed for treatment of numerous intestinal conditions, typhoid fever, gout and rheumatism, and for infections of the mouth, skin and urinary tract.

Economically, currant and gooseberry production around the world is based mainly in Asia, Europe and Oceania. In Asia, the Russian Federation represents more than 99% of the total production, while Poland and Germany together produce more than 70% of the European crop (Table 14.2). New Zealand represents almost the totality of the crop harvested in Oceania; 6110 t out of the 7110 t harvested during 2005. Currently, North American acreage for currants and gooseberries is increasing and this is due in part to both the lifting of the legislation that prohibited blackcurrant cultivation in several states and the release of new resistant varieties. This said, the crop still has not reached the popularity it currently has in Europe, and such is reflected by the paucity of research undertaken on these berries by the USA as compared with that on other berry fruits.

Table 14.2. Production (1000 t) of currants and gooseberries for the main producing countries within the European Union.

Country	1990	1995	2000	2005
Poland	165.3	196.9	175.3	203.5
Germany	230.7	252.3	246.5	186
UK	18.9	21.9	13.8	22.4
Austria	25.8	19.7	24.7	21.1
Czech Republic	–	32.5	24.9	18.6
Hungary	23.4	16	16.5	13.4
France	7.5	11.3	8.4	13
Total	*471.6*	*550.6*	*510.1*	*478*

Source: FAOSTAT (2008).

14.1.2 *Rubus*

Rubus is a broad genus of flowering plants in the family *Rosaceae*, subfamily *Rosoideae*. It is found worldwide, except in desert areas, but is present mainly in the northern hemisphere. The important cultivated species from the genus include the European red raspberry (*Ru. idaeus* ssp. *vulgatus*), the North American red raspberry (*Ru. strigosus* (Michx.) Maxim.), the eastern North American blackberry (*Ru. occidentalis* L.) and the Andean blackberry hybrid (*Ru. glaucus* Benth., *Ru. adenotrichus* Schlech.) (Mertz *et al.*, 2007). Blackberry fruit, for instance, tend to be first green and red to brown-red and hard when immature but turn into black-coloured and juicy fruit as the berry ripens. Most of the commercial blackberry production occurs in the USA, but with appreciable amounts also grown in the UK and New Zealand (Dai *et al.*, 2007) (Fig. 14.1). In the USA, the Pacific Coast region produces *c.*80% of the total national production. Raspberries are the most important species in the genus *Rubus*, although considered one of the most perishable fruit, with the risk of decay, colour darkening and changes in flavour occurring rapidly after harvest (Krüger *et al.*, 2003). Historically, raspberries and other *Rubus* species can be tracked to ancient times. However, the first written mention of raspberries can only be found in an English book on herbal medicine dated 1548. Juices and extracts from *Rubus* fruit were used extensively in the 16th century for the treatment of several conditions, but mainly for the treatment of infections (Dai *et al.*, 2007).

14.2 Identity and Role of Bioactives

Ribes and *Rubus* species are characterized by their high anthocyanin and phenolic contents, as well as other bioactives (e.g. ascorbic acid). Berries from these species have some of the highest antioxidant capacities of any common fruit. Generally, it is accepted that total phenolic content (TP) and total flavonoids (TF) are well correlated with antioxidant capacity, as determined by any of the standard assays such as FRAP or ORAC (see Chapters 18 and 19 of this volume for further information). For instance, TP and TF of four different raspberry cultivars were found to be strongly correlated with antioxidant capacity ($R^2 = 0.988$ and $R^2 = 0.996$, respectively) in a study conducted by Liu *et al.* (2002).

14.2.1 Polyphenolic compounds

Phenolic compounds are widely distributed in both *Ribes* and *Rubus* species, ranging from simple moieties with an individual hydroxylated aromatic ring to complex polymeric molecules (Harborne and Williams, 1995). Phenolics are plant secondary metabolites that are synthesized and accumulate in the plant via processes that are controlled endogenously or regulated by exogenous factors

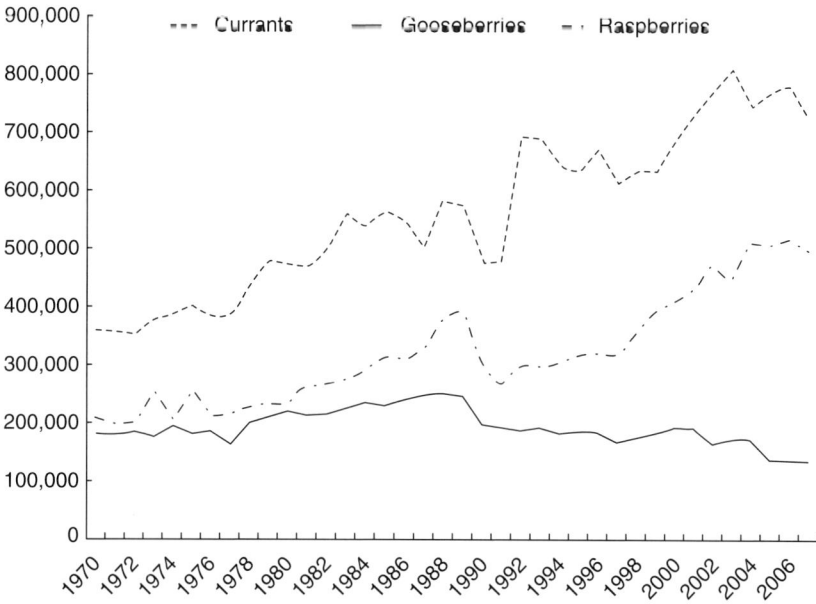

Fig. 14.1. Worldwide production (tonnes) of certain *Ribes* and *Rubus* species from 1970 to 2007 (data from FAOSTAT, 2008).

such as environmental conditions (namely, temperature and light) (Dixon and Paiva, 1995). Indeed, the diverse range of phenolics found in berries from *Ribes* or *Rubus* species are responsible for the fruit's astringency, bitterness, colour and flavour, and also for the oxidative stability of their derived products. Due in part to the high concentrations of phenolic compounds in *Ribes* and *Rubus* fruit, a great deal of research has investigated the different polyphenolic fractions of these berries. However, it is important to notice that most of this information is focused on blackberries, raspberries and blackcurrants (Zadernowski *et al.*, 2005; Mertz *et al.*, 2007; Giné Bordonaba and Terry, 2008), whereas little is known about other minor *Ribes* and *Rubus* berries.

Generally, polyphenol content may be estimated by adaptations of the standard Folin–Ciocalteu method. Briefly, this method is based on the reduction of a phosphowolframate–phosphomolibdate complex by phenolic compounds, resulting in blue reaction products (see Chapter 18 of this volume for further information), which are then measured spectrophotometrically. By using this method with any of its reported modifications, many papers refer to the high total phenolic content, expressed as gallic acid equivalents (GAE), of different *Ribes* and *Rubus* fruit, often in comparison with other fruit and vegetables (Fig. 14.2). That said, the values for TP content found in the literature are often controversial, since the reported variation in the content of total phenolics between berry types is due mainly to differences in cultivar, agroclimatic and growing conditions and, finally, to differences in the methods used in each study (Giné Bordonaba and Terry, 2008).

Flavonoids

Flavonoids are a group of polyphenolic compounds that can be divided into different subclasses such as flavanols, flavonols, flavones, flavanones, isoflavones and anthocyaninins (Pinent *et al.*, 2008). Most berries from the *Ribes* and *Rubus* genera are rich sources of these compounds, with blackcurrants, for instance, containing *c.*tenfold greater flavonol concentrations than other berries (Häkkinen *et al.*, 1999). Some of these flavonols (i.e.

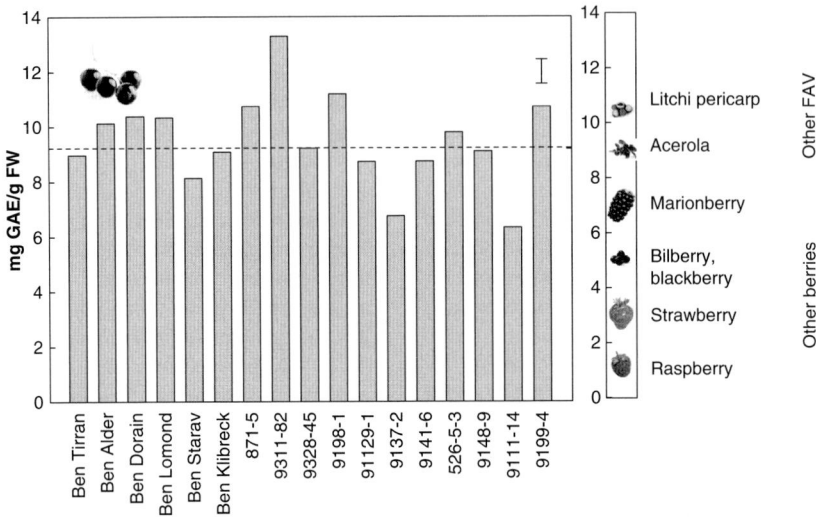

Fig. 14.2. Concentration of total phenolics (mg/GAE g), measured by the Folin–Ciocalteu method, in 17 UK grown blackcurrant cultivars, compared with those of other fruit and vegetables (FAV). Results are expressed on a fresh weight (FW) basis and the bar indicates LSD value ($P < 0.05$) (based on Giné Bordonaba and Terry, 2008).

quercetin) are found ubiquitously in most *Ribes* berries, accounting for 46.3, 39.6, 29.8, 14.3 and 10.1% of the total phenolic and flavonol fraction in green gooseberry, redcurrant, blackcurrant, green currant and white currant, respectively (Häkkinen *et al.*, 1999). In contrast, *Rubus* species, including red raspberry, arctic bramble and cloudberry, had no more than 2.5% of quercetin (Fig. 14.4). In the same study, relative concentrations of myricetin and kaempferol ranged from 0 to 9.4% for all the above-mentioned *Ribes* and *Rubus* species (Häkkinen *et al.*, 1999).

Anthocyanins are considered one of the main plant pigments visible to the human eye. They belong to the flavonoid class and they usually conjugate to form glycosides of polyhydroxy and polymetoxy derivatives of 2-phenylbenzopyrilium or flavylium salts. The differences between various anthocyanins relate to the number of hydroxyl groups, the nature and number of sugars attached to the molecule, the position of these sugars, usually C3 and less frequently at C5 or C7, and the nature and number of aliphatic or aromatic acids attached to the sugars (Fig. 14.3b). In fruit, anthocyanins are found generally in the external layers of the skin (hypodermis), and within the skin these compounds are

encountered in vacuoles of different sizes. Anthocyanins in berries including raspberry, blackberry and blackcurrant have been studied extensively during the past years, not only for their interest as natural colorants but also for their health-promoting properties. Indeed, blackcurrant extracts have hitherto acted as an important model for understanding anthocyanin absorption in both humans and animals (Netzel *et al.*, 2001; Nielsen *et al.*, 2003; Wu *et al.*, 2005; Matsumoto *et al.*, 2006). A simple survey of the literature, using any of the available search engines, reveals that the number of published articles referring to blackcurrant anthocyanins has increased exponentially during the last 15 years (from one in 1991 up to 15 articles in 2008), and similar results can be obtained if searching for other *Ribes* or *Rubus* species. All studies so far have concluded that four major anthocyanins (Fig. 14.3b) (namely, cyanidin-3-glucoside, cyanidin-3-rutinoside, delphinidin-3-glucoside and delphinidin-3-rutinoside) constitute almost 90% of the total anthocyanin content of blackcurrants (Häkkinen *et al.*, 1999; Anttonen and Karjalainen, 2006; Manhita *et al.*, 2006; Rubinskiene *et al.*, 2006; Jordheim *et al.*, 2007; Giné Bordonaba and Terry, 2008). Other anthocyanins, including peonidin-3-rutinoside and

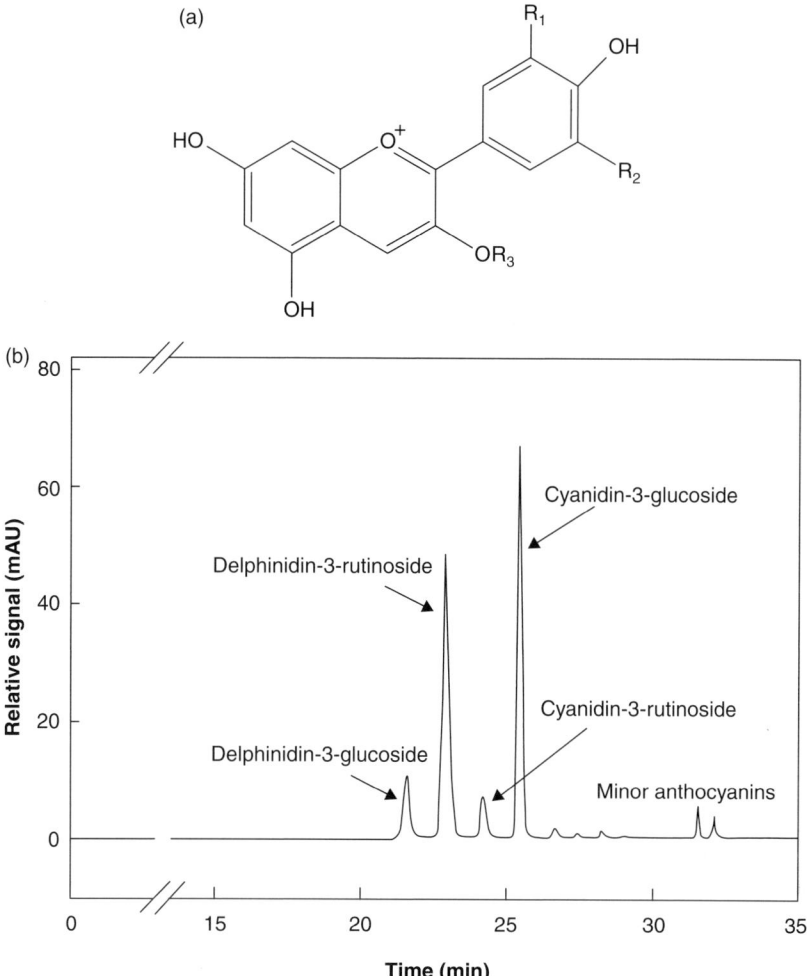

Fig. 14.3. (a) Chemical structure of *Ribes* and *Rubus* anthocyanins and their occurrence in selected berries. (b) Chromatographic profile of major blackcurrant anthocyanins identified by HPLC coupled to DAD (Giné Bordonaba *et al.*, 2010).

malvidin-3-glucoside, have been reported in blackcurrants, but in lesser amounts (Frøytlog *et al.*, 1998; Slimestad and Solheim, 2002; Fig. 14.3b). In raspberries, cyanidin-3-sophoroside, cyanidin-3-glucoside, cyanidin-3-rutinoside, cyanidin-3-glucorutinoside, pelargonidin-3-sophoroside and pelargonidin-3-glucoside have all been identified (De Ancos *et al.*, 2000; Fan-Chiang and Wrolstad, 2005). Marionberry anthocyanins include cyanidin-3-glucoside, cyanidin-3-rutinoside, pelargonidin-3-glucoside and acylated cyanidin-based anthocyanins (Wu *et al.*, 2004). Anthocyanin

concentration in raspberries, as determined by the pH differential method (see Chapters 18 and 19 of this volume), ranged from 1.7 to 576 μg/g FW, depending on the cultivar (Liu *et al.*, 2002). Blackberries contain cyanidin-3-galactoside, cyanidin-3-glucoside, cyanidin-3-arabinoside, pelargonidin-3-glucoside, cyanidin-3-xyloside and malvidin-3-glucoside, cyanidin-3-glucoside being the dominant anthocyanin (Goiffon *et al.*, 1991; Fan-Chiang and Wrolstad, 2005). Indeed, anthocyanin distribution in *Ribes* and *Rubus* is species dependent. Certain European gooseberry cultivars

Compound	R_1	R_2	R_3	Berry	Reference
Delphinidin-3-glucoside	OH	OH	D-Glucose	Blackcurrant, gooseberry	Frøytlog et al. (1998); Wu et al. (2004); Jordheim et al. (2007); Giné Bordonaba and Terry (2008)
Delphinidin-3-rutinoside	OH	OH	D-Glucose-L-rhamnose	Blackcurrant, gooseberry	Frøytlog et al. (1998); Wu et al. (2004); Jordheim et al. (2007); Giné Bordonaba and Terry (2008)
Cyanidin-3-glucoside	OH	H	D-Glucose	Blackcurrant, raspberry, blackberry, boysenberry, marionberry, gooseberry, redcurrant	Goiffon et al. (1991); Frøytlog et al. (1998); Cooney et al. (2004); Wu et al. (2004); Jordheim et al. (2007); Mertz et al. (2007); Giné Bordonaba and Terry (2008)
Cyanidin-3-rutinoside	OH	H	D-Glucose-L-rhamnose	Blackcurrant, raspberry, blackberry, boysenberry, marionberry, gooseberry, redcurrant	Goiffon et al. (1991); Frøytlog et al. (1998); Cooney et al. (2004); Wu et al. (2004); Jordheim et al. (2007); Mertz et al. (2007); Giné Bordonaba and Terry (2008)
Cyanidin-3-arabinoside	OH	H	D-Arabinose	Blackberry	Goiffon et al. (1991)
Pelargonidin-3-glucoside	H	H	D-Glucose	Marionberry, blackberry	Goiffon et al. (1991); De Ancos et al. (2000); Proteggente et al. (2002); Wu et al. (2004); Fan-Chiang and Wrolstad (2005)
Pelargonidin-3-rutinoside	H	H	D-Glucose-L-rhamnose	Blackcurrant	Wu et al. (2004)
Peonidin-3-rutinoside	OCH3	H	D-Glucose-L-rhamnose	Blackcurrant, gooseberry	Frøytlog et al. (1998); Slimestad and Soldheim (2002); Wu et al. (2004); Jordheim et al. (2007)
Peonidin-3-glucoside	OCH3	H	D-Glucose	Blackcurrant, gooseberry	Wu et al. (2004); Jordheim et al. (2007)
Malvidin-3-glucoside	OCH3	OCH3	D-Glucose	Blackcurrant, blackberry	Goiffon et al. (1991); Frøytlog et al. (1998); Slimestad and Soldheim (2002); Wu et al. (2004); Jordheim et al. (2007)
Cyanidin-3-sophoroside	OH	H	D-Glucose-L-glucose	Raspberry, redcurrant, boysenberry	Cooney et al. (2004); Wu et al. (2004)
Cyanidin-3-glucorutinoside	OH	H	D-glucose-L-rhamnosyl-D-glucose	Raspberry, blackberry	Goiffon et al. (1991); Gónzalez et al. (2003)
Pelargonidin-3-sophoroside	H	H	D-Glucose-L-glucose	Raspberry	De Ancos et al. (2000); Proteggente et al. (2002); Fan-Chiang and Wrolstad (2005)
Cyanidin-3-galactoside	OH	H	D-Galactose	Blackberry	Goiffon et al. (1991)
Cyanidin-3-arabinoside	OH	H	D-Arabinose	Blackberry	Goiffon et al. (1991)
Cyanidin-3-sambubioside	OH	H		Redcurrant	Wu et al. (2004)
Cyanidin-3-xyloside	OH	H	D-Xylose	Blackberry, gooseberry	Wu et al. (2004)
Petunidin-3-glucoside	OH	OCH3	D-Glucose	Blackcurrant	Wu et al. (2004)
Delphinidin-3-xyloside	OH	OH	D-Xylose	Blackcurrant	Wu et al. (2004)
Petunidin-3-rutinoside	OH	OCH3	D-Glucose-L-rhamnose	Blackcurrant	Wu et al. (2004)
Malvidin-3-rutinoside	OCH3	OCH3	D-Glucose-L-rhamnose	Blackcurrant	Frøytlog et al. (1998); Slimestad and Solheim (2002)

Fig. 14.3. *Continued*

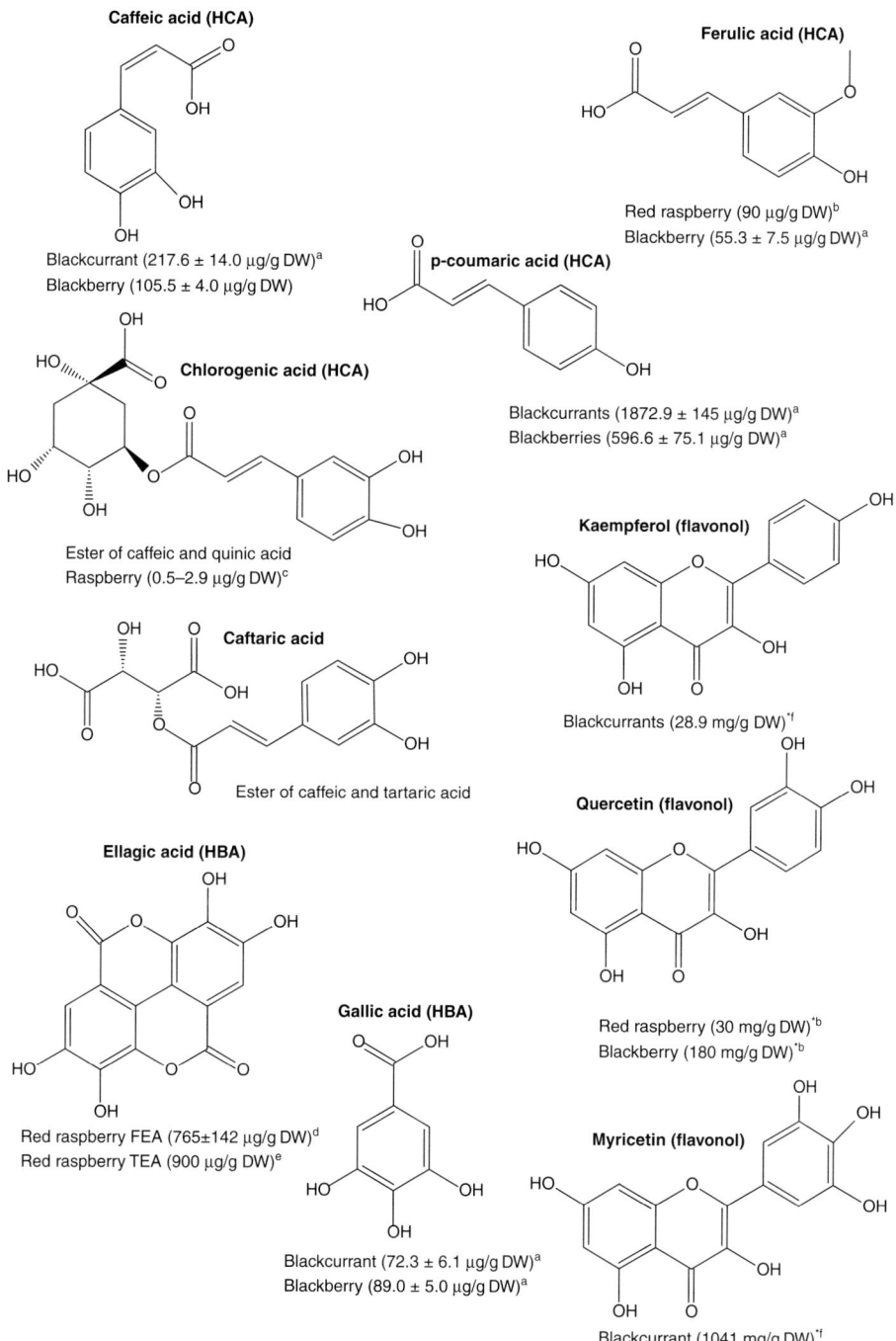

Fig. 14.4. Chemical structure of phenolic acids (hydroxycinnamates (HCA) and hydroxybenzoic acids (HBAs)) and flavonols and their reported occurrence in certain *Ribes* and *Rubus* berries. FEA = free ellagic acid; TEA = total ellagic acid after acid hydrolysis. [a]Zadernowsky *et al.*, 2005; [b]Jakobek *et al.*, 2009; [c]Zhang *et al.*, 2010; [d]Olsson *et al.*, 2004; [e]Daniel *et al.*, 1989; [f]Häkkinen *et al.*, 2000. *Values were transformed to dry weight (DW) basis based on 10% dry matter content. Values given for coumaric acid correspond to the *m*-isomer.

(*Ri. uva-crispa* L.) contained up to ten different anthocyanins, with a higher proportion of aromatic acylated anthocyanins, than seen in other commercially available berries (Jordheim *et al.*, 2007) (Fig. 14.4). In blackcurrants and blackberries, anthocyanin content is well correlated with berry colour, since the deeper the colour of the fruit, the higher the anthocyanin content. Recently, the anthocyanin profile has been proposed as a valuable tool to distinguish between different *Rubus* species (Mertz *et al.*, 2007). Similarly, by combining the anthocyanin profile with multivariate data analysis, recent work has been able to discriminate between different blackcurrant cultivars (Giné Bordonaba and Terry, 2008). Besides variations between cultivars and degrees of maturities (Rubinskiene *et al.*, 2006; Giné Bordonaba and Terry, 2008; Giné Bordonaba and Terry, unpublished data), anthocyanin content in *Ribes* and *Rubus* species depends on the harvest season and agroclimatic conditions (Fan-Chiang and Wrolstad, 2005; Rubinskiene *et al.*, 2006). Over the last decade, a vast number of publications have referred to the bioavailability or health-related properties of anthocyanins, including those of *Ribes* and *Rubus* fruit. So far, anthocyanins, including those commonly found in *Ribes* and *Rubus* fruit, have exhibited anti-inflammatory, antioxidant, vasomodulatory and antihaemostatic (Rechner and Kroner, 2005) activities when assessed *in vitro*. In addition, the beneficial effect of these compounds on the treatment of retinopathies has been known since the early 1980s (Scharrer and Ober, 1981). In earlier works, Matsumoto *et al.* (2001) observed that, despite the low bioavailability of these flavonoids, anthocyanins were absorbed directly, distributed to the blood and excreted in urine as their glycosilated forms. Similar findings have since been reported by other authors when working with *Ribes* or *Rubus* berry extracts as anthocyanin sources (Netzel *et al.*, 2001; Mülleder *et al.*, 2002; McGhie *et al.*, 2003; Hollands *et al.*, 2008).

As mentioned earlier, all flavonoids except flavanols from *Ribes* and *Rubus* species are found in glycosylated forms, which clearly affects their absorption (Scalbert and Williamson, 2000). Absorption in the stomach is possible for some flavonoids in their aglycone form, but not for their glycosides. It has been postulated that glycoside forms may resist gastric hydrolysis and therefore arrive in the duodenum as intact molecules. Similarly in the small intestine, absorption is limited to aglycones and some of their glucosides. As a result, most flavonoid molecules linked to rhamnose or other glycoside moieties need to be hydrolysed by the colon microflora prior to their absorption (Scalbert and Williamson, 2000). Anthocyanins, though, may represent an exception, since intact glycosides have been recovered from urine or identified as the main form in blood. In contrast, there is little evidence of anthocyanin aglycones in human blood or urine (Kay, 2006), which may be related to the poor stability of such compounds in neutral pH conditions. This dichotomy has resulted in the mechanisms involved in anthocyanin metabolism and absorption still not being understood fully. For instance, Passamonti *et al.* (2003) suggested that glycosides of anthocyanins might be transported by bilitranslocase at the gastric level, whereas Wu *et al.* (2004) proposed that these molecules might be converted into glucuronides by uridine 5′-diphosphate (UDP) glucose dehydrogenase. Generally, the urinary excretion of anthocyanins reported is very low, ranging from 0.016 to 0.13% of dosage within the first 2–8 h after consumption (Nielsen *et al.*, 2003). Nevertheless, recent evidence strongly suggests that anthocyanin metabolites may be overlooked with the current identification methods and hence the absorption of these compounds may have been underestimated dramatically (Felgines *et al.*, 2003). In the particular case of blackcurrants, Nielsen *et al.* (2003) studied the absorption and excretion of blackcurrant anthocyanins and found that the rutinoside forms were detected in urine, from both Watanabe heritable hyperlipidaemic rabbits and healthy humans, in higher concentrations (per cent excretion from 0 to 4 h; 0.058 ± 0.033) than the anthocyanin glucosides (0.046 ± 0.043). The authors suggested that this was due probably to the cleavage of the glucoside forms, but not of the rutinosides, in the small intestines, by β-glucosidases. Interestingly, blackcurrant berries are especially rich in both cyanidin and delphinidin rutinoside (Giné Bordonaba and Terry, 2008). Other studies have found larger proportions of ingested delphinidin glycosides than of ingested cyanidin glycosides in blood (Matsumoto *et al.*, 2001) or that the concentration of anthocyanin glycosides in plasma increases and decreases more rapidly compared with their respective rutinoside forms. Similarly, the plasma concentration:dose ratio in pigs after ingestion of marionberry freeze-dried powder was greater for cyanidin-3-rutinoside than for cyanidin-3-glucoside (Wu *et al.*, 2004). Bioavailability and fate of anthocyanins are, however, also influenced by the food matrix and few studies to date have focused on this issue. Nielsen *et al.* (2003) studied different food matrixes and found that rabbits fed blackcurrant juice showed higher plasma level of anthocyanins than animals fed with purified anthocyanins in an aqueous citric acid matrix. However, results from the same study showed that, in human subjects, the concentration of anthocyanins in plasma was not affected by the additional ingestion of a highly carbohydrate-rich meal (Nielsen *et al.*, 2003). Further research should address the role that the food matrix may have on the bioavailability of anthocyanin and other flavonoids.

Phenolic acids

Phenolic acids, including hydroxybenzoic or hydroxycinnamate acids, are non-flavonoid polyphenolic compounds that are of significant importance in berries from the *Ribes* and *Rubus* genera (Häkkinen *et al.*, 1998, 1999) (Fig. 14.4). Nevertheless, significant discrepancies exist between different published works found in the literature regarding phenolic concentrations within these berries. Most of these variations may be due not only to the different cultivars or agroclimatic conditions assessed, but also to the techniques used to extract and quantify the different phenolic fractions (Giné Bordonaba and Terry, 2008).

Hydroxybenzoic acids have a general structure derived directly from benzoic acid, with variations in the hydroxylation or methylation or the aromatic ring, whereas hydroxycinnamic acids tend to occur naturally as conjugated forms, being esters of hydroxy acids such as quinic, shikimic and tartaric acids or the corresponding sugar derivatives (Naczk and Shahidi, 2006). For instance,

ellagic acid (Fig. 14.4), a type of hydroxybenzoic acid derivative, is known to be present in both *Ribes* and *Rubus* berries, but particularly in raspberries, where it accounts for approximately 88% of the total phenolic acids (Häkkinen *et al.*, 1999; Amakura *et al.*, 2000; Olsson *et al.*, 2004). Raspberries and blackberries contain three times more ellagic acid than walnuts and 15 times more than other fruit and nuts (Tomás-Barberán and Clifford, 2000). Accordingly, ellagic acid concentration in raspberries was reported as $765 \pm 142\ \mu g/g$ DW, while this compound was not detected in any of the other berries analysed (Olsson *et al.*, 2004). Other studies have detected $c.400\ \mu g/g$ DW in freeze-dried raspberries when extracted with methanol (Daniel *et al.*, 1989). In the same study, after acid hydrolysis, concentrations rose up to $1900\ \mu g/g$ DW, indicating that most of the ellagic acid present in raspberries was encountered as ellagitannins (Daniel *et al.*, 1989). Similarly, in raspberry cvs. Zeva, Heritage and Williamet ellagic acid was detected together with six other ellagic acid derivatives (Tomás-Barberán and Clifford, 2000). The corresponding ellagic acid concentrations in blackberry pulp and seeds were reported as 2.43 mg/g DW and 3.37 mg/g DW (Wang *et al.*, 1996). Hydroxycinnamic acids ($113 \pm 43\ \mu g/g$ DW), quercetin ($28 \pm 9\ \mu g/g$ DW), quercetin-glycosides ($99 \pm 35\ \mu g/g$ DW) and other flavonols ($50 \pm 34\ \mu g/g$ DW) were detected by Olsson *et al.* (2004) in blackcurrant berries using an HPLC coupled to a diode array detector (DAD). Häkkinen *et al.* (1999) found large quantities of *p*-coumaric and caffeic acids in blackcurrant berries when a large set of different berries was screened for their phenolic content. When assessing variation in the phenolic content of different small, Polish grown berries, including blackcurrant and blackberries, Zadernowski *et al.* (2005) reported up to 14 different phenolic compounds. *m*-Coumaric acid derivatives were the principal phenolic compounds, with concentration over threefold higher in blackcurrants ($1872.9 \pm 145\ \mu g/g$ DW) than in blackberries ($596.6 \pm 75.1\ \mu g/g$ DW).

Although phenolic acids are the major polyphenols ingested by humans, the bioavailability of these compounds has not yet

received the same attention as that of flavonoids (Lafay and Gil-Izquierdo, 2008). The
limited information so far reveals that, for
instance, absorption of ferulic acid (Fig. 14.4)
takes place mainly in the small intestine,
with a urinary excretion of 40% of the
ingested dose, whereas ferulic acid conjugates are absorbed principally in the large
intestine (Kern *et al.*, 2003a,b) and with
lower recoveries. Gallic acid (Fig. 14.4) is
absorbed fairly well in the upper part of the
gut, with urinary excretions ranging from 36
to 40% of the ingested dose, depending very
much on the food source (Shahrzad *et al.*,
2001). Similarly, when hydroxycinnamic
acids are ingested, they are absorbed rapidly, which indicates an absorption in the
upper part of the gut (Lafay and Gil-Izquierdo, 2008).

14.2.2 Tannins and stilbenes

Tannins are also important components of
berry fruit (Szajdek and Borowska, 2008) and
are responsible for the astringent taste of some
fruit from different *Rubus* and *Ribes* species.
Basically, the tart taste of certain berries can be
attributed, in part, to the interactions between
this type of polyphenol and proteins. Tannins
include both condensed non-hydrolysable tannins (namely, proanthocyanidins) and hydrolysable tannins (namely, esters of ellagic and
gallic acids, also known as ellagitannins and
gallotannins, respectively). Although hydrolysable tannins are encountered more rarely in
berries (Szajdek and Borowska, 2008), Mertz
et al. (2007) has described two different ellagi-
tanins detected in blackberry extracts, the first
one consisting of lamberianin C and the second one identified tentatively as sanguiin H-6,
which has been previously identified in *Ribes*
species (Määttä *et al.*, 2003). Recently, McDougall *et al.* (2008) identified similar mixtures of
ellagitannin components and ellagic acid in a
tannin-enriched extract profile from raspberry
and cloudberry fruit. Similarly, Nohynek *et al.*
(2006) found similar concentrations of ellagitannins in both raspberry and cloudberry,
while these components were not detected in
blackcurrants.

In blackcurrants, all the cultivars investigated by Wu *et al.* (2004) had a similar
proanthocyanidin profile containing both
procyanidins and prodelphinidins. In the
same study, polymeric proanthocyanidins
with a degree of polymerization superior to
ten were the main proanthocyanidins
detected (80% of 1.21–1.66 mg/g FW) (Wu
et al., 2004). Similarly, earlier works established that the average degree of polymerization in blackcurrant proanthocyanidins was
38.7 (Gu *et al.*, 2003). In a range of gooseberry
cultivars and redcurrants cv. Red Lake, proanthocyanidins with a high degree of polymerization (> 10) also accounted for most of the
total concentrations of these compounds
(0.45–1.34 and 60.8 mg/g FW, respectively)
(Wu *et al.*, 2004).

Both condensed tannins and hydrolysable tannins show a greater free radical scavenging capacity than vitamin C or other
types of polyphenol (Szajdek and Borowska,
2008), and hence their potential role in ame-
liorating oxidative stress related to many
diseases. The bioavailability and metabolism
of these types of polyphenols have been
studied extensively during the past years
and the results indicate, for instance, that
ellagitannins are metabolized primarily by
the intestinal flora rather than being
absorbed directly in the human body (Cerdá
et al., 2005). Similarly, *in vitro* studies using
human colonic microflora have demonstrated, to a certain extent, that polymeric
proanthocyanidins are almost completely
degraded in 48 h (Déprez *et al.*, 2000).
Although initial studies showed that proanthocyanidins were metabolized and
absorbed in both mice and rats (Santos-
Buelga and Scalbert, 2000), more recent studies have failed to corroborate such findings
(for the interested reader, see the excellent
review by Beecher, 2004).

14.2.3 Ascorbic acid (vitamin C)

Ascorbic acid (AsA) is one of the most important water-soluble vitamins. Most plants and
animals are able to synthesize this compound;
however, apes and humans lack the enzymes

required and therefore AsA has to be supplemented, mainly through the consumption of fruit and vegetables (Naidu, 2003).

Similarly to that described for polyphenolic-type compounds, vitamin C concentration in berries from *Ribes* and *Rubus* species depends on several factors such as genotype, cultivation techniques, agroclimatic conditions, ripeness and postharvest storage and time (Hancock *et al.*, 2007; Giné Bordonaba and Terry, 2008; Chope, Giné Bordonaba and Terry, unpublished data). In blackcurrant, the synthesis and role of AsA have been elucidated recently (Hancock *et al.*, 2007). Variation in AsA content exists among blackcurrant cultivars (Viola *et al.*, 2000; Giné Bordonaba and Terry, 2008) and such variation has been suggested as being established during the initial development stages (Viola *et al.*, 2000). Recently, a wide range of UK grown blackcurrant cultivars was screened for several quality and health-related components (Giné Bordonaba and Terry, 2008) and the concentrations of AsA detected ranged from 1.922 to 5.415 mg/g FW, and therefore were far higher than those found in other common berry fruit, in which AsA content is commonly < 1 mg/g FW. Similarly, other studies also found that AsA content in blackcurrant samples was much higher than that in the other berries analysed (Remberg *et al.*, 2007). In other *Ribes* and *Rubus* fruit, concentration of AsA varies from 0.15 to 0.17 in blackberries, 0.15 to 0.32 in raspberries and 0.17 to 0.21 mg/g FW in redcurrants (Hägg *et al.*, 1995; De Ancos *et al.*, 2000; Haffner *et al.*, 2002; Benvenuti *et al.*, 2004).

Consumption of products naturally rich in AsA is associated with multiple health benefits. For instance, both ascorbate and dehydroascorbate delay the initiation of low-density lipoprotein (LDL) oxidation (Retsky and Frei, 1995), which is a process related to the formation of atherosclerosis. In addition, vitamin C plays an important role in the biosynthesis of certain vital constituents (namely, collagen, carnitine, neurotransmitters) and also stimulates immunological resistance, and can act as a detoxicant for certain mutagenic and carcinogenic compounds (Coulter *et al.*, 2006). In this context, extensive clinical, animal and *in vitro* studies have been conducted during the past decades, trying to elucidate such health-promoting

properties (for further information, see the review by Naidu, 2003). Nevertheless, a study by Olsson *et al.* (2004) failed to demonstrate any prevention of cancer cell proliferation using an ascorbate standard alone (Olsson *et al.*, 2004). In the same study, a correlation was found between inhibition of cancer cell growth and AsA content between the different *Ribes* and *Rubus* extracts analysed and, therefore, the authors speculated that such a phenomenon was most probably the result of a synergistic effect of vitamin C with other bioactives present in the extracts studied (Olsson *et al.*, 2004).

14.2.4 Fatty acids

Research over the past two decades has been carried out on the metabolism of polyunsaturated fatty acids (PUFAs) in general, with special emphasis on that of n-3 fatty acids (Simopoulos, 1999). This is due, in part, to the early evidence indicating that reducing the ratio of n-6 to n-3 fatty acids might play a role in decreasing the risk of heart disease and cancer. Nevertheless, numerous health-promoting properties are also reported for certain n-6 fatty acids (Ruiz del Castillo *et al.*, 2002) that are encountered relatively rarely in nature. Today, it is known that fatty acids are essential for normal growth and development, and also may have a crucial role in the prevention and treatment of coronary and metabolic diseases, as well as inflammatory and autoimmune disorders and cancer (Simopoulos, 1999). Within the 30 different blackcurrant genotypes studied, Ruiz del Castillo *et al.* (2002) found that γ-linolenic (n-6) acid (GLA) ranged from 11 to 19% of the total fatty acid fraction, whereas two other fatty acids, stearidonic and α-linolenic (n-3), varied from 2 to 4% and 10 to 19%, respectively (Ruiz del Castillo *et al.*, 2002). Few natural products are such rich sources of GLA as blackcurrant seeds. This fatty acid, in particular, is transformed to dihomo-γ-linolenic acid (DGLA; 20:3 n-6), the intermediate precursor of prostaglandin E_1, which is recognized for its anti-inflammatory and immunomodulating properties (Leventhal *et al.*, 1994). Supplementation with GLA has been shown to be a satisfactory remedy for a diverse range of conditions, including rheumatoid arthritis and atopic eczema. In the *Rubus* genus, blackberries are an exceptionally rich source of omega-3 (α-linolenic acid; n-3) and other PUFAs, owing in part to their numerous and large seeds (Bushman *et al.*, 2004). Seeds, however, tend to pass intact through the alimentary canal and hence any bioactives contained in this part of the fruit may not be assimilated. The preparation of berry extracts may overcome this limitation by enabling a better homogenization of the different components distributed in the whole fruit rather than specific tissues. Cold-pressed black raspberry, marionberry and boysenberry seed oil had 32.4, 15.8 and 19.5%, respectively, of α-linolenic acid and 53.0, 62.8 and 53.8%, respectively, of linoleic acid (Parry *et al.*, 2005). In the same study, boysenberry seed oil showed the highest scavenging activity against 2,2-diphenyl-1-picrylhydrazyl radical (DPPH•) and peroxyl radicals induced by 2,2'-azobis(2-amidinopropane) dihydrochloride (AAPH), followed by red raspberry and marionberry (Parry *et al.*, 2005).

14.3 Chemopreventive Activity and Bioavailability

14.3.1 Introduction

Generally, it is accepted that a correct balance between oxidants and antioxidants is synonymous with good health and that alterations to this balance are associated with certain pathologic conditions such as ageing, cancer and cardiovascular diseases. In this context, most of the health benefits associated with the intake of berries from *Ribes* and *Rubus* species have been linked largely with the high antioxidant capacity of these fruit, as assessed *in vitro* or in cell-free systems. However, the health-related properties of these berries may not be limited to the presence of antioxidant compounds. For example, several studies have reported the benefits derived from blackcurrant seed oil (BSO) due to its high content of GLA (Noli *et al.*, 2007), as well as the health benefits derived from the intake of blackcurrant

polysaccharide fractions (Takata *et al.*, 2005). Moreover, when considering recent studies and taking into account the low bioavailability of certain phytochemicals such as flavonoids, it appears that the health benefits associated with these berries may be the result of more complex biological processes rather than simply their capacity to scavenge free radicals. Williams *et al.* (2004) suggested that flavonoids might act as modulators of intracellular signalling processes, which could modify cellular redox status. Others (Seeram, 2008) suggested a synergistic effect among different berry bioactives as being responsible for many of the reported health-promoting properties.

Fruit and other parts from *Ribes* and *Rubus* plants have been used extensively as remedies for many diseases, and the relevant data can be traced back to as early as the 16th century (Dai *et al.*, 2007). Nowadays, there is a plethora of scientific reports available that describe the beneficiary role these berries may have on cardiovascular diseases, brain dysfunction and ageing, eye care, urinary tract health, and antimutagenic, anticarcinogenic, antibiotic and anti-inflammatory processes (Tables 14.3 and 14.4). Some of the most relevant information is summarized in the following sections.

14.3.2 Cancer studies

The anticarcinogenic effects derived from the intake of *Ribes* and *Rubus* species are well documented (Table 14.3). Bioactive compounds in the berries play different roles in cancer prevention, such as protection against oxidative DNA damage and the formation of DNA adducts, enhancement of DNA repair mechanisms and modulation of signalling pathways involved in different crucial cellular processes (namely, cell proliferation, apoptosis, inflammation, angiogenesis and arrest of the cell cycle) (Stoner *et al.*, 2008). ROS-induced DNA damage may be recognized as the possible first step involved in the complex process of carcinogenesis. Several studies (Table 14.3) have demonstrated *in vitro* or even *in vivo* (using animal models) the effects of *Ribes* and *Rubus* bioactives in

Table 14.3 Reported anticarcinogenic properties of fruit from *Ribes* and *Rubus* species.

Activity	Action	System	Dose	Extract type[a]	Reference
Anticarcinogenic	Inhibition of cancer cell proliferation in a dose-dependent manner	HepG2 human liver cancer cells	Extract equivalent to 50 mg/ml of raspberry extracts	Raspberry extracts from four different cultivars (Heritage, Kiwigold, Goldie and Anne)	Liu *et al.* (2002)
Anticarcinogenic	Inhibition of the growth of premalignant and malignant human oral cell lines	Human oral epithelial cell lines; malignant (83-01-82CA), premalignant (SCC-83-01-82)	50–200 µg/ml twice over 6 days	Different fractions from freeze-dried black raspberry extracts (namely, ferulic acid, β-sitosterol) from cv. Jewel	Han *et al.* (2005)
Anticarcinogenic	Preventing cell proliferation	Human colon cancer cells (CaCo-2) and human cervical cancer cells (Hela)	25–75 µg of GAE/ml	Digested[b] raspberry extracts (cv. Glen Ample)	McDougall *et al.* (2005)
Anticarcinogenic	Inhibition of tumour induction by N-nitroso-methylbenzylamine	Mouse epidermal (JB6 C1 41) cells	50 and 100 µg/ml of bioactive fractions	Different bioactive fractions from freeze-dried black raspberries cv. Jewel	Hecht *et al.* (2006)
Anticarcinogenic	Scavenge ultraviolet induced OH and O_2 radicals; Decrease the number of malignant and non-malignant skin tumours	*In vitro*: JB6 cells; *In vivo*: mouse model	3.5 µM C3G/mouse	Cyanidin-3-glucoside from blackberry	Ding *et al.* (2006)
Anticarcinogenic	Inhibition of human colon tumour cell growth; Supression of interleukin-12 release	HT 29 human cancer cells; Mouse bone marrow-derived dentritic cells	13.6–49.2 µg anthocyanins/ml; 0–40 µg anthocyanins/ml	Blackberry (cv. Hull) extracts	Dai *et al.* (2007)

(Continued)

Table 14.3 *Continued*

Activity	Action	System	Dose	Extract type[a]	Reference
Anticarcinogenic	Inhibition of cancer cell proliferation in a dose-dependent manner	Human cervical cancer (HeLa) *in vitro*	17.5 µg/ml GAE	Ellagitannin-rich fraction from raspberry cv. Glen Ample	Ross *et al.* (2007)
Anticarcinogenic	Preventing cell proliferation	Human colon cancer cells (CaCo-2) and human cervical cancer cells (Hela)	EC_{50} 25–40 µg polyphenols/ml	Different polyphenolic fractions from various berry extracts (lingonberry, raspberry, etc.)	McDougall *et al.* (2008)
Anticarcinogenic	Inhibition of N-nitrosomethylbenzy-lamine-induced tumours in the rat oesophagus	Sprague–Dawley male rats	Different treatments containing either 5% anthocyanin fractions of black raspberry at different concentrations or freeze-dried black raspberry extract	Freeze-dried black raspberry extracts	Wang *et al.* (2009)

Notes: [a]Whenever possible sample tissue or the cultivars used are specified; [b]samples were digested chemically, mimicking the conditions that occur in the gastrointestinal tract. GAE = gallic acid equivalents.

Table 14.4. Miscellaneous health-promoting properties of *Ribes* and *Rubus* species reported in the literature.

Activity	Action	System	Dose	Extract type	Reference
Antirheumatoid	Reduction in signs and symptoms of disease activity in rheumatoid arthritis patients	Human subjects	1.05 g/day	Blackcurrant seed oil (BSO)	Leventhal et al. (1994)
Anti-inflammatory	Supression of both cellular and fluid phases of inflammation as induced by monosodium urate crystals	Sprague–Dawley rats	ND	BSO (γ-linolenic and α-linolenic acid)	Tate and Zurier (1994)
Cardiovascular health	Inhibition of blood pressure (BP) over 40% and reduction in diastolic BP	Human subjects	6 g BSO/day over 8-week period	BSO	Deferne and Leeds (1996)
Cardiovascular health	Favourable blood pressure lowering effect of gammalinolenic acid	Spontaneous hypertensive rats	11% by weight of BSO	GLA-enriched BSO (17% GLA)	Engler and Engler (1998)
Eye health	Preventing myopic refractory shift during visual tasks and promoting visual recovery	Double-blind, placebo-controlled crossover study with healthy human subjects	12.5, 20 and 50 mg/subject	Blackcurrant anthocyanoside concentrate	Nakaishi et al. (2000)
Antiurolithiasis	Alkalizing effect in urine (greater pH and oxalic acid and citric acid in urine) which could support the metaphylaxis and treatment of urolithiasis	Human subjects	330 ml blackcurrant juice in three loading phases/person	Blackcurrant juice	Keβler et al. (2002)

(Continued)

protection against oxidative DNA damage. However, probably most of the information available pertains to *in vitro*-based studies showing the inhibitory effect of berry extracts on different types of cancer cell lines (Liu *et al.*, 2002; Olsson *et al.*, 2004; Han *et al.*, 2005; Ross *et al.*, 2007; McDougall *et al.*, 2008), as well as the ability of extracts to scavenge ROS (Jiao *et al.*, 2005; Hecht *et al.*, 2006).

Recently, McDougall *et al.* (2008), when assessing the inhibitory effect of a wide range of berry extracts on human cervical cancer and colon cancer cell lines, found that particularly those from the *Rubus* family were the most effective in preventing cell proliferation. Raspberry extracts from cv. Glen Ample, digested previously in conditions similar to those that occurr in the upper gastrointestinal tract, were shown to reduce the population of human HT29 cancer cells in the G1 phase of the cell cycle (Coates *et al.*, 2007). In the same study, the authors observed a protective effect against DNA damage in the HT29 cancer cells due to the same berry extract. In another study, raspberry extracts, from cvs. Heritage, Kiwigold, Goldie and Anne, inhibited HepG2 cell proliferation satisfactorily in a dose-dependent manner (> 10 mg/ml) (Liu *et al.*, 2002). In this case, the authors could not explain the inhibitory effect as a result of the phenolic/flavonoid fraction of the different extracts investigated, therefore suggesting that most probably other phytochemicals were involved. Similarly, Ross *et al.* (2007) found that ellagitannin content in raspberry extracts (cv. Glen Ample) was strongly correlated with the inhibition of cell proliferation and therefore concluded that antiproliferative activity from raspberries was associated predominantly with the content of these bioactives. In this, as in many other studies, anthocyanin and other polyphenolic fractions were purified from interfering compounds by solid-phase extraction (Fig. 14.5).

Seeram *et al.* (2006) also demonstrated the antiproliferative properties of different berry extracts, including blackberry and black and red raspberry, yet the applied dose (200 µg/ml) was probably far superior to that which could be supplied *in vivo* (Stoner *et al.*, 2008). That said, this seminal work clearly demonstrated the following: (i) that significant differences exist between the efficacy of different berry extracts on different cells (oral, breast, colon and prostate human cancer cell lines) and (ii) that extracts from black raspberry resulted in a significant induction of cell apoptosis (Seeram *et al.*, 2006). Others (Dai *et al.*, 2007; Wu *et al.*, 2007) also concluded that extracts from different berries, including

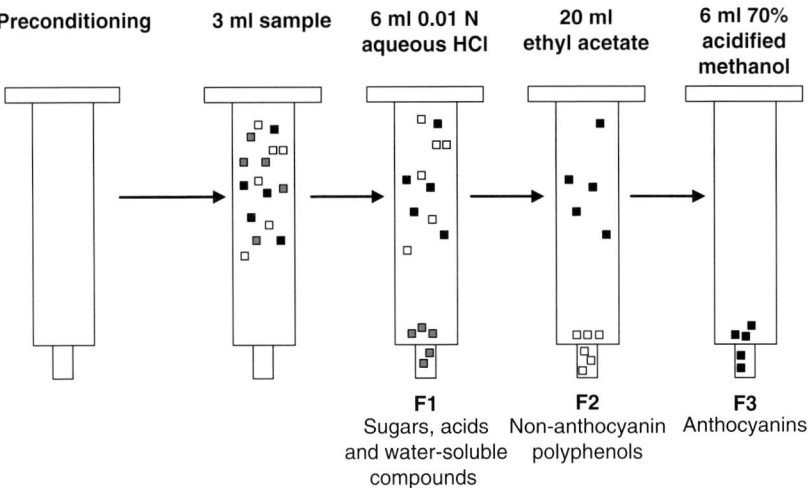

Fig. 14.5. Example of fractionation of polyphenolic compounds from berries into anthocyanins and other polyphenolic compounds using C18 cartridges (■ anthocyanins, □ non-anthocyanin polyphenols and ▪ sugars, acids and other water-soluble compounds).

blackberry (cv. Hull), inhibited cancer cell proliferation and, in specific cases, increased certain markers of cancer cell apoptosis. Certain bioactive fractions from black raspberry and blackberry extracts have also been reported as inhibitors of tumours induced by *N*-nitrosomethylbenzylamine (Hecht *et al.*, 2006) or potent scavengers of ultraviolet light-induced ·OH and O_2 radicals when assessed *in vitro*, in mouse epidermal JB6 cells.

The health benefits associated with the intake of berry-derived products have also been investigated with the aim of developing new foodstuffs with added nutritional value. Recently, raspberry seed flour was shown to inhibit cell proliferation in human colon cancer cells (Parry *et al.*, 2006a,b). The authors from the latest study highlighted the potential benefits of these berry-derived products in the formulation of new products. However, in an intervention study, blackcurrant seed press residue, a by-product without specific commercial value, consumed as part of a 250 g/day bread meal (containing 8% of the press residue), failed to reduce, and rather increased, oxidative stress markers in the stools and urine of 36 women (Helbig *et al.*, 2009). Despite the fact that serum and stool total tocopherol concentrations were increased as a result of the blackcurrant seed press residue, Helbig *et al.* (2009) pointed out that consumption of ground berry seed might not represent any health advantage.

Although there is still a paucity of information from *in vivo* trials or intervention studies, as compared with that from *in vitro* studies, work has also been conducted using both animals and human subjects to try to elucidate the effect of *Rubus* and/or *Ribes* extracts on cancer cell proliferation. For instance, freeze-dried black raspberry extracts (cv. not specified) inhibited tumour-induced development in the rat oesophagus, by inhibiting the formation of DNA adducts and reducing the proliferation of prenoplastic cells (Chen *et al.*, 2006). In the same study, the authors determined the possible mechanisms of action by which raspberry extracts inhibited tumour development. In addition, a polysaccharide-rich fraction from blackcurrant juice was shown to retard tumour growth when tested in Etarlich carcinoma-bearing

mice (Takata *et al.*, 2005). Clinical data from human studies also exist; for instance, Stoner's group showed that after consumption of lyophilized black raspberry extract (60 g/day) by 50 subjects with colorectal cancer and/or polyps, proliferation and angiogenesis biomarkers were diminished, while apoptosis was enhanced (Stoner *et al.*, 2007). Kresty *et al.* (2006) also showed that lyophilized black raspberry extract (32 or 45 g/day) diminished urine markers of oxidative stress in patients (*n* = 10) with Barrett's oesophagus. In contrast to all these positive and encouraging results, there are also cases of human intervention studies that failed to demonstrate any beneficial effects of berry intake (Møller *et al.*, 2004). This dichotomy or contradiction should be appreciated, as sometimes the positive benefits of fruit and vegetables are overexaggerated. Therefore, it is evident that further research is required at all levels (*in vitro*, *in vivo* and intervention studies) to understand the mechanisms by which consumption of *Ribes* and *Rubus* fruit may help against cancer.

14.3.3 Cardiovascular and metabolic diseases

In certain pathologic conditions (namely, hypertension, diabetes and atherosclerosis), the endothelium-dependent vasorelaxation caused by various vasodilator agonists is restrained considerably. Such a phenomenon is associated directly with a decrease in the release of NO, which certainly is crucial for the regulation of vasomotor tone and structure under certain physiological conditions. Given this, the development of vasodilator compounds with the ability to restore NO levels potentially could contribute to the treatments of some diseases. Nakamura *et al.* (2002) showed that blackcurrant concentrate could have an endothelium-dependent vasorelaxation effect when tested *in vitro* with rat thoracic aorta tissues. The authors found that increased levels of NO were one of the mechanisms involved in the vasorelaxation caused by the blackcurrant concentrate. In another study, purified anthocyanins from blackcurrant berries or blackcurrant juice demonstrated an antiatheroesclerotic effect when tested in Watanable heritable hyperlipidaemic rabbits (Nielsen *et al.*, 2005), suggesting the potential of these extracts in the prevention of certain cardiovascular conditions. A recent study in which an elderly population was given blackcurrant and other berry drinks showed a statistical significant improvement in oxidative status, as measured by plasma antioxidant capacity (McGhie *et al.*, 2007).

There is increasing attention over the positive effect that berry-derived bioactives, and specifically anthocyanins, have on blood vessel walls (namely, vasodilation, permeability, fragility, etc.) (Kähkönen *et al.*, 2003). Concomitant to this, anthocyanins from blackcurrant had considerable antioxidant activity when tested *in vitro* in lipid environments such as methyl linoleate and human LDL (Kähkönen *et al.*, 2003). Nevertheless, some of the positive effects from berries of *Ribes* and *Rubus* species on cardiovascular health (Table 14.4) may be associated with the fatty acid composition of the oil obtained from the seeds of those species. Certain disorders, and specifically hypertension, are related to abnormalities in tissue fatty acid metabolism, due in part to a reduction in desaturase activity. As discussed earlier in this chapter, BSO is a rich source of GLA, a PUFA with known health-related properties. Engler and Engler (1998) demonstrated more than a decade ago that oil enriched with GLA from blackcurrant has a significant blood pressure lowering effect when tested in spontaneous hypertensive rats.

Evidence from *in vitro* studies conducted with *Ribes* and *Rubus* species, as well as other sources of anthocyanin-rich compounds, suggests that certain compounds present in these fruit may mitigate certain metabolic diseases (Table 14.4), such as diabetes. Indeed, *in vitro* studies revealed that flavonoids modified the insulin-secreting capacity, reduced the NaF-induced apoptosis and modulated the cell proliferation of β cells (Pinent *et al.*, 2008). As β cells are the pancreatic cells responsible for producing and releasing insulin, they control blood glucose levels. McDougall *et al.* (2005) showed that, when tested *in vitro*, blackcurrant and raspberry polyphenol-rich extracts

(with phenolic concentrations ranging from 10 to 1500 μg) had an insulin-like effect since they inhibited both α-amylase and α-glucosidase significantly (these enzymes being responsible for hydrolysing complex carbohydrates into glucose and other simple sugars and hence elevating blood glucose levels). Whereas blackcurrant (cv. Ben Lomond) extracts inhibited the α-glucosidase better, α-amylase was inhibited more readily by the raspberry (cv. Glen Ample) extract (McDougall *et al.*, 2005). Jayaprakasam *et al.* (2005) found that specifically cyanidin-3-glucoside and delphinidin-3-glucoside, both anthocyanins commonly present in *Ribes* and *Rubus* species (Table 14.4), were the most effective insulin secretagogues among several anthocyanins tested *in vitro*. Sugimoto *et al.* (2003) studied the protective effects of major boysenberry anthocyanins against oxidative stress in streptozotocin induced diabetic rats. Elevated concentrations of oxidative substances in the plasma and also in the liver fell back to the levels of those observed in control rats when a diet with the berry anthocyanins was given to the diabetic animals. Accordingly, the authors pointed out that boysenberry anthocyanins were effective in protecting the development of *in vivo* oxidation involved with diabetes. Nevertheless, few *in vivo* studies and clinical data are available yet in order to validate the *in vitro* observations.

14.3.4 Urinary tract health and inhibition of intestinal pathogens

During the late 1950s and 1960s, several studies verified the role of anthocyanins and other polyphenols in altering microbial activity. The results from those studies demonstrated that, for instance, anthocyanins had stimulatory as well as inhibitory effects on microbial growth. More recently, the influence of blackcurrant concentrates or isolated anthocyanins from the same berry on the growth of microorganisms has been evaluated (Werlein *et al.*, 2005). The authors concluded that, while the anthocyanin fraction alone did not have significant effects on the growth of the microorganisms studied, blackcurrant extract inhibited *in vitro* the growth of certain microorganisms (*Staphylococcus aureus, Enterococcus faecium*), as well as stimulated the growth of *Saccharomyces cerevisiae* (Werlein *et al.*, 2005). Finnish researchers have demonstrated, in several *in vitro* studies (Puupponen-Pimiä *et al.*, 2001, 2005), the inhibitory effect of berry extracts, including raspberry, artic bramble and cloudberry, on the growth of both Gram-positive and Gram-negative intestinal pathogens. Recent research on this (Table 14.4) also supports the notion that proanthocyanindins commonly found in *Ribes* and *Rubus* berries prevent the adhesion of certain pathogenic bacteria to uroepithelial cells (Foo *et al.*, 2000).

14.3.5 Ageing and brain health

There are numerous motor and cognitive behavioural deficits that occur during ageing. Although many of the mechanisms involved still remain unclear, numerous researchers sustain that oxidative stress and inflammation are, in part, involved in the ageing process (Lau *et al.*, 2006). Indeed, Lau *et al.* (2006) suggested that combinations of antioxidants and anti-inflamatory polyphenols from berries might be key compounds to help prevent, suppress or inhibit age-related deficits. Studies conducted on animals showed that supplementation with dietary antioxidants improved cognitive function (Joseph *et al.*, 1998). Even though little research has been conducted in this regard with fruit from *Ribes* and *Rubus*, it is assumed that similar results to those obtained with blueberries or other berries may be observed following the consumption of these fruit. Shukitt-Hale *et al.* (2009) recently examined the effect of a 2% blackberry-supplemented diet in reversing the age-related deficits of rats. Results indicated that the blackberry diet not only improved motor performance on various tasks but that blackberry-fed rats had significantly greater working, or short-term, memory performance than the control rats (Shukitt-Hale *et al.*, 2009). Another of the few relevant studies conducted on *Ribes* berries is that by McGhie *et al.* (2007) in which the ability of blackcurrant-based

drinks to improve measures of oxidative stress and inflammation in an elderly population was assessed and an improvement in plasma antioxidant capacity was observed. Nevertheless, after the blackcurrant intake, plasma antioxidant capacity was the only indicator, from a wide range of oxidative stress markers studied, that improved. Anthocyanins and other polyphenolic fractions, at concentrations from 100 to 500 µg/ml, of *Rubus* (boysenberry cv. Riwaka Choice) and *Ribes* (blackcurrant cv. Ben Ard) species, have been reported to offer protection against the cytotoxic or neurotoxic effect of dopamine and amyloid β_{25-35} in M1 muscarinic receptor-transfected COS-7 brain cells (Ghosh *et al.*, 2007). Either dopamine or amyloid β_{25-35} can disrupt the Ca^{2+} buffer ability of brain cells, leading to further oxidative stress and cell degeneration associated with ageing. The mechanisms underlying the positive effects of berries and other fruit and vegetables on ageing have been reviewed recently by Shukitt-Hale *et al.* (2008) and Joseph *et al.* (2009).

14.3.6 Other health-promoting properties

BSO supplemented to patients suffering from rheumatoid arthritis during a 24-week trial resulted in a significant reduction of the signs and symptoms of disease activity (Leventhal *et al.*, 1994). The authors concluded that BSO was a potentially effective treatment for rheumatoid arthritis. Similarly, other studies conducted *in vivo* with Sprague–Dawley rats demonstrated that BSO suppressed both the cellular and fluid phases of inflammation significantly (Tate and Zurier, 1994). In this context, later studies showed that purified anthocyanins from blackcurrant and other berries were responsible for the inhibition of nuclear factor-κB, which controls the expression of many genes involved in the inflammatory response, as well as the reduction of proinflammatory mediators, when tested in healthy adults (Karlsen *et al.*, 2007).

Consumption of blackcurrant berries has also been associated with positive effects against kidney stone formation. Keßler *et al.* (2002) showed that blackcurrant juice could be used as a support treatment and metaphylaxis of uric acid stones due to its alkalizing effects. Crude extracts from wild blackcurrant berries had antiviral effects against influenza virus in a study conducted by Suzutani *et al.* (2003). This antiviral activity was speculated to be related to the interaction between the combining site on the viral envelope and certain constituents, not identified, in the crude extracts.

BSO administered to dogs suffering from atopic dermatitis resulted in increased concentrations of both GLA and dihomo-linolenic acid in the serum of the animals, but, more importantly, an improvement in the dermatitis was also observed (Noli *et al.*, 2007). Other health-promoting properties from *Ribes* or *Rubus* berries relate to eye vision. Nakaishi *et al.* (2000) demonstrated, in a double-blind placebo-controlled crossover study with healthy human subjects, that blackcurrant anthocyanins at doses of 12.5, 20 or 50 mg had a positive effect, preventing myopic shift during visual tasks and promoting visual recovery. In the same study, oral intake of blackcurrant anthocyanins was found to decrease the dark adaptation threshold in a dose-dependent manner.

A few years later, Matsumoto *et al.* (2006), revealed the ocular distribution of blackcurrant anthocyanins in rats and rabbits after the oral, intravenous or intraperitoneal administration of anthocyanins isolated from blackcurrants. This study revealed, for the first time, that blackcurrant anthocyanins were absorbed and distributed in ocular tissues as intact forms and passed through the blood–aqueous barriers and blood–retinal barriers in both of the animals investigated. In summary, the above-mentioned studies may have demonstrated that oral intake of purified anthocyanins or anthocyanin-rich extracts from *Ribes* and *Rubus* species may be used therapeutically for the treatment of certain ophthalmological conditions.

14.4 Effect of Preharvest, Postharvest and Processing

The level of secondary metabolites in plants from *Ribes* and *Rubus* is regulated by both environmental and genetic factors. Plants produce a wide range of bioactive compounds as a result of survival or adaptive strategies. These bioactive compounds are plant secondary metabolites produced for defence, protection and cell-to-cell signalling as a response to exposure to certain environmental stresses. Although the environmental mechanisms responsible for enhanced bioactive content in *Ribes* and *Rubus* species still remain unclear, cultivation of plants under certain stress conditions is one of the means by which the content of these berry bioactives can be enhanced. In addition, extensive research is being done, through several breeding programmes worldwide, to develop improved varieties with enhanced content of phytochemicals in combination with low-input cropping systems (Brennan *et al.*, 2008).

Blackberries, blackcurrants and raspberries, like other berries from the *Ribes* and *Rubus* genera, are not only available fresh but are distributed mainly as frozen and thermally processed products (namely, jams, jellies, juices, purees, cobblers and pies). For instance, most of the blackcurrant market in the UK is designated for the production of blackcurrant juice (e.g. Ribena®). After harvest, the quality of both *Ribes* and *Rubus* fruit declines dramatically, making postharvest storage at chilling temperatures (around 0°C) a requirement for the industry, generally for periods no longer than 3 weeks (Harb *et al.*, 2008). Controlled atmosphere (CAs) are also used occasionally to extend storage life, not only for blackcurrants but also for many other perishable berry fruit (Agar *et al.*, 1997; Terry *et al.*, 2009), when prolonged storage is required. However, both nutritional value and quality of berries are known to be affected negatively by postharvest storage conditions. For example, the concentration of AsA in berries tends to decrease with increased storage temperature and time (Roelofs *et al.*, 1993; Agar *et al.*, 1997; Kalt *et al.*, 1999; Häkkinen *et al.*, 2000; Viola *et al.*, 2000; Antunes *et al.*, 2003). In particular, Roelofs *et al.* (1993) showed that AsA content was reduced significantly when redcurrant berries were stored for 25 days at either 1°C or at fluctuating temperatures between 10 and 20°C. Similarly, a reduction of 40% in AsA content was observed in blackcurrant berries stored for 10 days at 10 or 20°C (Viola *et al.*, 2000). Even greater reductions in AsA, up to 50% of the initial content, were reported by Antunes *et al.* (2003) in blackberries stored at 20°C. Generally, the decline in AsA, and, indeed, in overall acid concentrations, during storage is accompanied by a darkening of the berry (Chope, Giné Bordonaba and Terry, unpublished data). This change in coloration has been related to an increase in anthocyanin concentration (Robbins *et al.*, 1989; Kalt *et al.*, 1999), which occurs in a temperature- and time-dependent manner. Raspberries stored at 0°C for 24 days contained 70% more anthocyanins than the initial values after harvest (Robbins *et al.*, 1989). Conversely, another study (Chanjirakul *et al.*, 2006) showed that, in raspberries stored for 7 or 10 days at 10°C, the concentration of anthocyanins was reduced considerably as compared with initial values before storage. Other bioactives are also affected by storage temperature. Ellagic acid content in red raspberry was reduced by 30% after 9 months of storage at −20°C in a study conducted by Häkkinen *et al.* (2000). As mentioned earlier, CAs, and in particular those with high CO_2 concentrations, may be used to extend the shelf life of many berries, including blackberry, raspberry and currants (Terry *et al.*, 2009). However, under these storage conditions, berries from *Ribes* (namely, black- and redcurrants) and *Rubus* (blackberry and raspberry) tend to suffer considerable reductions in their AsA content (Agar *et al.*, 1997). Little research has been conducted on elucidating the effects of CA storage on other common bioactives from *Ribes* and *Rubus* species.

The effect of postharvest processing treatment is also well documented. In all thermally processed blackberry-derived products, the concentration of monomeric anthocyanins, as well as the antioxidant activity of the products, declined dramatically compared with those seen in non-treated products (Hager *et al.*, 2008). In the same study, juice processing resulted in the greatest losses, whereas canned products were the least affected by processing. Most of the anthocyanin losses occurred during blanching and enzymatic treatment of blackberry juice (34% loss in total monomeric

anthocyanins). In contrast, total phenolic concentration of blackcurrant juices stored at 4°C tended to decline from day 0 (1919.8 ± 149.5 mg/ml GAE) to day 15 (1309.6 ± 107.8 mg/ml GAE), but returned to their initial values after 29 days of storage (Piljac-Žegarac *et al.*, 2009). However, antioxidant capacity, as measured by the Trolox equivalent antioxidant capacity (TEAC) assay, was diminished significantly. Postharvest storage of processed blackberry products also resulted in significant losses of monomeric anthocyanins, but had little or no significant effect on the antioxidant activity of most of the products. In another study, processing blackcurrant-derived products also reduced the content of total anthocyanins dramatically (to 0.05–10.3% of the levels in fresh fruit) but did not enhance the urinary yield in human subjects (Hollands *et al.*, 2008).

14.5 Conclusions and Future Research Needs

A considerable number of recent studies advocate that a high intake of *Ribes* and/or *Rubus* fruit may offer a number of health benefits against degenerative diseases and can promote longevity. Based on the survey of the literature presented herein, there is no doubt that most *Ribes* and *Rubus* berries are particularly rich sources of biologically active compounds (i.e. they have high levels of anthocyanins, proanthocyanidins, quercetin, myricetin, phenolic acids, etc.). In addition, blackcurrants are one of the richest sources of vitamin C, contributing, together with bioactive phenolics, to the high antioxidant activity of the berries. The array of health-promoting properties of these berries includes the inhibition of the development of certain cancers, cardiovascular and metabolic disorders and inflammation-related diseases. Besides, *Ribes* and *Rubus* fruit may be used therapeutically to treat urinary infections, ophthalmological diseases and even fight against ageing-related conditions. Blackcurrant was demonstrated recently to provide effective neuroprotection against oxidative stress-induced neuronal damage in human cell cultures. Among the bioactives of these berries, anthocyanins have received much more attention than the other polyphenols or non-polyphenol-type compounds, and hence further research should clarify the health-promoting properties of *Ribes* and *Rubus* bioactives other than anthocyanins.

As indicated, most *Ribes* and *Rubus* berries are consumed as derived products rather than fresh. However, most of the information related to health-promoting properties refers to fresh berries or purified berry fractions rather than the products that the consumer normally ingests. Only a limited number of studies have shown the detrimental effect of processing on the concentration and bioavailability of certain *Ribes* and *Rubus* bioactives (Hollands *et al.*, 2008).

Despite all the positive effects mentioned earlier, robust animal and human intervention trials are still necessary in order to substantiate any claims of human health benefits.

References

Agar, I.T., Streif, J. and Bangerth, F. (1997) Effect of high CO_2 and controlled atmosphere (CA) on the ascorbic and dehydroascorbic acid content of some berry fruits. *Postharvest Biology and Technology* 11, 47–55.

Amakura, Y., Okada, M., Tsuji, S. and Tonogai, Y. (2000) High-performance liquid chromatographic determination with photodiode array detection of ellagic acid in fresh and processed fruits. *Journal of Chromatography A* 896, 87–93.

Anttonen, M.J. and Karjalainen, R.O. (2006) High-performance liquid chromatography analysis of black currant (*Ribes nigrum* L.) fruit phenolics grown either conventionally or organically. *Journal of Agricultural and Food Chemistry* 54, 7530–7538.

Antunes, L.E., Duarte Filho, J. and de Souza, C.M. (2003) Postharvest conservation of blackberry fruits. *Pesquisa Agropecuaria Brasileira* 38, 413–419.

Barney, D.L. and Hummer, K.E. (2005) *Currants, Gooseberries, and Jostaberries: a Guide to Growers, Marketers, and Researchers in North America.* Food Products Press, Binghamton, New York.

Beecher, G.R. (2004) Proanthocyanidins: biological activities associated with human health. *Pharmaceutical Biology* 42, 2–20.

Benvenuti, S., Pellati, F., Melegari, M. and Bertelli, D. (2004) Polyphenols, anthocyanins, ascorbic acid, and radical scavenging activity of *Rubus*, *Ribes*, and *Aronia*. *Journal of Food Science* 69, 164–169.

Brennan, R. (2005) Currant and gooseberries (*Ribes* L.). In: Janick, J. (ed.) *The Encyclopedia of Fruit and Nut Crops.* CABI International, Wallingford, UK, pp. 191–295.

Brennan, R.M., Hunter, E.A. and Muir, D.D. (1997) Genotypic effects on sensory quality of blackcurrant juice using descriptive sensory profiling. *Food Research International* 30, 381–390.

Brennan, R., Jorgensen, L., Hackett, C., Woodhead, M., Gordon, S. and Russell, J. (2008) The development of a genetic linkage map of blackcurrant (*Ribes nigrum* L.) and the identification of regions associated with key fruit quality and agronomic traits. *Euphytica* 161, 19–34.

Bushman, B.S., Phillips, B., Isbell, T., Ou, B., Crane, J.M. and Knapp, S.J. (2004) Chemical composition of caneberry (*Rubus* spp.) seeds and oils and their antioxidant potential. *Journal of Agricultural and Food Chemistry* 52, 7982–7987.

Cerdá, B., Tomás-Barberán, F.A. and Espín, J.C. (2005) Metabolism of antioxidant and chemopreventive ellagitannins from strawberries, raspberries, walnuts, and oak-aged wine in humans: identification of biomarkers and individual variability. *Journal of Agricultural and Food Chemistry* 53, 227–235

Chanjirakul, K., Wang, S.Y., Wang, C.Y. and Siriphanich, J. (2006) Effect of natural volatile compounds on antioxidant capacity and antioxidant enzymes in raspberries. *Postharvest Biology and* Technology 40, 106–115.

Chen, T., Rose, M.E., Hwang, H., Nines, R.G. and Stoner, G.D. (2006) Black raspberries inhibit N-nitrosomethylbenzylamine (NMBA)-induced angiogenesis in rat esophagus parallel to the suppression of COX-2 and iNOS. *Carcinogenesis* 27, 2301–2307.

Coates, E.M., Popa, G., Gill, C.I.R., McCann, M.J., McDougall, G.J., Stewart, D., *et al.* (2007) Colon-available raspberry polyphenols exhibit anti-cancer effects on *in vitro* models of colon cancer. *Journal of Carcinogenesis* 6, 4–17.

Cooney, J.M., Jensen, D.J. and McGhie, T.K. (2004) LC-MS identification of anthocyanins in boysenberry extract and anthocyanin metabolites in human urine following dosing. *Journal of the Science of Food and Agriculture* 84, 237–245.

Coulter, I.D., Hardy, M.L., Morton, S.C., Hilton, L.G., Tu, W., Valentine, D., *et al.* (2006) Antioxidants vitamin C and vitamin E for the prevention and treatment of cancer. *Journal of General Internal Medicine* 21, 735–744.

Dai, J., Patel, J.D. and Mumper, R.J. (2007) Characterization of blackcberry extract and its antiproliferative and anti-inflamatory properties. *Journal of Medicinal Food* 10, 258–265.

Daniel, E.M., Krupnick, A.S., Heur, Y., Blinzler, J.A., Nims, R.W. and Stoner, G.D. (1989) Extraction, stability, and quantitation of ellagic acid in various fruits and nuts. *Journal of Food Composition and Analysis* 2, 338–349.

De Ancos, B., Ibañez, E., Reglero, G. and Cano, M.P. (2000) Frozen storage effects on anthocyanins and volatile compounds of raspberry fruit. *Journal of Agricultural and Food Chemistry* 48, 873–879.

Deferne, J.L. and Leeds, A.R. (1996) Resting blood pressure and cardiovascular reactivity to mental arithmetic in mild hypertensive males supplemented with blackcurrant seed oil. *Journal of Human Hypertension*, 10, 531–537.

Déprez, S., Brezillon, C., Rabot, S., Philippe, C., Mila, I., Lapierre, C., *et al.* (2000) Polymeric proanthocyanidins are catabolized by human colonic microflora into low-molecular-weight phenolic acids. *Journal of Nutrition* 130, 2733–2738.

Ding, M., Feng, R., Wang, S.Y., Bowman, L., Lu, Y., Qian, Y., *et al.* (2006) Cyanidin-3-glucoside, a natural product derived from blackberry, exhibits chemopreventive and chemotherapeutic activity. *Journal of Biological Chemistry* 281, 17359–17368.

Dixon, R.A. and Paiva, N.L. (1995) Stress-induced phenylpropanoid metabolism. *Plant Cell* 7, 1085–1097.

Engler, M.M. and Engler, M.B. (1998) The effects of dietary evening primrose, black currant, borage and fungal oils on plasma, hepatic and vascular tissue fatty acid composition in the spontaneously hypertensive rat. *Nutrition Research* 18, 1533–1544.

Fan-Chiang, H. and Wrolstad, R.E. (2005) Anthocyanin pigment composition of blackberries. *Journal of Food Science* 70, C198–C202.

FAOSTAT (2008) FAOSTAT statistics database-agriculture. Food and Agriculture Organization of the United Nations, Rome, Italy (http://faostat.fao.org/site/567/DesktopDefault.aspx?PageID=567#ancor, accessed 16 September 2008).

Felgines, C., Talavéra, S., Gonthier, M., Texier, O., Scalbert, A., Lamaison, J., *et al.* (2003) Strawberry anthocyanins are recovered in urine as glucuro- and sulfoconjugates in humans. *Journal of Nutrition* 133, 1296–1301.

Foo, L.Y., Lu, Y., Howell, A.B. and Vorsa, N. (2000) The structure of cranberry proanthocyanidins which inhibit adherence of uropathogenic P-fimbriated *Escherichia coli in vitro*. *Phytochemistry* 54, 173–181

Frøytlog, C., Slimestad, R. and Andersen, Q.M. (1998) Combination of chromatographic techniques for the preparative isolation of anthocyanins applied on blackcurrant (*Ribes nigrum*) fruits. *Journal of Chromatography A* 825, 89–95.

Ghosh D., McGhie, T.K., Zhang, J., Adaim, A. and Skinner, M. (2006) Effects of anthocyanins and other phenolics of boysenberry and blackcurrant as inhibitors of oxidative stress and damage to cellular DNA in SH-SY5Y and HL-60 cells. *Journal of the Science of Food and Agriculture* 86, 678–686.

Ghosh, D., McGhie, T.K., Fisher, D.R. and Joseph, J.A. (2007) Cytoprotective effects of anthocyanins and other phenolic fractions of boysenberry and blackcurrant on dopamine and amyloid β-induced oxidative stress in transfected COS-7 cells. *Journal of the Science of Food and Agriculture* 87, 2061–2067.

Giné Bordonaba, J. and Terry, L.A. (2008) Biochemical profiling and chemometric analysis of seventeen UK-grown black currant cultivars. *Journal of Agricultural and Food Chemistry* 56, 7422–7430.

Giné Bordonaba, J., Crespo, P. and Terry, L.A. (2011). A new acetonitrile-free mobile phase for HPLC-DAD determination of individual anthocyanins in blackcurrant and strawberry fruits: a comparison and validation study. *Food Chemistry* 129, 1265–1273.

Goiffon, J., Brun, M. and Bourrier, M. (1991) High-performance liquid chromatography of red fruit anthocyanins. *Journal of Chromatography* 537, 101–121.

Gu, L., Kelm, M.A., Hammerstone, J.F., Beecher, G., Holden, J., Haytowitz, D. and Prior, R.L. (2003) Screening of foods containing proanthocyanidins and their structural characterization using LC-MS/MS and thiolytic degradation. *Journal of Agricultural and Food Chemistry* 51, 7513–7521.

Haffner, K., Rosenfeld, H.J., Skrede, G. and Wang, L. (2002) Quality of red raspberry (*Rubus idaeus* L.) cultivars after storage in controlled and normal atmospheres. *Postharvest Biology and Technology* 24, 279–289.

Hager, T.J., Howard, L.R. and Prior, R.L. (2008) Processing and storage effects on monomeric anthocyanins, percent polymeric color, and antioxidant capacity of processed blackberry products. *Journal of Agricultural and Food Chemistry* 56, 689–695.

Hägg, M., Ylikoski, S. and Kumpulainen, J. (1995) Vitamin C content in fruits and berries consumed in finland. *Journal of Food Composition and Analysis* 8, 12–20.

Häkkinen, S.H., Kärenlampi, S.O., Heinonen, I.M., Mykkänen, H.M. and Törrönen, A.R. (1998) HPLC method for screening of flavonoids and phenolic acids in berries. *Journal of the Science of Food and Agriculture* 77, 543–551.

Häkkinen, S., Heinonen, M., Kärenlampi, S., Mykkänen, H., Ruuskanen, J. and Törrönen, R. (1999) Screening of selected flavonoids and phenolic acids in 19 berries. *Food Research International* 32, 345–353.

Häkkinen, S.H., Kärelampi, S.O., Mykkänen, H.M., Heinonen, I.M. and Törrönen, A.R. (2000) Ellagic acid content in berries: influence of domestic processing and storage. *European Food Research and Technology* 212, 75–80.

Han, C., Ding, H., Casto, B., Stoner, G.D. and D'Ambrosio, S.M. (2005) Inhibition of the growth of premalignant and malignant human oral cell lines by extracts and components of black raspberries. *Nutrition and Cancer* 51, 207–217.

Hancock, R.D., Walker, P.G., Pont, S.D.A., Marquis, N., Vivera, S., Gordon, S.L., *et al.* (2007) L-Ascorbic acid accumulation in fruit of *Ribes nigrum* occurs by *in situ* biosynthesis via the L-galactose pathway. *Functional Plant Biology* 34, 1080–1091.

Harb, J., Bisharat, R. and Streif, J. (2008) Changes in volatile constituents of blackcurrants (*Ribes nigrum* L. cv. 'Titania') following controlled atmosphere storage. *Postharvest Biology and Technology* 47, 271–279.

Harborne, J.B. and Williams, C.A. (1995) Anthocyanins and other flavonoids. *Natural Product Reports* 12, 639–657.

Hecht, S.S., Huang, C., Stoner, G.D., Li, J., Kenney, P.M.J., Sturla, S.J., *et al.* (2006) Identification of cyanidin glycosides as constituents of freeze-dried black raspberries which inhibit anti-benzo[a]pyrene-7,8-diol-9,10-epoxide induced NFκB and AP-1 activity. *Carcinogenesis* 27, 1617–1626.

Helbig, D., Wagner, A., Glei, M., Basu, S., Schubert, R. and Jahreis, G. (2009) Blackcurrant seed press residue increases tocopherol concentrations in serum and stool whilst biomarkers in stool and urine indicate increased oxidative stress in human subjects. *British Journal of Nutrition* 102, 554–562.

Hollands, W., Brett, G.M., Radreau, P., Saha, S., Teucher, B., Bennett, R.N. and Kroon, P.A. (2008) Processing blackcurrant dramatically reduces the content and does not enhance the urinary yield of anthocyanins in human subjects. *Food Chemistry* 108, 869–878.

Jakobek, L., Šeruga, M., Šeruga, B., Novak, I. and Medvidović-Kosanović, M. (2009) Phenolic compound composition and antioxidant activity of fruits of *Rubus* and *Prunus* species from Croatia. *International Journal of Food Science and Technology* 44, 860–868.

Jayprakasam, B., Vareed, S.K., Olson, L.K. and Nair, M.G. (2005) Insulin secretion by bioactive anthocyanins and anthocyanidins present in fruits. *Journal of Agricultural and Food Chemistry* 53, 28–31.

Jiao, Z., Liu, J. and Wang, S. (2005) Antioxidant activities of total pigment extract from blackberries. *Food Technology and Biotechnology* 43, 97–102.

Jordheim, M., Måge, F. and Andersen, O.M. (2007) Anthocyanins in berries of Ribes including gooseberry cultivars with high content of acylated pigments. *Journal of Agricultural and Food Chemistry* 55, 5529–5535.

Joseph, J.A., Denisova, N., Fisher, D., Shukitt-Hale, B., Bickford, P., Prior, R., *et al.* (1998) Age-related neurodegeneration and oxidative stress: putative nutritional intervention. *Neurologic Clinics* 16, 747–755.

Joseph, J.A., Cole, G., Head, E. and Ingram, D. (2009) Nutrition, brain aging, and neurodegeneration. *Journal of Neuroscience* 29, 12795–12801.

Kähkönen, M.P., Heinämäki, J., Ollilainen, V. and Heinonen, M. (2003) Berry anthocyanins: isolation, identification and antioxidant activities. *Journal of the Science of Food and Agriculture* 83, 1403–1411.

Kalt, W., Forney, C.F., Martin, A. and Prior, R.L. (1999) Antioxidant capacity, vitamin C, phenolics, and anthocyanins after fresh storage of small fruits. *Journal of Agricultural and Food Chemistry* 47, 4638–4644.

Karlsen, A., Retterstøl, L., Laake, P., Paur, I., Kjølsrud-Bøhn, S., Sandvik, L., *et al.* (2007) Anthocyanins inhibit nuclear factor-κB activation in monocytes and reduce plasma concentrations of pro-inflammatory mediators in healthy adults. *Journal of Nutrition* 137, 1951–1954.

Kay, C.D. (2006) Aspects of anthocyanin absorption, metabolism and pharmacokinetics in humans. *Nutrition Research Reviews* 19, 137–146.

Keβler, T., Jansen, B. and Hesse, A. (2002) Effect of blackcurrant, cranberry and plum juice consumption on risk factors associated with kidney stone formation. *European Journal of Clinical Nutrition* 56, 1020–1023.

Kern, S.M., Bennett, R.N., Mellon, F.A., Kroon, P.A. and García-Conesa, M.T. (2003a) Absorption of hydroxycinnamates in humans after high-bran cereal consumption. *Journal of Agricultural and Food Chemistry* 51, 6050–6055.

Kern, S.M., Bennett, R.N., Needs, P.W., Mellon, F.A., Kroon, P.A. and Garcia-Conesa, M. (2003b) Characterization of metabolites of hydroxycinnamates in the *in vitro* model of human small intestinal epithelium Caco-2 cells. *Journal of Agricultural and Food Chemistry* 51, 7884–7891.

Knox, Y.M., Suzutani, T., Yosida, I. and Azuma, M. (2003). Anti-influenza virus activity of crude extract of *Ribes nigrum* L. *Phytotherapy Research* 17, 120–122.

Kresty, L.A., Frankel, W.L., Hammond, C.D., Baird, M.E., Mele, J.M., Stoner, G.D., *et al.* (2006) Transitioning from preclinical to clinical chemopreventive assessments of lyophilized black raspberries: interim results show berries modulate markers of oxidative stress in barrett's esophagus patients. *Nutrition and Cancer* 54, 148–156.

Krüger, E., Schöpplein, E., Rasim, S., Cocca, G. and Fischer, H. (2003) Effects of ripening stage and storage time on quality parameters of red raspberry fruit. *European Journal of Horticultural Science* 68, 176–182.

Lafay, S. and Gil-Izquierdo, A. (2008) Bioavailability of phenolic acids. *Phytochemistry Reviews* 7, 301–311.

Lau, F.C., Shukitt-Hale, B. and Joseph, J.A. (2006) Beneficial effects of berry fruit polyphenols on neuronal and behavioral aging. *Journal of the Science of Food and Agriculture* 86, 2251–2255.

Leventhal, L.J., Boyce, E.G. and Zurier, R.B. (1994) Treatment of rheumatoid arthritis with blackcurrant seed oil. *British Journal of Rheumatology* 33, 847–852.

Liu, M., Li, X.Q., Weber, C., Lee, C.Y., Brown, J. and Liu, R.H. (2002) Antioxidant and antiproliferative activities of raspberries. *Journal of Agricultural and Food Chemistry* 50, 2926–2930.

Määttä, K.R., Kamal-Eldin, A. and Riitta Törrönen, A. (2003) High-performance liquid chromatography (HPLC) analysis of phenolic compounds in berries with diode array and electrospray ionization mass spectrometric (MS) detection: *Ribes* species. *Journal of Agricultural and Food Chemistry* 51, 6736–6744.

McDougall, G.J., Dobson, P., Smith, P., Blake, A. and Stewart, D. (2005) Assessing potential bioavailability of raspberry anthocyanins using an *in vitro* digestion system. *Journal of Agricultural and Food Chemistry* 53, 5896–5904.

McDougall, G.J., Ross, H.A., Ikeji, M. and Stewart, D. (2008) Berry extracts exert different antiproliferative effects against cervical and colon cancer cells grown in vitro. *Journal of Agricultural and Food Chemistry* 56, 3016–3023.

McDougall, G.J., Kulkarni, N.N. and Stewart, D. (2009) Berry polyphenols inhibit pancreatic lipase activity in vitro. *Food Chemistry* 115, 193–199.

McGhie, T.K., Ainge, G.D., Barnett, L.E., Cooney, J.M. and Jensen, D.J. (2003) Anthocyanin glycosides from berry fruit are absorbed and excreted unmetabolized by both humans and rats. *Journal of Agricultural and Food Chemistry* 51, 4539–4548.

McGhie, T.K., Walton, M.C., Barnett, L.E., Vather, R., Martin, H., Au, J., *et al.* (2007) Boysenberry and blackcurrant drinks increased the plasma antioxidant capacity in an elderly population but had little effect on other markers of oxidative stress. *Journal of the Science of Food and Agriculture* 87, 2519–2527.

Manhita, A.C., Teixeira, D.M. and da Costa, C.T. (2006) Application of sample disruption methods in the extraction of anthocyanins from solid or semi-solid vegetable samples. *Journal of Chromatography A* 1129, 14–20.

Matsumoto, H., Inaba, H., Kishi, M., Tominaga, S., Hirayama, M. and Tsuda, T. (2001) Orally administered delphinidin 3-rutinoside and cyanidin 3-rutinoside are directly absorbed in rats and humans and appear in the blood as the intact forms. *Journal of Agricultural and Food Chemistry* 49, 1546–1551.

Matsumoto, H., Nakamura, Y., Iida, H., Ito, K. and Ohguro, H. (2006) Comparative assessment of distribution of blackcurrant anthocyanins in rabbit and rat ocular tissues. *Experimental Eye Research* 83, 348–356.

Mertz, C., Cheynier, V., Günata, Z. and Brat, P. (2007) Analysis of phenolic compounds in two blackberry species (*Rubus glaucus* and *Rubus adenotrichus*) by high-performance liquid chromatography with diode array detection and electrospray ion trap mass spectrometry. *Journal of Agricultural and Food Chemistry* 55, 8616–8624.

Møller, P., Loft, S., Alfthan, G. and Freese, R. (2004) Oxidative DNA damage in circulating mononuclear blood cells after ingestion of blackcurrant juice or anthocyanin-rich drink. *Mutation Research – Fundamental and Molecular Mechanisms of Mutagenesis* 551, 119–126.

Mülleder, U., Murkovic, M. and Pfannhauser, W. (2002) Urinary excretion of cyanidin glycosides. *Journal of Biochemical and Biophysical Methods* 53, 61–66.

Naczk, M. and Shahidi, F. (2006) Phenolics in cereals, fruits and vegetables: occurrence, extraction and analysis. *Journal of Pharmaceutical and Biomedical Analysis* 41, 1523–1542.

Naidu, K.A. (2003) Vitamin C in human health and disease is still a mystery? An overview. *Nutrition Journal* 2, 1–10.

Nakaishi, H., Matsumoto, H., Tominaga, S. and Hirayama, M. (2000) Effects of black currant anthocyanoside intake on dark adaptation and VDT work-induced transient refractive alteration in healthy humans. *Alternative Medicine Review* 5, 553–562.

Nakamura, Y., Matsumoto, H. and Todoki, K. (2002) Endothelium-dependent vasorelaxation induced by black currant concentrate in rat thoracic aorta. *Japanese Journal of Pharmacology* 89, 29–35.

Netzel, M., Strass, G., Janssen, M., Bitsch, I. and Bitsch, R. (2001) Bioactive anthocyanins detected in human urine after ingestion of blackcurrant juice. *Journal of Environmental Pathology, Toxicology and Oncology* 20, 89–95.

Nielsen, I.L.F., Haren, G.R., Magnusen, E.L., Dragsted, L.O. and Rasmussen, S.E. (2003) Quantification of anthocyanins in commercial black currant juices by simple high-performance liquid chromatography. Investigation of their pH stability and antioxidative potency. *Journal of Agricultural and Food Chemistry* 51, 5861–5866.

Nielsen, I.L.F., Rasmussen, S.E., Mortensen, A., Ravn-Haren, G., Ma, H.P., Knuthsen, P., *et al.* (2005) Anthocyanins increase low-density lipoprotein and plasma cholesterol and do not reduce atherosclerosis in Watanabe heritable hyperlipidemic rabbits. *Molecular Nutrition and Food Research* 49, 301–308.

Nohynek, L.J., Alakomi, H., Kähkönen, M.P., Heinonen, M., Helander, I.M., Oksman-Caldentey, K., *et al.* (2006) Berry phenolics: antimicrobial properties and mechanisms of action against severe human pathogens. *Nutrition and Cancer* 54, 18–32.

Noli, C., Carta, G., Cordeddu, L., Melis, M.P., Murru, E. and Banni, S. (2007) Conjugated linoleic acid and black currant seed oil in the treatment of canine atopic dermatitis: a preliminary report. *Veterinary Journal* 173, 413–421.

Olsson, M.E., Gustavsson, K., Andersson, S., Nilsson, A. and Duan, R. (2004) Inhibition of cancer cell proliferation *in vitro* by fruit and berry extracts and correlations with antioxidant levels. *Journal of Agricultural and Food Chemistry* 52, 7264–7271.

Parry, J., Su, L., Luther, M., Zhou, K., Yurawecz, M.P., Whittaker, P. and Yu, Y. (2005) Fatty acid composition and antioxidant properties of cold-pressed marionberry, boysenberry, red raspberry, and blueberry seed oils. *Journal of Agricultural and Food Chemistry* 53, 566–573.

Parry J., Su, L., Luther, M., Zhou, K., Zhou, K., Yurawecz, M.P., Whittaker, P. and Yu, L. (2006a) Fatty acid composition and antioxidant properties of cold-pressed marionberry, boysenberry, red raspberry, and blueberry seed oils. *Journal of Agricultural and Food Chemistry* 53, 566–573.

Parry, J., Su, L., Moore, J., Cheng, Z., Luther, M., Rao, J.N., *et al.* (2006b) Chemical compositions, antioxidant capacities, and antiproliferative activities of selected fruit seed flours. *Journal of Agricultural and Food Chemistry* 54, 3773–3778.

Passamonti, S., Vrhovsek, U., Vanzo, A. and Mattivi, F. (2003) The stomach as a site for anthocyanins absorption from food. *FEBS Letters* 544, 210–213.

Piljac-Zegarac, J., Valek, L., Martinez, S. and Belščak, A. (2009) Fluctuations in the phenolic content and antioxidant capacity of dark fruit juices in refrigerated storage. *Food Chemistry* 113, 394–400.

Pinent, M., Castell, A., Baiges, I., Montagut, G., Arola, L. and Ardévol, A. (2008) Bioactivity of flavonoids on insulin-secreting cells. *Comprehensive Reviews in Food Science and Food Safety* 7, 299–308.

Proteggente, A.R., Pannala, A.S., Paganga, G., van Buren, L., Wagner, E., Wiseman, S., van de Put, F., Dacombe, C. and Rice-Evans, C.A. (2002). The antioxidant activity of regularly consumed fruit and vegetables reflects their phenolic and vitamin C composition. *Free Radical Research*, 36, 217–233.

Puupponen-Pimiä, R., Nohynek, L., Meier, C., Kähkönen, M., Heinonen, M., Hopia, A., *et al.* (2001) Antimicrobial properties of phenolic compounds from berries. *Journal of Applied Microbiology* 90, 494–507.

Puupponen-Pimiä, R., Nohynek, L., Hartmann-Schmidlin, S., Kähkönen, M., Heinonen, M., Määttä-Riihinen, K., *et al.* (2005) Berry phenolics selectively inhibit the growth of intestinal pathogens. *Journal of Applied Microbiology* 98, 991–1000.

Rechner, A.R. and Kroner, C. (2005) Anthocyanins and colonic metabolites of dietary polyphenols inhibit platelet function. *Thrombosis Research* 116, 327–334.

Remberg, S.F., Måge, F., Haffner, K. and Blomhoff, R. (2007) Highbush blueberries (*Vaccinium corymbosum* L.), raspberries (*Rubus idaeus* L.) and black currants (*Ribes nigrum* L.) – influence of cultivar on antioxidant activity and other quality parameters. *Acta Horticulturae* 744, 259–266.

Retsky, K.L. and Frei, B. (1995) Vitamin C prevents metal ion-dependent initiation and propagation of lipid peroxidation in human low-density lipoprotein. *Biochimica et Biophysica Acta – Lipids and Lipid Metabolism* 1257, 279–287.

Robbins, J., Sjulin, T.M. and Patterson, M. (1989) Postharvest storage characteristics and respiration rates in five cultivars of red raspberry. *HortScience* 24, 980–982.

Roelofs, F., Van-de-Waart, A., Smolarz, K. and Zmarlicki, K. (1993) Long-term storage of red currants under controlled atmosphere conditions. *Acta Horticulturae* 352, 217–222.

Ross, H.A., McDougall, G.J. and Stewart, D. (2007) Antiproliferative activity is predominantly associated with ellagitannins in raspberry extracts. *Phytochemistry* 68, 218–228.

Rubinskiene, M., Viskelis, P., Jasutiene, I., Duchovskis, P. and Bobinas, C. (2006) Changes in biologically active constituents during ripening in black currants. *Journal of Fruit and Ornamental Plant Research* 14, 237–246.

Ruiz del Castillo, M.L., Dobson, G., Brennan, R. and Gordon, S. (2002) Genotypic variation in fatty acid content of blackcurrant seeds. *Journal of Agricultural and Food Chemistry* 50, 332–335.

Santos-Buelga, C. and Scalbert, A. (2000) Proanthocyanidins and tannin-like compounds – nature, occurrence, dietary intake and effects on nutrition and health. *Journal of the Science of Food and Agriculture* 80, 1094-1117.

Scalbert, A. and Williamson, G. (2000). Dietary intake and bioavailability of polyphenols. *Journal of Nutrition* 130, 2073S–2085S.

Scharrer, A. and Ober, M. (1981) Anthocyanosides in the treatment of retinopathies. *Klinische Monatsblatter für Augenheilkunde* 178, 386–389.

Seeram, N.P. (2008) Berry fruits for cancer prevention: current status and future prospects. *Journal of Agricultural and Food Chemistry* 56, 630–635.

Seeram, N.P., Adams, L.S., Zhang, Y., Lee, R., Sand, D., Scheuller, H.S., *et al.* (2006) Blackberry, black raspberry, blueberry, cranberry, red raspberry, and strawberry extracts inhibit growth and stimulate apoptosis of human cancer cells in vitro. *Journal of Agricultural and Food Chemistry* 54, 9329–9339.

Shahrzad, S., Aoyagi, K., Winter, A., Koyama, A. and Bitsch, I. (2001) Pharmacokinetics of gallic acid and its relative bioavailability from tea in healthy humans. *Journal of Nutrition* 131, 1207–1210.

Shukitt-Hale, B., Lau, F.C. and Joseph, J.A. (2008) Berry fruit supplementation and the aging brain. *Journal of Agricultural and Food Chemistry* 56, 636–641.

Shukitt-Hale, B., Cheng, V. and Joseph, J.A. (2009) Effects of blackberries on motor and cognitive function in aged rats. *Nutritional Neuroscience* 12, 135–140.

Simopoulos, A.P. (1999) Essential fatty acids in health and chronic disease. *American Journal of Clinical Nutrition* 70, 560S–569S.

Slimestad, R. and Solheim, H. (2002) Anthocyanins from black currants (*Ribes nigrum* L.). *Journal of Agricultural and Food Chemistry* 50, 3228–3231.

Stoner, G.D., Wang, L., Zikri, N., Chen, T., Hecht, S.S., Huang, C., *et al.* (2007) Cancer prevention with freeze-dried berries and berry components. *Seminars in Cancer Biology* 17, 403–410.

Stoner, G.D., Wang, L. and Casto, B.C. (2008) Laboratory and clinical studies of cancer chemoprevention by antioxidants in berries. *Carcinogenesis* 29, 1665–1674.

Sugimoto, E., Igarashi, K., Kubo, K., Molyneux, J. and Kubomura, K. (2003) Protective effects of boysenberry anthocyanins on oxidative stress in diabetic rats. *Food Science and Technology Research* 9, 345–349.

Suzutani, T., Ogasawara, M., Yoshida, I., Azuma, M. and Knox, Y.M. (2003) Anti-herpesvirus activity of an extract of *Ribes nigrum* L. *Phytotherapy Research* 17, 609–613.

Szajdek, A. and Borowska, E.J. (2008) Bioactive compounds and health-promoting properties of berry fruits: a review. *Plant Foods for Human Nutrition* 63, 147–156.

Takata, R., Yamamoto, R., Yanai, T., Konno, T. and Okubo, T. (2005) Immunostimulatory effects of a polysaccharide-rich substance with antitumor activity of isolated from black currant (*Ribes nigrum* L.). *Bioscience, Biotechnology and Biochemistry* 69, 2042–2050.

Tate, G.A. and Zurier, R.B. (1994) Suppression of monosodium urate crystal-induced inflammation by black currant seed oil. *Agents and Actions* 43, 35–38.

Terry, L.A., Crisosto, C.H. and Forney, C.F. (2009) Small fruit and berries. In: Yahia, E.M. (ed.) *Modified and Controlled Atmosphere for the Storage, Transportation and Packaging of Horticultural Commodities.* CRC Press, Boca Raton, Florida, pp. 363–395.

Tomás-Barberán, F.A. and Clifford, M.N. (2000) Dietary hydroxybenzoic acid derivatives – nature, occurrence and dietary burden. *Journal of the Science of Food and Agriculture* 80, 1024–1032.

Viola, R., Brennan, R.M., Davies, H.V. and Sommerville, L. (2000) L-Ascorbic acid accumulation in berries of *Ribes nigrum* L. *Journal of Horticultural Science and Biotechnology* 75, 409–412.

Wang, H., Cao, G. and Prior, R.L. (1996) Total antioxidant capacity of fruits. *Journal of Agricultural and Food Chemistry* 44, 701–705.

Wang, L.S., Hecht, S.S., Carmella, S.G., Yu, N., Larue, B., Henry, C., McIntyre, C., Rocha, C., Lechner, J.F. and Stoner, G.D. (2009) Anthocyanins from raspberries prevent esophaegal tumours in rats. *Cancer Prevention Research* 2, 84–93.

Werlein, H., Kütemeyer, C., Schatton, G., Hubbermann, E.M. and Schwarz, K. (2005) Influence of elderberry and blackcurrant concentrates on the growth of microorganisms. *Food Control* 16, 729–733.

Williams, R.J., Spencer, J.P.E. and Rice-Evans, C. (2004) Flavonoids: antioxidants or signalling molecules? *Free Radical Biology and Medicine* 36, 838–849.

Wolfe, K.L., Kang, X., He, X., Dong, M., Zhang, Q. and Liu, R.H. (2008) Cellular antioxidant activity of common fruits. *Journal of Agricultural and Food Chemistry* 56, 8418–8426.

Wu, Q.K., Koponen, J.M., Mykkänen, H.M. and Törrönen, A.R. (2007) Berry phenolic extracts modulate the expression of p21WAF1 and bax but not bcl-2 in HT-29 colon cancer cells. *Journal of Agricultural and Food Chemistry* 55, 1156–1163.

Wu, X., Pittman, H.E. III and Prior, R.L. (2004) Pelargonidin is absorbed and metabolized differently than cyanidin after marionberry consumption in pigs. *Journal of Nutrition* 134, 2603–2610.

Wu, X., Pittman, H.E. III, Mckay, S. and Prior, R.L. (2005) Aglycones and sugar moieties alter anthocyanin absorption and metabolism after berry consumption in weanling pigs. *Journal of Nutrition* 135, 2417–2424.

Zadernowski, R., Naczk, M. and Nesterowicz, J. (2005) Phenolic acid profiles in some small berries. *Journal of Agricultural and Food Chemistry* 53, 2118–2124.

Zhang, L., Li, J., Hogan, S., Chung, H., Welbaum, G.E. and Zhou, K. (2010) Inhibitory effect of raspberries on starch digestive enzyme and their antioxidant properties and phenolic composition. *Food Chemistry* 119, 592–599.

15   Strawberry

Jordi Giné Bordonaba and Leon A. Terry

15.1 Introduction

Strawberry fruit (*Fragaria × ananassa* Duch.) is one of the most widely consumed fruit world-wide, as fresh fruit, processed products or even as dietary supplements, and represents an overall cultivated area greater than 200,000 ha (Liu *et al.*, 2007). As fresh fruit or derived products, strawberries constitute a rich source of diverse bioactives with an array of known health-promoting properties. Recently, within a large set of 1113 food samples obtained from the US Department of Agriculture National Food and Nutrient Analysis Program, strawberries were ranked among the top three regarding their antioxidant content (3.584 mmol/serving) (Halvorsen *et al.*, 2006). In a following study, Wolfe *et al.* (2008) reported that strawberry fruit were among the largest suppliers of cellular antioxidant activity from 25 different fruit and vegetables consumed by the American population. In agreement with others, strawberries were also the top source of antioxidants from fruit and vegetables (FAV) in the Scottish population studied by Haleem *et al.* (2008).

15.1.1 The strawberry fruit

The commercial strawberry fruit belongs to the *Rosoideae* order of the *Rosaceae* family.

Strawberry is a perennial plant with rooting runners that usually bears red fruit once it is developed. Botanically, a strawberry fruit is in fact a 'false fruit', being described as a modified receptacle with one-seeded fruits or achenes located on the outer surface (Perkins-Veazie *et al.*, 1995). Whereas most crops were domesticated some 10,000 years ago, the first strawberry species can only be tracked to Roman times (approximately 2200 years ago), when wild species were grown for their appealing flavour and fragrance. Cultivation of strawberry fruit in Europe did not start until many years later, in the 14th century, when first the French, followed rapidly by the English, started to see strawberry plants not only as ornamental plants but as an intriguing food source. Towards the end of the 16th century, three European *Fragaria* species were commonly referred to – *F. vesca* (diploid), *F. moschata* (hexaploid) and *F. viridis* (diploid) – and shortly after, a new wild strawberry, discovered in the eastern part of North America (*F. virginiana*; octoploid), was introduced in European gardens, due mainly to its intense fragrance. Later, another American species, *F. chilonensis* (octoploid), characterized by good-size fruit, reached Europe after its discovery on the Pacific coast of the American continent (Medina-Minguez, 2008).

Due to the sterility of *F. chilonensis* plants when introduced in Europe, *F. moschata* and *F. viginiana* grown rapidly in between *F. chilonensis* plants, leading to a new strawberry hybrid that was later named *F. × ananassa* by Duchesne, in about 1780. The generation of this octoploid hybrid only 230 years ago was the origin of most of the cultivated strawberries now grown worldwide. Since then, hundreds of strawberry cultivars (*F. × ananassa* Duch.) have been grown, with cv. Elsanta nowadays probably being the predominant cultivar in north–western Europe.

Through history, strawberries have been appreciated not only for their particular flavour but also for their medicinal properties. Indeed, in the 13th century, when medical books were filled with botanical remedies, a Greek doctor, named Nicholas Myrepsur, detailed the medicinal properties of strawberries for the treatment of several illnesses (Medina-Minguez, 2008). Nowadays, the popularity of the strawberry may be attributed to its characteristic flavour and taste, which is defined in part by the balance between sugars and acids within the fruit, and to its known potential health-promoting properties.

15.1.2 Economic importance

Worldwide, the production of strawberries has grown steadily during the past 40 years, with most of the production in the northern hemisphere (> 95%). The USA is, in official numbers, the leading producing nation, followed by Spain, Turkey and the Russian Federation (Fig. 15.1). However, no official statistics are available for the size of the strawberry industry in China, even though it is accepted that China is nowadays a direct competitor for most of the major strawberry producing regions, with estimated values for the period 2001–2003 of *c*.1.5 million tonnes (Mt) (Carter *et al.*, 2005). In addition, strawberry production is a major part of the European soft fruit industry, accounting in the UK, for instance, for a production value of £96 million in 2003 (Defra, 2003).

15.2 Identity, Role and Bioavailability of Strawberry Bioactives

15.2.1 Introduction

Certainly, strawberry fruit have long been recognized as one of the main sources of vitamin C, folic acid and dietary fibre, as well as an excellent source of polyphenols in the diet (Table 15.1). Of all the types of bioactives present in strawberry fruit, polyphenols are without doubt the ones that have received most attention. In strawberries the main polyphenols are ellagic acid, ellagic acid glycosides, ellagitannins, gallotannins, flavanols, flavonols, anthocyanins and coumaroyl glycosides (Hannum, 2004; Zhang *et al.*, 2008), which occur with non-polyphenol bioactives such as folate and vitamin C (Tulipani *et al.*, 2009). Up to 40 different phenolic compounds have been detected recently by Aaby *et al.* (2007) in strawberry fruit cv. Senga Sengana, using high performance liquid chromatography (HPLC) coupled to various detectors (namely, diode array, mass spectrometer and coulometric array).

The following sections describe in detail the main bioactive compounds present in strawberry fruit, as well as their bioavailability.

15.2.2 Flavanols

Flavanols, not to be confused with flavonols, are a type of flavonoid that may play an important role in the prevention of certain pathologies (Santos-Buelga and Scalbert, 2000; González-Paramás *et al.*, 2006). *In vitro* studies have endeavoured to reveal the different biological activities of these compounds, some of which are antioxidants, scavengers, of free radicals, inhibitors of tumour growth and development or antibacterial agents (Pascual-Teresa *et al.*, 2000). The content of flavanols in strawberries was described in detail by Pascual-Teresa *et al.* (2000), who identified ten different compounds, catechin-(4,8)-catechin (10.1 µg/g FW), catechin (15.7 µg/g FW) and epicatechin-3-*O*-gallate (6.6 µg/g FW) being the main ones. Studies on flavan-3-ols showed that their bioavailability was mainly

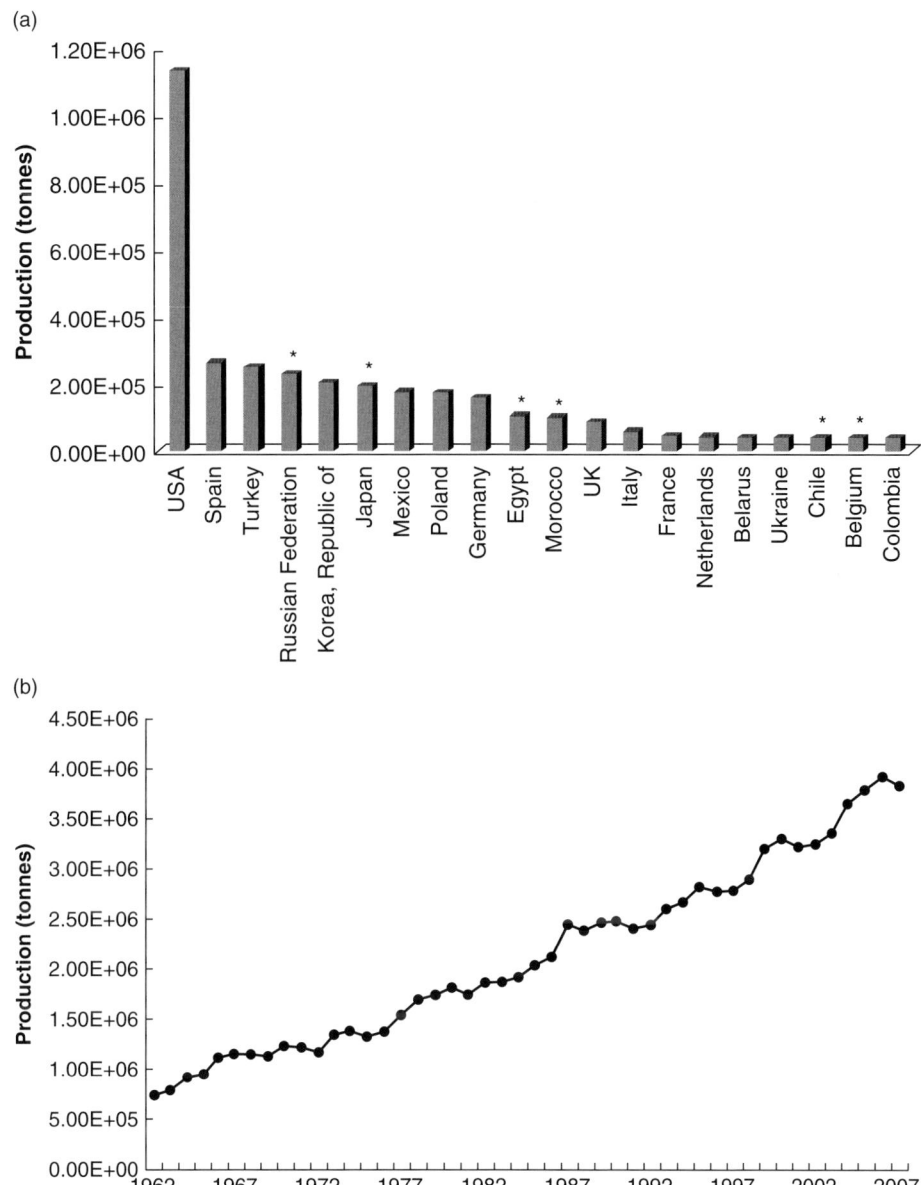

Fig. 15.1. (a) Twenty highest strawberry producing countries (tonnes) in 2007. * = FAO estimates or unofficial figures for production values. (b) Worldwide strawberry production over the past 45 years (source: FAO, 2009).

dependent on, and linked intimately to, their chemical structure. For instance, catechin and epicatechin monomers are supposed to be some of the most bioavailable polyphenols, with urinary excretions ranging from 1 to 30% of the ingested amounts (Tomás-Bar-

berán, 2008). None the less, further studies are required on other flavan-3-ols to reach any conclusion on their metabolism and differential adsorption in the human body.

Proanthocyanidins are better known as condensed tannins; they are mixtures of

Table 15.1. Nutrient and bioactive composition of strawberry fruit based on data available in the literature. Values are presented as mg or µg/g of fresh fruit (FW).

	Concentration	Reference
Dry matter (mg)	53–125	Terry et al. (2007); Tulipani et al. (2008); Giné Bordonaba and Terry (2009)
Fibre (mg)	23	ESHA food database
Sugars (mg)	61.95–110.45	Terry et al. (2007); Giné Bordonaba and Terry (2009)
Glucose	18.01–31.00	
Fructose	22.54–36.10	
Sucrose	21.40–43.35	
Organic acids (mg)	5.96–14.29	Terry et al. (2007); Giné Bordonaba and Terry (2009)
Ascorbate (vitamin C)	0.24–0.74	
Citrate	4.2–10.1	
Malate	1.52–3.45	
Proanthocyanidins (flavanols) (mg)	4.47	Hosseinian et al. (2007)
Monomers (epicatechin, B2)	0.31	
Dimers	0.39	
Oligomers	1.55	
Polymers (up to hexamer)	2.22	
Anthocyanins (µg)	66–571	Wang et al. (2003); Kosar et al. (2004); Määttä-Riihinen et al. (2004); Skupień anc Oszmiań
ski (2004); Terry et al. (2007)
Cyanidin-3-glucoside	4.5–34	
Pelargonidin-3-glucoside	53–441	
Pelargonidin derivatives	8.4–95.9	

Ellagic acid (µg)	19.9–522	Gil et al. (1997); Häkkinen and Törrönen (2000); Terry et al. (unpublished data)
Folate (vitamin B) (µg)	0.13–0.96	Strålsjö et al. (2003); Tulipani et al. (2008)
Resveratrol (µg)	0.12–2[a]	Wang et al. (2007)
Quercetin (µg)	3–40	Gil et al. (1997); Häkkinen and Törrönen (2000)
Kaempferol (µg)	2–13.7	Gil et al. (1997); Häkkinen and Törrönen (2000)
Total phenolics[b]	0.86–3.75	Terry et al. (2007); Giné Bordonaba and Terry (2009)
Antioxidant capacity[c]	6.2–17.8	Terry et al. (2007); Giné Bordonaba and Terry (2009)

Notes: [a]Values for pulp and achenes, respectively; [b]measured by the Folin–Ciocalteu assay and results expressed as mg gallic acid equivalents (GAE)/g FW; [c]measured by the FRAP assay and results expressed as µmol Fe^{2+}/g FW.

oligomers and polymers built up of flavan-3-ol units and principally held together by C4-C8 bonds (Gu et al., 2003). Recent evidence suggests that certain proanthocyanins occur not only in red wines but also in several food-stuffs, including strawberries (Pascual-Teresa et al., 2000; Fossen et al., 2004; González-Paramás et al., 2006). Proanthocyanidins are supposed to contribute to the 'French paradox phenomenon' by exerting several health-promoting properties, such as antioxidant, anti-inflammatory and anticarcinogenic activity (Santos-Buelga and Scalbert, 2000). The proanthocyanidin concentrations in strawber-ries and other berries have been described recently by Hosseinian et al. (2007). When the whole fruit was considered, strawberry (4.47 mg/g FW) ranked second after raspberry (5.05 mg/g FW) in terms of total proanthocy-anidin concentration. In the same study, 0.31, 0.39, 1.55 and 2.22 mg/g FW corresponded to monomers, dimers, oligomers and polymers of the total anthocyanidin fractions encoun-tered in the fruit.

15.2.3 Flavonoids

Flavonoids are a type of polyphenolic com-pound that can be divided into different subclasses, including flavanols, flavonols, fla-vones, flavanones, isoflavones and anthocy-anidins. Within the group of flavonoids, most attention has probably been paid to the antho-cyanins. These are water-soluble, flavonoid-type polyphenols widely expressed in the plant kingdom. Anthocyanins have been reported to have anticarcinogenic, anti-inflammatory, vasoprotective and anti-obesity properties (McGhie and Walton, 2007). In addition, research suggests that anthocyanins may play a role in enhancing vision and improving memory (Joseph et al., 1998). Recent reports estimate an anthocyanin con-sumption in the USA of c.12.5 mg/day (Wu et al., 2006), a value that is exceeded in certain European countries (Mullen et al., 2008). Yet, greater consumption does not always result in greater bioavailability and/or effect. Gener-ally, these pigments are extracted using aque-ous mixtures of ethanol, methanol or acetone (Terry et al., 2007; Giné Bordonaba and Terry,

2008); however, discrepancies exist in whether acetone- or methanol-based solvents appear more efficient for the extraction of these com-pounds from different FAV (Giné Bordonaba and Terry, 2008). Separation and quantifica-tion of anthocyanins are generally achieved using reversed phase HPLC coupled to vari-ous detection systems (usually a photo diode array (PDA) or mass spectrometry (MS)) and acetonitrile is generally as the mobile phase of choice, due to its elution strength, low viscos-ity and good miscibility with water.

Anthocyanin concentration in straw-berry fruit varies greatly between cultivars and is also influenced by growing conditions and maturity at harvest (Wang and Lin, 2000; Lopes da Silva et al., 2007; Terry et al., 2007; Crespo et al., 2010). In contrast to some other fruit, a vast array of information is available detailing the anthocyanin profile and concen-tration of strawberry fruit, all of the relevant studies identifying pelargonidin-3-glucoside as the main anthocyanin, representing over 80% of the total anthocyanin pool (Fig. 15.2).

Lopes da Silva et al. (2007) detected, through a detailed study of strawberry pig-ments (by means of HPLC coupled to PDA and MS detection), up to 25 different anthocy-anins within the five different strawberry culti-vars analysed. The authors highlighted the notable variability among anthocyanin con-tent in samples of the same variety and har-vest, and therefore pointed out the strong influence of degree of maturity, edaphic–climatic conditions and postharvest storage on the concentration of these pigments. Terry et al. (2007) identified three major anthocyanins (namely, cyanidin-3-glucoside, pelargonidin-3-glucoside derivative and pelargonidin-3-glucoside at concentrations of 2.165, 33.56 and 121.54 μg/g FW, respectively) in straw-berry cv. Elsanta fruit grown in a glasshouse and subjected to full or deficit irrigation (Fig. 15.3). In the same work, the authors dem-onstrated that differences existed in anthocy-anin content between primary and secondary fruit from the same primary truss (Fig. 15.3). Similar concentrations were later reported by Crespo et al. (2010) when studying anthocy-anin concentrations in fruit from four different cultivars grown at different Swiss production sites.

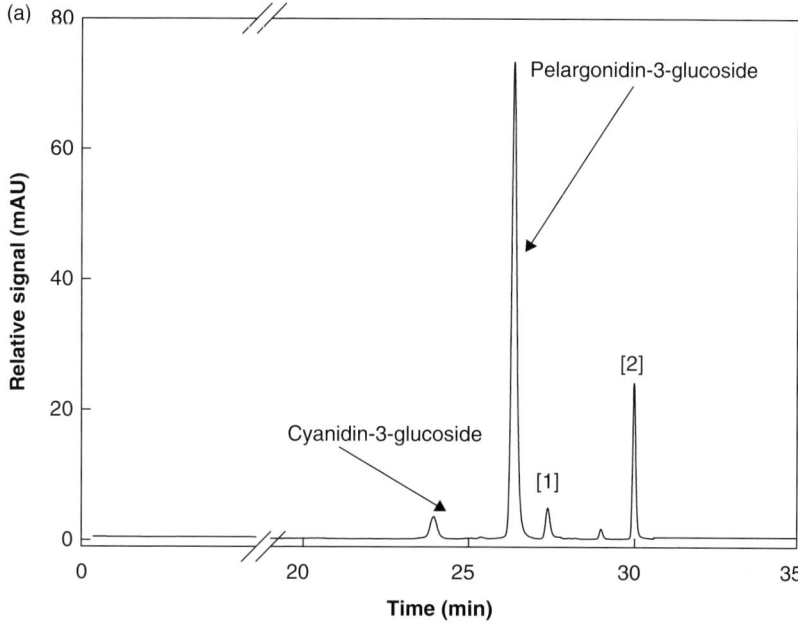

Fig. 15.2. Anthocyanin profile (a) and chemical structure (b) of the main anthocyanins identified in strawberry fruit (cv. Matis). Peaks [1] and [2] are pelargonidin derivatives.

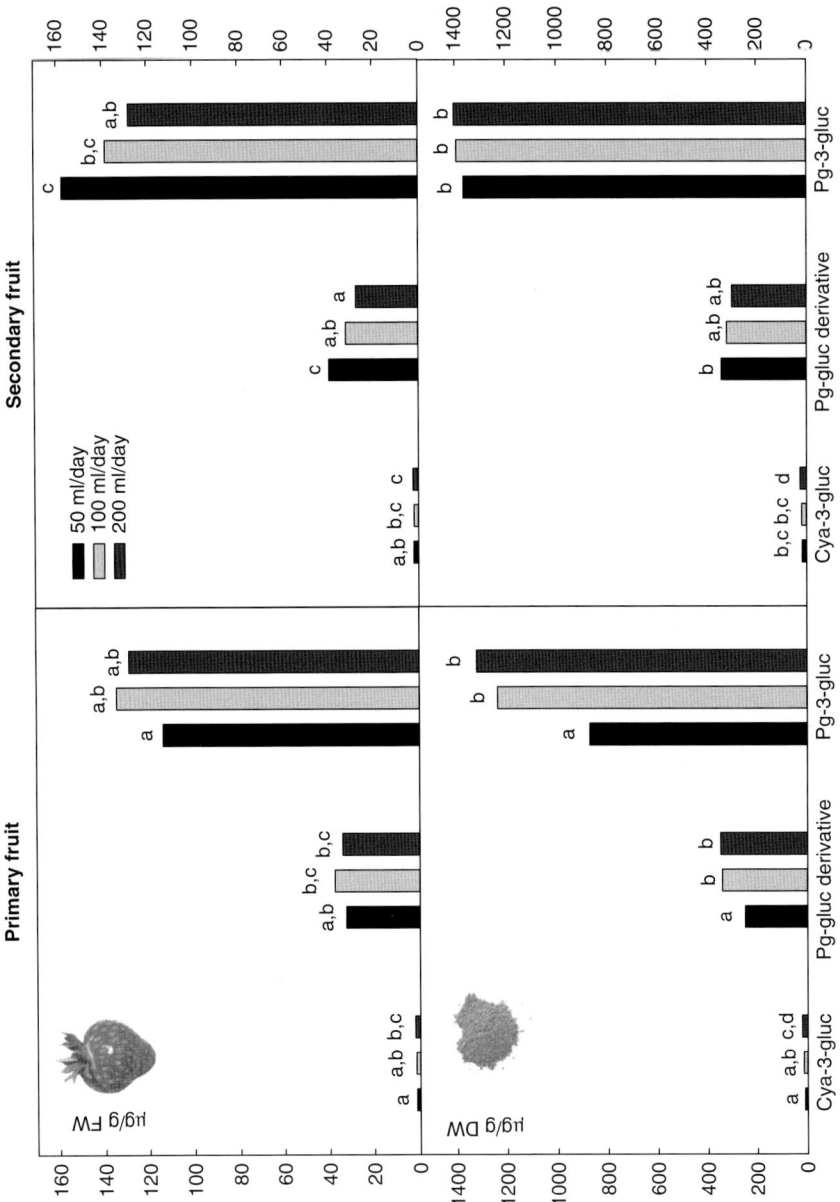

Fig. 15.3. Effect of different water irrigation regimes (ml/day) on the concentration of main anthocyanins (cyanidin-3-glucoside (Cya-3-gluc), pelargonidin-3-glucoside (Pg-3-glucoside) and pelargonidin glucoside derivative (Pg-gluc derivative)) in strawberry (cv. Elsanta) primary and secondary fruit. Results are expressed on a fresh weight (FW; upper panel) and dry weight (DW; lower panel) basis. The different letters above the columns indicate significant differences between fruits from different irrigation regimes.

In addition to fruit position, anthocyanin profiles differ spatially with different tissues/locations within the fruit. Aaby *et al.* (2005) showed the different anthocyanin profiles in receptacle tissue and achenes from two different strawberry cvs. (Totem and Puget Reliance). Both cultivars had pelargonidin-3-glucoside as the main anthocyanin in the flesh, whereas similar amounts of this anthocyanin and cyanidin-3-glucoside were detected in the achenes of both cultivars. Almeida *et al.* (2007) later demonstrated that anthocyanin levels increased during ripening, whereas flavan-3-ols decreased, in both cvs. Queen Elisa and Korona. Indeed, the authors demonstrated that, whereas, at the white stage, flavan-3-ol content was associated mainly with vascular epidermal tissue and pith, it was limited mostly to vascular tissues at more advanced development stages. Recently, Zhang *et al.* (2008), when studying the antioxidant properties of different strawberry fractions, found that anthocyanins including cyanidin-3-glucoside (7156 μM Trolox/mg), pelargonidin (4922 μM Trolox/mg) and pelargonidin-3-rutinoside (5514 μM Trolox/mg) gave greater antioxidant activities, as determined by the popular Trolox equivalent antioxidant capacity (TEAC) assay (see Chapters 18 and 19 of this volume), than other purified fractions from the same berries. In the same study, the authors highlighted that strawberry extracts and their purified compounds had antiproliferative activity in a dose-dependent manner when assessed in different lines of human cancer cells (oral, colon and prostate).

The biological activity of strawberry fruit is due in part to the biological activity of its polyphenols, including anthocyanins. Therefore, it is crucial to determine the bioavailability of these compounds by means of human and animal studies designed to: (i) obtain information on target organ and organ tissue distribution; (ii) assess the concentrations reached in the organs, tissues or fluids; and also (iii) assess the potential toxic effects, if any. Although not much information exists in this regard, it is evident that the body of literature regarding the bioavailability of strawberry bioactives is currently growing, and data are available on the bioavailability of

ellagic acid, ellagitannins, anthocyanins, procyanidins and flavonols (Tomás-Barberán, 2008). In the particular case of strawberry anthocyanins, it has been reported that consumption of 200 g of strawberries per person resulted in a mean concentration of pelargonidin-3-glucuronide in blood plasma of 274 nmol/l (Mullen *et al.*, 2008). Another human study, with six healthy volunteers, three from each gender, was that conducted by Felgines *et al.* (2003). In this particular case, results showed that after consumption of 200 g of strawberries by each volunteer, pelargonidin-3-glucoside, its aglycon and three monoglucuronides, as well as a sulfo-conjugate were detectable as urinary metabolites. Yet, the total urinary excretion of anthocyanin metabolites represented no more than 1.80 ± 0.29% of the total pelargonidin-3-glucoside ingested. Hollands *et al.* (2008) concluded that strawberry anthocyanins were partially bioavailable, with a linear relation between intake and urine excretion (each additional unit of dose ingested resulted in 0.0166 units excreted). Similar findings were shown by Carkeet *et al.* (2008) where the linear response doses were reported to be in the range of 15–60 μmol. In all cases, pelargonidin, the main anthocyanin detected in strawberry fruit, seems to be better absorbed and excreted (with recoveries in urine of 0.58% over 24 h) than any other anthocyanins commonly present in strawberry fruit (0.084–0.087%) (Carkeet *et al.*, 2008). Although anthocyanins are absorbed mainly in the duodenum, with very little or no absorption taking place in the small intestine (Tomás-Barberán, 2008), Andres-Lacueva *et al.* (2005) have shown that these compounds can even be detected in the brain. For the interested reader, an excellent review on the bioavailability and absorption of anthocyanins is that published by McGhie and Walton (2007). Regardless of the poor bioavailability, it is accepted that the little portion that is absorbed may be very biologically active (Carkeet *et al.*, 2008).

Other types of flavonoids are flavonols, which are present in strawberries as glucosides and glucuronides of quercetin and kaempferol aglycons (Seeram *et al.*, 2006c). Epidemiological data suggest an inverse relationship between intake of flavonols and the incidence of cardiovascular disease or certain cancer types (Knekt *et al.*, 2002). In this context, Häkkinen *et al.* (1999) studied the flavonol content of different berries, including those from strawberry cvs. Senga Sengana and Jonsok, and found both quercetin (8 µg/g FW) and kaempferol (5–8 µg/g FW) in relatively low concentrations in the strawberries. In another study, Seeram *et al.* (2006c) identified quercetin-rutinoside, quercetin-glucoside, quercetin-glucuronide and kaempferol-glucuronide as the main flavonols in strawberry fruit. However, according to the literature, flavonol content in strawberries varies drastically. These great variations may be explained, in part, by differences in the genotypes studied, the growing conditions and the sample preparation and extraction methodology used (Häkkinen, *et al.*, 1999). In addition, as observed for anthocyanins (Terry *et al.*, 2007), flavonol concentrations may also vary according to fruit position on the cymose inflorescence. The bioavailability of flavonols has been investigated too, and, so far, there seems to be enough scientific evidence to suggest that monoglucosides are absorbed better than their corresponding aglycones, which in part seems to be related to the presence of glucose transporters in the intestinal wall (Tomás-Barberán, 2008).

There is a paucity of research data available describing the effect that different food matrices may have on the absorption and metabolism of strawberry bioactives in general, and flavonoids in particular. However, Mullen and co-authors demonstrated that consumption of strawberries with 100 ml of double cream delayed absorption of pelargonidin-3-glucoside by more than 1 h, but had no effect on the C_{max}, as compared with fruit ingested without cream (Mullen *et al.*, 2008).

15.2.4 Ellagic acid, ellagitannins and derivatives

Ellagic acid is a polyphenol occurring naturally, in both the free form and esterified to glucose in water-soluble, hydrolysable tannins, in certain fruit and nuts. Although concentrations of ellagic acid for strawberry differ markedly according to the literature (19.9 and 522 µg/g FW; Gil *et al.*, 1997, and Häkkinen and Törrönen, 2000, respectively), strawberry fruit are one of the richest sources of this polyphenol, especially if compared with other commonly consumed fruit (Williner *et al.*, 2003). Fruit tissue distribution and variability among cultivars have been highlighted by several works (Maas *et al.*, 1991; Atkinson *et al.*, 2006). Ellagic acid content was shown to decrease during fruit development in all of the five cultivars studied by Williner *et al.* (2003). For instance, cv. Chandler had 2.07 ± 0.10 mg/g DW at white stage, which decreased to 1.08 ± 0.11 and 0.46 ± 0.07 mg/g DW, respectively, in 50% and 100% red stage fruit (Williner *et al.*, 2003). Ellagic acid, per se, has been associated with numerous health-promoting properties and up to now many data support the role of this phenolic compound as a chemopreventive agent (Hannum, 2004). Häkkinen *et al.* (2000) reported that ellagic acid in strawberries accounted for approximately 51% of the phenolic profile from this berry; besides, significant differences in ellagic acid concentrations existed between cultivars (396 µg/g FW in cv. Senga Sengana but 522 µg/g FW in cv. Jonsok). Genotypic differences in ellagic acid concentrations (60–341 µg/g FW) and in the ratio between conjugated ellagic acid and free ellagic acid were also seen by Atkinson *et al.* (2006). Häkkinen *et al.* (2000) detected other selected phenolic acids in strawberries at lower proportions than those described for ellagic acid (namely, kaempferol, 3.1%; quercetin, 6.0%; myrecitin, 1.6%; *p*-coumaric acid, 34.3%; *p*-hydroxybenzoic acid, 4%). Whereas, for most berries, ellagic acid seems to be concentrated in the seeds, Daniel *et al.* (1990) found that most of the ellagic acid in strawberries was to be found in the pulp (95.7%). In the same study, ellagic acid concentrations (~ 63 µg/g FW) in strawberry fruit were below the concentrations found in other berries (raspberry and blackberry: ~ 150 µg/g FW) but were far superior to those found in the other fruit analysed. The higher concentration in pulp than achenes was later highlighted as the possible explanation for the greater bioavailability of ellagic acid from strawberries compared with that from other fruit (Hannum, 2004). In contrast, in a more recent study conducted by Aaby and collaborators, the achenes from two different strawberry cultivars were the main source of ellagic acid (cv. Totem 87.3 ± 14.6 mg/100 g FW and cv. Puget 34.4 ± 3.7 mg/100 g FW) or ellagic acid derivatives as compared with the flesh (cv. Totem 0.3 mg/100 g FW and cv. Puget 0.2 mg/100 g FW) (Aaby *et al.*, 2005). In this context, the discrepancies encountered between the study of Aaby *et al.* (2005) and that of Daniel *et al.* (1990) are due most probably to variations in the methodologies used, such as hydrolysis conditions, extraction solvents used, etc.

Aaby *et al.* (2005) not only reported greater amounts of ellagic acid and anthocyanins in achenes than pulp but also found up to 20-fold greater total antioxidant activities and 10 times larger total phenolic values in achenes than in the flesh of the different strawberry cultivars investigated. In agreement, Terry (2002) and Terry *et al.* (2004) also showed, by thin layer chromatography (TLC), that strawberry achenes contained larger amounts of antifungal compounds, many of them being phenolic compounds, than other fruit tissues (Fig. 15.4).

Nevertheless, most of the achenes from the fruit tend to pass intact through the alimentary canal, ending up in the faeces, and hence any bioactives present in this fruit tissue are probably unavailable.

Other strawberry compounds derived from ellagic acid are ellagitannins, which are water-soluble polyphenols belonging to the hydrolysable tannins class. This type of compound can occur in complex polymeric forms of high molecular weight. Ellagitannins can be quantified directly or indirectly by hydrolysing the polymer with an acid or base, to yield ellagic acid (Häkkinen *et al.*, 2000), thus resulting in the concept of free ellagic acid or conjugated ellagic commonly found in the literature (Fig. 15.5).

Seeram *et al.* (2006c) identified sanguiin H-6 and ellagic acid glycosides among the major hydrolysable tannins present in strawberry fruit, whereas other types of ellagitannins were reported by Cerdà *et al.* (2005). Cerdà and collaborators studied the metabolism of ellagitannins from strawberries and identified urolithin B, a previously identified antiangiogenic and hyaluronidase inhibitor compound, as a suitable biomarker for ellagitannin consumption in healthy humans. After strawberry (cultivar not specified) intake, urolithin B excretion was c.2.8% of the ingested dose and varied considerably between individuals. Similar findings were observed by Seeram *et al.* (2006c) when studying the pharmacokinetic parameters of ellagic acid after ingestion of pomegranate juice. Despite the limited absorption of either ellagic acid or ellagitannins (Tomás-Barberán, 2008), these compounds and metabolites have been detected in the kidneys and liver of rats, as well as the prostate gland of mice (Cerdà *et al.*, 2005; Seeram *et al.*, 2006b; Seeram and Heber, 2007).

15.2.5 Ascorbate, folate and other bioactives

With constant advances in analytical sciences, new strawberry-derived compounds have been identified and associated with potential health-promoting properties. Resveratrol, a compound commonly found in grapes and synthesized from cinnamic acid derivatives, has been indentified in strawberry fruit by Wang *et al.* (2007). During the past decades, several studies have shown the potential health-promoting properties of resveratrol (namely, antioxidant, anticarcinogenic, anti-inflammatory, cardioprotection) (for the interested reader, see the review by Aggarwal *et al.*, 2004), and, as a result, increasing interest has been raised to identify new sources of this phytoalexin among fruit and vegetables. Even though there was evidence of this compound in other berries, the first detailed study on the compound in strawberries was probably that conducted by Wang *et al.* (2007). The authors not only studied resveratrol content of strawberry cvs. Kent and Earliglow grown in a glasshouse but also detailed the effect that different preharvest factors (namely, genotype, fruit maturity, cultural practices and environmental conditions) had on that content. Resveratrol concentration was greater in fully ripe than in early ripe berries, as well as being higher in

(a)

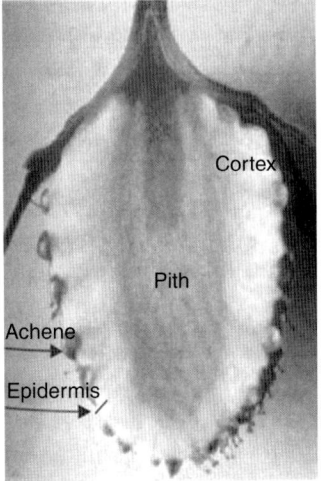

(b)

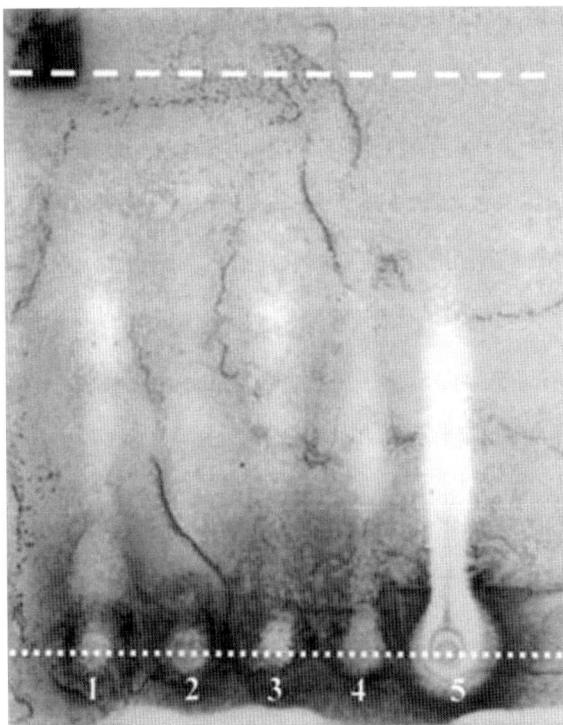

Fig. 15.4. (a) Section of a strawberry fruit and differentiation between main fruit tissues. (b) TLC *Cladosporium cladosporioides* bioassay of cv. Elsanta fruit crude ethanol extracts (100 ml, 0.2 µl/g FW), from different tissues of Green I strawberry fruit, run in hexane:ethyl acetate:methanol (60:40:10 v/v/v) (Terry, 2002; Terry *et al.*, 2004).

(a)

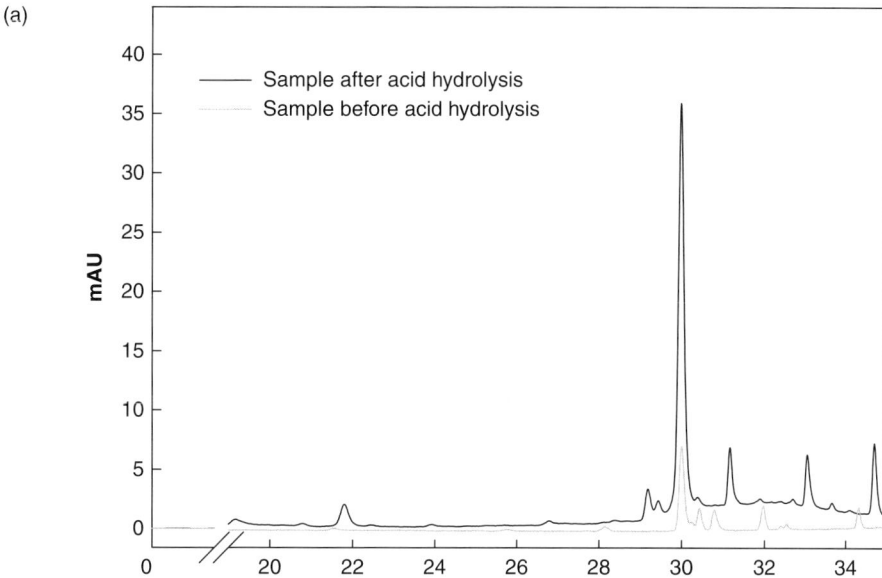

(b)

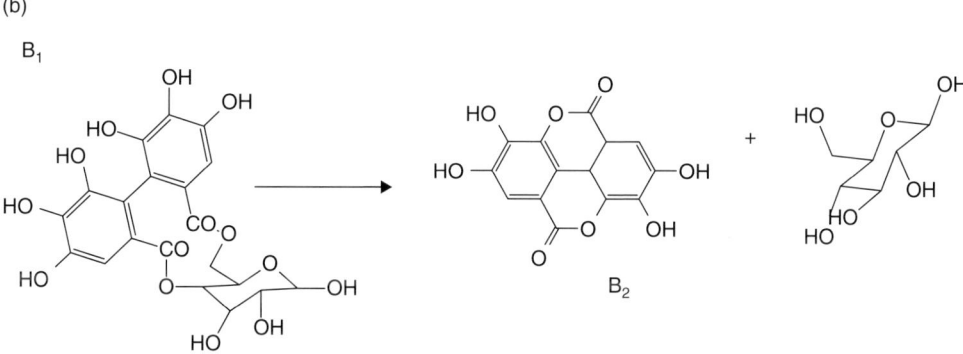

Fig. 15.5. (a) Chromatographic profile of ellagic acid in strawberry fruit (cv. Camarosa) prior to and after 90 min hydrolysis. (b) General scheme for the hydrolysis of ellagitannins from strawberry fruit; before hydrolysis the principal components correspond to ellagic acid glycosides, simple ellagitannins (B₁) and complex oligomeric ellagitannins that, when hydrolysed, give rise to ellagic acid (B₂), methyl gallate and methyl sanguisorboate.

achenes (~ 2 µg/g DW) than in pulp (~ 0.12 µg/g DW). Overall, values for resveratrol concentrations in strawberry fruit were far below the ~ 10 µg/g FW that could be found in the skin of red mature grapes cv. Napoleon (Cantos *et al.*, 2000).

Another compound found in strawberry fruit with reported anti-inflammatory, antiproliferative and antiangiogenic (Sung *et al.*, 2007) as well as memory enhancer properties (Maher *et al.*, 2006) is fisetin. Fisetin is a flavo-

noid occurring naturally in certain fruit and vegetables, with concentrations ranging from 2–160 µg/g (Arai *et al.*, 2000). The mechanism by which fisetin exerts its anticarcinogenic effects still remains unclear, although recent research has shown that it inhibits cyclin-dependent kinase 6 and downregulates nuclear factor-κB-regulated cell proliferation (Sung *et al.*, 2007).

In addition to all the polyphenols mentioned earlier, strawberry fruit are one of the richest natural sources of two water-soluble vitamins, folate and ascorbate. Inadequate folate status in humans has been associated with an increased risk of the chronic diseases that may affect the elderly population in particular (Rampersaud *et al.*, 2003; Tulipani *et al.*, 2009). Similarly, ascorbate deficiency is known to be related to detrimental health effects (see Chapter 14 of this volume). For the determination of folate content in strawberry fruit, both a microbial assay and a radio-protein binding assay have been developed and validated recently in berry fruits (Stråsljö *et al.*, 2003; Tulipani *et al.*, 2008). In a recent study, total folate content from nine Italian strawberry cultivars ranged from ~ 0.2 to ~ 1.0 µg/g FW (Tulipani *et al.*, 2008) and was in agreement with earlier studies conducted on Swedish grown strawberries (0.73–0.99 µg/g FW; Stråsljö *et al.*, 2003). In this context, it is generally accepted that 250–350 g of berries (~ 125 µg folate) can provide almost the totality of European daily intake recommendations (200–300 µg/day) (Bailey and Gregory, 1999), making strawberries one of the most appealing sources of this vitamin. Supplementation with folate to individuals with homocysteinaemia has been shown to reduce levels of homocystein in plasma, and therefore reduce the risk of heart disease (Spiller and Dewell, 2003).

Ascorbic acid content in strawberry fruit varies drastically among different genotypes (Fig. 15.6) and agroclimatic conditions (Cordenunsi *et al.*, 2005; Atkinson *et al.*, 2006; Terry *et al.*, 2007; Giné Bordonaba and Terry, 2009, 2010; Crespo *et al.*, 2010), with reported concentrations ranging from 0.2–0.9 mg/g FW (Atkinson *et al.*, 2006; Terry *et al.*, 2007; Giné Bordonaba and Terry, 2009; Crespo *et al.*, 2010).

15.3 Chemopreventive and Health-related Properties

15.3.1 Introduction

The antioxidant properties of strawberries are well documented (Wang and Lin, 2000; Terry *et al.*, 2007; Wolfe *et al.*, 2008). Since oxidative stress has been suggested to play an important role in the development of certain conditions, including cancers, it was expected initially that the antioxidant capacity of the fruit would correlate well with its antiproliferative properties. Nevertheless, Meyers *et al.* (2003) demonstrated that antioxidant activity from eight different strawberry cultivars was not related to their antiproliferative properties. Nowadays, a plethora of research studies has shown that berries, including strawberry, as well as berry purified phenolic compounds, inhibit cancer cell proliferation, regulate cell cycle arrest and, in some cases, induce apoptosis by multimechanistic means of actions beyond antioxidation (reviewed by Seeram and Heber, 2007). Besides, little or no cytotoxic effect was observed when strawberry extracts were tested on normal cells, resulting in no doubts about the potential anticancer effects of strawberry fruit (Seeram, 2008). As for many other fruit and vegetables, most of the reported anticancer activity of strawberry fruit is based on *in vitro* studies rather than *in vivo* or intervention trials. Among the limitations of *in vitro* tests, several authors (Roques *et al.*, 2002; Kern *et al.*, 2007) have pointed out that the results of *in vitro* assays may be an artefact from the generation of hydrogen peroxide in the culture media by the antioxidants tested, and hence it is evident that results from this type of assay should be interpreted cautiously.

In this context, the following sections aim to describe the latest scientific evidence obtained from *in vitro* and *in vivo* experiments or intervention studies regarding the beneficial effects that strawberry fruit or their extracts (namely, freeze-dried powders, concentrates, etc.) may have on preventing or fighting against cancer as well as other illnesses (Table 15.2).

15.3.2 Cancer studies

Several studies have demonstrated that strawberry extracts inhibit the growth of human carcinoma cells when tested *in vitro* (Table 15.2). For instance, strawberries effectively inhibited, by different mechanisms, the growth of oral, breast and prostate (Seeram *et al.*, 2006a) or liver cancer cell lines (Ramos *et al.*, 2005) in a dose-dependent manner. Ramos *et al.* (2005) proved that whole strawberry extracts arrested

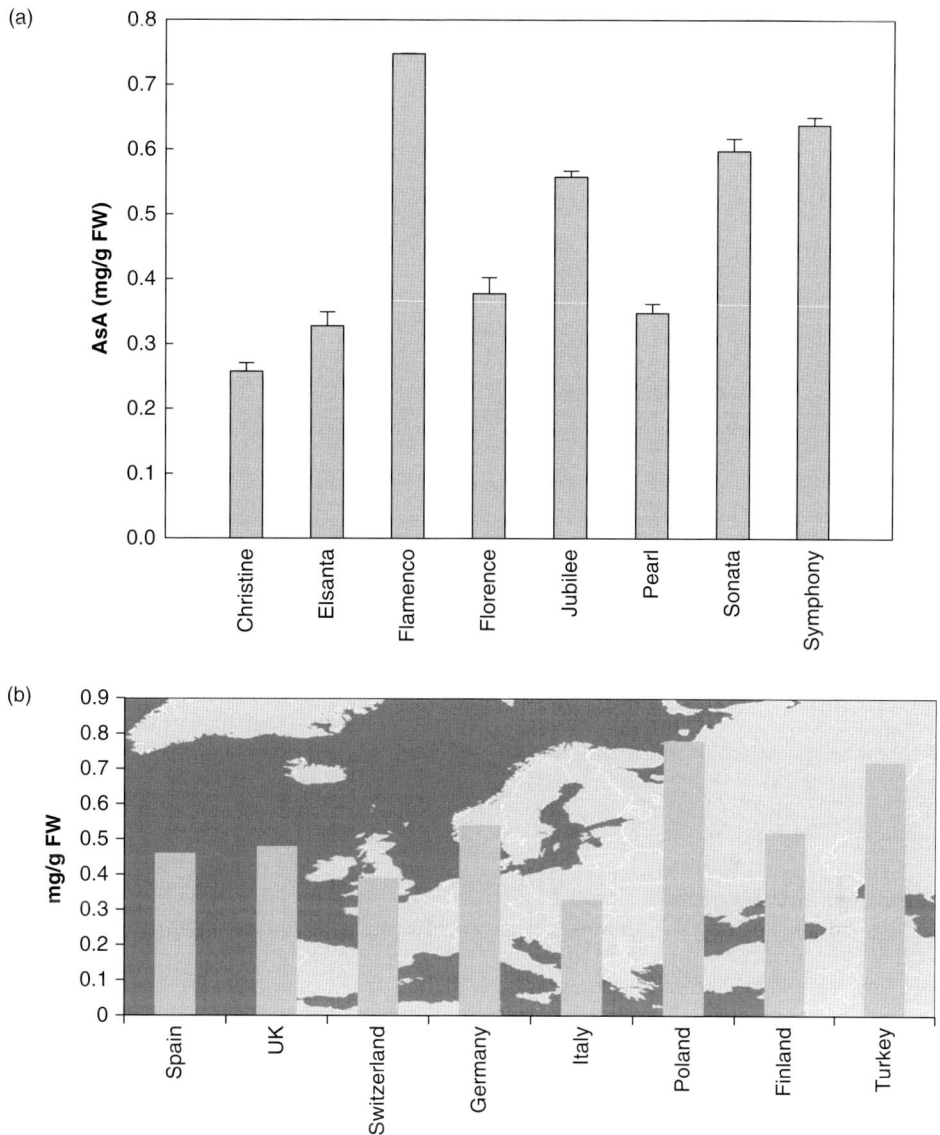

Fig. 15.6. (a) Ascorbic acid (AsA) concentrations (mg/g FW) in a range of UK grown strawberry cultivars (Giné Bordonaba and Terry, 2009). (b) Reported AsA concentrations (mg/g FW) in strawberry fruit from different locations in Europe (NB: values correspond to average concentrations from different cultivars and different harvest years obtained from various literature).

the G1 phase, therefore showing proapoptotic effects. In a more recent study, Wu and collaborators (2007) investigated whether strawberry and other berry extracts had any effect on cell viability and expression of apoptotic cell markers in human HT29 colon cancer cells. The results suggested that berry extracts inhibited

cell proliferation through the cyclin kinase inhibitor pathway, p21WAF1 (Wu *et al.*, 2007). Other *in vitro* studies with extracts from two different strawberry cultivars (namely, Sweet Charlie and Carlsbad) showed the potential of both cultivars to inhibit breast cancer and cervical cancer cell proliferation (Wedge *et al.*,

Table 15.2. Reported health-promoting properties of *Fragaria* × *ananassa* fruit or derived products.

Activity	Action	System	Dose	Extract type[a]	Reference
Anticarcinogenic	Inhibit the initiation and promotion of oesophageal cancer	Rat	Diet (AIN76) containing 5 and 10% of strawberries (~ ellagic acid concentration was 0.34 and 0.67 mg/kg diet, respectively)	Freeze-dried strawberry puree	Stoner et al. (1999)
Anticarcinogenic	Reduction of tumour multiplicity in oesophageal cancer	Rat	Diet (AIN76) containing 5 and 10% of strawberries	Freeze-dried strawberry (var. Comander)	Carlton et al. (2001)
Anticarcinogenic	Protective effect against endogenous generation of carcinogens	27 male and 13 female volunteers (ten healthy subjects in each group)	300 g on the 4th day of a 4-day trial	Fresh strawberries	Chung et al. (2002)
Anticarcinogenic	Inhibition of cancer cell proliferation by inhibiting transcription factors and activating protein-1 (AP-1) and nuclear factor kappa B (NFκB)	In vitro: human lung epithelial cancer A549 cells and mouse epidermal JB6 P+ cell lines	–	Filtered and diluted strawberry homogenates	Wang et al. (2005)
Anticarcinogenic	Growth inhibition of liver cancer cells	In vitro: HepG2 cell cultures	0.1–0.8 mg/ml	Lyophilized strawberry extract	Ramos et al. (2005)
Anticarcinogenic	Growth inhibition of oral, breast and prostate cancer cells	In vitro: cell cultures	25–200 µg/ml	Polyphenolic enriched strawberry extract (sugars and acids removed by C18)	Seeram et al. (2006a)
Anticarcinogenic	Reduced cell proliferation through the cyclin kinase inhibitor pathway p21WAF1	In vitro: HT29 colon cancer cells	0–60 mg/ml	Homogenized strawberry extract	Wu et al. (2007)
Anticarcinogenic	Inhibition of nuclear factor of activated T cells (NAFT) and tumour necrosis factor (TNF)	In vitro: mouse epidermal JB6 Cl 41 cell lines	1–100 µg/ml	Lyophilized strawberry extract	Li et al. (2008)
Anticarcinogenic	Inhibition of tumour formation of oral cancer cells	Hamster	Diet (AIN76) containing 10% of strawberries or strawberries enriched with selenium (0.5 ppm)	Lyophilized strawberry or selenium-enriched lyophilized strawberries	Warner et al. (2008)

Health effect	Effect	Model/system	Dose	Strawberry form	Reference
Anticardiovascular disease	Endothelium-dependent vasorelaxation through the activation of PI3 kinase/akt	Rabbit aorta	0.1–10 mg/ml	Freeze-dried strawberry powder	Edirisinghe et al. (2008)
Anticardiovascular disease	Greater reduction in oxidative damage to low-density lipoproteins	Human intervention study on 28 hyperlipidaemic subjects	454 g/day in a randomized 1-month crossover study with a 2-week washout	Fresh berries	Jenkins et al. (2008)
Anticardiovascular disease	Antiplatelet activity	Mice	~ 11 ml/kg	Strawberry filtrate	Naemura et al. (2008)
Anticardiovascular disease	Borderline significant, multi-variate 14% lower likelihood of an elevated C-reactive protein of ≥ 3 mg/l	Cohort study	At least two servings/week	–	Sesso et al. (2007)
Anticardiovascular disease	Hypocholesterolaemic effects and reduced lipid peroxidation	Women suffering metabolic syndrome	25 g/day	Freeze-dried strawberry powder	Basu et al. (2009)
Anti-inflammatory	Inhibition key inflammation enzymes (COX1)	Cell culture	125 µg/ml	Lyophilized crude anthocyanins from strawberries cv. Honeoye	Seeram et al. (2001)
Antineuro-degenerative	Reversion of the ageing process by protecting against decrease in mental performance	Rat	9.5 g/kg of standard diet for 6 months	Lyophilized strawberry powder	Joseph et al. (1998)
Antineurodegenera-tive	Reduction of oxidative stress-induced neurotoxicity	Neuronal cells	100–2000 µg/ml	Phenolics extracted from 10 g of dried sample	Heo and Lee (2005)
Antidiabetic	Limiting post-meal blood glucose levels by reducing α-amylase	In vitro assays	0–500 µg/assay	Phenolic-rich fractions from breeding variety 932034 and cv. Elsanta from local growers	McDougall et al. (2005)
Antiobesity	Increased weight gain / Reduced weight gain	Mice	Diet containing 10% freeze-dried strawberry powder / Purified anthocyanins from strawberry given in the water	Whole fruit / Purified anthocyanins from strawberries	Prior et al. (2008)

Note: [a]Most strawberry extracts used for *in vitro* studies are based on acidified methanol aqueous extraction followed by an acetone–water extraction, evaporation of the solvents and resuspension of the extracts in water prior to being deposited on to the cell cultures. Whenever data are available, the cultivars used are specified. (–) concentration dose not specified.

2001). Li *et al.* (2008) not only demonstrated the anticarcinogenic properties of strawberry extracts but also elucidated that they specifically inhibited nuclear factor of activated T cells (NFAT) and tumour necrosis factor (TNF). Results from the same study suggested that the chemopreventive properties of black raspberry and strawberry bioactives might be targeted through different signalling pathways. In an earlier study, Wang *et al.* (2005) showed that strawberry extracts suppressed cancer cell proliferation and transformation by means of inhibiting the transcription factors, activating protein-1 (AP-1) and nuclear factor kappa B (NFκB). In the same study, the authors postulated that the antioxidant properties and the ability to reduce oxidative stress most probably were related to the ability of the same extracts to block ultraviolet B (UVB) and TPA-induced AP-1 and NFκB activation.

Besides the earlier highlighted mechanisms of action, strawberry fruit are thought to possess anticarcinogenic effects by inhibiting possible mutations. In this context, Hope *et al.* (2004) pointed out that strawberry tannin fractions were very effective in inhibiting mutations caused by both methyl methanesulfonate and benzopyrene.

Whereas animal and human studies are still limited, many research groups are currently investigating the possible role of strawberry fruit in preventing not only cancer but also several other diseases. The dietary intake of products rich in ellagitannins, such as strawberries, has been shown to inhibit the initiation and promotion of oesophageal cancer (Stoner *et al.*, 1999). Stoner and collaborators included 5 or 10% of freeze-dried strawberries into the diet of rats prior to the induction of oesophagus cancer with *N*-nitrosomethylbenzylamine (NMBA) and observed an inhibition effect depending on the dose concentration. Using a similar approach, the chemopreventive effect of strawberry lyophilized extracts on NMBA-induced rat oesophagus carcenogenesis was studied by Carlton *et al.* (2001). They proved that although the berry extract had no effect on tumour incidence, it reduced tumour multiplicity significantly compared with that seen in control or non-treated cells. Yet an earlier study by Stoner *et al.* (1999) did not detect any effect of

strawberries in the reduction of lung cancer in rats, induced by other carcinogenic compounds. In a different study conducted by Chung *et al.* (2002), the effect of strawberries against the endogenous generation of carcinogens in healthy individuals consuming a diet with excessive nitrates was evaluated. Results from that study demonstrated that consumption of 300 g of strawberry resulted in a reduction of 70% in the urinary concentration of the carcinogen, NMBA. Recently, Warner *et al.* (2008) demonstrated that lyophilized strawberries or selenium-enriched lyophilized strawberries satisfactorily inhibited tumour formation by 43 and 59%, respectively. Overall, the authors concluded that, based on their results using a hamster cheek pouch model, strawberries and strawberries with selenium could prevent or delay the development of oral cancer (Warner *et al.* 2008).

15.3.3 Cardiovascular disease

Epidemiological data suggest that consumption of fruit and vegetables may lower the risk of cardiovascular disease (CVD). Again, antioxidant activity has been cited as the possible mechanisms by which strawberries or specific polyphenols found in strawberries may exert their beneficial effects. In this context, it is postulated that the antioxidant activity of strawberry is crucial in the prevention of atherosclerosis, since the oxidation of low-density lipoproteins (LDL) is a key phenomenon associated with the development of such conditions (Diaz *et al.*, 1997; Edirisinghe *et al.*, 2008).

In addition, specific compounds found in strawberries (anthocyanins for instance) are known vasodilators and help towards reducing the incidence of coronary diseases. Edirisinghe *et al.* (2008) showed, for the first time, that not only did freeze-dried strawberry powder from California strawberries cause endothelium-dependent relaxation in the rabbit aorta (EDR) but also that this was achieved through the activation of Pl3 kinase/akt (Table 15.2). The major phenolic compounds present in the strawberry extracts were pelargonidin-3-*O*-glucoside, coumaryl-3-*O*-glucoside and *trans*-cinnamoyl-*O*-glucoside.

Previous studies also demonstrated that ascorbate had EDR effects at concentrations similar to those found in strawberry fruit, and therefore the authors pointed out the possible synergistic effect between ascorbate and polyphenols. The loss of proper endothelial function is frequent in people suffering from diabetes mellitus, hypertension and other chronic conditions that can therefore increase the risk of heart disease. The role that certain polyphenols present in strawberry fruit (namely, kaempferol, catechin and anthocyanins) have in inhibiting the formation of atheroma plaque, and therefore reducing the incidence of thrombosis, has been demonstrated (Rein *et al.*, 2000). In a cohort study conducted by Sesso *et al.* (2007), higher strawberry intake (more than two servings/week) was associated with a reduced borderline but significant likelihood of having elevated C-reactive protein (CRP) levels. However, in the same study, no association was found between the risk of incidence of CVD, lipids, or CRP in middle-aged and older women and the consumption of strawberry fruit. The authors pointed out that additional epidemiological data were needed to clarify any role of strawberries in CVD prevention.

Jenkins and co-workers (2008) assessed the effect of adding strawberries as a source of antioxidants to improve the antioxidant effect of a cholesterol-lowering diet on 28 hyperlipidaemic subjects. Results from this study revealed that supplementation with strawberry fruit resulted in a greater reduction in oxidative damage to LDL, while preserving reductions in blood lipids and enhancing diet palatability (Jenkins *et al.*, 2008).

In another study conducted by Naemura *et al.* (2008), an *in vitro* platelet function test (haemostatometry) was used to screen different strawberry cultivars. In the same study, those cultivars showing significant antiplatelet function were further examined *in vivo* by means of a laser-induced thrombosis test in mice. Results suggested that strawberry varieties KYSt-4 (Nohime), KYSt-11 (Kurume IH-1) and KYSt-17 (Kurume 58) showed significant antiplatelet activity both *in vitro* and *in vivo*.

Recently, the effect of freeze-dried strawberry powder supplementation has been evaluated on women suffering metabolic syndrome (Basu *et al.*, 2009). Short-term supplementation of a strawberry drink, consisting of 25 g of freeze-dried strawberry powder (unspecified cultivar), one cup of water, artificial sweeteners and vanilla essence, resulted in hypocholesterolaemic effects and decrease in lipid peroxidation (Basu *et al.*, 2009) at 4 weeks as compared with baseline values. When a similar study was undertaken with liperlipidaemic subjects each ingesting 453 g of fresh strawberries daily for 4 weeks, no differences were observed in lipid levels (Jenkins *et al.*, 2008). In the same study, however, strawberry consumption was associated with reduction of lipid oxidative damage.

15.3.4 Other beneficial effects

Anti-inflammatory

It is possible that certain phenolic compounds from strawberry fruit may exert positive effects on the immune system. For instance, Seeram *et al.* (2001) reported on the inhibitory effect of berry anthocyanins on cyclooxygenase (COX). COX is a key enzyme in inflammation and its inhibition is the target mechanism of different drugs, including the common aspirin. Strawberries have been proven to be very effective in inhibiting COX2, though not so effective against COX1. Given that inflammation is a process involved in the aetiology of several pathologic conditions, including cancer, cardiovascular, Alzheimer's etc., the findings by Seeram *et al.* (2001) highlighted the potential of strawberries for the treatment of multiple conditions (Tomás-Barberán, 2008).

Anti-neurodegenerative

Scientific evidence suggests that strawberries (Table 15.2) and other berries may have a role in delaying or even overturning age-related degenerative diseases. Already, studies performed during the last decade have shown the ability of diets supplemented with strawberries to retard and reverse the ageing process in rats (Joseph *et al.*, 1998; Shukitt-Hale *et al.*, 1999; Bickford *et al.*, 2000). In the study carried out by Joseph and collaborators, rats of 6 months of age were fed with strawberries for 8 months. Results revealed the potential of strawberry fruit in preserving cerebral function and protecting against the decrease in mental performance associated with ageing (Joseph *et al.*, 1998). Other studies conducted on aged rats showed that strawberries improved the rodents' motor and learning skills, and reduced the deficit in the cognitive capacity of the animals (Bickford *et al.*, 2000). More recent data, generally obtained from *in vitro* studies, reveals the potential of strawberry fruit, as well as their anthocyanins or other polyphenols, to reduce oxidative stress-induced apoptosis in neuronal cells. In the study conducted by Heo and Lee (2005), strawberries reduced oxidative stress-induced neurotoxicity significantly in neuronal cells in a greater manner than observed for banana or orange (Heo and Lee, 2005). Accordingly, Shukitt-Hale *et al.* (2007) suggested that combinations of antioxidants and anti-inflammatory polyphenols from berries might be key compounds to help prevent, suppress or inhibit age-related deficits by several mechanisms. Recently, Shukitt-Hale *et al.* (2007) have investigated whether strawberry and other berry extracts can mitigate the oxidative stress in rats exposed to irradiation. Exposing young rats to irradiation enhances oxidative stress and disrupts the dopaminergic system in a way similar to that observed in aged rats (Shukitt-Hale *et al.* 2007). In this context, the authors demonstrated the ability of strawberry extracts to protect against spatial deficits.

Obesity and anti-metabolic disorders

Following earlier studies in which anthocyanins were hypothesized to have important implications in the prevention of obesity and diabetes, Prior and collaborators investigated whether whole blueberry and strawberry fruit, or their purified anthocyanin extracts, had any anti-obesity effect on a mouse model. The authors proved that, while feeding mice with the whole fruit did not prevent and even increased obesity, feeding purified anthocyanins from the same fruit had a crucial effect on reducing obesity (Prior *et al.*, 2008). The potential therapeutic effects of strawberry extracts (limiting post-meal blood glucose levels by reducing α-amylase) were also demonstrated by McDougall *et al.* (2005). The authors found that, from a range of berry extracts, strawberry and raspberry were the most effective in inhibiting α-amylase, which the authors hypothethized was due to their high concentrations of soluble tannins.

15.4 Effect of Preharvest and Postharvest Continuum

Strawberry growth and development are characterized by changes in colour, texture and flavour, with four or five different stages commonly described in the literature that are based on the development of non-ovarian receptacle tissue (Terry *et al.*, 2004). These stages include *small green*, *large green*, *white* and *full red*. The contents of certain bioactives that account for the potential health-promoting properties of the fruit vary markedly, depending on developmental stage. Changes in anthocyanins and other bioactives during strawberry fruit development have been described in detail recently (Carbone *et al.*, 2009). In addition, during fruit growth and development, exposure of the plant to certain abiotic and biotic conditions may result in enhancing oxidative stress, and therefore generation of reactive oxygen species (ROS). It is believed that under such conditions the plant responds by increasing bioactive-related gene expression, thus enhancing the production of ROS scavengers, mainly antioxidants that may counteract ROS at different levels. It has been demonstrated that growing strawberry plants under different agroclimatic conditions may result in fruit with different contents of health-promoting components (Atkinson *et al.*, 2006; Terry *et al.*, 2007).

Recent studies have elucidated that exposing the plant to different stress conditions (namely, deficit irrigation, salinity, etc.) resulted in enhanced content of specific bioactives (Terry *et al.*, 2007; Keutgen and Pawelzik, 2008; Table 15.3). Specifically, Terry *et al.* (2007) showed that anthocyanin content, total phenolics and antioxidant capacity were greater in cv. Elsanta plants irrigated with 50 ml/day than in the plants receiving greater amounts of water (100 or 200 ml/day) (Fig. 15.3).

Table 15.3. Effect of preharvest factors (a) and postharvest treatments (b) on the health-related composition of strawberries.

(a) Preharvest factors	Effect on bioactives	Reference
Conventional versus organic cultivation systems	No effect on total phenolics	Häkkinen and Törrönen (2000)
Growing temperature	Strawberry grown at higher temperatures (°C) showed higher concentrations of bioactive compounds	Wang and Zheng (2001)
Cultural system (hill plasticulture versus matted row)	Hill plasticulture systems resulted in higher content of phenolics, flavonoids and ascorbate	Wang et al. (2002)
Ozone exposure	No significant effect on antioxidant activity or bioactives	Keutgen and Pawelzik (2007)
Salinity stress	Moderate salinity resulted in increase antioxidant activity and bioactives	Keutgen and Pawelzik (2007)
Deficit irrigation	Higher content of certain anthocyanins, total phenolics and antioxidant activity	Terry et al. (2007)
Inoculation with Botrytis cinerea[a]	No effect on strawberry (cv. Elsanta) bioactives or antioxidant activity	Terry et al. (2007)
Organic and conventional nutrient amendments	No significant differences between treatments on antioxidant activity	Hargreaves et al. (2008)
Methyl jasmonate (MeJa) applied in fully or deficit irrigated plants	Higher concentrations of anthocyanins found in MeJa-treated plants and changes in fruit and leaves antioxidant capacity	Giné Bordonaba and Terry (unpublished data)

(b) Postharvest treatment	Effect on bioactives	Reference
1-MCP	Lower accumulation of phenolic anthocyanins in cv. Everest treated fruits	Jiang et al. (2001)
Heat treatment before refrigerated storage	Treated fruits (cv. Selva) showed less anthocyanin accumulation than controls when held at 20°C	Vicente et al. (2002)
Exogenous abscisic acid (ABA) application	Anthocyanin and phenolic contents and PAL activity increased during storage of ABA-treated strawberry fruit (cv. Everest) more rapidly than in non-treated fruit	Jiang and Joyce (2003)

(Continued)

Table 15.3. *Continued*

(b) Postharvest treatment	Effect on bioactives	Reference
Storage temperature	Higher temperatures (5° and 10°C) resulted in greater antioxidant capacity, total phenolics and anthocyanins of the fruit (cv. Chandler) as compared with fruit stored at 0°C	Ayala-Zavala *et al.* (2004)
UV-C (4.1 kJ/m²) and heat treatment (45°C, 3 h in air) either separately or combined	All treatments reduced the accumulation of anthocyanins in strawberries cv. Seascape	Pan *et al.* (2004)
Strawberry wrapping with polyvinyl chloride and stored at 1°C	Wrapped fruit (cv. Oso Grande) suffered lower water loss and maintained better anthocyanin and other soluble phenolics as compared with unwrapped fruit	Nunes *et al.* (2005)
Storage temperature	Lower temperatures affected anthocyanins and ascorbate negatively but had no effect on flavonols, ellagic acid and total phenolics and antioxidant capacity	Cordenunsi *et al.* (2005)
MeJA in combination with ethanol	Enhanced antioxidant capacity, volatile compounds and postharvest life of strawberries (cv. Chandler)	Ayala-Zavala *et al.* (2005)
Superatmospheric storage conditions	High oxygen atmospheres (> 40 kPa) resulted in fruit (cv. Chandler) with higher antioxidant capacity, total phenolics, less decay and longer postharvest life than those stored in air	Ayala-Zavala *et al.* (2007)

Note: [a]At anthesis of primary flower.

The antiproliferative effect of strawberry fruit grown conventionally or organically was assessed recently by Olsson et al. (2006). The organically grown strawberries showed greater antiproliferative activity, which the authors related to the higher content of secondary metabolites found in those berries and which in turn might be associated with the exposure of the plants to greater stress from pathogens when grown under organic cultivation systems. In contrast, others (Hargreaves et al., 2008) could not find differences in several bioactive constituents between organically and conventionally grown strawberries, and hence further research is required to elucidate whether or not organic production may result in 'healthier' berries.

Clearly, metabolism is known to continue beyond fruit harvest; however, this is often ignored or, more commonly, overlooked or not appreciated. The concentration of certain metabolites is expected to change during postharvest storage and through the supply chain. Given this, however, few works have studied the role that postharvest treatments have on strawberry bioactives. In addition, the effectiveness of any postharvest treatment is dependent on whether the treatment is focused on preserving appearance or maintaining the health-related composition of the fruit. Earlier works have shown that postharvest treatments focused on maintaining appearance do not correlate with, for instance, better maintenance of certain bioactives (Pelayo et al., 2003). Postharvest storage temperature (5 or 22°C) did not have a significant effect on ellagic acid content in strawberries stored for 24 h (Häkkinen et al., 2000). Under longer storage conditions, Gil et al. (1997) reported that content of ellagic acid increased over the course of 10 days in fruit stored at 5°C and pointed out that such a phenomenon probably was the result of degradation of ellagitannins (Fig. 15.5). In another study, Cordenunsi et al. (2005) evaluated the chemical composition and antioxidant activity changes of different strawberry cultivars (namely, Dover, Campineiro and Oso Grande) stored at 6, 16 and 25°C for 6 days. The authors concluded that low temperatures affected anthocyanins and ascorbic acid negatively,

whereas it had no significant effect on the content of flavonols, ellagic acid or total phenolics. Antioxidant activity, on the other hand, was similar between cultivars and in all cases decreased after harvest, independent of temperature. Interestingly, the observed increase in ascorbate in all three cultivars during storage (10% greater than initial values) was experienced only at 16°C and therefore was in disagreement with previous studies (Nunes et al., 1998; Cordenunsi et al., 2003), which highlighted that low temperatures and high humidity during storage might retard ascorbate degradation (Nunes, 2008). Nunes et al. (1998) showed that, in strawberry fruit stored for 8 days at 1 or 10°C, as well as 4 days at 20°C, ascorbic degradation was greater at the higher temperatures. In the same study, postharvest shelf life was enhanced and ascorbic acid degradation reduced by 7.5-fold in fruit stored at 1°C. In this context, Cordenunsi et al. (2005) postulated that ascorbic acid synthesis might occur during postharvest storage and that it was indeed affected by lowering temperature. Hakkinen et al. (2000) also demonstrated that quercetin content increased markedly in strawberries or strawberry jams stored for 9 months, whereas ellagic acid tended to decline in fresh fruit but not in jams from the same berry. Similarly, others also reported that flavonol content increased in strawberries or other fruit stored under refrigerated conditions (Gil et al., 1997). Controlled atmosphere (CA) during storage may reduce the rate of accumulation of anthocyanins normally observed in strawberry fruit after harvest (Zheng et al., 2007). Nevertheless, Gil et al. (1997) showed that enriched CO_2 atmospheres had a minimal effect on anthocyanin concentrations of cv. Selva fruit. In the particular case of anthocyanins, it is likely that changes in other components within the fruit (i.e. organic acids) may affect the pH, and hence the stability, co-pigmentation and spectra of these pigments within the fruit (Terry et al., 2009), thus accounting for the reported changes during postharvest storage. In addition, given the variability in colour within different genotypes (Giné Bordonaba and Terry, 2009), it may be feasible to speculate that changes in anthocyanins and other pigments, as a result of any postharvest treatment, may be genotype-dependent, and this should be investigated further.

The effects that certain non-conventional treatments have on preharvest or postharvest strawberry bioactives have also been studied increasingly over the last decade. As an example, Zabetakis *et al.* (2000) studied the effect that treatment with high hydrostatic pressure and further storage at different temperatures had on strawberry anthocyanins. High hydrostatic pressure may be an alternative to conventional heat treatments for preservation of strawberry-derived products. The authors observed that samples pressurized under 800 MPa for 15 min (the greatest of the different treatments applied) resulted in the lowest losses of anthocyanins. Furthermore, after high hydrostatic pressure treatments, storage at 4°C also resulted in the best maintenance of both pelargonidin-3-glucoside and pelargonidin-3-rutinoside. Accordingly, treatment with 400 and 600 MPa resulted in greater enzymatic activity involved in anthocyanin degradation than seen with 800 MPa (Zabetakis *et al.*, 2000).

In a different study, Wang *et al.* (2007) showed that strawberries treated postharvest with essential oils inhibited human HT-29 colon cancer cell proliferation better than those not treated.

The application of essential oils preharvest (methyl jasmonate for instance) resulted in marked changes in anthocyanin concentrations in different strawberry cultivars (Giné Bordonaba and Terry, unpublished data).

15.5 Future Research Needs and Conclusions

Unlike many other fruit and vegetables, there appears to be substantial scientific evidence to confirm that strawberries are one of the main sources of vitamin C, folic acid and dietary fibre, as well as an excellent source of dietary polyphenols. From the survey of the literature presented herein, anthocyanin pigments, together with hydrolysable and non-hydrolysable tannins, are among the main compounds with reported health-promoting properties (namely, anticarcinogenic, anti-cardiovascular and anti-neurodegenerative diseases, anti-inflammatory, etc). However, in order to understand fully the health-promoting properties of this berry, more studies are still required to elucidate further not only the heterogeneity in bioactive compounds among strawberry genotypes as affected by preharvest/postharvest continuum, but also to broaden investigative research on the bioavailability of specific strawberry bioactives in different food matrices.

In vitro studies must continue, since they provide vital information on the mechanisms and actions of specific bioactives, but it is clear that such studies may present certain limitations and the results cannot be translated *in vivo*. Consequently, further *in vivo* and intervention studies must be conducted to sustain the information obtained so far, as well as to study the long-term beneficial or toxic effect, if any, derived from strawberry consumption.

Daily consumption of dietary polyphenols can vary from a few hundred mg to almost 2 g per capita. As described throughout this chapter, only a very minor fraction of the polyphenols found in strawberry fruit is absorbed directly. Generally, most polyphenols pass the stomach and come across the gut microflora, resulting in a diverse range of metabolites, most of them still unknown, which undoubtedly may have potential health-related properties. Further work should aim to clarify whether the metabolites generated by the interaction between gut microflora and polyphenols exert health-promoting properties.

References

Aaby, K., Skrede, G. and Wrolstad, R.E. (2005) Phenolic composition and antioxidant activities in flesh and achenes of strawberries (*Fragaria* × *ananassa*). *Journal of Agricultural and Food Chemistry* 53, 4032–4040.

Aaby, K., Ekeberg, D. and Skrede, G. (2007) Characterization of phenolic compounds in strawberry (*Fragaria* × *ananassa*) fruits by different HPLC detectors and contribution of individual compounds to total antioxidant capacity. *Journal of Agricultural and Food Chemistry* 55, 4395–4406.

Aggarwal, B.B., Bhardwaj, A., Aggarwal, R.I., Seeram, N.P., Shishodia, S. and Takada, Y. (2004) Role of resveratrol in prevention and therapy of cancer. Precliniclal and clinical studies. *Anticancer Research* 24, 2783–2840.

Almeida, J.R.M., D'Amico, E., Preuss, A., Carbone, F., de Vos, C.H.R., Deiml, B., Mourgues, F., Perrota, G., Fischer, T.C., Bovy, A.G., Martens, S. and Rosati, C. (2007) Characterization of major enzymes and genes involved in flavonoid and proanthocyanidin biosynthesis during fruit development in strawberry (*Fragaria* × *ananassa*). *Archives of Biochemistry and Biophysics* 465, 61–71.

Andres-Lacueva, C., Shukitt-Hale, B., Galli, R.L., Jauregui, O., Lamuela-Raventos, R.M. and Joseph, J.A. (2005) Anthocyanins in aged blueberry-fed rats are found centrally and may enhance memory. *Nutritional Neuroscience* 8 , 111–120.

Arai, Y., Watanabe, S., Kimira, M., Shimoi, K., Mochizuki, R. and Kinae, N. (2000) Dietary intakes of flavonols, flavones and isoflavones by Japanese women and the inverse correlation between quercetin intake and plasma LDL cholesterol concentration. *Journal of Nutrition* 130, 2243–2250.

Atkinson, C.J., Doods, P.A.A., Ford, Y.Y., Le Mière, J., Taylor, J.M., Blake, P.S. and Paul, N. (2006) Effect of cultivar, fruit number and reflected phosynthetically active radiation on *Fragaria* × *ananassa* productivity and fruit ellagic acid and ascorbic acid concentrations. *Annals of Botany* 97, 429–441.

Ayala-Zavala, J.F., Wang, S.Y., Wang, C.Y. and González-Aguilar, G.A. (2004) Effect of storage temperatures on antioxidant capacity and aroma compounds in strawberry fruit. *LWT – Food Science and Technology* 37, 687–695.

Ayala-Zavala, J.F., Wang, S.Y., Wang, C.Y. and González-Aguilar, G.A. (2005) Methyl jasmonate in conjunction with ethanol treatment increases antioxidant capacity, volatile compounds and postharvest life of strawberry fruit. *European Food Research and Technology* 221, 731–738.

Ayala-Zavala, J.F., Wang, S.Y., Wang, C.Y. and González-Aguilar, G.A. (2007) High oxygen treatment increases antioxidant capacity and postharvest life of strawberry fruit. *Food Technology and Biotechnology* 45, 169–173.

Bailey, L.B. and Gregory, J.F. (1999) Folate metabolism and requirements. *Journal of Nutrition* 129, 779–782.

Basu, A., Wilkinson, M., Penugonda, K., Simmons, B., Betts, N.M. and Lyons, T.J. (2009) Freeze-dried strawberry powder improves lipid profile and lipid peroxidation in women with metabolic syndrome: baseline and post intervention effects. *Nutrition Journal* 8, 1–7.

Bickford, P., Gould, T., Briederick, L., Chadman, K., Pollock, A., Young, D., Shukitt-Hale, B. and Joseph, J. (2000) Antioxidant-rich diets improve cerebellar physiology and motor learning in aged rats. *Brain Research* 866, 211–217.

Cantos, E., Garcia-Viguera, C., de Pascual-Teresa, S. and Tomas-Barberan, F.A. (2000) Effect of postharvest ultraviolet irradiation on resveratrol and other phenolics of cv. Napoleon table grapes. *Journal of Agricultural and Food Chemistry* 48, 4606–4612.

Carbone, F., Preuss, A., de Vos, R.C.H., D'amico, E., Perrotta, G., Bovy, A. G., Martens, S. and Rosati, C. (2009) Developmental, genetic and environmental factors affect the expression of flavonoid genes, enzymes and metabolites in strawberry fruits. *Plant Cell and Environment* 32, 1117–1131.

Carkeet, C., Clevidence, B.A. and Novotny, J.A. (2008) Anthocyanin excretion by humans increases linearly with increasing strawberry dose. *Journal of Nutrition* 138, 897–902.

Carlton, P.S., Kresty, L.A., Siglin, J.C., Morse, M.A., Lu, J., Morgan, C. and Stoner, G.D. (2001) Inhibition of N-nitrosomethylbenzylamine-induced tumorigenesis in the rat esophagus by dietary freeze-dried strawberries. *Carcinogenesis* 22, 441–446.

Carter, C.A., Chalfant, J.A. and Goodhue, R.E. (2005) China's strawberry industry: an emerging competitor for California? Giannini Foundation of Agricultural Economics (http://www.agecon.ucdavis.edu/extension/update/articles/v9n1_3.pdf, accessed 20 September 2008).

Cerdà, B., Tomás-Barberán, F.A. and Espín, J.C. (2005) Metabolism of antioxidant and chemopreventive ellagitannins from strawberries, raspberries, walnuts, and oaked-aged wine in humans: identification of biomarkers and individual variability. *Journal of Agricultural and Food Chemistry* 53, 227–235.

Chung, M.J., Lee, S.H. and Sung, N.J. (2002) Inhibitory effect of whole strawberries, garlic juice or kale juice on endogenous formation of N-nitrosodimethylamine in humans. *Cancer Letters* 182, 1–10.

Cordenunsi, B.R., Nascimento, J.R.O. and Lajolo, F.M. (2003) Physico-chemical changes related to quality of five strawberry fruit cultivars during cool-storage. *Food Chemistry* 83, 167–173.

Cordenunsi, B.R., Genovese, M.I., Nascimento, J.R.O., Hassimoto, N.M.A., dos Santos, R.J. and Lajolo, F.M. (2005) Effect of temperature on the chemical composition and antioxidant activity of three strawberry cultivars. *Food Chemistry* 91, 113–121.

Crespo, P., Giné Bordonaba, J., Terry, L.A. and Carlen, C. (2010) Characterisation of major taste and health-related compounds of four strawberry genotypes grown at different Swiss production sites. *Food Chemistry* 122, 16–24.

Daniel, E.M., Krupnick, A.S., Hew, Y., Blinzler, J.A., Nems, R.M. and Stoner, G.D. (1990) Extraction, stability, and quanlitation of ellagic acid in various fruits and nuts. *Journal of Food Composition and Analysis* 2, 338–349.

Defra (2003) *Basic Horticultural Statistics for the United Kingdom, Calendar and Crop Years 1992/93–2002/03.* National Statistics, Defra PB 8889. Department for Environment, Food and Rural Affairs, London.

Diaz, M.N., Frei, B., Vita, J.A. and Keaney, J.F. (1997) Mechanisms of disease: antioxidants and atherosclerotic heart disease. *New England Journal of Medicine* 337, 408–416.

Edirisinghe, E., Burton-Freeman, B., Varelis, P. and Kappagoda, T. (2008) Strawberry extract caused endothelium-dependent relaxation through the activation of Pl3 kinase/akt. *Journal of Agricultural and Food Chemistry* 56, 9383–9390.

FAO (2009) FAOSTAT statistics database. Food and Agriculture Organization of the United Nations. Rome, Italy (http://faostat.fao.org, accesed 12 December 2009).

Felgines, C., Talavera, S., Gonthier, M.P., Texier, O., Scalbert, A., Lamaison, J.L. and Remesy, C. (2003) Strawberry anthocyanins are recovered in urine as glucuro- and sulfoconjugates in humans. *Journal of Nutrition* 133, 1296–1301.

Fossen, T., Rayyan, S. and Andersen, Ø.M. (2004) Dimeric anthocyanins from strawberry (*Fragaria ananassa*) consisting of pelargonidin 3-glucoside covalently linked to four flavan-3-ols. *Phytochemistry* 65, 1421–1428.

Gil, M.I., Holdcroft, D.M. and Kader, A.A. (1997) Changes in strawberry anthocyanins and other polyphenols in response to carbon dioxide treatments. *Journal of Agricultural and Food Chemistry* 45, 1662–1667.

Giné Bordonaba, J. and Terry, L.A. (2008) Biochemical profiling and chemometric analysis of 17 UK-grown blackcurrant cultivars. *Journal of Agricultural and Food Chemistry* 56, 7422–7430.

Giné Bordonaba, J. and Terry, L.A. (2009) Development of a glucose biosensor for rapid assessment of strawberry quality: relationship between biosensor response and fruit composition. *Journal of Agricultural and Food Chemistry* 57, 8220–8226.

Giné Bordonaba, J. and Terry, L.A. (2010) Manipulating the taste-related composition of strawberry fruits (*Fragaria ananassa*) from different cultivars using deficit irrigation. *Food Chemistry* 122, 1020–1026.

González-Paramás, A.M., Lopes da Silva, F., Martin-López, P., Macz-Pop, G., González-Manzano, S., Alcalde-Eon, C., Pérez-Alonso, J.J., Escribano-Bailón, M.T., Rivas-Gonzalo, J.C. and Santos-Buelga, C. (2006) Flavanol-anthocyanin condensed pigments in plant extracts. *Food Chemistry* 94, 428–436.

Gu, L., Kelm, M.A., Hammerstone, J.F., Beecher, G., Holden, J., Haytowitz, D. and Prior, R.L. (2003) Screening of foods containing proanthocyanidins and their structural characterization using LC-MS/MS and thiolytic degradation. *Journal of Agricultural and Food Chemistry* 51, 7513–7521.

Häkkinen, S.H. and Törrönen, A.R. (2000) Content of flavonols and selected phenolic acid in strawberries and vaccinum species: influence of cultivar, cultivation site and technique. *Food Research International* 33, 517–524.

Häkkinen, S.H., Kärelampi, H.M., Heinonen, S.O., Mykkänen, I.M. and Törrönen, A.R. (1999) Content of the flavonols quercetin, myricetin, and kaempferol in 25 edible berries. *Journal of Agricultural and Food Chemistry* 47, 2274–2279.

Häkkinen, S.H., Kärelampi, S.O., Mykkänen, H.M., Heinonen, I.M. and Törrönen, A.R. (2000) Ellagic acid in berries: influence of domestic processing and storage. *European Food Research and Technology* 212, 75–80.

Haleem, M.A., Barton, K.L., Borges, G., Crozier, A. and Anderson, A.S. (2008) Increasing antioxidant intake from fruits and vegetables: practical strategies for the Scottish population. *Journal of Human Nutrition and Dietetics* 21, 539–546.

Halvorsen, B.L., Carlsen, M.H., Philips, K.M., BØhn, S.K., Holte, K., Jacobs, D.R. and Blomhoff, R. (2006) Content of redox-active compounds (i.e. antioxidants) in foods consumed in the United States. *American Journal of Clinical Nutrition* 84, 95–135.

Hannum, S.M. (2004) Potential impact of strawberries on human health: a review of the science. *Critical Reviews in Food Science and Nutrition* 44, 1–17.

Hargreaves, J.C., Adl, M.S., Warman, P.R. and Rupasinghe, H.P.V. (2008) The effects of organic and conventional nutrient amendments on strawberry cultivation: fruit yield and quality. *Journal of the Science of Food and Agriculture* 88, 2669–2675.

Heo, H.J. and Lee, C.Y. (2005) Strawberry and its anthocyanins reduce oxidative stress-induced apoptosis in PC12 cells. *Journal of Agricultural and Food Chemistry* 53, 1984–1989.

Hollands, W., Brett, G.M., Dainty, J.R., Teucher, B. and Kroon, P.A. (2008) Urinary excretion of strawberry anthocyanins is dose dependent for physiological oral doses of fresh fruit. *Molecular Nutrition and Food Research* 52, 1097–1105.

Hope, S.S., Tate, P.L., Huang, G., Magee, J.B., Meepagala, K.M., Wedge, D.M. and Larcom, L.L. (2004) Antimutagenic activity of berry extracts. *Journal of Medicinal Food* 7, 450–455.

Hosseinian, F.S., Li, W., Hydamaka, A.W., Tsopmo, A., Lowry, L., Friel, J. and Beta, T. (2007) Proanthocyanidin profile and ORAC values of manitoba berries, chokecherries, and seabuckthorn. *Journal of Agricultural and Food Chemistry* 55, 6970–6976.

Jenkins, D.J.A., Nguyen, T.H., Kendall, C.W.C., Faulkner, D.A., Bashyam, B., Kim, I.J., *et al.* (2008) The effect of strawberries in a cholesterol-lowering dietary portfolio. *Metabolism: Clinical and Experimental* 57, 1636–1644.

Jiang, Y. and Joyce, D.C. (2003) ABA effects on ethylene production, PAL activity, anthocyanin and phenolic contents of strawberry fruit. *Plant Growth Regulation* 39, 171–174.

Jiang, Y., Joyce, D.C. and Terry, L.A. (2001) 1-Methylcyclopropene treatment affects strawberry fruit decay. *Postharvest Biology and Technology* 23, 227–232.

Joseph, J.A., Shukitt-Hale, B., Denisova, B., Prior, R.L., Cao, G., Martin, A., Tagliatela, G. and Bickford, P.C. (1998) Long-term dietary strawberry, spinach, or vitamin E supplementation retards the onset of age-related neuronal signal-transduction and cognitive behavioral deficits. *Journal of Neuroscience* 18, 8047–8055.

Kern, M., Fridrich, D., Reichert, J., Skrbek, S., Nussher, A., Hofem, S., *et al.* (2007) Limited stability in cell culture medium and hydrogen peroxide formation affect the growth inhibitory properties of delphinidin and its degradation product gallic acid. *Molecular Nutrition and Food Research* 51, 1163–1172.

Keutgen, A. and Pawelzik, E. (2007) Modifications of taste-relevant compounds in strawberry fruit under NaCl salinity. *Food Chemistry* 105, 1487–1494.

Keutgen, A.J. and Pawelzik, E. (2008) Quality and nutritional value of strawberry fruit under long term salt stress. *Food Chemistry* 107, 1413–1420.

Knekt, P., Kumpulainen, J., Järvinen, R., Rissanen, H., Heliövaara, M., Reunanen, A., *et al.* (2002) Flavonoid intake and risk of chronic diseases. *American Journal of Clinical Nutrition* 76, 560–568.

Kosar, M., Kafkas, E., Paydas, S. and Baser, K.H.C. (2004) Phenolic composition of strawberry genotypes at different maturation stages. *Journal of Agricultural and Food Chemistry* 52, 1586–1589.

Li, J., Zhang, D., Stoner, G.D. and Huang, C. (2008) Differential effects of black raspberry and strawberry extracts on BaPDE-induced activation of transcription factors and their target genes. *Molecular carcinogenesis* 47, 286–294.

Liu, F., Savić, S., Jensen, C.R., Shahnazari, A., Jacobsen, S.E., Stikić, R. and Andersen, M.N. (2007) Water relations and yield of lysimeter-grown strawberries under limited irrigation. *Scientia Horticulturae* 111, 128–132.

Lopes da Silva, F., Escribano-Bailón, M.T., Pérez Alonso, J.J., Rivas-Gonzalo, J.C. and Santos-Buelga, C. (2007) Anthocyanins pigments in strawberry. *LWT – Food Science and Technology* 40, 374–382.

Maas, J.L., Wang, S.Y. and Galleta, G.J. (1991) Evaluation of strawberry cultivars for ellagic acid content. *HortScience* 26, 66–68.

Määttä-Riihinen, K.R., Kamal-Eldin, A. and Törrönen, A.R. (2004) Identification and quantification of phenolic compounds in berries of *Fragaria* and *Rubus* species (family *Rosaceae*). *Journal of Agricultural and Food Chemistry* 52, 6178–6187.

McDougall, G.J., Shapiro, F., Dobson, P., Smith, P., Blake, A. and Stewart, D. (2005) Different polyphenolic components of soft fruits inhibit α-amylase and α-glycosidase. *Journal of Agricultural and Food Chemistry* 53, 2760–2766.

McGhie, T.K. and Walton, M.C. (2007) The bioavailability and absorption of anthocyanins: towards a better understanding. *Molecular Nutrition and Food Research* 51, 702–713.

Maher, P., Akaishi, T. and Abe, K. (2006) Flavonoid fisetin promotes ERK-dependent long-term potentiation and enhances memory. *Proceedings of the National Academy of Sciences of the United States of America* 103, 16568–16573.

Medina-Mínguez, J.J. (2008) Origin of the crop: a pioneer. In: de Andalucía, J. (ed.) *The Strawberry Crop at Huelva*. Consejería de Agricultura y Pesca, Sevilla, pp. 17–46.

Meyers, K.J., Watkins, C.B., Pritts, M.P. and Liu. R.H. (2003) Antioxidant and antiproliferative activities of strawberries. *Journal of Agriculture and Food Chemistry* 51, 6887–6892.

Mullen, W., Edwards, C.A., Serafini, M. and Crozier, A. (2008) Bioavailability of pelargonidin-3-O-glucoside and its metabolites in humans following the ingestion of strawberries with and without cream. *Journal of Agricultural and Food Chemistry* 56, 713–719.

Naemura, A., Mitani, T., Ijiri, Y., Tamura, Y., Yamashita, T., Okimura, M. and Yamamoto, J. (2008) Anti-thrombotic effect of strawberries. *Blood Coagulation and Fibrinolysis* 16, 501–509.

Nunes, M.C.N. (2008) Strawberry. In: Nunes, M.C.N. (ed.) *Color Atlas of Postharvest Quality of Fruits and Vegetables*. Blackwell Publishing, Ames, Iowa, pp. 175–185.

Nunes, M.C.N., Brecht, J.K., Morais, A.M.M.B. and Sargent, S.A. (1998) Controlling temperature and water loss to maintain ascorbic acid levels in strawberries during postharvest handling. *Journal of Food Science* 63, 1033–1036.

Nunes, M.C.N., Brecht, J.K., Morais, A.M.M.B. and Sargent, S.A. (2005) Possible influences of water loss and polyphenol oxidase activity on anthocyanin content and discoloration in fresh ripe strawberry (cv. Oso Grande) during storage at 1°C. *Journal of Food Science* 70, S79–S84.

Olsson, M.E., Andersson, C.S., Oredsson, S., Berglund, R.H. and Gustavsson, K.E. (2006) Antioxidant levels and inhibition of cancer cell proliferation *in-vitro* by extracts from organically and conventionally cultivated strawberries. *Journal of Agricultural and Food Chemistry* 54, 1248–1255.

Pan, J., Vicente, A.R., Martínez, G.A., Chaves, A.R. and Civello, P.M. (2004) Combined use of UV-C irradication and heat treatment to improve postharvest life of strawberry fruit. *Journal of the Science of Food and Agriculture* 84, 1831–1838.

Pascual-Teresa, S., Santos-Buelga, C. and Rivas-Gonzalo, J.C. (2000) Quantitative analysis of flavn-3-ols in Spanish foodstuffs and beverages. *Journal of Agricultural and Food Chemistry* 48, 5331–5337.

Pelayo, C., Ebeler, S.E. and Kader, A.A. (2003) Postharvest life and flavor quality of three strawberry cultivars kept at 5°C in air or air +20 kPa CO_2. *Postharvest Biology and Technology* 27, 171–183.

Perkins-Veazie, P.M., Huber, D.J. and Brecht, J.K. (1995) Characterization of ethylene production in developing strawberry fruit. *Plant Growth Regulation* 17, 33–39.

Prior, R. L., Wu, X., Gu, L., Hager, T., Hager, A. and Howard, L. R. (2008) Whole berries vs. berry anthocyanins: interactions with dietary fat levels in the C57BL/6J mouse model of obesity. *Journal of Agriculture and Food Chemistry* 56, 647–658.

Ramos, S., Alia, M., Bravo, L. and Goya, L. (2005) Comparative effects of food-derived polyphenols on the viability and apoptosis of a human hepatoma cell line HepG2. *Journal of Agricultural and Food Chemistry* 53, 1271–1280.

Rampersaud, G.C., Kauwell, G.P.A. and Bailey, L.B. (2003) Folate: a key to optimising health and reducing disease risk in the elderly. *Journal of the American College of Nutrition* 22, 1–8.

Rein, D., Paglieroni, T.G., Pearson, D.A., Wan, T., Schmitz, H.H., Gosselin, R. and Keen, C.L. (2000) Cocoa and wine polyphenols modulate platelet activation and function. *Journal of Nutrition* 130, 2120S–2126S.

Roques, S.C., Landrault, N., Teissèdre, P.L., Laurent, C., Besançon, P., Rouane, J.M. and Caporiccio, B. (2002) Hydrogen peroxide generation in Caco-2 cell culture medium by addition of phenolic compounds: effect of ascorbic acid. *Free Radical Research* 36, 593–599.

Santos-Buelga, C. and Scalbert, A. (2000) Proanthocaynidins and tannin-like compounds-nature, occurrence, dietary intake, and effects on nutrition and health. *Journal of the Science of Food and Agriculture* 80, 1094–1117.

Seeram, N.P. (2008) Berry fruits for cancer prevention: current status and future prospects. *Journal of Agricultural and Food Chemistry* 56, 630–635.

Seeram, N.P. and Heber, D. (2007) Impact of berry phytochemicals on human health: effects beyond antioxidation. *ACS Symposium Series* 956, 326–336.

Seeram, N.P., Momin, R.A., Nair, M.G. and Bourquin, L.D. (2001) Cyclooxygenase inhibitory and antioxidant cyanidin glycosides in cherries and berries. *Phytomedicine* 8, 362–369.

Seeram, N.P., Adams, L.S., Zhang, Y., Lee, R., Sand, D., Scheuller, H.S., *et al.* (2006a) Blackberry, black raspberry, blueberry, cranberry, red raspberry, and strawberry extracts inhibit growth and stimulate apoptosis of human cancer cells in vitro. *Journal of Agricultural and Food Chemistry* 54, 9329–9339.

Seeram, N.P., Henning, S.M., Zhang, Y., Suchard, M., Li, Z. and Heber, D. (2006b) Pomegranate juice ellagitannin metabolites are present in human plasma and some persist in urine for up to 48 hours. *Journal of Nutrition* 136, 2481–2485.

Seeram, N.P., Lee, R., Scheuller, H.S. and Heber, D. (2006c) Identification of phenolic compounds in strawberries by liquid chromatography electrospray ionization mass spectroscopy. *Food Chemistry* 97, 1–11.

Sesso, H.D., Gaziano, J.M., Jenkins, D.J.A. and Buring, J.E. (2007) Strawberry intake, lipids, C-reactive protein, and the risk of cardiovascular disease in women. *Journal of the American College of Nutrition* 26, 303–310.

Shukitt-Hale, B., Smith, D.E., Meydani, M. and Joseph, J.A. (1999) The effects of dietary antioxidants on psychomotor performance in aged mice. *Experimental Gerontology* 34, 797–808.

Shukitt-Hale, B, Carey, A.N., Jenkins, D., Rabin, B.M. and Joseph, J.A. (2007) Beneficial effects of fruit extracts on neuronal function and behavior in a rodent model of accelerated aging. *Neurobiology of Aging* 28, 1187–1194.

Skupienń, K. and Oszmianński, J. (2004) Comparison of six cultivars of strawberries (*Fragaria* × *ananassa* Duch.) grown in northwest Poland. *European Food Research and Technology* 219, 66–70.

Spiller, G.A. and Dewell, A. (2003) Report on health and strawberries study. Californian Strawberry Commission (www.calstrawberry.com, accessed 13 November 2008).

Stoner, G.D., Kresty, L.A., Carlton, P.S., Siglin, J.C. and Morse, M.A. (1999) Isothiocyanates and freeze-dried strawberries as inhibitors of esophageal cancer. *Toxicological Science* 52, 95–100.

Strålsjö, L., Ảhlin, H., Witthöft, C.M. and Jastrebova, J. (2003) Folate determination in Swedish berries by radioprotein-binding assay (RPBA) and high performance liquid chromatography (HPLC). *European Food Research Technology* 216, 264–269.

Sung, B., Pandey, M.K. and Aggarwal, B.B. (2007) Fisetin, an inhibitor of cyclin-dependent kinase 6, down-regulates nuclear factor-kB-regulated cell proliferation, antiapoptotic and metastatic gene products through the suppression of TAK-1 abd receptor-interacting protein regulated IκBα kinase activation. *Molecular Pharmacology* 71, 1703–1714.

Terry, L.A. (2002) Natural disease resistance in strawberry fruit and Geraldton waxflower flowers. PhD thesis, Institute of Bioscience and Technology, Cranfield University, Bedfordshire, UK.

Terry, L.A., Joyce, D.C., Adikaram, N.K.B. and Khambay, B.P.S. (2004) Preformed antifungal compounds in strawberry fruit and flower tissues. *Postharvest Biology and Technology* 31, 201–212.

Terry, L.A., Chope, G.A. and Bordonaba, J.G. (2007) Effect of water deficit irrigation and inoculation with *Botrytis cinerea* on strawberry (*Fragaria* × *ananassa*) fruit quality. *Journal of Agricultural and Food Chemistry* 55, 10812–10819.

Terry, L.A., Crisosto, C.H. and Forney, C.F. (2009) Small fruit and berries. In: Yahi, E.M. (ed.) *Modified and Controlled Atmosphere for the Storage, Transportation and Packaging of Horticultural Commodities*. CRC Press, Boca Raton, Florida, pp. 119–158.

Tomás-Barberán, F.A. (2008) The strawberry. A very healthy food. In: de Andalucía, J. (ed.) *The Strawberry Crop at Huelva*. Conserjería de Agricultura y Pesca, Sevilla, pp. 307–333.

Tulipani, S., Mezzetti, B., Capocasa, F., Bompadre, S., Beekwilder, J., de Vos, C.H.R., *et al.* (2008) Antioxidants, phenolic compounds, and nutritional quality of different strawberry genotypes. *Journal of Agricultural and Food Chemistry* 56, 696–704.

Tulipani, S., Mezzetti, B. and Battino, M. (2009) Impact of strawberries on human health: insight into marginally discussed bioactive compounds for the Mediterranean diet. *Public Health Nutrition* 12, 1656–1662.

Vicente, A.R., Martínez, G.A., Civello, P.M. and Chaves, A.R. (2002) Quality of heat-treated strawberry fruit during refrigerated storage. *Postharvest Biology and Technology* 25, 59–71.

Wang, S.Y. and Lin, H.S. (2000) Antioxidant activity in fruit and leaves of blackberry, raspberry and strawberry varies with cultivar and development stage. *Journal of Agricultural and Food Chemistry* 48, 140–146.

Wang, S.Y. and Zheng, W. (2001) Effect of plant growth temperature on antioxidant capacity in strawberry. *Journal of Agriculture and Food Chemistry* 49, 4977–4982.

Wang, S.Y., Zheng, W. and Galletta, G.J. (2002) Cultural system affects fruit quality and antioxidant capacity in strawberries. *Journal of Agricultural and Food Chemistry* 50, 6534–6542.

Wang, S.Y., Bunce, J.A. and Maas, J.L. (2003) Elevated carbon dioxide increases contents of antioxidant compounds in field-grown strawberries. *Journal of Agricultural and Food Chemistry* 51, 4315–4320.

Wang, S.Y., Feng, Y.R., Bowman, L. and Ding. (2005) Inhibitory effect on activator protein-1, nuclear factor-kappa B, and cell transformation by extracts of strawberries (*Fragaria* × *ananassa* Duch). *Journal of Agricultural and Food Chemistry* 53, 4187–4193.

Wang, S.Y., Chen, C., Wang, C.Y. and Chen, P. (2007) Resveratrol content in strawberry fruit is affected by preharvest conditions. *Journal of Agricultural and Food Chemistry* 55, 8269–8274.

Warner, B., Casto, B., Knobloch, T., Accurso, B., Galioto, R., Tieche, S., Funt, R. and Weghorst, C. (2008) Preclinical evaluation of strawberries and strawberries with selenium on the chemoprevention of oral cancer. *Cancer Prevention Research* 1, B96.

Wedge, D., Meepagala, K.M., Magee, J.B., Smith, S.H., Huang, G. and Larcom, L.L. (2001) Anticarcinogenic activity of strawberry, blueberry and raspberry extracts to breast and cervical cancer cells. *Journal of Medicinal Foods* 4, 49–51.

Williner, M.R., Pirovani, M.E. and Güemes, D.R. (2003) Ellagic acid content in strawberries of different cultivars and ripening stages. *Journal of the Science of Food and Agriculture* 83, 842–845.

Wolfe, K.L., Kang, X., He, X., Dong, M., Zhang, Q. and Liu, R.H. (2008) Cellular antioxidant activity of common fruits. *Journal of Agricultural and Food Chemistry* 56, 8418–8426.

Wu, Q.K., Koponen, J.M., Mykkänen, H.M. and Törrönen, A.R. (2007) Berry phenolics extracts modulate the expression of p21 (WAF1) and Bax but not Bcl-2 in HT-29 colon cancer cells. *Journal of Agricultural and Food Chemistry* 55, 1156–1163.

Wu, X. Beecher, G., Holden, J., Haytowitz, D.B., Gebhardt, S.E. and Prior, R.L. (2006) Concentration of anthocyanins in common foods in the United States and estimation of normal consumption. *Journal of Agricultural and Food Chemistry* 54, 4069–4075.

Zabetakis, I., Leclerc, D. and Kajda, P. (2000) The effect of high hydrostatic pressure on the strawberry anthocyanins. *Journal of Agricultural and Food Chemistry* 48, 2749–2754.

Zhang, Y., Seeram, N.P., Lee, R., Feng, L. and Heber, D. (2008) Isolation and identification of strawberry phenolics with antioxidant and human cancer cell antiproliferative properties. *Journal of Agricultural and Food Chemistry* 56, 670–675.

Zheng, Y., Wang, S.Y., Wang, C.Y. and Zheng, W. (2007) Changes in strawberry phenolics, anthocyanins, an antioxidant capacity in response to high oxygen treatments. *LWT – Food Science and Technology* 40, 49–57.

16 Tomato and Other Solanaceous Fruits

Amarat H. Simonne, Cecilia do Nascimento Nunes and Jeffrey K. Brecht

16.1 Main Introduction

The *Solanaceae* or nightshade plant family is one of the major families of plants supplying vegetable and staple food crops in the world (Swiader and Ware, 2001). Tomato (*Solanum lycopersicum* L.) is the second most important crop in this family, after potato (*S. tuberosum* L.) (see Chapter 12 of this volume). Tomatoes and peppers (*Capsicum annuum* L.) are the Solanaceous fruit crops representing the most diverse fruit morphology (Paran and van der Knaap, 2007). Tomatoes, peppers and aubergines (eggplants; *S. melongena* L.) are grown worldwide, while other *Solanaceae* such as tomatillo (*Physalis philadelphica* Lam) and uchuva, or cape gooseberry (*P. peruviana* L.), are grown in limited areas of the world. The current tomato world production is 126 million tonnes (Mt)/year, while the annual world production of other significant Solanaceous fruit, aubergines and peppers, is 32 and 28 Mt, respectively. The production statistics for peppers include dried chilli peppers as well as fresh chilli peppers and bell peppers (FAOSTAT, 2009). Although tomatillo and uchuva or cape gooseberry are also listed among some other commercially grown Solanaceous fruit, due to their small production volumes, current world production figures for these crops are not readily available. However, production of tomatillo has been reported to be significant in Mexico

and Central America (Moriconi *et al.*, 1990; Hernándo Bermejo and Leon, 1994), while the main production areas for uchuva or cape gooseberry include Asia, Africa and South America (Morton, 1987). The fruit from *Solanaceae* plants are important food sources as they provide important bioactive nutrients to the human diet, as well as many secondary metabolites with reported medicinal properties.

16.2 Tomato

16.2.1 Introduction

Cultivated tomato fruit come in many sizes, shapes and colours, resulting from a long history of genetic improvement of the plant for various usages, as well as for improving disease resistance and other desirable traits (Figs. 16.1 and 16.2). Despite the diverse genotypes, the majority of domesticated tomatoes are the results of crosses among the cultivated species and their wild relatives. Although experts agree that the genetic base of the cultivated tomato is narrow, more than 75,000 *S. lycopersicum* germplasm accessions are maintained by countries around the world (Robertson and Labate, 2007). Per capita consumption of tomatoes varies around the world, with countries considered as having high, average or low per capita consumption

Fig. 16.1. Large tomato types. Top from left to right: red round vine ripe green house tomatoes, red round field grown tomatoes; red heirloom ugly tomato cv Mortgage Lifter grown in a shaded condition, and OSU P20 (*AftAft atvatv*) with anthocyanin expression (photo is a courtesy of James R. Myers from Oregon State University); bottom from left to right: orange round field grown tomatoes, yellow round field grown tomatoes, red heirloom field grown ugly tomatoes.

Fig. 16.2. Small tomato types. Top from left to right: heirloom cherry tomatoes cv Brown Berry, red grape tomatoes cv Chiquita and Honey Bunch, yellow grape tomatoes cv. Morning Light; bottom from left to right: heirloom yellow plum tomatoes cv. Cream Sausage, red plum tomatoes and cherry tomatoes.

(Bieche and Covis, 1992). Countries with high tomato per capita consumption include the USA, Italy, Canada and Algeria, while others with average tomato per capita consumption include the UK, France, the former USSR and Germany, and those with low per capita consumption include Japan and Brazil (Bieche and Covis, 1992). Tomato consumption patterns in recent years show increases in both processed and fresh consumption, with mostly processed tomato products being consumed in the USA and mostly fresh tomatoes being consumed in the EU (Harvey *et al.*, 2003), but these changes are by no means monolithic. In Europe, for example, annual per capita tomato consumption in 2007–2008 varied from 30 kg or less in Germany and the Netherlands, mostly as processed products, to 82 kg in Greece, of which 75% was fresh tomatoes (Eurostat, 2006).

16.2.2 Identity and role of bioactive compounds

Tomato contributes significant dietary components to human health due to its popularity, availability and high per capita consumption (Stommel, 2007). A 100 g portion of fresh or raw tomatoes supplies 93–95 g of water, 15–23 calories, 1–2 g of fibre, 9–23 mg of ascorbic acid, 9–30 µg of folate, less than 1 g of fat and no cholesterol (USDA, 2008). In addition to traditional nutritional components such as vitamins and minerals, tomato is rich in other bioactive components including carotenoids (lycopene, β-carotene, α-carotene, lutein, zeaxanthin, unique lycopene metabolite) (Burri *et al.*, 2009), and phenolic compounds (Jones *et al.*, 2003). Several other tomato components may also influence human health, including flavonoids, folic acid and the tocopherols, or vitamin E (Dorais *et al.*, 2001).

Lycopene is the pigment responsible for the red colour of ripe tomato fruit, while other carotenoids in tomato fruit are either yellow or colourless. The physiological function of carotenoids in plants relates primarily to their photoprotective and chemoprotective role as antioxidant free radical scavengers, counteracting the damaging effects of reactive oxygen species (Demmig-Adams *et al.*, 1996). Carotenoids are also precursors of some important aroma volatile compounds in tomato (Lewinsohn *et al.*, 2005).

Phenolic compounds have numerous roles in plants, including cell wall structure, pigmentation and protection against microbial pathogens (von Roepenack-Lahaye *et al.*, 2003), and, like carotenoids, they also play roles as antioxidants and free radical scavengers. Phenolic compounds in Solanaceous plants (as well as in other plants) are classified into many groups depending on the complexity of their structures, including the numbers of aromatic rings (one, two or more) and the substitutional groups on those rings. To date, there is no one perfect way to group them and, therefore, phenolic compounds such as flavonoids, anthocyanins, ascorbic acid or quercetin may not be classified under the same heading, despite their general classification as phenolic compounds (Shahidi, 2005; Shahidi and Ho, 2005; Slimestad and Verheul, 2009). Although phenolic compounds in tomatoes are present in lower concentrations than other phytonutrients, such as lycopene in mature and ripe fruit, the concentrations of these compounds depend on genotype, maturity and location within the fruit, with the highest concentrations found in the skin and placental tissues (Buta and Spaulding, 1997; Slimestad and Verheul, 2009). These phenolic compounds serve to protect the fruit from oxidative and other stresses (Winkel-Shirley, 2002). A recent comprehensive review on flavonoids and other phenolic compounds in fruit of different tomato cultivars revealed fast-growing interest in the subject, and the authors concluded that choices of cultivar as well as growing environment affected the quantity and quality of phenolics in tomatoes (Slimestad and Verheul, 2009).

Carotenoids (lycopene, β-carotene, lutein)

Tomato fruit are generally known to contain high levels of lycopene, which is responsible for the typical characteristic deep-red colour of ripe tomato fruit and the red colour of tomato products. Lycopene comprises 80–90% of the pigments in red tomatoes and a typical red tomato fruit may contain 0.01–0.2 mg of lycopene/g on a fresh weight (FW) basis (Shi and Le Maguer, 2000). Furthermore, tomato varieties with the crimson gene tend to have extremely high levels of lycopene (Thompson *et al.*, 2000), but the distribution of different types of carotenoids in fact determines the fruit colour (Fig. 16.3). In addition to lycopene, red tomatoes also contain very small amounts of other carotenoids such as β-carotene, α-carotene, β-cryptoxanthin and lutein (Shi and Le Maguer, 2000). The content of these other carotenoids is relatively low in comparison with lycopene, but β-carotene is an important precursor of vitamin A. The major carotenoids in yellow (and some orange) tomatoes are β-carotene and lutein, but the contents of these compounds are often very low in comparison with lycopene content in red tomatoes. For example, Simonne *et al.* (2007) reported the β-carotene and lutein contents to vary from 0.6–0.9 and from 0.1–0.4 µg/g FW, respectively, in yellow tomato cv. Honey Bunch.

Tomato fruit constitute a generous source of vitamin C as well as other important bioactive compounds. At the red ripe stage, tomatoes contain on average 0.13 mg of vitamin C/g FW (USDA, 2008). However, depending on the type or cultivar, weather conditions, agricultural practices and postharvest environmental conditions, vitamin C content can range from 0.15–0.95 mg/g FW of tomato

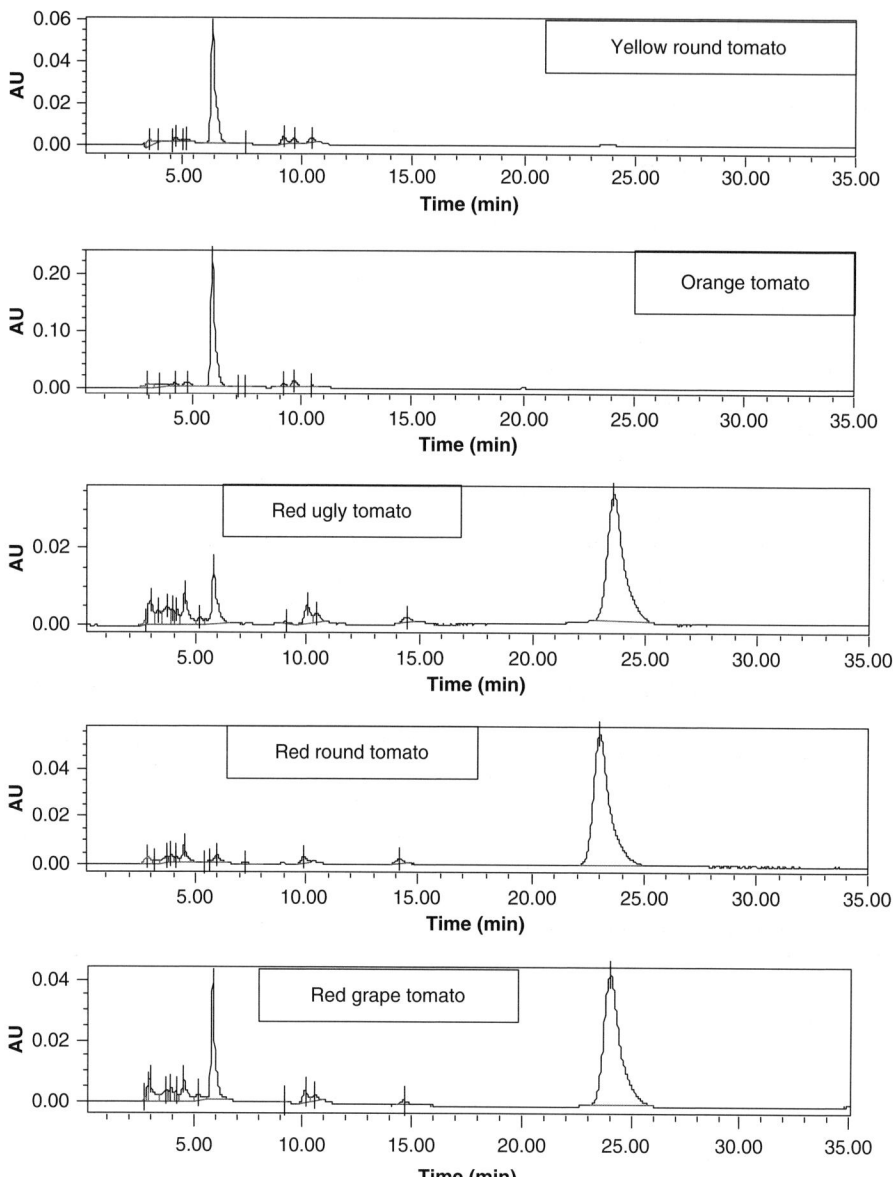

Fig. 16.3. Typical chromatograms of tomatoes as a function of colours and types. First and second from the top are chromatograms of yellow and orange round tomatoes, respectively, with a prominent peak of β-carotene at six minutes retention time. The third, fourth, and fifth from the top are chromatograms of red ugly, red round and red grape tomatoes, respectively, all with a prominent lycopene peak at 25 minutes retention time and small β-carotene peak at six minutes retention time for red ugly and red grape tomatoes. These chromatograms were obtained from A.H. Simonne research laboratory with HPLC conditions in Simonne et al. (2007).

fruit (Abushita et al., 2000; Yahia et al., 2001; Dumas et al., 2003). Vitamin C content in tomatoes was, however, reported to be more reliant on the cultivar and maturity stage of the fruit than on seasonal variations (Shinohara et al., 1982; Raffo et al., 2006).

16.2.3 Chemopreventive activity and bioavailability

Tomatoes have been documented as foods with potential chemopreventive activities against many chronic diseases because of the high levels of lycopene and other bioactive compounds (Giovannucci et al., 1995; Giovannucci, 1999, 2002). Experts agreed that, in order for the bioactive compounds from foods or supplements to exert any health effects, they must be consumed, absorbed and metabolized by the body and remain at certain concentrations in various tissues; however, data on absorption and metabolism of tomato bioactive compounds remain incomplete (Porrini et al., 1998). Therefore, the subjects of absorption and pharmacokinetic properties of bioactive compounds such as lycopene have been investigated by many researchers (Hadley et al., 2002; Diwadkar-Navsariwala et al., 2003; Cohn et al., 2004; Gustin et al., 2004; Basu and Imrhan, 2007; Lindshield et al., 2006; Unlu et al., 2007a,b; Devaraj et al., 2008; Burri et al., 2009). It is well accepted that uptake of lycopene progresses from intestinal mucosa into lymph, then the liver before deposition in various tissues (Schmitz et al., 1991; Stahl et al., 1992; Clinton et al., 1996; Bramley, 2000). Furthermore, lycopene may undergo changes in the human body post-consumption and absorption, due to the biochemical processes in the human body (Lindshield et al., 2006).

Overall, based on the current research, bioavailability of bioactive compounds depends on the type of compounds, food matrix (processing) and interactions with other food components, as well as the stage of gastrointestinal physiology. For example, bioavailability of lycopene in cooked or processed tomatoes is greater than in raw tomatoes because the processing treatments increase the bioaccessibility of the plant tissue (Gartner et al., 1997; Dewanto et al., 2002; Richelle et al., 2002). Also, bioavailability and absorption of lycopene is further dependent on the type of isomers, with cis-isomers being more bioavailable than the all-trans form (Failla et al., 2007; Unlu et al., 2007a), lipid levels (Gustin et al., 2004) and tomato variety (Unlu et al., 2007b; Burri et al., 2009).

Cancer studies

After the research publication by Giovannucci et al. (1995) suggesting that tomato-based food may be beneficial in reducing prostate cancer risk, and because of the anti oxidant properties of lycopene, tomato products containing high levels of lycopene have received much attention as potential cancer fighting foods. A comprehensive review of the epidemiological literature in English by Giovannucci (1999, 2002) revealed that intake of tomatoes and tomato-based products and lycopene levels in plasma were strongly negatively correlated with the risk of some types of cancers, such as lung, stomach and prostate gland, but only weakly correlated, if at all, with the risk of cancers of the cervix, breast, oral cavity, pancreas, colorectum and oesophagus. Another review by Miller et al. (2002), summarizing accumulated research on tomato products, lycopene and prostate cancer risk, recommended that consumption of one serving/day or five servings/week of tomato products, as a part of a health dietary pattern, might reduce the risk of prostate cancer or other chronic diseases. These reviews (Giovannucci, 1999, 2002; Miller et al., 2002) also revealed that, although the benefit of tomato and tomato products for reducing risk of certain cancers was often attributed to lycopene, there was no proof of direct benefit from ingesting lycopene alone. Giovannucci (1999) suggested ultimately that diets rich in a variety of fruit and vegetables, including tomatoes and tomato-based products, would provide health benefits.

After these reviews were published, many more research studies reported positive, neutral (inconclusive) or negative outcomes in attempts to link lycopene intakes and cancer. Among the positive outcomes, many studies have shown an inverse association between consumption of tomato and/or lycopene supplements and the risk of certain types of cancers, but others have only suggested that increased consumption of lycopene from tomatoes and tomato products may provide protection, while not enough information is available on the therapeutic use of lycopene (Chen et al., 2001; Stacewicz-Sapuntzakis and Bowen, 2005). The US Food and Drug

Administration evaluated two health claim petitions submitted in 2004 and found very limited evidence to support the association between tomato consumption and reduced risks of prostate, ovarian, gastric and pancreatic cancers (Kavanaugh *et al.*, 2007). Another study, by Peters *et al.* (2007), examining the association between prediagnostic serum carotenoids (including lycopene) and the risk of prostate, lung, colorectal and ovarian cancers, revealed that high serum β-carotene increased risk for aggressive prostate cancer and that lycopene and other carotenoids were unrelated to prostate cancers. The authors further suggested that lycopene or tomato-based products would not be effective for prostate cancer prevention. Authors of other reviews (Etminan *et al.*, 2004; Seren *et al.*, 2008) examining the use of lycopene in cancer prevention and treatment came to the conclusion that, because there were not enough data regarding the benefit of lycopene supplementation, the best potential benefits could be obtained from consumption of lycopene-rich fruit and vegetables, including tomatoes. A comprehensive review of dietary lycopene in relation to its properties and anticarcinogenic effects revealed inconsistencies in the epidemiological data related to disease prevention by lycopene, and the pharmacokinetic properties of lycopene still remain poorly understood (Singh and Goyal, 2008). Another recent review (Amin *et al.*, 2009), on the potential use of many natural products such as lycopene in cancer prevention, has shown that lycopene may decrease growth of some cancer cells, but could not define clear mechanisms of how this compound prevents cancer. Another recent study on lycopene and health claims (Cámara and Fernánadez-Ruiz, 2009) also concluded that more research was needed.

Because of the inconsistent outcomes in regard to consumption of tomatoes or tomato products (as a source of lycopene and other bioactive compounds) and different types of cancers, many researchers continue to focus on understanding the modes of action or basic mechanisms of action or roles of lycopene and bioactive compounds in oxidative stress (Porrini and Riso, 2000; Chen *et al.*, 2001; Basu and Imrhan, 2007), carcinogenesis (Porrini and Riso, 2000; Sharoni *et al.*, 2004;

Wertz *et al.*, 2004) and specific gene regulation (Zhang *et al.*, 1992). Among the reported modes of action of lycopene are: DNA protection (Porrini and Riso, 2000), increased communication of gap-junction (Wertz *et al.*, 2004), inhibition of IGF-I (insulin-like growth factor I) signal transduction (Wertz *et al.*, 2004), inhibited or reduced gene expression (Herzog *et al.*, 2004; Wertz *et al.*, 2004) and receptors of specific molecules (Wertz *et al.*, 2004), gene transcription (Sharoni *et al.*, 2004) and cell cycle regulation (Karas *et al.*, 2000; Rao and Rao, 2007). Lindshield *et al.* (2006) suggested that tomato carotenoid metabolites (lycopenoids) may be responsible for reduced risks of prostate cancer in men who have consumed high levels of tomato products, but more research is needed.

Cardiovascular diseases

A comprehensive review of tomatoes and cardiovascular health (Wilcox *et al.*, 2003) revealed that tomatoes and tomato products contained nutrients essential for cardiovascular health, namely lycopene and β-carotene, as well as other vitamins and bioactive compounds. Earlier studies revealed a strong association between diets rich in fruit and vegetables and reduced risk of cardiovascular diseases (CVD) (Clarke and Armitage, 2002; Bazzano *et al.*, 2003; Hung *et al.*, 2004; Omoni and Aluko, 2005; Dauchet *et al.*, 2006). An extensive study on lycopene and myocardial infarction risk from ten European countries (Kohlmeier *et al.*, 1997) eliminated the association of α- and β-carotene with myocardial infarction risk, but correlated it with lycopene, a carotenoid that is more common in food sources. Subsequently, basic research has revealed that the oxidation of low-density lipoproteins (LDL) increases the risk of CVD. A study by Fuhrman *et al.* (1997) showed the hypocholesterolaemic effect of lycopene and β-carotene, which suppressed cellular cholesterol synthesis from acetate, and that the inhibition occurred at the same time as the stimulation of LDL receptor activity on macrophages, leading to clearance of LDL from the plasma. Epidemiological observations also suggest that antioxidant vitamins (i.e. vitamins E and C) and carotenoids such as lycopene and β-carotene may have protective effects against CVD (Kris-Etherton *et al.*, 2002; Rao, 2002). These trends led to years of research to evaluate the effects of both pure supplements and specific foods on the prevention of CVD and cancers, but comprehensive reviews of the results have failed to confirm any protective effects of these supplements/foods against either cancer or CVD (Clarke and Armitage, 2002). Recent research results, however, continue to show that fruit and vegetable consumption is associated inversely with CVD (Dauchet *et al.*, 2006), with tomato being one of the top nutrient-rich fresh produce types on the list. Possible mechanisms of action of tomatoes, tomato products and tomato constituents on the prevention of CVD have been attributed to antioxidative (e.g. reduced LDL oxidation) as well as non-oxidative effects (e.g. reduced HMG-CoA-reductase activity, reduced homocysteine levels in the blood and reduced platelet aggregation); however, many research reports continue to suggest that additional examination of the health benefits of tomato consumption is required (Wilcox *et al.*, 2003).

Other beneficial/detrimental effects

Intake of tomatoes, cooked vegetables and fruit has been documented as being protective against wheezing in children (Farchi *et al.*, 2003). Another review by Sies and Stahl (2004) revealed that dietary lycopene could protect against skin damage (erythema) due to sunlight. Shao and Hathcock (2006) systematically evaluated the risk of high intake of lutein and lycopene and found that intakes of up to 20 mg/day for lutein and 75 mg/day for lycopene did not result in any clear adverse effects, but they did not have enough data to determine long-term safety.

16.2.4 Effect of preharvest and postharvest continuum

Genotypes

Commercial tomatoes include several different types, including round tomatoes, which are most commonly used for fresh consumption,

oblong 'roma' or 'plum' types, typically grown for processing, and small 'cherry' and 'grape' tomatoes, usually eaten fresh. Furthermore, while most ripe tomatoes are red, there are also cultivars that become yellow or orange during ripening. Research has shown that nutritional quality and bioactive compound levels may be different for various genotypes of tomatoes (Leonardi *et al.*, 2000). For example, the different colours found in various tomato genotypes represent the relative amounts of red lycopene and yellow β-carotene pigments. The lycopene content of most red tomato cultivars is about 30–50 mg/kg, while deep red cultivars contain more than 150 mg/kg and yellow cultivars only about 5 mg/kg (Hart and Scott, 1995); the ranges of lycopene content found in fresh market tomato cultivars are similar to the levels measured in processing tomatoes (Garcia and Barrett, 2005). In recent years, efforts have also been made to enhance further the levels of bioactive compounds in tomatoes by examining tomato genotypes with the capacity to produce anthocyanins in the fruit (Jones *et al.*, 2003) and ways to express genes related to anthocyanin production (Butelli *et al.*, 2008). Other efforts have also been made to understand the upregulation of a number of tomato mutants with high pigment levels, as well as those with high antioxidant contents (Long *et al.*, 2006; Kolotilin *et al.*, 2007).

Production conditions

In general, as tomato fruit mature on the plant, vitamin C content tends to increase. For example, vitamin C content increased from about 0.13 mg/g FW at 18 days after fruit set to 0.95 mg/g FW at 74 days after fruit set, at the red ripe stage (Yahia *et al.*, 2001). Consequently, vitamin C levels are usually higher in vine-ripe tomatoes than in fruit harvested at the mature-green stage (Yahia *et al.*, 2001; Dumas *et al.*, 2003; Slimestad and Verheul, 2005).

Dumas *et al.* (2003) and Dorais (2007) have reviewed the effects of environment and production conditions on the antioxidant content of tomato fruit. The amount of lycopene, the predominant antioxidant compound in tomato fruit, is influenced by the environmental conditions under which the fruit develop and ripen. Red colour development and vitamin C content are limited by extreme high and low temperatures and solar radiation occurring during the time when the tomato fruit are ripening on the plant (Brandt *et al.*, 2006; Dorais, 2007; Dorais *et al.*, 2008). Furthermore, light exposure also contributes to vitamin C accumulation in tomato fruit (Dumas *et al.*, 2003). While vitamin C synthesis in tomato fruit is stimulated by light exposure, lycopene synthesis is influenced negatively by light (Passam *et al.*, 2007). The effect of UV-B irradiation on antioxidant potential in tomato fruit has been found to be either negligible or detrimental, depending on the tomato genotype; although depletion of UV-B radiation in the light growth conditions induced a significant increase in the vitamin C content of DRW 5981 and Esperanza tomato cultivars, vitamin C levels in HP 1 fruit tended to decrease (Giuntini *et al.*, 2005). Similarly, Giuntini *et al.* (2005) reported that vitamin C accumulation was either negligible or was promoted, and carotenoid accumulation was either inhibited or promoted, in different tomato genotypes under such radiation. Luthria *et al.* (2006) found that reduced UV radiation, with no difference in photosynthetically active radiation, resulted in lower total phenolics, as well as lower concentrations of individual phenolic compounds in ripe tomato cvs. Oregon Spring and Red Sun fruit. In a season-long study in southern Spain, using greenhouse cherry tomatoes, Rosales *et al.* (2006) showed that lycopene and antioxidant levels in cv. Naomi tomato fruit were correlated negatively with temperature and overall solar radiation. In another greenhouse tomato study in New Zealand, Toor *et al.* (2006) showed similar trends for cvs. Excell, Tradiro and Flavourine, with the total phenolics and antioxidant activity (AOX) in the three cultivars being higher in summer than in spring and the lycopene contents being lower in the summer months; any effects of light intensity or temperature on vitamin C were unclear.

Water deficit during fruit development is well known to increase tomato soluble solids content, as well as increasing lycopene and vitamin C contents (Dumas *et al.*, 2003). Lycopene content in tomato fruit can also be enhanced by the fertilization regime, with moderate N and increasing P and K levels resulting in greater carotenoid development (Dumas *et al.*, 2003). A report by Taber *et al.* (2008) indicated that the response of tomato to a high K fertilization rate was genotype dependent, with a high-lycopene (crimson gene) variety developing greater carotenoid content in response to increasing K, while a normal variety showed no response. Excess N has also been reported to result in lower vitamin C in tomatoes (Mozafar, 1993; Kobryn′ and Hallmann, 2005). A report by Subbiah and Perumal (1990) indicated that Ca sprays applied to tomatoes during fruit growth increased the vitamin C content. Tomato fruit grown on organic substrates were reported to have higher vitamin C levels than fruit grown in hydroponic substrates (Premuzic *et al.*, 1998).

Another recent review by Dorais *et al.* (2008) concluded that producing tomatoes with specific health attributes might have to be accomplished under unique growing conditions that might not result in the highest yield, and thus much more work will be needed to achieve any given specific benefit.

Postharvest treatments and storage

Lycopene content in ripening tomato fruit is influenced strongly by storage temperature, as lycopene synthesis is inhibited at temperatures at or above 30°C (Goodwin and Jamikorn, 1952; Tomes, 1963). Synthesis of β-carotene, however, continues up to at least 35°C (Hamauzu *et al.*, 1998). Tomato ripening does not cease completely at 30°C, however, and high temperature inhibition of lycopene synthesis is reversible if the fruit are returned to a lower temperature to the extent that they remain viable to continue ripening.

Postharvest exposure of unripe or partially ripe tomato fruit to temperatures below 12°C also inhibits carotenoid synthesis due to chilling injury, which either inhibits ripening or results in abnormal colour development during ripening (Wang, 1993). Exposure of green tomatoes to ethylene initiates and accelerates ripening, including red colour development, but does not influence the final lycopene content of ripe fruit. Conflicting reports of differences in lycopene development in tomato fruit ripened on the plant versus postharvest (Giovanelli *et al.*, 2001; Wold *et al.*, 2004) may be due to temperature differences, with the temperature regime that is more conducive to lycopene development resulting in greater levels of lycopene.

Vitamin C content of tomato fruit generally increases during ripening on the plant and declines during postharvest storage and handling (Bisogni *et al.*, 1976; Betancourt *et al.*, 1977; Soto-Zamora *et al.*, 2000). The vitamin C content may increase during storage of tomatoes harvested at earlier stages of ripeness, but it never attains the levels found in vine-ripened fruit (Scott and Kramer, 1949). Although tomatoes ripened on the plant tend to accumulate more vitamin C than fruit ripened off the plant, the accumulation patterns in vine-ripened and postharvest-ripened fruit differ (Giovanelli *et al.*, 1999). In vine-ripened tomatoes, vitamin C accumulation takes place during the first stages of fruit ripening and then decreases, whereas in postharvest-ripened tomatoes (20°C) vitamin C shows an initial decrease, followed by a significant increase in the last stages of fruit ripening (Giovanelli *et al.*, 1999). Tomato fruit harvested at the breaker stage and ripened at 20°C contained only 43.6–62.9% of their potential vitamin C content if ripened on the plant to the red ripe stage (Betancourt *et al.*, 1977). Tomatoes harvested green and ripened at 20°C contained about 55–65% of the vitamin C content relative to those harvested at the table-ripe stage (Kader *et al.*, 1977).

Postharvest ethylene treatment to accelerate ripening results in higher vitamin C levels in the ripened fruit than seen in tomatoes ripened without ethylene, due to less vitamin C loss during the shorter time required for ethylene-treated fruit to reach the full ripe stage (Watada *et al.*, 1976; Kader *et al.*, 1978).

Environmental conditions during the postharvest period, namely temperature, may also contribute to increased or decreased vitamin C content of tomato fruit. For example, in tomatoes (cvs. Roma, Marglobe, Sioux, Best of All, Red Plum, Pusa Ruby, Ponderosa and H.S. 102) stored at 20°C, vitamin C content showed a tendency to increase during the first 8 days of storage but decreased afterwards (Syamal, 1990). In general, postharvest conditions or treatments such as ethylene or controlled atmospheres that either accelerate or delay ripening have indirect rather than direct effects on those tomato bioactive compounds that change in concentration during ripening.

Storage of tomatoes at chilling temperatures (i.e. below 10°C) increases the rate of vitamin C loss. Tomatoes harvested at the colour-break stage and stored for 10 days at 21°C showed the highest content of vitamin C with greatest retention during storage, compared with storage at the chilling temperatures of 1.6 or 10°C (Scott and Kramer, 1949). In tomato fruit stored at 4 or 10°C, vitamin C content increased initially but then decreased (Soto-Zamora *et al.*, 2005).

Ripening of tomato fruit at high temperatures leads to a decrease in vitamin C content due to ascorbate oxidation (Dumas *et al.*, 2003); however, heating cv. Rhapsody tomato fruit in air at 34°C for 24 h prior to storage at 10°C for up to 30 days reduced losses in antioxidant content compared with unheated fruit, and fruit colour developed adequately (Soto-Zamora *et al.*, 2005). Exposure of mature green cv. Rhapsody tomato fruit to 34°C and 95% RH for 24 h promoted the tomato antioxidant system during ripening at 20°C (Yahia *et al.*, 2007).

16.3 Peppers

16.3.1 Introduction

Although more than 30 species of peppers exist, commercial production around the world consists of the five main domesticated species: *C. annuum* (bell peppers, paprika, cayenne, jalapeños and the chiltepin), *C. frutescens* (tabasco peppers), *C. chinense* (naga, habanero, datil and Scotch bonnet), *C. pubescens* (South American rocoto peppers) and *C. baccatum* (wax peppers and berry-like South American aji peppers) (McLeod *et al.*, 1982; Pickersgill, 1997; Lefebvre *et al.*, 2001; Votava *et al.*, 2005; Moscone *et al.*, 2007; Paran and van der Knaap, 2007; Jarret and Berke, 2008). Among the five domesticated pepper species, *C. annuum* represents the largest group of peppers grown worldwide. The fresh non-pungent peppers, which are known as bell peppers, are more economically and nutritionally important, because of their higher consumption level (Paran and van der Knaap, 2007), than those used as seasonings (paprika, dried pepper and chilli powder) (Pradeep *et al.*, 1992; Wall *et al.*, 2001).

16.3.2 Identity and role of bioactive compounds

Peppers are rich in traditional nutrients such as vitamin C and are a moderate source of bioactive compounds such as carotenoids, phenolics, flavonoids, vitamin E and capsaicin, to name just a few. While bioactive compounds in peppers may have some physiological function in the plants or fruit themselves, and serve as functional food for humans, some researchers have concluded that peppers are not as good a source of carotenoids (zea xanthin, lutein, β-carotene), vitamin E (α-tocopherol) or the flavonoids quercetin and luteolin as some other fruit and vegetables (Lee *et al.*, 2005).

Carotenoids

The fruit of the genus *Capsicum*, which includes both sweet (*C. annum*) and hot (*C. frutescens*) peppers, are rich in many typical carotenoids (e.g. provitamin A carotenoids, lutein and lycopene) and unique keto-carotenoids as well as allylic apo-carotenols. Pepper carotenoids have been well investigated in relation to paprika, which is commonly used as a natural colourant, as well as in terms of their vitamin A precursor activity (Gross, 1987). Pepper carotenoid profiles have been summarized comprehensively by Gross (1987). Carotenoid content of peppers varies in different cultivars (Simonne *et al.*, 1997; Howard *et al.*, 2000; Breithaupt and Bamedi, 2001; Wall *et al.*, 2001; Deepa *et al.*, 2006; Ha *et al.*, 2007; Suzuki *et al.*, 2007) and, in general, the total carotenoid and β-carotene contents increase with maturation and ripening (Howard *et al.*, 2000; Deepa *et al.*, 2006; Ha *et al.*, 2007). Also, it was found that red peppers accumulated relatively high amounts of total carotenoids during ripening,

while non-red peppers accumulated lower levels of total carotenoids of varying compositions (Ha *et al.*, 2007). Many hot pepper varieties have been documented to have extremely high amounts of β-carotene (up to 1.2 mg/g dry weight) and total carotenoids (up to 10 mg/g dry weight) (Wall *et al.*, 2001). However, given the extreme pungency of many hot peppers, the amount of consumption is quite low. Thus, in a practical sense, high levels of carotenoids in very hot peppers may be of limited nutritional interest because normally a large amount cannot be consumed.

Phenolic compounds and vitamin C

Bell peppers are one of the most important sources of vitamin C in the human diet, with some cultivars contributing almost 500% of the RDA (recommended daily allowance) for this vitamin per 100 g serving (Howard *et al.*, 2000). Coloured bell peppers tend to have higher contents of vitamin C and carotenoids than green peppers; however, green, red and orange peppers have all been reported to have much higher contents of vitamin C and carotenoids than more unusually coloured peppers, such as black, purple or white (Simonne *et al.*, 1997). Green bell peppers contain on average 0.15–0.80 mg of vitamin C/g FW, whereas red and yellow peppers may contain on average 127.7 and 315.3 mg of vitamin C/100 g FW, respectively (Osuna-García *et al.*, 1998; Yahia *et al.*, 2001; Geleta and Labuschagne, 2006; USDA, 2008). Green *C. annuum* bell peppers and green hot peppers contain much higher levels of most polyphenols than the ripe fruit, but the amounts of some phenolic compounds increase during ripening (Marín *et al.*, 2004; Materska and Perucka, 2005). Coloured peppers have, in general, a higher nutritional value than green fruit because they may contain three times more vitamin C and four times more vitamin E.

Vitamin C levels increase during ripening and peak at the full ripe stage of development (Howard *et al.*, 1994, 2000; Luning *et al.*, 1994; Simonne *et al.*, 1997; Osuna-García *et al.*, 1998; Márkus *et al.*, 1999; Yahia *et al.*, 2001). At the full ripe stage, peppers may contain 95% more vitamin C than at the mature green stage (Howard *et al.*, 1994). Minimum vitamin

C levels (0.52 mg/g FW) were detected in bell peppers at 22 days after fruit set, increasing rapidly to 1.36 mg/g FW at 51 days after fruit set (Yahia *et al.*, 2001). This trend was also observed in 43 pepper cultivars and breeding lines grown in Texas, which showed a substantial increase in vitamin C concentrations with maturation (Crosby *et al.*, 2008). Similarly, vitamin E (α-tocopherol), also present in different chilli pepper varieties, increases during ripening, from the mature green to the fully red stages (Osuna-García *et al.*, 1998; Márkus *et al.*, 1999).

Flavonoids

Peppers are documented to have moderate to high levels of flavonoids as per the definition provided by Peterson and Dwyer (1998), in which low, moderate and high flavonoid concentration ranges are 0.01–0.4, 0.04–0.1 and > 0.1 mg/g, respectively. Total flavonoid contents reported for various pepper species (*C. annuum*, *C. frutescens* and *C. chinense*) ranged from 0.002 mg/g (*C. chinense* habanero type cv. Red Savina) to 0.85 mg/g (*C. annuum*, long yellow-type pepper cv. Inferno), while the flavonoid content of a yellow bell pepper (*C. annuum*) was 0.32 mg/g (Howard *et al.*, 2000). The levels of flavonoids in the peppers used in the latter study could not be predicted from maturity, as the flavonoid content of some fruit increased while that of others others decreased during maturation, and it was suggested that extremely low levels of flavonoids in very hot peppers might reflect the interchange between flavonoids and capsaicin (Howard *et al.*, 2000).

In another study, Saga and Sato (2003) examined immature and ripe Japanese hot and sweet peppers and found that the amount of total phenolics was higher in hot and ripe peppers than in sweet and immature peppers, but there was no difference in quercetin content between hot and sweet peppers during the fruit growing season. In a later study by Marín *et al.* (2004), of antioxidant constituents of sweet peppers (hydroxycinnamic acids, flavonoids, carotenoids and ascorbic acid), clear differences in individual and total phenolic contents were demonstrated for different maturity stages;

immature green peppers exhibited the highest content of polyphenols, while red ripe fruit had the highest contents of ascorbic acid and provitamin A.

Based on many studies, it appears that flavonoid contents in peppers may depend on genotype (types of peppers), maturity stage, postharvest treatment, cooking and processing, as well as use. In general, the flavonoids reported most often in peppers are quercetin, luteolin (Howard *et al.*, 2000), coumaric acid, caffeic acid derivatives, apigenin and luteolin derivatives (Materska *et al.*, 2003; Materska and Perucka, 2005). However, specific information on the effects of cooking on these bioactive components remains conflicting; while some have reported no significant changes (Turkmen *et al.*, 2005; Ornelas-Paz *et al.*, 2010), others have suggested that losses are highly variable (Rickman *et al.*, 2007).

Capsaicin

Capsaicin (*trans*-8-methyl-*N*-vanillyl-6-nonenamide) is a pungent (perceived as 'hot') compound found in various species of *Capsicum* and is classified as part of the alkaloid family along with other capsaicinoids (Thompson *et al.*, 2006; Hayman and Kam, 2008; NPIC, 2009). Documented uses of capsaicin include as a flavouring agent (Mortensen and Mortensen, 2009), a pesticide and insect repellent (NPIC, 2009) and an analgesic (Hayman and Kam, 2008), as well as a tool in neurobiological research (Buck and Burks, 1986; Franco-Cereceda *et al.*, 1987; Merck Index, 1996). Capsaicin has also been widely documented for use in topical pain management, diabetes and obesity control, cancer treatment, headache control and as an anti-inflammatory agent (Mortensen and Mortensen, 2009), as well as for control of overactive bladders (a condition of frequent urination) (Cronin, 2002). Other uses of capsaicin include pepper spray for personal protection (Mortensen and Mortensen, 2009). Although capsaicin is widely used topically and is well absorbed via the skin for various conditions including pain (Gilbert *et al.*, 2007; Singh and Nulu, 2008), this chapter will focus only on the food and oral use of this compound. Despite a long historic use of chilli peppers, the chemical investigation of these fruit did not begin until the 19th century and the chemical structure of capsaicin was identified only in the early 20th century (Barceloux, 2009). Furthermore, although the capsaicinoid family consists of more than 20 compounds, only three (6-ene-8-methyl capsaicin (6,8-C), 8-methyl dihydrocapsaicin (8-DC) and *N*-vanillylnonanamide (NVN)) are well characterized and commercially available in pure form (Thompson *et al.*, 2006).

The highest level of capsaicin is present in the fruit placental tissues (which hold the seeds) of hot pepper fruit. Based on the pungency rating using Scoville heat units, bell peppers score 0–100, habanero peppers score 200,000–575,000, while pure capsaicin registers as 16 million Scoville units (Hayman and Kam, 2008; Mortensen and Mortensen, 2009). Characterization and quantification of capsaicin in hot peppers have been subjects of much research, and in a recent study by Garcés-Claver *et al.* (2006) it was found that two forms of capsaicin (capsaicin and dihydrocapsaicin) were the predominant contributors to pungency in hot peppers. To date, information concerning the levels of capsaicin in hot peppers is somewhat fragmented; however, Garcés-Claver *et al.* (2006) reported that the concentrations of capsaicin in different pepper genotypes ranged from 0.002–6.6 mg/g.

Other phytonutrients

Other phytonutrients, such as non-pungent capsaicinoids (capsiate) in sweet peppers, have been evaluated for chemopreventive and anticancer potential (Macho *et al.*, 2003). Others have reported that the capsaicinoids cause an increase in body temperature (Ohnuki *et al.*, 2001b), suppress fat accumulation (Ohnuki *et al.*, 2001a) and block pathologic angiogenesis and vascular permeability caused by vascular endothelial growth factor (Pyun *et al.*, 2008). Some efforts have been made to identify and characterize additional capsaicinoids (Kozukue *et al.*, 2005; Thompson *et al.*, 2005) for further assessment of their biological activities.

16.3.3 Chemopreventive activity and bioavailability

Chemopreventive properties of some pungent ingredients in red peppers have been described by Surh *et al.* (1998). For peppers, many of the chemopreventive properties identified have been attributed to capsaicin (Surh, 2002). Although red and green *C. annuum* peppers have high levels of antioxidative activity, a recent study has shown that methanolic extracts of pepper fruit tissue do not exhibit any antiproliferative activity in contrast to many other commonly consumed raw vegetables (Park *et al.*, 2000). Bioavailability or absorbability of some bioactive components in peppers is limited compared with other vegetables or fruit such as tomatoes. Suresh and Srinivasan (2007) compared the absorbability of three spice active principles, curcumin, piperine and capsaicin, in an *in vitro* rat intestine model system. Although these three compounds were similar in structure, the researchers found that capsaicin (10–500 µg/10 ml) was absorbed the least for a given concentration and the amount absorbed did not increase proportionally with the concentration applied; however, absorption increased when the same compound was present in micelles rather than in its native form.

Cancer studies

Capsaicin in peppers has been at centre stage when it comes to cancer studies. Capsaicin has been documented to have both negative and positive effects on various types of cancers, but so far the results have not been conclusive. A case–control study that took place in Mexico City from 1989 to 1990, examining the relation between chilli peppers and gastric cancer, found that chilli pepper consumption was considered a possible risk factor for gastric cancer, but more studies are needed to prove this assumption (López-Carrillo *et al.*, 1994). These authors later investigated an association of gastric cancer with capsaicin consumption and *Helicobacter pylori* infection, but they concluded that, in Mexico, chilli pepper consumption might be independent of gastric cancer (López-Carrillo *et al.*,

2003). Another case–control study in Korea (Lee *et al.*, 1995) revealed an increased risk of stomach cancer among those people with high consumption of soybean paste stew and hot peppers. Another study (Serra *et al.*, 2002) revealed an association of chilli pepper consumption, low socio-economic status and longstanding gallstones with gallbladder cancer in a Chilean population, but the researchers also suggested additional studies needed to be conducted to examine the risk factors further.

Laboratory studies suggest that capsaicin in hot chilli peppers may be a carcinogen. In reviews by Surh and Lee (1995, 1996) examining capsaicin as a carcinogen, co-carcinogen or anticarcinogen, it was revealed that capsaicin might have dual effects on chemically induced carcinogenesis and mutagenesis, but the results were conflicting. Metabolism of capsaicin appeared to involve microsomal cytochrome P450-dependent monooxygenases, which are involved in both activation and detoxification of many chemical carcinogens and mutagens. Another review of both experimental and clinical data on capsaicin and stomach disease revealed that capsaicin at low concentrations protected rat gastric mucosa from injury by ulcerogenic agents via stimulation of the local defence system; however, in humans, the higher concentrations of capsaicin obtained from hot foods produced the opposite effects. The authors of this review also concluded that more studies were needed to support their findings (Abdel-Salam *et al.*, 1997). Based on these reviews, it is possible that the positive or negative effects of capsaicin may depend on dosage (Surh and Lee, 1995, 1996; Abdel-Salam *et al.*, 1997; Surh *et al.*, 1998). Yet, it remains unclear whether capsaicin is carcinogenic or anticarcinogenic (Surh *et al.*, 1998).

A review by Archer and Jones (2002) examining the association between capsaicin and ethnicity strengthened further the results of prior case–control studies on the association of stomach cancer with capsaicin in peppers; they found elevated stomach cancer incidence among five cultural groups with high usage of peppers in their cooking. In this study, a reduced colon cancer rate was found among high capsaicin pepper users, but the

authors also suggested additional studies to test their findings further. Laohavechvanich *et al.* (2006) reported that four Thai hot chilli peppers (*C. frutescens*) contained mixtures of antimutagens, but could not establish an association between the antigenotoxicity and glutathione transferase activity tests with larvae of *Drosophila melanogaster*. The anti-tumour or anticancer action of capsaicin and luteolin in peppers may be due to apoptosis of tumour or cancer cells. Such results may have potential application for delaying cancer growth or for cancer treatment (Khan *et al.*, 2006). Although many studies have revealed the anticancer effects of capsaicin and luteolin (Table 16.1), based on current knowledge there are not enough data to make a final conclusion because some of the negative or positive effects are dose dependent.

Cardiovascular diseases

Direct studies testing the potential benefits of peppers on CVD have not been documented. Many studies have, however, included peppers as one of the vegetables with potential properties against CVD due to the high content of many antioxidative compounds that may help prevent the oxidation of cholesterol and other blood lipids (Perucka and Materska, 2001; Suhaj, 2006; Sun *et al.*, 2007; Antonious *et al.*, 2009).

Other beneficial and detrimental effects

Peppers and their bioactive compounds have been documented to have antimicrobial properties against some foodborne pathogens and and chemoprotective effects against ethanol-induced injury of gastric mucosa, to be potential remedies for functional dyspepsia, to have antioxidative properties in brain tissues and to increase carbohydrate oxidation and metabolic rate in humans. Furthermore, capsaicin has also been reported to have a thermoregulatory effect in animals in both cold and warm environments (Szikszay *et al.*, 1982) and thermogenesis (Mahmmoud, 2008).

Antimicrobial properties of chilli peppers as well as their uses in Mayan medicine have been recorded, with capsaicin and dihydro-capsaicin showing varying degrees of growth

Table 16.1. Health-promoting action of Solanaceous fruit.

Activity	Action	System	Dose	Extract type	References
Anticancer	Induced programmed cell death by activating the peroxisome proliferator-activated receptor	HT-29 human colon cancer cells	0–300 μm	Capsaicin	Kim *et al.* (2004)
Anticancer	Inhibit growth of prostate cancer cells	Prostate cancer cells (LNCap, PC-3 and Du-145 cells)	1×10^{-4} – 5×10^{-4} mol/l for PC-3 cells and 1×10^{-4} – 5×10^{-4} mol/l LNCaP cells	Pure capsaicin	Mcri *et al.* (2006)
Anticancer	Induce programmed cell death and inhibit xenograft prostate tumour growth	Prostate tumour PC-3 cells	$IC_{50} = 20$ μm	Capsaicin	Sánchez *et al.* (2006)
Anticancer	Block STAT3 activation pathway	Multiple myeloma cells	Varied	Capsaicin	Bhutani *et al.* (2007)
Antimicrobial	Inhibit growth of *Helicobacter pylori*	*In vitro*, 16 clinical isolates of *H. pylori*	25–50 μg/ml	Capsaicin	Zeyrek and Oguz (2005)
Antioxidative	Protect against ethanol-induced oxidative injury by inhibiting cyclooxygenase-2	Gastric mucosa of rat	0.5–10 mg/kg	Capsaicin	Par< *et al.* (2000)
Increased metabolism	Increase carbohydrate oxidation	Human subjects	10 g	Red hot peppers	Lim *et al.* (1997)
Increased body temperature	Increase O_2 consumption	Human subjects	0.1 g/kg body weight	Sweet peppers: CH-19 and California-Wandar	Ohnuki *et al.* (2001a)
Increased metabolism	Increased energy metabolism and suppressed body fat accumulation	Mice	10 and 50 mg/kg body weight	Capsiate	Ohnuki *et al.* (2001b)
Cancer prevention	Antiproliferative and apoptotic activity	Human colon cancer cells (Hepa-1c1 × 7 and Sw480 human colon cancer cells	1–20 μg/ml	Ixocarpalactone A (IxoA) from tomatillo	Choi *et al.* (2006)

inhibition of *Bacillus cereus, B. subtilis, Clostridium sporogenes, C. tetani* and *Streptococcus pyogenes* (Cichewicz and Thorpe, 1996). Pepper extracts (*C. annum,* such as habanero, Serrano and pimiento morron peppers) were documented to be most effective against growth of *Listeria* and least effective against *Salmonella,* but the inhibitory effects were attributed to cinnamic and *m*-coumaric acids, not to capsaicin (Dorantes *et al.,* 2000).

Red peppers were more effective than a placebo in decreasing the intensity of dyspeptic symptoms for people with functional dyspepsia (without gastro-oesophageal reflux disease and irritable bowel syndrome), and the decrease in symptoms was attributed to desensitization of gastric nociceptive C-fibres induced by capsaicin (Bortolotti *et al.,* 2002). Water extracts from *C. pubescens* (tree peppers) were also found to inhibit lipid peroxidation induced by different pro-oxidant agents in rat brain, in an *in vitro* study (Oboh *et al.,* 2007; Oboh and Rocha, 2008), and the antioxidative effect was attributed to the high phenolic content in the peppers.

In addition to beneficial effects, pepper capsaicin has been reported to have some adverse effects on humans, including a fatal case of pepper poisoning of an 8-month-old infant (Snyman *et al.,* 2001). In recent years, capsaicin has been linked to arterial hypertensive crisis and acute myocardial infarction (with high thyroid-stimulating hormones), including the case of a 59-year-old male after ingestion of a large quantity of pepper and chilli peppers the day before (Patanè *et al.,* 2008) and a case of arterial hypertensive crisis in a 19-year-old male (Patanè *et al.,* 2009).

Additional information on some beneficial properties is provided in Table 16.1.

16.3.4 Effect of preharvest and postharvest continuum

Genotypes

The types and amounts of antioxidants in peppers vary by genotype and maturity and are influenced by growing conditions (Davies *et al.,* 1970; Mejia *et al.,* 1988; Lee *et al.,* 1995;

Daood *et al.,* 1996; Simonne *et al.,* 1997; Osuna-García *et al.,* 1998; Márkus *et al.,* 1999; Deepa *et al.,* 2006). Howard *et al.* (2000) measured the carotenoids, flavonoids, phenolic acids, ascorbic acid and AOX in bell pepper and tabasco- and habanero-type peppers. The habanero peppers contained much lower amounts of carotenoids and flavonoids than the bell pepper and tabasco types, but total soluble phenolic content, vitamin C and AOX were similar in all three pepper types. Capsaicinoid content is also affected by the types and varieties of peppers (Cruz-Perez *et al.,* 2007; Hayman and Kam, 2008; Mortensen and Mortensen, 2009), as well as by the stages of development (Cruz-Perez *et al.,* 2007); the capsaicinoids tended to decrease at later stages of ripeness (after 58–96 days) in most varieties of manzano hot peppers used in the latter study. Konisho *et al.* (2005) investigated the specific variation of capsaicinoid concentration in cultivated *Capsicum* species (145 accessions) from around the world and found negative correlations between capsaicinoid concentration and fruit size. Furthermore, the researchers found high levels of variation in capsaicinoid content and composition between the domesticated *Capsicum* species. A recent study by Antonious and Jarret (2006), examining 90 *Capsicum* accessions selected from the USDA *Capsicum* germplasm collection, revealed that capsaicin concentrations in peppers were typically higher than those of dihydrocapsaicin, and that the concentrations of total capsaicinoids were non-detectable in *C. chinense* and highest in the *C. frutescens* group.

Production conditions

The growing environment was shown by Lee *et al.* (2005) to affect the phenolic and carotenoid compositions of different pepper types produced in a greenhouse at College Station and in field plots at Uvalde and Weslaco, Texas, USA. The three growing environments differed in terms of soil or growth medium, temperature, humidity, rainfall and evapotranspiration. The best sources of β-carotene were mature, greenhouse-grown fruit of cvs. Fidel and C127. Mature, greenhouse-grown fruit of cvs. Tropic Bell and PI357509 had high

continuously until the full ripe stage. Cape gooseberries contain on average 0.02–0.03 mg of β-carotene and 0.30–0.54 mg of vitamin C/g fruit FW (Mazumdar and Basu, 1979; Morton, 1987; Sarkar and Chattopadhyay, 1993; Fischer *et al.*, 2000; Ramadan and Morsel, 2005).

To date, no extensive research has been reported on cape gooseberry fruit and limited information can be found in the literature regarding fruit quality and composition, particularly when handled under different environmental conditions; however, compared with other temperatures, cape gooseberries maintain the longest postharvest life and best visual quality when stored at 10°C. Storage at 0 or 5°C reduced the postharvest life of the fruit due to the development of skin injury and discoloration on transfer to ambient temperature, which was probably the result of chilling injury, while storage at 15 or 20°C resulted in accelerated shrivelling and development of decay (Nunes, 2008).

Little was found in the literature regarding tomatillo or husk tomato (*P. ixocarpa* Brot.) quality and composition, namely vitamin C content during fruit development. An early study reported that vitamin C content of tomatillo was relatively low (0.03–0.04 mg/g FW) at early stages of development and remained unchanged during further development (Cantwell *et al.*, 1992). At the ripe stage, tomatillo fruit may contain on average 0.12 mg of vitamin C/g FW (USDA, 2008).

16.4.3 Chemopreventive activity and bioavailability

Plant phenolics have been well documented as having multiple chemopreventive properties, but limited information is available specifically for aubergine, cape gooseberry or tomatillo. Until recent years, very little information on the health benefits of tomatillo had been published. Kennelly *et al.* (1997) revealed an induction of quinone reductase by withanolides isolated from tomatillo; withanolides are classes of steroids first isolated from *Withania somnifera* in the late 1960s (Su *et al.*, 2004). According to a review

by Su *et al.* (2004), several new withanolides have been isolated and characterized as inducers of quinone reductase enzymes, and previously withanolides have been documented as having many potential chemopreventive properties, including the antiproliferative and apoptotic properties of ixocarpalactone A in tomatillo (Choi *et al.*, 2006). With regard to the bioavailability of bioactive compounds, one study revealed that steam cooking improved *in vitro* bile acid binding significantly in many vegetables, including aubergines, in comparison with uncooked ones (Kahlon *et al.*, 2007).

Cancer studies

According to some folk medicines, some plants of the *Solanaceae* family have been implicated in anticancer effects (Kennelly *et al.*, 1997; Su *et al.*, 2004; Choi *et al.*, 2006), but limited information is available on reasons why those members of the plant family have these effects. Historically, atropine (derived from (–)-hyoscyamine and -hyoscine) from *Solanaceae* species has been used clinically (Kanto and Klotz, 1987; Phillipson, 2001). A recent study by Lee *et al.* (2004) revealed that glycoalkaloids and metabolites found in potato, aubergine and tomato inhibited growth of human colon and liver cancer cells. A study by Yeh and Yen (2005) suggested that fruit and vegetables, including aubergines, containing components that had the ability to induce a detoxifying enzyme called phenol-sulfotransferase could play a role in the prevention of chronic diseases such as cancer. Azevedo *et al.* (2007) revealed that a purified anthocyanin (delphinin) from aubergines might have potential as a natural colorant that could prevent mutation, but the researchers suggested additional studies to confirm their findings.

Cardiovascular diseases

Phenolic compounds have been well documented as effective antioxidants and, although aubergines have been used in many areas of the world to control cholesterol, very little information is available on this specific subject. A study by Botelho *et al.* (2004)

revealed that aubergine extract increased oxidative stress in LDL receptor knockout mice (LDLR$^{-/-}$), which meant that aubergines could be a risk factor for atherosclerosis. However, another study revealed an antiangiogenic activity of nasunin (an anthocyanin in aubergine peel) in *in vitro* models based on endothelial cells from human umbilical vein (Matsubara *et al.*, 2005). From existing information, it is not possible to reach any conclusion on the relationship between CVD and this plant family.

Other beneficial and detrimental effects

Because of their high content of carotene, lutein and zeaxanthin, the consumption of cape gooseberry may help reduce the incidence of cataracts and macular degeneration (Lyle, 2006). Another study revealed that *P. peruvtana* fruit juice exhibited a mild anti-inflammatory activity and this activity might relate to its ability to inhibit fibroblast growth (Pardo *et al.*, 2008). However, a recent evaluation of some native Peruvian fruit for antihyperglycaemia and antihypertension potential, using *in vitro* models, did not show any significant potential for cape gooseberry for these two conditions; the researchers correlated high phenolic contents in some of the fruit to antioxidant activity (Da Silva Pinto *et al.*, 2009). Kwon *et al.* (2008) reported that aubergine phenolics inhibited key enzymes (α-glucosidase and angiotesin I converting enzyme) for type II diabetes and hypertension in an *in vitro* study.

In addition to beneficial effects, aubergines have been reported recently to have some negative effects, including anaphylaxis in a patient with latex allergy (Lee *et al.*, 2004) and other allergy reactions (Pramod and Venkatesh, 2004; Harish Babu *et al.*, 2008).

16.5 Future Research Needs

Further research is needed on the potential effects of preharvest factors, such as irrigation, chemicals and light, on the bioactive components of Solanaceous fruit, as well as on the effectiveness of specific postharvest treatments, such as temperature, humidity or modified atmospheres, as a way to boost the nutritional value and bioactive compounds in fruit from Solanaceous plants.

Further research is also needed to confirm the bioavailability of specific bioactive compounds, namely lycopene, anthocyanins and other minor bioactive compounds, when consumed as part of a 'fruit or vegetable cocktail', as in real life, rather than individually. However, the effects of lycopene and other bioactive compounds used individually also need further investigation, in order to understand how they act in protecting the human body against some degenerative diseases.

Study in more detail is also needed for some specialty crops, such as cape gooseberry and tomatillo, in terms of their composition, postharvest requirements and potential health benefits when part of a balanced diet.

16.6 Conclusion

More collaborative research should be conducted that includes all those involved in the study of the *Solanaceae* family – namely, food scientists, postharvest physiologists and nutritionists – in an effort to understand better the characteristics, pre- and postharvest requirements and potential health benefits of Solenanaceous fruit.

References

Abdel-Salam, O.M.E., Szolcsanyáyi, J. and Mózsik, G. (1997) Capsaicin and the stomach: a review of experimental and clinical data. *Journal of Physiology* 91, 151–171.

Abushita, A.A., Daood, H.G. and Biacs, P.A. (2000) Changes in carotenoids and antioxidant vitamins in tomato as a function of varietal and technological factors. *Journal of Agricultural and Food Chemistry* 48, 2075–2081.

Amin, R., Kucuk, O., Khuri, F.R. and Shin, D.M. (2009) Perspectives for cancer prevention with natural compounds. *Journal of Clinical Oncology* 27, 2712–2725.

Antonious, G.F. and Jarret, R.L. (2006) Screening *Capsicum* accessions for capsaicinoids content. *Journal of Environmental Science and Health: Part B. Pesticides Food Contaminants and Agricultural Wastes* 41, 717–729.

Antonious, G.F., Lobel, L., Kochhar, T., Berke, T. and Jarret, R.L. (2009) Antioxidants in *Capscicum chinenses*: variation among countries of origins. *Journal of Environmental Science and Health* 44, 621–626.

Archer, V.E. and Jones, D.W. (2002) Capsaicin pepper, cancer and ethnicity. *Medical Hypothesis* 59, 450–457.

Arvanitoyannis, I.S., Khah, E.M., Christakou, E.C. and Bletsos, F.A. (2005) Effect of grafting and modified atmosphere packaging on eggplant quality parameters during storage. *International Journal of Food Science and Technology* 40, 311–322.

Aubert, S. (1971) L'aubergine (*Solanum melongena* L.). *Annales de Technologie Agricole* 20, 241–264.

Azevedo, L., Alves de Lima, P.L., Gomes, J.C., Stringheta, P.C., Ribeiro, D.A. and Salvadori, D.M.F. (2007) Differential response related to genotoxicity between eggplant (*Solanum melongena*) skin aqueous extract and its main purified anthocyanin (delphinin) *in vitro*. *Food and Chemical Toxicology* 45, 852–858.

Barceloux, D.G. (2009) Pepper and capsaicin (*Capsicum* and *Piper* Species). In: Barceloux, D.G. (ed.) *Medical Toxicology of Natural Substances: Foods, Fungi, Medicinal Herbs, Toxic Plants, and Venomous Animals*. John Wiley and Sons, Hoboken, New Jersey, pp. 71–76.

Basu, A. and Imrhan, V. (2007) Tomatoes versus lycopene in oxidative stress and carcinogenesis: conclusions from clinical trials. *European Journal of Clinical Nutrition* 61, 295–303.

Bazzano, L.A., Serdula, M.K. and Liu, S. (2003) Dietary intake of fruits and vegetables and risk of cardiovascular disease. *Current Atherosclerosis Reports* 5, 492–499.

Betancourt, L.A., Stevens, M.A. and Kader, A.A. (1977) Accumulation and loss of sugars and reduced ascorbic acid in attached and detached tomato fruit. *Journal of the American Society for Horticultural Science* 102, 721–723.

Bhutani, M., Pathak, A.K., Nair, A.S., Kunnumakkara, A.G., Guha, S., Sethi, G. and Aggarwal, B.B. (2007) Capsaicin is a novel blocker of constitutive and interleukin-6-inducible STAT3 activation. *Clinical Cancer Research* 13, 3024–3032.

Bieche, B. and Covis M. (1992) Worldwide dynamics of tomato product consumption. *Acta Horticulturae* 301, 23–31.

Bisogni, C.A., Armbruster, G. and Brecht, P.E. (1976) Quality comparisons of room ripened and field ripened tomato fruits. *Journal of Food Science* 41, 333–338.

Bortolotti, M., Coccia, G., Grossi, G. and Miglioli, M. (2002) The treatment of functional dyspepsia with red pepper. *New England Journal of Medicine* 346, 947–948.

Botelho, F.V., Eneas, L.R., Cesar, G.C., Bizzotto, C.S., Tavares, E., Olivera, F.A., Gloria, M.B.A., Silvestre, M.P.C., Arantes, R.M.E. and Alvarez-Leite, J.I. (2004) Effects of eggplant (*Solanum melongena*) on the atherogenesis and oxidative stress in LDL receptor knockout mice (LDLR$^{-/-}$). *Food and Chemical Toxicology* 42, 1259–1267.

Bramley, P.M. (2000) Is lycopene beneficial to human health? *Phytochemistry* 54, 233–236.

Brandt, S., Pek, Z., Barna, E., Lugasi, A. and Helyes, L. (2006) Lycopene content and colour of ripening tomatoes as affected by environmental conditions. *Journal of the Science of Food and Agriculture* 86, 568–572.

Breithaupt, D.E. and Bamedi, A. (2001) Carotenoid esters in vegetables and fruits: a screening with emphasis on β-cryptoxanthin esters. *Journal of Agricultural and Food Chemistry* 49, 2064–2070.

Buck, S.H. and Burks, T.F. (1986) The neuropharmarcology of capsaicin: review of some recent observations. *Pharmacological Reviews* 38, 179–226.

Burri, B.J., Chapman, M.H., Neidlinger, T.R., Seo, J.S. and Ishida, B.K. (2009) Tangerine tomatoes increase total and tetra-*cis*-lycopene isomer concentrations more than red tomatoes in healthy adult humans. *International Journal of Food Sciences and Nutrition* 60(S1), 1–16.

Buta, J.G. and Spaulding, D.W. (1997) Endogenous levels of phenolics in tomato fruit during growth and maturation. *Journal of Plant Growth Regulation* 16, 43–46.

Butelli, E., Titta, L., Giorgio, M., Mock, H.P., Matros, A., Peterek, S., Schijlen, E.G.W.M., Hall, R.D., Bovy, A.G., Luo, J. and Martin, C. (2008) Enrichment of tomato fruit with health-promoting anthocynins by expression of selected transcription factors. *Nature Biotechnology* 26, 1301–1308.

Cámara, M. and Fernánadez-Ruiz, V. (2009) European nutrition and health claims on foods: the case of lycopene. *Acta Horticulturae* 823, 243–248.

Cantwell, M., Flores-Minutti, J. and Trejo-Gonzalez, A. (1992) Developmental changes and postharvest physiology of tomatillo fruits (*Physalis ixocarpa* Brot.). *Scientia Horticulturae* 50, 59–70.

Chanasut, U. and Rattanapanone, N. (2008) Screening methods to measure antioxidant activities of phenolic compound extracts from some varieties of Thai eggplants. *Acta Horticulturae* 765, 291–296.

Chen, L., Stacewicz-Sapuntzakis, M., Duncan, C., Sharifi, R., Ghosh, L., van Breemen, R., Ashton, D. and Bowen, P.E. (2001) Oxidative DNA damage in prostate cancer patients consuming tomato sauce-based entrees as a whole-food intervention. *Journal of the National Cancer Institute* 93, 1872–1879.

Choi, J.K., Murillo, G., Su, B.N., Perzzuto, J.M., Kinghorn, A.D. and Mehta, R.G. (2006) Ixocarpalactone A isolated from the Mexican tomatillo shows potent antiproliferative and apoptotic activity in colon cancer cells. *FEBS Journal* 273, 5714–5723.

Cichewicz, R.H. and Thorpe, P.A. (1996) The antimicrobial properties of chile peppers (*Capsicum* species) and their uses in Mayan medicine. *Journal of Ethnopharmacology* 52, 61–70.

Clarke, R. and Armitage, J. (2002) Symposium: Vitamin therapy and ischemic heart disease. *Cardiovascular Drugs and Therapy* 16, 411–415.

Clinton, S.K., Emenhiser, C. and Schwartz, S.J. (1996) *Cis-trans* lycopene isomers, carotenoids and retinol in human prostate. *Cancer Epidemiology Biomarker and Prevention* 5, 823–833.

Cohn, W., Thürmann, P., Tenter, U., Aebischer, C., Schierle, J. and Schalch, W. (2004) Comparative multiple dose plasma kinetics of lycopene administered in tomato juice, tomato soup or lycopene tablets. *European Journal of Nutrition* 43, 304–312.

Concellón, A., Añón, María, C. and Chaves, A.R. (2007) Effect of low temperature storage on physical and physiological characteristics of eggplant fruit (*Solanum melongena* L.). *LWT – Food Science and Technology* 40, 389–396.

Conforti, F., Statti, G.A. and Menichini, F. (2007) Chemical and biological variability of hot pepper fruits (*Capsicum annuum* var. *acuminatum* L.) in relation to maturity stage. *Food Chemistry* 102, 1096–1104.

Cronin, J.R. (2002) The biochemistry of alternative medicine: the chili pepper's pungent principle capsaicin delivers diverse health benefits. *Alternative and Complementary Therapies* 8, 110–113.

Crosby, K., Jifon, J. and Leskovar, D. (2008) Agronomy and the nutritional quality of vegetables. In: Tomás-Barberán, F.A. and Gil, M.I. (eds) *Improving the Health-promoting Properties of Fruit and Vegetable Products*. CRC Press, Inc, Woodhead Publishing Limited, Cambridge, UK, pp. 392–411.

Cruz-Perez, A.B., Gonzalez-Hernandez, V.A., Soto-Hernandez, R.M., Gutierrze-Espinosa, M.A., Gardea-Bejar, A.A. and Perez-Grajalez, M. (2007) Capsaicinoids, vitamin C and heterosis during fruit development of manzano hot pepper. *Agrociencia* 41, 627–635.

Da Silva Pinto, M., Ranilla, L.G., Apostolidis, E., Lajolo, F.M., Genovese, M.I. and Shetty, K. (2009) Evaluation of antihyperglycemia and antihypertension potential of native Peruvian fruits using *in vitro* models. *Journal of Medicinal Foods* 12, 278–291.

Daood, H.G., Vinkler, M., Markus, F., Hebshi, E.A. and Biacs, P.A. (1996) Antioxidant vitamin content of spice red pepper (paprika) as affected by technological and varietal factors. *Food Chemistry* 55, 365–372.

Dauchet, L., Amouyel, P., Herchberg, S. and Dallongeville, J. (2006) Fruit and vegetable consumption and risk of coronary heart disease: a meta-analysis of cohort studies. *Journal of Nutrition* 136, 2588–2593.

Davies, B.H., Matthews, S. and Kirk, J.T.O. (1970) The nature and biosynthesis of the carotenoids of different colour varieties of *Capsicum annuum*. *Phytochemistry* 9, 797–805.

Deepa, N., Kaur, C., Singh, B. and Kapoor, H.C. (2006) Antioxidant activity in some red sweet pepper cultivars. *Journal of Food Composition and Analysis* 19, 572–578.

Demmig-Adams, B., Gilmore, A.M. and Adams, W.W. III (1996) *In vivo* functions of carotenoids in higher plants. *FASEB Journal* 10, 403–412.

Devaraj, S., Mathur, S., Basu, A., Aung, H.H., Vasu, V.T., Meyers, S. and Jialal, I. (2008) A dose–response study on the effects of purified lycopene supplementation on biomarkers of oxidative stress. *Journal of American College of Nutrition* 27, 267–273.

Dewanto, V., Wu, Z., Adom, K.K. and Liu, R.H. (2002) Thermal processing enhances the nutritional value of tomatoes by increasing total antioxidant activity. *Journal of Agricultural and Food Chemistry* 50, 3010–3014.

Diwadkar-Navsariwala, V., Novotny, J.A., Gustin, D.M., Sosman, J.A., Rodvold, K.A., Crowell, J.A., Stacewicz-Sapuntzakis, M. and Bowen, P.E. (2003) A physicological pharmokinetic model describing the disposition of lycopene in healthy men. *Journal of Lipid Research* 44, 1927–1939.

Dorais, M. (2007) Effect of cultural management on tomato fruit health qualities. *Acta Horticulturae* 744, 279–294.

Dorais, M., Papadopoulos, A.P. and Gosselin, A. (2001) Greenhouse tomato fruit quality. *Horticultural Reviews* 26, 239–319.

Dorais, M., Ehret, D.L. and Papadopoulos, A.P. (2008) Tomato (*Solanum lycopersicum*) health components: from seed to the consumer. *Phytochemistry Review* 7, 231–250.

Dorantes, L., Colmenero, R., Hernandez, H., Mota, L., Jaramillo, M.E., Fernandez, E. and Solano, C. (2000) Inhibition of growth of some foodborne pathogenic bacteria by *Capsicum annum* extracts. *International Journal of Food Microbiology* 57, 125–128.

Dorji, K., Behboudian, M.H. and Zegbe-Dominguez, J.A. (2005) Water relations, growth, yield, and fruit quality of hot pepper under deficit irrigation and partial root zone drying. *Scientia Horticulturae* 104, 137–149.

Dumas, Y., Dadomo, M., Di Lucca, G. and Grolier, P. (2003) Effects of environmental factors and agricultural techniques on antioxidant content of tomatoes. *Journal of the Science of Food and Agriculture* 83, 369–382.

Esteban, R.M., Molla, E., Villaroya, M.B. and Lopez-Andreu, F.J. (1989) Changes in the chemical composition of eggplant fruits during storage. *Scientia Horticulturae* 41, 19–25.

Esteban, R.M., Mollá, E.M., Robredo, L.M. and López-Andréu, J. (1992) Changes in the chemical composition of eggplant fruits during development and ripening. *Journal of Agricultural and Food Chemistry* 40, 998–1000.

Etminan, M., Takkouche, B. and Caamano-Isorna, F. (2004) The role of tomato products and lycopene in the prevention of prostate cancer: a meta analysis of observational studies. *Cancer Epidemiology Biomarkers and Prevention* 13, 340–345.

Eurostat (2006) *Agricultural Statistics – Quarterly Bulletin No. 04/2005.* Eurostat, ISBN 1607-2308.

Failla, M.L., Chitchumroonchokchai, C. and Ishida, B.K. (2007) *In vitro* micellarization and intestinal cell uptake of *cis* isomers of lycopene exceed those of all-*trans* lycopene. *Journal of Nutrition* 138, 482–486.

FAOSTAT (2009) Food and Agriculture Statistics Division (http://faostat.fao.org/site/567/DesktopDefault.aspx?PageID=567#ancor, accessed 14 May 2009).

Farchi, S., Forastiere, F., Agabiti, N., Corbo, G., Pistelli, R., Fortes, C., Dell'Orco, V. and Perucci, C.A. (2003) Dietary factors associated with wheezing and allergic rhinitis in children. *European Respiratory Journal* 22, 772–780.

Fischer, G., Ebert, G. and Lüdders, P. (2000) Provitamin A carotenoids, organic acids and ascorbic acid content of cape gooseberry (*Physalis peruviana* L.) ecotypes grown at two tropical altitudes. *Acta Horticulturae* 531, 263–267.

Flick, G.J., Ory, R.L. and Angelo, A.J. St (1977) Comparison of nutrient composition and of enzyme activity in purple, green, and white eggplants. *Journal of Agricultural and Food Chemistry* 25, 117–120.

Floyd, W.W. and Fraps, G.S. (1939) Vitamin C content of some Texas fruits and vegetables. *Journal of Food Science* 4, 87–91.

Franco-Cereceda, A., Henke, H., Lundberg, J.M., Petermann, J.B., Hökfelt, T. and Fisher, J.A. (1987) Calcitonin gene-related peptide (CGRP) in capsaicin-sensitive substance P-immunoreactive sensory neurons in animals and man: distribution and release by capsaicin. *Peptides* 8, 399–410.

Fuhrman, B., Elis, A. and Aviram, M. (1997) Hypocholesterolemic effect of lycopene and β-carotene is related to suppression of cholesterol synthesis and augmentation of LDL recptor activity in macrophages. *Biochemical and Biophysical Research Communications* 233, 658–662.

Garcés-Claver, A., Arnedo-Andrés, M.S., Abdia, J., Gil-Ortega, R. and Alverez-Fernandez, A. (2006) Determination of capsaicin and dihydrocapsaicin in *Capsicum* fruit by liquid chromatography – electrospray/time-of-flight mass spectrometry. *Journal of Agricultural and Food Chemistry* 54, 9303–9311.

Garcia, E. and Barrett, D. (2005) Assessing lycopene content in California processing tomatoes. *Journal of Food Processing and Preservation* 30, 56–70.

Gartner, C., Stahl, W. and Sies, H. (1997) Lycopene is more bioavailable from tomato paste than from fresh tomatoes. *Journal of Clinical Nutrition* 66, 116–122.

Geleta, L.F. and Labuschagne, M.T. (2006) Combining ability and heritability for vitamin C and total soluble solids in pepper (*Capsicum annuum* L.). *Journal of the Science of Food and Agriculture* 86, 1317–1320.

Gilbert, D., Funk, K., Dekowski, B., Lechler, R., Keller, S., Möhrlen, F., Frings, S. and Hagen, V. (2007) Caged capsaicins: new tools for the examination of TRPV1 channels in somatosensory neurons. *ChemBioChem* 8, 89–97.

Giovanelli, G., Lavelli, V., Peri, C. and Nobili, S. (1999) Variation in antioxidant components of tomato during vine and post-harvest ripening. *Journal of the Science of Food and Agriculture* 79, 1583–1588.

Giovanelli, G., Lavelli, V., Peri, C. and Nobili, S. (2001) The antioxidant activity of tomato. II. Effects of vine and post-harvest ripening. *Acta Horticulturae* 542, 211–216.

Giovannucci, E. (1999) Tomatoes, tomato-based products, lycopene, and cancer: review of the epidemiologic literature. *Journal of the National Cancer Institute* 91, 317–331.

Giovannucci, E. (2002) A review of epidemiologic studies of tomatoes, lycopene, and prostate cancer. *Experimental Biology and Medicine* 227, 852–859.

Giovannucci, E., Asherio, A., Rimm, E.B., Stampfer, M.J., Colditz, G.A. and Willet, W.C. (1995) Intake of carotenoids and retinol in relation to risk of prostate cancer. *Journal of National Cancer Institute* 87, 1767–1776.

Giuntini, D., Graziani, G., Lercari, B., Fogliano, V., Soldatini, G.F. and Ranieri, A. (2005) Changes in carotenoid and ascorbic acid contents in fruit of different tomato genotypes related to the depletion of UV-B radiation. *Journal of Agricultural and Food Chemistry* 53, 3174–3181.

González, M., Centurión, A., Sauri, E. and Latournerie, L. (2005) Influence of refrigerated storage on the quality and shelf life of 'Habanero' chili peppers (*Capsicum chinense* Jacq.). *Acta Horticulturae* 682, 1297–1302.

González-Aguilar, G.A. (2004) Pepper. In: Gross, K.C., Wang, C.Y. and Saltveit, M. (eds) *The Commercial Storage of Fruits, Vegetables, and Florist and Nursery Stock. Agriculture Handbook No. 66*. US Department of Agriculture, Agricultural Research Service, Washington, DC. (Available at: http://usna.usda.gov/hb66/108pepper.pdf. Accessed 13 April 2011)

Goodwin, T.W. and Jamikorn, M. (1952) Biosynthesis of carotenes in ripening tomatoes. *Nature* 170, 104–105.

Gross, J. (1987) *Pigments in Fruits*. Alden Press, Oxford, UK.

Gustin, D.M., Rodvold, K.A., Sosman, J.A., Diwadkar-Navsriwala, V., Stacewicz-Spuntzakis, M., Viana, M., Crowel, J.A., Murray, J., Tiller, P. and Bowen, P. (2004) Single-dose pharmacokinetic study of lycopene delivered in a well-defined food-based lycopene delivery system (tomato paste–oil mixture) in healthy adult male subjects. *Cancer Epidemiology, Biomarkers and Prevention* 13, 850–860.

Ha, S.-W., Kim, J.-B., Park, J.-S., Lee, S.-W. and Cho, K.-J. (2007) A comparison of carotenoid accumulation in *Capsicum* varieties that show different ripening colors: deletion of the capsanthin-capsorubin synthase gene is not a prerequisite for the formation of a yellow pepper. *Journal of Experimental Botany* 58, 3135–3144.

Hadley, C.W., Clinton, S.K. and Schwartz, S.J. (2002) The consumption of processed tomato products enhances plasma lycopene concentrations in association with a reduced liproprotein sensitivity to oxidative damage. *Journal of Nutrition* 133, 727–732.

Hamauzu, Y., Chachin, K. and Ueda, Y. (1998) Effect of postharvest temperature on the conversion of 14C-mevalonic acid to carotenes in tomato fruit. *Journal of Japanese Society of Horticultural Science* 67, 549–555.

Hanson, P.M., Yang, R.Y., Tsou, S.C.S., Ledesma, D., Engle, L. and Lee, T.C. (2006) Diversity in eggplant (*Solanum melongena*) for superoxide scavenging activity, total phenolics, and ascorbic acid. *Journal of Food Composition and Analysis* 19, 594–600.

Harish Babu, B.N., Mahesh, P.A. and Venkatesh, Y.P. (2008) A cross-sectional study on the prevalence of food allergy to eggplant (*Solanum melongena* L.) reveals female predominance. *Clinical Experimental Allergy* 38, 1795–1802.

Hart, D.J. and Scott, K.J. (1995) Development and evaluation of an HPLC method for the analysis of carotenoids in foods, and the measurement of the carotenoid content of vegetables and fruits commonly consumed in the UK. *Food Chemistry* 54, 101–111.

Harvey, M., Quilley, S. and Beynon, H. (2003) *Exploring the Tomato: Transformations of Nature, Society, and Economy*. Edward Elgar Publishing, Cheltenham, UK and Northampton, Massachusetts.

Hayman, M. and Kam, P.C.A. (2008) Capsaicin: a review of its pharmarcology and clinical applications. *Current Anaesthesia and Critical Care* 19, 338–343.

Hernándo Bermejo, J.E. and Leon, J. (1994) Neglected crops: 1492 from a different perspective. *Plant Production and Protection Series No. 26*. FAO, Rome, pp. 117–122.

Herzog, A., Siler, U., Spitzer, V., Seifert, N., Denelavas, A., Hunziker, P.B., Hunziker, W., Goralczyk, R. and Wertz, K. (2004) Lycopene reduced gene expression of steroid targets and inflammatory markers in normal rat prostate. *FASEB Journal* 19, 272–274.

Hewett, E.W. (1993) New horticultural crops in New Zealand. In: Janick, J. and Simon, J.E. (eds) *New Crops*. John Wiley and Sons, Inc., New York, pp. 57–64.

Howard, L.R., Smith, R.T., Wagner, A.B., Villalon, B. and Burns, E.E. (1994) Provitamin A and ascorbic acid content of fresh pepper cultivars (*Capsicum annuum*) and processed jalapeños. *Journal of Food Science* 59, 362–365.

Howard, L.R., Talcott, S.T., Brenes, C.H. and Villalon, B. (2000) Changes in phytochemical and antioxidant activity of selected pepper cultivars (*Capsicum* species) as influenced by maturity. *Journal of Agricultural and Food Chemistry* 48, 1713–1720.

Hung, H.-C., Joshipura, K.J., Jiang, R., Hu, F.B., Hunter, D., Smith-Warner, S.A., Colditz, G.A., Rosner, B., Spiegelman, D. and Willett, W.C. (2004) Fruit and vegetable intake and risk of major chronic disease. *Journal of the National Cancer Institute* 96, 1577–1584.

Ichiyanagi, T., Kashiwada, Y., Shida, Y., Ikeshiro, Y., Kaneyuki, T. and Konishi, T. (2005) Nasunin from eggplant consists of *cis-trans* isomers of delphinidin 3-[4-(*p*-coumaroyl)-L-rhamnosyl (1 6) gluco-pyranoside]-5-glucopyranoside. *Journal of Agricultural and Food Chemistry* 53, 9472–9477.

Jarret, R.L. and Berke, T. (2008) Variation for fruit morphological characteristics in a *Capsicum chinense* Jacq. germplasm collection. *HortScience* 43, 1694–1697.

Jiménez, A., Romojaro, F., Gómez, J.M., Llanos, M.R. and Sevilla, F. (2003) Antioxidant systems and their relationship with the response of pepper fruits to storage at 20°C. *Journal of Agricultural and Food Chemistry* 51, 6293–6299.

Jones, C.M., Mes, P. and Myers, J.R. (2003) Characterization and inheritance of the anthocyanin fruit (Aft) tomato. *Journal of Heredity* 94, 449–456.

Kader, A.A., Stevens, M.A., Albright-Holton, M., Morris, L.L. and Algazi, M. (1977) Effect of fruit ripeness when picked on flavor and composition of fresh market tomatoes. *Journal of the American Society for Horticultural Science* 102, 724–731.

Kader, A.A., Morris, L.L., Stevens, M.A. and Albright-Holton, M. (1978) Composition and flavor quality of fresh market tomatoes as influenced by some postharvest handling procedures. *Journal of the American Society for Horticultural Science* 103, 6–13.

Kahlon, T.S., Chiu, M.C.M. and Chapman, M.H. (2007) Steam cooking significantly improves *in vitro* bile acid binding of beets, eggplant, asparagus, carrots, green beans, and cauliflower. *Nutrition Research* 27, 750–755.

Kanto, J. and Klotz, U. (1987) Pharmacokinetic implications for the clinical use of atropine, scopolamine and glycopyrrolate. *Acta Anesthesiologica Scandinavica* 32, 69–78.

Karas, M., Amir, H., Fishman, D., Danilenko, M., Segal, S., Nahum, A., Koifmann, A., Giat, Y., Levey, J. and Sharoni, Y. (2000) Lycopene interferes with cell cycle progression and insulin-like growth factor I signalling in mammary cancer cells. *Nutrition and Cancer* 36, 101–111.

Kavanaugh, C.J., Trumbo, P.R. and Ellwood, K.C. (2007) The US Food and Drug Administration's evidence-based review for qualified health claims: tomatoes, lycopene, and cancer. *Journal of National Cancer Institute* 99, 1074–1085.

Kaynas, K., Özelkök, S., Sürmeli, N. and Abak, K. (1995) Controlled and modified atmosphere storage of eggplant (*Solanum melongena* L.) fruits. *Acta Horticulturae* 412, 143–151.

Kennelly, E.J., Gerhäuser, C., Song, L.I., Graham, J.G., Beacher, C.W.W., Pezzuto, J.M. and Kinhorn, A.D. (1997) Induction of quinine reductase by withanolides isolated from *Physalis philadelphica* (tomatillos). *Journal of Agricultural and Food Chemistry* 45, 3771–3777.

Khan, N., Afaq, F. and Mukhtar, H. (2006) Apoptosis by dietary factors: the suicide solution for delaying cancer growth. *Carcinogenesis* 28, 233–239.

Kim, C.-S., Park, W.-H., Park, J.Y., Kang, J.H., Kim, M.O., Kawada, T., Yoo, H., Han, I.S. and Yu, R. (2004) Capsaicin, a spicy component of hot pepper, induces apoptosis by activation of the peroxisome proliferator-activated receptor gamma in HT-29 human colon cancer cells. *Journal Medicinal Food* 7, 267–273.

Kobryń, J. and Hallmann, E. (2005) The effect of nitrogen fertilization on the quality of three tomato types cultivated on rockwool. *Acta Horticulturae* 691, 341–348.

Kohlmeier, L., Kark, J.D., Gomez-Garcia, E., Martin, B.C., Steck, S.E., Kardinaal, A.F.M., Ringstad, J., Thamm, M., Maserv, V., Rimersma, R., Martin-Moreno, J.M., Huttunen, J.K. and Kok, F.J. (1997) Lycopene and myocardial infarction risk in the EURAMIC study. *American Journal of Epidemiology* 146, 616–626.

Kolotilin, I., Koltai, H., Tadmor, Y., Bar-Or, C., Reuveni, M., Meir, A., Nahon, S., Shlomo, H., Chen, L. and Levin, I. (2007) Transcriptional profiling of high pigment-2dg tomato mutant links early fruit plastic biogenesis with its overproduction of phytonutrients. *Plant Physiology* 145, 389–401.

Konisho, K., Minami, M., Matsushima, K. and Nemoto, K. (2005) Inter- and intra-specific variation of capsicinoid concentration in chili pepper (*Capsicum* spp). *Horticultural Research (Japan)* 4, 153–158.

Kozukue, N., Kozukue, E. and Kishiguchi, M. (1979) Changes in the contents of phenolic substances, phenylalanine ammonia-lyase (PAL) and tyrosine ammonia-lyase (TAL) accompanying chilling-injury of eggplant fruit. *Scientia Horticulturae* 11, 51–59.

Kozukue, N., Han, J.S., Kozukue, E., Lee, S.J., Kim, J.A., Lee, K.R., Leven, C.E. and Friedman, M. (2005) Analysis of eight capsaicinoids in peppers and pepper-containing foods by high performance liquid chromatography and liquid chromatography-mass spectrometry. *Journal of Agricultural and Food Chemistry* 53, 9172–9181.

Kris-Etherton, P.M., Hecker, K.D., Bonanome, A., Coval, S.M., Binkoski, A.E., Hilpert, K.F., Griel, A.E. and Etherton, T.D. (2002) Bioactive compounds in foods: their role in prevention of cardiovascular disease and cancer. *American Journal of Medicine* 113, 71S–88S.

Kwon, Y.I., Apostolidis, E. and Shetty, K. (2008) *In vitro* studies of eggplant (*Solanum melongena*) phenolics as inhibitors of key enzymes relevant for type 2 diabetes and hypertension. *Bioresource Technology* 99, 2981–2988.

Laohavechvanich, P., Kangsadalampai, K., Tirawanchai, N. and Ketterman, A.J. (2006) Effect of different Thai traditional processing of various hot chilli peppers on urethane-induced somatic mutation and recombination in *Drosophila melanogaster*: Assessment of the role of glutathione transferase activity. *Food and Chemical Toxicology* 44, 1348–1354.

Lee, J., Cho, Y.S., Park, S.Y., Lee, Ch.K., Yoo, B., Moon, H.B. and Park, H.S. (2004) Eggplant anaphylaxis in a patient with latex allergy. Letter to the Editor. *Journal of Allergy and Clinical Immunology* 113, 995–996.

Lee, J.J. Crosby, K.M., Pike, L.M., Yoo, K.S. and Leskovar, D.I. (2005) Impact of genetic and environmental variation on development of flavonoids and carotenoids in pepper (*Capsicum* spp.). *Scientia Horticulturae* 106, 341–352.

Lee, J.K., Park, B.-J., Yoo, K.-Y. and Ahn, Y.O. (1995) Dietary factors and stomach cancer: a case control study in Korea. *International Journal of Epidemiology* 24, 33–41.

Lee, K.R., Kozukue, N., Han, J.S., Park, J.H., Chang, E.Y., Baek, E.J., Chang, J.S. and Friedman, M. (2004) Glycoalkaloids and metabolites inhibit the growth of human colon (HT 29) and liver (HepG2) cancer cells. *Journal of Agricultural and Food Chemistry* 52, 2832–2839.

Lee, Y., Howard, L.R. and Villalon, B. (1995) Flavonoids and antioxidant activity of fresh pepper (*Capsicum annuum*) cultivars. *Journal of Food Science* 60, 473–476.

Lefebvre, V., Goffinet, B., Chauvet, J.C., Caromel, B., Signoret, P., Brand, B. and Palloix, A. (2001) Evaluation of genetic distances between pepper inbred lines for cultivar protection purposes: comparison of AFLP, RAPD and phenotypic data. *Theoretical and Applied Genetics* 102, 741–750.

Leonardi, C., Ambrosino, P., Esposito, F. and Foglinano, V. (2000) Antioxidative activity and carotenoid and tomatine contents in different typologies of fresh consumption tomatos. *Journal of Agricultural and Food Chemistry* 48, 4723–4727.

Lewinsohn, E., Sitrit, Y., Bar, E., Azulay, Y., Meir, A., Zamir, D. and Tadmor, Y. (2005) Carotenoid pigmentation affects the volatile composition of tomato and watermelon fruits, as revealed by comparative genetic analyses. *Journal of Agricultural and Food Chemistry* 53, 3142–3148.

Lim, K., Yoshioka, M., Kikuzato, S., Kiyonaga, A., Tanaka, H., Shindo, M. and Suzuki, M. (1997) Dietary red pepper ingestion increases carbohydrate oxidation at rest and during exercise in runners. *Sports and Exercise* 29, 355–361.

Lindshield, B.L., Canene-Adams, K. and Erdman, J.W. Jr (2006) Lycopenonids: are lycopene metabolites bioactive? *Archives of Biochemistry and Biophysics* 458, 136–140.

Long, M., Millar, D.J., Kimura, Y., Donovan, G., Rees, J., Fraser, P.D., Bramley, P.M. and Bolwell, G.P. (2006) Metabolite profiling of carotenoid and phenolic pathways in mutant and transgenic lines of tomato: identification of a high antioxidant fruit line. *Phytochemistry* 67, 1750–1757.

López-Carrillo, L., Avila, M.H. and Dubrow, R. (1994) Chili pepper consumption and gastric cancer in Mexico: a case-control study. *American Journal of Epidemiology* 139, 263–271.

López-Carrillo, L., López-Cervantes, M., Robles-Diaz, G., Ramirez-Espitta, A., Mohar-Betancourt, A., Meneses-Garcia, A., López-Vidal, Y. and Blair, A. (2003) Capsaicin consumption, *Helicobacter pylori* positivity and gastric cancer in Mexico. *International Journal of Cancer* 106, 277–282.

Luning, P.A., Vries, R.V., Yuksel, D., Ebbenhorst-Seller, T., Wichers, H.J. and Roozen, J.P. (1994) Combined instrumental and sensory evaluation of flavour of fresh bell peppers (*Capsicum annuum*) harvested at three maturation stages. *Journal of Agricultural and Food Chemistry* 42, 2855–2861.

Luthria, D.L., Mukhopadhyay, S. and Krizek, D.T. (2006) Content of total phenolics and phenolic acids in tomato (*Lycopersicon esculentum* Mill.) fruits as influenced by cultivar and solar UV radiation. *Journal of Food Composition and Analysis* 19, 771–777.

Lyle, S. (2006) *Physalis peruviana*. In: *Fruits and Nuts. A Comprehensive Guide to the Cultivation, Uses and Health Benefits of Over 300 Food-Producing Plants*. Timber Press, Portland, Oregon, pp. 323–325.

McCain, R. (1993) Goldenberry, passionfruit, and white sapote: potential fruits for cool subtropical areas. In: Janick, J. and Simon, J.E. (eds) *New Crops*. John Wiley and Sons, Inc, New York, pp. 479–486.

Macho, A., Lucena, C., Sancho, R., Daddario, N., Minassi, A., Minassi, A., Munoz, E. and Appendino, G. (2003) Non-pungent capsaicinoids from sweet pepper; synthesis and evaluation of the chemopreventive and anticancer potential. *European Journal of Nutrition* 42, 2–9.

McLeod, M.J., Guttman, S.I. and Eshbaugh, W.H. (1982) Early evolution of chili peppers (*Capsicum*). *Economic Botany* 36, 361–368.

Mahmmoud, Y.A. (2008) Capsaicin stimulates uncoupled ATP hydrolysis by the sarcoplasmic reticulum calcium pump. *Journal of Biological Chemistry* 283, 21418–21426.

Marín, A., Ferreres, F., Tomás-Barberán, F.A. and Gil, M.I. (2004) Characterization and quantitation of antioxidant constituents of sweet pepper (*Capsicum annuum* L.). *Journal of Agricultural and Food Chemistry* 52, 3861–3869.

Márkus, F., Daood, H.G., Kapitány, J. and Biacs, P.A. (1999) Change in the carotenoid and antioxidant content of spice red pepper (paprika) as a function of ripening and some technological factors. *Journal of Agricultural and Food Chemistry* 47, 100–107.

Materska, M. and Perucka, I. (2005) Antioxidant activity of the main phenolic compounds isolated from hot pepper fruit (*Capsicum annuum* L.). *Journal of Agricultural and Food Chemistry* 53, 1750–1756.

Materska, M., Piacente, S., Stochmal, A., Pizza, C., Oleszek, W. and Perucka, I. (2003) Isolation and structure elucidation of flavonoid and phenolic acid glycosides from pericarp of hot pepper fruits *Capsicum annuum* L. *Phytochemistry* 63, 893–898.

Matsubara, K., Kaneyuki, T., Miyake, T. and Mori, M. (2005) Antiangiogenic activity of nasunin, and antioxidant anthocynin, in eggplant peels. *Journal of Agricultural and Food Chemistry* 53, 6272–6275.

Mazumdar, B.C. and Basu, T.K. (1979) Analysis of cape gooseberry fruits. *Plant Science* 11, 101.

Mejia, L.A., Hudson, E., Gonzalez de Mejia, E. and Vasquez, F. (1988) Carotenoid content and vitamin A activity of some common cultivars of Mexican peppers (*Capsicum annuum*) as determined by HPLC. *Journal of Food Science* 53, 1448–1451.

Merck Index (1996) *An Encyclopedia of Chemicals, Drugs, and Biologicals*, 12th edn. Merck Research Laboratories Division of Merck & Co., Inc, Whitehouse Station, New Jersey.

Miller, E.C., Giovannucci, E., Erdman, J.W., Bahnson, R., Schwartz, S.J. and Clinton, S.K. (2002) Tomato products, lycopene, and prostate cancer risk. *Urologic Clinics of North America* 29, 83–93.

Mori, A., Lehmann, S., O'Kelly, J., Kumagai, T., Desmond, J.C., Pervan, M., McBride, W.H., Kizaki, M. and Koeffler, H.P. (2006) Capsaicin, a component of red peppers, inhibits the growth of androgen-independent p53 mutant prostate cancer cells. *Cancer Research* 66, 3222–3229.

Moriconi, D.N., Rush, D.N. and Flores, H. (1990) Tomatillo: a potential vegetable crop for Louisiana. In: Janick, J. and Simon, J.E. (eds) *Advances in New Crops*. Timber Press, Portland, Oregan, pp. 407–413 (http://www.hort.purdue.edu/newcrop/proceedings1990/v1-407.html, accessed 18 May 2009).

Mortensen, J.M. and Mortensen J.E. (2009) The power of capsaicin. *Journal of Continuing Education* 11, 8–12.

Morton, J.F. (1987) Cape gooseberry. In: Morton, J.F. (ed.) *Fruits of Warm Climates*. Creative Resource Systems, Inc, Winterville, North Carolina, pp. 430–434 (http://www.hort.purdue.edu/newcrop/morton/cape_gooseberry.html, accessed 19 December 2007).

Moscone, E.A., Scaldaferro, M.A., Garbiele, M., Cecchini, N.M., Garcia, Y.S., Jarret, R., Davina, J.R., Ducasse, D.A., Barboza, G.E. and Ehrendorfer, F. (2007) The evolution of chili pepper (*Capsicum – Solanaceae*): a cytogenetic perspective. *Acta Horticulturae* 745, 137–170.

Mozafar, A. (1993) Nitrogen fertilizers and the amount of vitamins in plants: a review. *Journal of Plant Nutrition* 16, 2479–2506.

Navarro, J.M., Flores, P., Garrido, C. and Martinez, V. (2006) Changes in the contents of antioxidant compounds in pepper fruits at different ripening stages, as affected by salinity. *Food Chemistry* 96, 66–73.

Noda, Y., Kneyuki, T., Igarashi, K., Mori, A. and Packer, L. (2000) Antioxidant activity of nasunin, an anthocyanin in eggplant peels. *Toxicology* 148, 119–123.

Nothmann, J. (1986) Eggplant. In: Monselise, S.P. (ed.) *Handbook of Fruit Set and Development*. CRC Press, Inc, Boca Raton, Florida, pp. 145–152.

Nothmann, J., Rylski, I. and Spigelman, M. (1976) Color and variations in color intensity of fruit of eggplant cultivars. *Scientia Horticulturae* 4, 191–197.

NPIC (2009) Capsaicin: technical fact sheet (http://npic.orst.edu/factsheets/Capsaicintech.pdf, accessed 30 September 2009).

Nunes, M.C.N. (2008) *Color Atlas of Postharvest Quality of Fruits and Vegetables*. Wiley-Blackwell Publishing, Ames, Iowa.

Oboh, G. and Rocha, J.B.T. (2008) Water extractable phytochemicals from *Capsicum pubescens* (tree pepper) inhibit lipid peroxidation induced by different pro-oxidant agents in brain: *in vitro. European Food Research and Technology* 226, 707–713.

Oboh, G., Puntel, R.L. and Rocha, J.B.T. (2007) Hot pepper (*Capsicum annum*, tepin and *Capsicum chinese*, habanero) prevents Fe2+ -induced lipid peroxidation in brain – *in vitro. Food Chemistry* 102, 178–185.

Ohnuki, K., Haramizu, S., Oki, K., Watanabe, T., Yazawa, S. and Fushiki, T. (2001a) Administration of capsiate, a non-pungent capsaicin analog, promotes energy metabolism and suppress body fat accumulation in mice. *Bioscience Biotechnology and Biochemistry* 65, 2735–2740.

Ohnuki, K., Niwa, S., Maeda, S., Inoue, N., Yazawa, S. and Fushiki, T. (2001b) CH-19 Sweet, a non-pungent cultivar of red pepper, increased body temperature and oxygen consumption in humans. *Bioscience Biotechnology and Biochemistry* 65, 2033–2036.

Omoni, A.O. and Aluko, R.E. (2005) The anticarcinogenic and anti-atherogenic effects of lycopene: a review. *Trends in Food Science and Technology* 16, 344–350.

Ornelas-Paz, J.J., Martinez-Burrola, M.M., Ruiz-Cruz, S., Santana-Rodriguez, V., Ibarra-Junquera, V., Olivas, G.I. and Perez-Martinez, J.D. (2010) Effect of cooking on the capsaicinoids and phenolics contents of Mexican peppers. *Food Chemistry* 119, 1619–1625.

Osuna-García, J.A., Wall, M.M. and Waddell, C.A. (1998) Endogenous levels of tocopherols and ascorbic acid during fruit ripening of New Mexican-type chile (*Capsicum annuum* L.) cultivars. *Journal of Agricultural and Food Chemistry* 46, 5093–5096.

Paran, I. and van der Knaap, E. (2007) Genetic and molecular regulation of fruit and plant domestication traits in tomato and pepper. *Journal of Experimental Botany* 58, 3841–3852.

Pardo, J.M., Fontanilla, M.R., Ospina, L.F. and Espinosa, L. (2008) Determining the pharmacological activity of *Physalis peruviana* fruit juice on rabbit eyes and fibroblast primary cultures. *Investigative Ophthalmology and Visual Science* 49, 3074–3079.

Park, J.S., Choi, M.A., Kim, B.S., Han, I.S., Kurata, T. and Yu, R. (2000) Capsaicin protects against ethanol-induced oxidative injury in the gastric mucosa of rats. *Life Science* 67, 3087–3093.

Passam, H.C., Karaponos, I.C., Bebeli, P.J. and Savvas, D. (2007) A review of recent research on tomato nutrition, breeding and post-harvest technology with reference to fruit quality. *European Journal of Plant Science and Biotechnology* 1, 1–21.

Patanè, S., Marte, F., Di Bella, G., Cerrito, M. and Coglitore, S. (2008) Capsaicin, arterial hypertensive crisis and acute myocardial infarction associated with high levels of thyroid stimulating hormone. *International Journal of Cardiology* 134, 130–132.

Patanè, S., Marte, F., La Rosa, F.C. and La Rocca, R. (2009) Capsaicin and arterial hypertensive crisis. *International Journal of Cardiology* 144, e26–e27.

Perucka, I. and Materska, M. (2001) Phenylalanine ammonia-lyase and antioxidant activities of lipophilic fraction of fresh pepper fruits *Capsicum annum* L. *Innovative Food Science and Emerging Technologies* 2, 189–192.

Peters, U., Leitzmann, M.F., Chatterjee, N., Wang, Y., Albanes, D., Gelmann, E.P., Friesen, M.D., Riboki, E. and Hayes, R.B. (2007) Serum lycopene, other carotenoids, and prostate cancer risk: a nested case–control study in the prostate, lung, colorectal, and ovarian cancer screening trial. *Cancer Epidemiology Biomarkers Prevention* 16, 962–968.

Peterson, J. and Dwyer, J. (1998) Taxonomic classification helps identify flavonoid-containing foods on a semiquantitative food frequency questionnaire. *Journal of American Dietetic Association* 98, 677–682.

Phillipson, J.D. (2001) Phytochemistry and medicinal plants. *Phytochemistry* 56, 237–243.

Pickersgill, B. (1997) Genetic resources and breeding of *Capsicum* spp. *Euphytica* 96, 129–133.

Pordesimo, L.O., Li, H., Lee, J.H. and Reddick, B.B. (2004) Effects of drying procedure, cultivar, and harvest number on capsaicin levels in dried jalapeno peppers. *Applied Engineering in Agriculture* 20, 35–38.

Porrini, M. and Riso, P. (2000) Lymphocyte lycopene concentration and DNA protection from oxidative damage is increased after a short period of tomato consumption. *Journal of Nutrition* 130, 189–192.

Porrini, M., Riso, P. and Testolin, G. (1998) Absorption of lycopene from single or daily portions of raw and processed tomato. *British Journal of Nutrition* 80, 353–361.

Pradeep, K.M., Geervani, P. and Eggum, B.O. (1992) Common Indian spices: nutrient composition, consumption and contribution to dietary value. *Plant Foods for Human Nutrition (formerly Qualitas Plantarum)* 44, 137–148.

Pramod, S.N. and Venkatesh, Y.P. (2004) Allergy to eggplant (*Solanum melongena*). *Journal of Allergy and Clinical Immunology* 113, 171–173.

Premuzic, Z., Bargiela, M., Garcia, A., Rendina, A. and Iorio, A. (1998) Calcium, iron, potassium, phosphorus, and vitamin C content of organic and hydroponic tomatoes. *HortScience* 33, 255–257.

Prohens, J., Rodriguez-Burruezo, A., Raigon, M.D. and Nuez, F. (2007) Total phenolic concentration and browning susceptibility in collection of different varietal types and hybrids of eggplant: implications of breeding for higher nutritional quality and reduced browning. *Journal of the American Society for Horticultural Science* 132, 638–646.

Prohens, J. Munoz-Falcon, J.E., Rodriguez-Burruezo, A. and Nuez, F. (2008) Strategies for breeding of eggplants with improved nutritional quality. *Acta Horticulturae* 767, 285–291.

Pyun, B.-J., Choi, S., Lee, Y., Kim, T.W., Min, J.K., Kim, Y., Kim, B.D., Kim, J.H., Kim, T.Y., Kim, Y.M. and Kwan, Y.G. (2008) Capsiate, a nonpungent capsaicin-like compound, inhibits angiogenesis and vascular permeability via a direct inhibition of Src kinase activity. *Cancer Research* 68, 227–235.

Raffo, A., La Malfa, G., Fogliano, V., Maiani, G. and Quaglia, G. (2006) Seasonal variations in antioxidant components of cherry tomatoes (*Lycopersicon esculentum* cv. Naomi F1). *Journal of Food Composition and Analysis* 19, 11–19.

Raffo, A., Baiamonte I., Nardo, N. and Paoletti, F. (2007) Internal quality and antioxidant content of cold-stored red sweet peppers as affected by polyethylene bag packaging and hot water treatment. *European Food Research and Technology* 225, 395–405.

Raigón, M.D., Prophen, J., Munoz-Falcon, J.E. and Nuez, F. (2008) Comparison of eggplant landraces and commercial varieties for fruit content of phenolics, minerals, dry matter and protein. *Journal of Food Composition and Analysis* 21, 370–376.

Ramadan, F.M. and Morsel, J.T. (2005) Cape gooseberry. A golden fruit of golden future. *Fruit Processing* 6, 396–400.

Rao, A.V. (2002) Lycopene, tomatoes, and the prevention of coronary heart disease. *Experimental Biology Medicine* 227, 908–913.

Rao, A.V. and Rao, L.G. (2007) Invited review: carotenoids and human health. *Pharmacological Research* 55, 207–216.

Richelle, M., Bortlik, K., Liardet, S., Hagar, C., Llambelet, P., Baur, M., Applegate, L.A. and Offord, E.A. (2002) A food-based formulation provides lycopene with the same bioavailability to humans as that from tomato paste. *Journal of Nutrition* 132, 404–408.

Rickman, J.C., Barrett, D.M., and Bruhn, C.M. (2007) Nutritional comparison of fresh frozen and canned fruits and vegetables. Part 1. Vitamin C and B and phenolic compounds. *Journal of the Science of Food and Agriculture* 87, 930–944.

Robertson, L.D. and Labate, J.A. (2007) Genetic resources of tomato (*Lycopersicon esculentum* Mill.) and wild relatives. In: Razdan, M.K. and Matto, A.K. (eds) *Genetic Improvement of Solanaceous Crops. Volume 2: Tomato.* Science Publishers, Enfield, New Hampshire, pp. 25–75.

Rosales, M.A., Ruiz, J.M., Hernández, J., Soriano, T., Castilla, N. and Romero, L. (2006) Antioxidant content and ascorbate metabolism in cherry tomato exocarp in relation to temperature and solar radiation. *Journal of the Science of Food and Agriculture* 86, 1545–1551.

Saga, K. and Sato, G. (2003) Varietal differences in phenolic, flavonoid and capsaicinoid contents in pepper fruits (*Capsicum annuum* L). *Journal of the Japanese Society for Horticultural Science* 72, 335–341.

Sakamura, S. and Obata, J. (1963) The structure of the major anthocyanins in eggplant. *Agricultural and Biological Chemistry* 23, 663–665.

Sánchez, A.M., Sánchez, M.G., Malagarie-Cazenava, S., Olea, N. and Diaz-Laviada, I. (2006) Induction of apoptosis in prostate tumor PC-3 cells and inhibition of xenograft prostate tumor growth by the vanilloid capsaicin. *Apoptosis* 11, 89–99.

Sarkar, T.K. and Chattopadhyay, T.K. (1993) Correlation studies on cape gooseberry (*Physalis peruviana* L.). *Annals of Agricultural Research* 14, 211–214.

Schmitz, H.N., Poor, C.L., Wellman, R.B. and Erdman, J.W. Jr (1991) Concentrations of selected carotenoids and vitamin A in human liver, kidney, and lung tissues. *Journal of Nutrition* 121, 1613–1621.

Scott, L.E. and Kramer, A. (1949) The effect of storage upon the ascorbic acid content of tomatoes harvested at different maturity stages. *Proceedings of the American Society for Horticultural Science* 54, 277–280.

Seren, S., Leberman, R., Bayraktar, U.D., Heath, E.H., Sahin, K., Andic, F. and Kucuk, O. (2008) Lycopene in cancer prevention and treatment. *American Journal of Therapeutics* 15, 66–81.

Serra, I., Yamamoto, M., Calvo, C., Cavada, G., Baez, S., Endoh, K., Watanabe, H. and Tajma, K. (2002) Association of chili pepper consumption, low socioeconomic status and longstanding gallstones with gallbladder cancer in a Chilean population. *International Journal Cancer* 102, 407–411.

Shahidi, F. (2005) Nutraceuticals and functional foods in health promotion and disease prevention. *Acta Horticulturae* 680, 13–24.

Shahidi, F. and Ho, C.T. (2005) Phenolics in foods and natural health products: an overview. In: Shahidi, F. and Ho, C.T. (eds) *ACS Symposium Series 909: Phenolic Compounds in Foods and Natural Health Products*. American Chemical Society, Washingtion, DC, pp. 1–8.

Shao, A. and Hathcock, J.N. (2006) Risk assessment for the carotenoids lutein and lycopene. *Regulatory Toxicology and Pharmacology* 45, 289–298.

Sharoni, Y., Danilenko, M., Dubi, N., Ben-Dor, A. and Levy, J. (2004) Carotenoids and transduction. *Archives of Biochemistry and Biophysics* 430, 89–96.

Shi, J. and Le Maguer, M. (2000) Lycopene in tomatoes: chemical and physical properties affected by food processing. *Critical Reviews in Food Science and Nutrition* 40, 1–42.

Shinohara, Y., Suzuki, Y. and Shibuya, M. (1982) Effects of cultivation methods, growing season and cultivar on the ascorbic acid content of tomato fruits. *Journal of Japanese Society for Horticultural Science* 51, 338–343.

Sies, H. and Stahl, W. (2004) Nutritional protection against skin damage from sunlight. *Annual Review of Nutrition* 24, 173–200.

Simonne, A.H., Simonne, E.H., Eitenmiller, R.R., Mills, H.A. and Green, N.R. (1997) Ascorbic acid and provitamin A contents in unusually colored bell peppers (*Capsicum annuum* L.). *Journal of Food Composition and Analysis* 10, 299–311.

Simonne, A.H., Fuzere, J.M., Simonne, E.H., Hochmuth, R.C. and Marshall, M.R. (2007) Effects of nitrogen rates on chemical composition of yellow grape tomatoes grown in a sub-tropical climate. *Journal of Plant Nutrition* 30, 927–935.

Singh, A.P., Luthria, D., Wilson, T., Vorsa, N., Singh, V., Banuelos, G.S. and Pasakdee, S. (2009) Polyphenols content and antioxidant capacity of eggplant pulp. *Food Chemistry* 114, 955–961.

Singh, C.U. and Nulu, J.R. (2008) Esters of capsaicin for treating pain. United States Patent Application Publication. Pub. No: US 2008/0020996 A1.

Singh, P. and Goyal, G.K. (2008) Dietary lycopene: its properties and anticarcinogenic effects. *Comprehensive Reviews in Food Science and Food Safety* 7, 255–270.

Slimestad, R. and Verheul, M.J. (2005) Content of chalconaringenin and chlorogenic acid in cherry tomatoes is strongly reduced during postharvest ripening. *Journal of Agricultural and Food Chemistry* 53, 7251–7256.

Slimestad, R. and Verheul, M. (2009) Review of flavonoids and other phenolics from fruits of different tomato (*Lycopersicon esculentum* Mill.) cultivars. *Journal of the Science of Food and Agriculture* 89, 1255–1270.

Snyman, T., Stewart, M.J. and Steenkamp, V. (2001) A fatal case of pepper poisoning. *Forensic Science International* 124, 43–46.

Soto-Zamora, G., Yahia, E.M. and Steta-Gandara, M. (2000) Changes in ascorbic acid and relation with ascorbate oxidase and ascorbate peroxidase in vine-ripe or ethylene-ripened tomato fruit. In: Florkowski, W.J., Prussia, S.E. and Shewfelt, R.L. (eds) *Proceedings International Multidisciplinary Conference: Integrated View of Fruit and Vegetable Quality*. Technomic Publishing Co, Inc, Lancaster, Basel, pp. 81–90.

Soto-Zamora, G., Yahia, E.M., Brecht, J.K. and Gardea, A. (2005) Effects of postharvest hot air treatments on the quality and antioxidant levels in tomato fruit. *LWT – Food Science and Technology* 38, 657–663.

Stacewicz-Sapuntzakis, M. and Bowen, P.E. (2005) Role of lycopene and tomato products in prostate health. *Biochemica et Biophysica Acta* 1740, 202–205.

Stahl, W., Schwarz, W., Sundquist, A.R. and Sies, H. (1992) Cis-trans isomers of lycopene and β-carotene in human serum and tissues. *Archives of Biochemistry and Biophysics* 430, 89–96.

Stommel, J.R. (2007) Genetic enhancement of tomato fruit nutritive values. In: Razdan, M.K. and Mattoo, A.K. (eds) *Genetic Improvement of Solanaceous Crops. Volume 2: Tomato*. Science Publishers, Enfield, New Hampshire, pp. 193–238.

Stommel, J.R. and Whitaker, B.D. (2003) Phenolic content and composition of eggplant fruit in a germplasm core subset. *Journal of the American Society for Horticultural Science* 128, 704–710.

Su, B.N., Gu, J.Q., Kang, Y.H., Park, E.J., Pezzuto, J.M. and Kinghorn, A.D. (2004) Induction of the phase II enzyme, quinone reductase, by withanolides and norwithanolides from Solanaceous species. *Mini-Review in Organic Chemistry* 1, 115–123.

Subbiah, K. and Perumal, R. (1990) Effect of calcium sources, concentrations, stages and number of sprays on physicochemical properties of tomato fruits. *South Indian Horticulture* 38, 20–27.

Suhaj, M. (2006) Spice antioxidants isolation and their antiradical activity: a review. *Journal of Food Composition and Analyses* 19, 531–537.

Sun, T., Xu, Z., Wu, C.T., Janes, M., Prinyawiwatkul, W. and No, H.K. (2007) Antioxidative activities of different colored sweet bell peppers (*Capsicum annum* L.). *Journal of Food Science* 72, 98–102.

Suresh, D. and Srinivasan, K. (2007) Studies on the *in vitro* absorption of spice principles – curcumin, capsaicin and piperine in rat intestines. *Food and Chemical Toxicology* 45, 1437–1442.

Surh, Y.J. (2002) Anti-tumor promoting potential of selected spice ingredients with antioxidative and anti-inflammatory activities: a short review. *Food and Chemical Toxicology* 40, 1091–1097.

Surh, Y.J. and Lee, S.S. (1995) Capsaicin, a double-edged sword: toxicity, metabolism, and chemopreventive potential. *Life Sciences* 56, 1845–1855.

Surh, Y.-J. and Lee, S.S. (1996) Capsaicin in hot chili pepper: carcinogen, co-carcinogen or anticarcinogen? *Food Chemistry Toxicology* 34, 313–316.

Surh, Y.-J., Lee, E. and Lee, J.M. (1998) Chemopreventive properties of some pungent ingredients present in red pepper and ginger. *Mutation Research* 402, 259–267.

Sutoh, K., Kobata, K. and Watanabe, T. (2001) Stability of capsinoid in various solvents. *Journal of Agricultural and Food Chemistry* 49, 4026–4030.

Suzuki, K., Mori, M., Ishikawa, K., Takizawa, K. and Nunomura, O. (2007) Carotenoid composition in mature *Capsicum annum*. *Food Science and Technology Research* 13, 77–80.

Swiader, J.M. and Ware, G.W. (2001) *Producing Vegetable Crops*. Interstate Publishers, Inc, Danville, Illinois.

Syamal, M.M. (1990) Biochemical composition of tomato fruits during storage. *Acta Horticulturae* 287, 369–374.

Szikszay, M., Obál, F. Jr and Obál, F. (1982) Dose–response relationship in the thermoregulatory effects of capsaicin. *Archives of Pharmacology* 320, 97–100.

Taber, H., Perkins-Veazie, P., Li, S.S., White, W., Rodermel, S. and Xu, Y. (2008) Enhancement of tomato fruit lycopene by potassium is cultivar dependent. *HortScience* 43, 159–165.

Thompson, K.A., Marshall, M.R., Sims, C.A., Wei, C.I., Sargent, S.A. and Scott, J.W. (2000) Cultivar, maturity, and heat treatment on lycopene content in tomatoes. *Journal of Food Science* 65, 791–795.

Thompson, R.Q., Phinney, K.W., Sander, L.C. and Welch, M.J. (2005) Reversed-phase liquid chromatography and argentation chromatography of the minor capsaicinoids. *Analytical and Bioanalytical Chemistry* 381, 1432–1440.

Thompson, R.Q., Pennino, M.J., Brenner, M.J. and Mehta, M.A. (2006) Isolation of individual capsaicinoids from a mixture and their characterization by ^{13}C NMR spectrometry. *Talanta* 70, 315–322.

Todaro, A., Cimino, F., Rapisard, P., Catano, A.E., Barbagallo, R.N. and Spagna, G. (2009) Recovery of anthocyanins from eggplant peel. *Food Chemistry* 114, 434–439.

Tomes, M.L. (1963) Temperature inhibition of carotene biosynthesis in tomato. *Botanical Gazette* 124, 180–185.

Toor, R.K., Savage, G.P. and Lister, C.E. (2006) Seasonal variations in the antioxidant composition of greenhouse grown tomatoes. *Journal of Food Composition and Analysis* 19, 1–10.

Turkmen, N., Sari, F. and Veioglu, Y.S. (2005) The effect of cooking methods on total phenolics and antioxidant activity of selected green vegetables. *Food Chemistry* 93, 713–718.

Unlu, N.Z., Bohn, T., Francis, D.M., Nagaraja, H.N., Clinton, S.K. and Schwartz, S.J. (2007a) Lycopene from heat-induced *cis*-isomer-rich tomato sauce is more bioavailable than from all-*trans*-rich tomato sauce in human subjects. *British Journal of Nutrition* 98, 140–146.

Unlu, N.Z., Bohn, T., Francis, D., Clinton, S.K. and Schwartz, S.J. (2007b) Carotenoid absorption in humans consuming tomato sauces obtained from tangerine or high-β-carotene varieties of tomatoes. *Journal of Agricultural and Food Chemistry* 55, 1597–1603.

USDA (2008) USDA National Nutrient Database for Standard Reference, Release 21. US Department of Agriculture, Agricultural Research Service, Nutrient Data Laboratory (http://www.nal.usda.gov/fnic/foodcomp/search/, accessed 15 October 2009).

van Wyk, B.E. (2005) *Physalis peruviana*, cape gooseberry. In: *Food Plants of the World*, Timber Press, Portland, Oregon, p. 294.

von Roepenack-Lahaye, E., Newman, M.A., Schornack, S., Hammond-Kosack, K.E., Lahaye, T., Jones, J.D.G., Daniels, M.J. and Dow, J.M. (2003) *p*-Coumaroylnoradrenaline, a noval plant metabolite implicated in tomato defense against pathogens. *Journal of Biological Chemistry* 278, 43373–43383.

Votava, E.J., Baral, J.B. and Bosland, P.W. (2005) Genetic diversity of chile (*Capsicum annuum* var. *annuum* L.) landraces from northern New Mexico, Colorado, and Mexico. *Economic Botany* 59, 8–17.

Wall, M.M., Waddell, C.A. and Bosland, P.W. (2001) Variation in β-carotene and total carotenoid content in fruits of Capsicum. *HortScience* 36, 746–749.

Wang, C.Y. (1993) Approaches to reduce chilling injury of fruits and vegetables. *Horticultural Reviews* 15, 83–95.

Watada, A.E., Aulenbach, B.B. and Worthington, J.T. (1976) Vitamin A and C in ripe tomatoes affected by stage of ripeness at harvest and supplementary ethylene. *Journal of Food Science* 41, 856–858.

Wertz, K., Siler, U. and Gorlczyk, R. (2004) Lycopene: mode of action to promote prostate health. *Archives of Biochemistry and Biophysics* 430, 127–134.

Whitaker, B.D. and Stommel, J.R. (2003) Distribution of hydroxycinnamic acid conjugates in fruit of commercial eggplant (*Solanum melongena* L.) cultivars. *Journal of Agricultural and Food Chemistry* 51, 3448–3454.

Wilcox, J.K., Catignani, G.L. and Lazarus, S. (2003) Tomatoes and cardiovascular health. *Critical Reviews in Food Science and Nutrition* 43, 1–18.

Winkel-Shirley, B. (2002) Biosynthesis of flavonoids and effects of stress. *Current Opinion in Plant Biology* 5, 218–223.

Wold, A.B., Rosenfeld, H.J., Holte, K., Baugerød, H., Blomhoff, R. and Haffner, K. (2004) Colour of post-harvest ripened and vine ripened tomatoes (*Lycopersicon esculentum* Mill.) as related to total antioxidant capacity and chemical composition. *International Journal of Food Science and Technology* 39, 295–302.

Yahia, E.M., Contrera-Padilla, M. and Gonzalez-Aguilar, G. (2001) Ascorbic acid content in relation to ascorbic acid oxidase activity and polyamine content in tomato and bell pepper fruits during development, maturation and senescence. *LWT – Food Science and Technology* 34, 452–457.

Yahia, E.M., Soto-Zamora, G., Brecht, J.K. and Gardea, A. (2007) Postharvest hot air treatment effects on the antioxidant system in stored mature-green tomatoes. *Postharvest Biology and Technology* 44, 107–115.

Yeh, C.T. and Yen, G.C. (2005) Effects of vegetables on human phenolsulfotransferases in relation to their antioxidant activity and total phenolics. *Free Radical Research* 39, 893–904.

Zeyrek, F.Y. and Oguz, E. (2005) *In vitro* activity of capsaicin against *Helicobacter pylori*. *Annals of Microbiology* 55, 125–127.

Zhang, L.X., Cooney, R.V. and Bertran, J.S. (1992) Carotenoids up-regulate connexin43 gene expression independent of their provitamin A or antioxidant properties. *Cancer Research* 52, 5707–5712.

17 Tropical Fruit
[Banana, Pineapple, Papaya and Mango]

Thiruchelvam Thanaraj and Leon A. Terry

17.1 Introduction

Tropical fruit crops (banana, pineapple, papaya and mango) are a group of botanically unrelated plants. They are grouped together for ease in this chapter merely because of their tropical adaptation and global importance. Tropical crops form a substantial part of the export economy of several developing countries. Banana, pineapple and mango are categorized as major tropical fruit crops, while papaya is considered a minor crop (Galán Saúco, 1996).

Tropical fruit are commonly consumed fresh. Despite their generally relatively low calorific value (banana and plantain are the exceptions), tropical fruit play a major role in the human diet, mainly because of their high and diverse concentrations of vitamins, minerals, carotenoids and other bioactive components. The antioxidant activities of tropical fruit have been discussed in a few studies only, even though some of these fruits are rich in dietary antioxidants (Jimenez-Escrig et al., 2001; Bashir et al., 2003; Bennett et al., 2010). Mango and papaya are good sources of carotenoids (β-carotene) and vitamin C, while banana is rich in polyphenols (Arora et al., 2008; Vijayakumar et al., 2008; Bennet et al., 2010) and pineapple is rich in vitamin C (Cordenunsi et al., 2010). However, pineapple and papaya contain proteolytic enzymes, namely

bromelain and papain, respectively. These enzymes have significant chemopreventive activities and are used in several industrial applications. Tropical fruit also contain high levels of pectin, fibre and cellulose, which are believed to promote intestinal motility. The relatively high organic acid content of many tropical fruit may also stimulate appetite and aid digestion (Martin et al., 1987).

In the main, tropical fruit are rich in health-promoting properties and are associated with several health benefits to humans; however, the published information is still very limited. Therefore, it is necessary to highlight the health benefits of tropical fruit, which may improve awareness among people, especially in the developing world, in selecting and consuming these fruit.

17.2 Banana

17.2.1 Introduction

Banana is a common name for the fruit of the genus *Musa*, which bears edible fruit of the family *Musaceae*. Edible fruit-bearing banana cultivars belong to the species endemic to South–east Asia (*Musa acuminata* and *M. balbisiana*) (Zhang et al., 2005). Banana usually refers to a soft, sweet dessert fruit but there are also green, firm and starchy fruit from a group referred to as the plantains, which are also valued as a staple crop. Dessert banana fruit are imported in large quantities from the tropics to temperate regions. Banana cv. Cavendish is still the most important among commercial cultivars and accounts for a huge bulk of the bananas imported from the tropics (Arora et al., 2008). Total banana production was about 81.3 million tonnes (Mt) in 2007. India, Ecuador, Brazil and China are the leading banana producers; however, Ecuador, along with Costa Rica, the Philippines and Colombia are the leading banana exporting countries (FAO, 2005). Because of the dominance of cv. Cavendish, less is known about the vast plethora of other banana varieties (Arora et al., 2008).

Banana fruit dry matter consists mainly of sugars (glucose, fructose and sucrose), starch and fibre, making it an ideal immediate and slightly prolonged source of energy. Due to the fibre content, banana can help to restore normal bowel function. Banana fruit also contains pectin, a soluble fibre (hydrocolloid) that can help normalize movement through the digestive tract. Other than these beneficial effects, banana fruit is also a rich source of health-related compounds such as vitamins, phenolic compounds, carotenoids, minerals and certain amino acids. Though banana is low in primary antioxidants, it is rich in secondary antioxidants (Lim et al., 2007).

17.2.2 Identity and role of bioactives

Phenolic compounds

Total phenolic (TP) content of banana peel varies greatly among cultivars; for example, cv. Kluai Hom Thong, which has 3.0 mg gallic acid equivalents (GAE)/g fresh weight (FW), contains higher TP than cv. Kluai Khai (0.9 mg GAE/g FW) (Nguyen et al., 2003). The apparent TP content of Malaysian banana cv. Pisang-mas varies between 0.24 and 0.72 mg GAE/g FW depending on the extraction method (Lim et al., 2007; Alothman et al., 2009). The TP and flavonoids contents correlate positively with antioxidant capacities, as measured using ferric reducing antioxidant power (FRAP) or 2,2-diphenyl-1-picrylhydrazyl (DPPH) (Alothman et al., 2009).

Generally, flavonoids are believed to exhibit anti-inflamatory, antineoplastic and hepatoprotective activities and to reduce acid secretion from gastric parietal cells (Havsteen, 1983; Beil et al., 1995). Leucodelphinidin (a flavonoid) is more abundant in plantain than in banana pulp and has been reported to have an antiulcerogenic effect in humans, and also to have a protective effect against aspirin-induced damage of the gastric mucosa (Lewis et al., 1999). Gallocatechin is more abundant in banana peel (1.58 mg/g dry weight (DW)) than in pulp (0.3 mg/g DW) and thus is probably not of importance (Someya et al., 2002). Bennett et al. (2010) showed recently that catechin, gallocatechin, epicatechin and condensed tannins were present in soluble extracts of banana fruit pulp, and reviewed the chemical structures commonly found in various Musa spp.

Banana bracts are abundant edible residues of banana production and are consumed as a vegetable in most banana producing countries. Bracts contain various anthocyanins (~ 0.32 mg/g FW). Cyanidin-3-rutinoside is the prominent anthocyanin and contributes about ~ 80%; however, 3-rutinoside derivatives of delphinidin, pelargonidin, peonidin and malvidin also contribute in considerable quantities.

Vitamins

Both ascorbic acid (AsA) and dehydroascorbic acid (DHA) contribute to the total vitamin C concentration. Banana is a moderate source of vitamin C (0.33 mg/g FW) (Cano et al., 1997). Dwarf Brazilian ('apple') banana fruit contain about threefold higher levels of vitamin C (0.13 mg/g FW) than cvs. Williams and Pisang-mas (0.05 mg/g FW) (Wall, 2006; Lim et al., 2007). However, the vitamin C content of banana cv. Cavendish ranges between 0.02 and 0.19 mg/g FW (Leong and Shui, 2002; USDA-ARS, 2004). Vitamin A content of Dwarf Brazilian ('apple') banana fruit is around 12.4 µg retinol activity equivalents (RAE)/100 g FW, while cv. Williams has 8.2 µg RAE/100 g FW (Wall, 2006). These values are higher than the vitamin A content of cv. Cavendish fruit (4.5 µg RAE/100 g FW) (Wenkam, 1990). However, vitamin A content of banana is based mainly on the β-carotene (most active provitamin A pigment) and α-carotene (less active provitamin A pigment) concentrations (Wall, 2006).

Carotenoids

Banana fruit are generally recommended for young children and pregnant/lactating mothers since they are rich in carotenoids that protect against vitamin A deficiency (Englberger *et al.*, 2003). Banana fruit with orange-coloured flesh have higher concentrations of bioavailable carotenoids, such as β-carotene (9.4–27.8 µg/g FW), α-carotene (6.1–9.5 µg/g FW), lutein (0.4–1.0 µg/g FW) and zeaxanthin (0.1–0.2 µg/g FW), than yellow and more creamy-fleshed banana pulp (Englberger *et al.*, 2003). Since the banana matrix is digestible, the bioavailability of β-carotene in banana fruit is relatively high (Englberger *et al.*, 2003). The average lutein concentrations of Dwarf Brazilian banana fruit and cv. Williams were found to be 1.6 and 1.1 µg/g FW, respectively, which exceed the corresponding values for the provitamin A pigments such as β-carotene and α-carotene (Englberger *et al.*, 2003; Wall, 2006). Arora *et al.* (2008) detailed the variation in β-carotene content in the pulp and peel of a number of Indian-derived banana cultivars and showed that there might be an opportunity to exploit these as a by-product as the higher carotenoid levels are found in the peel. The Red banana was ranked as one of the highest for total carotenoid levels found in both pulp and peel (Arora *et al.*, 2008).

Minerals

Potassium (5.09 mg/g FW) is the prominent mineral found in banana, followed by phosphorus (0.59 mg/g FW), magnesium (0.38 mg/g FW) and calcium (0.38 mg/g FW) (Hardisson *et al.*, 2001). The peel of some Cameroon banana cultivars has a relatively high content of minerals, namely potassium (50.0 mg/g DW), phosphorus (22.0 mg/g DW), magnesium (11 mg/g DW) and calcium (18 mg/g DW) (Emaga *et al.*, 2007). Generally, calcium content is low in banana; however, Micronesian cv. Karat contains relatively high calcium content (Englberger *et al.*, 2003). Banana contributes about 2.7% of the total potassium and fibre consumed by an average adult (USDA-HHS, 2004).

Essential amino acids

Dopamine is a strong, water-soluble antioxidant found in the peel (0.8–5.6 mg/g FW) and pulp (0.03–0.1 mg/g FW) of banana cv. Cavendish, and is one of the catecholamines that suppress the oxygen uptake of linoleic acid. Dopamine has similar antioxidant potency to strong antioxidants such as gallocatechin gallate and AsA (Kanazawa and Sakakibara, 2000). Bioactive amines such as putrescine, spermidine and serotonin have been identified in high concentrations from banana cv. Prata. The content of serotonin, which is responsible for regulating a number of important functions in humans, i.e. sleeping, thirst, hunger, mood and sexual activity, reduces during ripening, while that of some other amino acids is maintained (Coutts *et al.*, 1986; Adão and Glória, 2005). Vetorazzi (1974), Marriott and Palmer (1980) and Coutts *et al.* (1986) reported that bioactive amines such as serotonin, dopamine, noradrenaline, octopamine, histamine, 2-phenylethylamine and tyramine have been identified in various banana cultivars.

17.2.3 Chemopreventive activity and bioavailability

Banana contains relatively high iron content. Frequent consumption of banana (four to six times/week) reduces the risk of kidney cancer and other disorders related to the kidney and urinary tract. Banana has relatively high potassium (3.5 mg/g FW) and very low sodium content, a good ratio for preventing high blood pressure and stroke. Consuming bananas regularly in the diet can reduce significantly the risk of death caused by strokes, by about 40%. Phytonutrients from the fruit generally stimulate natural detoxifying enzymes in the body; these enzymes reduce the risk of atherosclerosis and cancer (Ames *et al.*, 1993). The ability of banana (9%) to bind to bile acids *in vitro* is higher than that for other fruit such as pineapple (5.9%), grape (5%), peach (6%) and pear (4.7%). Preventing the recirculation of bile acids reduces fat absorption, increases the excretion of cancer-causing toxic metabolites and increases cholesterol conversion to more bile acids (Kahlon and Smith, 2007). Banana can cause a natural anti-acid effect in the human body; therefore, eating banana can provide a soothing relief in people suffering from heartburn. Plantain has long been recommended to treat digestive disorders in humans since it is believed to contain active antiulcerogenic agents (Goel *et al.*, 1989; Dunjic *et al.*, 1993).

17.2.4 Effect of preharvest and postharvest continuum

Variety, harvest maturity, state of ripeness, soil type, soil condition, fertilization, irrigation and weather are all important preharvest factors that affect the quality of banana fruit (Tahvonen, 1993). Bananas for export are transported in a preclimacteric state and thus need to be ripened in specialist ripening rooms near to the point of sale. Even though the banana dominates the fresh fruit export market, there has been little research conducted on detailing the temporal changes in health-promoting compounds after harvest and linking these with effects on human health. This is surprising considering the dominance of the banana compared with other frequently consumed fruit and may reflect the commoditization of the product.

17.3 Pineapple

17.3.1 Introduction

The pineapple (*Ananas comosus* L.) is the most economically important member of the family *Bromeliaceae* and the only bromeliad bearing edible fruit. The family *Bromeliaceae* includes about 45 genera and over 2000 cultivars of pineapple, which have been cultivated around the world. Smooth Cayenne, Natal Queen, Red Spanish and Kona Sugarloaf are some of the prominent commercial cultivars. Among these cultivars, cv. Smooth Cayenne is popular around the world for its excellent processing characteristics, such as regularity of shape and size, high sugar and acid content, ability to withstand rough handling and an acceptable shelf life (Kelly and Bagshaw,

1993; Paull, 1993). Pineapple is ranked third among world tropical crops, preceded only by banana and citrus (Uriza-Avila, 2005). Although pineapple is thought to be native to the South Americas (Brazil, Hawaii, Paraguay, etc.), it was first discovered by Europeans, in 1493, in the Caribbean. The pineapple is now grown extensively in Hawaii, the Philippines, Ivory Coast, the Caribbean, Malaysia, Costa Rica, Taiwan, Thailand, Australia, Mexico, Kenya and South Africa. Pineapple has long been considered as one of the most popular non-citrus tropical and subtropical crops owing to its attractive flavour and refreshing sugar–acid balance (Bartolomé *et al.*, 1996; Morse, 2008). Global production of pineapples was about 18.87 Mt in 2007. Thailand is the largest producer, accounting for 16% of global output, followed by the Philippines (12%) and Brazil (10%) (Rebolledo-Martínez *et al.*, 2005; Fold and Gough, 2008).

17.3.2 Identity and role of bioactives

Bromelain

Pineapple has been used extensively as a folk remedy for several health ailments, including digestive problems. Bromelain is a complex mixture of substances (collectively named as proteolytic enzymes or proteases) found mainly in the stem and core of the pineapple fruit. Bromelain is considered to be one of the best vegetable proteases, with numerous applications in the food industries, medicine and pharmacology (Devakate *et al.*, 2009). Bromelain obtained from the stem is widely used in industries; fruit bromelain (from the core) is not available commercially but is believed to have possible digestion-related and anti-inflammatory benefits (Devakate *et al.*, 2009).

Vitamins and minerals

The average vitamin C concentration in fresh pineapple flesh (0.15 mg/g FW) is higher than that in its juice (0.11 mg/g FW) (USDA, 1992; Morse, 2008). The manganese concentration (15–20 mg/l) of commercial pineapple juice is much higher than the concentrations of chromium, iron and copper. However, the nutritional bioavailability of the manganese is uncertain (Beattie and Quoc, 2000). In addition to manganese, pineapple is also a good source of potassium, calcium, iron and magnesium (Nakasone and Paull, 1998). A combination of glucosamine, chondroitin sulfate and manganese may offer significant improvement of symptoms for those with mild to moderate osteoarthritis of the knee (Orlando, 2006).

Carotenoids and phenolic compounds

Total carotenoid concentration is proportional to the degree of yellow colour in pineapple flesh. Carotenoid content is higher in the flesh than in the juice of cvs. Del Monte Hawaii Gold (1.36 µg/g in flesh and 0.25 µg/g in juice) and Smooth Cayenne (0.45 µg/g in flesh and 0.07 µg/g in juice). β-Carotene is a primary provitamin A found in pineapple and contributes to about 35% of total carotenoids, its content also being considerably lower in juice than flesh, irrespective of cultivar (Paull, 1993; Ramsaroop and Saulo, 2007).

Phenolic compounds are associated with several health benefits. Total phenolics are one of the important antioxidants in pineapple and their concentration varies between 0.32 mg GAE/g FW and 0.52 mg GAE/g FW. Flavonoid content of pineapple ranges between 0.01 mg catechin equivalents (CEQ)/g FW and 0.04 mg CEQ/g FW (Alothman *et al.*, 2009).

17.3.3 Chemopreventive activity and bioavailability

Therapeutic doses of bromelain are believed to reduce excessive inflammation, coagulation of the blood and certain types of tumour growth. Protein molecules from bromelain, such as CCZ and CCS, have been identified as powerful anticancer agents and could lead to a new class of cancer-fighting drugs. Antioxidant capacity of pineapple varied between 1.72 µmol Fe/g and 5.3 µmol Fe/g, while DPPH inhibition ranged between 12.7 and 90.8% (Alothman *et al.*, 2009). Since pineapple is rich in fibre, fresh pineapple prevents constipation and also relieves constipation in

those who already have it (Morse, 2008). Bromelain is also effective in treating sore throat pain, upper respiratory conditions and acute sinusitis. Fresh pineapple juice also speeds up the natural healing of warts (Orlando, 2006).

17.3.4 Effect of preharvest and postharvest continuum

Adequate soil nutrition (minerals) is crucial for good quality and productivity of fruits. Addition of potassium (K) to the soil can play a significant role in producing good quality pineapple, increasing total solids and producing fruit with a larger diameter. However, high concentrations of K in the soil may lead to very acidic fruit with pale and rigid pulps (Dull, 1971). Adequate K in the soil may improve pineapple flavour, increase stalk diameter and increase ascorbic acid concentrations. The ascorbic acid may prevent some degree of enzymatic browning, by inhibition of polyphenoloxidase activity (Soares *et al.*, 2005). Acidic soil condition, although often associated with higher manganese levels, may lead to iron deficiency (Beattie and Quoc, 2000).

Natural flowering has been a longlasting agronomic problem in pineapple cultivation, since it causes considerable postharvest losses and irregular market supply. Therefore, flowering is usually induced using various external induction agents (ethephon, ethylene gas, etc.) (Bartholomew *et al.*, 2003; Wang *et al.*, 2007). Pineapple fruit is commercially important and a major export crop for some countries. However, occurrence of internal browning (black heart or endogenous brown spots) presents challenges in maintaining adequate fruit quality standards and export potential. The quality of the fruit essentially depends on planting, harvesting and pre- and postharvest factors (Selvarajah and Herath, 1997; Selvarajah *et al.*, 2001).

As the external colour of pineapple is not an exact predictor of ripeness, selecting the optimum harvest maturity or ripeness is a challenging task. Ripeness has traditionally been judged by sniffing at the stem end of the fruit or by looking for fresh deep green leaves.

Since carotenoids are very sensitive to heat and light, heat treatment during preparation of pineapple juice reduces the carotenoid concentration (Hodgson and Hodgson, 1993). Since bromelain is deactivated by high temperatures, the bromalain content may be lower in pineapple juice and canned pineapple than in fresh fruit.

17.4 Papaya

17.4.1 Introduction

Papaya (*Carica papaya* L.) is the major economically important species out of the 21 species within the genus *Carica* of the small dicotyledonous family *Caricaceae*. Most of the papaya cultivars are grown in tropical and subtropical countries, while cvs. Solo and Sunrise have widespread distribution beyond the tropics. Papaya is called by different names around the globe, namely 'tepayas' in East Malaysia, 'betik' in Peninsular Malaysia, 'lechosa' in Venezuela, 'pawpaw' in Sri Lanka and 'papali' in India (Fasihuddin and Ghazally, 2003). Tropical Central America (Mexico and Costa Rica) is believed to be the origin of papaya; however, it is now cultivated throughout the tropics and subtropics. The main papaya producing countries are Brazil, Nigeria, India, Mexico and Indonesia (Banerjee-Bhattacharya, 2002). Global production of papaya has increased continuously over the past decade; production reached around 6.94 Mt in 2007. Brazil is the major producing country and contributes about 24.6% of total production. The European Union countries are the most prominent papaya importers, with imports increasing by 50% between 2001 and 2003. Since papaya contains specialized cells called laticifers, it is referred to as a laticiferous plant (Azarkan *et al.*, 2003). The papaya fruit is a melon-like, oval to nearly round, 15–20 cm long fleshy berry. A rich orange-coloured flesh with either yellow or pink hues is deliciously sweet, with musky undertones and a soft, butter-like consistency. The papaya is available year-round and the ripe fruit is a favourite breakfast and dessert. It can also be used to make fruit salad, refreshing drinks, jam, jelly, marmalade, candies and crystallized fruit, etc. (Banerjee-Bhattacharya, 2002).

Papaya fruit is highly accepted worldwide and is sought after for its health-promoting properties, flavour and digestive properties (Fernandes *et al.*, 2006). Papaya fruit not only offers luscious taste and the sunlit colour of the tropics, but also is a good source of energy and antioxidant nutrients such as carotenoids, vitamins, minerals, flavonoids, anthocyanins, etc. Some of these nutrients improve the cardiovascular system and provide protection against colon cancer (Banerjee-Bhattacharya, 2002). The fruit and other parts of the papaya plant contain papain, an enzyme that is reported to help in the digestion of protein (Banerjee-Bhattacharya, 2002). However, this enzyme is especially concentrated in unripe fruit. Papain is extracted commercially to make digestive enzyme dietary supplements and is also used as an ingredient in some chewing gums.

17.4.2 Identity and role of bioactives

Carotenoids and phenolics

β-Carotene (2.32–5.98 µg/g FW) and β-caryophyllene (5.94 µg/g FW) are the prominent carotenoids in papaya; however, β-carotene is the more active provitamin A pigment. Red-fleshed papaya cultivars (Sunrise and SunUp) contain significant concentrations of lycopene (13.50–42.81 µg/g FW); however, lycopene has not been detected in yellow-fleshed papaya cvs. Formosa, Kapoho, Laie Gold, Maradol and Rainbow (Chandrika *et al.*, 2003; Wall, 2006; de Oliveira *et al.*, 2010; Rivera-Pastrana *et al.*, 2010). Lycopene has the highest free radical scavenging ability, followed by β-caryophyllene and β-carotene (Miller *et al.*, 1996). The concentration of β-carotene is also higher in red-fleshed papaya fruit (7.0 µg/g DW) than in yellow-fleshed fruit (1.4 µg/g DW); however, β-cryptoxanthin is in a more or less similar quantity (16.0 µg/g DW) in both red- and yellow-fleshed fruit (de Oliveira *et al.*, 2010).

Papaya has a comparatively high primary antioxidant potential, similar to those of guava and starfruit (see Chapter 8 of this volume). Total phenolic concentrations (0.54 mg GAE/g) of papaya are similar in the peel and flesh and across maturity stages (Kondo *et al.*, 2005) and are lower than those of most other tropical crops, such as mango, guava and pineapple (Patthamakanokporn *et al.*, 2008). However, papaya cv. Soursop contains relatively high concentrations of flavonoids (10–30 μg/g). Rivera-Pastrana *et al.* (2010) reported that ferulic acid, caffeic acid and rutin were the most abundant phenolic compounds identified in mesocarp tissue.

Vitamins and minerals

Vitamin A concentration of papaya fruit ranges between 0.19 and 0.74 μg RAE/g FW. Papaya is a very rich source of vitamin C, yet the content may vary between 0.46 and 1.45 mg/g FW depending on the cultivar, ripening stage and handling after harvest (Firmin, 1997; Proulx, 2002; de Oliveira *et al.*, 2010). According to Lim *et al.* (2007), papaya has a higher content of ascorbic acid (1.08 mg/g FW), than most other tropical fruit apart from guava. Ascorbic acid increases with maturity in both the peel and pulp (Kondo *et al.*, 2005). In addition to vitamins A, C and E, folate is also found in considerable quantities in papaya.

Papaya fruit can also contribute a significant amount of minerals to the human diet, but the mineral composition of the fruit reflects the trace mineral content of soil and also varies with climate, maturity, cultivars, agricultural practices, etc. (Forster *et al.*, 2002). Phosphorus (0.05–0.09 mg/g FW), potassium (0.90–2.21 mg/g FW), calcium (0.1–0.32 mg/g FW), magnesium (0.19–0.33 mg/g FW) and sodium (0.05–0.24 mg/g FW) are the important minerals identified in significant quantities from the prominent papaya cvs. Kapoho, Laie Gold, Rainbow, Sunrise and SunUp (Wall, 2006).

Papain

Unripe papaya fruit contains an enzyme called papain, which is a cysteine protease with an action similar to the pepsin in gastric juice. Papain is derived, for commercial use, from the latex of the unripe papaya fruit immediately after harvest. Chymopapain, caricain and papaya proteinase VI are the other cysteine proteases isolated from papaya latex (Morton, 1987a; Ohara *et al.*, 1995). Papain has been used in several industrial applications, such as in beer as a clarifier, in meat as a tenderizer, in the preparation of protein hydrolysates and in the pharmaceutical industry (in the treatment of osteoporosis, arthritis, vascular diseases and cancer) (Siewinsky *et al.*, 1996; Brömme *et al.*, 2004). Unripe papaya peel has high levels of papain and chymopapain (Emeruwa, 1982; Osato *et al.*, 1993) and has been used as an effective external treatment for skin wounds.

17.4.3 Chemopreventive activity and bioavailability

Preparation method, processing and cooking affect the total antioxidant capacity (TAC) of fruit and vegetables, as does genotype. According to Prior and Cao (2000), a study has revealed that cooking generally reduces the antioxidant capacity in fruit and vegetables by 15%. Antioxidant activities of papaya (measured as oxygen radical absorbance capacity or FRAP) are lower than for other tropical crops such as mango, guava and pineapple (Patthamakanokporn *et al.*, 2008). However, Lako *et al.* (2007) reported that Hawaiian papaya cv. Soursop contained higher TAC than other fruit grown in Fiji, and moderate levels of total polyphenol (TPP).

The bioavailability of carotenoids in red-fleshed papaya is under investigation. It has been noted that excess intake of red-fleshed papaya can cause a yellow-orange discoloration of the skin of the palm (lycopenaemia) among the peoples of Sri Lanka. No other yellow or red fruit in Sri Lanka causes this symptom (Chandrika *et al.*, 2003). Lycopene has no provitamin activity as it misses a β-ionone ring; however, conjugated and non-conjugated double bonds make lycopene highly reactive towards oxygen and free radicals. Perspective and retrospective epidemiological studies have indicated that oral intake of lycopene is bioavailable and reduces prostrate cancer risk. In addition, *in vitro* experiments also indicate that lycopene can induce apoptosis in cancer cells and inhibit their proliferation by producing cell cycle arrest (van Breemen and Pajkovic, 2008). Since lycopene is bound tightly to macromolecules within the food matrix, its bioavailability is relatively poor. Cooking or processing lycopene-rich food like tomato and papaya, however, releases lycopene from protein complexes and enhances its bioavailability. Consumption with lipids can also increase the bioavailability of lycopene, as it is highly lipophilic (Erdman *et al.*, 1993; van Breemen and Pajkovic, 2008).

17.4.4 Other beneficial and negative effects

Different parts of the papaya fruit, such as the peel, flesh, seeds and latex, traditionally have been used to treat various ailments in humans. Papaya seed is a rich source of biologically active isothiocyanate (Nakamura *et al.*, 2007); the seeds have also been used as an emmenagogue, thirst quencher, or carminative to alleviate pain from insect bites and stings (Wiart, 2006). Papaya latex is a rich source of four cysteine endopeptidases (papain, chymopapain, glycyl endopeptidase and caricain), the content of these enzymes varying in different parts of the plant. Since latex concentration decreases during ripening, it is commercially extracted from unripe papaya fruit (Azarkan *et al.*, 2003). Latex is very efficient against gastrointestinal nematodes (Stepek *et al.*, 2007) and is also used to treat eczema and psoriasis in Cambodia, Laos and Vietnam (Amenta *et al.*, 2000). Fresh papaya latex is smeared on boils, warts, scalds and freckles, and is given as a vermifuge. It has also been used as an anthelmintic in livestock and to cure dyspepsia (Satrija *et al.*, 1995). Papain has been used to treat ulcers, dissolve membranes in diphtheria, and reduce swelling, fever and adhesions after surgery.

Unripe papaya, pineapple, banana and jackfruit are considered harmful to pregnant women in Indian traditional medicine. The unripe papaya fruit and seeds are consumed purposely to cause abortion and sometimes ingested hazardously in some parts of the world (Tiwari *et al.*, 1982; Adebiyi *et al.*, 2002). These beliefs have been supported scientifically, as crude papaya latex and its proteonases have been found to be strong uterine contractants, which enhance the abortifacient properties of papaya (Cherian, 2000; Adebiyi *et al.*, 2002). Adebiyi *et al.* (2002) further described how the phytochemicals of papaya may suppress the effect of progesterone, which leads to the contraction of uterine smooth muscle and consequently causes abortion. This may be the most probable reason for the abortifacient properties of unripe papaya (progesterone plays a major role during pregnancy) (Dewick, 2002). Papaya is also used to treat constipation, kidney problems, intestinal disorders, piles, dyspepsia, liver, spleen and digestive disorders and urology problems.

Extract of unripe papaya peel has beneficial effects in the treatment of wounds; topical application of unripe fruit also promotes desloughing, granulation and healing, with reduced odour, in chronic skin ulcers (Hewitt *et al.*, 2000; Anuar *et al.*, 2008). Papaya fruit also contains carpanie, which is an alkaloid with a strong depressant action on the heart (Hornick *et al.*, 1978). The content of arginine, which is known to be essential for male fertility, and is very effective in treating nematodes such as roundworms (*Ascaridilla galli*) in humans, may explain why the latex of young fruit has an anthelmintic activity. Intestinal nematode infections may cause intestinal disorders, discomfort and loss of productivity through direct or indirect interference with host nutrition and metabolism in humans and animals (Satrija *et al.*, 1995).

17.4.5 Effect of preharvest and postharvest continuum

The nutritional composition of a papaya fruit varies widely based on cultivar, maturity, climate, soil type and fertility. Carotenoids and ascorbic acid increase with maturation and ripening. Ascorbic acid concentration is also influenced by the availability of light to the crop and fruit (Wenkam, 1990; Lee and Kader, 2000). The reported concentrations of β-carotene in papaya vary with genotype, quantification methods, etc. For example, cv. Khakdahm ripe fruit were reported to contain 4.71 µg/g of β-carotene along with 21.69 µg/g of lycopene (Charoensiri *et al.*, 2009); however, β-carotene concentrations of 2.76 µg/g (Holden *et al.*, 1999), 2.28 µg/g (Tee and Lim, 1991) and 4.40 µg/g (Setiawan *et al.*, 2001) have also been estimated for the same cultivar using different methods. The ascorbic acid content of papaya is higher in first-harvested fruit (7.50 mg/g DW) than in second-harvested fruit (4.00 mg/g DW) (Lee and Kader, 2000; Proulx *et al.*, 2005).

The quality of papaya fruit may be reduced by the adverse environmental and physical conditions encountered during transportation, distribution and retailing. Poor appearance, flavour and nutritional value may result from extreme or fluctuating temperatures and/or mechanical damage, combined with improper harvesting and handling practices. Papaya may have a shelf life of 4–6 days under ambient tropical conditions (25–28°C) or up to 3 weeks at lower temperatures (10–12°C) if it is handled properly after harvest (Paull *et al.*, 1997). Chilling injury symptoms occur in mature green papaya after 14 days, and in 60% of yellow fruit after 21 days, at 7°C. Rivera-Pastrana *et al.* (2010) demonstrated that carotenoids (except β-carotene) in cv. Maradol papaya fruit were affected detrimentally by low temperature storage at 1°C and suggested that storage at 25°C had a lesser negative effect on these metabolites.

17.5 Mango

17.5.1 Introduction

Mango (*Mangifera indica* L.) is known as the king of fruit and belongs to genus *Mangifera*, which consists of numerous species of tropical fruit in the family *Anacardiaceae*. Mango is considered indigenous to Eastern Asia, Myanmar and Assam state in India; however, it is widely cultivated as a fruit tree in frost-free tropical and warmer subtropical regions (Morton, 1987b). Despite most ripe mango fruit being consumed as a dessert, mature mangoes are also eaten as pickles, sliced or grated in fresh salads, soaked in water or sugar, salted and dried, sliced in vinegar or fish sauce, etc. Mango has special significance in Pakistan, India, Bangladesh, Sri Lanka and the Philippines, as its leaves are used spiritually for floral decorations at Hindu marriages and religious ceremonies (Kim *et al.*, 2009). Total global mango production was forecast to reach 30.7 Mt/year by 2010 (FAO, 2003). Though mango fruit is well known for its characteristic aroma and taste, it is also an excellent source of carbohydrates and health-promoting dietary antioxidants (Kauer and Kapoor, 2001) such as ascorbic acid (Franke *et al.*, 2004; de Oliveira *et al.*, 2010), carotenoids (Godoy and Rodriguez-Amaya, 1989; de Oliveira *et al.*, 2010), phenolic compounds (Berardini *et al.*, 2004, 2005), vitamin E (α-tocopherol) (Xianli *et al.*, 2004) and minerals (Ribeiro *et al.*, 2007). Many of these compounds possess not only antioxidant (Berardini *et al.*, 2004; Talcott *et al.*, 2005; Mahattanatawee *et al.*, 2006) but also immunomodulatory (Naved *et al.*, 2005), antimutagenic (Botting *et al.*, 1999) and/or anticancer activity (Percival *et al.*, 2006).

17.5.2 Identity and role of bioactives

Phenolic compounds

Studies have demonstrated that polyphenolic compounds of mango fruit include various flavonoids, xanthones, phenolic acids and gallotannins (Schieber *et al.*, 2003; Berardini *et al.*, 2005). Among these compounds, gallic acid and hydrolysable tannins are the major antioxidant polyphenolics in mango. Mango peel has generally high TP content but the reported level varies widely with cultivar and assay method. Sri Lankan mango cv. Willard was found to have higher TP than cvs. Karutha Colomban and Malgova (Table 17.1) (Thanaraj, 2010).

About 12 flavonoids and xanthones were identified in Brazilian mango cvs. Haden (0.48 mg/g DW), Tommy Atkins, Palmer and Uba; however, cv. Uba (2.09 mg/g DW) contained a significantly higher concentration of these compounds than the other cultivars (Ribeiro *et al.*, 2008). Phenolic compounds

Table 17.1. Total phenolic content in the peel of ripe mango fruit.

Mango cultivars	Total phenolics (mg GAE/g DW)	References
Haden	52.3	Larrauri *et al.* (1996)
Raspuri	46.3	Ajila *et al.* (2007)
Badami	33.3	Ajila *et al.* (2007)
Uba	57.2	Ribeiro *et al.* (2008)
Willard	84.6	Thanaraj (2010)
Karutha Colomban	74.3	Thanaraj (2010)
Malgova	31.9	Thanaraj (2010)
Anwar Rotale	48.9	Rajwana *et al.* (2010)
Faiz Kareem	29.8	Rajwana *et al.* (2010)
Chaunsa	51.2	Rajwana *et al.* (2010)

such as flavonol and xanthone glycosides, gallotannins and benzophenone derivatives (mangiferin, quercetin 3-*O*-galactoside, quercetin 3-*O*-glucoside, kaempferol-3-*O*-glucoside, quercetin-3-*O*-rhamnoside, quercetin 3-*O*-diglycoside, quercetin, quercetin 3-*O*-arabinofuranoside, quercetin-3-D-xyloside, isomangiferin and rhamnetinare) are present mainly in mango peels and seeds. Mangiferin (xanthone-*C*-glycoside) is the prominent flavonoid in Sri Lankan mango cvs. Willard (5620 µg/g DW), Karutha Colomban (3920 µg/g DW) and Malgova (2350 µg/g DW), followed by quercetin 3-*O*-galactoside and quercetin 3-*O*-glucoside (Thanaraj, 2010). Mangiferin is also the main flavonoid in mango cvs. Uba (199 µg/g DW) and Tommy Atkins (1690 µg/g DW). Phenolic compounds generally help protect humans against some chronic degenerative diseases related to oxidative stress (Manach *et al.*, 2005). Flavonols have strong antioxidant (Pannala *et al.*, 2001), anticarcinogenic (Peng *et al.*, 2006), antiatherogenic (Kim *et al.*, 2006) and antitumour and antiviral (Guha *et al.*, 1996) activities. The antioxidant activity of phenolic acids generally depends on their chemical structure, the more hydroxyl groups present in the phenol the higher the antioxidant capacity (Heo *et al.*, 2007).

Vitamins

Both the peel and pulp of mango fruit are an excellent source of AsA. However, the concentration varies extensively with the cultivar, tissue, stage of maturity, postharvest ripening and storage, climatic conditions, cultural practices and pre- and postharvest factors (Lee and Kader, 2000). Vinci *et al.* (1995) reported that the AsA concentration of mango pulp ranges from 0.1 to 1.0 mg/g FW, although, again, reported values vary with cultivar and quantification method (Table 17.2). Sri Lankan mango cv. Willard (5.8 mg/g DW) contains exceptionally high levels of AsA, while cvs. Karutha Colomban (1.1 mg/g DW) and Malgova (0.6 mg/g DW) have moderate content (Thanaraj, 2010). Mango fruit (4.42 µg/g FW in cv. Ataulfo) is also a good source of vitamins A and E (α-tocopherol) (Corral-Aguayo *et al.*, 2008).

Carotenoids

Mango is a rich source of carotenoids, which are responsible for the yellow-to-orange colour in ripe fruit, and contributes substantially to β-carotene supply in tropical countries. All-*trans*-β-carotene (> 50%) and violaxanthins and their isomers (xantophylls) are the most common carotenoids in mango cultivars (Chen *et al.*, 2004). The β-carotene provides higher provitamin A value and antioxidant capacity (Godoy *et al.*, 1994; Pott *et al.*, 2003). Ripe mango fruit generally contains a higher level of provitamin A than other tropical fruit; however, the level depends on cultivar, pulp colour and stage of ripening (West and Poortvliet, 1993;

Table 17.2.   Concentration of ascorbic acid (AsA) in different mango cultivars.

Mango cultivars	Plant tissue	AsA (mg/g FW)	References
Tommy Atkins	Pulp	0.23	Mansour *et al.* (2006)
Keitt	Pulp	0.33	Mansour *et al.* (2006)
Kent	Pulp	0.32	Mansour *et al.* (2006)
Raspuri	Peel	0.35	Ajila *et al.* (2007)
Badami	Peel	0.39	Ajila *et al.* (2007)
Haden	Pulp	0.18	Ribeiro *et al.* (2007)
Tommy Atkins	Pulp	0.1	Ribeiro *et al.* (2007)
Palmer	Pulp	0.1	Ribeiro *et al.* (2007)
Uba	Pulp	0.72	Ribeiro *et al.* (2007)
Langra	Peel	5.2	Thomas and Oke (2007)
Langra	Pulp	0.98	Thomas and Oke (2007)
Ataulfo	Pulp	1.0	Corral-Aguayo *et al.* (2008)
Tainong	Pulp	0.45	Wang *et al.* (2009)

Englberger *et al.*, 2003). Miller *et al.* (1996) reported that, among the various carotenoids present in mango fruit, lycopene has the higher antioxidant activity, followed by β-caryophyllene, β-carotene, lutein and zeaxanthin. Total carotenoid content usually varies from 9–92 µg/g FW in most of the mango cultivars; however, Indian mango cv. Alphonso (110 µg/g FW) (Padmini and Prabha, 1997) and Sri Lankan mango cvs. Willard (63 µg/g FW) and Karutha Colomban (76 µg/g FW) (Thanaraj, 2010) contain exceptionally high concentrations.

17.5.3 Chemopreventive activity and bioavailability

Percival *et al.* (2006) reported that whole mango fruit juice has the ability to inhibit cell proliferation in the leukaemic cell line HL-60 and also inhibits the neoplastic transformation of BALB/3T3 cells. The presence of antioxidant and antimutagenic (Botting *et al.*, 1999) activities in mango and its antineoplastic effects detected using mammalian *in vitro* systems support the anticancer activity seen *in vivo* (García-Solís *et al.*, 2008). Studies in humans show that accumulation of carotenoids by breast adipose tissue reduces the risk of breast cancer (Yeum *et al.*, 1998). Carotenoids are also correlated with activities against different types of cancer and heart diseases and are a precursor of provitamin A

(Yahia *et al.*, 2006). However, the bioavailability of β-carotene depends on the ripeness of the fruit, although it can be improved by consuming the fruit with other, fat-containing food products (Ornelas-Paz *et al.*, 2007). The mango triterpene (lupeol) is an effective inhibitor in laboratory models of prostate and skin cancers. Mango peel also contains the oil, urushiol, which can trigger a skin rash called urushiol-induced contact dermatitis (Nigam *et al.*, 2007; Chaturvedi *et al.*, 2008). Mango extract (Vimang) could be a useful new (natural) drug for preventing oxidative damage during hepatic injury associated with free radical generation (Sanchez *et al.*, 2003). It is also believed that mango fruit may offer better protection against radical scavenger activity than many common food additives (Saleem *et al.*, 2004).

17.5.4 Other beneficial effects

In India, mango fruit sap has been used to treat the pain of bee and scorpion stings. Many of the traditional Indian medicinal uses of mango involve eating unripe fruit. It should be noted that unripe fruit contains much toxic sap, which can cause throat irritation, indigestion, dysentery and colic. Unripe mango juice is also used as a remedy for exhaustion and heat stroke. Half-ripe fruit is usually eaten with salt and honey and is also used for the treatment of gastrointestinal, bilious and blood disorders and scurvy in many tropical countries (Nigam *et al.*, 2007; Prasad *et al.*, 2008).

17.5.5 Effect of preharvest and postharvest continuum

The concentration of antioxidants varies greatly with genotype and pre- and postharvest factors such as climatic conditions, agricultural practices, harvest maturity, ripening stage, storage and processing (Lee and Kader, 2000). The total phenolic concentration of mango fruit decreases consistently during ripening and has positive correlation with antioxidant capacity (Kim *et al.*, 2007). However, during low temperature storage, total phenolics decrease but this trend has no significant correlation with antioxidant capacity (Shivashankara *et al.*, 2004). Therefore, low temperature (20°C) ripening of mango fruit is preferred over higher temperature (30°C) ripening as far as health-promoting properties are concerned (Thanaraj, 2010). The shelf life and external quality of mango fruit are improved greatly by low O_2 and/or high CO_2 storage conditions and hot water treatments. However, the colour, aroma and nutritional composition of mango fruit are reported to be reduced by controlled atmospheres (CAs). Despite this, hot water treatment plus CA storage is considered an effective treatment combination to extend the postharvest life of mangoes without greatly affecting the nutritional profile and overall quality adversely (Kim *et al.*, 2007). Gallic acid and other hydrolysable tannins and their resultant antioxidant capacity were not affected by hot water treatment (Kim *et al.*, 2007). It has been suggested, however, that major

polyphenolic compounds in mango fruit may decrease during prolonged hot water treatment (Kim *et al.*, 2009).

Hancock and Viola (2005) and Wang *et al.* (2009) reported that AsA content of mango decreases by 50% during ripening. However, Thanaraj (2010) found that there was no significant variation in AsA content during ripening in Sri Lankan mango cvs. Willard and Malgova, but that it decreased by 50% in cv. Karutha Colomban. He has also observed that high (30°C) temperature ripening enhanced the loss of AsA during ripening compared with a lower ripening temperature (20°C). Thomas and Oke (2007) made similar observations using mango cv. Alphonso. AsA losses are minimal during the storage of fresh-cut mangoes.

17.6 General Conclusion

Tropical fruit are eaten for their unique taste and aroma attributes; however, they are also a good source of dietary antioxidants and contribute significantly to daily dietary requirements, especially in developing countries. Given their importance, it is perhaps surprising that there is a paucity of information detailing the variation in health-promoting compounds between varieties and as affected by postharvest handling and storage. This may reflect the commoditization of many of these products and the concomitant narrowing of commercially available genotypes. Tropical fruit arguably have the greatest potential to improve the health of more people across the world than any other fresh produce category, given that they are dominant in many developing countries. A refocusing of research activity is thus required.

References

Adão, A.C. and Glória, M.B.A. (2005) Bioactive amines and carbohydrate changes during ripening of 'Prata' banana (*Musa acuminata* × *M. balbisiana*). *Food Chemistry* 90, 705–711.

Adebiyi, A., Adaikan, P.G. and Prasad, R.N.V. (2002) Papaya (*Carica papaya*) consumption is unsafe in pregnancy: fact or fable? Scientific evaluation of a common belief in some parts of Asia using a rat model. *British Journal of Nutrition* 88, 199–203.

Ajila, C.M., Bhat, S.G. and Prasada Rao, U.J.S. (2007) Valuable components of raw and ripe peels from two Indian mango varieties. *Food Chemistry* 102, 1006–1011.

Alothman, M., Bhat, R. and Karim, A.A. (2009) Antioxidant capacity and phenolic content of selected tropical fruits from Malaysia extracted with different solvents. *Food Chemistry* 115, 785–788.

Amenta, R., Camardab, L., di Stefanob, V., Lentinia, F. and Venza, F. (2000) Traditional medicine as a source of new therapeutic agents against psoriasis. *Fitoterapia* 71, S13–S20.

Ames, B.M., Shigena, M.K. and Hagen, T.M. (1993) Oxidants, antioxidants and the degenerative diseases of aging. *Proceedings of the National Academy of Sciences of the United States of America* 90, 7915–7922.

Anuar, N.S., Zahari, S.S., Taib, I.A. and Rahman, M.T. (2008) Effect of green and ripe *Carica papaya* epicarp extracts on wound healing and during pregnancy. *Food and Chemical Toxicology* 46, 2384–2389.

Arora, A., Choudhary, D., Agarwal, G., Singh, V.P. (2008) Compositional variation in β-carotene content, carbohydrate and antioxidant enzymes in selected banana cultivars. *International Journal of Food Science and Technology* 43, 1913–1921.

Azarkan, M., El Moussaoui, A., Wuytswinkel, V.D., Dehon, G. and Looze, Y. (2003) Fractionation and purification of the enzymes stored in the latex of *Carica papaya*. *Journal of Chromatography B* 790, 229–238.

Banerjee-Bhattacharya, J. (2002) Tissue culture and transformation studies in Indian cultivars of papaya (*Carica papaya* L.). Ph.D Thesis. University of Pune, India.

Bartholomew, D.P., Malézieux, E., Sanewski, G.M. and Sinclair, E. (2003) Inflorescence and fruit development and yield. In: Bartholomew, D.P., Paull, R.E. and Rohrbach, K.G. (eds) *The Pineapple: Botany, Production and Uses*. CAB International, Wallingford, UK, pp. 167–202.

Bartolomé, A.P., Rupérez, P. and Fúster, C. (1996) Non-volatile organic acids, pH and titratable acidity changes in pineapple fruit slices during frozen storage. *Journal of the Science of Food and Agriculture* 70, 475–480.

Bashir, H.A., Abu-Bakr, A. and Abu-Goukh, A. (2003) Compositional changes during guava fruit ripening. *Food Chemistry* 80, 557–563.

Beattie, J.K. and Quoc, T.N. (2000) Manganese in pineapple. *Food Chemistry* 68, 37–39.

Beil, W., Birkholz, C. and Sewing, K.F. (1995) Effects of flavonoids on parietal cell acid secretion, gastric mucosal prostaglandin production and *Helicobacter pylori* growth. *Drug Research* 45, 697–700.

Bennett, R.N., Shiga, T.M., Aymoto Hassimotto, N.M., Rosa, E.A.S., Lajolo, F.M. and Cordenunsi, B.R. (2010) Phenolics and antioxidant properties of fruit pulp and cell wall fractions of postharvest banana (*Musa acuminata* Juss.) cultivars. *Journal of Agricultural and Food Chemistry* 58, 7991–8003.

Berardini, N., Carle, R. and Schieber, A. (2004) Characterization of gallotannins and benzophenone derivatives from mango (*Mangifera indica* L. cv. Tommy Atkins) peels, pulp and kernels by high-performance liquid chromatography/electrospray ionization mass spectrometry. *Rapid Communications in Mass Spectrometry* 18, 2208–2216.

Berardini, N., Knödler, M., Schieber, A. and Carle, R. (2005) Utilization of mango peels as a source of pectin and polyphenolics. *Innovative Food Science and Emerging Technologies* 6 442–452.

Botting, K.J., Young, M.M., Pearson, A.E., Harris, P.J. and Ferguson, L.R. (1999) Antimutagens in food plants eaten by Polynesians: micronutrients, phytochemicals and protection against bacterial mutagenicity of the heterocyclic amine 2-amino-3-methylimidazo(4,5-f)quinoline. *Food and Chemical Toxicology* 37, 95–103.

Brömme, D., Nallaseth, F.S. and Turk, B. (2004) Production and activation of recombinant papain-like cysteine proteases. *Methods* 32, 199–206.

Cano, M.P., de Ancos, B., Matallana, M.C., Cámara, M., Reglero, G. and Tabera, J. (1997) Differences among Spanish and Latin-American banana cultivars: morphological, chemical and sensory characteristics. *Food Chemistry* 59, 411–419.

Chandrika, U.G., Jansz, E.R., Wickramasinghe, N.S.M.D. and Warnasuriya, N.D. (2003) Carotenoids in yellow- and red-fleshed papaya (*Carica papaya* L). *Journal of the Science of Food and Agriculture* 83, 1279–1282.

Charoensiri, R., Kongkachuichai, R., Suknicom, S. and Sungpuag, P. (2009) Beta-carotene, lycopene, and alpha-tocopherol contents of selected Thai fruits. *Food Chemistry* 113, 202–207.

Chaturvedi, P.K., Bhui, K. and Shukla, Y. (2008) Lupeol: connotations for chemoprevention. *Cancer Letters* 8, 1–13.

Chen, J.P., Tai, C.Y. and Chen, B.H. (2004) Improved liquid chromatographic method for determination of carotenoids in Taiwanese mango (*Mangifera indica* L.). *Journal of Chromatography A* 1054, 261–268.

Cherian, T. (2000) Effect of papaya latex extract on gravid and non-gravid rat uterine preparations *in-vitro*. *Journal of Ethnopharmacology* 70, 205–212.

Conn, P.F., Schalch, W. and Truscott, T.G. (1991) The singlet oxygen and carotenoid interaction. *Journal of Photochemistry and Photobiology B: Biology* 11, 41–47.

Cordenunsi, B., Saura-Calixto, F., Diaz-Rubio, M.E., Zuleta, A., Tine, M.A., Buckeridge, M.S., de Silva, G.B., Carpio, C., Giuntini, E.B., de Menezes, E.W. and Lajolo, F. (2010) Carbohydrate composition of ripe pineapple (cv. Perola) and the glycemic response in humans. *Ciência e Tecnologia de Alimentos (Food Science and Technology)*, Campinas 30(1), 282–288.

Corral-Aguayo, R.D., Yahia, E.M., Carrillo-Lopez, A. and Gonzalez-Aguilar, G. (2008) Correlation between some nutritional components and the total antioxidant capacity measured with six different assays in eight horticultural crops. *Journal of Agricultural and Food Chemistry* 56, 10,498–10,504.

Coutts, R.T., Baker, G.B. and Pasutto, F.M. (1986) Foodstuffs as sources of psychoactive amines and their precursors: content, significance and identification. *Advances in Drug Research* 15, 69–232.

de Oliveira, E.J., dos Santos Silva, A., de Carvalho, F.M., dos Santos, L.F., Costa, J.L., de Oliveira Amorim, V.B., *et al.* (2010) Polymorphic microsatellite marker set for *Carica papaya* L. and its use in molecular-assisted selection. *Euphytica* 173, 279–287.

Devakate, R.V., Patil, V.V., Waje, S.S. and Thorat, B.N. (2009) Purification and drying of bromelain. *Separation and Purification Technology* 64, 259–264.

Dewick, P.M. (2002) Progestogen. In: *Medicinal Natural Products*, 2nd edn. John Wiley, New York, pp. 175–176.

Dull, G.G. (1971) The pineapple: general. In: Hulme, A.C. (ed.) *The Biochemistry of Fruits and their Products*. Academic Press, London, pp. 303–324.

Dunjic, B.S., Svensson, I., Axelson, J., Adlercreutz, P., Ar'rajab, A., Larsson, K. and Bengmark, S. (1993) Green banana protection of gastric mucosa against aspirin induced injuries in rats. A multicomponent Mechanism? *Scandinavian Journal of Gastroenterology* 28, 894–898.

Emaga, T.H., Andrianaivo, R.H., Wathelet, B., Tchango, J.T. and Paquot, M. (2007) Effects of the stage of maturation and varieties on the chemical composition of banana and plantain peels. *Food Chemistry* 103, 590–600.

Emeruwa, A. (1982) Antibacterial substance from *Carica papaya* fruit extract. *Journal of Natural Products* 45, 132–137.

Englberger, L., Darnton-Hill, I., Coyne, T., Fitzgerald, M.H. and Marks, G.C. (2003) Carotenoid-rich bananas: a potential food source for alleviating vitamin A deficiency. *Food and Nutrition Bulletin* 24, 303–318.

Erdman, J.W. Jr, Bierer, T.L. and Gugger, E.T. (1993) Absorption and transport of carotenoids. *Annals of the New York Academy of Sciences* 691, 76–85.

FAO (2003) FAOSTAT statistics database. Food and Agriculture Organization of the United Nations, Rome, Italy (http://faostat.fao.org, accessed 10 April 2009).

FAO (2005) FAOSTAT statistics database. Food and Agriculture Organization of the United Nations. Rome, Italy (http://faostat.fao.org, accessed 11 December 2008).

FAO (2007) FAOSTAT statistics database. Food and Agriculture Organization of the United Nations, Rome, Italy (http://faostat.fao.org, accessed 10 April 2009).

Fasihuddin, A. and Ghazally, I. (2003) *Medicinal Plants Used by Kadazan Dusun Communities around Crocker Range*. ASEAN Review of Biodiversity and Environmental Conservation (ARBEC), University of Malaysia Sarawak, Malaysia.

Fernandes, F.A., Rodrigues, S., Gaspareto, O.C. and Oliveira, E.L. (2006) Optimization of osmotic dehydration of papaya followed by air-drying. *Food Research International* 39, 492–498.

Firmin, A. (1997) Physicochemical changes in papaya during storage. *Tropical Science* 37, 49–51.

Fold, N. and Gough, K.V. (2008) From smallholders to transnationals: the impact of changing consumer preferences in the EU on Ghana's pineapple sector. *Geoforum* 39, 1687–1697.

Forster, M.P., Rodriguez, E., Martin, J.D. and Romero, C.D. (2002) Statistical differentiation of bananas according to their mineral composition. *Journal of Agricultural and Food Chemistry* 50, 6130–6135.

Franke, A.A., Custer, L.J., Arakaki, C. and Murphy, S.P. (2004) Vitamin C and flavonoid levels of fruits and vegetables consumed in Hawaii. *Journal of Food Composition and Analysis* 17, 1–35.

Galán Saúco, V. (1996) Current situation, trends, and future of agronomic research on tropical fruits. In: *Proceedings of the International Conference on Tropical Fruits* Malaysian Agricultural Research and Development Institute (MARDI), Kuala Lumpur, Malaysia, 23–26 July 1996.

García-Solís, P., Yahia, E.M. and Aceves, C. (2008) Study of the effect of 'Ataulfo' mango (*Mangifera indica* L.) intake on mammary carcinogenesis and antioxidant capacity in plasma of *N*-methyl-*N*-nitrosourea (MNU)-treated rats. *Food Chemistry* 111, 309–315.

Godoy, H.T. and Rodriguez-Amaya, D.B. (1989) Carotenoid composition of commercial mangoes from Brazil. *LWT – Food Science and Technology* 22, 100–103.

Godoy, H.T. and Rodriguez-Amaya, D.B. (1994) Occurrence of *cis*-isomers of provitamin A in Brazilian fruits. *Journal of Agricultural and Food Chemistry* 42, 1306–1313.

Goel, R.K., Tavares, I.A. and Bennett, A. (1989) Stimulation of gastric and colonic mucosal eicosonoid synthesis. *Journal of Pharmacy and Pharmacology* 41, 747–750.

Guha, S., Ghosal, S. and Chattopadhay, U. (1996) Antitumor, immunomodulatory and anti-HIV effect of mangiferin, a naturally occurring glucosylxanthone. *Chemotherapy* 42, 443–451.

Halliwell, B. (1996) Vitamin C: antioxidant or pro-oxidant *in vivo. Free Radical Research* 25, 439–454.

Hancock R.D. and Viola, R. (2005) Improving the nutritional value of crops through enhancement of L-ascorbic acid (vitamin C) content: rationale and biochemical opportunities. *Journal of Agricultural and Food Chemistry* 53, 5248–5257.

Hardisson, A., Rubio, C., Baez, A., Martin, M., Alvarez, R. and Mineral, E.D. (2001) Composition of the banana (*Musa acuminata*) from the island of Tenerife. *Food Chemistry* 73, 153–161.

Havsteen, B. (1983) Flavonoids, a class of natural compounds of high pharmacological potency. *Biochemical Pharmacology* 32, 1141–1148.

Heo, H.J., Kim, Y.J., Chung, D. and Kim, D.O. (2007) Antioxidant capacities of individual and combined phenolics in a model system, *Food Chemistry* 104, 87–92.

Hewitt, H., Whittle, S., Lopez, E.B. and Weaver, S. (2000) Topical use of papaya in chronic skin ulcer therapy in Jamaica. *West Indian Medical Journal* 49, 32–33.

Hodgson, A.S. and Hodgson, L.R. (1993) Pineapple juice. In: Nagy, S., Chen, C.S. and Shaw, P.E. (eds) *Fruit Juice Processing Technology*. Agscience, Inc, Auburndale, Florida, pp. 378–435.

Holden, J.M., Eldrige, A.L., Beecher, G.R., Buzzard, M., Bhagwat, S., Carol, S., *et al.* (1999) Carotenoid content of US foods: an update of database. *Journal of Food Composition Analysis* 12, 169–196.

Hornick, C.A., Sanders, L.I. and Lin, Y.C. (1978) Effect of carpaine, a papaya alkaloid, on the circulatory function in the rat. *Research Communications in Chemical Pathology and Pharmacology* 22, 277–289.

Jimenez-Escrig, A., Rincon, M., Pulido, R. and Saura-Calixto, F. (2001) Guava fruit (*Psidium quajava* L.) as a new source of antioxidant dietary fibre. *Journal of Agricultural and Food Chemistry* 51, 5489–5493.

Kahlon, T.S. and Smith, G.E. (2007) *In vitro* binding of bile acids by bananas, peaches, pineapple, grapes, pears, apricots and nectarines. *Food Chemistry* 101, 1046–1051.

Kanazawa, K. and Sakakibara, H. (2000) High content of dopamine, a strong antioxidant, in Cavendish banana. *Journal of Agricultural and Food Chemistry* 48, 844–848.

Kauer, C. and Kapoor, H.C. (2001) Antioxidants in fruits and vegetables – the millennium's health. *International Journal of Food Science and Technology* 36, 703–725.

Keijer, J., Bunschoten, A., Palou, A. and Franssen-van Hal, N.L.W. (2005) Beta-carotene and the application of transcriptomics in risk–benefit evaluation of natural dietary components. *Biochemica et Biophysica Acta* 1740, 139–146.

Kelly, D.E.S. and Bagshaw, J. (1993) Effect of fruit handling and fruit coatings on Blackheart (internal brown spot) and other aspects of fresh pineapple quality. *Acta Horticulturae* 334, 305–316.

Kim, J.D., Liu, L., Guo, W. and Meydani, M. (2006) Chemical structure of flavonols in relation to modulation of angiogenesis and immune-endothelial cell adhesion. *The Journal of Nutritional Biochemistry* 17, 165–176.

Kim, Y., Brecht, J.K. and Talcott, S.T. (2007) Antioxidant phytochemical and fruit quality changes in mango (*Mangifera indica* L.) following hot water immersion and controlled atmosphere storage. *Food Chemistry* 105, 1327–1334.

Kim, Y., Lounds-Singleton, A.J. and Talcott, S.T. (2009) Antioxidant phytochemical and quality changes associated with hot water immersion treatment of mangoes (*Mangifera indica* L.). *Food Chemistry* 105, 1327–1334.

Kondo, S., Kittikorn, M. and Kanlayanarat, S. (2005) Preharvest antioxidant activities of tropical fruit and the effect of low temperature storage on antioxidants and jasmonates. *Postharvest Biology and Technology* 36, 309–318.

Lako, J., Trenerry, V.C., Wahlqvist, M., Wattanapenpaiboon, N., Sotheeswaran, S. and Premier, R. (2007) Phytochemical flavonols, carotenoids and the antioxidant properties of a wide selection of Fijian fruit, vegetables and other readily available foods. *Food Chemistry* 101, 1727–1741.

Larrauri, J.A., Ruperez, P., Borroto, B. and Saura-calixto, F. (1996) Mango peel as a new tropical fibre: preparation and characterization. *LWT – Food Science and Technology* 29, 729–733.

Lee, S.K. and Kader, A.A. (2000) Pre-harvest and postharvest factors influencing vitamin C content of horticultural crops. *Postharvest Biology and Technology* 20, 207–220.

Leong, L.P. and Shui, G. (2002) An investigation of antioxidant capacity of fruits in Singapore markets. *Food Chemistry* 76, 69–75.

Lewis, D.A., Fields, W.N. and Shaw, G.P. (1999) A natural flavonoid present in unripe plantain banana pulp (*Musa sapientum* L. var. *paradisiaca*) protects the gastric mucosa from aspirin-induced erosions. *Journal of Ethnopharmacology* 65, 283–288.

Lim, Y.Y., Lim, T.T. and Tee, J.J. (2007) Antioxidant properties of several tropical fruits: a comparative study. *Food Chemistry* 103, 1003–1008.

Mahattanatawee, K., Manthey, J.A., Luzio, G., Talcott, S.T., Goodner, K. and Baldwin, E.A. (2006) Total antioxidant activity and fibre content of select Florida-grown tropical fruits. *Journal of Agricultural and Food Chemistry* 54, 7355–7363.

Manach, C., Williamson, G., Morand, C., Scalbert, A. and Rémésy, C. (2005) Bioavailability and bioefficacy of polyphenols in humans. I. Review of 97 bioavailability studies. *American Journal of Clinical Nutrition* 81, 230S–242S.

Mansour, F.S., Abd-El-Aziz, S.A. and Helal, G.A. (2006) Effect of fruit heat treatment in three mango varieties on incidence of postharvest fungal disease. *Journal of Plant Pathology* 88, 141–148.

Marriott, J. and Palmer, J.K. (1980) *Bananas – Physiology and Biochemistry of Storage and Ripening for Optimum Quality.* CRC Critical Reviews in Food Science and Nutrition 13(1). CRC Press, Boca Raton, Florida, pp. 41–88.

Martin, F.W., Campbell, C.W. and Ruberté, R.M. (1987) *Perennial Edible Fruits of the Tropics: an Inventory.* Washington, D.C., U.S. Department of Agriculture, Agricultural Research Service.

Miller, N.J., Sampson, J., Candeias, L.P., Bramley, P.M. and Rice-Evans, C.A. (1996) Antioxidant activities of carotenes and xanthophylls. *FEBS Letters* 384, 240–242.

Morse, J. (2008) *Health Benefits of Pineapple.* (http://www.associatedcontent.com/article, accessed on 12 November 2008).

Morton, J.F. (1987a) Papaya. In: Morton, J.F. (ed.) *Fruits of Warm Climates.* Media Incorporate, Florida, pp. 336–346.

Morton, J.F. (1987b) Mango. In: Morton, J.F. (ed.) *Fruits of Warm Climates.* Media Incorporate, Florida, pp. 221–239.

Nakamura, Y., Yoshimoto, M., Murata, Y., Shimoishi, Y., Asai, Y., Park, E.Y. and Sato, K. (2007) Papaya seed represents a rich source of biologically active isothiocyanate. *Journal of Agricultural and Food Chemistry* 55, 4407–4413.

Nakasone, H.Y. and Paull, R.E. (1998) Pineapple. In: *Tropical Fruits.* CAB International, Wallingford, UK, pp. 292–327.

Naved, T., Siddiqui, J.I., Ansari, S.H., Ansari, A.A. and Mukhtar, H.M. (2005) Immunomodulatory activity of *Mangifera indica* L. *Journal of Natural Remedies* 5, 137–140.

Nguyen, T.B.T., Ketsa, S. and van Doorn, W.G. (2003) Relationship between browning and the activities of polyphenoloxidase and phenylalanine ammonia lyase in banana peel during low temperature storage. *Postharvest Biology and Technology* 30, 187–193.

Nigam, N., Prasad, S. and Shukla, Y. (2007) Preventive effects of lupeol on DMBA induced DNA alkylation damage in mouse skin. *Food Chemistry and Toxicology* 45, 2331–2335.

Ohara, B.P., Hemming, A.M., Buttle, D.J. and Pearl, L.H. (1995) Crystal structure of glycil endopeptidase from *Carica papaya*: a cysteine endopeptidase of unusual substrate specificity. *Journal of the American Chemical Society* 34, 13190–13195.

Orlando, L. (2006) The amazing pineapple, Hawaii's natural health booster (http://www.buzzle.com/editorials/4-2-2006-92514.asp, accessed 20 November 2008).

Ornelas-Paz, J.D.J., Yahia, E.M. and Gardea-Bejar, A. (2007) Identification and quantification of xanthophyll esters, carotenes and tocopherols in fruit of seven Mexican mango cultivars by liquid chromatography-atmospheric pressure chemical ionization-time-of-flight mass spectrometry [LC-(APcI(+))-MS]. *Journal of Agricultural and Food Chemistry* 55, 6628–6635.

Osato, J.A., Santiago, L.A., Remo, G.M., Cuadra, M.S. and Mori, A. (1993) Antimicrobial and antioxidant activities of unripe papaya. *Life Science* 53, 1383–1389.

Padmini, S. and Prabha, T.N. (1997) Biochemical changes during acetylene-induced ripening in mangoes (var. Alphonso). *Tropical Agriculture* 74, 265–271.

Pannala, A.S., Chan, T.S., O'Brien, P.J. and Rice-Evans, C.A. (2001) Flavonoid B-ring chemistry and antioxidant activity: fast reaction kinetics. *Biochemical and Biophysical Research Communications* 282, 1161–1168.

Patthamakanokporn, O., Puwastien, P., Nitithamyong, A. and Sirichakwal, P.P. (2008) Changes of antioxidant activity and total phenolic compounds during storage of selected fruits. *Journal of Food Composition and Analysis* 21, 241–248.

Paull, R.E. (1993) Postharvest handling of Smooth Cayenne pineapple in Hawaii for the fresh fruit market. *Acta Horticulturae* 334, 273–285.

Paull, R.E., Nishijima, W., Marcelino, R. and Cavaletto, C. (1997) Postharvest handling and losses during marketing of papaya (*Carica papaya* L.). *Postharvest Biology and Technology* 11, 165–179.

Peng, G., Dixon, D.J., Muga, S.J., Smith, T.J. and Wargovich, M.J. (2006) Green tea polyphenol (–)-epigallocatechin-3-gallate inhibits cyclooxygenase-2-expression in colon carcinogenesis. *Molecular Carcinogenesis* 45, 309–319.

Percival, S.S., Talcott, S.T., Chin, S.T., Mallak, A.C., Lounds-Singleton, A. and Pettit-Moore, J. (2006) Neoplastic transformation of BALB/3T3 cells and cell cycle of HL-60 cells are inhibited by mango (*Mangifera indica* L.) juice and mango juice extracts. *Journal of Nutrition* 136, 1300–1304.

Pott, I., Breithaupt, D.E. and Carle, R. (2003) Detection of unusual carotenoid esters in fresh mango (*Mangifera indica* L. cv. 'Kent'). Institute for Agricultural Engineering in the Tropics and Subtropics, Hohenheim Uv. Stuttgart, Germany. *Phytochemistry* 64, 825–829.

Prasad, S., Kalra, N., Singh, M. and Shukla, Y. (2008) Protective effects of lupeol and mango extract against androgen induced oxidative stress in Swiss albino mice. *Asian Journal of Andrology* 10, 313–318.

Prior, R.L. and Cao, G. (2000) Antioxidant phytochemicals in fruits and vegetables; diet and health implications. *HortScience* 35, 588–592.

Proulx, E. (2002) Étude de la qualité de la papaya (*Carica papaya* L.) et des haricots verts (*Phaseoulus vulgaris* L.) en fonction de la température d'entreposage. MSc thesis, Université Laval, Québec, Canada.

Proulx, E., Cecilia, M., Nunes, N., Emond, J.P. and Brecht, J.K. (2005) Quality attributes limiting papaya postharvest life at chilling and non-chilling temperatures. *Proceedings of the Florida State Horticultural Society* 118, 389–395.

Rajwana, I.A., Thanaraj, T., Malik, A.U. and Terry, L.A. (2010) Changes in taste-related compounds, total phenolics and antioxidant capacity in peel and pulp of Pakistani mango (*Mangifera indica* L.) cultivars during high temperature storage. *Journal of Horticultural Science and Biotechnology* (In press).

Ramsaroop, R.E.S. and Saulo, A.A. (2007) Comparative consumer and physicochemical analysis of Del Monte Hawaii Gold and Smooth Cayenne pineapple cultivars. *Journal of Food Quality* 30, 135–159.

Rebolledo-Martínez, L., Uriza, D.E., Rebolledo-Martínez, A. and Zágada, G. (2005) Slip production of MD-2 hybrid pineapple by three methods: gaullin, leaf pruning and a growth regulator. *Acta Horticulturae* 666, 277–285.

Ribeiro, S.M.R., Queiroz, J.H., Queiroz, M.E.R.L., Campos, F.M. and Pinheiro-Sant'Ana, H.M. (2007) Antioxidants in mango (*Mangifera indica* L.) pulp. *Plant Foods for Human Nutrition* 62, 13–17.

Ribeiro, S.M.R., Barbosa, L.C.A., Queiroz, J.H., Knödler, M. and Schieber, A. (2008) Phenolic compounds and antioxidant capacity of Brazilian mango (*Mangifera indica* L.) varieties. *Food Chemistry* 110, 620–626.

Rivera-Pastrana, D.M., Yahia, E.M. and González-Aguilar, G.A. (2010) Phenolic and carotenoid profiles of papaya fruit (*Carica papaya* L.) and their contents under low temperature storage. *Journal of the Science of Food and Agriculture* 90(14), 2358–2365.

Saleem, M., Afaq, F., Adhami, V.M. and Mukhtar, H. (2004) Lupeol modulates NF-kappaB and PI3K/Akt pathways and inhibits skin cancer in CD-1 mice. *Oncogene* 1(23), 5203–5214.

Sanchez, G.M., Rodríguez, H.M.A., Giuliani, A., Núñez Sellés, A.J., Rodríguez, N.P., León-Fernández, O.S. and Re, L. (2003) Protective effect of *Mangifera indica* L. extract (Vimang) on the injury associated with hepatic ischaemia reperfusion. *Pharmacological Research* 17, 197–201.

Satrija, F., Nansen, P., Murtini, S. and He, S. (1995) Anthelmintic activity of papaya latex against patent *Heligmosomoides polygyrus* infections in mice. *Journal of Ethnopharmacology* 48, 161–164.

Schieber, A., Berardini, N. and Carle, R. (2003) Identification of flavonol and xanthone glycosides from mango (*Mangifera indica* L. cv. 'Tommy Atkins') peels by high-performance chromatography–electrospray ionization mass spectrometry. *Journal of Agricultural and Food Chemistry* 51, 5006–5011.

Selvarajah, S. and Herath, H.M.W. (1997) Effect of edible coating on some quality and physico-chemical parameters of pineapple. *Tropical Agricultural Research* 9, 77–89.

Selvarajah, S., Bauchot, A.D. and John, P. (2001) Internal browning on cold-store pineapples is suppressed by a postharvest application of 1-methycyclopropene. *Postharvest Biology and Technology* 23, 167–170.

Setiawan, B., Sulaeman, A., Giraud, D.W. and Driskell, J.A. (2001) Carotenoid content of selected Indonesian fruits. *Journal of Food Composition Analysis* 14, 169–196.

Shivashankara, K.S., Isobe, S., Al-Haq, M.I., Takenaka, M. and Shina, T. (2004) Fruit antioxidant activity, ascorbic acid, total phenol, quercitin, and carotene of Irwin mango fruits stored at low-temperature after high electric field treatment. *Journal of Agricultural of Food Chemistry* 52, 1281–1286.

Siewinsky, M., Gutowicz, J., Zarzycki, A. and Mikvlewicz, W. (1996) Role of cysteine endopeptidases in cancerogenesis. *Cancer Biotherapy and Radiopharmaceuticals* 11, 169–176.

Soares, A.G., Trugob, L.C., Neide Botrela, N. and Souzac, L.F.D.S. (2005) Reduction of internal browning of pineapple fruit (*Ananas comusus* L.) by preharvest soil application of potassium. *Postharvest Biology and Technology* 35, 201–207.

Someya, S., Yoshiki, Y. and Okubo, K. (2002) Antioxidant compounds from bananas (M*usa* Cavendish). *Food Chemistry* 79, 351–354.

Stepek, G., Lowe, A.E., Buttle, D.J., Duce, I.R. and Behnke, J.M. (2007) The anthelmintic efficacy of plant-derived cysteine proteinases against the rodent gastrointestinal nematode, *Heligmosomoides polygyrus*, in vivo. *Parasitology* 134, 1409–1419.

Tahvonen, R. (1993) Contents of selected elements in some fruits, berries and vegetables on the Finnish market in 1987–1989. *Journal of Food Composition Analysis* 6, 75–86.

Talcott, S.T., Pettit-Moore, J., Lounds-Singleton, A. and Percival, S.S. (2005) Ripening associated phyto-chemical changes in mangos (*Mangifera indica*) following thermal quarantine and low temperature storage. *Journal Food Science* 70, 337–341.

Tee, E.S. and Lim, C.L. (1991) Carotenoid composition and content of Malaysian vegetables and fruits by the AOAC and HPLC methods. *Food Chemistry* 41, 309–339.

Thanaraj, T. (2010) Understanding the postharvest changes of Sri Lankan mango cultivars. PhD thesis. Cranfield University, UK.

Thomas, T. and Oke, M.S. (2007) Technical note: vitamin C content and distribution in mangoes during ripening. *International Journal of Food Science and Technology* 15, 669–672.

Tiwari, K.C., Majumder, R. and Bhattacharjee, S. (1982) Folklore information from Assam for family planning and birth control. *International Journal of Crude Drug Research* 20, 133–137.

Uriza-Avila, D. (2005) Foreword and Preface. IV International Pineapple Symposium. *Acta Horticulturae*, 666.

USDA (1992) Composition of foods: fruits and fruit juices – raw, processed, prepared. In: *Agriculture Handbook No 8–9*. US Government, Printing Office, Washington, DC, p. 225.

USDA-ARS (2004) USDA National Nutrient Database for Standard Reference, Release 17. Nutrient Data Laboratory. US Department of Agriculture, Agricultural Research Service (http://www.nal.usda.gov/fnic/foodcompS, accessed 8 January 2009).

USDA-HHS (2004) Dietary Guidelines Advisory Committee Report. US Department of Agriculture, Health and Human Services (www.health.gov/dietaryguidelines/dga2005/reportS, accessed 8 January 2009).

van Breemen, R.B. and Pajkovic, N. (2008) Multitargeted therapy of cancer by lycopene. *Cancer Letters* 269, 339–351.

Vetorazzi, G. (1974) 5-Hydroxytryptamine content of bananas and banana products. *Food and Cosmetics Toxicology* 12, 107–113.

Vijayakumar, S., Presannakumar, G. and Vijayakumar, N.R. (2008) Antioxidant activity of banana flavonoids. *Fitoterapia* 79, 279-282.

Vinci, G., Botré, F., Mele, G. and Ruggieri, G. (1995) Ascorbic acid in exotic fruits: a liquid chromatographic investigation. *Food Chemistry* 53, 211–214.

Wall, M.M. (2006) Ascorbic acid, vitamin A, and mineral composition of banana (*Musa* sp.) and papaya (*Carica papaya*) cultivars grown in Hawaii. *Journal of Food Composition and Analysis* 19, 434–445.

Wang, B., Wang, J., Feng, X., Lin, L., Zhao, Y. and Jiang, W. (2009) Effects of 1-MCP and exogenous ethylene on fruit ripening and antioxidants in stored mango. *Plant Growth Regulators* 57, 185–192.

Wang, R.H., Hsu, Y.M., Bartholomew, D.P., Maruthasalam, S. and Lin, C.H. (2007) Delaying natural flowering in pineapple through foliar application of aviglycine, an inhibitor of ethylene biosynthesis. *HortScience* 42, 1188–1191.

Wenkam, N.S. (1990) *Food of Hawaii and the Pacific Basin, Fruits and Fruit Products: Raw, Processed, and Prepared. Vol. 4: Composition.* Hawaii Agricultural Experiment Station Research and Extension Series 110, p. 96.

West, C.E. and Poortvliet, E.J. (1993) *The Carotenoid Content of Foods with Special Reference to Developing Countries.* International Science and Technology Institute, Arlington, Virginia.

Wiart, C. (2006) *Family Caricaceae. Medicinal Plants of the Asia-Pacific: Drugs for the Future?* World Scientific Publishing, Singapore, pp. 183–186.

Xianli, W., Beecher, G.R., Holden, J.M., Haytowitz, D.B., Gebhardt, S.E. and Prior, R.L. (2004) Lipophilic and hydrophilic antioxidant capacities of common foods in the United States. *Journal of Agricultural and Food Chemistry* 52, 4026–4037.

Yahia, E.M., Ornelas-Paz, J.J. and Gardea, A. (2006) Extraction, separation and partial identification of 'Ataulfo' mango fruit carotenoids. *Acta Horticulturae* 712, 333–338.

Yeum, K.J., Ahn, S.H., Rupp de Pavia, S.A., Lee-Kim, Y.C., Krinsky, N.I. and Russell, R.M. (1998) Correlation between carotenoids concentrations in serum and normal breast adipose tissue of women with benign breast tumor or breast cancer. *Journal of Nutrition* 128, 920–1926.

Zhang, P., Whistler, R.L., BeMiller, J.N. and Hamaker, B.R. (2005) Banana starch: production, physico-chemical properties, and digestibility – a review. *Carbohydrate Polymers* 59, 443–458.

18 Methodologies for Extraction, Isolation, Characterization and Quantification of Bioactive Compounds

Katherine Cools, Ariel Vicente and Leon A. Terry

18.1 Introduction

Since it is widely recognized that the beneficial influence of fruit on human health is linked with the presence of specific phytochemicals, determining the nature of these compounds in different commodities, and the influence of preharvest, postharvest and processing treatments, have been major areas of research (Espín *et al.*, 2007; Frankel, 2007). Antioxidants can be measured as individual compounds or as a total capacity, yet some methods used to quantify total antioxidant activity are thought to take into account compounds not classed as antioxidants, for instance, reducing sugars (Huang *et al.*, 2005). This chapter summarizes the wide variety of extraction and characterization techniques used to quantify individual or total antioxidants, as well as other bioactive compounds.

18.2 Sample Preparation

Fruit and vegetables contain a complex profile of bioactive compounds, of which most are labile to heat, air and/or light (Brat, 2008). To extract bioactive compounds effectively from fresh tissue, samples are usually cut into small pieces and then snap-frozen immediately in liquid nitrogen, since chopped material is much more unstable. Failure to prepare samples in this way can result in enzymatic browning, as well as undesirable molecular, biochemical and physiological changes. Samples are often freeze-dried before extraction takes place, to remove water. Lyophilization aids sample grinding, which benefits extraction by increasing solvent penetration due to greater sample surface area. Puupponen-Pimiä *et al.* (2003) investigated the effects of freezer storage on fresh vegetables. After being frozen quickly to –40°C (method not stated), frozen vegetables were then stored at –20°C for 6, 12 and 18 months. The authors found that different bioactives were affected by long-term freezer storage in different vegetable groups. A decrease in α- and β-carotene was observed in processed carrots, yet not in peas, between 6 and 12 months at –20°C. Similarly, over twofold reductions in quercetin and kaempferol content were observed in cauliflower and broccoli over the same period. It is therefore important, especially when comparing different fruit and vegetable groups, that analyses are undertaken as soon as possible after sample preparation and, ideally, that samples are kept at or below –40°C. Differences among the literature may indeed be due to sample preparation and subsequent storage.

18.3 Bioactive Extraction

A factor that certainly differs among the literature is extraction procedure. An example of this is the extraction of fructans, which are thought to have prebiotic effects whereby they promote the beneficial bacteria of the gut; their health-benefiting properties are further described in Chapter 2 of this volume. Studies have shown that methanol is a superior extraction solvent to ethanol when extracting fructans from fresh produce (Davis *et al.*, 2007; Downes and Terry, 2010). Downes and Terry (2010) found that extraction of fructans from onion using 62.5% (v/v) methanol (O'Donoghue *et al.*, 2004) yielded approximately 40 mg/g DW more fructans than when extracted using 80% (v/v) ethanol (Vågen and Slimestad, 2008) (Fig. 18.1). Similarly, anthocyanins are another group of health-benefiting compounds that have been extracted using various solvent mixtures. Initial work by Giné Bordonaba and Terry (2008) identified acidified aqueous methanol as a superior extraction solvent to acidified aqueous ethanol when extracting anthocyanins from blackcurrant berries. However, acetone has also been used for the extraction of anthocyanins (Awika *et al.*, 2004; Anttonnen and Karjalainen, 2006); therefore, recent work (Giné Bordonaba and Terry, in press) has compared the methanol-based method with extraction using acidified 70% (v/v) acetone. Results from this work revealed that acidified aqueous methanol extracted higher concentrations of anthocyanins, specifically pelargonadin derivatives, from strawberry.

18.4 Bioactive Isolation, Separation and Quantification

Various methods are available to measure ascorbic acid (AsA), carotenoids and phenolic compounds (Tsao and Deng, 2004) individually

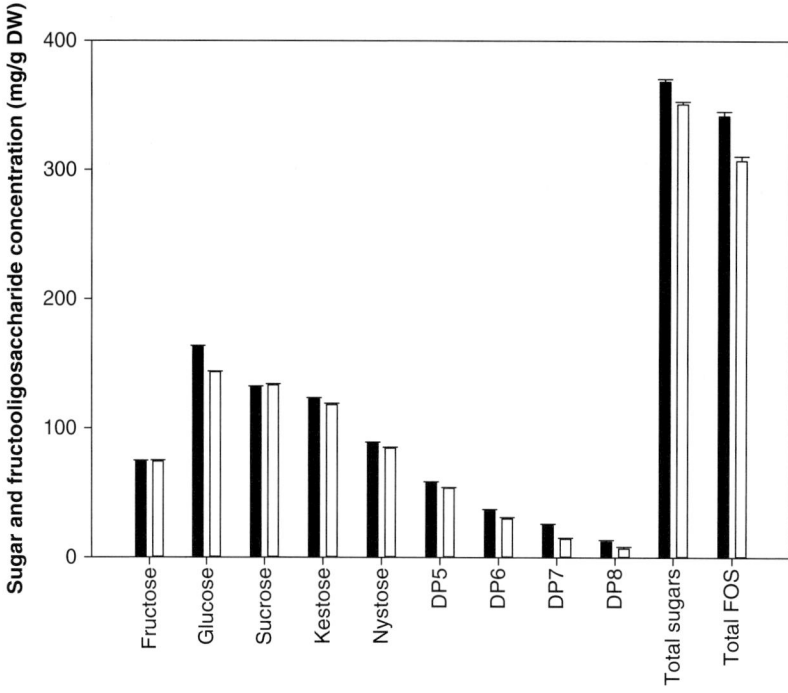

Fig. 18.1.   Mean sugar and fructooligosaccharide concentrations (mg/g DW) in onion samples extracted with either methanol (MeOH ▬) or ethanol (EtOH ▭). Total sugars = sum of fructose, glucose and sucrose; total FOS = sum of DP3 – DP8 (Downes and Terry, 2010).

in foods. High performance liquid chromato-graphic (HPLC) techniques are now most widely used for both separation and quantification. For phenolic acids and flavonoid analysis, the chromatographic conditions of the HPLC methods may include the use of reverse phase C18 columns, a UV–visible diode array detector (DAD) and a binary solvent system containing acidified water and a polar organic solvent such as methanol or acetonitrile (for review, see Robards, 2003; Naczk and Shahidi, 2006). Separation of saponified carotenoids can be carried out on silica columns using gradient elution from 95% of light petroleum to 95% acetone (Almela et al., 1990). For AsA, samples must be prepared carefully to prevent oxidation. This usually involves extraction in citric or metaphosphoric acid at low temperature and in the dark, followed by rapid filtration and/or centrifugation. After that, the supernatant is injected into the HPLC. Most HPLC methods involve an isocratic method using a reversed phase column. AsA and dehydro-ascorbic acid (DHA) can be monitored with a UV detector. AsA can also be measured using test kits in which increase in absorbance at 578 nm, following the reduction of MTT (3-(4,5-dimethylthiazolyl-2)-2,5-diphenyltetrazolium bromide) to a formazan compound by AsA, is measured (Megazyme, 2010).

Mass spectrometers are extremely useful for studying phytochemicals, either as highly selective detectors or as powerful tools for metabolite identification and profiling. When quantifying a bioactive compound using HPLC, peak identification and concentration are calculated according to a pure standard of known concentration. For compounds with high degrees of polymerization, for instance procyanidins and fructans (Fig. 18.2), standards are either scarce or expensive. Mass spectrometry can be used to identify unknown peaks according to the ion mass-to-charge ratio; a compound represented by such a peak can then be purified for future use as a standard.

Thin layer chromatography (TLC) can also be used to isolate bioactive compounds; extracts are spotted on to thin layer chromatography plates coated with silica gel, which can then be developed using an appropriate running solvent. This method is particularly useful for determining the antifungal activity of bioactive compounds. Once the compounds have been separated, the plates are dried and spores/conidia can then be sprayed on the surface. Compounds with antifungal activity can be identified as the area with reduced mycelial growth (Adikaram and Ratnayake Bandara, 1998; Terry et al., 2003, 2004). The identity of these antifungal compounds can be determined by developing spots of known standards.

18.5 Measurement of Total Antioxidants in Fruit

Single metabolite analysis might be the best choice for certain objectives but, given the variety of compounds present in fruit and the limited number of assays available in a given laboratory, such an approach might, in some cases, be tedious and impractical. In addition, the antioxidant properties of a pure compound might be different to those observed in most real samples, in which many other antioxidants are also present and/or in different matrices. Ideally, it might be desirable to have simple and accurate methods for the rapid quantification of food antioxidant capacity that could prevent diseases (Huang et al., 2005). Unfortunately, these methods are not yet available. Total antioxidant capacity (TAC; defined as the ability of a given sample to prevent the action of pro-oxidants) has been assumed to be related to the chemoprotective effect of foods. However, it has to be taken into account that studies comparing in vitro and in vivo testing of antioxidants generally have shown divergent results (Frankel, 2007; Frankel and Finley, 2008). The association between the TAC and the health effects of fruit and vegetables needs to be studied further. Multiple protocols have been used to evaluate the TAC of foods and these have employed a wide variety of free radical generating systems and different methods of inducing oxidation and final detection (Frankel and Meyer, 2000; Guiselli et al., 2000; Antolovich et al., 2002; Sánchez Moreno, 2002). There is evidence that in vitro assays might not reflect in vivo antioxidant capacity, especially if the compounds are metabolized to other derivatives (Cerdá et al., 2004, 2005). The results from such assays should not be considered a

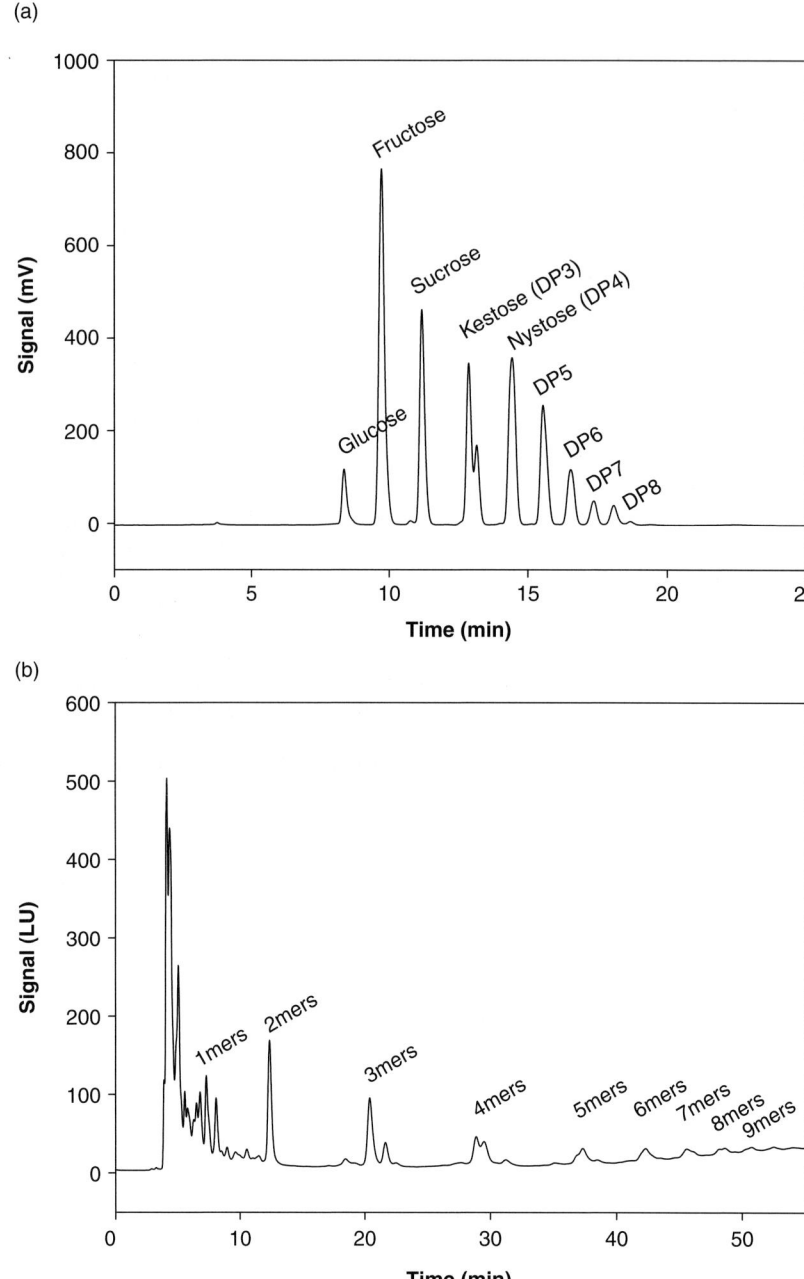

Fig. 18.2. Chromatograms of fructans (a) and procyanidins (b) with degrees of polymerization ranging from one to eight and one to nine, respectively.

reflection of the effect on the human body, as sometimes has been the case. *In vitro* antioxidant assays might be used as indicators of relative capacity to quench specific radicals and might be of some value for general comparisons of different products or treatments. There seems to be no consensus of opinion about a single method that should be used as a reference (Prior and Cao, 1999; Huang *et al.*, 2005; Roginsky and Lissi, 2005). The detailed analysis and comparison of the different assays are beyond the scope of this chapter (reviewed by Prior and Cao, 1999; Huang *et al.*, 2005). Here, we summarize briefly the main characteristics of the most commonly used methods: oxygen radical absorbance capacity (ORAC), Trolox equivalence antioxidant capacity (TEAC), ferric ion reducing antioxidant power (FRAP), the 2,2-diphenyl-1-picrylhydrazyl (DPPH•) assay, and Folin–Ciocalteu reagent reducing substances (FC).

18.5.1 ORAC assay

The oxygen radical absorbance capacity (ORAC) was developed by Cao *et al.* (1995) in order to measure antioxidant scavenging activity. It was designed originally to determine the antioxidant status in biological systems. Later, the ORAC protocol was widely used to test antioxidants in food samples (Guo *et al.*, 1997; Caldwell, 2001). Measurements are based on the quenching of phycoerytrin (PE) fluorescence in a phosphate buffer (excitation and emission at 540 nm and 565 nm, respectively), as caused by peroxyl radicals generated by thermolysis of 2,2'-azobis(2-amidinopropane) dihydrochloride (AAPH). In the presence of antioxidant-containing samples, loss of PE fluorescence would be reduced. For a given sample, fluorescence is measured over time and the area under the curve of the sample is compared to a blank. The ORAC is usually expressed in Trolox equivalents (Trolox being a water-soluble antioxidant analogue to vitamin E). Modifications to this protocol were proposed later by Ou *et al.* (2001). The latter authors showed that fluorescein (FL) (3',6'-dihydroxyspiro[isobenzofuran-1[3H],9' [9H]-

xanthen]-3-one) (excitation at 493 nm and emission at 515 nm) might be a better probe than PE, which gave inconsistent behaviour between different lots, might be photobleached and could bind polyphenols. The value of some other assays used to assess food antioxidant capacity (e.g. ABTS+•(see below) and DPPH•) has been questioned because the radicals used are not present in biological systems. In contrast, the ORAC assay is based on quenching of peroxyl radicals. It is considered by some to be a preferable method because it directly estimates the chain-breaking antioxidant activity, while most other assays measure the specific reducing power.

18.5.2 TEAC assay

The Trolox equivalent antioxidant capacity (TEAC) assay was first reported by Miller *et al.* (1993). The assay should not be confused with other techniques that also use Trolox as a reference antioxidant to express the final results. TEAC measurements are based on antioxidant-mediated reduction of the absorbance of a radical cation (2,2'-azinobis(3-ethylbenzothiazoline 6-sulfonate)) (ABTS+•). The radical can be produced from the commercially available ammonium salt of ABTS using potassium persulfate (Alonso *et al.*, 2002). ABTS+• has a strong absorption in the range of 600–750 nm and it is moderately stable. Conversion of ABTS+• into a non-coloured form, as caused by the addition of antioxidants, can be determined spectrophotometrically. Results are then expressed as Trolox equivalents. One advantage of this test is its simplicity, which makes it suitable for routine determinations of antioxidants.

18.5.3 FRAP assay

The FRAP (ferric reducing antioxidant power) assay was introduced by Benzie and Strain (1996) and was based on the ability of antioxidants to reduce iron from the ferric to the ferrous state. When this occurs in the presence of 2,4,6-trypyridyl-*s*-triazine, the reduction is accompanied by the formation of a blue complex, with Fe^{2+} increasing absorption at 593 nm. Other reagents able to form specific coloured complexes with Fe^{2+}, such as 1,10-fenantrolin (ferroin), may also be used for the determination. In this case, a red complex is formed and measured at 510 nm. In both cases, stronger absorption indicates a higher reducing power of the phytochemical, and thus a higher reducing capacity.

18.5.4 DPPH• assay

The 2,2-diphenyl-1-picrylhydrazyl (DPPH•) method was first reported by Blois (1958). DPPH• is one of a few stable and commercially available free radicals. The test is based on the reaction of DPPH• with antioxidants. DPPH• shows a very intensive absorption in the visible region and its concentration can be determined by visible spectroscopy (515 nm). Sample antioxidants cause a decrease in the initial DPPH• (Brand-Williams *et al.*, 1995). Several aliquots of the sample to analyse are added to test tubes containing a DPPH• solution and left to react until the change in absorbance reaches a plateau. Then, the EC_{50} (amount of sample required to reduce the initial concentration of DPPH• by 50%) is determined. The antioxidant capacity can then be expressed as EC_{50}^{-1}. This parameter does not take into account the time required for the antioxidants to consume 50% of the radical (t50), which might be quite different depending on the compound measured. Ascorbic acid reacts rapidly with DPPH•, while in samples rich in phenolic compounds the time needed for the reaction to reach the end point is longer (30–60 min). Sánchez-Moreno *et al.* (1998) suggested using the combination of kinetic (t50) and static parameters (EC_{50}) to characterize the antiradical efficiency (AE) as $EC_{50}^{-1} \times t50$. The DPPH• assay is simple and has the advantage that the reagent can be purchased directly. However, variations in initial DPPH• concentration would result in different EC_{50} values for the same sample, making comparisons difficult between different lots. In addition, the radical has no similarity with peroxyl radicals involved in lipid peroxidation and many antioxidants reacting quickly with peroxyl radicals only react slowly with DPPH•.

18.5.5 The Folin–Ciocalteu assay (FC)

The FC assay was used initially to analyse proteins, taking advantage of the reagent's reaction with aromatic amino acids (Folin and Ciocalteu, 1927). Singleton and Rossi (1965) proposed extending the use of this procedure in order to measure total phenolic compounds (Singleton *et al.*, 1999). The reagent, containing phosphomolybdic and phosphotungstic heteropoly acids, is added to aliquots of the antioxidant-containing samples, and subsequently the pH is raised by addition of Na_2CO_3. To prevent potential precipitation of carbonates, a mixture of NaOH and more diluted Na_2CO_3 may be used as an alkalizing reagent. Under basic conditions, phenolics can form fenolate anions, which in turn can reduce molybdenum, yielding a blue oxide showing a high extinction coefficient near 750 nm. Results are commonly expressed in gallic acid equivalents. The FC assay actually measures a sample's reducing capacity and not necessarily 'total phenolics', as is usually stated. Apart from flavonoids, groups also classed as polyphenols include stilbenes, lignans and tannins, many of which contain compounds not detected by the Folin–Ciocalteu method (Brat, 2008). AsA reacts significantly with the FC reagent. Depending on the fruit considered, the use of the term 'total phenolics' would be misleading to different extents. For instance, in blueberries the content of phenolic compounds is so far beyond the content of other reductants that calling results from FC assays as total phenolics might be reasonably accurate. One advantage of the FC assay over the ABTS and the DPPH• tests is that it is associated with increased absorbance rather than with a reduction. However, there is uncertainty over the compounds that are quantified by the FC assay since, in addition to phenolics, other reducing agents such as reducing sugars and possibly some metal chelators may be included (Prior *et al.*, 2005). The assay is straightforward and reproducible, and has become a routine protocol for characterizing antioxidants in foods.

18.6 Conclusions

To prepare samples for quantification of bioactive compounds, care should be taken to avoid degradation or oxidation, by minimizing prolonged exposure to light, high or ambient temperatures and air. Extraction procedures should be chosen based on the literature, which has compared multiple techniques to identify the most effective method, as well as correct solvent choice. Methods used for the quantification of total bioactives, for instance the FC method, are less likely to differ among authors, since these methods are well documented and relatively easy to carry out. Yet many compounds exist that have reducing potential and are therefore quantified but may not necessarily be classed as antioxidants. The ORAC method is becoming the preferred assay for determination of

antioxidant activity, as its mechanism is not through reducing power but an alternative mechanism. In addition, the ORAC method is favoured since antioxidant status can be determined in both lipophilic and hydrophilic fractions. Caution should be taken in comparing *in vitro* measurements and extrapolating these to prove a real effect *in vivo*.

The quantification of individual bioactive analytes using methods such as HPLC is more likely to differ between studies due to the multiple factors (mobile phases, stationary phases, column temperatures, etc.) that can differ between quantification techniques. Continued research comparing methodologies for the extraction, separation, isolation and quantification of bioactive compounds can only help to standardize methods further and reduce discrepancies among the literature.

References

Adikaram, N.K.B. and Ratnayake Bandara, B.M. (1998) Methodology for studying defence mechanisms against fungal pathogens: an overview. In: Johnson, G.I., Highley, E. and Joyce, D.C. (eds) *Disease Resistance in Fruit*. ACIAR Proceedings, Australian Centre for International Agricultural Research, Australia, pp. 177–185.

Almela, L., Lopez-Roca, J.M., Candela, M.E. and Alcazar, M.D. (1990) Separation and determination of individual carotenoids in a *Capsicum* cultivar by normal-phase high-performance liquid chromatography. *Journal of Chromatography A* 502, 95–106.

Alonso, A.M., Domínguez, C., Guillén, D.A. and Barroso, C.G. (2002) Determination of antioxidant power of red and white wines by a new electrochemical method and its correlation with polyphenolic content. *Journal of Agricultural and Food Chemistry* 50, 3112–3115.

Antolovich, M., Prenzler, P.D., Tatsalides, E., McDonald, S. and Robards, K. (2002) Methods for testing antioxidant activity. *Analyst* 127, 183–198.

Anttonen, M.J. and Karjalainen, R.O. (2006) High performance liquid chromatography analysis of black currant (*Ribes nigrum* L.) fruit phenolics grown either conventionally or organically. *Journal of Agricultural and Food Chemistry* 54, 7530–7538.

Awika, J.M., Rooney, L.W. and Waniska, R.D. (2004) Properties of 3-deoxyanthocyanins from Sorghum. *Journal of Agricultural and Food Chemistry* 52, 4388–4394.

Benzie, I.F.F. and Strain, J.J. (1996) The ferric reducing ability of plasma (FRAP) as a measure of "Antioxidant Power": the FRAP assay. *Analytical Biochemistry* 239, 70–76.

Blois, M.S. (1958) Antioxidant determination by the use of a stable free radical. *Nature* 181, 1199–1200.

Brand-Williams, W., Cuvelier, M.E. and Berset, C. (1995) Use of a free radical method to evaluate antioxidant activity. *LWT – Food Sicence and Technology* 28, 25–30.

Brat, P. (2008) Rapid analysis of phytochemicals in fruits and vegetables In: Tomás-Barberán, F.A. and Gil, M.I. (eds) *Improving the Health-promoting Properties of Fruits and Vegetable Products*. Woodhead Publishing Limited and CRC Press LLC, Cambridge, UK, pp. 248–273.

Caldwell, C.R. (2001) Oxygen radical absorbance capacity of the phenolic compounds in plant extracts fractionated by high performance liquid chromatography. *Analytical Biochemistry* 293, 232–238.

Cao, G., Verdon, C.P., Wu, A.H.B., Wang, H. and Prior, R.L. (1995) Automated assay of oxygen radical absorbance capacity with the COBAS FARA II. *Clinical Chemistry* 41, 1738–1744.

Cerdá, B., Espín, J.C., Parra, A., Martínez, P. and Tomás-Barberán, F.A. (2004) The potent *in vitro* antioxidant ellagitannins from pomegranate juice are metabolized into bioavailable but poor antioxidant hydroxy-6H-dibenzopyran-6-one derivatives by the colonic microflora of healthy humans. *European Journal of Nutrition* 43, 205–220.

Cerdá, B., Tomás-Barberán, F. and Espín, J.C. (2005) Metabolism of chemopreventive and antioxidant ellagitannins from strawberries, raspberries, walnuts and oak-aged wines in humans: identification of biomarkers and individual variability. *Journal of Agricultural and Food Chemistry* 53, 227–235.

Davis, F., Terry, L.A., Chope, G.A. and Faul, C.F.J. (2007) Effect of extraction procedure on measured sugar concentrations in onion (*Allium cepa* L.) bulbs. *Journal of Agricultural and Food Chemistry* 55, 4299–4306.

Downes, K. and Terry, L.A. (2010) A new acetonitrile-free mobile phase method for LC-ELSD quantification of fructooligosaccharides in onion (*Allium cepa* L.). *Talanta* 82, 118–124.

Espín, J.C., García-Conesa, M.T. and Tomás-Barberán, F.A. (2007) Nutraceuticals: facts and fiction. *Phytochemistry* 68, 2986–3008.

Folin, O. and Ciocalteu, V. (1927) Tyrosine and tryptophan determinations proteins. *Journal of Biological Chemistry* 73, 627.

Frankel, E.N. (2007) *Antioxidants in Food and Biology. Facts and Fiction*. The Oily Press, Bridgwater, UK.

Frankel, E. and Finley, J.W. (2008) How to standardize the multiplicity of methods to evaluate natural antioxidants. *Journal of Agricultural and Food Chemistry* 56, 4901–4908.

Frankel, E.N. and Meyer, A.S. (2000) The problems of using one dimensional methods to evaluate multi-functional food and biological antioxidants. *Journal of the Science of Food and Agriculture* 80, 1925–1941.

Giné Bordonaba, J., Crespo, P., and Terry, L.A. (2008) Biochemical profiling and chemometric analysis of seventeen UK-grown black currant cultivars. *Journal of Agricultural and Food Chemistry* 56, 7422–7430.

Giné Bordonaba, J. and Terry, L.A. (in press) A new acetonitrile-free mobile phase for HPLC-DAD determination of anthocyanins in blackcurrant and strawberry fruits: a comparison and validation study. *Food Chemistry* 129, 1265–1273.

Guiselli, A., Serafini, M., Natella, F. and Scaccini, C. (2000) Total antioxidant capacity as a tool to assess redox status: critical view and experimental data. *Free Radical Biology and Medicine* 11, 1106–1114.

Guo, C., Cao, G., Sofic, E. and Prior, R.L. (1997) High-performance liquid chromatography coupled with colometric array detection of electroactive components in fruits and vegetables: relationship to oxygen radical absorbance capacity. *Journal of Agricultural and Food Chemistry* 45, 1787–1796.

Huang, D., Ou, B. and Prior, R.L. (2005) The chemistry behind antioxidant capacity assays. *Journal of Agricultural and Food Chemistry* 53, 1841–1856.

Megazyme (2010) L-*Ascorbic Acid (L-Ascorbate) Assay Procedure*. Megazyme International Ireland Limited K-ASCO 06/10.

Miller, N.J., Rice-Evans, C., Davies, M.J., Gopinathan, V. and Milner, A. (1993) A novel method for measuring antioxidant capacity and its application to monitoring the antioxidant status in premature neonates. *Clinical Science* 84, 407–412.

Naczk, M. and Shahidi, F. (2006) Phenolics in cereals, fruits and vegetables: occurrence, extraction and analysis. *Journal of Pharmaceutical and Biomedical Analysis* 41, 1523–1542.

O'Donoghue, E.M., Somerfield, S.D., Shaw, M., Bendall, M., Hedderly, D., Eason, J. and Sims, I. (2004) Evaluation of carbohydrates in Pukekohe Longkeeper and Grano cultivars of *Allium cepa*. *Journal of Agricultural and Food Chemistry* 52, 5383–5390.

Ou, B., Hampsch-Woodill, M. and Prior, R.L. (2001) Development and validation of an improved oxygen radical absorbance capacity using fluorescein as the fluorescent probe. *Journal of Agricultural and Food Chemistry* 49, 4619–4626.

Prior, R.L. and Cao, G. (1999) *In vivo* total antioxidant capacity: comparison of different analytical methods. *Free Radical Biology and Medicine* 27, 1173–1181.

Prior, R.L., Wu, X. and Schaich, K. (2005) Standardized methods for the determination of antioxidant capacity and phenolics in foods and dietary supplements. *Journal of Agricultural and Food Chemistry* 53, 4290–4302.

Puupponen-Pimiä, R., Häkkinen, S.T., Aarni, M., Suortti, T., Lampi, A.-M., Eurola, M., Piironen, V., Nuutila, A.M. and Oksman-Caldentey, K.-M. (2003) Blanching and long-term freezing affect various bioactive compounds of vegetables in different ways. *Journal of Agricultural and Food Chemistry* 83, 1389–1402.

Robards, K. (2003) Strategies for the determination of bioactive phenols in plants, fruit and vegetables. *Journal of Chromatography A* 1000, 657–691.

Roginsky, V. and Lissi, E.A. (2005) Review of methods to determine chain-breaking antioxidant activity in food. *Food Chemistry* 92, 235–254.

Sánchez-Moreno, C. (2002) Review: Methods used to evaluate the free radical scavenging activity in foods and biological systems. *Food Science and Technology International* 8, 121–137.

Sánchez-Moreno, C., Larrauri, J.A. and Saura Calixto, F. (1998) A procedure to measure the antiradical efficiency of polyphenols. *Journal of the Science of Food and Agriculture* 76, 270–276.

Singleton, V.L. and Rossi, J.A. (1965) Colorimetry of total phenolics with phosphomolybdic-phosphotungstic acid reagents. *American Journal of Enology and Viticulture* 16, 144–158.

Singleton, V.L., Orthofer, R. and Lamuela-Raventos, R.M. (1999) Analysis of total phenols and other oxidation substrates and antioxidants by means of Folin–Ciocalteu reagent. *Methods in Enzymology* 299, 152–178.

Terry, L.A., Joyce, D.C. and Khambay, B.P.S. (2003) Antifungal compounds in Geraldton waxflower tissues. *Australian Plant Pathology* 32, 411–420.

Terry, L.A., Joyce, D.C., Adikaram, N.K.B. and Khambay, B.P.S. (2004) Preformed antifungal compounds in strawberry fruit and flower tissues. *Postharvest Biology and Technology* 31, 201–212.

Tsao, R. and Deng, Z. (2004) Separation procedures for naturally occurring antioxidant phytochemicals. *Journal of Chromatography A* 812, 85–99.

Vågen, I.M. and Slimestad, R. (2008) Amount of characteristic compounds in 15 cultivars of onion (*Allium cepa* L.) in controlled field trials. *Journal of the Science of Food and Agriculture* 88, 404–411.

19 Methodologies for Evaluating *In Vitro* and *In Vivo* Activities of Bioactive Compounds

Paul J. Thornalley, Mingzhan Xue and Naila Rabbani

19.1 Introduction

The health-promoting effects of bioactive compounds found in fruit and vegetables are typically associated with decreasing risk of cancer, maintaining good vascular health and suppression of ageing-related disorders. Evaluations of related pharmacological activities of bioactive compounds are made in chemical and biochemical cell-free systems, cell systems (including use of inducible expression reporter systems), preclinical animal models and clinical dietary intervention and diet supplement trials. These methods are reviewed succinctly in this chapter. There is a continuing tendency to view the health effects of dietary bioactive compounds as being linked inextricably and sometimes only with antioxidant activity (Gorinstein *et al.*, 2009). This is probably a too narrow, blinkered approach (Stevenson and Hurst, 2007). Bioactive compounds are not essential nutrients for growth and development but rather are mainly non-nutrients improving health – particularly in adult and later life. They have been called 'lifespan essentials' (Holst and Williamson, 2008). The emerging consensus view on metabolic control for healthy ageing is a requirement to prevent not only oxidative damage in tissues but also to prevent and repair glycation damage, to regain or reset glycolytic and lipogenic control when lost in metabolic syndrome, diabetes and obesity (Kitano *et al.*, 2004), to remove damaged proteins and nucleotides and to maintain efficient metabolism and clearance of damaging metabolites and xenobiotic compounds (oxidizing and non-oxidizing) (Kwak *et al.*, 2003; Leahy, 2006; Xue *et al.*, 2008a). Indeed, bioactives may induce a weak damaging response to activate a stronger antistress gene response that is eventually protective. Hormesis is the beneficial effect of a treatment that, at a higher intensity, is harmful (Calabrese *et al.*, 1999). Agents inducing hormesis are called 'hormetins' and many bioactive compounds may be of this type. This might explain the observations, which currently are judged controversial, that bioactive compounds at high doses have adverse toxic effects, whereas at lower doses they are beneficial (Lambert *et al.*, 2007). For hormesis, this dose-dependent behaviour is expected – and will be accepted providing there are plausible mechanistic explanations for beneficial and adverse effects. Hormesis has been proposed to play a role in healthy ageing (Gems and Partridge, 2008). Improved evaluation of bioactive compounds in the future may have to take hormetic considerations into account.

19.2 Cell-free Assessment of Bioactivity

19.2.1 Antioxidant activity

Several protocols have been proposed to compare the antioxidant activity of bioactive compounds. These assays have been classified into two types, depending on the reactions involved: assays depending on hydrogen atom transfer (HAT) reactions and assays based on electron transfer (ET). Most HAT-based assays use a competitive assay system in which the test antioxidant and a standard substrate compete for thermally generated peroxyl radicals through the decomposition of azo compounds. These assays include: oxygen radical absorbance capacity (ORAC), total radical trapping antioxidant parameter (TRAP), crocin bleaching assays, and inhibition of induced low-density lipoprotein autoxidation. ET-based assays measure the propensity of an antioxidant to reduce an oxidant – the reaction typically monitored by chromophoric change in the oxidant accompanying reduction. ET-based assays include: total phenols assay based on the Folin–Ciocalteu reagent (FCR), Trolox equivalence antioxidant capacity (TEAC), ferric ion reducing antioxidant power (FRAP), 'total antioxidant potential' assay using a Cu(II) complex as an oxidant, and the 2,2-diphenyl-1-picrylhydrazyl radical (DPPH•) assay. Other assays assess the scavenging capacity of test compounds of biologically relevant oxidants such as singlet oxygen, superoxide anion, peroxynitrite, and hydroxyl radical. It has been suggested that a HAT-based test of free radical scavenging activity and an ET-based test of reducing activity should be performed for bioactive compounds (Huang *et al.*, 2005).

19.2.2 Oxygen radical absorbance capacity (ORAC) assay

This was developed by Cao *et al.* (1993) and employed as the standard substrate B-phycoerythrin (B-PE) – a protein product isolated from *Porphyridium cruentum*. B-PE is a fluorescence protein that loses fluorescence in reaction with peroxyl radicals. 2,2'-Azinobis (2-amidinopropane) dihydrochloride (AAPH) was used as the peroxyl radical generator. Use of B-PE suffered several disadvantages: (i) commercial sources of B-PE had marked variability in ORAC; (ii) B-PE was photobleached under microplate-reader conditions; and (iii) polyphenols and other bioactives bound to B-PE and lost fluorescence without peroxyl radical involvement. To avoid these problems, B-PE was replaced by fluorescein (FL) (Naguib, 2000). This improved the ORAC assay, provided a direct measure of the hydrophilic and lipophilic chain-breaking antioxidant capacity versus peroxyl radicals (Huang *et al.*, 2002a) and was adapted for high-throughput application (Huang *et al.*, 2002b). Further modification includes an organic solvent-based ORAC assay for lipophilic samples. The ORAC assay has been applied as a method of choice to quantify antioxidant capacity, and an antioxidant database has been generated applying the ORAC assay in combination with the total phenols assay (Wu *et al.*, 2004b).

19.2.3 Crocin bleaching assay

This assay measures the ability of an antioxidant to protect bleaching of the carotenoid derivative crocin by peroxyl radicals generated by AAPH (Bors *et al.*, 1984). The progress of the reaction is monitored spectrophotometrically at 443 nm. The bleaching rate becomes linear about 1 min after the addition of AAPH and is monitored for 10 min. By varying the antioxidant concentration, the ratio of the rate constants for reaction of the antioxidant and crocin with peroxyl radical is deduced and this is the measurement of antioxidant activity. Crocin is a mixture of natural pigments extracted from saffron, and preparations vary in composition, which limits the reproducibility of this procedure.

19.2.4 Total peroxyl radical-trapping antioxidant parameter (TRAP) assay

This assay uses R-phycoerythrin (R-PE) as a fluorescent substrate, with AAPH as peroxyl radical generator. The reaction progress of

R-PE with AAPH is monitored fluoro-metrically (at an excitation wavelength of 495 nm and an emission wavelength of 575 nm). The antioxidant capacity of test anti-oxidants is normalized to that of 6-hydroxy-2,5,7,8-tetramethylchroman-2-carboxylic acid (Trolox) – a water-soluble derivative of vita-min E, and is therefore expressed in 'Trolox equivalents' (Ghiselli *et al.*, 2000). A further modification has been to use dichlorodihy-drofluorescein diacetate (DCFH-DA) as a flu-orogenic reporter. In this case, peroxyl radical generation by AAPH oxidizes DCFH-DA to fluorescent dichlorofluorescein (DCF) (Valkonen and Kuusi, 1997).

19.2.5 Total phenols assay by Folin–Ciocalteu reagent

Folin–Ciocalteu reagent (FCR) was used ini-tially in the Lowry protein assay. Singleton applied this reagent for the analysis of total phenols in wine (Singleton *et al.*, 1999). The FCR-based assay is known commonly as the total phenols (or phenolic) assay. It measures the reducing activity of the sample. The exact chemical nature of the FCR is thought to be heteropolyphosphotungstates–molybdates, which undergo reversible one- or two-electron reduction reactions, leading to formation of a blue chromophore, possibly $(PMoW_{11}O_{40})^{4-}$ (Huang *et al.*, 2005). The FCR is not specific to phenolic compounds and it can be reduced by many non-phenolic compounds: vitamin C, Cu(I) and many others. Phenolic com-pounds react with FCR only under basic conditions (in sodium carbonate buffer, pH 10). The phenolate anion reduces FCR. The total phenols assay by FCR is convenient, simple and apparently reproducible. It has become a routine assay in studies of phenolic antioxidants (Jimenez-Alvarez *et al.*, 2008).

19.2.6 Trolox equivalent antioxidant capacity assay (TEAC)

This assay was first reported by Miller *et al.* (1993), with later improvements (Re *et al.*, 1999). In the improved version, persulfate oxidation

of 2,2′-azino-*bis*(3-ethylbenzothiazoline-6-sulfonic acid) forms a radical cation 2,2′-azino bis-(3-ethylbenzothiazoline-6-sulfonate), which is chromophoric and becomes colour-less when reduced. The amount of antioxi-dant required to produce the same decrease in absorbance as Trolox is deduced. TEAC values of many compounds and food sam-ples are reported (Seeram *et al.*, 2008).

19.2.7 Ferric ion reducing antioxidant power (FRAP) assay

In the FRAP assay, a ferric complex with 2,4,6-tripyridyl-*s*-triazine (TPTZ), $FeTPTZ_2Cl_3$, is reduced by the test antioxidant (Huang *et al.*, 2002a). The FRAP assay is performed under acidic conditions (pH 3.6). Interfer-ences in the assay arise from chelators in food extracts binding to Fe(III) and forming com-plexes. FRAP values of many compounds and food samples are reported (Seeram *et al.*, 2008).

19.2.8 2,2-Diphenyl-1-picrylhydrazyl radical scavenging capacity assay

2,2-Diphenyl-1-picrylhydrazyl (DPPH) is a stable, chromophoric and commercially avail-able free radical. The chromophoric proper-ties are decreased when DPPH is reduced by antioxidants. The time and concentration of antioxidant required to decrease the concen-tration of DPPH by 50% is recorded (Sharma and Bhat, 2009).

19.2.9 Problems of antioxidant capacity measurements of bioactive compounds

The major weaknesses and problems of the antioxidant assays considered above (and oth-ers) are: (i) assessment on the basis of scaveng-ing a free radical or use of free radical generator that is not of physiological relevance; (ii) inter-ference of non-antioxidant effects on the reporter chromophore or fluorophore; (iii) variability in composition and instability of the reporter molecule; (iv) antioxidant capacity estimation under non-physiological conditions; and (v) sample matrix effects. The design of these assays has compromised quantitation and specificity for ease of procedure, accessibility to reagents and ease of understanding and comparison of outcomes. When comparing these methods in fruit juices and beverages, the TEAC assay gave lower estimates than the ORAC assay (Zulueta *et al.*, 2009). Comparison of antioxidant capacity of 104 plant foods, beverages and oils consumed in Italy by the TEAC, TRAP and FRAP assays showed good correlations of estimates by these different methods (Pellegrini *et al.*, 2003). Comparison of antioxidant capacity assays is intractable, however, in human serum. There was only a weak linear correlation between serum ORAC and serum FRAP activities and no correlation between serum ORAC and TEAC or serum FRAP and serum TEAC outcomes (Cao and Prior, 1998). Confounding interferences, masking effects or other antioxidant activities appear to be operating *in vivo*, such that the simple assays listed above appear unable to provide a coherent report on antioxidant capacity.

19.2.10 Oxidation of low-density lipoprotein

A potential target of bioactive compounds to decrease atherosclerosis and cardiovascular disease is low-density lipoprotein (LDL). A decrease in oxidation of LDL is linked to concomitant decrease in atherogenicity (Fraley and Tsimikas, 2006). The ability of isolated bioactive compounds to delay the oxidation of LDL initiated by copper sulfate (10 µM) in physiological saline can be assessed. Oxidation is followed by spectrophotometric detection of conjugated dienes at 234 nm. The protective effect is assessed from the increase in time required before the oxidation enters the rapid propagation phase (Souza *et al.*, 2008). The weakness of this approach is that the kinetics of LDL oxidation employed therein are far higher than those found *in vivo*, where even small, dense atherogenic LDL has low or minimal modification (Navab *et al.*, 2004).

19.3 *In Vitro* Assessments: Cell Culture Systems

The pharmacological activities of bioactive compounds and fruit and vegetable extracts are achieved mostly within the body, where exposure of cells to the proinflammatory effects of bacterial endotoxins is suppressed by endotoxin-binding proteins (Munford, 2005). It is important, therefore, in the evaluation of pharmacological activities of bioactive compounds and extracts in cell culture cells to avoid inadvertent introduction of endotoxin (or lipopolysaccharide) into the test system with the added bioactive compound or extract. The normal concentration of endotoxin in human plasma is *c*.0.1 endotoxin unit/ml (Hiki *et al.*, 1999), and endotoxin levels should not be increased significantly beyond this by addition of the test compound. Endotoxin testing is performed conveniently by the chromogenic limulus assay (Piotrowicz *et al.*, 1985). Endotoxin may be removed from test solutions by ultrafiltration, elution through columns of Detoxigel™ (agarose-immobilized polymyxin B) or neutralized by addition of polymyxin B. Care is required if Detoxigel™ or the convenient addition of polymyxin B is the chosen strategy. Polymyxin B does not inactivate all types of endotoxin (Kluger *et al.*, 1985). Polycationic polymyxin B binds to polyanionic endotoxin. If the test compound or extract contains polyanionic substances, they may displace endotoxin from polymyxin B and activate an endotoxin-mediated response.

19.3.1 Cellular antioxidant activity assay

Bioactive compounds absorbed from dietary foodstuffs may derive part of their antioxidant activity by interaction with the metabolic capacity of cells. An attempt to quantify antioxidant activity of bioactive compounds in a cellular matrix is made in the cellular antioxidant assay (CAA) (Wolfe and Liu, 2007). The CAA of bioactive food extracts and dietary supplements uses DCFH-DA as a fluorogenic reporter probe. This probe enters the human hepatoma HepG2 cells used in the assay and is de-esterified by non-specific esterase and trapped therein as the fluorogenic reporter dichlorodihydrofluorescein. The assay measures the ability of compounds to prevent the formation of DCF by the free radical generator AAPH in HepG2 cells. The results have been expressed in micromoles of quercetin equivalents/100 μmol of phytochemical or in micromoles of quercetin equivalents/100 g of fresh fruit; quercetin had the highest CAA value (Wolfe and Liu, 2007). Of the selected fruit tested by the CAA assay, blueberry had the highest CAA value, followed by cranberry > apple ≈ red grape > green grape (Wolfe and Liu, 2007). The CAA assay has been applied to flavonoids, common fruit and other bioactive compounds and foodstuffs (Wolfe and Liu, 2007, 2008; Wolfe *et al.*, 2008).

The CAA assay has greater biological relevance than cell-free antioxidant capacity activity assays, as potentially it is influenced by cellular uptake, metabolism and location of antioxidant compounds within the HepG2 cells. As bioactive compounds influence the antioxidant capacity of cells by indirect methods – activation of antioxidant-linked gene expression, for example (see below) – there are often potential confounding and masking effects. To explore these, the CAA assay could be performed after varied periods of pre-incubation of HepG2 cells with bioactive compounds or food extract. The assay retains the weakness of using a non-physiological stressor, AAPH, to produce oxidizing free radicals. Use of cultured cells also requires that activators of cellular oxidative stress are not introduced into the assay inadvertently – such as endotoxin, for example. It is not clear from current use of the CAA assay that endotoxin contamination has been controlled and excluded.

19.3.2 Activation of NF-E2-related factor-2 and the antistress gene response

One of the key transcriptional systems in which dietary bioactive compounds are thought to have influence – leading to beneficial health effects – is transcription factor NF-E2-related factor-2 (Nrf2) and its interaction with promoter antioxidant response elements (AREs) (Fig. 19.1). This factor coordinates the antistress gene response. ARE-linked genes code for a battery of protective and metabolic enzymes: γ-glutamylcysteine ligase (GCL), glutathione reductase (GSHRd), aldo-keto reductase (AKRd), glutathione transferases (GSTs), quinone reductase (NQO1), Nrf2 itself (Kwak *et al.*, 2002) and others (Thimmulappa *et al.*, 2002). Nrf2-linked gene expression has a key role in the protection of cells against oxidative stress, carbonyl compounds and electrophilic agents. Key enzymes of the pentosephosphate pathway – the source of reducing equivalents for several protective enzymes, transketolase (TK) and transaldolase (TALDO) – are also coded by ARE-linked genes (Thimmulappa *et al.*, 2002). Recent studies have shown that the expression of TK is particularly important in resisting mechanisms underlying vascular disease induced by the hyperglycaemia associated with diabetes (Xue *et al.*, 2008b). Components of the ubiquitin-independent 20S proteasome that degrade damaged proteins also have ARE-linked expression (Woods *et al.*, 1995). All of these genes have expression increased by activation of Nrf2, tempered by inhibitory or activatory effects of small Maf proteins, F, G and K. Some other genes linked to lipogenesis – those coding for sterol regulatory element binding protein 1 (SREBP1) and lipoprotein lipase (LPL) – have expression repressed through the ARE promoters (Thimmulappa *et al.*, 2002). Repression of SREBP1 decreased expression and activities of diacylglycerol acyltransferase-1 and -2 activity and fatty acid synthase (FASN), leading to decreased synthesis of triglycerides (TG) and cholesterol esters (CE) and secretion of apolipoprotein B100 (apoB) in very low-density lipoprotein (VLDL) (Maiyoh *et al.*, 2007). Activation of Nrf2, therefore, has the potential to protect against proteome and lipidome damage, enhance removal of residual damage and decrease lipogenesis – all important for maintaining good vascular health (Fig. 19.2).

Under basal conditions, Nrf2 is complexed with Kelch-like ECH-associated protein 1 (Keap1) – a BTB-Kelch protein. Keap1 is a substrate adaptor protein for Cullin-3 (Cul3)-dependent E2 ubiquitin ligase complex, directing Nrf2 for proteasomal degradation (Kobayashi *et al.*, 2004). In oxidative stress, lipid peroxidation products, 4-hydroxynonenal and

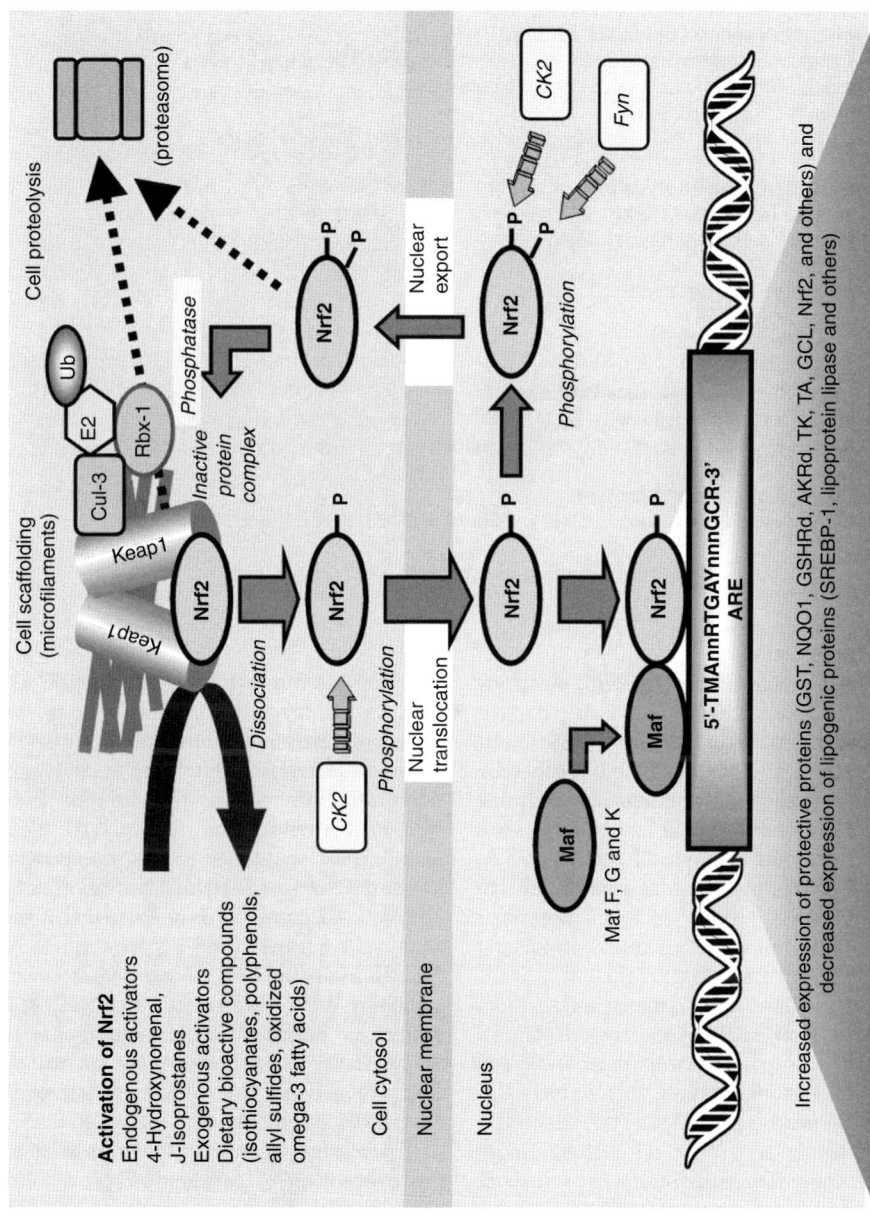

Fig. 19.1. Activation of Nrf2 and dynamic nuclear–cytoplasmic shuttling of Nrf2 for expression of antioxidant response element-linked genes.

Antistress gene response:
Transcription factor Nrf2 / antagonist Keap1 (iNrf2)

Antioxidant response element (ARE)-linked genes

ARE

Protection against oxidative damage
Glutathione synthesis and GSH- dependent enzymes GCL, GSHHd, GSHPx and GSTs
Superoxide dismutase-1 (SOD1) and catalase (CAT)
Peroxiredoxins (PRDXs)
Thioredoxin (TXN) and thioredoxin reductase (TXNRd1) … and others

Protection against glycation
Aldoketo reductase (AKRd)
Aldehyde dehydrogenase (AldDH)

Protection against metabolic stress
Transketolase (TKT)
Transaldolase (TALDO)

Protection against lipogenic stress
SREBP-1 (down regulation)
Lipoprotein lipase and others (downregulation)

Removal of damaged proteins
20S Proteasome induction (PSMA1, PSMA4, PSMB3, PSMB5, and PSMB6)

Fig. 19.2.   Antistress gene response: a multi-layered protective response.

J$_3$-isoprostanes, may disrupt the Nrf2–Keap1 or Keap1–Cul3 interactions. These are current candidate physiological activators of Nrf2 signalling; there may be others. The Nrf2 thereby liberated or stabilized from degradation translocates to the nucleus and, combining with small Maf protein (Motohashi *et al.*, 1997), in Maf–Nrf2 heterodimer or (Maf)$_2$–Nrf2 homodimer complexes (Kimura *et al.*, 2007), induces or represses ARE-linked gene expression (Zhang *et al.*, 2005) (Fig. 19.1). MafF, MafG and MafK isoforms have differential potency and specificity for subsets of ARE-linked genes (Katsuoka *et al.*, 2005). Keap1 has concurrent increased susceptibility to degradation, but also has ARE-linked gene expression and may be induced by Nrf2 activation providing an autoregulatory feedback loop for post-stimulation return to Nrf2 homeostasis (Lee *et al.*, 2007). Nrf2 also undergoes nuclear export, establishing cytoplasmic/nuclear dynamic shuttling (Jain *et al.*, 2005). Some studies suggest Keap1 undergoes nuclear translocation during the shutdown phase, binding Nrf2 in the nucleus and escorting it to the cytosol (Ahmed *et al.*, 2005). This is controversial; recent studies with improved specificity immunoblotting suggest there is minimal nuclear Keap1 in basal conditions and little nuclear translocation during Nrf2 activation (Watai *et al.*, 2007). Recent research has suggested the serine/threonine kinase CK2 has a role in nuclear import of Nrf2 and both CK2 and tyrosine kinase Fyn influence nuclear export and degradation of Nrf2 (Jain and Jaiswal, 2007; Pi *et al.*, 2007; Sun *et al.*, 2007) (Fig. 19.1). There is an active Nrf2–Keap1 system in human vascular endothelial cells (Chen *et al.*, 2003), hepatocytes (Keum *et al.*, 2006), related cell lines (Qian *et al.*, 2006; Lee *et al.*, 2007) and many other cell types.

Activation of Nrf2 has been implicated in the prevention of cancer (Surh *et al.*, 2008), cardiovascular disease (Zhu *et al.*, 2005), vascular complications of diabetes (Xue *et al.*, 2008b) and renal failure (Thornalley and Rabbani, 2009), dementia (Singh *et al.*, 2008), arthritis (Mahajan *et al.*, 1994), other diseases and ageing. In this regard, health benefit may be achieved by dietary bioactive compounds enhancing the endogenous activation of the Keap1–Nrf2 system and provide a supra-normal defence pathogenic mechanism. Key dietary bioactive activators of this system are: glucosinolate-derived dietary isothiocyanates (Thornalley and IARC Workgroup, 2004) and indoles (Maiyoh *et al.*, 2007), thioethers and disulfides (Chen *et al.*, 2004), polyphenols (Tanigawa *et al.*, 2007), flavonoids (Mann *et al.*, 2007) and carotenoids (Ben Dor *et al.*, 2005), as well as omega-3 fatty acids via formation of J_3-isoprostanes (Gao *et al.*, 2007). The activation is thought to occur by different mechanisms: isothiocyanates such as sulforaphane (SFN) release Nrf2 from Keap1 by modification of critical cysteine thiol residues (Dinkova-Kostova *et al.*, 2002); some polyphenols induce downregulation of Keap1 expression (Tanigawa *et al.*, 2007) and/or induce mild oxidative/nitrosative stress (Mann *et al.*, 2007), and J_3-isoprostanes disrupt the Keap1–Cul3 complex, preventing Keap1–Nrf2 targeting to the proteasome (Gao *et al.*, 2007). Increased levels of ARE-linked gene products provide for increased protection against reactive oxidizing, nitrating and glycating species – preserving the integrity of the vascular cell proteome and lipidome and enhancing the activity of enzymes of detoxification (Thimmulappa *et al.*, 2002). Activation of Nrf2 also suppresses the expression of key regulators of lipogenesis (Kwak *et al.*, 2003).

The ability of bioactive compounds to increase inducible expression of a particular protective enzyme via the ARE promoter may be assessed by the design of reporter constructs in expression vectors containing the ARE nucleotide sequence of interest linked to a luciferase or similar reporter system (Table 19.1). Transfection efficiency is controlled by incorporation of a different constitutive reporter either in the same vector or in a similar vector included in co-transfection. Additional controls include transfection with empty vector and with vector containing a mutant non-functional ARE promoter.

Alternatively, the ability of bioactive compounds to induce ARE-linked gene expression may be assessed by gene microarray analysis, real time RT-PCR array analysis, immunoblotting of ARE-linked gene products and measurement of the activities of related enzymes. Typical marker enzymes of the anti-stress gene response are quinone reductase (NQO1) and isozymes of glutathione transferase (GST). Other ARE-linked gene expression may be assessed for particular applications; for example, TK involved in countering metabolic dysfunction in hyperglycaemia (Xue *et al.*, 2008b) and SREBP1 in reversal of dyslipidaemia (Maiyoh *et al.*, 2007) (Table 19.2).

The effects of bioactive compounds on the Nrf2 system may increase the cellular concentration of GSH by increased expression of the GCL involved in GSH synthesis and GSHRd. The effect may be time dependent; for example, dietary isothiocyanates tend to induce initially a decrease in GSH after 3 h exposure, with a rebound to increased

Table 19.1. Reporter systems for antioxidant linked gene expression.

Gene	Reporter system	Expression vector/construct	Bioactives	Reference
Quinone reductase (NQO1)	Luciferase	NQO1hARE-tk-luc	Lycopene, flavonoids	Mulcahy *et al.* (1997); Ben Dor *et al.* (2005); Lee-Hilz *et al.* (2006)
γ-Glutamylcysteine ligase – heavy chain (GCL$_h$)	Luciferase	GCShARE4-tk-luc	Lycopene	Mulcahy *et al.* (1997); Ben Dor *et al.* (2005)
Human GSTP1	Luciferase	−336-GSTP1-luc	Curcumin	Nishinaka *et al.* (2007)
Thioredoxin reductase	Luciferase		SFN	Hintze *et al.* (2003)
Glutathione *S*-transferase	Chloramphenicol amino transferase	pARE-CAT construct	SFN	Yu *et al.* (1999)

Table 19.2. Evaluation of antioxidant response element linked gene expression in cell systems *in vitro*.

Target disease/abnormal physiological state	Cell type	Gene expression and metabolites	Bioactives	Reference
Oxidative stress	Human mammary cancer MCF-7 cells and hepatoma HepG2 cells	γ-Glutamylcysteine ligase – heavy chain (GCL$_h$)	Lycopene	Ben Dor *et al.* (2005)
Oxidative stress	Rat hepatocytes	CAT, SOD, GSHPx, GSR, GST, NQO1 activities	Resveratrol (25–75 μm)	Rubiolo *et al.* (2008)
	Human HepG2 cells	Nrf2	SFN, AITC, I3C	Jeong *et al.* (2005) Xu and Thornalley (2000a); Ye and Zhang (2001)
	Human leukaemia HL60 and ML-1 cells; HepG2 cells	NQO1, GST and cellular GSH		
Hyperglycaemia (diabetes)	Microvascular endothelial cells	TK, GSRd, NQO1	SFN	Xue *et al.* (2008b)
Dyslipidaemia	Human hepatoma HepG2 cells	Apolipoprotein B100 secretion, LDL uptake, triglycerides, cholesterol, cholesterol esters and lipogenic gene expression	Naringenin (citrus flavonoid); I3C; red grape juice	Borradaile *et al.* (2003); Davalos *et al.* (2006); Maiych *et al.* (2007)
Arthritis	Human chondrocytes	NQO1, GST, Nrf2 and others	SFN	Healy *et al.* (2005)

cellular GSH concentrations thereafter (Xu and Thornalley, 2000a; Ye and Zhang, 2001).

Activation of the Nrf2 system by vegetable and fruit extracts in cell culture has also been studied. Broccoli sprout extract processed with myrosinase to convert glucosinolates to related isothiocyanates induced both GST and NQO1 significantly in cultured bladder cells (Zhang *et al.*, 2006). Aqueous extracts of black radish (*Raphanus sativus* L. var. niger) increased the activity of NQO1 in HepG2 cells *in vitro* (Hanlon *et al.*, 2007). Anthocyanin-enriched bilberry extracts modulated pre- or post-translational levels of oxidative stress defence enzymes heme-oxygenase-1 and GST in cultured human retinal pigment epithelial cells (Milbury *et al.*, 2007).

Innovative high-throughput screening methods are in development; for example, an ARE sequence-specific, fluorescence-quenching hybridization probe (Wang *et al.*, 2008). However, the Nrf2 system has complex time course activation characteristics influenced by multiple positive and negative feedback mechanisms. All of these influences are potential factors affecting bioactive compound stimulatory activity.

19.3.3 Inhibition of malignant and non-malignant cell growth

Bioactive compounds have been assessed for their ability to inhibit the growth of malignant and non-malignant cells *in vitro*; for example, inhibition of human leukaemia cells and human lymphocyte growth (Xu and Thornalley, 2000b). Inhibition of malignant tumour cell growth has been linked to the cancer preventive activity of many bioactive compounds via inhibition of growth and induction of cell death of preclinical tumours. This remains controversial and it is often unclear how concentrations can be achieved and maintained *in vivo* to produce this response. Glucosinolate-derived isothiocyanates (ITCs) inhibited the growth of human tumour cells *in vitro*. The median growth inhibitory concentration (GC_{50}) values were in the range 0.7–40 µM (Table 19.3). Apoptotic death was a characteristic of dose-limiting toxicity; at much higher concentrations, ITCs

induced necrotic cell death. ITCs may also be toxic to non-malignant cells *in vitro* (Table 19.4). Mixtures of dietary ITCs induced similar effects (Fimognari *et al.*, 2005). There is relatively low toxicity of ITCs to colonic, prostate, bronchial and oral mucosal epithelial cells, but significant toxicity to keratinocytes and renal tubular, cervical and mammary epithelial cells (reviewed by Thornalley and IARC Workgroup, 2004).

Assessment of inhibition of cell growth is made with greatest security by counting viable cell numbers in a haemocytometer or by automated low-cost image analysis. Viability is assessed by exclusion of the dye, Trypan blue (Maruhashi *et al.*, 1994). Other indirect methods have been used: reduction of the redox dye 3-(4,5-dimethylthiazol-2-yl)-2,5-diphenyl tetrazolium bromide (MTT), (3-(4,5-dimethylthiazol-2-yl)-5-(3-carboxymethoxyphenyl)-2-(4-sulfophenyl)-2H-tetrazolium (MTS) or related compounds (Wang *et al.*, 1996), uptake of neutral red (Repetto *et al.*, 2008) and sulforhodamine B detection of protein content (Vichai and Kirtikara, 2006). The colorimetric methods relate metabolic reduction activity or protein dye binding to cell number. If treatment of cells with bioactive compounds interferes in the metabolic reduction activity or cell protein content without change in cell number, then false positive outcomes will be registered. Viable cell number counts are the most secure method for assessing effect on cell growth and viability.

19.3.4 Assessment of cytotoxicity

Toxicity of bioactive compounds is usually undesirable unless it is selective to malignant cells or invading microorganisms. In pharmacological studies of bioactive compounds and fruit and vegetable extracts in cell culture models, initial study of the effect of concentration dependence on cell viability, at least up to the peak plasma concentration, is advisable to ensure that other pharmacological effects are not compromised by decreased cell viability. As for all cell culture studies, bioactive compounds and fruit and vegetable extracts should be endotoxin-free before

Table 19.3. Inhibition of human tumour cell growth and induction of cytotoxicity by dietary isothiocyanates *in vitro*.

Compound	Cell	GC_{50} (µM)	TC_{50} (µM)	Reference
4-Methylsulfinylbutyl isothiocyanate (SFN)	Jurkat	4.9	–	Fimognari *et al.* (2002b)
	HT29	15	–	Gamet-Payrastre *et al.* (1998)
	LS-174	55	–	Bonnesen *et al.* (2001)
	Caco-2	40	–	Bonnesen *et al.* (2001)
	P-3	20	–	Nastruzzi *et al.* (2000)
	K562	11	–	Tang *et al.* (2006)
	UM-UC-3	6.8	–	
Allyl isothiocyanate	HL60	2.6	11	Xu and Thornalley (2000b)
	ML-1	2.6	7.7	Xu and Thornalley (2000b)
	HeLa	20	–	Hasegawa *et al.* (1993)
Benzyl isothiocyanate	K562	1.5	–	Nastruzzi *et al.* (2000)
	LS-174	15	–	Bonnesen *et al.* (2001)
	Caco-2	2	–	Bonnesen *et al.* (2001)
	HT29	3	–	Musk *et al.* (1995)
	Jurkat	2	–	Chen *et al.* (1998)
	HeLa	2	–	Hasegawa *et al.* (1993)
Phenethyl isothiocyanate	HL60	1.5	5.0	Xu and Thornalley (2000b)
	ML-1	2.7	3.3	Xu and Thornalley (2000b)
	LS-174	20	–	Bonnesen *et al.* (2001)
	Caco-2	12	–	Bonnesen *et al.* (2001)
	HeLa	2	–	Hasegawa *et al.* (1993)
	Jurkat	3	–	Chen *et al.* (1998)
	PC-3	6	–	Xiao and Singh (2002)
Broccoli isothiocyanates (extract)	UM-UC-3	6.6	–	Tang *et al.* (2006)

Notes: Cell lines: Caco-2, human colonic adenocarcinomas; HCEC, SV40 T-antigen immortalized human colonic epithelial cells; HeLa, human cervical carcinoma; HL60, human acute myeloblastic leukaemia; HT29, human colonic adenocarcinoma cells; Jurkat, human T-cell leukaemia; K562, human erythroleukaemia; LS-174, human colorectal adenocarcinomas; ML-1, human myeloblastic leukaemia; PC-3, human prostate cancer; UM-UC-3, bladder carcinoma.

Table 19.4. Toxicity of dietary isothiocyanates to non-malignant cells *in vitro*.

Compound	Species origin	Cell type	Cell growth TC_{50} (µM)	Reference
SFN	Human	Lymphocytes	30	Fimognari *et al.* (2002a)
AITC	Rat	RL-4 liver epithelial	175	Bruggeman *et al.* (1986)
BITC	Human	Colon epithelial	22	Bonnesen *et al.* (2001)
	Rat	RL-4 liver epithelial	25	Bruggeman *et al.* (1986)
PEITC	Human	Colon epithelial	20	Bonnesen *et al.* (2001)
	Human	Lymphocytes	53	Xu and Thornalley (2000b)
	Human	Keratinocytes	0.5	Elmore *et al.* (2001)
	Human	Renal tubular epithelial	4.5	Elmore *et al.* (2001)
	Human	Mammary epithelial	1.3	Elmore *et al.* (2001)
	Human	Cervical epithelial	3.2	Elmore *et al.* (2001)
	Human	Prostate epithelial	12	Elmore *et al.* (2001)
	Human	Bronchial epithelial	15	Elmore *et al.* (2001)
	Human	Oral mucosal epithelial	19	Elmore *et al.* (2001)
	Human	Liver epithelial (Chang)	20	Elmore *et al.* (2001)
Resveratrol	Bovine	Vascular smooth muscle cells	c.10	Poussier *et al.* (2005)

being added to cultured cells for evaluation of cell viability. Cell death may be induced by several mechanisms: apoptosis, autophagy, necrosis, anoikis (detachment stimulated apoptosis) and other mechanisms. The generally agreed indications of cell death are when one or more of the following molecular or morphological criteria are met: (i) the cell has lost the integrity of the plasma membrane, as defined by vital dyes *in vitro*; (ii) the cell including its nucleus has undergone complete fragmentation into discrete bodies (which are frequently referred to as 'apoptotic bodies'); and/or (iii) its corpse (or its fragments) have been engulfed by an adjacent cell *in vivo* (Kroemer *et al.*, 2005).

One of the most commonly used methods to assess cell death is exclusion of the dye, Trypan blue. This has been automated recently for low-cost image analysis (Maruhashi *et al.*, 1994). In a cytotoxic response, some cells may die and fragment prior to visualization and hence the total cell number seen (viable plus non-viable cells) may be less than in control cultures. The commitment to cell death or point-of-no-return may have been reached many hours before dying cells become permeable to Trypan blue. This can be investigated by examining the minimum period of exposure to bioactive compound (washing it away and replacing with fresh medium thereafter) required for later development of decreased cell viability.

Surrogate markers of cell death relate to biochemical detection of changes thought to be characteristic of commitment to cell death. In this regard, a profound increase in caspase activity, complete permeabilization of the outer mitochondrial membrane, or exposure of phosphatidylserine residues are markers of commitment to cell death by apoptosis. There are some examples, however, where these phenomena are found without cell death, and hence evidence of the harder end points of cell death should be sought (Kroemer *et al.*, 2005). Other methods used for assessment of apoptosis are: the terminal deoxynucleotidyl transferase (TDT)-mediated 2'-deoxyuridine-5'-triphosphate (dUTP) nick end labelling (TUNEL) assay (which utilizes TDT to add a poly-U tail on to DNA strand breaks, which are then detected using immunocytochemis-

try) (Heatwole, 1999), and propidium iodide staining of cellular DNA and DNA fragments (Riccardi and Nicoletti, 2006). Dual staining with Hoechst 33342 and propidium iodide can discriminate between cells dying by necrosis and apoptosis: necrotic cells show red fluorescence due to the uptake of propidium iodide, apoptotic cells show bright blue fluorescence and normal cells show low blue, low red fluorescence (Ormerod *et al.*, 1993). Anoikis is detected by analysis for apoptosis in cells floating free of the extracellular matrix to which they are normally adherent. All methods require care in interpretation and the reader is directed to the expert references cited herein for further guidance.

19.3.5 Anti-inflammatory activity

The anti-inflammatory properties of some bioactive compound ITCs have been investigated in cellular systems. The inhibition of bioactive compounds on formation of superoxide in the respiratory burst of neutrophils stimulated by phorbol ester – an activator of protein kinase C – has been studied. The median inhibitory concentration (EC_{50}) value was 3.5 µM for phenethyl isothiocyanate (PEITC) and 0.26 µM for hydroxytyrosol (Visioli *et al.*, 1998; Gerhauser *et al.*, 2003). The inhibition of lipopolysaccharide (LPS)-induced inflammatory responses by bioactive compounds has also been investigated via the formation of nitric oxide in a macrophage cell line – an inducible nitric oxide synthase response, the secretion of tumour necrosis factor (TNF) and the expression of COX-2 and associated formation of prostaglandin E_2 (Heiss *et al.*, 2001; Ippoushi *et al.*, 2002; Gerhauser *et al.*, 2003; Allen and Walker, 2003).

19.3.6 Antibacterial activity

Antibacterial activity of bioactive compounds has been investigated in relation to countering gastrointestinal infection by *Helicobacter pylori*, which has been linked to gastritis, peptic ulcer disease and gastric cancer. Bioactive compounds are also potentially health beneficial in countering dermatological infections and foodborne bacterial pathogens. Both glucosinolate-derived ITCs, the flavonoids quercetin and cranberry juice extract rich in polyphenols have been found to have antibacterial activity (Fahey *et al.*, 2002; Ibrahim *et al.*, 2007; Wu *et al.*, 2008).

19.3.7 Other effects: effect on endogenous conversion of *cis*- to *trans*-fatty acids

Consumption of unsaturated fatty acids with at least one double bond in the *trans* configuration has been linked to increased cardiovascular disease. This is thought to be mediated by effects on membrane structure, function and cell signalling (Mozaffarian *et al.*, 2006). It has emerged recently that *trans*-fatty acid content of cells can also be increased by *in situ* geometric isomerization of *cis*-fatty acids of membrane lipids. This process is mediated by thiyl free radicals formed in oxidative stress (Ferreri *et al.*, 2005). It will be of future interest to study the effects of dietary bioactive compounds for their ability to prevent the *cis*-/*trans*-isomerization of fatty acids in membrane lipids and assess the link to risk of cardiovascular disease.

19.4 Preclinical Assessment of Bioactive Compounds

The pharmacological activities of bioactive compounds are evaluated in preclinical studies typically in laboratory animals (mice, rats, hamsters and others). Initially, bioactive compounds are administered to healthy animals and pharmacological effects assessed. Any adverse effects are also recorded. Traditionally, the effects of bioactives compounds on target enzyme activities are assessed. With modern transcriptomic, proteomic and metabolomic techniques, a different approach may be taken. Firstly, gene microarray analysis can give a qualitative assessment of global changes in expression of c.30,000 genes. This is followed by quantitative analysis of expression of a selection of genes of particular interest, by real-time RT-PCR array. Changes in

gene product levels may then be investigated by quantitative Western blotting and also assay of relative enzyme activities. Finally, changes in metabolites of interest may also be measured. Changes in gene expression will indicate the type of the most potent pharmacological activity: cancer chemopreventive activity, antioxidant activity, antilipogenic activity or other. The bioactive compounds may then be taken forward for evaluation in specific disease models in further preclinical studies. Examples are given in Table 19.5.

19.4.1 Chemoprevention of cancer

Experimental models used in preclinical assessment of chemoprevention of cancer are typified by those employed in assessment of cancer preventive activity of dietary-derived isothiocyanates and *Brassica* vegetable consumption – reviewed in the International Agency for Research on Cancer (IARC) handbook (Thornalley and IARC Workgroup, 2004). The standard protocol involves administration of a bioactive compound of dietary intervention before, during and after administration of a chemical carcinogen and assessing tumour incidence, tumour mass or related biomarkers of neoplasma (e.g. aberrant crypt foci). The IARC report concluded there was sufficient evidence to indicate that consumption of cruciferous vegetables and intake of aromatic dietary isothiocyanates and indole-3-carbinol (I3C) decreased the risk of cancers of the colon, mammary gland and liver in experimental animal models; there was insufficient evidence of similar effects of glucosinolates or SFN.

19.4.2 Prevention of cardiovascular disease

Prevention of cardiovascular disease in experimental animals commonly targets suppression of atherosclerotic plaque formation in the atherosclerosis-prone apolipoprotein E knockout mouse or hamster (Zadelaar *et al.*, 2007). Prevention of atherosclerosis by quercetin and catechin has been studied by Hayek *et al.* (1997). The effect of extracts has also

Table 19.5. Evaluation of bioactive compounds in preclinical animal models.

Target disease	Animal model	Specimen dosing	Primary end point	Other	Reference
None	Mouse, Balb/c	SFN, c.160 mg/kg	Microarray gene expression		Thimmulappa et al. (2002)
None	Rat, Fisher F344	SFN, c.60 mg/kg	Microarray gene expression		Hu et al. (2004)
Cancer	Mouse, rat – various strains	0.075–1.25 μmol/g PEITC	Tumour incidence	Tumour mass. Biomarkers – aberrant crypt foci, etc.	Thornalley and IARC Workgroup (2004)
Cardiovascular disease	Apolipoprotein E deficient mice	Quercetin and catechin (50 μg/day) and red wine (0.5 ml/day)	Atherosclerotic plaque area	Serum paraoxonase and LDL aggregation	Hayek et al. (1997)
Hypertension	Spontaneously hypertensive stroke-prone rat	Broccoli sprouts, dried (200 mg/day)	Blood pressure – decreased by 20 mmHg	Improved endothelial-dependent relaxation, increased heart and kidney GSH, GSRd and GSHPx	Wu et al. (2004a)
Longevity	Caenorhabditis elegans	Resveratrol, 100 μM in food	Median lifespan		Viswanathan et al. (2005); Bass et al. (2007)
	Drosophila melanogaster	Resveratrol, 1 μM–1 mM in food	Median lifespan	Fecundity	Baur and Sinclair (2006); Bass et al. (2007)
	Mice, male C57BL/6NIA	Resveratrol, 0.01 and 0.04% (w/w) diet	Median lifespan	Vascular function, histopathology, other	Fraley and Tsimikas (2006)
Arthritis	Rat, adjuvant-induced arthritis	5 × 30 mg/kg quercetin	Clinical score (swelling, immobility)	Inflammatory mediators concentrations	Mamani-Matsuda et al. (2006)

been studied: green and black tea, pomegranate juice, grape extracts, red wine, dealcoholized wine and wine polyphenols. The effects observed may be due to the antioxidant and anti-inflammatory activities of the bioactives. The doses of bioactive compound applied for significant beneficial effect are, however, often much higher than typically seen in dietary exposure of human subjects (reviewed in Manach *et al.*, 2005).

19.4.3 Renal disease

Renal disease is a major health concern as it is associated with markedly decreased life expectancy and increased risk of cardiovascular disease. One of the major causes of renal disease is diabetes. Bioactive compounds may be evaluated for ability to prevent diabetic nephropathy in a streptozotocin-induced diabetic rat model (McNeil, 1999). Early stage nephropathy develops over 12–24 weeks, as indicated by increased urinary albumin excretion rate. Tea catechins and other bioactive compounds suppress the development of diabetic nephropathy (Hase *et al.*, 2006). A model of chronic renal failure can be produced in rats by partial nephrectomy, where one whole kidney and part of the other kidney is removed surgically. Bioactive compounds were found to improve residual renal function in this model (Yokozawa *et al.*, 1996).

19.4.4 Survival and longevity studies

Bioactive compounds are often evaluated for effects on survival and lifespan in experimental models of ageing – such as the nematode *Caenorhabditis elegans*, the fruitfly *Drosophila melanogaster* and laboratory mice. The compounds are often thought to act as antioxidants or as dietary restriction mimetics – in the latter case, seeking to mimic the life extension effect achieved by caloric restriction regimes (Lane *et al.*, 2007). Many claims for life-extending effects have not been confirmed by later studies. For example, resveratrol was claimed to increase the lifespan of *C. elegans* (Viswanathan *et al.*, 2005) and *Drosophila* (Baur

and Sinclair, 2006), but these claims could not be verified in later studies (Bass *et al.*, 2007). Recent translation of these findings to mice found some beneficial effects of resveratrol on general health but no significant increase in lifespan (Fraley and Tsimikas, 2006). Longevity studies have often been poorly designed – particularly underpowered and susceptible to false positive outcomes. In response to these and other concerns, the US National Institute on Ageing has formulated standard operating procedures for conducting such studies (Nadon *et al.*, 2008).

19.4.5 Other effects

The anti-inflammatory activity of bioactive compounds is evaluated in preclinical studies against experimental models of arthritis and endotoxaemia. Quercetin decreased inflammation in rat adjuvant-induced arthritis (Mamani-Matsuda *et al.*, 2006).

SFN (*c*.2 mg/kg, daily) inhibited angiogenesis – as judged by decease of blood content of a subcutaneous implant in mice of an extracellular matrix Matrigel™ plug impregnated with recombinant murine vascular endothelial growth factor (VEGF) and heparin (Jackson *et al.*, 2007). This dose, however, is *c*.20-fold higher than achieved maximally by consumption of 100 g broccoli (Song and Thornalley, 2007).

19.5 Clinical Assessment of Bioactive Compounds

Clinical assessments of the health benefits of bioactive compounds are often judged by dietary interventions – change in the dietary content of compounds known to be rich in the bioactive compound of interest. The effects on clinical end points or biomarkers indicative of health benefit are often predicted to be small and therefore studies must be designed, randomized and powered for significant effect. Where possible, environmental factors should be controlled for and participant and investigator awareness of the study group allocation of each subject should be prevented (by participant and investigator blinding). Increased statistical power may be achieved by expanding subject or patient numbers in the study, by crossover study design in which participant groups are switched between periods of normal and intervention dietary regime, and by repeated observations over time or 'repeated measures' analysis. A simple and effective way to control for environmental factors and provide for participant and investigator blinding is to identify two varieties of dietary component (fruit or vegetable, for example), one relatively poor and one relatively rich in the bioactive compound of interest, and produce a dietary intervention with these in a randomized, participant- and investigator-blinded design. In addition, periods of abstaining from the dietary component of interest, or washout period before and after the dietary intervention – providing it is ethical to do so (with no significant and long-term health impairment likely) – will assist in assessing the health benefit of the dietary bioactive compound and whether the benefits achieved are maintained or reversed with change in diet.

Given an appropriate study design and protocol, there is also assessment and monitoring of compliance to the dietary regime to be considered. Dietary compliance may be assessed by questionnaire, interaction with the subject by calls and visits by the research clinic, and by urinary metabolite analysis. Steps to improve compliance have been: diet counselling (at the research clinic and remotely by calls and electronic communication media), direct supply of dietary supplement, written goals and instructions, portion size guides, study cookbook containing appropriate recipes, skills training in a teaching kitchen, subject motivation assessment in pre-screening, diet diaries with periodic review with a researcher, and dietician-led group sessions (Fowke *et al.*, 2006).

19.5.1 Biomarkers

The use of biomarkers in the surrogate assessment of clinical end points is advantageous because: (i) they often indicate early, preclinical stages or risk of chronic disease and can be accessed in subjects in otherwise good health;

(ii) they are present in a larger population, which decreases subject recruitment problems; (iii) they may be measured with non-invasive or minimally invasive procedures (sampling blood and urine); (iv) they are often more responsive to interventions than established disease; and (v) they may be relatively inexpensive to measure. Clinical studies using biomarkers depend on the validity of the biomarker employed. The biomarker should be a validated surrogate end point for the disease or abnormal metabolic state of interest.

One of the most common biomarker assessments used is that of oxidative stress. The Biomarkers of Oxidative Stress Study (BOSS) was designed to validate markers of oxidative stress in specific animal models of oxidative stress: (i) carbon tetrachloride poisoning; (ii) environmental exposure to ozone; (iii) systemic exposure to endotoxin; and (iv) continuous skin exposure to cumene hydroperoxide. It remains unclear how these models translate to oxidative stress in the clinical setting; initial conclusions indicate that measurements of malondialdehyde (MDA) and isoprostanes in plasma and urine and 8-hydroxydeoxyguanosine in urine are potential biomarkers of oxidative stress (Kadiiska *et al.*, 2005).

A further area of bioactive compound evaluation where biomarkers offer critical insights is in the chemoprevention of cancer. The IARC Working Party assessing chemoprevention of cancer by cruciferous vegetables accepted evidence of the following biomarkers: (i) detectable precancerous changes in tissue assessed by histology; (ii) change in gene expression thought to play a causal role; (iii) DNA damage; (iv) exposure to known carcinogen; and (v) effects on metabolic factors thought to be involved in cancer aetiology. Intervention studies with cruciferous vegetables were reviewed (Thornalley and IARC Workgroup, 2004).

19.5.2 Clinical evaluation of bioactive compounds

A selection of clinical studies on the effect of bioactive compounds derived from fruit and vegetable is presented in Table 19.6. The reader is also directed to several detailed

Table 19.6. Evaluation of bioactive compounds in clinical intervention studies.

Target disease/abnormal physiological state	Study design (participant number)	Dietary intervention	Primary end point	Outcome	Reference
Colon adenoma patients	Randomized, crossover (n = 20)	Brassica vegetables (>160 g/day) for 4 weeks	Urinary isoprostanes	20% decrease with intervention	Fowke et al. (2006)
Normal healthy subjects	Three-phase crossover (n = 16)	Broccoli soup (normal and high glucosinolate)	Gastric mucosa gene expression	Induction of expression of GCLm TrRd	Gasper et al. (2007)
Normal healthy subjects	Open trial (n = 12)	Vegetable soup, 'gazpacho', 500 ml/day	Biomarkers of oxidative stress and vascular inflammation	28% decrease in plasma isoprostanes	Sanchez-Moreno et al. (2004)
Normal healthy subjects	Three-phase crossover (n = 59)	Baseline diet, then sunflower oil diet for 25 days, followed by rapeseed oil for 25 days	Markers of lipid metabolism	Decreased total cholesterol and LDL	Sanchez-Moreno et al. (2004)
Normal healthy subjects	Double-blind, randomized, placebo controlled study (n = 173)	Oil supplements for 8 months: placebo, sunflower oil, evening primrose oil, soyabean oil, tuna fish oil and tuna/evening primrose oil mix	Endothelium-dependent and -independent vascular responses	Acetylcholine responses were improved significantly after tuna oil supplementation	Khan et al. (2003)
Normal healthy subjects	Randomized open trial (n = 175)	β-Carotene, lutein, lycopene (15 mg/day) or placebo/day for 12 weeks	LDL oxidation, blood GSH, plasma protein thiols and antioxidant enzyme activities	No significant effects	Hininger et al. (2001)

reviews of these studies – including analysis of combinations of studies: cruciferous vegetables (Thornalley and IARC Workgroup, 2004), lycopene and tomato paste (Basu and Imrhan, 2007), flavonoids (Erlund, 2004), cranberries (Neto, 2007), blackcurrants (Lister *et al.*, 2002), n-3 polyunsaturated fatty acids (Hartweg *et al.*, 2007), tea polyphenols (Sajilata *et al.*, 2008) and other chapters in this book. There is often insufficient and inadequate clinical evidence of the health benefit of bioactive compounds derived from fruit and vegetables, and a tendency to claim clinical benefits of bioactive compounds before sufficient evidence has been obtained to sustain such claims.

19.6 Conclusions

The wealth and diversity of bioactive compounds present in and extracted from fruit, vegetables and other plants provide a rich palate on which to draw to produce valuable pharmacology for health benefit. Many compounds appear to exert their strongest effects by enhancing the antistress gene response, strengthening the body's innate response to stress and tissue damage. Working by this transcriptional and enzymatic gene product mechanism, the potential potency of action of bioactive compounds is much greater than small molecule antioxidants with stoichiometric action. Nonantioxidant effects – such as antilipogenic activities – will also likely provide future valuable therapeutics for cardiovascular disease. There remains much further preclinical and clinical evaluation to be done to reap the potential of dietary bioactive compounds.

19.7 Acknowledgements

The authors thank the Biotechnology and Biological Sciences Research Council (UK), the Wellcome Trust, British Heart Foundation (BHF) and Diabetes UK for support for their bioactive compound research. NR is a BHF Intermediate Research Fellow.

References

Ahmed, N., Babaei-Jadidi, R., Howell, S.K., Thornalley, P.J. and Beisswenger, P.J. (2005) Glycated and oxidized protein degradation products are indicators of fasting and postprandial hyperglycemia in diabetes. *Diabetes Care* 28, 2465–2471.

Allen, K.V. and Walker, J.D. (2003) Microalbuminuria and mortality in long-term duration type 1 diabetes. *Diabetes Care* 26, 2389–2391.

Bass, T.M., Weinkove, D., Houthoofd, K., Gems, D. and Partridge, L. (2007) Effects of resveratrol on lifespan in *Drosophila melanogaster* and *Caenorhabditis elegans*. *Mechanisms of Ageing and Development* 128, 546–552.

Basu, A. and Imrhan, V. (2007) Tomatoes versus lycopene in oxidative stress and carcinogenesis: conclusions from clinical trials. *European Journal of Clinical Nutrition* 61, 295–303.

Baur, J.A. and Sinclair, D.A. (2006) Therapeutic potential of resveratrol: the *in vivo* evidence. *Nature Reviews Drug Discovery* 5, 493–506.

Ben Dor, A., Steiner, M., Gheber, L., Danilenko, M., Dubi, N., Linnewiel, K., Zick, A., Sharoni, Y. and Levy, J. (2005) Carotenoids activate the antioxidant response element transcription system. *Molecular Cancer Therapeutics* 4, 177–186.

Bonnesen, C., Eggleston, I.M. and Hayes, J.D. (2001) Dietary indoles and isothiocyanates that are generated from cruciferous vegetables can both stimulate apoptosis and confer protection against DNA damage in human colon cell lines. *Cancer Research* 61, 6120–6130.

Borradaile, N.M., de Dreu, L.E. and Huff, M.W. (2003) Inhibition of net HepG2 cell apolipoprotein B secretion by the citrus flavonoid naringenin involves activation of phosphatidylinositol 3-kinase, independent of insulin receptor substrate-1 phosphorylation. *Diabetes* 52, 2554–2561.

Bors, W., Michel, C. and Saran, M. (1984) Inhibition of the bleaching of the carotenoid crocin – a rapid test for quantifying antioxidant activity. *Biochimica et Biophysica Acta* 796, 312–319.

Bruggeman, I.M., Temmink, J.H.M. and Van Bladeren, P.J. (1986) Glutathione- and cysteine-mediated cyto-toxicity of allyl and benzyl iothiocyanate. *Toxicology and Applied Pharmacology* 83, 349–359.

Calabrese, E.J., Baldwin, L.A. and Holland, C.D. (1999) Hormesis: a highly generalizable and reproducible phenomenon with important implications for risk assessment. *Risk Analysis* 19, 261–281.

Cao, G. and Prior, R.L. (1998) Comparison of different analytical methods for assessing total antioxidant capacity of human serum. *Clinical Chemistry* 44, 1309–1315.

Cao, G.H., Alessio, H.M. and Cutler, R.G. (1993) Oxygen-radical absorbency capacity assay for antioxidants. *Free Radical Biology and Medicine* 14, 303–311.

Chen, C., Pung, D., Leong, V., Hebbar, V., Shen, G., Nair, S., Li, W. and Kong, A.N.T. (2004) Induction of detoxifying enzymes by garlic organosulfur compounds through transcription factor Nrf2: effect of chemical structure and stress signals. *Free Radical Biology and Medicine* 37, 1578–1590.

Chen, X.L., Varner, S.E., Rao, A.S., Grey, J.Y., Thomas, S., Cook, C.K., Wasserman, M.A., Medford, R.M., Jaiswal, A.K. and Kunsch, C. (2003) Laminar flow induction of antioxidant response element-mediated genes in endothelial cells. *Journal of Biological Chemistry* 278, 703–711.

Chen, Y.-R., Wang, W., Kong, T. and Tan, T.-H. (1998) Molecular mechanism of c-Jun N-terminal kinase-mediated apoptosis induced by anticarcinogenic isothiocyanates. *Journal of Biological Chemistry* 273, 1769–1775.

Davalos, A., Fernandez-Hernando, C., Cerrato, F., Martinez-Botas, J., Gomez-Coronado, D., Gomez-Cordoves, C. and Lasuncion, M.A. (2006) Red grape juice polyphenols alter cholesterol homeostasis and increase LDL-receptor activity in human cells *in vitro*. *Journal of Nutrition* 136, 1766–1773.

Dinkova-Kostova, A.T., Holtzclaw, W.D., Cole, R.N., Itoh, K., Wakabayashi, N., Katoh, Y., Yamamoto, M. and Talalay, P. (2002) Direct evidence that sulfhydryl groups of Keap1 are the sensors regulating induction of phase 2 enzymes that protect against carcinogens and oxidants. *Proceedings of the National Academy of Sciences of the United States of America* 99, 11,908–11,913.

Elmore, E., Luc, T.-T., Steele, A.E. and Redpath, J.L. (2001) Comparative tissue-specific toxicities of 20 cancer preventive agents using cultured cells from 8 different normal human epithelia. *In Vitro and Molecular Toxicology* 14, 191–207.

Erlund, I. (2004) Review of the flavonoids quercetin, hesperetin naringenin. Dietary sources, bioactivities, and epidemiology. *Nutrition Research* 24, 851–874.

Fahey, J.W., Haristoy, X., Dolan, P.M., Kensler, T.W., Scholtus, I., Stephenson, S.S., Talalay, P. and Lozniewski, A. (2002) Sulforaphane inhibits extracellular, intracellular, and antibiotic-resistant strains of *Helicobacter pylori* and prevents benzo[a]pyrene-induced stomach tumours. *Proceedings of the National Academy of Sciences of the United States of America* 99, 7610–7615.

Ferreri, C., Kratzsch, S., Brede, O., Marciniak, B. and Chatgilialoglu, C. (2005) Trans lipid formation induced by thiols in human monocytic leukemia cells. *Free Radical Biology and Medicine* 38, 1180–1187.

Fimognari, C., Nusse, M., Berti, F., Iori, R., Cantelli-Forti, G. and Hrelia, P. (2002a) Cyclin D3 and p53 mediate sulforaphane-induced cell cycle delay and apoptosis in non-transformed human T lymphocytes. *Cellular and Molecular Life Sciences* 59, 2004–2012.

Fimognari, C., Nussse, M., Cesari, R., Iori, R., Cantelli-Forti, G. and Hrelia, P. (2002b) Growth inhibition, cell-cycle arrest and apoptosis in human T-cell leukemia by the isothiocyanate sulforaphane. *Carcinogenesis* 23, 581–586.

Fimognari, C., Berti, F., Iori, R., Cantelli-Forti, G. and Hrelia, P. (2005) Micronucleus formation and induction of apoptosis by different isothiocyanates and a mixture of isothiocyanates in human lymphocyte cultures. *Mutation Research-Genetic Toxicology and Environmental Mutagenesis* 582, 1–10.

Fowke, J.H., Morrow, J.D., Motley, S., Bostick, R.M. and Ness, R.M. (2006) Brassica vegetable consumption reduces urinary F2-isoprostane levels independent of micronutrient intake. *Carcinogenesis* 27, 2096–2102.

Fraley, A.E. and Tsimikas, S. (2006) Clinical applications of circulating oxidized low-density lipoprotein biomarkers in cardiovascular disease. *Current Opinion in Lipidology* 17, 502–509.

Frydoonfar, H.R., McGrath, D.R. and Spigelman, A.D. (2003) The effect of indole-3-carbinol and sulforaphane on a prostate cancer cell line. *Australian and New Zealand Journal of Surgery* 73, 154–156.

Gamet-Payrastre, L., Lumeau, S., Gasc, N., Cassar, G., Rollin, P. and Tulliez, J. (1998) Selective cytostatic and cytotoxic effects of glucosinolates hydrolysis products on human colon cancer cells *in vitro*. *Anti-Cancer Drugs* 9, 141–148.

Gao, L., Wang, J.K., Sekhar, K.R., Yin, H.Y., Yared, N.F., Schneider, S.N., Sasi, S., Dalton, T.P., Anderson, M.E., Chan, J.Y., Morrow, J.D. and Freeman, M.L. (2007) Novel n-3 fatty acid oxidation products activate Nrf2 by destabilizing the association between Keap1 and Cullin3. *Journal of Biological Chemistry* 282, 2529–2537.

Gasper, A.V., Traka, M., Bacon, J.R., Smith, J.A., Taylor, M.A., Hawkey, C.J., Barrett, D.A. and Mithen, R.F. (2007) Consuming broccoli does not induce genes associated with xenobiotic metabolism and cell cycle control in human gastric mucosa. *Journal of Nutrition* 137, 1718–1724.

Gems, D. and Partridge, L. (2008) Stress–response hormesis and aging: 'that which does not kill us makes us stronger'. *Cell Metabolism* 7, 200–203.

Gerhauser, C., Klimo, K., Heiss, E., Neumann, I., Gamal-Eldeen, A., Knauft, J., Liu, G.-Y., Sitthimonchai, S. and Frank, N. (2003) Mechanism-based *in vitro* screening of potential cancer chemopreventive agents. *Mutation Research* 523–524, 163–172.

Ghiselli, A., Serafini, M., Natella, F. and Scaccini, C. (2000) Total antioxidant capacity as a tool to assess redox status: critical view and experimental data. *Free Radical Biology and Medicine* 29, 1106–1114.

Gorinstein, S., Jastrzebski, Z., Leontowicz, H., Leontowicz, M., Namiesnik, J., Najman, K., Park, Y.S., Heo, B.G., Cho, J.Y. and Bae, J.H. (2009) Comparative control of the bioactivity of some frequently consumed vegetables subjected to different processing conditions. *Food Control* 20, 407–413.

Hanlon, P.R., Webber, D.M. and Barnes, D.M. (2007) Aqueous extract from Spanish black radish (*Raphanus sativus* L. var. niger) induces detoxification enzymes in the HepG2 human hepatoma cell line. *Journal of Agricultural and Food Chemistry* 55, 6439–6446.

Hartweg, J., Farmer, A.J., Perera, R., Holman, R.R. and Neil, H.A.W. (2007) Meta-analysis of the effects of n-3 polyunsaturated fatty acids on lipoproteins and other emerging lipid cardiovascular risk markers in patients with type 2 diabetes. *Diabetologia* 50, 1593–1602.

Hase, M., Babazono, T., Karibe, S., Kinae, N. and Iwamoto, Y. (2006) Renoprotective effects of tea catechin in streptozotocin- induced diabetic rats. *International Urology and Nephrology* 38, 693–699.

Hasegawa, T., Nishino, H. and Iwashima, A. (1993) Isothiocyanates inhibit cell cycle progression of HeLa cells at G_2/M phase. *Anti-Cancer Drugs* 4, 273–279.

Hayek, T., Fuhrman, B., Vaya, J., Rosenblat, M., Belinky, P., Coleman, R., Elis, A. and Aviram, M. (1997) Reduced progression of atherosclerosis in apolipoprotein E-deficient mice following consumption of red wine, or its polyphenols quercetin or catechin, is associated with reduced susceptibility of LDL to oxidation and aggregation. *Arteriosclerosis, Thrombosis, and Vascular Biology* 17, 2744–2752.

Healy, Z.R., Lee, N.H., Gao, X.Q., Goldring, M.B., Talalay, P., Kensler, T.W. and Konstantopoulos, K. (2005) Divergent responses of chondrocytes and endothelial cells to shear stress: cross-talk among COX-2, the phase 2 response, and apoptosis. *Proceedings of the National Academy of Sciences of the United States of America* 102, 14,010–14,015.

Heatwole, V.M. (1999) TUNEL assay for apoptotic cells. *Methods in Molecular Biology* 115, 141–148.

Heiss, E., Herhaus, C., Klimo, K., Bartsch, H. and Gerhauser, C. (2001) Nuclear factor kappa B is a molecular target for sulforaphane-mediated anti-inflammatory mechanisms. *Journal of Biological Chemistry* 216, 32008–32015.

Hiki, N., Berger, D., Dentener, M.A., Mimura, Y., Buurman, M.A., Prigl, C., Seidelmann, M., Tsuji, E., Kaminishi, M. and Beger, H.G. (1999) Changes in endotoxin-binding proteins during major elective surgery: important role for soluble CD14 in regulation of biological activity of systemic endotoxin. *Clinical and Diagnostic Laboratory Immunology* 6, 844–850.

Hininger, I.A., Meyer-Wenger, A., Moser, U., Wright, A., Southon, S., Thurnham, D., Chopra, M., Van Den Berg, H., Olmedilla, B., Favier, A.E. and Roussel, A.M. (2001) No significant effects of lutein, lycopene or {beta}-carotene supplementation on biological markers of oxidative stress and LDL oxidizability in healthy adult subjects. *Journal of the American College of Nutrition* 20, 232–238.

Hintze, K.J., Wald, K.A., Zeng, H., Jeffery, E.H. and Finley, J.W. (2003) Thioredoxin reductase in human hepatoma cells is transcriptionally regulated by sulforaphane and other electrophiles via an antioxidant response element. *Journal of Nutrition* 133, 2721–2727.

Holst, B. and Williamson, G. (2008) Nutrients and phytochemicals: from bioavailability to bioefficacy beyond antioxidants. *Current Opinion in Biotechnology* 19, 73–82.

Hu, R., Hebbar, V., Kim, B.R., Chen, C., Winnik, B., Buckley, B., Soteropoulos, P., Tolias, P., Hart, R.P. and Kong, A.N.T. (2004) *In vivo* pharmacokinetics and regulation of gene expression profiles by isothiocyanate sulforaphane in the rat. *Journal of Pharmacology and Experimental Therapeutics* 310, 263–271.

Huang, D., Ou, B., Hampsch-Woodill, M., Flanagan, J.A. and Deemer, E.K. (2002a) Development and validation of oxygen radical absorbance capacity assay for lipophilic antioxidants using randomly methylated beta-cyclodextrin as the solubility enhancer. *Journal of Agricultural and Food Chemistry* 50, 1815–1821.

Huang, D., Ou, B., Hampsch-Woodill, M., Flanagan, J.A. and Prior, R.L. (2002b) High-throughput assay of oxygen radical absorbance capacity (ORAC) using a multichannel liquid handling system coupled with a microplate fluorescence reader in 96-well format. *Journal of Agricultural and Food Chemistry* 50, 4437–4444.

Huang, D., Ou, B. and Prior, R.L. (2005) The chemistry behind antioxidant capacity assays. *Journal of Agricultural and Food Chemistry* 53, 1841–1856.

Ibrahim, E.S.A., Hassan, M.A., El Mahdy, M.M. and Mohamed, A.S. (2007) Formulation and evaluation of quercetin in certain dermatological preparations. *Journal of Drug Delivery Science and Technology* 17, 431–436.

Ippoushi, K., Itou, H., Azuma, K. and Higashio, H. (2002) Effect of naturally occurring organosulfur compounds on nitric oxide production in lipopolysaccharide-activated macrophages. *Life Science* 71, 411–419.

Jackson, S.J.T., Singletary, K.W. and Venema, R.C. (2007) Sulforaphane suppresses angiogenesis and disrupts endothelial mitotic progression and microtubule polymerization. *Vascular Pharmacology* 46, 77–84.

Jain, A.K. and Jaiswal, A.K. (2007) GSK-3beta acts upstream of Fyn kinase in regulation of nuclear export and degradation of NF-E2 related factor 2. *Journal of Biological Chemistry* 282, 16502–16510.

Jain, A.K., Bloom, D.A. and Jaiswal, A.K. (2005) Nuclear import and export signals in control of Nrf2. *Journal of Biological Chemistry* 280, 29158–29168.

Jeong, W.S., Keum, Y.S., Chen, C., Jain, M.R., Shen, G.X., Kim, J.H., Li, W.G. and Kong, A.N.T. (2005) Differential expression and stability of endogenous nuclear factor E2-related factor 2 (Nrf2) by natural chemopreventive compounds in HepG2 human hepatoma cells. *Journal of Biochemistry and Molecular Biology* 38, 167–176.

Jimenez-Alvarez, D., Giuffrida, F., Vanrobaeys, F., Golay, P.A., Cotring, C., Lardeau, A. and Keely, B.J. (2008) High-throughput methods to assess lipophilic and hydrophilic antioxidant capacity of food extracts *in vitro*. *Journal of Agricultural and Food Chemistry* 56, 3470–3477.

Kadiiska, M.B., Gladen, B.C., Baird, D.D., Germolec, D., Graham, L.B., Parker, C.E., Nyska, A., Wachsman, J.T., Ames, B.N., Basu, S., Brot, N., FitzGerald, G.A., Floyd, R.A., George, M., Heinecke, J.W., Hatch, G.E., Hensley, K., Lawson, J.A., Marnett, L.J., Morrow, J.D., Murray, D.M., Plastaras, J., Roberts, L.J. II, Rokach, J., Shigenaga, M.K., Sohal, R.S., Sun, J., Tice, R.R., Van Thiel, D.H., Wellner, D., Walter, P.B., Tomer, K.B., Mason, R.P. and Barrett, J.C. (2005) Biomarkers of Oxidative Stress Study II: are oxidation products of lipids, proteins, and DNA markers of CCl4 poisoning? *Free Radical Biology and Medicine* 38, 698–710.

Katsuoka, F., Motohashi, H., Ishii, T., Aburatani, H., Engel, J.D. and Yamamoto, M. (2005) Genetic evidence that small Maf proteins are essential for the activation of antioxidant response element-dependent genes. *Molecular and Cellular Biology* 25, 8044–8051.

Keum, Y.S., Han, Y.H., Liew, C., Kim, J.H., Xu, C.J., Yuan, X.L., Shakarjian, M.P., Chong, S.H. and Kong, A.N. (2006) Induction of heme oxygenase-1 (HO-1) and NAD[P]H: quinone oxidoreductase 1 (NQO1) by a phenolic antioxidant, butylated hydroxyanisole (BHA) and its metabolite, tert-butylhydroquinone (tBHQ) in primary-cultured human and rat hepatocytes. *Pharmaceutical Research* 23, 2586–2594.

Khan, F., Elherik, K., Bolton-Smith, C., Barr, R., Hill, A., Murrie, I. and Belch, J.J.F. (2003) The effects of dietary fatty acid supplementation on endothelial function and vascular tone in healthy subjects. *Cardiovascular Research* 59, 955–962.

Kimura, M., Yamamoto, T., Zhang, J., Itoh, K., Kyo, M., Kamiya, T., Aburatani, H., Katsuoka, F., Kurokawa, H., Tanaka, T., Motohashi, H. and Yamamoto, M. (2007) Molecular basis distinguishing the DNA binding profile of Nrf2-Maf heterodimer from that of Maf homodimer. *Journal of Biological Chemistry* 282, 33681–33690.

Kitano, H., Oda, K., Kimura, T., Matsuoka, Y., Csete, M., Doyle, J. and Muramatsu, M. (2004) Metabolic syndrome and robustness tradeoffs. *Diabetes* 53, S6–S15.

Kluger, M.J., Singer, R. and Eiger, S.M. (1985) Polymyxin B use does not ensure endotoxin-free solution. *Journal of Immunological Methods* 83, 201–207.

Kobayashi, A., Kang, M.I., Okawa, H., Ohtsuji, M., Zenke, Y., Chiba, T., Igarashi, K. and Yamamoto, M. (2004) Oxidative stress sensor Keap1 functions as an adaptor for Cul3-based E3 ligase to regulate proteasomal degradation of Nrf2. *Molecular and Cellular Biology* 24, 7130–7139.

Kroemer, G., El Deiry, W.S., Golstein, P., Peter, M.E., Vaux, D., Vandenabeele, P., Zhivotovsky, B., Blagosklonny, M.V., Malorni, W., Knight, R.A., Piacentini, M., Nagata, S. and Melino, G. (2005) Classification of cell death: recommendations of the Nomenclature Committee on Cell Death. *Cell Death and Differentiaton* 12, 1463–1467.

Kwak, M.K., Itoh, K., Yamamoto, M. and Kensler, T.W. (2002) Enhanced expression of the transcription factor Nrf2 by cancer chemopreventive agents: role of antioxidant response element-like sequences in the Nrf2 promoter. *Molecular and Cellular Biology* 22, 2883–2892.

Kwak, M.K., Wakabayashi, N., Itoh, K., Motohashi, H., Yamamoto, M. and Kensler, T.W. (2003) Modulation of gene expression by cancer chemopreventive dithiolethiones through the Keap1-Nrf2 pathway. Identification of novel gene clusters for cell survival. *Journal of Biological Chemistry* 278, 8135–8145.

Lambert, J.D., Sang, S. and Yang, C.S. (2007) Possible controversy over dietary polyphenols: benefits vs risks. *Chemical Research in Toxicology* 20, 583–585.

Lane, M.A., Roth, G.S. and Ingram, D.K. (2007) Caloric restriction mimetics: a novel approach for bio-gerontology. *Methods in Molecular Biology* 371, 143–149.

Leahy, D.E. (2006) Integrating *in vitro* ADMET data through generic physiologically based pharmacokinetic models. *Expert Opinion on Drug Metabolism and Toxicology* 2, 619–628.

Lee, O.H., Jain, A.K., Papusha, V. and Jaiswal, A.K. (2007) An auto-regulatory loop between stress sensors INrf2 and Nrf2 controls their cellular abundance. *Journal of Biological Chemistry* 282, 36412–36420.

Lee-Hilz, Y.Y., Boerboom, A.M.J.F., Westphal, A.H., van Berkel, W.J.H., Aarts, J.M.M.J. and Rietjens, I.M.C.M. (2006) Pro-oxidant activity of flavonoids induces EpRE-mediated gene expression. *Chemical Research in Toxicology* 19, 1499–1505.

Lister, C.E., Wilson, P.E., Sutton, K.H. and Morrison, S.C. (2002) Understanding the health benefits of black-currants. *Acta Horticulturae* 585, 443–449.

McNeil, J.H. (1999) *Experimental Models of Diabetes*. CRC Press Inc, Boca Raton, Florida.

Mahajan, M.A., Acara, M. and Taub, M. (1994) Uptake and phosphorylation of thiamine in rabbit primary proximal tubule cells and Madin Darby canine kidney cells. II. Effect of ethanol. *Journal of Pharmacology And Experimental Therapeutics* 268, 1316–1320.

Maiyoh, G.K., Kuh, J.E., Casaschi, A. and Theriault, A.G. (2007) Cruciferous indole-3-carbinol inhibits apo-lipoprotein B secretion in HepG2 cells. *Journal of Nutrition* 137, 2185–2189.

Mamani-Matsuda, M., Kauss, T., Al Kharrat, A., Rambert, J., Fawaz, F., Thiolat, D., Moynet, D., Coves, S., Malvy, D. and Mossalayi, M.D. (2006) Therapeutic and preventive properties of quercetin in experimental arthritis correlate with decreased macrophage inflammatory mediators. *Biochemical Pharmacology* 72, 1304–1310.

Manach, C., Mazur, A. and Scalbert, A. (2005) Polyphenols and prevention of cardiovascular diseases. *Current Opinion in Lipidology* 16, 77–84.

Mann, G.E., Rowlands, D.J., Li, F.Y.L., de Winter, P. and Siow, R.C.M. (2007) Activation of endothelial nitric oxide synthase by dietary isoflavones: role of NO in Nrf2-mediated antioxidant gene expression. *Cardiovascular Research* 75, 261–274.

Maruhashi, F., Murakami, S. and Baba, K. (1994) Automated monitoring of cell concentration and viability using an image analysis system. *Cytotechnology* 15, 281–289.

Milbury, P.E., Graf, B., Curran-Celentano, J.M. and Blumberg, J.B. (2007) Bilberry (*Vaccinium myrtillus*) anthocyanins modulate heme oxygenase-1 and glutathione S-transferase-pi expression in ARPE-19 cells. *Investigative Ophthalmology and Visual Science* 48, 2343–2349.

Miller, N.J., Riceevans, C., Davies, M.J., Gopinathan, V. and Milner, A. (1993) A novel method for measuring antioxidant capacity and its application to monitoring the antioxidant status in premature neonates. *Clinical Science* 84, 407–412.

Motohashi, H., Shavit, J.A., Igarashi, K., Yamamoto, M. and Engel, J.D. (1997) The world according to Maf. *Nucleic Acids Research* 25, 2953–2959.

Mozaffarian, D., Katan, M.B., Ascherio, A., Stampfer, M.J. and Willett, W.C. (2006) Trans fatty acids and cardiovascular disease. *New England Journal of Medicine* 354, 1601–1613.

Mulcahy, R.T., Wartman, M.A., Bailey, H.H. and Gipp, J.J. (1997) Constitutive and beta-naphthoflavone-induced expression of the human gamma-glutamylcysteine synthetase heavy subunit gene is regulated by a distal antioxidant response element/TRE sequence. *Journal of Biological Chemistry* 272, 7445–7454.

Munford, R.S. (2005) Invited review: detoxifying endotoxin: time, place and person. *Journal of Endotoxin Research* 11, 69–84.

Musk, S.R.R., Astley, S.B., Edwards, S.M., Stephenson, P., Hubert, R.B. and Johnson, I.T. (1995) Cytotoxic and clastogenic effects of benzyl isothiocyanate towards cultured mammalian cells. *Food Chemistry and Toxicology* 33, 31–37.

Nadon, N., Strong, R., Miller, R., Nelson, J., Javors, M., Sharp, Z., Peralba, J. and Harrison, D. (2008) Design of aging intervention studies: the NIA interventions testing program. *Age* 30, 187–199.

Naguib, Y.M.A. (2000) A fluorometric method for measurement of oxygen radical-scavenging activity of water-soluble antioxidants. *Analytical Biochemistry* 284, 93–98.

Nastruzzi, C., Cortesi, R., Esposito, E., Menegatti, E., Leoni, O., Iori, R. and Palmieri, S. (2000) *In vitro* antiproliferative activity of isothiocyanates and nitriles generated by myrosinase-mediated hydrolysis of glucosinolates from seeds of cruciferous vegetables. *Journal of Agriculture and Food Chemistry* 48, 3572–3575.

Navab, M., Ananthramaiah, G.M., Reddy, S.T., Van Lenten, B.J., Ansell, B.J., Fonarow, G.C., Vahabzadeh, K., Hama, S., Hough, G., Kamranpour, N., Berliner, J.A., Lusis, A.J. and Fogelman, A.M. (2004) Thematic review series: the pathogenesis of atherosclerosis. The oxidation hypothesis of atherogenesis: the role of oxidized phospholipids and HDL. *Journal of Lipid Research* 45, 993–1007.

Neto, C.C. (2007) Cranberry and its phytochemicals: a review of *in vitro* anticancer studies. *Journal of Nutrition* 137, 186S–193S.

Nishinaka, T., Ichijo, Y., Ito, M., Kimura, M., Katsuyama, M., Iwata, K., Miura, T., Terada, T. and Yabe-Nishimura, C. (2007) Curcumin activates human glutathione S-transferase P1 expression through antioxidant response element. *Toxicology Letters* 170, 238–247.

Ormerod, M.G., Sun, X.M., Brown, D., Snowden, R.T. and Cohen, G.M. (1993) Quantification of apoptosis and necrosis by flow-cytometry. *Acta Oncologica* 32, 417–424.

Pellegrini, N., Serafini, M., Colombi, B., Del Rio, D., Salvatore, S., Bianchi, M. and Brighenti, F. (2003) Total antioxidant capacity of plant foods, beverages and oils consumed in Italy assessed by three different *in vitro* assays. *Journal of Nutrition* 133, 2812–2819.

Pi, J.B., Bai, Y.S., Reece, J.M., Williams, J., Liu, D.X., Freeman, M.L., Fahl, W.E., Shugar, D., Liu, J., Qu, W., Collins, S. and Waalkes, M.P. (2007) Molecular mechanism of human Nrf2 activation and degradation: role of sequential phosphorylation by protein kinase CK2. *Free Radical Biology and Medicine* 42, 1797–1806.

Piotrowicz, B., Watt, I., Edlin, S. and McCartney, A. (1985) A micromethod for endotoxin assay in human plasma using limulus amoebocyte lysate and a chromogenic substrate. *European Journal of Clinical Microbiology and Infectious Diseases* 4, 52–54.

Poussier, B., Cordova, A.C., Becquemin, J.P. and Sumpio, B.E. (2005) Resveratrol inhibits vascular smooth muscle cell proliferation and induces apoptosis. *Journal of Vascular Surgery* 42, 1190–1197.

Qian, Q.W., Babaei-Jadidi, R., Ahmed, N. and Thornalley, P.J. (2006) Reversal of biochemical dysfunction in endothelial cells in hyperglycemia by induction of antioxidant response element-linked gene expression by sulforaphane. *Diabetes* 55, A184–A185.

Re, R., Pellegrini, N., Proteggente, A., Pannala, A., Yang, M. and Rice-Evans, C. (1999) Antioxidant activity applying an improved ABTS radical cation decolorization assay. *Free Radical Biology and Medicine* 26, 1231–1237.

Repetto, G., del Peso, A. and Zurita, J.L. (2008) Neutral red uptake assay for the estimation of cell viability/cytotoxicity. *Nature Protocols* 3, 1125–1131.

Riccardi, C. and Nicoletti, I. (2006) Analysis of apoptosis by propidium iodide staining and flow cytometry. *Nature Protocols* 1, 1458–1461.

Rubiolo, J.A., Mithieux, G. and Vega, F.V. (2008) Resveratrol protects primary rat hepatocytes against oxidative stress damage: activation of the Nrf2 transcription factor and augmented activities of antioxidant enzymes. *European Journal of Pharmacology* 591, 66–72.

Sajilata, M.G., Bajaj, P.R. and Singhal, R.S. (2008) Tea polyphenols as nutraceuticals. *Comprehensive Reviews in Food Science and Food Safety* 7, 229–254.

Sanchez-Moreno, C., Cano, M.P., de Ancos, B., Plaza, L., Olmedilla, B., Granado, F. and Martin, A. (2004) Consumption of high-pressurized vegetable soup increases plasma vitamin C and decreases oxidative stress and inflammatory biomarkers in healthy humans. *Journal of Nutrition* 134, 3021–3025.

Seeram, N.P., Aviram, M., Zhang, Y., Henning, S.M., Feng, L., Dreher, M. and Heber, D. (2008) Comparison of antioxidant potency of commonly consumed polyphenol-rich beverages in the United States. *Journal of Agricultural and Food Chemistry* 56, 1415–1422.

Sharma, O.P. and Bhat, T.K. (2009) DPPH antioxidant assay revisited. *Food Chemistry* 113, 1202–1205.

Singh, M., Arseneault, M., Sanderson, T., Murthy, V. and Ramassamy, C. (2008) Challenges for research on polyphenols from foods in Alzheimer's disease: bioavailability, metabolism, and cellular and molecular mechanisms. *Journal of Agricultural and Food Chemistry* 56, 4855–4873.

Singleton, V.L., Orthofer, R. and Lamuela-Raventos, R.M. (1999) Analysis of total phenols and other oxidation substrates and antioxidants by means of Folin–Ciocalteu reagent. *Methods in Enzymology* 299, 152–178.

Song, L. and Thornalley, P.J. (2007) Effect of storage, processing and cooking on glucosinolate content of Brassica vegetables. *Food and Chemical Toxicology* 45, 216–224.

Souza, J.N.S., Silva, E.M., Loir, A., Rees, J.F., Rogez, H. and Larondelle, Y. (2008) Antioxidant capacity of four polyphenol-rich Amazonian plant extracts: a correlation study using chemical and biological *in vitro* assays. *Food Chemistry* 106, 331–339.

Stevenson, D. and Hurst, R. (2007) Polyphenolic phytochemicals – just antioxidants or much more? *Cellular and Molecular Life Sciences* 64, 2900–2916.

Sun, Z., Zhang, S., Chan, J.Y. and Zhang, D.D. (2007) Keap1 controls postinduction repression of the Nrf2-mediated antioxidant response by escorting nuclear export of Nrf2. *Molecular and Cellular Biology* 27, 6334–6349.

Surh, Y.J., Kundu, J.K. and Na, H.K. (2008) Nrf2 as a master redox switch in turning on the cellular signaling involved in the induction of cytoprotective genes by some chemopreventive phytochemicals. *Planta Medica* 74, 1526–1539.

Tang, L., Zhang, Y., Jobson, H.E., Li, J., Stephenson, K.K., Wade, K.L. and Fahey, J.W. (2006) Potent activation of mitochondria-mediated apoptosis and arrest in S and M phases of cancer cells by a broccoli sprout extract. *Molecular Cancer Therapeutics* 5, 935–944.

Tanigawa, S., Fujii, M. and Hou, D.X. (2007) Action of Nrf2 and Keap1 in ARE-mediated NQO1 expression by quercetin. *Free Radical Biology and Medicine* 42, 1690–1703.

Thimmulappa, R.K., Mai, K.H., Srisuma, S., Kensler, T.W., Yamamoto, M. and Biswal, S. (2002) Identification of Nrf2-regulated genes induced by the chemopreventive agent sulforaphane by oligonucleotide array. *Cancer Research* 62, 5196–5203.

Thornalley, P.J. and IARC (International Agency For Research on Cancer) Workgroup (2004) *Cruciferous Vegetables, Isothiocyanates and Indoles – IARC Handbook on Chemoprevention of Cancer.* IARC Press, Lyon, France.

Thornalley, P.J. and Rabbani, N. (2009) Highlights and hotspots of protein glycation in end stage renal disease. *Seminars in Dialysis* 22, 400–404.

Valkonen, M. and Kuusi, T. (1997) Spectrophotometric assay for total peroxyl radical-trapping antioxidant potential in human serum. *Journal of Lipid Research* 38, 823–833.

Vichai, V. and Kirtikara, K. (2006) Sulforhodamine B colorimetric assay for cytotoxicity screening. *Nature Protocols* 1, 1112–1116.

Visioli, F., Bellomo, G. and Galli, C. (1998) Free radical-scavenging properties of olive oil polyphenols. *Biochemical and Biophysical Research Communications* 247, 60–64.

Viswanathan, M., Kim, S.K., Berdichevsky, A. and Guarente, L. (2005) A role for SIR-2.1 regulation of ER stress response genes in determining *C. elegans* life span. *Developmental Cell* 9, 605–615.

Wang, L., Sun, J., Horvat, M., Koutalistras, N., Johnston, B. and Ross Sheil, A.G. (1996) Evaluation of MTS, XTT, MTT and 3HTdR incorporation for assessing hepatocyte density, viability and proliferation. *Methods in Cell Science* 18, 249–255.

Wang, Z., Gidwani, V., Sun, Z., Zhang, D.D. and Wong, P.K. (2008) Development of a molecular assay for rapid screening of chemopreventive compounds targeting Nrf2. *Journal of the Association for Laboratory Automation* 13, 243–248.

Watai, Y., Kobayashi, A., Nagase, H., Mizukami, M., McEvoy, J., Singer, J.D., Itoh, K. and Yamamoto, M. (2007) Subcellular localization and cytoplasmic complex status of endogenous Keap1. *Genes to Cells* 12, 1163–1178.

Wolfe, K.L. and Liu, R.H. (2007) Cellular antioxidant activity (CAA) assay for assessing antioxidants, foods, and dietary supplements. *Journal of Agricultural and Food Chemistry* 55, 8896–8907.

Wolfe, K.L. and Liu, R.H. (2008) Structure–activity relationships of flavonoids in the cellular antioxidant activity assay. *Journal of Agricultural and Food Chemistry* 56, 8404–8411.

Wolfe, K.L., Kang, X.M., He, X.J., Dong, M., Zhang, Q.Y. and Liu, R.H. (2008) Cellular antioxidant activity of common fruits. *Journal of Agricultural and Food Chemistry* 56, 8418–8426.

Woods, A.S., Buchsbaum, J.C., Worrall, T.A., Berg, J.M. and Cotter, R.J. (1995) Matrix-assisted laser desorption/ionization of non-covalently bound compounds. *Analytical Chemistry* 67, 4462–4465.

Wu, L.Y., Ashraf, M.H.N., Facci, M., Wang, R., Paterson, P.G., Ferrie, A. and Juurlink, B.H.J. (2004a) Dietary approach to attenuate oxidative stress, hypertension, and inflammation in the cardiovascular system. *Proceedings of the National Academy of Sciences of the United States of America* 101, 7094–7099.

Wu, V.C.-H., Qiu, X., Bushway, A. and Harper, L. (2008) Antibacterial effects of American cranberry (*Vaccinium macrocarpon*) concentrate on foodborne pathogens. *LWT – Food Science and Technology* 41, 1834–1841.

Wu, X., Beecher, G.R., Holden, J.M., Haytowitz, D.B., Gebhardt, S.E. and Prior, R.L. (2004b) Lipophilic and hydrophilic antioxidant capacities of common foods in the United States. *Journal of Agricultural and Food Chemistry* 52, 4026–4037.

Xiao, D. and Singh, S.V. (2002) Phenethyl isothiocyanate-induced apoptosis in p53-deficient PC-3 human prostate cancer cell line is mediated by extracellular signal-regulated kinases. *Cancer Research* 62, 3615–3619.

Xu, K. and Thornalley, P.J. (2000a) Involvement of GSH metabolism in the cytotoxicity of the phenethyl isothiocyanate and its cysteine conjugate to human leukaemia cells *in vitro*. *Biochemical Pharmacology* 61, 165–177.

Xu, K. and Thornalley, P.J. (2000b) Studies on the mechanism of the inhibition of human leukaemia cell growth by dietary isothiocyanates and their cysteine adducts *in vitro*. *Biochemical Pharmacology* 60, 221–231.

Xue, M., AntonySunil, A., Rabbani, N. and Thornalley, P.J. (2008a) Protein damage by glycation, oxidation and nitration in the ageing process. Advances in quantitation of protein damage and the emerging importance of decline in enzymatic defences as the ageing phenotype develops. In: Foyer, C.H., Faraghar, R. and Thornalley, P.J. (eds) *Redox Metabolism and Longevity Relationships in Animals and Plants*. Garland Science, London, pp. 227–265.

Xue, M., Qian, Q., Adaikalakoteswari, A., Rabbani, N., Babaei-Jadidi, R. and Thornalley, P.J. (2008b) Activation of NF-E2-related factor-2 reverses biochemical dysfunction of endothelial cells induced by hyperglycemia linked to vascular disease. *Diabetes* 57, 2809–2817.

Ye, L. and Zhang, Y. (2001) Total intracellular accumulation levels of dietary isothiocyanates determine their activity in elevation of cellular glutathione and induction of phase 2 detoxification enzymes. *Carcinogenesis* 22, 1987–1992.

Yokozawa, T., Chung, H.Y., He, L.Q. and Oura, H. (1996) Effectiveness of green tea tannin on rats with chronic renal failure. *Bioscience Biotechnology and Biochemistry* 60, 1000–1005.

Yu, R., Lei, W., Mandelkar, S., Weber, M.J., Der, C.J., Wu, J. and Kong, A.N.T. (1999) Role of a mitogen-activated protein kinase pathway in the induction of phase II detoxifying enzymes by chemicals. *Journal of Biological Chemistry* 274, 27545–27552.

Zadelaar, S., Kleemann, R., Verschuren, L., de Vries-Van der Weij, J., van der Hoorn, J., Princen, H.M. and Kooistra, T. (2007) Mouse models for atherosclerosis and pharmaceutical modifiers. *Arteriosclerosis Thrombosis and Vascular Biology* 27, 1706–1721.

Zhang, D.D., Lo, S.C., Sun, Z., Habib, G.M., Lieberman, M.W. and Hannink, M. (2005) Ubiquitination of Keap1, a BTB-kelch substrate adaptor protein for Cul3, targets Keap1 for degradation by a proteasome-independent pathway. *Journal of Biological Chemistry* 280, 30,091–30,099.

Zhang, Y.S., Munday, R., Jobson, H.E., Munday, C.M., Lister, C., Wilson, P., Fahey, J.W. and Mhawech-Fauceglia, P. (2006) Induction of GST and NQO1 in cultured bladder cells and in the urinary bladders of rats by an extract of broccoli (*Brassica oleracea italica*) sprouts. *Journal of Agricultural and Food Chemistry* 54, 9370–9376.

Zhu, H., Itoh, K., Yamamoto, M., Zweier, J.L. and Li, Y.B. (2005) Role of Nrf2 signaling in regulation of antioxidants and phase 2 enzymes in cardiac fibroblasts: protection against reactive oxygen and nitrogen species-induced cell injury. *FEBS Letters* 579, 3029–3036.

Zulueta, A., Esteve, M.J. and Frigola, A. (2009) ORAC and TEAC assays comparison to measure the antioxidant capacity of food products. *Food Chemistry* 114, 310–316.

Index